P9-DEL-205

A Dictionary of **Euphemisms**

A Dictionary of Euphemisms

R. W. HOLDER

OXFORD UNIVERSITY PRESS
1995

Oxford University Press, Walton Street, Oxford OX2 6DP

Oxford New York
Athens Auckland Bangkok Bombay
Calcutta Cape Town Dar es Salaam Delhi
Florence Hong Kong Istanbul Karachi
Kuala Lumpur Madras Madrid Melbourne
Mexico City Nairobi Paris Singapore
Taipei Tokyo Toronto

and associated companies in
Berlin Ibadan

Oxford is a trade mark of Oxford University Press

Published in the United States by
Oxford University Press Inc., New York

First published as A Dictionary of American and British Euphemisms *by Bath University Press 1987*

Revised edition published by Faber and Faber Limited 1989

This edition first published by Oxford University Press 1995

British Library Cataloguing in Publication Data
Data available

Library of Congress Cataloging in Publication Data
Data available
ISBN 0-19-869275-7

10 9 8 7 6 5 4 3 2 1

Typeset by Barbers Ltd, Wrotham, Kent
Printed in Great Britain
at The Bath Press, Avon

Contents

An Explanation

When I started gathering euphemisms in 1977 with a dictionary in mind, nothing of the kind had previously been published and, although two American works were to appear before my first edition in 1987, I had already decided on the form the collection should take – what to include or omit, when it might be helpful to look at etymology, to what extent supporting quotations were needed and how to deal with the numerous sexual and lavatorial examples. I knew that certain of the entries would give offence to some readers but I wanted to make the whole readable (which is why I was pleased when one critic called it a 'browser's delight') and to draw attention to the inherent silliness of some of the euphemistic words and expressions we all use. It was not, I thought, a project to be taken too seriously.

I based my selection mainly on Fowler's definition: *Euphemism means (the use of a) mild or vague or periphrastic expression as a substitute for blunt precision or disagreeable truth* (*Modern English Usage*, 1957). I eventually evolved a second test, that the euphemistic word or phrase once meant, or prima facie still means, something else. This latter refinement led to my leaving out many entries which other lexicographers have classified as euphemistic. For example, I do not include *golly* as a corruption of *God*, but as a mild swearword, and I resist the temptation to add a note about Florence Upton's delightful character, the original Golliwog whose resource and bravery a century ago saved The Dutch Dolls from so many perils. Conversely I do include many ways of speaking allusively of the devil.

Because many euphemisms have become so much part of standard English that we think only of the current usage, I sometimes draw the reader's attention to what the word means literally, or used to mean. You will also find a few foreign words or phrases which we use as euphemisms in English, most of which are euphemisms in the original language.

As I increased my collection, it became apparent that we use euphemism not for a single purpose, but for four: it is the language of evasion, of hypocrisy, of prudery, and of deceit. Some of the entries might be classified as dysphemisms by other writers and I am sure that Allan and Burridge must be right when they tell us that *Euphemistic dysphemisms and dysphemistic euphemisms have locutions that are at odds with their illocutionary point* (*Euphemism and Dysphemism*, 1991). I would only add that one man's euphemism can be another woman's dysphemism.

I decided to leave out anything which was not in literary or common use, unless it struck me as interesting, and to omit words or phrases which I have seen only in another dictionary. Inevitably, because I live in England and work in Ireland, the selection reflects the speech of those countries more than that of the United States, which I visit regularly for a few days only, or other English-speaking places to which I seldom go. Happily English literature is universal, with Indian, South African, and Australian writers as available as those from North America and the British Isles. That fertile ground for picking up the development of language, the newspaper, is unfortunately only regularly available to me in England.

Euphemisms are by their nature closely associated with taboo, and the taboos of one generation are not necessarily those of the next, although those connected with sexual behaviour or defecation have shown remarkable staying powers. We are more open than the Victorians about brothels and prostitution, less prudish about courtship and childbirth, less terrified of bankruptcy. In our turn we have created fresh taboos, relating to skin pigmentation, charity, education, and commercial practice, for example. We have also banished the word *fart* from polite speech. Our use of language reveals social attitudes and mores in a manner which is particularly useful to historians because it is contemporary evidence which has passed down to us unnoticed and which we in turn will pass on to our descendants. Henry Mayhew tells us more about Victorian London than he supposed.

Very often the derivation of a euphemism is obvious, as with resting and death or bathrooms and urination, and I have not bothered to spell it out. In other cases I have spent enjoyable hours tracking down etymologies, occasionally finding two or three possible sources. As with all detective work, after a while a pattern emerges. Sometimes I have had to speculate or guess and occasionally I admit defeat. I also note occasional differences between usages on either side of the Atlantic, which are much more frequent and insidious than the commonly quoted *rubber* or *fanny*.

Those of us for whom English is our first language use subtly different euphemisms to express shades of meaning, especially in sexual matters but also in other topics such as death or dismissal from employment. These nuances may differ from place to place, or even between generations, but I have tried to take them into account in framing definitions. It seemed a denial of what I was trying to achieve if I had to define one euphemism with another, but this could not be entirely avoided for a variety of reasons. There is no word in English for *fuck* except 'fuck', but the constant repetition of that noun and verb is not conducive to comfortable reading. My use of *copulate* and *copulation* is unsatisfactory, but better than the alternatives. So too with *shit* and defecation. We have no non-euphemistic word for a woman who regularly copulates and cohabits with a man outside wedlock, and so I use *sexual mistress*. Her counterpart, or *lover*, causes similar problems. *Lavatory* is a euphemism which I use as a definition, in default of anything better. In my earlier editions I used *cripple* and *whore*, both of which are thought to have acquired offensive overtones (although *crippled* has not), and as there are non-euphemistic alternatives like *lame* and *prostitute*, I have accepted the changes.

The celebrated lecturer on criminal law at Cambridge, the late Henry Barnes, chose scandalous or sexual illustrations from case law on the basis that these would help students to remember both them and the principles involved: I certainly have never forgotten *This is illustrated by rex versus Biggar: change one letter, you have the facts*. While unable to achieve a comparable standard of wit or entertainment, I have sought where possible to choose illustrative quotations which are of interest in themselves. As most of the quotations after 1900 come from my own reading, the scope for this has been limited. Many of the earlier quotations come from Dr Joseph Wright's majestic *English Dialect Dictionary* but where I have lifted a quotation from another compiler, I say so.

You will find a Bibliography and an Index. I occasionally give a talk about euphemisms (although it always seems quickly to turn into a seminar, however large the audience) and am asked what other books on euphemism are enjoyable to read. I can recommend with confidence *Fair of Speech*, edited by the poet and critic D. J. Enright, and Hugh Rawson's *Dictionary of Euphemisms & Other Doubletalk*. I will not tell you what I do not recommend. As to the Index, which may seem an odd thing for a dictionary, you will find an attempt to classify the alphabetical entries, partly to enable the reader to pick out everything relating to a chosen subject but also to avoid excessive cross-referencing in the text.

I know readers, especially critics, usually skip introductions and those who do not, fall away at an author's list of benefactors. I must nonetheless acknowledge again the many parts played by Dr Michael Allen of Bath University in publishing the original hardback in 1987 and in everything which has happened since. This edition has greatly benefited from suggestions, corrections, advice, and other assistance from those wise lexicographers at the Oxford University Press, Sara Tulloch and Patrick Hanks, while Julia Elliott has been the most thoughtful and sympathetic of editors. I have also received encouragement and information from many people on both sides of the ocean, to whom I am deeply indebted.

My labours are not dissimilar to those of Sisyphus. The language continues to evolve and it is a poor week when I fail to note two or three new euphemisms. As I complete this Explanation, I hope the stone is near the top of the hill but I also know that from this moment on it will inexorably roll down until, with good fortune, I am asked to prepare another edition.

R. W. Holder

West Monkton
1995

Editorial Note

Labels such as *American* and *Scottish* indicate that usage is restricted to the regional English specified. Those words or phrases for which no evidence of current usage is available are identified by the label *obsolete*.

As informal language is a common characteristic of euphemism, and to avoid superfluous labelling, labels such as *informal*, *slang*, or *colloquial* have not been used in this dictionary.

Cross references are printed in small capitals.

Bibliography

Quotations have been included in the text to show how each word or phrase was or is used, and when. As with all libraries, mine is subject to continual depredations and in a handful of cases I have been unable to verify the title of a work from which I have quoted.

The date given for each title refers to the first publication or to the edition which I have used. Where an author has deliberately used archaic language, I mention this in the text.

In addition to the sources of the quoted material the bibliography lists relevant dictionaries and reference books, and some further titles which are useful background reading. Some of these are referred to in the text by an abbreviation, and these abbreviations are listed first.

BDPF	*The Dictionary of Phrase and Fable* (Brewer, 1978)
CBSLD	*The 'Official' CB Slanguage Language Dictionary* (Dills, 1976)
DAS	*Dictionary of American Slang* (Wentworth and Flexner, 1975)
DDE	*The Dictionary of Diseased English* (Hudson, 1977)
DHS	*A Dictionary of Historical Slang* (Partridge and Simpson, 1972)
DNB	*The Dictionary of National Biography* (1978)
Dr Johnson	*A Dictionary of the English Language* (1755)
DRS	*A Dictionary of Rhyming Slang* (Franklin, 1961)
DSUE	*A Dictionary of Slang and Unconventional English* (Partridge, 1970)
EDD	*The English Dialect Dictionary* (1898–1905)
FDMT	*The Fontana Dictionary of Modern Thought* (Bullock and Stallybrass, 1977)
Grose	*Dictionary of the Vulgar Tongue* (1811)
ODEE	*The Oxford Dictionary of English Etymology* (Onions, 1978)
ODEP	*The Oxford Dictionary of English Proverbs* (Smith and Wilson, 1970)
OED	*The Oxford English Dictionary* (1989)
SOED	*The Shorter Oxford English Dictionary* (1993)
WNCD	*Webster's New Collegiate Dictionary* (1977)

Adams, J. (1985) *Good Intentions*
Agnus, Orme (1900) *Jan Oxber*
'Agrikler' (1872) *Rhymes in the West of England Dialect*
Ainslie, Hew (1892) *A Pilgrimage to the Land of Burns*
Aldiss, Brian (1988) *Forgotten Life*
Alexander, William (1875–82 edition) *Sketches of Life among my Ain Folk*
Allan, Keith and Burridge, Kate (1991) *Euphemism and Dysphemism*
Allbeury, Ted (1975) *Palomino Blonde*
—— (1976) *The Only Good German*
—— (1976) *Moscow Quadrille*
—— (1977) *The Special Connection*
—— (1978) *The Lantern Network*
—— (1980) *The Twentieth Day of January*
—— (1980) *The Reaper*
—— (1981) *The Secret Whispers*
—— (1982) *All our Tomorrows*
—— (1983) *Pay Any Price*
Allen, Charles (1975) *Plain Tales from the Raj*
Allen, Paula Gunn (1992) *The Sacred Hoop: Recovering the Feminine in American Indian Tradition*
Allen, Richard (1971) *Suedehead*
Alter (1960) *The Exile*
Amis, Kingsley (1978) *Jake's Thing*
—— (1980) *Russian Hide-and-Seek*
—— (1986) *The Old Devils*
—— (1988) *Difficulties with Girls*
—— (1990) *The Folks That Live on the Hill*
Anderson, David (1826) *Poems Chiefly in the Scottish Dialect*
Anderson, R. (1805–8 edition) *Ballads in the Cumberland Dialect*
Anderson, William (1851–67 edition) *Rhymes, Reveries and Reminiscences*
Andrews, William (1899) *Bygone Church Life in Scotland*
Antrobus, C. L. (1901) *Wildersmoor*
Archer, Jeffrey (1979) *Kane and Abel*
Armstrong, Andrew (1890) *Ingleside Musings and Tales*
Armstrong, Louis (1955) *Satchmo*
Atkinson, J. C. (1891) *Forty Years in a Moorland Parish*
Aubrey, John (1696) *Collected Works*
Axon, W. E. A. (1870) *The Black Knight of Ashton*
Ayto, John (1993) *Euphemisms*

Bacon, Francis (1627) *Essays*
Bagley, Desmond (1977) *The Enemy*
—— (1982) *Windfall*
Bagnall, Jos (1852) *Songs of the Tyne*
Baldwin, William (1993) *The Hard to Catch Mercy*
Ballantine, James (1869) *The Miller of Deanhaugh*
Banim, John (1825) *O'Hara Tales*
Barber, Lyn (1991) *Mostly Men*
Barber, Noel (1981) *Taramara*
Barham, R. H. (1840) *Ingoldsby Legends*
Barlow, Jane (1892) *Bogland Studies*
Barnard, Howard and Lauwerys, Joseph (1963) *A Handbook of British Educational Terms*
Baron, Alexander (1948) *From the City, From the Plough*
Barnes, Julian (1989) *A History of the World in 10½ Chapters*
—— (1991) *Talking it Over*
Barr, John (1861) *Poems and Songs*
Bartram, George (1897) *The People of Clopton*
—— (1898) *The White-Headed Boy*
Beard, Henry and Cerf, Christopher (1992) *The Official Politically Correct Dictionary & Handbook*
Beattie, Ann (1989) *Picturing Will*
Beattie, William (1801) *Fruits of Time Parings*
Beatty, W. (1897) *The Secretar*
Behr, Edward (1978) *Anyone here been raped & speaks English*
—— (1989) *Hirohito: Beyond the Myth*
Bence-Jones, Mark (1987) *Twilight of the Ascendancy*
Benet, Stephen (1943) *A Judgment in the Mountains*
Besant, Walter and Rice, James (1872) *Ready Money Mortiboy*
Binchy, Maeve (1985) *Echoes*
Binns, Æthelbert (1889) *Yorkshire Dialect Words*
Blacker, Terence (1992) *The Fame Hotel*
Blackhall, Alex (1849) *Lays of the North*
Blackmore, R. D. (1869) *Lorna Doone*
Blair, Emma (1990) *Maggie Jordan*
Blanch, Leslie (1954) *The Wilder Shores of Love*
Block, Thomas (1979) *Mayday*
Blythe, Ronald (1969) *Akenfield*
Bogarde, Dirk (1972) *A Postilion Struck by Lightning*
—— (1978) *Snakes and Ladders*
—— (1981) *Voices in the Garden*
—— (1983) *An Orderly Man*
Boldrewood, Rolf (1890) *A Colonial Reformer*
Bolger, Dermot (1990) *The Journey Home*
Book of Common Prayer (1662)
Boswell, Alexander (1803) *Songs*
—— (1871 edition) *Poetical Works*
Boswell, James (c. 1792–3) *London Journal*
Boyd, William (1981) *A Good Man in Africa*
—— (1982) *An Ice-Cream War*
—— (1983) *Stars and Bars*
—— (1987) *The New Confessions*
—— (1993) *The Blue Afternoon*
Boyle, Andrew (1979) *The Climate of Treason*
Bradbury, Malcolm (1959) *Eating People is Wrong*
—— (1965) *Stepping Westward*
—— (1975) *The History Man*
—— (1976) *Who Do You Think You Are*
—— (1983) *Rates of Exchange*
Bradley, Edward (1853) *The Adventures of Mr Verdent Green*
Brand, John (1789) *The History and Antiquities of Newcastle-upon-Tyne*
Brewer, E. Cobham (1978 edition) *The Dictionary of Phrase and Fable*
Brierley, Benjamin (1854) *Treadlepin Fold and Other Tales*
—— (1865) *Irkdale*
—— (1886) *The Cotters of Mossburn*
Brown, Harry (1944) *A Walk in the Sun*
Brown, Ivor (1958) *Words in our Time*
Browning, D. C. (1962) *Everyman's Dictionary of Literary Biography*
Bryce, J. B. (1888) *The American Constitution*
Bryson, Bill (1989) *The Lost Continent*
—— (1991) *Neither Here Nor There*
—— (1994) *Made in America*
Buchan, John (1898) *John Burnet of Barns*
Buckman, S. S. (1870) *John Darke's Sojourn in the Cotswolds*
Bullock, Alan and Stallybrass, Oliver (1977) *The Fontana Dictionary of Modern Thought*
Bunyan, John (1678–84) *The Pilgrim's Progress*
Burgess, Anthony (1959) *Beds in the East*
—— (1980) *Earthly Powers*
Burmester, F. G. (1902) *John Lot's Alice*
Burnet, Gilbert (1714) *History of the Reformation of the Church of England*
Burnley, James (1880) *Poems and Sketches*
Burns, Robert (1868) *Poems, Songs and Letters*
Burroughs, William (1959) *Naked Lunch*
—— (1984) *The Place of Dead Roads*
Burton, Anthony (1989) *The Great Days of the Canals*
Burton, Robert (1621) *The Anatomy of Melancholy*
Butcher, Harry C. (1946) *Three Years with Eisenhower*
Butler, Samuel (1903) *The Way of All Flesh*
Byrnes, J. H. (1974) *Mrs Byrnes's Dictionary of Unusual, Obscure and Preposterous Words*
Byron, G. G. N. (1809–24) *Works*
Bywater, Abel (1839) *The Sheffield Dialect*
—— (1853) *The Shevvild Chap's Annual*

Caine, T. H. H. (1885) *The Shadow of a Crime*
Carleton, William (1836) *Fordorougha the Miser*
Carrick, J. D. (1835) *The Laird of Logan*
Carter, Angela (1984) *Nights at Circus*
Caufield, Catherine (1990) *Multiple Exposures*
Cauthery, Philip and Stanway, Andrew (1983) *Complete Book of Love and Sex*
Chambers, Robert (1870) *Popular Rhymes of Scotland*
Chamier, F. quoted in *The Oxford Dictionary of English Proverbs*
Chandler, Raymond (1934) *Finger Man*
—— (1939) *Trouble is my Business*
—— (1940) *Farewell my Lovely*
—— (1943) *The High Window*
—— (1944) *The Lady in the Lake*
—— (1950) *The Big Sleep*
—— (1951) *The Little Sister*
—— (1953) *The Long Goodbye*
—— (1958) *Playback*
Chase, C. David (1987) *Mugged on Wall Street*
Cheng, Nien (1984) *Life and Death in Shanghai*
Christie, Agatha (1939) *Evil Under the Sun*

—— (1940) *Ten Little Niggers*
Clancy, Tom (1986) *Red Storm Rising*
—— (1987) *Patriot Games*
—— (1988) *The Cardinal in the Kremlin*
—— (1989) *Clear and Present Danger*
Clare, John (1827) *The Shepherd's Calendar*
Clark, Alan (1993) *Diaries*
Clark, Charles (1839) *John Noakes and Mary Styles*
Clark, Miles (1991) *High Endeavours*
Cleland, John (1749) *Memoirs of a Woman of Pleasure* (Fanny Hill)
Cobbett, William (1830) *Rural Rides*
Coghill, James (1890) *Poems, Songs and Sonnets*
Cole, John (1995) *As it Seemed to Me*
Collins English Dictionary (1979 edition)
Collins, Jackie (1981) *Chances*
Colodny, Lee and Gettlin, Robert (1991) *Silent Coup*
Colvil, Samuel (1796) *The Whig's Supplication*
Colville, John (1967) *Fringes of Power*
—— (1976) *Footprints in Time*
Commager, Henry (1972) *The Defeat of America*
Condon, Richard (1966) *Any God Will Do*
Congreve, William (1695) *Love for Love*
Cookson, Catherine (1967) *Slinky Jane*
—— (1969) *Our Kate*
Cork, Kenneth (1988) *Cork on Cork*
Corley, T. A. B. (1961) *Democratic Despot*
Cornwell, Bernard (1993) *Rebel*
Cosgrave, Patrick (1989) *The Lives of Enoch Powell*
Coyle, Harold (1987) *Team Yankee*
Crews, Harry (1990) *Body: A Tragicomedy*
Crisp, N. J. (1982) *The Brink*
Crockett, S. R. (1894) *The Raiders*
—— (1896) *The Grey Man*
Croker, T. C. (1862) *Fairy Legends & Traditions of South Ireland*
Cromwell, Oliver (1643) *Letter*
Cross, William (1844) *The Disruption*
Crossman, Richard (1981) *Backbench Diaries*
Cussler, Clive (1984) *Deep Six*

Davidson, Lionel (1978) *The Chelsea Murders*
de la Billière, Peter (1992) *Storm Command*
de Mille, Nelson (1988) *Charm School*
Defoe, Daniel (1721) *Moll Flanders*
Deighton, Len (1972) *Close-up*
—— (1978) *SS-GB*
—— (1981) *XPD*
—— (1982) *Goodbye Mickey Mouse*
—— (1985) *London Match*
—— (1987) *Winter*
—— (1988) *Spy Hook*
—— (1989) *Spy Line*
—— (1990) *Spy Sinker*
—— (1991) *City of Gold*
—— (1994) *Faith*
Desai, Boman (1988) *The Memory of Elephants*
Dickens, Charles (1840) *The Old Curiosity Shop*
—— (1853) *Bleak House*
—— (1861) *Great Expectations*
Dickinson, William (1866) *Scallow Beck Boggle*
Dickson, Paul (1978) *The Official Rules*
Dictionary of Cautionary Words and Phrases (1989)
Dictionary of National Biography (1978 edition)
Diehl, William (1978) *Sharky's Machine*
Dills, Lanie (1976) *The `Official' CB Slanguage Language Dictionary*
Dixon, J. H. (1846) *Ancient Poems, Ballads and Songs of the Peasantry of England*
Doherty, Austen (1884) *Nathan Barlow*
Donaldson, Frances (1990) *Yours Plum. The Letters of P. G. Wodehouse*
Douglas, George (1901) *The House with the Green Shutters*
Doyle, Arthur Conan (1917) *His Last Bow*
Doyle, Ezra (1855) *Polly's Gaon*
Doyle, Roddy (1993) *Paddy Clarke Ha Ha Ha*
Dryden, John (1668–98) *Poetical Works*
du Maurier, Daphne (1938) *Rebecca*
Dunning, Robert (1993) *Somerset One Hundred Years Ago*
Egerton, J. C. (1884) *Sussex Folks and Sussex Ways*
Eliot, George (1871–2) *Middlemarch*
Ellis, William (1750) *The Modern Husbandman*
Ellman, Lucy (1988) *Sweet Desserts*
Emerson, P. H. (1890) *Wild Life on a Tidal Water*
—— (1892) *A Son of the Fens*
Enright, D. J. (editor) (1985) *Fair of Speech*
Erdman, Paul (1974) *The Silver Bears*
—— (1981) *The Last Days of America*
—— (1986) *The Panic of '89*
—— (1987) *The Palace*
Etherege, George (1676) *The Man of Mode*
Evans, Bergen (1962) *Comfortable Words*
Evelyn, John *Diary* (published posthumously in 1818)

Faderman, Lillian (1991) *Odd Girls and Twilight Lovers*
Faith, Nicholas (1990) *The World the Railways Made*
Farmer, J. S. and Henley, W. E. (1890–4) *Slang and its Analogues*
Farran, Roy (1948) *Winged Dagger*
Farrell, J.G. (1973) *The Siege of Krishnapur*
Fergusson, Bernard (1945) *Beyond the Chindwin*
Fergusson, Robert (1773) *Poems on Various Subjects*
Fielding, Henry (1729) *The Author's Farce*
—— (1742) *The History and Adventures of Joseph Andrews*
Fine, Anne (1989) *Goggle-Eyes*
Fingall, Elizabeth (Countess of) (1937) *Seventy Years Young*
Flanagan, Thomas (1979) *The Year of the French*
—— (1988) *The Tenants of Time*
Fleming, Lionel (1965) *Head or Harp*
Fletcher, John (1618) *Valentinian*
Follett, Ken (1978) *The Eye of the Needle*
—— (1979) *Triple*
—— (1991) *Night over Water*
Forbes, Brian (1972) *The Distant Laughter*
—— (1983) *The Rewrite Man*
—— (1986) *The Endless Game*
—— (1989) *A Song at Twilight*
Forbes, Colin (1983) *The Leader and the Damned*
—— (1985) *Cover Story*
—— (1987) *The Janus Man*
—— (1992) *By Stealth*
Ford, Robert (1891) *Thistledown*
Forsey, Eugene (1990) *A Life on the Fringe*
Forsyth, Frederick (1984) *The Fourth Protocol*
—— (1994) *The Fist of God*

Foster, Brian (1968) *The Changing English Language*
Foster, R. F. (1988) *Modern Ireland 1600–1972*
—— (1993) *Paddy & Mr Punch*
Fowler, H. W. (1957) *Modern English Usage*
Fowles, John (1977) *The Magus (revised)*
—— (1985) *A Maggot*
Fox, James (1982) *White Mischief*
Francis, Dick (1962) *Dead Cert*
—— (1978) *Trial Run*
—— (1981) *Twice Shy*
—— (1982) *Banker*
—— (1985) *Break In*
—— (1987) *Hot Money*
—— (1988) *The Edge*
Francis, M. E. (1901) *Pastorals of Dorset*
Franklin, Benjamin (1757) *The Way to Wealth*
Franklyn, Julian (1960) *A Dictionary of Rhyming Slang*
Fraser, George MacDonald (1969) *Flashman*
—— (1970) *Royal Flash*
—— (1971) *Flash for Freedom*
—— (1973) *Flashman at the Charge*
—— (1975) *Flashman in the Great Game*
—— (1977) *Flashman's Lady*
—— (1982) *Flashman and the Redskins*
—— (1983) *The Pyrates*
—— (1985) *Flashman and the Dragon*
—— (1992) *Quartered Safe out Here*
Freemantle, Brian (1977) *Charlie Muffin*
Fry, Stephen (1991) *The Liar*
—— (1994) *The Hippopotamus*
Funk, Charles E. (the elder) (1955) *Heavens to Betsy and Other Curious Sayings*

Galloway, George (1810) *Poems*
Galsworthy, John (1924) *The White Monkey*
Galt, John (1821) *The Ayrshire Legatees*
—— (1823) *The Entail*
—— (1826) *The Last of the Lairds*
Gannel-Milne, G. N. (1933) *Wind in the Wires*
Gardner, J. (1983) *Elephants in the Attic*
Garmondsway, George and Simpson, Jacqueline (1969) *The Penguin English Dictionary*
Garner, James F. (1994) *Politically Correct Bedtime Stories*
Gascoigne, George (1576, 1907–10 edition) *Works*
Gaskell, E. C. (1863) *Sylvia's Lovers*
Genet, Jean (1969) *Funeral Rites* (in translation)
Gissing, Algernon (1890) *A Village Hampden*
Goebbels, Josef (1945) *Diaries—The Last Days* (translated by Richard Barry)
Goldman, William (1984) *The Colour of Light*
—— (1986) *Brothers*
Gordon, Alexander (1894) *Northward Ho!*
Gordon, Frank (1885) *Pyotshaw*
Gordon, J. F. S. (1880) *The Book of the Chronicles of Keith*
Grinnell-Milne, Duncan (1933) *Wind in the Wires*
Gordon, Lyndall (1994) *Charlotte Brontë*
Gores, Joseph N. (1975) *Hammett*
Gosling, John and Warner, Douglas (1960) *The Shame of a City*
Graham, Dougal (1883) *The Collected Writings*
Grant, David (1884) *Lays and Legends of the North*
Graves, Robert (1940) *Sergeant Lamb of the Ninth*
—— (1941) *Proceed Sergeant Lamb*
Grayson, H. (1975) *The Last Alderman*
Greeley, Andrew M. (1986) *God Games*
Green, Jonathon (1991) *Neologisms—New Words since 1960*
Green, Shirley (1979) *Rachman*
Greene, Graham (1932) *Stamboul Train*
—— (1934) *It's a Battlefield*
—— (1967) *May We Borrow Your Husband*
—— (1978) *The Human Factor*
Greene, G. A. (1599) *Works*
Grisham, John (1992) *The Pelican Brief*
Grose, Francis (1811 edition) *Dictionary of the Vulgar Tongue*
Guinness, Alec (1985) *Blessings in Disguise*

Hackett, John (1978) *The Third World War*
Hailey, Arthur (1973) *Wheels*
—— (1975) *The Money-Changers*
—— (1979) *Overlord*
—— (1984) *Strong Medicine*
Hall, Adam (1969) *The Ninth Directive*
—— (1979) *The Scorpion Signal*
—— (1988) *Quiller's Run*
Hallam, Reuben (1866) *Wadsley Jack*
Hamilton, Ernest (1897) *The Outlaws of the Marshes*
—— (1898) *The Mawkin of the Flow*
Hardy, Thomas (1874) *Far From the Madding Crowd*
—— (1888) *Wessex Tales*
Harland, John and Wilkinson, T. T. (1867) *Folk Lore*
Harris, Frank (1925) *My Life and Loves*
Harris, Robert (1992) *Fatherland*
Harris, Thomas (1988) *The Silence of the Lambs*
Hartley, John (1870) *Heart Broken*
Harvey, William (1628) *Anatomica de Motu Cordis etc.*
Hastings, Max (1987) *The Korean War*
Hastings, Selina (1994) *Evelyn Waugh*
Hayden, Eleanor (1902) *From a Thatched Cottage*
Heath, Robert (1650) *Clarastella: together with poems ocassionall etc.*
Hector, William (1876) *Selections from the Judicial Records of Renfrewshire*
Hemingway, Ernest (1941) *For Whom the Bell Tolls*
Henderson, George (1856) *The Popular Rhymes, Sayings and Proverbs of the County of Berwick*
Henderson, William (1879) *Notes on the Folk Lore of the Northern Counties etc.*
Herd, David (1771) *Ancient and Modern Scottish Songs*
Herr, Michael (1977) *Dispatches*
Herriot, James (1981) *The Good Lord Made Them All*
Hetrick, Robert (1826) *Poems and Songs*
Hewett, Sarah (1892) *The Peasant Speech of Devon*
Heywood, John (1546) *Works*
Hibbert, Samuel (1822) *A Description of the Shetland Islands*
Higgins, Jack (1976) *Storm Warning*
Hogg, James (1822) *Perils of Man*
—— (1866) *Tales and Sketches*
Holder, R. W. (1992) *Thinking About Management*
Holmes, Richard (1993) *Dr Johnson and Mr Savage*
Holt, Alfred (1961) *Phrase and Word Origins*
Hood, Thomas (c. 1830) *Poems*
Horne, Alastair (1969) *To Lose a Battle*
Horrocks, Brian (1960) *A Full Life*
Horsley, J. W. (1887) *Jottings from Jail*
Housman, A. E. (1896) *A Shropshire Lad*

Howard, Anthony (1977) *New Words for Old*
—— (1978) *Weasel Words*
Howat, Gerald (1979) *Who Did What*
Hudson, Bob and Pickering, Larry (1986) *First Australian Dictionary of Vulgarities and Obscenities*
Hudson, Kenneth (1977) *The Dictionary of Diseased English*
—— (1978) *The Jargon of the Professions*
Hughes, Robert (1987) *The Fatal Shore*
Hughes, Thomas (1856) *Tom Brown's Schooldays*
Hunt, Holman (1854) *letter*
Hunt, Robert (1865–96 edition) *Popular Romances of the West of England*
Hutchinson, Lucy (c. 1650) *letter*
Hynd, (1949)

Iacocca, Lee (1984) *Iacocca*
Ingelo (1830) *Reminiscences*
Inglis, James (1895) *Oor Ain Folk*
Innes, Hammond (1991) *Isvik*
Irvine, Lucy (1986) *Runaway*

James, Haddy (Surgeon) (1816) *Journal*
James, P. D. (1962) *Cover Her face*
—— (1972) *An Unsuitable Job for a Woman*
—— (1975) *The Black Tower*
—— (1980) *Innocent Blood*
—— (1986) *A Taste for Death*
Jane, Fred (1897) *The Lordship, the Passen and We*
Jennings, Gary (1965) *Personalities of Language*
Jefferies, Richard (1880) *Hodge and his Masters*
Johnson, Samuel (Dr) (1755) *A Dictionary of the English Language*
Johnston, Henry (1891) *Kilmallie*
Joliffe and Mayle (1984) *Man's Best Friend*
Jolly, Rick (1988) *Jackspeak—The Pusser's Rum Guide*
Jones, R. V. (1978) *Most Secret War*
Jonson, Ben (1598–1633) *Works* (edited by Herford and Simpson, 1925–52)
Joyce, James (1922) *Ulysses*
Kay, Valerie and Strevens, Peter (1974) *Beyond the Dictionary in English*
Kee, Robert (1993) *The Laurel and the Ivy*
Keegan, John (1989) *The Second World War*
Keith, Leslie (1896) *The Indian Uncle*
Kelly, James (1721) *A Complete Collection of Scottish Proverbs*
Keneally, Tom (1979) *Confederates*
—— (1982) *Schindler's Ark*
—— (1985) *A Family Madness*
—— (1987) *The Playmaster*
Kennedy, Patrick (1867) *The Banks of the Boro*
Kersh, Gerald (1936) *Night and the City*
Kinloch, George R. (1827) *The Ballad Book*
Kirkton, James (1817) *The Secret and True History of the Church of Scotland etc.*
Kyle, Duncan (1975) *The Semenov Impulse*
—— (1983) *The King's Commander*
—— (1988) *The Honey Ant*
Lacey, Robert (1986) *Ford*
Lauderdale, John (1796) *A Collection of Poems*
Lavine, Emanual (1930) *The Third Degree*
Lawless, Emily (1892) *Grania*
Lawrence, Karen (1990) *Springs of Living Water*
le Carré, John (1962) *A Murder of Quality*
—— (1980) *Smiley's People*
—— (1983) *Little Drummer Girl*
—— (1986) *A Perfect Spy*
—— (1989) *Russia House*
—— (1991) *The Secret Pilgrim*
Lee, John Alexander (1937) *Civilian into Soldier*
Lee, Joseph (1989) *Ireland, 1912–1985*
Lewis, Matthew (1795) *The Monk*
Lewis, Nigel (1989) *Channel Firing*
Liddle, William (1821) *Poems on Different Occasions*
Lingemann, Richard (1969) *Drugs from A to Z*
Linton, E. Lynn (1866) *Lizzie Lorton of Greyrigg*
Lochhead, Liz (1985) *True Confessions and New Clichés*
Lodge, David (1962) *Ginger You're Barmy*
—— (1975) *Changing Places*
—— (1980) *How Far Can You Go*
—— (1988) *Nice Work*
Londres, Albert (1928) *The Road to Buenos Ayres* (translated by Eric Sutton)
Longstreet, Stephen (1956) *The Real Jazz Old and New*
Lowson, Alexander (1890) *John Guidfollow*
Ludlum, Robert (1979) *The Materese Circle*
—— (1984) *The Aquitaine Progression*
Lumsden, James (1892) *Sheep-Head and Trotters*
Lyall, Gavin (1965) *Midnight Plus One*
—— (1969) *Venus with Pistol*
—— (1972) *Blame the Dead*
—— (1975) *Judas Country*
—— (1980) *The Secret Servant*
—— (1982) *The Conduct of Major Maxim*
—— (1985) *The Crocus List*
Lyly, John (1579) *Euphues, the Anatomy of Wit*
Lynn, Jonathan and Jay, Antony (1981) *Yes Minister*
—— (1986) *Yes Prime Minister*
—— (1989) *The Complete Yes Prime Minister*

Maas, Peter (1986) *Man Hunt*
McBain, Ed (1981) *Heat*
McCarthy, Mary (1963) *The Group*
—— (1967) *Vietnam*
McCrum, Robert, Cran, William and McNeil, Robert (1986) *The Story of English*
MacDonagh, Michael (1898) *Irish Life and Character*
Macdonald, Ross (1952) *The Ivory Grin*
—— (1971) *The Doomsters*
—— (1976) *The Blue Hammer*
McInerney, Jay (1992) *Brightness Falls*
Mackenzie, George Stewart, of Coul, quoted in Prebble, John (1963)
Mackie, Marlene (1983) *Gender Relations in Canada*
Maclaren, Ian (1895) *Beside the Bonnie Brier Bush*
MacManus, Seumas (1898) *The Bend of the Road*
—— (1899) *In Chimney Corners*
McNab, Andy (1993) *Bravo Two Zero*
McNair, Tom (1973) *A Guide to Hip Language and Culture*
Mactaggart, John (1824–76 edition) *Encyclopaedia*
Maidment, James (1844–5) *Spottiswoode Miscellany*
—— (1868) *A Book of Scotch Pasquils, 1568-1715*
Mailer, Norman (1965) *An American Dream*
Major, Clarence (1970) *Black Slang—A Dictionary of Afro-American Slang*
Manchester, William (1968) *The Arms of Krupp*

Mann, Mary (1902) *The Fields of Dulditch*
Manning, Olivia (1960) *The Great Fortune*
—— (1962) *The Spoilt City*
—— (1965) *Friends and Heroes*
—— (1977) *The Danger Tree*
—— (1978) *The Battle Lost and Won*
Mantle, Jonathan (1988) *In For a Penny*
Marmur, Jacland (1955) *The Kid in Command*
Marsh, Ngaio (1940) *Surfeit of Lampreys*
Marshall, William H. (1811/17/18) *Review and Abstract of the County Reports to the Board of Agriculture etc.*
Marvell, Andrew (c. 1670) *Poems*
Mason, A. E. W. (1927) *No Other Tiger*
Mason, David (1993) *Shadow over Babylon*
Mason, William S. (1815) *A Statistical Account or Parochial Survey of Ireland*
Massie, Robert (1992) *Dreadnought*
Masters, John (1976) *The Himalayan Concerto*
Mather, Joseph (1862) *Songs*
Matthew, Christopher (1978) *The Diary of a Somebody*
—— (1983) *How to Survive Middle Age*
Mayhew, Henry (1851) *London Labour and the London Poor*
—— (1861) *Mayhew's London*
—— (1862) *London's Underground*
Mayle, Peter (1993) *Hotel Pastis*
Mencken, Henry L. (1940–8) *The American Language*
Miller (1879)
Mitford, Jessica (1963) *The American Way of Death*
Mitford, Nancy (1945) *The Pursuit of Love*
—— (1949) *Love in a Cold Climate*
—— (1956) *Noblesse Oblige*
—— (1960) *Don't Tell Alfred*
Mockler, Anthony (1984) *Haile Selassie's War*
Moir, David (1828) *The Life of Mansie Wauch*
Moncrieff, William T. (1821) *Tom and Jerry, or Life in London*
Monkhouse, Bob (1993) *Crying with Laughter*
Monsarrat, Nicholas (1978) *The Master Mariner*
Moore, L. W. (1893) *His Own Story*
Morison, David (1790) *Poems*
Morley, Robert (1976) *Pass the Port*
Mort, Simon (1986) *Original Selection of New Words*
Mortimer, Geoffrey (1895) *Tales from the Western Moors*
Moss, Robert (1985) *Moscow Rules*
—— (1987) *Carnival of Space*
Moss, W. S. (1950) *Ill Met by Moonlight*
Moyes, P. (1980) *Angel Death*
Moynahan, Brian (1983) *Airport International*
Mucklebackit, Samuel (1885) *Rhymes*
Muggeridge, Malcolm (1972) *Chronicles of Wasted Time*
Muir, Frank (1990) *The Oxford Book of Humorous Prose*
Muir, George (1816) *The Clydesdale Minstrelsy*
Murdoch, Alexander (1873) *Lilts on the Doric Lyre*
—— (1895) *Scotch Readings*
Murdoch, Iris (1974) *The Sacred and Profane Love Machine*
—— (1977) *Henry and Cato*
—— (1978) *The Sea, the Sea*
—— (1980) *Nuns and Soldiers*
—— (1983) *The Philosopher's Pupil*
—— (1985) *The Good Apprentice*
Murray, C. S. (1989) *Crosstown Traffic*
Murray, D. Christie (1886) *Rainbow Gold*
—— (1890) *John Vale's Guardian*
Murray, Elisabeth (1977) *Caught in the Web of Words*

Nabokov, Vladimir (1968) *King, Queen, Knave*
Naipaul, V. S. (1964) *An Area of Darkness*
—— (1990) *India—A Million Mutinies Now*
Nares, Robert (1820) *A Glossary or Collection of Words etc.*
Neaman, Judith N. and Silver, Carole S. (1983) *Kind Words: A Thesaurus of Euphemisms*
New Larousse Encyclopaedia of Mythology (1968)
Nicholson, William (1814) *Poetical Works*
Nicholson and Burn (1777) *The History and Antiquities of the Counties of Westmoreland and Cumberland*

O'Connor, Joseph (1991) *Cowboys and Indians*
O'Donoghue, Maureen (1988) *Winner*
O'Reilly, R. (Mrs) (1880) *Sussex Stories*
Oakley, Ann (1984) *Taking it like a Woman*
Ogg, James (1873) *Willie Waly; and other Poems*
Olivier, Laurence (1982) *Confessions of an Actor*
Ollard, Richard (1974) *Pepys*
Ollivant, Alfred (1898) *Owd Bob, the Grey Dog of Kenmuir*
Onions, C. T. (1975) *The Oxford Dictionary of English Etymology*
Oxford English Dictionary (1989 edition)

Pae, David (1884) *Eustace the Outcast*
Parker, Dorothy (1944) *The Portable Dorothy Parker*
Partridge, Eric (1947) *Shakespeare's Bawdy*
—— (1959) *Origins*
—— (1969) *A Dictionary of Slang and Unconventional English*
—— (1972) *A Dictionary of Clichés*
——(1972) *A Dictionary of Historical Slang* (with Jacqueline Simpson)
—— (1973) *Usage and Abusage*
—— (1977) *A Dictionary of Catch Phrases*
Patterson, A. (1895) *Man and Nature on the Broads*
Payn, James (1878) *By Proxy*
Peacock, Edward (1870) *Ralf Skirlaugh, the Lincolnshire Farmer*
Peacock, F. M. (1890) *A Soldier and a Maid*
Pearsall, Ronald (1969) *The Worm in the Bud*
Pease, Howard (1894) *The Mark o' the Deil*
Peck, M. Scott (1987) *The Different Drum: Community-Making and Peace*
—— (1990) *A Bed by The Window: A Novel of Mystery and Redemption*
Pegge, Samuel (1803) *Anecdotes of the English Language*
Pei, Mario (1969) *Words in Sheep's Clothing*
—— (1978) *Weasel Words*
Pennecuik, Alexander (1715) *Description of Tweeddale and Poems*
Pepys, Samuel (1660–9) *Diary*
Pereira, M. (1972) *Singing Millionaire*
Perelman, S. J. (1937) *Strictly From Hunger*
Peshall (1773) *Ancient and Present State of the City of*

Oxford
Phillips, Julia (1991) *You'll Never Eat Lunch in this Town Again*
Picken, Ebenezer (1813) *Miscellaneous Poems and Songs*
Pilcher, Rosamund (1988) *The Shell Seekers*
Pincher, Chapman (1987) *The Spycatcher Affair*
'Pindar, Peter' (1816) *Works*
Pinnock, John (1895) *Tom Brown's Black Country Annual*
Playboy's Book of Limericks (1972) (edited by Clifford Crist)
Pope, Alexander (1735) *Poetical Works*
Pope-Hennessy, James (1967) *The Sins of the Fathers*
Poyer, Joe (1978) *The Contract*
Praed, Campbell (Mrs) (1890) *Romance Station*
Prebble, John (1963) *The Highland Clearances*
Price, Anthony (1970) *The Labyrinth Makers*
—— (1971) *The Alamut Ambush*
—— (1972) *Captain Butler's Wolf*
—— (1974) *Other Paths to Glory*
—— (1975) *Our Man in Camelot*
—— (1978) *The '44 Vintage*
—— (1979) *War Games*
—— (1982) *The Old Vengeful*
—— (1985) *Here Be Monsters*
—— (1987) *A New Kind of War*
Proudlock, Lewis (1896) *The Borderland Muse*
Pythiam, B. A. (1979) *A Concise Dictionary of Current English*

Quiller-Couch, Arthur (1890) *I Saw Three Ships*
—— (1891) *Naughts and Crosses*
—— (1893) *The Delectable Duchy*

Rabelais, François (1532) *Pantagruel* (in translation)
—— (1534) *Gargantua* (in translation)
Radford, Edwin and Smith, Alan (1973) *To Coin a Phrase*
Ramsay, Allan (1737) *Collection of Scots proverbs*
—— (1800 edition) *Poems*
Ramsay, E. B. (1858–61) *Reminiscences of Scottish Life and Character*
Rawson, Hugh (1981) *A Dictionary of Euphemisms and Other Doubletalk*
Ray, John (1678) *A Collection of English Proverbs*
Read, Piers Paul (1979) *A Married Man*
—— (1986) *The Free Frenchman*
Reeman, Douglas (1994) *Sunset*
Rees, Nigel (1980) *Graffiti*
Richards, David Adams (1988) *Nights Below Station Street*
Richards, Frank (1933) *Old Soldiers Never Die*
—— (1936) *Old Soldier Sahib*
Ritchie, A. I. (1883) *The Churches of St Baldred*
Robbins, Harold (1981) *Goodbye Jannette*
Roberts, Michael (1951) *The Estate of Man*
Rock, William F. (1867) *Jim an' Nell*
Rodger, Alexander (1838) *Poems and Songs*
Roethke, Theodore (1941) *'Open House' (a poem, in various collections)*
Roget's Thesaurus (1966 edition)
Ross, Alan (1956) *Noblesse Oblige*
—— (1971) *Words in Sheep's Clothing*
Russell, S. C. (c. 1900) *A Strange Voyage*
St Pierre, Paul (1983) *Smith and Other Events: Tales of Chilcotin*
Salinger, J. D. (1951) *The Catcher in the Rye*
Sale, Charles (1930) *The Specialist*
Sanders, Laurence (1970) *The Anderson Tapes*
—— (1973) *The First Deadly Sin*
—— (1977) *The Second Deadly Sin*
—— (1977) *The Tangent Objective*
—— (1979) *The Sixth Commandment*
—— (1980) *The Tenth Commandment*
—— (1980) *Caper*
—— (1981) *The Third Deadly Sin*
—— (1982) *The Case of Lucy Bending*
—— (1983) *The Seduction of Peter S.*
—— (1984) *The Passion of Molly T.*
—— (1985) *The Fourth Deadly Sin*
—— (1986) *The Eighth Commandment*
—— (1987) *The Timothy Files*
—— (1990) *Sullivan's Sting*
'Saxon' (1878) *Galloway Gossip Sixty Years Ago*
Scott, Andrew (1805) *Poems*
Scott, Paul (1968) *The Day of the Scorpion*
—— (1971) *The Towers of Silence*
—— (1973) *The Jewel in the Crown*
—— (1975) *A Division of the Spoils*
—— (1977) *Staying On*
Scott, Walter (1803) *Minstrelsy of the Scottish Border*
—— (1814) *Waverley*
—— (1815) *Guy Mannering*
—— (1816) *The Antiquary*
—— (1817) *Rob Roy*
—— (1818) *The Heart of Midlothian*
—— (1819) *The Battle of Lammermoor*
—— (1820) *The Abbot*
—— (1824) *Redgauntlet*
Service, John (1887) *The Life and Recollections of Dr Duguid of Kilwinning*
—— (1890) *The Notandums*
Seth, Vikram (1993) *A Suitable Boy*
Seymour, Gerald (1977) *Kingfisher*
—— (1980) *The Contract*
—— (1982) *Archangel*
—— (1984) *In Honour Bound*
—— (1989) *Home Run*
—— (1992) *The Journeyman Tailor*
Shakespeare, William: plays and sonnets as noted
Shankland (1980)
Sharpe, Tom (1974) *Porterhouse Blue*
—— (1975) *Blott on the Landscape*
—— (1976) *Wilt*
—— (1977) *The Great Pursuit*
—— (1978) *The Throwback*
—— (1979) *The Wilt Alternative*
—— (1982) *Vintage Stuff*
Shaw, Irwin (1946) *Short Stories; Five Decades*
Sheppard, Harvey (1970) *A Dictionary of Railway Slang*
Shipley, Joseph (1945) *A Dictionary of Word Origins*
Shirer, William (1984) *The Nightmare Years 1930–1940*
Shorter Oxford English Dictionary (1993 edition)
Sidney, Philip (1586) *Works*
Simon, Ted (1979) *Jupiter's Travels*
Simpson, John (1991) *From the House of War*

Sims (1902)
Skelton, John (1533) *Magnyfycence*
Slang Dictionary (The)—Etymological, Historical and Analytical (1874)
Slick, Samuel (1836) *Clockmaker*
Smith, Martin Cruz (1981) *Gorky Park*
Smith, Murray (1993) *The Devil's Juggler*
Smith, Tony (1986) *Family Doctor, Home Adviser*
Smith, Wilbur (1979) *Wild Justice*
Smith, William G. and Wilson, F. P. (1970) *The Oxford Dictionary of English Proverbs*
Smollett, Tobias (1748) *Roderick Random*
—— (1751) *Peregrine Pickle*
—— (1771) *Humphrey Clinker*
Sohmer, Steve (1988) *Favorite Son*
Solzhenitsyn, Alexander (1974) *The Gulag Archipelago* (translated by Thomas Whitney)
Somerville, A. E. and Ross, Martin (1894) The Real Charlotte
—— (1897) *Some Experiences of an Irish RM*
—— (1908) *Further Experiences of an Irish RM*
Spears, Richard (1981) *Slang and Euphemism*
Spence, Charles (1898) *From the Braes of the Carse*
Stegner, Wallace (1940) *The Women on the Wall*
Steinbeck, John (1961) *The Winter of our Discontent*
Stewart, George E. (1892) *Shetland Fireside Tales*
Stoker, Bram (1895) *The Watter's Mou'*
Stowe (1633)
Strachey, Lytton (1918) *Eminent Victorians*
Strain, E. H. (1900) *Elmslie's Drag-net*
Styron, William (1976) *Sophie's Choice*
Sullivan, Frank (1953) *The Night the Old Nostalgia Burned Down*
Sutcliffe, Halliwell (1899) *By Moor and Fell*
—— (1900) *Shameless Wayne*
—— (1901) *Mistress Barbara Cunliffe*
Sutherland, James (1975) *The Oxford Book of Literary Anecdotes*
Sutherland, William (1821) *Poems and Songs*
Swift, Jonathan (1723–38) *Works*

Tarras, William (1804) *Poems*
Taylor, Mary (1890) *Miss Miles*
Taylor, William (1787) *Scots Poems*
Teisser du Croix, Janet (1962) *Divided Loyalties*
Tennyson, Alfred (1859) *The Idylls of the King*
—— (1885) *The Spinster's Sweet-Arts*
Tester, William (1865) *Poems*
Thackeray, William (1837–55) *Works*
Theroux, Paul (1971) *Jungle Lovers*
—— (1973) *Saint Jack*
—— (1974) *The Black House*
—— (1975) *The Great Railway Bazaar*
—— (1976) *The Family Arsenal*
—— (1977) *The Consul's File*
—— (1978) *Picture Palace*
—— (1979) *The Old Patagonian Express*
—— (1980) *World's End and Other Stories*
—— (1981) *The Mosquito Coast*
—— (1982) *The London Embassy*
—— (1983) *The Kingdom by the Sea*
—— (1988) *Riding the Red Rooster*
—— (1989) *My Secret History*
—— (1990) *Chicago Loop*
—— (1992) *The Happy Isles of Oceania*
—— (1993) *Millroy the Magician*
Thom, Robert (1878) *The Courtship and Wedding of Jock o' the Knowe*
Thomas, Clive (1993) *Playing with Cobras*
Thomas, Hugh (1961) *The Spanish Civil War*
—— (1986) *Armed Truce*
—— (1993) *The Conquest of Mexico*
Thomas, Leslie (1977) *Bare Nell*
—— (1978) *Ormerod's Landing*
—— (1979) *That Old Gang of Mine*
—— (1981) *The Magic Army*
—— (1986) *The Adventures of Goodnight and Loving*
—— (1989) *Orders for New York*
Thomas, Michael (1980) *Green Monday*
—— (1982) *Someone Else's Money*
—— (1985) *Hard Money*
—— (1987) *The Ropespinner Conspiracy*
Thomson, David (1881) *Musings among the Heather*
Thwaite, Anthony (1992) *Selected Letters of Philip Larkin 1940–1985*
Torriano, Giovanni (1642) *A Common Place of Italian Proverbs And Proverbial Phrases*
Townsend, Sue (1982) *The Secret Diary of Adrian Mole Aged 13 ¾*
—— (1984) *The Growing Pains of Adrian Mole*
Train, John (1983) *Preserving Capital and Making it Grow*
'Treddlehoyle' (Rogers, Charles) (1846/92/93) *Bairnsla Annals* (reprinted in Leeds *Mercury*)
'Trevanian' (1972) *The Eiger Sanction*
—— (1973) *The Loo Sanction*
Trevor-Roper, Hugh (1977) *Introduction to Goebbels' Diaries*
Trollope, Anthony (1885) *The Land-Leaguers*
Trollope, Joanna (1992) *The Man and the Girls*
Tulloch, Sara (1991) *The Oxford Dictionary of New Words*
Turner, E. S. (1952) *The Shocking History of Advertising*
Turner, Graeme (1968) *Our Secret Economy*
Turner, Graham and Pearson, John (1965) *The Persuasion Industry*
Turow, Scott (1987) *Presumed Innocent*
—— (1990) *The Burden of Proof*
Tweeddale, John (1896) *Moff*
Tyrrell, Syd (1973) *A Countryman's Tale*

Upfield, A. (1932) *Royal Abduction*
Ustinov, Peter (1966) *The Frontiers of the Sea*
—— (1971) *Krumnagel*

Vachell, Horace (1934) *The Disappearance of Martha Panny*
van Druten, J. (1954) *I am a Camera*
van Lustbaden, Eric (1983) *Black Heart*
Vedder, David (1832) *Orcadian Sketches*
Verney (Lady) (1870) *Lettice Lisle*

Wainright, J. (1979) *Duty Elsewhere*
Wallace, James (1693) *A Description of the Isles of Orkney*
Wambaugh, Joseph (1972) *The Blue Knight*
—— (1975) *The Choirboys*
—— (1981) *The Glitter Dome*
—— (1983) *The Delta Star*
Ward, Geoffrey C. (1990) *The Civil War*

Ward, Mary (Mrs Humphrey) (1895) *The Story of Bessie Costrell*
Ward, T. (1708) *Some Queries to the Protestants etc.*
Wardrop, Alex (1881) *Johnnie Mathison's Courtship and Marriage*
Waugh, Auberon (from *Private Eye* and *Daily Telegraph* diaries as noted)
Waugh, Evelyn (1930) *Labels*
—— (1932) *Remote People*
—— (1933) *Scoop*
—— (1955) *Officers and Gentlemen*
—— (1956) *Noblesse Oblige*
Webster, John (1623) *The Duchess of Malfi*
Webster, Noah (1977 Merriam edition) *New Collegiate Dictionary*
Wentworth, Harold and Flexner, Stuart B. (1975 edition) *Dictionary of American Slang*
West, Morris (1979) *Proteus*
West, Nigel (1982) *M.I.5 1945–72*
Westall, William (1885) *The Old Factory*
Weverka, Robert (1973) *The Sting*
Wheeler, Ann (1790) *The Westmoreland Dialect*
Whitehead, Anthony (1896) *Legends of Penrith*
Whitehead, S. R. (1876) *Daft Davie*
Willock, A. Dewar (1886) *Rosetty Ends*
Wilson, Harry L. (1915) *Ruggles of Red Gap*
Wilson, John Mackay (1836) *Tales*
Wilson, Thomas (1843) *The Pitman's Pay*
Wilson (1603) *The Bachelor's Banquet*
Wodehouse, P. G. (1922) *Girl on Boat*
—— (1930) *Very Good, Jeeves!*
—— (1934) *Right Ho, Jeeves*
—— Letter (1930) in Donaldson (1990)
Wodrow, Robert (1721) *The History of the Sufferings of the Church of Scotland etc.*
Wood, Frederick (1962) *Current English Usage*
—— (1979) *Dictionary of English Colloquial Idioms* (with Robert Hill)
Woodward, Bob (1987) *Veil*
Wouk, Herman (1951) *The Caine Mutiny*
Wright, Joseph (1897) *Scenes of Scottish Life*
—— (1898–1905) *The English Dialect Dictionary*
Young, Edward (1721) *The Revenge*
Yule, Henry and Burnell, A. C. (1886) *Hobson-Jobson*
'Zack' (Keats, Gwendoline) (1901) *Tales of Dustable Wear*

A Dictionary of **Euphemisms**

A amphetamine used illegally.

A is also used by addicts to refer to LSD (see ACID). The punning *A bomb* is a combination of illicit narcotics, from the effect on the user. A *B bomb* is a benzedrine inhaler used illegally.

AC/DC indulging in both homosexual and heterosexual practices.

The reference is to alternating and direct electric current and the suggestion is that these are interchangeable:

> Young attractive housewife, AC/DC would like to meet married AC/DC people to join well-endowed husband for threesomes or moresomes. (*Daily Telegraph*, May 1980)

Occasionally spelt phonetically as *acey-deecy*:

> So, he was acey-deecy... Lots of old altar boys play hide-the-weenie when they shouldn't. (Sohmer, 1988)

AMW *American* a prostitute.

An abbreviation of *actress-model-whatever*, from the less disreputable avocations which some prostitutes profess:

> ... the Fleiss scandal has brought de luxe hooking out of the closet and encouraged more AMWs (actress-model-whatever) to go on the game. (*Daily Telegraph*, 6 December 1994, of a prosecution for procuring)

See also ACTRESS and MODEL.

à trois in a sexual relationship involving three people.

From *ménage à trois*. Often of a married couple and the outside sexual partner of one of them:

> I've been living *à trois* with a married couple. Do I shock you? (Murdoch, 1977)

Aaron's rod the penis.

Formerly an ornamental rod with a serpent entwined around it and now a popular name for several tall plants with flowering stems, especially the mullein, *Verbascum thapsus*. Less often as *Aaronic staff* or *baston*:

> ... the lips that were agape in a wordless prayer of gratitude now closed about the head and flower of the boy's Aaronic baston. (Burgess, 1980; a *baston* is a staff or cudgel)

According to the Bible, Aaron, the brother of Moses, wielded a rod which turned into a serpent (Exodus 2:10). On another occasion a rod bearing his name miraculously bore blossom and almonds (Numbers 17:8).

abandoned *obsolete* working as a prostitute.

Literally, forsaken, and epitomizing the formal 19th-century attitude to prostitution:

> The foolish idea... that once abandoned she must always be profligate. (Mayhew, 1862, of a prostitute)

The punning *abandoned habits* were the flashy clothes prostitutes wore when riding in London's Hyde Park.

abbess *obsolete* a female bawd.

Partly humorous and partly based on the supposition that nunneries were not solely occupied by chaste females:

> ... who should come in but the venerable mother Abbess herself. (Cleland, 1749, of a bawd)

abdomen a man's genitalia.

Literally, the lower cavity of the trunk. A convenient evasion for sports commentators when a player has been incapacitated by a blow to his testicles. Also as *lower abdomen*. An *abdominal protector* in genteel language is a box or shield worn by cricketers and others to protect the genitalia.

aberration an act of sex which is not heterosexual.

Another kind of deviation from the norm:

> There's a great deal of tolerance for, well, aberrations. (Burgess, 1980)

ableism insensitivity towards lame or injured people.

Coined in a world where all others are *temporarily abled*, on the basis that their turn will come:

> Likewise 'ableism' or 'oppression of the differently abled ('disabled' is discriminatory) by the temporarily abled', is firmly proscribed. (*Daily Telegraph*, 23 February 1991, of the Smith College, Massachusetts, list of POLITICALLY CORRECT attitudes)

ablutions a lavatory.

Originally, the religious rite of washing, whence washing the body on any occasion and the place in which you washed. An army usage:

> We were told to choose a bed site... shown where the Ablutions were. (Bogarde, 1978, of being drafted into the army—here and elsewhere note the capital letter for a word used euphemistically)

abnormal homosexual.

To be heterosexual is to be normal:

> ... lived an institutional life with other men in uniform without ever seriously arousing the suspicion that he was what is called abnormal. (P. Scott, 1975)

Of either sex. *Abnormality* is homosexuality:

> The fact that he revealed a hatred of 'abnormality' was only to be expected. 'What a filthy Lesbian trick,' he said. (McCarthy, 1963)

aboard[1] *?obsolete* drunk.

The imagery is from loading a ship. In dialect use, *something aboard*:

He's sum'uts aboard today; he could nobud just sit e' his gig. (EDD)

aboard[2] copulating with.

Male usage, from getting above the female:

I was aboard Lily Langtry long before he was. (Fraser, 1970—'he' was King Edward VII)

abode of love a brothel.

Where *love* imports 'copulation':

These abodes of love seen from the other side are strangely transfigured. All is order, cleanliness and respectability. (Londres, 1928, in translation)

above ground respectable or legitimate.

Usually of activities or behaviour after a period of shameful or illicit conduct:

She starred in dozens of blue movies before coming above ground. (Deighton, 1972)

The converse *underground* has implications of illegality or impropriety in many phrases.

Above, like *no*, is a prefix in many euphemistic expressions; thus an *above critical* nuclear reactor is very unsafe.

absent parents fathers who do not live with their infant children.

Not fathers and mothers who are temporarily away from home, but specifically divorced, single, or separated men who have fathered children for whose upbringing society rather than the progenitor is paying, in addition to the normal state provision of free schooling, free medical care, weekly cash allowance, etc.:

We must be careful that we do not empty our surgeries of angry absent parents only to fill them with angry lone parents instead. (British Social Security Secretary quoted in *Daily Telegraph*, 5 July 1994—the government had established an agency to try to reduce the cost to the state by seeking greater contributions from the fathers)

See also LONE PARENT.

abuse[1] to use for taboo sexual purposes.

Literally, to misuse or maltreat in any way. To *abuse* a woman or child is to engage in an unwanted or illegal sexual activity; *abuse* is also found as a noun in the same sense:

If Mayhew's figures for the abuse of children are suspect, so are his figures for rape. (Pearsall, 1969)

Carnal abuse is explicit:

During this period, Mayhew maintains, the cases for 'carnally abusing' girls between the ages of ten and twelve were a mere fifty-six. (ibid.)

To *abuse a bed* was not to leap about on it but to copulate extramaritally:

See the hell of having a false woman! My bed shall be abused. (Shakespeare, *Merry Wives of Windsor*)

To *abuse yourself* is to masturbate (of either sex); see also SELF-ABUSE. The rare *mutual abuse* is masturbation by each other of any combination of males and females.

abuse[2] the ingestion of any substance which is the subject of a taboo.

Usually illegal narcotics, solvents, etc. and also alcohol when taken excessively. The use *tout court* means illegal drugs:

...both now dead...Anthony from drink and 'abuse' in Dublin. (A. Clark, 1993)

Abyssinian medal *obsolete* an undone and visible trouser fly-button.

Various *medal* phrases found favour after successive military campaigns until the zip replaced buttons. The Abyssinian campaign lasted from 1893 to 1896, to be followed by a similar adventure in Egypt with its attendant medals, real and euphemistic. See also the general MEDAL SHOWING (A).

academic dismissal expulsion from college.

Not just the end of classes for the day:

No student ever gets expelled any more, though he may suffer 'academic dismissal'. (Jennings, 1965)

academy *obsolete* a brothel.

Literally, a school, from the original garden where Plato taught:

...the show of a shop was shut, the academy open'd; the mask of mock-modesty was completely taken off. (Cleland, 1749)

In the same sense prostitutes were styled *academicians*.

Acapulco gold marijuana.

The plant is grown near the Mexican city. *Gold* puns on the colour, the quality, and the cost:

Charters produced very strong, very good Acapulco Gold. He had two joints already rolled. (Collins, 1981)

This is an example of many similar evasions using geographical names.

access course a class to enable those lacking suitable qualifications to enter higher education.

Nothing so ordinary as an entrance examination or test which candidates may pass or fail on merit. The device may assist those from a MINORITY GROUP without attracting the attention of those who oppose *race-norming* (see at RACIAL) or facilitate the recruitment of the *mature student* (see at MATURE[1]), allowing the institution concerned to make up its numbers without appearing to lower standards of entry. It also provides us with the perfect euphemism, combining hypocrisy with prudery, prejudice, evasion, and deceit.

accident[1] involuntary urination or defecation.

Literally, anything which happens, whence in common use anything undesirable:

> I've never punished him, the way our mothers and nurses did, when he has an 'accident'. (McCarthy, 1963—here and elsewhere note the use of quotation marks for a word or phrase used euphemistically)

Mainly used of young children.

accident[2] an unwanted pregnancy.

To treat impregnation as though it were an unforeseen happening may seem unduly innocent or optimistic:

> I have the means to prevent any... accident. I promise I'll be very careful. (Styron, 1976)

The child born under these circumstances may also be called an *accident*.

accident[3] a collision which damages a vehicle or causes injury.

The usage seeks to transfer the blame for the mishap from ourselves, especially when driving:

> Threequarters of the road accidents are not accidents at all... Most of the deaths and injuries are due to 'carelessness, speeding, improper overtaking, not obeying signs, or effects of drink or drugs'. (*Daily Telegraph*, May 1981—the list is not exhaustive)

accommodate *obsolete* to copulate with extramaritally.

The female *accommodates* the male, usually for a consideration:

> ... nothin' but the best gentlemen to accommodate. (Fraser, 1982—a 19th-century bawd was speaking of her classy clientele)

Whence the *accommodation house* or 19th-century brothel:

> ... take him along to one of the accommodation houses in Haymarket, and get him paired off with a whore. (Fraser, 1973, writing in 19th-century style)

See also *house of accommodation* at HOUSE[1].

accommodate yourself to urinate.

Not seeking lodging as well:

> ... our guide stopped on the path and accommodated himself in a way that made me think that his reverence for the spot was far from fanatical. (E. Waugh, 1932, on being guided to a holy place)

accost to ask a stranger to copulate with you for payment.

The original meaning of lying alongside re-emerged in late 19th-century legal jargon, used of an approach to a potential client by a prostitute. It is now standard English of both prostitution and begging.

accouchement the period of childbirth.

What was a euphemism in French becomes doubly so in standard English use:

> Queen Victoria had taken a personal interest in the Empress's accouchement and had sent ... one of her ladies-in-waiting, to be present at the birth. (W. H. C. Smith, 1991)

See also *lying-in* at LIE IN.

account for to kill legally.

Of animals by hunters and humans by soldiers. It might imply a reckoning of the numbers slain but you can *account for* a single victim. Sir Walter Scott used *go on the account* for engaging in piracy.

ace *American* to kill.

From taking a trick at cards, although *trump* might seem more appropriate:

> The gaunt man, his hands enclosed in blood-covered surgeon's plastic gloves, looked up at him. 'Somebody's aced the lady.' (Diehl, 1978)

acid lysergic acid diethylamide.

Better known as *LSD* and only to chemists as $C_{20}H_{25}N_3O$. To *drop acid* is illegally to ingest LSD:

> ... he was dropping acid and bombed out of his gourd. (Sanders, 1977)

An *acid-freak* is someone addicted to LSD. *Acid Fascism* was the violence induced by Charles Manson, the Californian cult leader, and others while under its influence.

acorn academy *American* an institution for lunatics.

Where you might expect to find a NUT:

> 'Your Honor, were these the acts of a sane man?'—and Dan would be hidden away in an acorn academy for a period of years. (Sanders, 1973)

acorns testicles.

A less common American variant of NUTS:

> ... shrieked as the spray hit him in the acorns. (Wambaugh, 1975)

acquire to steal.

Literally, to obtain possession of legally. Partridge (DSUE) suggests this was originally British army use. Whence *acquisition*, obtaining by stealing or subterfuge:

> It said that Lafarge was 'at present furthering arrangements for the acquisition of one hundred Slingshots'. (Hall, 1988—he was trying to steal them)

See also GAIN and WIN[1].

act (the) copulation.

From any of the phrases which follow, but occasionally also used alone:

> My prepuce contracted so that the act would have been difficult. (Harris, 1925)

Act of shame referred to copulation outside marriage:

...she with Cassio hath the act of shame
A thousand times committed.
(Shakespeare, *Othello*)

Specifically, *act of generation*:

The embrace of the sexes in the act of generation (EDD)

act of intercourse:

An act of intercourse took place, in the course of which both partners achieved climax (Amis, 1978)

act of love:

It was the time after the act of love (West, 1979)

and *the sexual act*:

The sexual act is fully covered, but not in these pages. (Longstreet, 1956)

However, *a sexual act* implies something short of copulation.

act like a husband to copulate with a female to whom you are not married.

But not of a transaction with a prostitute:

Jessie confessed that her sister accused her of letting me 'act like a husband'. She must have seen a stain on my chemise. (Harris, 1925)

Actaeon *literary* one who cuckolds another.

In the legend Actaeon was no more than a casual observer of Artemis's nakedness, and she had no husband to take offence. Nevertheless she turned him into a stag and set his own pack on him:

Divulge Page himself for a secure and wilful Actaeon. (Shakespeare, *Merry Wives of Windsor*)

action[1] vice or illegal activity, or its proceeds.

Usually illegal gambling, narcotics, or prostitution:

...one waits, while Federal authorities, mayors, and the Mafia decide... how much of the action they want. (Allbeury, 1976)

A *piece* or *slice of the action* may also be an unmerited share in a dubious enterprise started by someone else:

He has claimed a piece of the action in the video production of operas at Covent Garden. (*Private Eye*, May 1981)

action[2] the brutal harassment of supposed opponents.

The *Aktion* of the Nazis:

Schindler did not dare believe that this red child had survived the Aktion process. (Keneally, 1982)

action[3] **(the)** activity which offers the prospect of casual copulation.

The ambience, rather than the proceeds of criminal activity as in ACTION[1]:

Then he stared around to check the action. (Sanders, 1982—a man had gone to a bar to pick up a woman)

active[1] not crippled by illness or age.

Denoting those American geriatrics who have retained a measure of mobility:

Active Adult Golf Community. (Advertisement at Gainesville, Florida, November 1987, for houses close to a golf course)

active[2] engaging regularly in sexual activity.

Perhaps only of homosexuals:

They say Willie Maugham had them, too, and he was still active, if you know what I mean, the day he died. (B. Forbes, 1972, writing of the use of 'youth pills')

active (air) defence see DEFENCE.

activist a political zealot.

No longer merely a supporter of the philosophical concept of activism:

On the few occasions when Chinese people supposedly demonstrated outside foreign embassies, activists had always been there among them to direct everything. (Cheng, 1984)

A western *activist* is often prepared to break the law for political purposes.

actress a prostitute.

Until a liberating decree of Charles II female roles on stage were played by males. Therefore until the late 19th century acting was not considered a respectable profession for a woman:

The actress and the singer were considered nothing much more than prostitutes with a sideline. (Longstreet, 1956, of New Orleans prior to 1917)

The connection between the stage and the couch persists:

Miss Keeler, 20, a freelance model, was visiting Miss Marilyn Rice-Davies, an actress. (*Daily Telegraph*, December 1962)

acute environmental reaction *American* an inability to continue fighting.

Vietnam jargon for the nervous illness induced by the shocks of war:

Most Americans would rather be told that their son is undergoing acute environmental reaction than to hear that he is suffering from shell shock. (Herr, 1977)

Adam a public lavatory for males.

Found in restaurants etc., with EVE as its counterpart.

Adam's arsenal the male genitalia.

Viewed sexually:

It wasn't just that she was unusually partial to Adam's arsenal. (Fraser, 1971, of a female who liked copulating)

Of the same tendency is *Eve's custom-house*, the vagina, where Adam was supposed to have made his first entry (Grose).

adapt to dye.

Of a woman's hair:

She 'mutates' or 'adapts' or 'colour-corrects' her hair. (Jennings, 1965)

additional means illegal drugs taken for body-building purposes.

One of the methods used under the former Communist regime in East Germany and elsewhere to achieve athletic success:

What is certain is that a large number of GDR sportsmen used 'additional means'. (*Sunday Telegraph*, 27 February 1994)

adjourn to urinate.

From the formal suspension of a meeting. Usually in the form *May we adjourn?* meaning 'I want to urinate'; or *I suggest we have a short adjournment*—this gives everyone a chance to urinate. See also STRETCH YOUR LEGS.

adjustment[1] an increase.

Commercial jargon used when prices go up. When they go down, the reduction needs no glossing over:

Price adjustment adds £5m to Carsington bill. (*Waterbulletin*, August 1983)

But see CURRENCY ADJUSTMENT.

adjustment[2] the concealment of an illegality.

In particular, the avoidance of justice through bribery or influence:

They caught him molesting a child in a public school in Queens. The desk sergeant had enough sense not to book him. The final adjustment cost about eighteen thousand dollars. (Condon, 1966)

adjustment[3] the cure of the mad or the punishment of the prisoner.

To correct the deviation from what is accepted as the norm:

Lucy is a very disturbed child, and a long way from adjustment. (Sanders, 1982)

An *adjustment centre* in an American prison is a cell for solitary confinement.

adjustment[4] a military defeat.

The front, defensive position, or line is *adjusted* under the compulsion of the enemy.

adjustment[5] a theft.

Military usage, when you *adjust* the rights of ownership:

No sirree, captain, there will be no theft. There will be adjustments. (Keneally, 1979—the conscripts then had their clothing taken from them)

adjustment[6] the subjective alteration of published accounts.

Usually to record an improvement on what actually transpired but sometimes, in a private company, to conceal profit and thereby reduce tax:

The purpose of the 'adjustments' was to put the bank in the best possible light when the year-end figures ultimately appeared in the annual report. (Erdman, 1986)

With most trading corporations the only verity is cash, which, short of detectable fraud, can seldom be misstated; this does not prevent inventories being commonly manipulated. Banks, whose inventory is cash, are allowed to play the game by special rules.

admirer *?obsolete* a man with whom a woman regularly copulates outside marriage.

Jane Austen uses *admirer* and *admire* of someone who has no more than proper heterosexual desire for another, but by the mid 19th century the euphemistic use had developed:

...met her admirer at a house in Bolton Row that she was in the habit of frequenting. (Mayhew, 1862)

Adriatic tummy diarrhoea.

Caught by foreign tourists in Italy and the Balkans:

In the end a bout of Adriatic tummy had persuaded them to cut their losses and fly home a week early. (Sharpe, 1982)

adult[1] pornographic.

Things so described are certainly unsuitable for children. An *adult movie* is a pornographic film; an *adult novelty* is a mechanism supposed to enhance sexual pleasure; an *adult book store* is a shop which sells explicit pornography:

...nothing but taverns, junkyards, and adult book stores. (Sanders, 1980)

It is difficult to believe that these are the normal tastes of the fully grown human.

adult[2] *American* excluding children.

As in the *adult trailer park*, which is not for grown-up trailers, nor a trailer park used for pornographic purposes, but one which bans residents with children.

adult[3] adulterous.

Or of regular copulation outside marriage, usually in the phrase *adult relationship*:

The Duchess has never made any secret of her adult relationships in the years before she was married. She had affairs with... (*Daily*

Telegraph, 14 December 1994, of a daughter-in-law of the British queen)

adventure[1] a war.

Originally, a chance happening. Used usually of a conflict in which the aggressor sees only easy gains:

Stalin will [not] allow himself to be dragged into the Pacific adventure. (Goebbels, 1945, in translation)

adventure[2] an act of copulation with someone other than your usual partner.

Again from the original sense, a chance or exciting event:

I cannot have an adventure with Martin. He would boast of me. (Theroux, 1980)

Adventure is also used of a series of such acts. An *adventuress* is a female who regularly so copulates:

... she was also an adventurer, in the precise sense of the word—one who has adventures, as opposed to an adventuress... one who has lovers. (Blanch, 1954)

adviser the representative of an imperial power in a client state.

Often put there in an attempt by an unpopular government to stay in power:

The Spanish Communist leaders moved out in the wake of their Russian 'advisers'. (Boyle, 1979)

aerated drunk.

Literally, of a liquid, charged with gas rather than, of a body, charged with liquid:

Now they know Master Frank; they know he's apt to get a bit aerated (or merry as other people might say). (Tyrrell, 1973)

Aerated may also mean 'angry' or 'agitated'.

aesthete a male homosexual.

Literally, one who affects a higher appreciation of beauty than others:

... aesthetes—you know—those awful effeminate creatures—pansies. (N. Mitford, 1949)

Whence *aestheticism*, male homosexuality:

He had been at the House, but remarked with a shade of regret that he had not found any aestheticism in his day. (E. Waugh, 1930, of a homosexual)

aesthetic procedure cosmetic surgery.

Not the sexual practices of a homosexual but an operation to enhance beauty:

They were concerned that my teeth never showed, even when I smiled, but they said the cure was simple. They had what they called an aesthetic procedure. (Iacocca, 1984)

affair a relationship which involves extramarital copulation.

Originally of the wife so involved but now of either sex:

... having a vigorous and even dangerous wife, and an affair problem. (Bradbury, 1975)

In the same sense further disguised as the French *affaire*:

He comes only to see the singer Floriana. He's her latest *affaire*. (Manning, 1960)

Some homosexual use, as:

His affairs with men had been few. (P. Scott, 1971)

This should not be confused with the *man of affairs*, a businessman. For Fanny Hill an *affair* was the penis:

... draws out his affair, so shrunk and diminish'd. (Cleland, 1749)

affinity *American, obsolete* a sexual mistress.

Literally, a natural liking for anything, whence a relationship which is like that of marriage:

Q: Why is a Model T like an affinity? (*Affinity* was the vogue euphemism for *mistress* in the 1920s) A: Because you hate to be seen on the streets with one. (Lacey, 1986)

affirmative action preferential treatment for particular classes of people when making appointments.

Originally, in America, denoting attempts to promote less qualified black people:

And of course, there's Affirmative Action. Apparently there aren't too many black or Hispanic Mesterwomen... How many spics and jigaboos you got working here, anyway? (M. Thomas, 1982, highlighting the prejudice which the practice seeks to overcome)

Now also used of attempts to give such preference to females, homosexuals, and lame people.

afflicted[1] subject to physical or mental abnormality.

Not just labouring under the effects of a temporary disease or ailment:

It's wrong to laugh at the afflicted. (Catchphrase of the English comedian Frankie Howerd)

In obsolete British use *afflictions* were mourning clothes.

afflicted[2] *?obsolete* drunk.

On a single occasion and not diseased through constant excess.

African-American a black person.

Another twist in the path of evasion:

Black people *may* be black, but many now prefer 'African American' or 'people of colour' —though *never* 'coloured people'. (*Daily Telegraph*, 23 February 1991)

There is also said to be a phenomenon in the United States known as an *African-American worldview*, in which protection of the interests of the black minority plays a prominent part.

after lusting for.

Of either sex, homosexually or heterosexually, from the pursuit:

She's after you, or was. Probably moved on to somebody else by now. (Amis, 1978)

after part the buttocks.

A rare alternative for BEHIND.

afterlife death.

Used especially by Quakers, spiritualists, and other bodies who have confidence that death is not the end:

'It is the smell of afterlife.' 'It smells more like that of afterdeath,' said Jessica. (Sharpe, 1978)

afternoon man a debauchee.

He is supposed not to get up in the morning:

They are a company of giddy-heads, afternoon men. (Burton, c. 1621)

Probably obsolete despite its use by Anthony Powell in the title of his 1931 novel.

aftershave a perfume used by males.

The ostensible justification was the alleviation of smarting following a shave with a blunt razor:

His sweet-whiskey fragrance of after-shave lotion stung my eyes. (Theroux, 1982)

I suppose the macho brand names relieve the doubts of those who affect these cosmetics.

afterthought a child born within wedlock following an unplanned conception.

Of the processes associated with the event, thought is usually the least prominent:

Being the youngest in the family—what is commonly called an 'afterthought'—she was also a little spoilt. (Read, 1986)

ageful old or geriatric.

Coined by the POLITICALLY CORRECT, for whom growing old is taboo and *ageism* is noticing that the process has occurred. In legal jargon to *be of full age* is to be an adult.

agent a participant in taboo employment or activities.

In espionage, a spy, and specifically a *secret agent*; in buggery, the donor—the recipient is the *patient*; in warfare, ordnance (a *chemical warfare agent* being a noxious poison, as for example the notorious *agent orange* of Vietnam). We also use *agent* to refer to our jobs, to enhance our standing. Thus the British *estate agent* is in law the agent of neither the buyer nor the seller. There is an infinite variety of American *agents*, often no more than junior employees.

aid[1] military repression.

From the phrase *aid to the civil power*, a common usage in the British Empire, when individual Britons commanded quite small native police forces with responsibility over large and often hostile populations:

Such a request for aid was, in a sense, really only a call to stand by. (P. Scott, 1973, of India—when the military became actively involved in riot control, you issued live rounds only to a sniper, who fired at the ringleaders)

aid[2] a gift from a rich country to a poor country.

Seldom made with as much altruism as the donor would have you believe:

As Lord Bauer taught us, aid is the transfer of money from poor people in rich countries to rich people in poor countries. (Christopher Fildes in *Daily Telegraph*, 11 February 1995)

Tied aid means that the donor is arranging the credits or spending cash to assist its exporters.

air (the) *mainly American* peremptory dismissal from employment or courtship.

Both the giver and the receiver use the term:

If Victoria wants to give Jamie the air, it's no business of ours. (Deighton, 1982, of a broken engagement)

Perhaps invoking the image of ejection from a building in which you were working.

air support attack on the enemy from aircraft.

Military jargon for the dropping of bombs to help your own troops. The usage is so common that we seldom think of the logical meaning of the phrase, including the phenomenon whereby a laminar flow of air maintains an aircraft in flight.

Ajax see JAKES.

alcohol an intoxicant.

The standard English use is shortened from the phrase *alcohol of wine*, from the meaning 'a condensed spirit', which in turn was derived from kohl, 'a fine powder produced by grinding or esp. by sublimation' (SOED).

Alderman Lushington see LUSH.

alienate to pilfer or steal.

I suspect this comes from the meaning 'to make less close or affectionate' rather than from the legal jargon, 'to transfer ownership':

You can 'alienate' as much pineboard as that? (Keneally, 1982, of stealing from a pile of lumber)

all-nighter a contract with a prostitute to stay with her all night.

Prostitutes' jargon of obvious derivation:

The price of a short-time with massage stayed the same, and an all-nighter cost only an extra three-fifty. (Theroux, 1973)

In obsolete British use an *all-night man* dug up recently buried corpses to sell for medical dissection. Wide belief in the resurrection of the dead led to an understandable reluctance to donate your body for medical purposes in case you might be faced in due course with a piecemeal return to earth.

all right prepared to engage in casual copulation.

Of a female, often described by males as *a bit of all right* despite her manifest moral turpitude. In modern use, it generally refers only to sexual attractiveness.

all-rounder a person of both homosexual and heterosexual tastes.

The reference is to sport, where it denotes ability in various aspects of a game, or in many games:

> She was a bit of an all-rounder. Both sexes, general fun and games. (Davidson, 1978)

all the way (of sexual activity) with full penetration.

Usually teenage use, to distinguish from the intermediate stages of caressing:

> 'Have you had sex together?' He blushed. 'Well, ah, not exactly. I mean, we've done... things. But not, you know all the way'. (Sanders, 1981, of a man's relationship with a schoolgirl)

In rare adult use, the phrase may mean 'without contraception':

> I *want* you, Nimrod. *All the way*. Now. (Hailey, 1979, said by a wife to her impotent husband's friend)

all up with about to die.

> It's all up with him, poor lad... His bowels is mortified. (Fraser, 1971)

From the slang *all up*, ended.

alley apple a horse turd dropped on the highway.

A perhaps obsolete American version of HORSE APPLES.

alley cat a prostitute.

Both tend to frequent narrow lanes:

> These alley cats pluck at your sleeve as you pick your way along the steep cobbled footpath. (Theroux, 1975)

To *alley-cat* means 'to copulate promiscuously with many partners':

> ... couldn't stand the thought of the guy alley-catting around. (Sanders, 1977)

alleyed dead.

First World War army usage. Probably from *tossing in the alley*, to acknowledge defeat in the game of marbles, but First World War soldiers did wonderful things with French verbs, including *aller*, to go.

along the passage to the lavatory.

The place to which a guest may be asked if he wishes to go, even though it could as well apply to the kitchen or the fuel store in most houses.

altar *American* a lavatory.

The kind of place you may claim to visit, from time to time, although THRONE may share the imagery, especially in the version *altar room*.

alternative different from existing social arrangement or convention.

The use implies that the methods proposed can lead to a better life without corresponding hardship; as by wind farms which do not involve unsightly towers and persistent noise replacing normal methods of electrical production; homoeopathy replacing tested drugs; etc.:

> Eva Wilt's... Alternative Medicine alternated with Alternative Gardening and Alternative Nutrition and even various Alternative Religions. (Sharpe, 1979)

Politically it may imply the acceptance of violence to achieve change:

> I'm into... Marxist aesthetics. I'm interested in alternative education. (Bradbury, 1976)

Militarily it may mean being opposed to nuclear weapons and countering your enemy's arsenal of hydrogen bombs with gunpowder:

> ... an 'alternative defence workshop' led by Mrs Joan Ruddock, CND Chairman. (*Daily Telegraph*, November 1983)

Alternative sexuality is homosexuality:

> Homosexuality, with the inevitable personal disorientations it generates, was shrugged off as 'alternative sexuality'. (*Daily Telegraph*, November 1979)

And a homosexual may be said to have an *alternative proclivity*:

> His relations with the women he photographed appear to have remained professional and friendly and—even though he never married—scandal never fastened on an alternative proclivity. (*Daily Telegraph*, August 1990, in the obituary of a fashion photographer)

amateur a female who copulates promiscuously without payment.

Literally, a person who loves doing something, whence a performer who does it without payment:

> ... stark except for her riding boots. That took me aback, for it ain't usual among amateurs. (Fraser, 1971)

In the 19th century it meant a prostitute who had other employment:

> ... working before losing their virtue, at some trade or other... called the 'amateurs' to contra-distinguish them from the professionals. (Mayhew, 1862)

amatory rites acts of copulation.

The physical procedures of love within or outside marriage:

> ... my two friends soon transferred both their sleeping arrangements and their deafening amatory rites to the bed in Nathan's quarters. (Styron, 1976)

amber fluid/liquid/nectar lager.

As described in television advertising on behalf of an Australian brand brewed in Britain.

ambidextrous having both homosexual and heterosexual tastes.

Of men or women, from the facility to use either hand with equal skill.

ambiguous homosexual.

Or bisexual, from having more than one meaning:

> By associating herself with the free love movement, by marrying a man with ambiguous sexual interests... (Pearsall, 1969)

ambivalent having both homosexual and heterosexual tastes.

Literally, entertaining two opposite emotions at the same time:

> Sexually I'd say some of the company was on the ambivalent side. (P. Scott, 1975)

ambrosia an intoxicant.

Originally, the food, and less often the drink, of the gods:

> Bring your own ambrosia or take pot luck. (Sharpe, 1976)

Nectar would be more accurate except that it implies sweetness.

ambulance chaser *American* a disreputable lawyer who touts for clients.

He likes, figuratively at least, to follow an ambulance to hospital in the hope of being briefed in a contingency suit which he may persuade the victim to bring:

> Madder was a shyster in the Quorn Building. An ambulance chaser, a small time fixer, an alibi builder-upper. (Chandler, 1939)

amenity the purchase of priority or privacy during a stay in hospital.

In British hospitals other accommodation not so described was free. The *amenity bed* or *ward* became less common with the emergence of commercial hospitals.

America first isolationism.

Very common before Pearl Harbor:

> Sloan did not care if Hitler gobbled up the whole of Europe—he was for America First. (McCarthy, 1963)

Whence *America Firster*:

> FDR has said, 'all aid short of war'. And the America Firsters would raise an almighty hue and cry. (Archer, 1979)

American disease (the) AIDS.

A limited British usage—we have to blame someone for these misfortunes:

> When the famous popular singer Freddie Mercury died two years ago, everyone said he had died of the American disease. (A. Waugh in *Daily Telegraph*, 3 November 1993)

ammunition[1] lavatory paper.

Of the same tendency as the obsolete jocular *bum fodder*, which now hides its origins in the short form *bumf*, masked still further as *bumph*, meaning no more than an excess of paperwork:

> Astounding how the bumph accumulated during even a short absence... Minutes! More like bloody hours. (Grayson, 1975)

ammunition[2] a towel worn during menstruation.

Possibly adverting to the red danger sign on explosives, or on its expendability. In obsolete British naval use an *ammunition wife* was a prostitute, lively in performance but expendable.

amorous favours copulation.

Of either sex:

> It had become embarrassingly and sickeningly plain that the fickle Kim was bestowing amorous favours simultaneously on Melinda. (Boyle, 1979, of the spy Kim Philby and Donald Maclean's wife)

See also AMATORY RITES and FAVOURS. For *amorous sport*, see SPORT (THE). An *amorous tie* is a commitment to another which involves copulation outside marriage:

> I have few friends and no 'amorous ties', I am alone and free. (Murdoch, 1978)

amour a sexual partner to whom you are not married.

Literally, love or affection. An *amour* may also less often be a single act of copulation outside marriage:

> Those women who live in apartments, and maintain themself by the product of their vagrant amours. (Mayhew, 1862)

Occasionally in this sense in the singular:

> ... the jolly athletic amour so obviously and exquisitely enjoyed. (Styron, 1976, of noisy copulation)

ample fat.

Only of women. Literally, wide and commodious:

> ... a generous figure. 'Ample', she used to call it, or, in a kinder manner, 'my Edwardian body'. (Bogarde, 1978)

amply endowed see WELL ENDOWED

amusing pornographic.

Jargon of the world of fine art which professes to ignore vulgarity:

> Pictures medium only, but some amusing ('amusing' means 'erotic', doesn't it, in an auctioneer's catalogue description). (A. Clark, 1993)

anarchistic groups those opposed to Communism.

Communist jargon. If you are a Marxist, those who advocate an alternative to the authoritarian rule which is implicit in the system must be advocating no rule at all:

> The word 'strike' has continuously featured alongside the customary poisonous euphemisms ('anti-Socialist elements', 'anarchistic groups') in the columns and even the headlines of the party Press. (*Sunday Telegraph*, August 1980, of Poland)

angel a male homosexual playing the masculine role.

He tries to secure devotion with gifts, which indicates derivation from the *angel* who backs plays.

angel dust an illegal narcotic or hallucinogenic drug.

From the heavenly feeling supposedly induced:

> And that shooting... wasn't just some kid on angel dust. (Deighton, 1981)

The phrase, successively used for various individual drugs or blends, is now usually applied to phenocyclidine hydrochloride. The obsolete *angel foam*, champagne, used the same imagery.

angle with a silver hook to indicate willingness to offer a bribe.

This meaning, which I have heard spoken but not read, replaces the 19th-century use, to pretend to have caught a fish which you have bought. See also CATCH FISH WITH A SILVER HOOK.

Anglo-Saxon (of language) crude or vulgar.

From the supposition that most obscenities in English have that ancestry.

animal rights the attribution to selected mammals of human characteristics.

What is for some concern that chosen mammals should not be treated with cruelty becomes for others a kind of anthropomorphism, a belief that largely the same groups of animals share characteristics normally attributed only to humans. This in turn may lead to intolerance and lawbreaking. (The animal predators of my vegetable garden may be used to indicate the selectivity. The badger (root crops and broad beans) has more 'rights' than the rabbit (brassica) or grey squirrel (strawberries, green peas), which in turn are preferred to the mouse (bean and pea seeds) or rat. So too for the birds, of which the most 'disadvantaged' is the wood pigeon. I have not yet categorized gastropods, caterpillars, or insects.)

anoint a palm see PALM[1].

anointed *Irish* expected to die soon.

From the practice of so treating the bodies of mortally ill Roman Catholics:

> ... sure there isn't a winter since her daughter wint to America that she wasn't anointed a couple of times. I'm thinkin' the people th' other side o' death will be throuncin' her for keepin' them waitin' on her this way! (Somerville and Ross, 1894)

Anschluss a military conquest.

Literally, the German word means 'connection'. The word was used by the Nazis with reference to their 1938 annexation of Austria and so became a euphemism in both German and English.

answer the call[1] to die.

Usually of those killed in war, called to arms and then to life eternal.

answer the call[2] to urinate.

The *call* is a CALL OF NATURE with perhaps jocular reference to dying:

> ... was answering an urgent call behind bushes when they stopped close by. (Cookson, 1967)

anti- avoiding a statement of your allegiance.

When the cause being promoted is likely to have few adherents, you declare yourself as being against something which well-meaning and gullible people are likely to abhor. Thus in the 5th century, Athenasius set himself up as an anti-Aryan, and millions have repeated his doctrinal niceties each Sunday. The linguistic device is still widely used.

anti-fascist Communist.

As in the *anti-fascist protection barrier*, the Berlin wall built in 1961 to prevent the inhabitants of that part of Germany then under Russian domination fleeing to the west. It was so described by its architects in the pretence that it was intended to keep people out:

> The anti-fascist protection barrier is particularly deep and formidable where the railway crosses the Alexander Ufer. (Deighton, 1988)

When it came down in 1989, so too did the Communist regime and spy writers had to look elsewhere for plots.

anti-freeze a spirituous intoxicant.

It warms in cold weather. Occasionally, too, used of heroin, but with less justification.

anti-personnel designed to kill.

It could merely mean opposed to people:

> 'Anti-personnel weapon' is a sophisticated euphemism for 'killer weapon'. (Pei, 1969)

Weapons, too, can be sophisticated, which means they are complicated, cause more destruction, and kill more people.

anti-social criminal.

Literally, no more than reclusive or self-centred. *Anti-social behaviour* may be a tendency to crime:

> ... he was 'jointed' for 'anti-social behaviour', the IRA's euphemism for petty crime. (*Sunday Telegraph*, January 1990)

anti-socialist elements supposed dissidents in a Communist country.

A convenient catch-all phrase for anyone the government thinks might challenge its autocratic rule. See also ANARCHISTIC GROUPS.

anticipating pregnant.

An American version of EXPECTANT.

antics in bed copulation.

The posture adopted may well be described as grotesque:

> There were even married women in my time who did not connect their husbands' antics in bed with the conception of children. (Fraser, 1969, writing of the 19th century)

any extramarital copulation.

Of either sex and of a single or repeated encounters. A prurient person may ask another known to be courting whether he or she is getting or has had *any*.

apartheid see SEPARATE DEVELOPMENT.

ape *mainly American* mad.

Usually of a temporary condition, from the supposed simian behaviour:

> Victor had something Jake will never have. It drove him ape. (Sanders, 1977)

appearance money *British* an improper payment for participation in an event.

The bribe openly paid for competing in field and track events to athletes who wished to retain their amateur status.

appeasement the avoidance of war at the expense of others.

To *appease* is literally to pacify, whence to seek to pacify by conceding demands. Specifically used of British and French behaviour in relation to Hitler prior to the Second World War and the abandonment of the Czechoslovaks.

appendage the penis.

Literally, something attached or hung on, but men don't so speak of their big toes:

> ... her mean little hand ready to perform its spiritless operation on my equally jaded appendage. (Styron, 1976, of masturbation)

appetite a desire to copulate.

Literally, desire of any kind and usually only to eat food. It acquires greater sexual significance in the plural:

> ... consigned to an early grave by his wife's various appetites. (Sharpe, 1974)

Also of homosexual desire.

apple-polish *American* to seek favour or advancement by flattery.

You rub the skin to make it shine and look more palatable:

> Why try to apple-polish the dinge downstairs? (Chandler, 1939—*dinge* was an offensive term for a black)

Whence an *apple-polisher*, who so acts:

> ... he thought Cutter was a shallow, self-serving apple-polisher with delusions of grandeur. (Clancy, 1989)

appliance medical equipment worn on the body.

Literally, anything which is applied for a specific purpose. A shortening of *surgical appliance*, which is a euphemism in its own right, because a scalpel might as well be so described. Usually denoting a truss but also wooden legs, hearing aids, or anything else you don't want to be too precise about.

appropriate[1] to steal.

Originally it meant to take for your own use, without taint of impropriety:

> All old *mali* had actually ever done, though, was appropriate his fair share of what he had hoed and sweated to grow. (P. Scott, 1977 —the mali, or 'gardener', had been dismissed for theft)

The now redundant 19th-century *misappropriate* was introduced to cover the illegal conversion of property.

appropriate[2] conforming with your dogmatic prejudices.

How the POLITICALLY CORRECT describe any action or expression which corresponds with their convictions:

> Freedom of speech is still guaranteed by the Constitution, but it can be exercised only so long as it is judged 'appropriate'. (A. Waugh, *Daily Telegraph*, 13 August 1994, commenting on the refusal of an American publisher to publish writings by the Pope because they were considered anti-feminist)

appropriate technology torture.

A choice of technology gives you the answers you want. Used by the Home Affairs minister in Mugabe's Zimbabwe:

> In the House of Assembly, Harare's Commons, he called it 'appropriate technology', a euphemism for electric shock treatment that drew appreciative nods from his colleagues. (*Daily Telegraph*, September 1983, referring to Ushekowokunza and the torture of white officers)

Appropriate and *appropriately* are described as the 'favourite words in the bureaucrat's lexicon—the grease for sliding round unpleasantness, the funk-hole for avoiding specifics' (R. Harris, 1992).

approved school a penal institute for children.

The *approval* was by the British Home Office as being suitable for young criminals. You would be wrong to assume that educational establishments not so described lack the blessing of society.

apron-string-hold *obsolete* the occupation by a man of his wife's property.

The use satirizes British land tenures, freehold, leasehold, and copyhold:

> A man being possessed of a house and large orchard by apron-string-hold, felled almost all his fruit trees, because he expected the death of his sick wife. (Ellis, 1750)

I include this obsolete entry because it indicates what people in the 18th century thought of a man who lived off the estate of his wife, whose property in those days vested automatically in her husband, for her lifetime or beneficially.

Arctic explorer *American* a person who illegally ingests cocaine.

From the use that he makes of SNOW[1].

ardent spirits spirituous intoxicants.

An older version perhaps of FIREWATER:

> He had committed the sin of lust, he had drunk ardent spirits. (Cornwell, 1993, writing in 19th-century style)

Arkansas toothpick a dagger.

This is a sample entry, many weapons being given geographical attributions, either mocking the uncouthness of the local inhabitants or applauding their manliness:

> ...the Kentucky abolitionist Cassius Marcellus Clay, wearing 'three pistols and an Arkansas tooth pick'. (G. C. Ward, 1990, quoting an 1861 source)

See also GLASGOW KISS.

arm a policeman.

An American use, probably from the cliché 'the long arm of the law'. In New York, perversely, the *arm* is an organization of criminals. The perhaps obsolete British *arm-bend* is not to assault or bribe the constabulary but to drink intoxicants:

> He was busy arm-bending in the public house when the tattoo sounded. (EDD)

armed struggle terrorism.

The language of the Provisional Irish Republican Army:

> ...you go saying that I'm in the Armed Struggle, then you've got real trouble. (Seymour, 1992—the speaker was a terrorist)

armour a contraceptive sheath.

As worn, or not, by Boswell:

> I took out my armour, but she begged that I might not put it on, as the sport was much pleasanter without it. (Boswell, c. 1792)

Before the 19th-century American invention of a latex version, sheep's intestines were used.

army form blank lavatory paper.

Some military use and also by former soldiers, to let you know that they, too, have served. It is said to be the only paper found in the British army without a form number on it.

around the Horn see RUN ROUND THE HORN.

arouse to cause sexual excitement in another.

Literally, to waken from sleep. It is used of either sex, heterosexually and homosexually:

> ...he aroused her in a way that her husband had never done. (Allbeury, 1976, and not by a new alarm clock)

Whence *arousal*, such excitement and specifically an erection of the penis:

> ...the muted talk of women made him excited and he had to roll onto his stomach to conceal his arousal. (Boyd, 1982)

arrange to do something underhand or the subject of a taboo.

For example, preparing accounts or reports in a misleading manner; bribing officials; or castrating domestic cats:

> You always ought to have tom cats arranged, you know—it makes 'em more companionable. (Noel Coward—reported speech)

In Ireland the marriage of a pregnant bride can be *arranged by circumstances*:

> We had our share... of marriages arranged by circumstances. (Flanagan, 1988, referring to such events)

An *arrangement* is what ensues. Thus an *arrangement with your creditors* means that they are not going to bankrupt you but will accept in full satisfaction a payment of less than you owe them, evidenced by a *Deed of Arrangement*.

The American *arrangements conference* is where an order is placed for a funeral:

> 'The arrangements conference' (in which the sale of the funeral to the survivors is made). (J. Mitford, 1963)

In former times, an *arrangement* might be a pot for urine, being *arranged* in a commode or under a bed for use in the night, partly because there were few lavatories in private houses and hotels and partly because wandering down dark corridors by males at night was discouraged. *Make arrangements*, to copulate extramaritally, is also obsolete:

> Give a gal a drop of beer, and make her half tipsy, and then they makes their arrangements. I've often heard the boys

boasting of having ruined girls. (Mayhew, 1851)

arse a person viewed sexually.

Literally, the buttocks, but because they were the subject of taboo, the old name for donkey was called in aid, until *ass* acquired its own connotations. Thus in obsolete British use, polite ladies were reduced to calling a jack-ass a *Johnny Bum*, *Jack* and *ass* being vulgar while *bum* was still respectable. (And if you think that odd, see *roosterswain* at ROOSTER.) The commoner use is by the male of the female, as *a bit* or *piece* of *ass* or *arse*:

> Am I to believe you would risk something like this for a piece of ass? (Diehl, 1978)

Less often by a female of a male:

> The stewardesses all agreed he was a piece of ass. (Follett, 1978—they desired him sexually)

Male homosexual use too, and of male and female prostitutes, who may be called *arse pedlars*. An *arse man* may be a homosexual or a man who copulates promiscuously:

> ... sexy as he smiled at the girl who was one of Engineering's assistants. He was the house ass-man. (M. Thomas, 1982)

article any object that is the subject of a taboo.

Thus it may be a pot for urine, shortened from *article of furniture*, which is also euphemistic:

> Article (meaning 'chamber-pot') is non-U. (Ross, 1956)

Also, anything which has been stolen; or, in white slavery jargon, a potential prostitute:

> Some 'articles' are from seventeen to twenty kilos, i.e. women from seventeen to twenty years old. (Londres, 1928, in translation)

artillery *American* a hypodermic needle used for illicit narcotics injection.

From landing its charge and the explosive effect:

> ... a piece of community artillery passed from junkie to junkie. (Wambaugh, 1975)

Aryan without Jewish ancestry.

Nazi jargon in their anti-Jewish dogma, although a pure German's ancestors' blood was permitted to have been so mingled prior to 1750. Originally an *Aryan* denoted 'a native or inhabitant of Ariana, the eastern part of ancient Iran' or 'a member of any of the peoples who spoke the parent language of the Indo-European (or esp. Indo-Iranian) family' (SOED), which would include Slavs, Persians, and Celts as well as Teutons.

as Allah made him naked.

The way he was born:

> Recognizably not wearing anything ... as Allah made him. (Davidson, 1978)

In the same sense others ascribe the manufacture to God.

asbestos drawers an imagined concomitant of female lust.

The figurative *asbestos drawers* are to contain the fire generated by HOT PANTS:

> Needs asbestos drawers, I hear. Another little number from the sticks with a rich husband and hot pants. (M. Thomas, 1982)

ashes hauled see HAUL YOUR ASHES.

Asian levy a bribe.

British shipowners employing British crews on British ships with British trade union practices at British pay rates found that they were often uncompetitive. The owners therefore agreed to pay the National Union of Seamen £30 annually for each lowly paid Asian employed on a British-registered ship and the Union agreed to keep quiet about it:

> The old NUS had a history of controversial financial deals including the now notorious 'Asian levy'. (*Daily Telegraph*, 28 September 1990)

asleep dead.

A tombstone favourite, often amplified to *asleep in Jesus*, etc. See also *fall asleep* at FALL[3].

ass see ARSE.

assault to attack sexually.

Literally, to use any force against another without consent. Of children, males homosexually, and females heterosexually:

> If I'd been assaulted by men of my own race I would have been an object of pity. (P. Scott, 1973, said by a woman raped by a gang of Indians)

Explicitly as *indecent assault* and see SEXUAL ASSAULT.

assembly centre a prison.

American Second World War name for the place of confinement of Americans of Japanese descent. They spent most of the war where they were assembled. See also RELOCATION CAMP.

asset a spy.

Literally, anything useful or valuable. Now common espionage jargon, according to the spy novelists:

> No, an asset we have in Norway. (Clancy, 1986—the speaker was answering a question as to the source of information)
>
> The memo stated that the CIA had played a direct role in placing underwater mines in these Nicaraguan harbors. This, according to the memo, had all been done by 'unilaterally controlled Latino assets'—the UCLA's. (Woodward, 1987)

assignation a meeting for extramarital copulation.

Literally, the allotment of something, whence a tryst, which remains the meaning in standard English:

> I have never really seriously thought of marriage... What suits me best is the drama of separation, of looking forward to assignations and rendezvous. (Murdoch, 1978)

Also of the deed itself:

> Palmerstone died there on the billiard table, reputedly after an assignation with one of the maids. (*Daily Telegraph*, 11 February 1995, referring to Brocket Hall)

assist the police see HELP THE POLICE (WITH THEIR INQUIRIES).

assistance a regular payment to the poor from public funds.

Literally, help of any kind, even across a busy street. Often as *public* or *national assistance.* To be *on assistance* carries the implication of unemployment and poverty. This usage has had an unusually long life in the sequence of expressions which try to cloak the charitable and perhaps degrading nature of state payments to the indigent.

assistant see PERSONAL ASSISTANT.

association[1] regular copulation with someone other than your spouse.

Literally, the action of combining for any purpose, but the housewife who forms an *association* with the milkman is doing more than confirming her requirement for dairy produce. To *associate intimately with* is specific:

> As in Hispaniola, many native women became associated intimately with the conquerors. (H. Thomas, 1993)

association[2] a cartel.

Commonly formed by traders selling a similar range of products. The impressive titles and impressions of public benevolence may mask the objectives of spying on each other, keeping up prices, and restricting competition. Because many of their discussions are illegal under the laws of some countries in which they operate, Zurich zoo became a favoured venue for delicate meetings.

astride copulating with.

Equine imagery and normally used of the male:

> 'Harry—you are *sure* you have not been astride Mrs Lade?'... 'Eh? Good God, girl, what d'you mean?' 'Have you mounted her?' (Fraser, 1977)

asylum an institution for the mad.

Originally, a place where pillage was sacrilegious, which is why there was so much fuss about Henry II's murder of Becket, who, as Chancellor, had supported his conquests. Then it became any safe place or benevolent institution. Now a short form of the standard English expression *lunatic asylum.*

at a rope's end dead by being hanged.

With your head in the noose:

> I'll have thee at a rope's end yet. (Fowles, 1985, writing in archaic style)

at half mast with a trouser zip undone.

From a flag incorrectly hoisted. Normally used as a hint from one male to another in mixed company.

at Her Majesty's pleasure indefinitely.

The wording when a convicted person cannot in the judge's view have any term placed on the confinement due to madness or other factors. It has nothing to do with keeping corgies or staying at Balmoral.

at it engaged in some taboo activity.

In appropriate circumstances, the phrase can apply to anything from picking your nose to bestiality. In the East End of London, it usually refers to being a villain:

> At least one of his uncles is 'at it', as they say, and drives around in a silver-grey Mercedes. (Read, 1979—the uncle was a habitual criminal)

at liberty involuntarily unemployed.

Actors' jargon; it is certainly correct that they are free to take another role:

> 'Laurence Olivier' (very careful checking every time for correct spelling) 'at liberty'. (Olivier, 1982, of an advertisement when he was out of work)

See also BETWEEN SHOWS.

at-need *American* after death.

Funeral jargon for doing business when you have a corpse to dispose of:

> In 1960 pre-needs of graves and crypts outnumbered 'at-need' sales by four to one. (J. Mitford, 1963)

at rest dead.

A tombstone favourite which might seem to discount the prospects of an afterlife. On the other hand a regimen of eternal harp-playing and hymn-singing might be restful, if tedious. Also as *at peace.*

athlete a male profligate.

Copulation is thought to provide the male with good exercise:

> Erroll was the greatest 'athlete' in Kenya... and was undoubtedly the love of Diana's life. (Fox, 1982)

athletic supporter a brief tight undergarment worn by males.

Not a football fan:

> The speaker stumbled sleepily past him towards the Silex, dressed in nothing but an athletic supporter. (Wouk, 1951)

athwart your hawse copulating with you.

A *hawse* is a rigid cable and, in this naval usage, the female is astride it:

> I was near crazy, with that naked alabaster beauty squirming athwart my hawse, as the sailors say. (Fraser, 1973)

attend to to copulate with.

Literally, to see to the needs of, which in this usage and the judgement of the male, is a need to copulate with him:

> I'm goin to 'tend to this li'l beauty right here an' now. (Fraser, 1971, of a man about to copulate with a woman)

attendance centre *British* a place to which young criminals are required to report for disciplinary training.

They have to go as part of their sentence, although, taken literally, the term might equally apply to a discotheque or a skating rink.

attention deficit disorder idleness or stupidity.

The latest educational jargon in a sequence of excuses for inattentive, stupid, ill-disciplined, or idle children, coined in America but now crossing the ocean:

> They said I had a learning disorder. ADD. Attention Deficit Disorder. (Theroux, 1993)

and:

> ADD was just another word for stupid. (ibid.)

attentions sexual activities with someone other than a regular partner.

What may be no more than a mark of respect, interest, or good manners in the singular assumes sexual overtones in the plural:

> ... dancing partners of wealthy widows and lonesome wives, ready to pay for attentions accorded to them. (Lavine, 1930)

It is usually the male who 'presses' *attentions* and the female who 'enjoys' them:

> Jack Profumo ... had become involved with a young lady who was also enjoying the attentions of the Soviet Military attaché. (A. Clark, 1993—the coincidence would have been less noteworthy if Profumo had not also been the British minister responsible for the defence of the realm)

au naturel naked.

American rather than British borrowing from the French to describe a condition that is still more subject to taboos in parts of the New World than in the Old. In many parts of the United States they still don't like you changing swimming trunks on a beach under the cover of a towel.

auction of kit one of the consequences of death.

British naval usage, when messmates bid inflated prices in the knowledge that the proceeds will go to the dependants of the deceased. Formerly referred to as the punning *sale before the mast*.

auld

The Scottish form of 'old' appears in many names for the devil—see OLD A' ILL THING. The *auld kirk* is whisky, from the name given to the established church of Scotland:

> Whisky for me—a dram o' guid Auld Kirk. (Coghill, 1890)

The ecclesiastical derivation is unclear, except to recall that after a Scottish sermon in winter, you feel in need of a restorative.

aunt[1] an elderly prostitute.

The modern American use recalls Shakespeare:

> ... summer songs for me and my aunts,
> While we lie tumbling in the hay.
> (*Winter's Tale*)

aunt[2] a lavatory.

To whom mostly women pay a mythical visit; in Victorian days it was your *Aunt Jones*, or *Mrs Jones*. If the punning *Aunt Flo* is visiting you, you are menstruating. See also UNCLE.

aunt[3] an elderly male homosexual.

Those so described are normally a generation or more older than those whose company they seek. Also as *auntie*:

> Some mincing auntie in a cell with flowered curtains. (Ustinov, 1971)

auto-da-fé killing by burning.

Literally (translated from the Portuguese) the *act of faith* of the Inquisition, itself no more in its own eyes than an 'inquiry'. In Spanish *auto de fe* and no more palatable, but there was no other way both to destroy a heretic's soul and deny him a body for resurrection.

auto-erotic habits masturbation.

By either sex and not just thinking sexy thoughts or watching pornographic videos. Also as *auto-erotic practices*:

> When the first menstruation coincided with the discovery of sex and possibly auto-erotic practices, this alarm combined with guilt feelings often created a climate for all kinds of neuroses. (Pearsall, 1969, of Victorian females. A problem in educating girls about menstruation was that its origin was not scientifically explained until 1878.)

avail yourself of to copulate with casually.

Usually of a male:

> ... any man who availed himself of the 'tree rats' or 'grass *bidis*' was properly dealt with. (Allen, 1975; see also TREE-RAT and BIBI)

available willing to copulate.

Mainly of females and outside a regular partnership, with or without payment:

> Aileen was the only girl who had ever turned him down. The rest were always available —however nice—however respectable. (Collins, 1981)

Thus for the American government an *available casual indigenous female companion* is a prostitute:

> Even now the US State Department cannot bring itself to use the word *prostitute*. Instead it refers to 'available casual indigenous female companions'. (Bryson, 1994)

aversion therapy torture.

Literally, curing a craving etc. by inducing nausea or distaste.

away[1] dead.

The use is obsolete but the concept survives in *gone*:

> Rachel mournynge for hir children and wolde not be comforted, because they were awaye. (Coverdale Bible, Jeremiah 31:15; the Authorized Version says 'because they were not')

away[2] in prison.

The use was more common when the stigma of imprisonment was greater. In obsolete Scottish usage *away the trip* meant pregnant, from the vacation far from home which young unmarried females took under those circumstances. There were also many regional euphemisms, some alluding to landmarks or towns which a pregnant girl might pass or go to for a discreet childbirth. Thus from Cheshire comes *She has given Lawton Gate a clap*, 'Spoken of one got with child and going to London to conceal it. Lawton is on the way to London from several parts of Cheshire' (Ray, 1678, quoted in ODEP).

awful experiment the prohibition of sale and consumption of intoxicants in the USA from 1920 to 1933.

Awful for those denied intoxicants or forced into illegality to obtain them; much more awful for the impetus that it gave to organized crime, which bedevils parts of the country still:

> A generation or so has come between us and the Awful Experiment. (Longstreet, 1956)

awkward pregnant.

Of women, from the clumsiness which is a consequence of the extra weight and size. When used of quadrupeds it denotes being on their backs and unable to get to their feet without help.

ax(e) to kill after judicial process.

Originally by beheading, then by any other form of judicial killing:

> They were brought to Berlin and axed. (Shirer, 1984, of two German socialist leaders handed over by Pétain in 1940 and executed by the Nazis. The Jews he and Laval handed over later died more slowly.)

Much figurative use, as:

> You were out to ax me. (Turow, 1987, of an attorney's assault on a hostile witness)

Because of the clean break which this form of execution achieved, given a sharp instrument and a good eye, the imagery is much used of peremptory dismissal from courtship or employment, especially by sub-editors seeking snappy headlines.

Aztec-twostep diarrhoea.

An affliction of visitors to Mexico—you have to keep dancing to a lavatory. Also as *Aztec hop* and see also MONTEZUMA'S REVENGE.

B anything taboo beginning with the letter B.

The common candidates are benzedrine, bitch, bloody, and bugger. A *B-pill* is benzedrine used illegally; a *silly B*, if a woman, is being insulted as a silly bitch; a *B fool* is a bloody fool; and you may be told to *B off* which is not *be off* but *bugger off*. The military spoken alphabet *baker* is sometimes used in the same way, as in *baker flying* (see at BAKER). For *B bomb*, see A; for *B girl*, see BAR; for *BO*, see BODY ODOUR; for *B Special*, see SPECIAL[3].

baby-sitting undisclosed telephone monitoring.

Presumably from the presence of a watchful third party in the home:

> Thomasson reports that Buzhardt made reference to 'baby-sitting people', a reference the reporters did not understand. (Colodny and Gettlin, 1991)

baby-snatcher a person with a much younger sexual partner.

The imagery is of a child taken from the cradle, as in CRADLE-SNATCHER. Some homosexual use, but mainly heterosexual for anything outside the normal male/female age pattern. Thus the woman may be very little older but the man has to transgress the rule of half his age plus seven years:

> He had been living with an older woman... baby-snatching, as everybody called it. (Murdoch, 1978)

The punning *baby-farmer*, used in the same sense, is rare.

bacchanalian drunken.

Literally, anything to do with Bacchus or Dionysus, who was the god of wine:

> Burgess fell from grace again at the Foreign Office as a result of another bacchanalian holiday trip. (Boyle, 1979)

A *devotee, son*, or *priest of Bacchus* is a drunkard and *bacchanals—Bag o'Nails* in English pub signs—a carouse.

bachelor's wife a prostitute.

One who has only transitory relationships with any man. A *bachelor girl*, on the other hand, is a single woman who places her career before marriage.

back (on your) see ON YOUR BACK.

back door see FRONT DOOR (THE).

back-door man *American* a woman's extramarital sexual partner.

He figuratively leaves by this exit when the husband arrives at the front door.

back-door trot *English, obsolete* diarrhoea.

Punning on the anus and the location of the lavatory in the yard. You could *get, be given*, or *be on the back-door trot*:

> Are teh poorly?—Ay, ah've been on t' back-door trot this mony a day. (EDD from the late 19th century)

back-gate parole the natural death of a prisoner.

American prison use—I suppose the corpse is taken out that way. Major in *Black Slang* gives *back-gate parlor*, which looks like a misprint.

back passage the anus.

Medical jargon which has passed into common vulgar use. Until this meaning became explicit, it also described a corridor to the rear of a house. The *back garden* or *back way* for the anus are less common.

back teeth afloat drunk.

You have raised your bodily liquid level to that extent. It can also mean a consequent need to urinate:

> I've got to go to the john first. My back teeth are floating. (Sanders, 1973)

backfire to fart.

The extra-cylindrical combustion of a motor coupled with the location of the exhaust pipe provides this male vulgarism.

backhander a bribe.

Literally, a blow with the back of the hand. The giver of a bribe figuratively rotates his palm to make payment. In obsolete British use a *backhander* was also a glass of wine taken out of turn, when the bottle had in error circulated in the wrong direction.

backside the buttocks.

This standard English use ignores the other parts of the body, from the back of the skull to the heels, which would equally qualify, if they were the subject of taboos:

> But then it was just my... backside was at risk. (Price, 1978—spoken by a cultured woman)

Backseat is rarer and more explicit. In obsolete British use a *backside* was a lavatory, from its location behind the house.

backward[1] very dull.

In educational jargon, it indicates an IQ of between 70 and 84. Lay people use it of idiots. It might have meant merely doing poorly in a class of normal children.

backward[2] poor and uncivilized.

Of sovereign states. Perhaps the first of the series of post-colonial euphemisms which mask the differing abilities of people when it came to governing themselves again:

> ... countries which have progressively and with increasing euphemism been termed backward, underdeveloped, less-developed, and developing. (FDMT)

See also SOUTH (THE) and THIRD WORLD.

backward[3] *obsolete* through the anus.

I give this entry as a lone example in these pages of Pepys's vulgar frankness:

> ... and so to Mrs Martin and there did what je voudrais avec her, both devante and backward, which is also muy bon plazer. (Pepys, 1660–9)

bad working as a prostitute.

A judgement on morals rather than job proficiency:

> ... lost her place for staying out one night with the man who seduced her; he afterwards deserted her and then she became bad. (Mayhew, 1862)

Bad girls are prostitutes:

> 'Bad girls here,' said the tonga driver when he dropped me in a seedy district of the old city; but I saw none, and nothing resembling a Lahore house. (Theroux, 1975)

bad lad (the) the devil.

Mainly Scottish usage and perhaps obsolete. Also as *the bad man*:

> The gite has a drop o' the bad man's bluid on it (Johnston, 1891—a *gite* is a dress)

and as *the black lad*:

> The auld black lad may hae my saul, if I ken but o' ae Macnab. (Ford, 1891)

See also OLD A' ILL THING.

bad-mouth to denigrate.

Originally an American use in the context of personal relationships; now common commercial jargon:

> This legendary trio were busy bad-mouthing the Segal/Fitzwalter management. (*Private Eye*, May 1981)

The American *potty mouth* is a habitual user of foul language.

bad news menstruation.

For those wishing to copulate:

> 'Shall we go to bed, Maggie?' ... 'Yes I'd like that ... But I'm afraid I have some rather bad news for you.' (Trevanian, 1973)

But it can be good news for those fearing an unwanted pregnancy.

bad powder a fart.

My assumption that the use comes from the slow and smelly combustion of a faulty charge in a firearm is reinforced by hearing people say it has been *burnt* or *let off*. Mainly used by men.

bad week menstruation.

From its normal duration. A woman normally uses the phrase of herself—*my bad week*.

badge *American* a policeman.

Of the same provenance as BUTTON[3] and BLUE[1]:

> You gonna go walkin round Center City with a stiff, better have a badge along. (Turow, 1990)

A *badge bandit* is a policeman on highway patrol who may, or may not, pocket your fine.

badger a prostitute.

Formerly a licensed huckster who had to wear a badge, from which the standard English meaning, to importune excessively. The use survives in the *badger game*, where the victim is led by a prostitute into a sexually compromising position and then blackmailed:

> Any man who accompanies a night-club or dance-hall hostess to her apartment ... runs a risk of being robbed or subjected to the well-known badger game. (Lavine, 1930)

bag[1] to steal.

From concealing and taking away the loot in a bag. Still particularly a schoolroom favourite, as in Tom Brown's schooldays:

> The idea of being had up to the Doctor for bagging fowls, quite unmans him. (Hughes, 1856)

The American *bag job* is burglary by a government agency. Private documents are removed without judicial approval for later examination, to secure evidence of treason or to discredit a political opponent.

bag[2] **(the)** dismissal from employment or courtship.

A rare form of the *sack*; see SACK (THE).

bag[3] to kill by hunting.

Standard English, referring to the birds and small mammals which are put in the hunter's bag; but you may figuratively *bag* a rhinoceros, despite its bulk and threatened extinction. A *bag* of partridges etc. indicates how many were killed on a particular occasion.

bag[4] a supply of illicit narcotics.

Originally it indicated a certain quantity of heroin, although that quantity seems to have differed from place to place. Now used to refer to any narcotic wrapped in paper.

bag[5] a prostitute. See BAGGAGE.

bag (in the) see IN THE BAG[1].

bag job see BAG[1].

baggage a prostitute.

Formerly in standard English a worthless person, male or female. Shakespeare uses the euphemism in one of his more complex sexual puns:

No barricado for a belly; know't;
It will let in and out the enemy,
with bag and baggage.
(*Winter's Tale—bag and baggage* is the impedimenta of any 16th-century army, with prostitutes among the other female camp-followers)

The white slavery jargon use came from the pimps taking the girls with them to South America etc. as so much merchandise:

The 'baggage' consisted of two 'underweights' who had travelled as stowaways. (Londres, 1928, in translation—an *underweight* would be a girl barely past puberty)

A *bag*, an unprepossessing prostitute, may be short for *baggage*, or may come from the phrase *bag and baggage*, or from the current slang meaning, 'an ill-favoured woman of any age'. In obsolete British use a *bag-shanty* was a brothel.

bagged drunk.

An American use, perhaps from BAG[3], as you often feel like death afterwards:

Al Mackey giggled. He was more than half bagged. (Wambaugh, 1981)

The obsolete British meaning 'pregnant' probably puns on the girl having being caught by a hunter, and on the resulting swelling:

Well, Venus shortly bagged, and ere long was Cupid bred. (Nares, c. 1820)

bagman *American* someone involved in an illegal or taboo occupation.

Originally, a tramp, with his bag over his shoulder. It now denotes a seller of illicit narcotics, from the container in which he carries them, or an intermediary who passes bribes, and can differentiate, between gangsters and politicians.

bagnio *obsolete* a brothel.

The common bathing imagery. A *bagnio* is literally a bath-house:

...he had seen her some two or three months since entering a bagnio behind St James. (Fowles, 1985, of an 18th-century prostitute going into a brothel)

bags trousers.

An abbreviation of *leg-bags*, which came from the 19th-century taboo on trousers:

The shapeless flannels which he called his 'bags'. (Manning, 1965)

Now dated slang *bags of* anything denotes an excess and not a form of packaging. The phrase can be used euphemistically as:

She knew 'bags of stamina' was a euphemism for 'not much finishing speed'. (Francis, 1985, writing about horse racing)

bait see JAIL BAIT.

baker *American* a taboo starting with the letter B.

From the military spoken alphabet. *Baker flying* means menstruating, from the red quartermaster (or baker) flag which is flown when a ship is loading fuel or ammunition, warning other craft to stand well clear.

balance of the mind disturbed a temporary insanity.

British legal jargon, referring to suicides in cases where people want to bury the corpse in consecrated ground or merely to reject the probability that someone had been driven to suicide as a rational choice:

The verdict at the inquest was that he took his life while the balance of his mind was disturbed. I know little of my son's mind but I reject that comfortable euphemism. (James, 1972)

bale out[1] to urinate.

A British naval usage, from the removal of surplus water from small boats.

bale out[2] to jump by parachute from a damaged aircraft.

Again from the action of throwing water over the side of a small boat:

I once baled out of a Provost. (Price, 1970—a Provost was a jet trainer)

Slang as much as euphemism, but it is taboo to talk about aircraft failures in flight. A Second World War pilot might also have been *on the silk* (see SILK (THE)).

ball to copulate with.

Probably punning on the slang meaning 'orgy' and on the testicles (see BALLS), although it is used of copulation by either sex:

Sure I balled Victor. I wish he had bathed more often, but sometimes that can be fun, too. (Sanders, 1977)

The obsolete British *ball money* was extortion at a wedding, being not a kind of copulation-fee but money ostensibly taken from guests by onlookers to pay for a football for the parish. In practice the recipients pocketed what they obtained.

ball bearing a term of male abuse.

For the more common *pillock* (see PILL[1]):

Terrible as that little ball-bearing is, he is less dangerous to us than Herbert Morrison. (Crossman, 1981)

balloon room *American* a place where marijuana is commonly smoked.

An addict usage, perhaps punning on the practice of storing the narcotic in rubber balloons and the feeling of levitation sometimes induced.

balls the testicles.

Men use the word more than women. There are also two common figurative usages: as a derisive riposte, when some suppose it is an abbreviation of *balderdash*; and to denote courage, the testicles being a feature of manliness:

I hate to admit it, but I got to admire him for that. The balls. (Sanders, 1980)

However I begin to doubt my derivation when I find the phrase used of a female:

> Maybe Mama even hustles him right here: she's got the balls for it. (Sanders, 1977)

balmy drunk.

From the mild or fragrant sensation which may come over you at one stage. The slang meaning 'insane' is usually spelt *barmy*.

bamboozled *American* drunk.

Literally, hoaxed, and perhaps adverting to the old excuse that you can be 'deceived in liquor'.

banana an act of copulation.

Perhaps from the slang meaning 'the penis'. Oddly then a man *has a banana with* a woman—never the reverse. If he has his banana *peeled*, he ejaculates.

banana skin a potentially embarrassing or dangerous situation.

From the supposed tendency of pedestrians to fall over after slipping on those discarded in the street. Journalistic jargon, mainly used of politicians but sometimes of other threatened species:

> Townsend, the Irish captain, is aware of the potential banana skin that awaits his side. (*Daily Telegraph*, 24 June 1994—it had to be a large specimen to threaten the entire soccer team, which did indeed go on to lose its match against the Mexicans)

See also OWN GOAL.

bananas mad.

The derivation has escaped me, although I suspect it comes in corrupted form from another language. It is often used to refer to mild hysteria:

> ... there's a poor cop called Captain Salvatore going bananas. (L. Thomas, 1979)

bandwagon a cause or chance for profit which attracts opportunists.

Literally, a vehicle carrying musicians in a circus parade:

> I'm on the bandwagon with him. (N. Mitford, 1960, of someone who had joined a scheme in which easy profits were made)

A *band-wagoner* is an opportunist:

> ... sufficiently politically confused to rank either as bandwagoner or a half-baked pain in the neck. (P. Scott, 1973, of Gandhi in 1942)

bang[1] to copulate with.

Literally, to beat, and using the common violent imagery for the male role:

> It'd be amusing to bang her under all those ducal Gainsboroughs (M. Thomas, 1980)

and, as a noun, denoting a single act of copulation by a male:

> Did you ever give the maid a bang? (Mailer, 1965)

The obsolete *bang-tail*, shortened to *banger*, was a prostitute:

> ... that little brown banger ... she was a lissome little wriggler. (Fraser, 1975, writing in 19th-century style)

A *gang-bang*, also in the American forms of *gang-shay* or *gang-shag*, is successive copulation with a woman by a number of males in each other's company. *Bang and biff* is rhyming slang for *syph*, 'syphilis', and perhaps punning on the copulation and the unhappy outcome.

bang[2] an illicit narcotic.

Usually taken by injection and so having an immediate result. See also SHOT[3] and ARTILLERY.

banish your bed *obsolete* to refuse to copulate with.

Usually a female sanction but also a kingly gesture:

> I banish her my bed and company. (Shakespeare, *2 Henry VI*)

bank *obsolete* to fail in business.

It was the bench on which Italian money-lenders conducted their business, which was turned over—*rupted*—if they failed to meet their commitments. In the late 19th century banks were still failing regularly and the phrase was still in use:

> Dunnot ye know at Turner's is banked. (Taylor, 1890)

A *banker* was a bankrupt, which sounds odd to us today as it is the bankers who do most of the bankrupting.

bar a place for the sale and consumption of intoxicants.

A plank was used both as a counter and a barrier, giving the world perhaps its most multinational word; even the French Academy doesn't claim it comes from the sea-perch via fish restaurants. There are many derivatives and compounds, such as *bar-fly*, a drunkard, and *bar girl*, a prostitute:

> Paul was getting anxious to fix me up with a bar girl. (Simon, 1979)

Bar girls started out looking for trade in bars but are now any casual or inferior prostitute and *B girl* puns on *bar* and *B* for second-rate:

> He's got a finger in the B-girl rackets. (Theroux, 1973)

Bar steward is used for *bastard* as a term of abuse:

> He has nobody, poor old bar steward, to lerve him. (Burgess, 1980)

Barclays an act of masturbation by a male.

From the rhyming slang *Barclays Bank*, wank. Used by the comedian Kenneth Williams in his diaries.

bareback copulating without a contraceptive sheath.

The common equine imagery, but this time without a saddle:

I always ride bareback myself. (Wambaugh, 1981, of copulation)

Men or women can be *bareback riders*:

...no females except the local bareback riders. (Fraser, 1971—the implication is that they had venereal disease)

barker *American* a handgun.

Neither a fairground tout nor a dog, but from the noise:

You knew you'd have to carry a barker on this job? (Sanders, 1970)

barley cap a person who habitually drinks whisky.

From the grain used in the manufacture. *Barley-fever* denoted drunkenness:

This was the first time he ever had fallen a victim to the barley-fever. (Moir, 1828)

John Barleycorn, sometimes knighted, was whisky:

Leeze me on thee, John Barleycorn,
Thou King o' grain.
(Burns, 1786)

I suspect all these uses are obsolete.

barrack-room lawyer an opinionated but well-informed know-all.

Usually an old soldier who has refused promotion and combines experience in the ways of the army with knowledge of military regulations and bloody-mindedness:

'Who says that, now?' cries this barrack-room lawyer. (Fraser, 1982, of such a person)

In America a *barracks lawyer* and in the navy a *sea lawyer* or *ship's lawyer* play the same part.

barrel-house a brothel.

Originally, a cheap American saloon, where the intoxicants were served from barrels:

The cribs, saloons, sinister dancing-schools, the barrel-houses... (Longstreet, 1956)

basement *American* a lavatory.

From its frequent location there in shopping malls, public rooms, etc. There is no connection with the slang anatomical meaning 'the stomach'. Usually in the query, 'Where's the basement?'

baser needs the need for copulation.

From the dated assumption that regular copulation improves your health but is at the same time reprehensible:

What you need is a sensible wife to take care of your baser needs. (Sharpe, 1982)

Having cooked meals is even more basic.

bash to work as a prostitute.

The slang *bash* means 'to walk', as in the army *square bashing*:

Lettin' a woman bash on the bloody streets. (Kersh, 1936)

On the bash is so working:

Anybody would think that I was asking you to go on the bash. (ibid.)

bash the bishop to masturbate.

Of a male, from the likeness of the penis to the chessman. Also as *flog the bishop* but see also SHAKE HANDS WITH THE BISHOP which means 'to urinate'.

basket[1] a bastard.

The two words sound alike and this is only used figuratively and often jocularly. In obsolete British use the punning *basket-making* was extramarital copulation by a male.

basket[2] the male genitalia seen through tight trousers.

Homosexual jargon:

The movement arched his entire body and made his basket bulge under the cloth of his trousers. (Genet, in 1969 translation)

Occasionally also, in heterosexual use, the vagina.

basket case a person who is unable to provide for himself.

A soldier who suffered the amputation of arms and legs in the First World War was so described:

The other part of him couldn't understand why a nurse in a nursing home should be so grief-stricken over the death... of an obvious basket case. (Peck, 1990)

Now used figuratively of the feckless or poor, whether countries or individuals:

Poland, which is economically a basket case. (*Daily Telegraph*, February 1982)

It is also used allusively:

You cock teasers have turned millions... into a generation of sexual basket cases. (Styron, 1976)

Basra belly diarrhoea.

Another of the many alliterative phrases, this time for those passing through the Persian Gulf.

basted drunk.

The common culinary imagery, which in this American instance likens the subject to a joint of meat being roasted and periodically covered with molten fat.

bat[1] a prostitute.

No doubt from flitting around at night. The obsolete 19th-century English use has re-emerged in 20th-century America. A *bat-house* is a brothel.

bat[2] a drunken carouse.

Possibly from an inning(s) at cricket or baseball, except that a *bat* was a habitual drunkard. *Over the bat* means 'drunk'.

bath-house a chamber for mass killing.

The Nazis stripped their victims and herded them into gas chambers on the pretence that they were going for decontamination:

> Of four thousand in the next four trainloads, two and a half thousand went at once to the 'bath-houses'. (Keneally, 1982, of Auschwitz)

bathroom *American* a lavatory.

In the long line of euphemisms which associate washing with urination and defecation; in this instance from the common location of lavatories in bathrooms:

> ... asked where the bathroom was. The restroom was filthy. (Diehl, 1978—and what was the lavatory like?)

This is a standard American use which leads to untold confusion in Europe. Whence *bathroom paper* or *tissue*, lavatory paper:

> Mommy they have a lovely house, *but their bathroom paper hurts*! (E. S. Turner, 1952, quoting from an American advertisement of the period)

bats (in the belfry) mental abnormality.

It covers anything from absent-mindedness through eccentricity to madness, when the wild ideas may circle in your head like the mammals in the church tower at twilight:

> Dear man, you've got bats in the belfry. (Mason, 1927)

Often used as an adjective, shortened to *batty* and *bats*:

> Told him he was bats. (C. Forbes, 1992)

battered drunk.

American culinary imagery again, but also punning on a drunk person's feeling that he has been roughly handled.

batting and bowling having both homosexual and heterosexual tastes.

As this is a British usage, the imagery has to come from the game of cricket, in which most players tend to specialize in one or the other. See also ALL-ROUNDER.

battle fatigue the inability to continue fighting.

In standard usage, acute neurosis induced by periods of fighting and even in that sense denoting much more than mere tiredness resulting from constant combat:

> ... wondering suddenly how much guilty truth and how much honest battle fatigue there had been. (Price, 1978, of an officer who had admitted being a coward)

The Second World War equivalent of the First World War SHELL SHOCK, in which the implication of cowardice is less often present. The American COMBAT FATIGUE can have the same overtones.

battle of the bulge a desire to slim.

The *bulge* is the evidence of obesity around the waist or hips:

> The 'battle of the bulge' became a corsetier's problem. (E. S. Turner, 1952)

The original battle was Eisenhower's, when the Germans broke through in the Ardennes in December 1944. The modern campaigns seldom last as long or achieve comparable losses.

bawd the keeper of a brothel.

It means no more than dirt:

> ... like sanctified and pious bawds,
> The better to beguile.
> (Shakespeare, *Hamlet*)

Bawdy has many of the moral meanings of DIRTY, of which *bawdy house*, a brothel, survives:

> I would not wreck it, turn it into a bawdy house, or receive any members of the press here. (Bogarde, 1978, commenting on the conditions on his lease)

bay window a fat person's stomach.

Normally used kindly of a male, from the architectural feature which often protrudes from the lower floor only.

be excused to urinate or defecate.

Literally, politely to obtain release from the company of others. Perhaps the first thing we all learnt when we started school.

be nice to to copulate with extramaritally.

Prostitutes' jargon and see also NICE TIME:

> ... wouldn't you like to be nice to Dasha? (Amis, 1980—Dasha was asking a man to copulate with her)

be with to copulate with extramaritally.

Of either sex:

> The girl talked. We know you've been with her. (Mailer, 1965—the police had evidence of recent copulation)

It is a reflection on human proclivities that mere proximity imports such a conclusion.
Been into refers almost explicitly to male copulation:

> He had never been into a girl either. (Bradbury, 1975)

An unmarried female said to have *been there* is not a virgin and a male who says he has *been there* is claiming to have copulated with a specified female.

beach to dismiss from employment.

Usually of a sailor, from the former practice of leaving one who was needed no more on a foreign shore, to find another ship or his own way home. *On the beach* is out of work as a sailor:

> You hear that, you Port Mahon bumboatman, you? You ought to be on the beach! (Fraser, 1971)

bear[1] to be pregnant or to give birth.

The standard English use makes us forget that anyone who lifts an infant thereby bears a child and is of child-bearing age:

> Asses are made to bear, and so are you. (Shakespeare, *Taming of the Shrew*, using both meanings)

Conversely when a mother *bears* a child, she no longer bears it; it is born but no longer borne.

bear[2] a policeman.

The common American use probably comes from the threat and the violence, characteristics which the quadruped and the officer of the law are thought to have in common. There are many derivatives, some of which come from Citizens' Band radio. *Bear bait* is a speeding car without Citizens' Band and therefore likely to be pulled in; a *bear cage* is a police station; a *bear in the air* is a police helicopter, especially one on traffic duty; *bear food* is any speeding vehicle; a *bear bite* is a ticket for speeding; a *bear trap* is police radar, illogical in that it is the *bears* who do the trapping; *the bears are crawling* means that the police are switching from one direction of the highway to another, to counter warnings of their presence on Citizens' Band; a *state bear* is a state trooper; the punning *bearded buddy* is any policeman, probably from the cartoon character, Yogi Bear; and a *lady bear* is a policewoman.

beard[1] *British, obsolete* the female pubic hair.

A male who parted it in the act of copulation was a *beard splitter*. A *bearded clam* is the female genitals viewed sexually by a male:

> I've ate close to three hundred bearded clams in my time. (Wambaugh, 1975)

beard[2] a male who falsely adopts the role of a woman's sexual partner.

The purpose is to protect another man performing that role and the derivation is from the false beard worn as a disguise:

> He's the 'beard'. That's what they call the other man who pretends to be the lover. (Sanders, 1981)

beast with two backs copulation.

From the facing posture of the parties. The *beast* may be *made*:

> Your daughter and the Moor are now making the beast with two backs (Shakespeare, *Othello*)

or *played*:

> She had the goods on me; in an idle moment I played the beast with two backs with her. (McCarthy, 1963)

Also found as the *two-backed beast*:

> I... know what it had been like with Deborah and him, what a hot burning two-backed beast (Mailer, 1965)

or as the *two-backed game*:

> She had a hearty appetite for the two-backed game. (Fraser, 1977)

I have traced one homosexual use:

> This was an embrace, and as the girls on stage made the beast with two backs, the man in the film hopped into the missionary position. (Theroux, 1975)

beastliness male masturbation.

Hardly fair, as most animals don't masturbate:

> ... the detrimental effects on sportsmen of masturbation, referred to in the sermon as beastliness. (Sharpe, 1982)

In the 19th century it meant 'copulation', of the male, when it was thought proper for women so to regard the occupation:

> While you were at your beastliness... (Fraser, 1971, writing of something which happened during copulation, in 19th-century style)

beat on (have a) see RAISE A BEAT.

beat the gong to smoke opium.

From an American association of Oriental ideas. However, a *beat pad* is a place where marijuana is smoked communally, usually characterized by the impurity of the product.

beat the gun to copulate with a proposed spouse before marriage.

The *gun* is the starter's pistol. Used specifically of conception before marriage even if only evident afterwards. See also CHEAT THE STARTER.

beat the mattress to copulate.

Punning on the domestic chore of beating a flock mattress to make it more even and the movements of the copulating pair:

> ... we beat the mattress regularly as far as Council Grove. (Fraser, 1982)

beat your meat to masturbate.

Of a male:

> 'To see that you don't beat your meat,' said the constable coarsely. (Sharpe, 1976, explaining why a prisoner was kept under observation)

See also MEAT. Less often as *beat your dummy*—a *dummy* is a penis. Also as *beat off*:

> Twenty minutes, he'll beat off and save the money. (Diehl, 1978—a male was waiting impatiently for a prostitute)

beau a woman's male sexual partner.

Not necessarily beautiful, but paying court to her and, if she is married, implying that she is copulating with him:

> 'You don't do it famously,'... 'I haven't heard a word of complaint from any new beau.' (Mailer, 1965)

beauty care the attempted concealment of signs of ageing.

Cosmetic and similar routines for women are thus described. The object is to persuade the subject that she is thus preserving her looks rather than donning a deceptive mask. Such routines may take place in a *beauty salon* or *parlour* and involve the concealment or imposition of a *beauty spot*, a facial blemish.

beaver the female genitals.

From the slang meaning 'a beard', and thence the pubic hair:

> He gobbles one beaver and gets promoted. (Wambaugh, 1975, referring to cunnilingus)

A *beaver-shot* is a photograph which shows the exposed genitals.

Bechuana tummy diarrhoea.

An African variant of the enteric indisposition common to places with impure water and poor sanitation. It is also used of the stomach pains which may follow drunkenness.

bed[1] childbirth.

The bed is the symbol of birth, marriage, and copulation. To be *brought to bed* is standard English for being delivered of a child:

> At the height of the gale a soldier's wife was brought to bed. (Graves, 1940, referring to childbirth)

bed[2] to copulate with.

From the normal location of the act. In certain societies and times the marriage ceremony is not complete until the witnesses have seen the bride and groom tucked up together, or *bedded*:

> Woo her, wed her and bed her. (Shakespeare, *Taming of the Shrew*)

Also in modern use, and referring not only to the marriage bed:

> ... what made the English 'Doctor X' bed as many young girls. (Lyall, 1980)

In bed means 'copulating':

> Afghan women were always hungry for men ... and very active in bed. (Fraser, 1969)

Bed down usually implies extramarital copulation, with the male taking the active part:

> Did you really think of bedding me down right there in the middle of that cocktail party? (Diehl, 1978)

In obsolete use, a *bed-faggot* was a prostitute and a *bed-house* an establishment to which you might take a female for extramarital copulation, paying corkage, as it were. *Bed-hopping* is promiscuous copulation:

> Given more privacy, some bed-hopping might have developed. (Hailey, 1979)

Bed and breakfast is a single act of extramarital overnight copulation, punning on what a guest-house offers:

> No mention of any bed-and-breakfast work, setting up ex-military members of parliament for possible blackmail. (Lyall, 1980)

Beddable means 'nubile':

> I'm wary of strong, clever women, however beddable they may be. (Fraser, 1973)

Bedwork, copulation, puns on the normal slang meaning 'a job so easy you could do it in bed'. *Bedworthy* describes a woman who is sexually attractive or who it is thought might benefit from having a child born to her. A *bedfellow* is a person with whom you copulate extramaritally:

> I've had better bedfellows, mistresses more given to the art of love. (Harris, 1925)

A woman with *bedroom eyes* invites copulation:

> A redheaded number with bedroom eyes. (Chandler, 1939)

To be *privy to a bed* was not to be a commode but to copulate with a woman outside marriage:

> Now, before you knew Dick was privy to her bed, had you marked any understanding between them? (Fowles, 1985)

bedewed sweating.

Of a female, who is not supposed to sweat in public or, in some circles, at all. The liquid excreted is likened to the pure morning dew:

> ... a lady might get 'bedewed', but she didn't sweat. (Jennings, 1965)

Bedfordshire bed.

An ancient pun on the English county:

> ... sleepy knaves, we pulled them out of Bedfordshire. (Thomas of Woodstock, 16th century, quoted in ODEP)

Used in talking to children who want to stay up; and in referring to adults who should have got up earlier.

bedpan see PAN.

bedwetting urinating in bed.

Usually of children involuntarily. This standard English use makes us forget that there are many other ways of making a bed damp.

bee *American* a quantity of illicit narcotics.

An inexact unit of measurement, probably based on the size of the insect rather than its ability to sting.

beef copulation.

Just as Africans may refer to any meat as *beef*, so in euphemism the two words are largely interchangeable:

> ... feeding him beef like a shogun in a geisha house. (Wambaugh, 1975)

Beef may also mean the penis, any woman viewed sexually by a male, or a prostitute. *Beefcake*, copying CHEESECAKE, is a photograph of a naked man for female erotic gratification.

been having urinated.

Polite usage and effectively the past tense of GO[3]:

> Hari's realization that I hadn't 'been' rather cast a blight on his evening. (P. Scott, 1973 —an Indian youth was entertaining a white girl)

Occasionally too of defecation.

beer and sandwiches (at Number 10) political appeasement of a trade union in an industrial dispute.

The location was the residence of the British Prime Minister—Number 10, Downing Street—and the fare was that thought to be agreeable to the trade union leaders whom he was trying to propitiate:

> Healey, like Roy Jenkins, favoured a 'proper dinner', so instead of beer and sandwiches at Number 10, there was more substantial fare at Number 11. (Cole, 1995)

beg a child of *obsolete* to copulate with.

Male usage, perhaps from a wish to generate an heir:

> I think he means to beg a child of her. (Shakespeare, *3 Henry VI*)

beggar a bugger.

As used in the expression *silly beggar*, a stupid person, or as *beggar it*, the mild expletive much favoured by my beloved maternal grandfather.

behind the buttocks.

It could be any part of you, from your head to your heels:

> ... reference to a female's buttocks as her 'behind'. (Jennings, 1965—but it is used equally of males)

Occasionally used for the anus:

> It was a serious insult, because that was the hand they used to wipe their behinds. (Simon, 1979)

There is also some male homosexual use:

> This bee-hind is for sale, boy. (Mailer, 1965)

behind a bush defecating.

In the open air:

> It certainly slowed us down a bit today, and Matt was behind a bush a lot as well. (Barnes, 1989—Matthew had eaten monkey meat)

behind the eight ball *American* in severe difficulty.

From a potentially losing position in the game of pool:

> Verdi would get the message that he could find himself behind the eight ball. (Deighton, 1994)

behind the wire in prison.

Especially describing prisoners of war, who were held in prisons whose defences included barbed or electrified wire. I suppose to *take something to the wire*, to carry it as far forward as you think possible, may come from the restrictions imposed on the inmates, whose approach to the fence would not be greeted with enthusiasm by the guards.

bell money *Scottish, obsolete* money extracted from a bridegroom at a wedding.

Not a corruption of *ball money*—see BALL—but from payment demanded by the ringers:

> At a wedding, the boys and girls of the neighbourhood assemble in front of the house, calling out 'Bell money, bell money, shabby waddin, canna spare a bawbee.' Money is then given to them. (N & Q 1855, quoted in EDD—such rudeness hardly deserved rewarding)

I include this entry to remind us that weddings have long been the occasion for extortion and wilful delay, even before the photographer made his appearance.

belle mère a mother-in-law.

A French euphemism doubly euphemistic in English. In the same vein as calling the Furies 'the Kindly Ones', the devil 'the Goodman', etc.

belly plea a claim that the accused is pregnant.

A pregnant woman could not be hanged and therefore so advised the judge if she were convicted on a capital charge:

> My mother pleaded her belly, and being found quick with child... (Defoe, 1721)

To *wear the belly high* was to be pregnant and to *slink a great belly away* was to have an abortion:

> Lady Castlemayne, who he believes hath lately slunk a great belly away. (Pepys, 1664—at least it saved us another dukedom)

belly to belly *American* copulating.

From the position adopted by the participants. In obsolete English use a *belly-piece* was a prostitute, punning on the meaning 'an apron'.

belly up bankrupt.

Usually of American companies, with piscine imagery:

> ... no government on earth in the mid-1980s let a company like MDC go belly up. (Erdman, 1981)

bellyful of lead see LEAD.

below medium height of unusually short stature.

Tallness short of gigantism is seen to be an attribute of manliness and shortness the converse:

> He was below medium height. (Obituary in *Daily Telegraph*, December 1989)

In more recent use he would have been VERTICALLY CHALLENGED.

below stairs employed as a domestic servant.

The construction of town houses afforded day accommodation for the servants in cellars or semi-basements and sleeping space in the attics, communication taking place through the *backstairs* of gossip fame. Servants might also be described in general terms as *downstairs*.

below the salt socially inferior.

The salt, being a scarce commodity needed by all, was put in the middle of the table with the more important diners in ascending order on one side of it, the less important in descending order on the other:

> ...in comparison with other professions—the Church, Education, the Law, the higher levels of journalism, and the BBC—I am afraid that it must be admitted that advertising sits rather below the salt. (E. S. Turner, 1952, quoting from a 1951 speech)

The saline distinction usually works only against you, but:

> ...it's a big dinner and you'll be well above the salt. (N. Mitford, 1960)

belt[1] to take any violent or wrong action.

Of many taboos, such as copulating extramaritally, smoking marijuana, or becoming very drunk. To *belt your batter* is to masturbate, perhaps a vulgar culinary pun on the slang *batter*, the penis. To *belt* is to drive a vehicle at an excessive speed, but beware of doing so on an American *belt-way*, or ring road. A *belter* is a cheap prostitute, not necessarily one with any special penchant for bondage.

belt[2] a drink of spirituous intoxicant.

Usually drunk fast and to excess:

> Dundee and Spencer had a couple of belts on the drive into Manhattan. (Sanders, 1984)

bench[1] a judicial authority.

Standard English for the office and for those who are appointed to and sit on it.

bench[2] to cause to withdraw from active participation.

From being relegated to the substitutes' bench while others do the playing:

> ...if I say you're benched, you're benched. (Deighton, 1982, of a commander grounding a flier)

A *bench-warmer* in America is a less competent athlete not picked as first choice for the team.

bend to drink intoxicants to excess.

Probably a shortened form of the euphemistic *bend the elbow*, from the locomotion of the drinking vessel:

> Bend well to the Madeira at dinner. (Ramsay, 1859)

In modern use an *elbow bender* is a drunkard and *elbow bending* the act of drinking intoxicants:

> Alrazi was a major leaguer at elbow-bending. (M. Thomas, 1980)

Bent in this context means 'drunk', *on the bend* is being engaged in an alcoholic debauch, and a *bender* is a drunken carouse. The 19th-century New England *benders* were the human legs in that age of prudery.

bend sinister an imputation of bastardy.

The heraldic *bend sinister* runs from the upper right to the lower left corner of a coat of arms. To suggest that someone, whether or not entitled to a coat of arms, has a *bend sinister* is to imply that he is actually or figuratively a bastard.

bend the rules to act illegally.

The implication is that the law has not really been broken:

> ...if he sometimes 'bent the rules'...he believed that the end justified the means. (P. Scott, 1973)

Bent here means 'stolen' or 'dishonest':

> Having sold a stolen or *bent* car to a complainant... (Lavine, 1930)

The punning *bent copper* is a dishonest policeman.

bends (the) menstruation.

Literally decompression sickness and its painful symptoms:

> She was having her damned period, she said, a real bastard, cramps, the bends, you name it. (le Carré, 1986)

benefit state aid paid to the poor.

Literally, an advantage, whence the specific advantage of being a member of a fund from which you could draw if you were ill. If the illness lasted too long, or you failed to keep up your contributions, you went *out of benefit*. The modern use is of regular or ad hoc payments:

> Jobless CSE candidates 'should be given £13 benefit'. (*Daily Telegraph*, December 1980)

benevolence an arbitrary tax.

Literally, the word means 'generosity'. English monarchs extracted it from their rich subjects under the guise of loans which were never repaid, until the 1689 Bill of Rights brought this method of royal fund-raising to an end. The British Second World War *post-war credit* was a similar device, except that it was repaid decades later in a depreciated currency.

bent[1] stolen; dishonest. See BEND THE RULES.

bent[2] drunk. See BEND.

bent[3] homosexual.

Usually of a male—he deviates from *straight*, or heterosexual, inclinations:

> ...he's bent as a tin spoon. (Bogarde, 1981, of a male homosexual)

besom *obsolete* a prostitute.

Originally the word meant 'broom', whence the aged menial who wielded it, with connotations of the preferred method of transportation of witches. In the 19th century 'A girl described as "a besom" without a qualifying adj. would imply unchastity' (EDD) and a *besomer* was a promiscuous person of either sex. If in northern England you decided to *hang out the besom*, you lived riotously during your wife's absence from home; probably because inn signs were poles with tufts on them, which looked like besoms, and punning on the meaning 'a prostitute'. But a woman *hanging out the broomstick* was scheming to get herself a husband, again from the sign advertising that the place was open for business.

bespattered *?obsolete* a mild oath.

For those who wanted to imply but avoid the once taboo *bloody*.

bestiality copulation of a human with an animal.

Literally, appertaining to a beast, but see also BEASTLINESS. Legal jargon for such a relationship with a mammal quadruped by either sex. The English case of *r. vs Brown* established the unlikely principle that you can be convicted of attempting an offence which it is impossible to commit. As *bestiality* embraced only such a relationship with cattle (which includes geese because they too graze) and Brown's amorous attentions were directed towards a duck, he was convicted of an 'attempt', and left to reflect on the axiom that 'hard cases make bad law'.

bestride to copulate with.

Of a male, with the common equine imagery:

> The tools of the fools who bestrode her. (*Playboy's Book of Limericks*)

bestseller a book of which the first impression is not remaindered.

Publishers' puff. An *international bestseller* is one in which the spellings 'honor', 'traveler', etc. are used to avoid re-setting the American edition. An *instant bestseller* indicates an expensive pre-release advertising campaign.

betray to copulate with a third party while married.

From the sense of proving false. There are two nuances—the married male may *betray* the unmarried female by copulating with her, despite her acquiescence in the transaction:

> ... servant girls ceased to be seduced and began to be betrayed (Mencken, 1940)

and either spouse may *betray* the other by so acting with a third party.

better country life after death.

The belief or hope of those who profess certain religions:

> ... strive to take it with faith and patience, looking to a 'better country'. (Sir James Murray in a 1915 letter)

Also as *better place*, *better* or *next world*, etc.

between shows involuntarily unemployed.

Theatrical jargon, not used of those rehearsing for a new part:

> 'I worked on lots a pictures a his over the years.' 'Did he know you were between shows?' (Wambaugh, 1981, of an out-of-work dancer)

Less often as *between jobs*, especially for those who do not tread the boards:

> 'What do you do?' ... 'I'm between jobs.' 'Are you an actor?' ... 'No.' (Hall, 1988)

between the legs on or around the genitals.

The term may be used to denote the location of anything from prickly heat to amorous fondling:

> ... her left arm around my neck stroking the back of my hair, her right hand still stroking me between the legs. (Barber, 1981)

between the sheets copulating.

From the bedlinen:

> We still suited very well between the sheets. (Fraser, 1970, of a wife)

Shakespeare used *twixt the sheets*:

> ... twixt my sheets,
> Has done my office.
> (*Othello*)

between the thighs of copulating with.

Used of extramarital copulation:

> A man can learn more between the thighs of a good woman than he ever needs to know. (Sharpe, 1974—academically and anatomically incorrect, for all its vivid imagery)

beverage an intoxicant.

Originally, any kind of drink. Now standard American use for any alcoholic drink served —'Would you like a beverage, sir?'—in a bar or *beverage room* by a bartender or *beverage host*.

beyond salvage condemned to death.

Espionage jargon for a defector who cannot be safely brought back to his former allegiance:

> The case against me is beyond salvage ... I'll be executed at the first opportune moment. (Ludlum, 1979)

bibi a prostitute.

In Hindi, it means 'lady', whence the 19th-century British Indian use, denoting a white woman married to a white man:

> The *bibi*, or white wife, was a great rarity; but the *bubu*, or native wife, was an accepted institution. (Blanch, 1954)

Later a *bibi* in British Indian army use became any Indian prostitute:

> Sahib, you want nice Bibi, me drive you to bungalow of nice half-caste, plenty clean, plenty cheap. (Richards, 1936)

You may also find it spelt *bidi* and as *grass bibi* or *bidi*. The army or colonial service carried the use into Africa:

> ... African mistress, also known as 'native comfort', '*bibi*' (Anglo-Indian) or '*bint*' —young woman (Arabic). (Allen, 1975)

biddy a woman with whom a man might copulate casually.

In 19th-century England it meant 'a young prostitute', in Ireland 'a chicken', and in both countries as well as America it was a short form of the Irish name Bridget, at a time when many maidservants were Irish. Whence general use:

> ...for a pound of sausages you could find a biddy who would actually chuck her old man out of bed and send him to sit downstairs till you'd finished. (Seymour, 1980, of Germany immediately after the Second World War)

big pregnant.

A shortened form of *big with child*, but used also before the swelling is visible:

> They said shoo'r big, but doctor said 'twor nought at all but cowld. (Doyle, 1855)

More explicitly, a *big belly* means 'pregnancy', without any suggestion of overeating:

> ...the consequence of which was a big belly, and the loss of place. (Cleland, 1749—a pregnant servant might expect summary dismissal)

big animal *American, obsolete* a bull.

From the days when the word *bull* was taboo, from its sexual overtones. Sometimes simply as *animal*. The fastidious had a plentiful choice of synonyms, with *big animal*, BRUTE, COW BRUTE, GENTLEMAN COW, HE-COW, MALE BEAST, MALE COW, MAN COW, and *seed-ox* (see SEED).

big-boned fat.

Of adults and children, suggesting that their frame needs that excess covering:

> ...in his beefy adolescence his mother had tactfully described him as 'big-boned', though 'burly' was how he now liked to see himself. (Boyd, 1981)

big brother the autocrat in a totalitarian tyranny.

Figurative use from the character in Orwell's *1984*. Technical progress in communication systems and computers has kept pace with the passing years since he wrote, and it now all seems possible. In American Citizens' Band slang, *big brother* is a policeman.

big C cancer.

An American usage for the dread disease:

> ...ailments are apt to be called by their own names or by superstitious shortened forms: arthritis, emphysema, cancer (or the Big C). (Diane Johnson and John Murray in Enright, 1985)

big D death.

Not necessarily from the BIG C. For military men, the *big stand-easy*, from the order to relax after being stood at ease.

big girl a large and sexually unattractive young woman.

Males use the term in a derogatory sense, because they need some physical dominance over a female, who therefore should not be larger or stronger than her master. A mother so describing her exceptionally large adolescent daughter does so kindly, to imply that she has merely grown quickly and all will be well in due course. Unfairly a *big man* is assumed to have the virtues of generosity and virility.

big house a prison.

American usage, from the size of the building:

> She has other worries besides trying to keep her ex-lover out of the Big House. (Lavine, 1930)

Also as *big pasture* or, for male convicts, *big school*, the *little school* being a prison for women or children. In obsolete British use, a *big house* was an institution for the destitute, or the common name for the residence of the squire.

big jobs defecation.

Nursery usage in several forms:

> ...done our bigs and wiped our bottoms. (Amis, 1978)

Little jobs or *littles* is urination.

big prize (the) copulation.

A male wins it after lesser awards during courtship:

> ...allowing moist liberties but with steel-trap relentlessness withholding the big prize. (Styron, 1976)

bijou inconveniently small.

British estate agents' jargon which seeks to persuade you that a minute apartment is a jewel:

> Now she lived in a tiny house off the King's Road...the sort of house agents called 'bijou'. (Deighton, 1988)

bikini wax the removal of women's pubic hair.

To prevent hair showing outside the scanty lower garment:

> Her fag hairdresser gives a great bikini wax. (Sanders, 1982)

It is the skin which is waxed, not the bikini.

bill a policeman.

Perhaps from the weapon once carried by constables:

> 'Eyes front,' said Murf. 'It's Bill.' A policeman in a helmet and gleaming rain-cape was coming towards them. (Theroux, 1976)

Also as *old bill*.

Billingsgate foul language.

It was used by women sellers of fish rather than by male porters in the London market closed in 1982. According to Dryden 'Parnassus spoke the cant of Billingsgate' and in modern use:

> ...his ears had surely overflowed with such billingsgate. (Styron, 1976)

bin an institution for the mad.

Literally, a container; short for *loony bin*:

> We shall be found stark staring mad with horror and live sixty more years in an expensive bin. (N. Mitford, 1949)

Now less used because of our heightened understanding of taboos about lunacy.

bind to cause to suffer from constipation.

Literally, to tie fast, whence to be constipated:

> Up and took phisique... only to loose me, for I am bound. (Pepys, 1662)
>
> Then the water will be madly binding. (N. Mitford, 1945—a hypochondriac was making the prediction)

binge a drunken carouse.

The verb means 'to soak' and in 19th-century use referred to drinking too much beer:

> A man goes to the ale-house to binge himself. (EDD)

bint a prostitute.

The British army picked up the Arabic word for 'girl' and carried it across the world:

> The women put it down to the rations we got, and the men down to the bint, as they called it. (Bogarde, 1978)

biographic leverage blackmail.

You exert the force through knowledge of the victim's past:

> Jonathan smiled at the cryptic jargon... 'biographic leverage' meant blackmail. (Trevanian, 1972)

biological reaction an erection of the penis.

It could have been no more than sweating in hot weather, or coming out in a rash:

> She could tell from her husband's biological reaction in the small of her back... (Clancy, 1987—they were occupying a double bed at the time)

bird[1] a prostitute.

When Holman Hunt wrote in 1854 of his *bird in a gilded cage*, the British meanings covered the spectrum from girl through sweetheart and sexual mistress to prostitute. An American *birdcage* is still a brothel and a *bird circuit* takes you, if so inclined, around the saloons frequented by male homosexuals in a vicinity where there is a choice of several. In most modern uses, a *bird* is no more than a young woman viewed sexually.

bird[2] imprisonment.

From the cage. To do *bird* is to be incarcerated. A *birdcage* is a prison cell.

bird[3] the vagina.

A male use, and the *bird's nest* is a female's pubic hair:

> This bitch wears these short shorts... when I'm down on my knees... I kneel there and look right at her bird. (Wambaugh, 1975)

bird circuit see BIRD[1].

bird dog[1] *American* a small gambler who follows the betting pattern of heavy gamblers.

From the animal that retrieves the quarry brought down by its master. But he is not that stupid; those who habitually bet large sums are likely on occasion to be party to criminal interference with the runners.

bird dog[2] an unsuspected accomplice to a criminal.

Again from picking up the carcass and bringing it back to the master:

> Your bird dog, the State Senator. (Chandler, 1939, of such a relationship)

To *bird-dog* is so to act:

> So he would be bird-dogging occasionally and bring you things? (Colodny and Gettlin, 1991, of a naval yeoman who stole secret documents and passed them to his superior)

bird in the air *American* a police helicopter watching traffic.

From the common slang meaning 'an aircraft'. It may be a corruption or variant of the Citizens' Band *bear in the air* (see at BEAR[2]).

bird's nest see BIRD[3].

birth control the prevention of conception during copulation.

The phrase would better describe stratagems by midwives to prevent the arrival of babies at weekends or other times inconvenient to themselves. A more accurate term would be *conception prevention*.

birth quota the system whereby the state restricts the size of families of married couples and orders abortions of unlicensed foetuses.

A feature of Chinese Communist rule:

> The organization is responsible for the weekly political indoctrination of the residents, running the day-care centres, distributing ration coupons, allocating birth quotas and arbitrating in disputes between neighbours. (Cheng, 1984, of the Residents' Committees who work under and report to the police)

birthday suit nakedness.

What you were born in:

> I went in the morning to a private place, along with the housemaid, and we bathed in our birthday soot. (Smollett, 1771—I am sure he meant they had one each)

Also as *birthday attire, gear*, or the obsolete *finery*:

> ...the figure I made...outshone all other *birthday* finery. (Cleland, 1749, of a naked woman)

bisexual having both homosexual and heterosexual tastes.

In biology, it means having both sexes in the same plant or animal. Often shortened to *bi*.

bit a woman viewed sexually by a male.

A synonym of PIECE[1] but not used of your own wife:

> 'Opale', said Cicero. 'Whose bit is that?' (Londres, 1928, in translation, of a prostitute)

Often as *a bit of all right* (or 'alright' for the illiterate), *arse, ass, crumpet, fluff, goods, hot stuff, how's your father, jam, meat, muslin, skirt, stuff, you-know-what*, etc., most of which are elaborated under those headings, so one illustration should suffice:

> One of them...was his own bit of goods. She was a married woman whose husband was away working. (Richards, 1936)

A *bit* can also be casual copulation by either sex:

> ...taking a little bit now and then from her husband's valet. (Condon, 1966)

A *bit on the side* is regular copulation with other than your professed sexual partner, adverting to the *side*, or additional, plate served with a formal meal:

> She had been used, had been just the fun you can't get when you're married, a bit on the side. (Bradbury, 1976)

A *bit of the other* is not a homosexual encounter but may describe a single act of extramarital copulation:

> ...off they go to this girlie restaurant...Duffy's not averse to a bit of the other. (le Carré, 1991)

bit missing (have a) to be of low intelligence.

> A bit missing by the way she asked. Someone took advantage of her I suppose. (James, 1962)

bit of meat (a) see MEAT.

bit on a bet.

Racing jargon, short for a *bit of money bet on*, and seeking to suggest that the wager was only a modest one. I don't think there is any pun on the *bit* the horse has in its mouth.

bitch *American* a male homosexual.

Homosexual jargon, used of someone thought to be as spiteful or vindictive as a woman so described. In obsolete use a *bitch* was a prostitute and to *bitch* was to be a regular user of brothels.

bite *obsolete* a swindler.

'A cheat; a trick; a fraud; in low and vulgar language' says Dr Johnson, and of *biter*, 'A tricker; a deceiver'.

> Is this wench an idiot, or a bite? (Fielding, 1742)

Although these senses are obsolete, we still say *the biter bit*. The American *bite* meant 'to borrow' before its present criminal connotations, as in *put the bite on*, to extort by threats:

> ...put the bite on you and you paid him a little now and then to avoid scandal. (Chandler, 1951)

All uses have dental imagery as their origin.

bite the dust to die.

> ...Jerry will unleash some devil's device and another brilliant novelist will bite the dust. (Letter written by Philip Larkin in 1944, in Thwaite, 1992)

See also LICK THE DUST.

black and whites *American* the police.

From the car colouring:

> ...didn't even notice the cops gliding up in the black-and-white. (Wambaugh, 1981)

black-bag (connected with) illicit enquiry.

Usually of telephone tapping or robbery, from the holdall in which tools are carried:

> I'd like you to authorise a black-bag job on Rathbone's town house. (Sanders, 1990)

black beauty an illegal narcotic pill.

Usually biphetamine and punning on the horse in the story:

> Librium and Valium in fives and tens; Quaaludes; Dexamyls and Benzedrines; Nembutols and Amyl Nitrate; black beauties... (M. Thomas, 1982)

black economy the sum of goods and services provided without tax or other official cognizance.

The international phenomenon of rich nations where all wish to avoid paying taxes and some wish to continue to draw unemployment pay despite working. In the USA it is *underground production, Schwarzarbeit* in Germany, *sotto governo* in Italy, etc.

> All public-spirited citizens will want to help the Inland Revenue in its battle against this 'black economy' of untaxed income and benefits. (A. Waugh, *Private Eye*, March 1981 —the satirist meant that the tax collector would get no help from the public)

black-eyed Susan *?obsolete* a prostitute.

Probably the song came first. Of prostitutes working sea ports:

> ...the sailors' wives and black-eyed Susans would...haunt the quays in sailing days. (Fowles, 1977)

black gentleman *?obsolete* the devil.

The Prince of Darkness was also thought to enter a house by the chimney in the days of coal fires and soot. Also as the *black lad, man, prince, Sam, spy*, etc.:

> The Black Man would gi'e her power... to kep the butter frae gatherin' in the kirn. (Service, 1890)

But *black George* was a poacher in standard English and in Scotland *black fishing* was poaching fish by night:

> Blackfishers, poachers and smugglers are a sort of gentry that will not be much checked. (W. Scott, 1824)

Most of these uses are obsolete as we don't have to refer allusively to the devil these days.

black hole *obsolete* a prison.

Most British towns had one, so called because it was insanitary, unlit, and below ground:

> Naething but law and vengeance, blackhole and fining. (Cross, 1844)

Also shortened to *hole*:

> They'l other foin us, or else send us to t'oil. (Bywater, 1839)

I include this obsolete British usage because generations of schoolchildren think the only *black hole* was the 1756 Calcutta version in which the native ruler had the effrontery to incarcerate some foreigners of whose activities he disapproved.

black ivory see BLACKBIRD.

black market illegal dealing in rationed goods.

Here, as elsewhere, *black* indicates illegality:

> A black market is beginning to appear, in sharp contrast to the orderly arrangements... in the food markets. (Goebbels, 1945, in translation)

In the black meant trading in such commodities, punning on *black market* and the profitability, as bank statements used to show credit statements in black ink. Debit accounts were shown in red and despite modern monochrome print-outs, we still talk about being *in the black*—in credit—or *in the red*—overdrawn.

black money cash gained illegally.

It can be the excess profits from a gambling establishment, which are concealed to avoid tax, the proceeds of any vice, a bribe received by a politician, etc. In each case there is a need to pass the funds through a lawful enterprise before spending it:

> Hasn't the wily oriental got black money tucked away? (Davidson, 1978)

Also as *black pounds, dollars*, etc.

> ...to do with black dollars... after returns from the orient. (ibid.)

But *black marks* are not illegally acquired German banknotes.

black operator *American* a secret agent or spy.

This CIA jargon does not mean he isn't a white man. In obsolete English use *black work* was the burial of corpses:

> ...employed in black work, or who, in other words, worked for an undertaker. (Egerton, 1884)

Similarly the punning *black art* was the business of arranging funerals, a *black job* was a funeral, and *blacks* were mourning clothes.

black order (the) Himmler's military Nazi organization.

The *Schutzstaffel*, better known as the *SS*, consisted of men 'racially pure since 1750'. Also known as *blackshirts* from their dress, as were other non-German Fascists who aped them and their uniform.

black pox syphilis caught from a black woman.

White males so infected thought that the disease took a more virulent form:

> John Lawrence, who got black pox, as he called syphilis, caught from a Negress... (Harris, 1925)

See also POX (THE).

black stuff opium.

Also as *black pills* or *smoke*:

> Imagine a clergyman peddling the black smoke. (Fraser, 1985, writing in archaic style of an opium smuggler)

The *white stuff* is cocaine. See also BLACK BEAUTY.

black velvet a dark-skinned prostitute.

Originally used by British soldiers in India, but the pun is now general:

> In sophisticated circles Black Velvet is a mix of champagne and Guinness. But in the outback the phrase has a different meaning derived from an obscure Ugandan dialect. (*Private Eye*, January 1982—see UGANDA for the equally obscure in-joke)

blackbird a negro slave.

The jargon of whites engaged in the slave trade or owning such slaves. Trading in slaves was *blackbirding*:

> I had recently forsaken African blackbirding in favour of river dealing. (Fraser, 1971)

The trade was also the *blackbird trade*:

> Things were making life more difficult in the blackbird trade (ibid.)

or *blackbird pie*:

> The old swine had his fingers in the blackbird pie. (ibid., of a British businessman)

A *blackbirder* was either a transporter of or a dealer in slaves:

> When the stinking ships of the blackbirders crossed the bars below the delta. (Longstreet, 1956)

The slaves in the mass were known as *black cattle*:

> ...wi' a full load of black cattle (Fraser, 1971, of a ship full of slaves)

or as *black hides*:

> There won't be a black hide aboard to show against us (ibid.—when slavery became illegal you might avoid conviction by pushing weighted slaves overboard to drown, before being boarded by an enforcement vessel)

or as *black sheep*:

> You're tender of black sheep (ibid.)

or as *black pigs*:

> I was shipping black pigs when you were hanging at your mother's teat (ibid.)

or, because of their value, as *black ivory*:

> ... a cargo of black ivory going over the Middle Passage. (ibid.)

Moral condemnation may be expected, of all the white people involved through several generations, directly and indirectly, and of the black chiefs and agents who provided the cargoes. I confine my comment to quoting von Longau in Longfellow's translation:

> Though the mills of God grind slowly, yet they grind exceeding small

and to referring you to Pope-Hennessy's 1967 *Sins of the Fathers*, a title aptly taken from Exodus 20:5 except that the ultimate price will be payable long after the third and fourth generations.

blackmail extortion by threats.

Mail was a tribute or tax, becoming *black* when paid by a Lowland Scot to a Highlander:

> And what is black-mail? A sort of protection money that Low-country gentlemen ... pay to some Highland chief that he may neither do them harm nor suffer it be done to them by others. (W. Scott, 1814)

Dr Wright in EDD was so rash as to call the use obsolete but it remains with us in standard English. See also GREENMAILER.

blackshirt see BLACK ORDER (THE).

blank[1] a mild oath.

A blank space may be left in print for the taboo swear word, which means that the euphemism is never written. The adjectival *blanking* is both spoken and written.

blank[2] *American* to kill.

From the void into which the victim is sent:

> ... none of whom seemed particularly distressed by the sudden blanking of Victor Maitland. (Sanders, 1977)

blanked drunk.

As a First World War usage, and in the light of what British soldiers did to the French language, it probably came from drinking too much *vin blanc* rather than from rendering yourself senseless.

blanket drill masturbation.

In some tropical stations an afternoon siesta on your charpoy was compulsory, the period being called *blanket drill*, although the last thing you needed was a blanket. In the heat and boredom there was ample opportunity for masturbation and the phrase was also used of copulation.

blast[1] *American* to kill.

By shooting, and so it probably refers to the explosion:

> We just got a message for the guy. We don't blast him. Not today. (Chandler, 1939)

blast[2] a mild oath.

The etymology is uncertain. It could come from the meanings 'a puff of air', or 'lightning'. Partridge in DSUE says of *blasted*, 'a euphemism for bloody' and of *bleating*, 'Among the lower classes a euphemism for bloody', but he didn't explain what led him to these conclusions.

blast[3] to take narcotics illegally.

Perhaps from the feeling of levitation induced:

> I'm higher than a giraffe's toupee. I started blasting when I was 13. (Longstreet, 1956)

An American *blast party* is one at which illegal narcotics are ingested communally. Occasionally, also, a *blast* may be an intoxicant:

> ... get me another blast, will you? Easy on the ice. (Sanders, 1982)

blasted drunk.

From the shattering effect and also no doubt used of narcosis:

> Takes a real sailorman to know how to get blasted. (Clancy, 1986)

blazes hell.

The fires burned you, without consuming your body or rendering you insensible to pain:

> You can count on J. B. to blazes and beyond. (Fraser, 1977)

Old blazes was the devil, who found the environment not uncongenial. Heretics had to be burned on earth without access to the eternal fires because that was the only way of destroying their soul and ensuring they would have no body for subsequent resurrection. It was very logical, if you believed in a medieval hell.

bleach to dye (hair).

Normally denoting the artificial colouring of women's hair. Dyeing your hair may mean that you are vain or trying to conceal your age; natural bleaching in the sun would not be anything to be ashamed of or even remark on.

bleed to extort money from, on a regular basis.

American underworld slang, of obvious derivation. The obsolete British *bleed the monkey* was to steal rum, taking a little at a time through a straw from the mess tub or *monkey*.

bleeding a mild oath.

Used for the slightly more taboo BLOODY[2].

bleeding heart a person who ostentatiously expresses concern about the suffering of others.

Not as tedious perhaps as a DO-GOODER. The victims for whom the heart metaphorically bleeds may be unaware of or unconcerned about their own supposed distress:

> The warden of the prison was a well-known bleeding-heart, little better than a Red. (Moss, 1987)

bleep an obscenity or swear word.

From the electronic noise introduced by broadcasters etc., to replace obscene or offensive speech in a recording.

blighty a serious but not fatal wound.

It comes from the Hindi *bilayati*, foreign, which effectively meant Great Britain for the soldiers and subjects of the British Indian empire. In First World War usage it denoted a wound serious enough for you to be sent home and so released, for a time at least, from the horror of the Western Front:

> What we used to call 'a nice blighty one'; sent me back to England. (Price, 1974)

blind a drunken carouse.

It should be because it makes you *blind drunk* but the use seems to predate both the cliché and its Scottish form *blind-fou*. Now often called a *blinder*.

blind copy a document of which a copy is circulated to third parties without informing the person to whom it is addressed.

Good manners suggest that the person to whom the document is sent should be told of other recipients. Occasionally as *silent copy* and not to be confused with a *leak* (see LEAK[2]).

blind pig *American* an unlicensed place for the consumption of intoxicants.

Hidden from the police, perhaps? Other unsocial activities such as the sale of illegal narcotics and prostitution might be found, the customers being usually blacks:

> Howison had raided it as long ago as February, 1966, and had discovered that it was, in fact, the front for a blind pig. (Lacey, 1986)

blip off *American* to kill.

Blips indicate that an oscilloscope or other monitoring equipment is working. They vanish if the instrument malfunctions, or is switched off.

blitzed drunk.

The consequent feeling of devastation is not much like the Second World War German *Blitz(krieg)*, or lightning war, still less the air raids on London which the word came to denote in English. Oddly more American than British use but:

> Miller has no health or weight problems and furthermore plans to get blitzed on February 1. (*Daily Telegraph*, 10 January 1995 of someone who had foresworn alcohol for the month of January)

block to copulate with.

Of a male, with obvious imagery:

> There was a young lady of Thun,
> Who was blocked by the man in the moon.
> (*Playboy's Book of Limericks*)

The punning *block her passage* is more explicit.

blockbuster *American* a real estate dealer who induces whites to sell their homes through threat of other racial groups moving into the area.

The use puns on the Second World War bomb and the breaking up of *blocks* of property occupied solely by whites, for which higher prices might be obtained on redevelopment. Now also publishers' jargon for a paperback novel which is not remaindered soon after publication; see also BESTSELLER.

blocked under narcotic influence.

I suppose from the sense 'hemmed in' or 'shut in'.

bloke *American* cocaine taken illegally.

Drug users' jargon, which could be rhyming slang for *coke*, cocaine—or merely rhyming with it coincidentally.

blood disease syphilis.

The condition was doubly taboo as being both incurable and contracted in a shameful manner. This American use just survives but *blood poison* for syphilis is obsolete:

> *Syphilis* became transformed into *blood-poison*, *specific blood poison* and *secret disease*. (Mencken, 1940, writing on 19th-century American evasions)

blood money extortion.

In standard usage, a reward for bringing about another's death, or compensation paid to surviving relatives in respect of a killing:

> ...collecting 'blood money', that is, shaking down prostitutes, poor peddlars, &c. (Lavine, 1930)

bloody[1] menstruating.

Female usage and in various expressions such as *the bloody flag is up*.

bloody[2] an oath.

OED says 'Prob. from the adv. use in its later phase' but there are proponents of derivation from obsolete popular oaths such as *God's blood*, or from a corruption of *By Our Lady*. *Bloody-minded* means 'cussed' or 'awkward'.

bloody shame a drink of tomato juice.

For a former alcoholic who used to like vodka with it, as a *bloody Mary*, after the British queen whose reign was marked by religious fervour and the execution of Protestants:

> 'Just tomato juice for me,' the sergeant said stolidly. 'A Virgin Mary,' the waiter nodded wisely. 'Also known as a Bloody Shame.' (Sanders, 1977)

blooming a mild oath.

I suppose it is for *bloody*, as most etymologists say so.

blot out to kill.

From the extinction of life:

> You can even blot me out suddenly so that I don't know about it. (Fraser, 1977)

Occasionally shortened to *blot*:

> The Emperor left here for Ethiopia today, flying to the frontier and then in by ground. I hope he doesn't get blotted. (Mockler, 1984 —in January 1941 Haile Selassie was more at risk from his compatriots than from the Italians)

To *blot your copybook* means merely to do or say something wrong, from the days when ink needed blotters, pens leaked ink when improperly used, and schoolchildren needed copybooks because they were taught to write properly.

blow[1] to copulate.

This use is perhaps obsolete, although you may still hear that a male may *blow through*, *blow the loose corns*, or *blow the groundsels*, which last is done on the floor. To *blow off* is to ejaculate semen, during copulation or otherwise, with imagery from the whale:

> Blew off all over the booth. (*Playboy's Book of Limericks*)

In obsolete use a *blow*, or the American *blowen*, was a prostitute. These noun meanings have today been almost entirely submerged by the verb sense 'orally to excite the genitals', homosexually or heterosexually:

> He was cruising down the interstate and his daughter's husband is blowing him. (Diehl, 1978)

A *blow job* is fellatio:

> 'You want for me to give you a blow job?' She got off the bed and came towards him. (Sharpe, 1976)

But for thieves a *blow job* may still be the use of explosives to gain access to a building or safe.

blow[2] to fart.

And as *blow off*—both uses are common. *Blow a raspberry* is to make a similar noise with the lips:

> The bank man blew the Marseilles equivalent of a raspberry and went home. (L. Thomas, 1977)

blow[3] to boast.

This perhaps obsolete use comes from puffing yourself up like a frog. A *blow* was a boast:

> Bonapart, loud vaunting smart,
> It was a fearfu' blaw that.
> (A. Scott, 1805)

In modern imagery you may *blow your own horn* or *trumpet*, or *blow smoke*, like a distant steamship advertising its presence:

> You think I'm blowing smoke? (Sanders, 1984, of someone accused of exaggeration)

blow[4] a mild oath.

Brewer says 'A play on the word Dash me, which is a euphemism for a more offensive oath' (BDPF) —how I wish those Victorians were more explicit. I suspect it is of the same tendency as BLAST[2].

blow[5] to betray to authority.

Espionage jargon, from *blowing*, or revealing, secret information:

> Did you tell the man to blow me? (Hall, 1979, of a betrayed spy)

In the underworld of crime, to *blow the whistle on someone* is to inform the authorities about his activities, from a referee's action in stopping a game:

> He was a number one hitman for the Cosa Nostra and he blew the whistle on them. (Diehl, 1978)
>
> —and we hate to ruin Kealy's career... So we'd feel bad about blowing the whistle on him. (Sanders, 1984)

The British *blow the gaff* has the same meaning. See also WHISTLEBLOWER.

blow[6] an ingestion of illicit narcotics.

Through the mouth, and I suppose you blow some of it out again. It is also used as a verb: you may *blow a stick*, *Charlie*, *horse*, *snow*, etc.

blow away *American* to kill.

Usually by gunfire at short range, although the corpse is left for disposal:

> He got blown away. I went to his funeral. (Sanders, 1977)

blow-in a foreigner taking part in Irish affairs.

More used in the south than the north:

> ... he fumed against 'blow-ins'—a jibe apparently aimed at Bruce Arnold, the English-born reporter of the Irish Independent. (J. J. Lee, 1989, of Cosgrave in May 1977; it was through lying about the tapping of Arnold's telephone that Haughey eventually fell from power)

blow me one *American* give me a draught beer.

The summary request is to the bartender, who then scoops or sometimes blows off the froth which has resulted from pouring bottom-fermented beer under pressure direct into a glass.

Blown, drunk, is not from drinking the beer—indeed *blown out* would be more appropriate—but from the 19th-century slang expression meaning 'exhausted', used of a horse.

blow off see BLOW[1, 2].

blow your cool to do something irrational or unexpected.

An imprecise American phrase which can confuse foreigners. If you *blow* your *cool, cork, lump, noggin, roof, stack, top*, etc. you may be angry, talk indiscreetly, drink intoxicants to excess, ingest illegal narcotics, commit suicide, or indulge in any other offbeat behaviour.

blue[1] a policeman.

Mainly American use, from the normal colour of the uniform:

> Okay, it was on the sixth floor when the first blues got to the Kipper townhouse. (Sanders, 1980, of the elevator)

A policeman is also a *bluebottle, bluebird, blue-belly, blue man, blue jeans*, or *blue and white* and the *men in blue* are the police. In the British Isles a *man in blue* or a *bluecoat* is a policeman and the *blue lamp*, a police station, from the standard exterior sign. Internationally the *blue police* are those in uniform and operating to known rules as distinct from more sinister varieties:

> ...speak to the SS man, to the Ukrainian auxiliary, to the Blue Police and to OD details. (Keneally, 1982, of Nazi-occupied Poland; German control rested in the army and these four organizations, the OD being Jews in authority over other Jews)

blue[2] erotic.

Probably from the French *bibliothèque bleu*, a collection of seamy works of literature, rather than from the colour of the brimstone fires which await the evil:

> She starred in dozens of blue movies before coming above ground. (Deighton, 1972)

blue[3] drunk.

The dialect *blue* was a jug of ale but this American use probably comes from post-alcoholic depression. Also as the punning *blue-eyed*.

blue box an electronic device to avoid paying for toll telephone calls.

A *black box*, or electronic control, of a different hue. The complexity of modern telecommunications coupled with the power of personal computers means that cheats and hackers are able to discover the codes which are reserved for line testing or the transfer of charges to a third party. As soon as one scam is blocked, another is developed and the information disseminated among aficionados, of whom there are tens of thousands.

blue cheer an illegal narcotic.

From the colour of the pill, or from its supposed ability to drive away the *blues*—depression. Also as *blue devil, flags, heaven, jay* (sodium amytal), *velvet*, etc. In obsolete northern British dialect *blue stone* was inferior whisky, punning on the word for vitriol. *Blue ruin* was any cheap intoxicant—usually gin—from the colour and the consequences:

> My ole man and me want some blue ruin to keep our spirits up. (Mayhew, 1862)

The obsolete *blue-devil factory* was not an illegal distillery but a mad-house, from one of the common delusions of lunatics:

> I saw him off for the last time to the blue-devil factory in the country where he died bawling with delirium tremens. (Fraser, 1975, writing in 19th-century style)

blue hair an old woman.

From the dye, or *blue rinse*, applied to the hair:

> This joint is where you find busloads a blue-hairs when they get off the freaking cruise ships. (Wambaugh, 1981)

blue-on-blue shelling or bombing your own troops.

From the colour used on chinagraph maps to indicate your own positions:

> I could see all the more clearly the potential for blue-on-blues (accidental attacks on friendly forces), particularly from the air. (de la Billière, 1992, writing of the Gulf War)

blue room a lavatory on an aircraft.

Crew jargon, perhaps from the subdued lighting:

> ...a passenger deliberately burnt himself to death in the right aft 'blue room' or toilet. (Moynahan, 1983)

blue veiner an erection of the penis.

From the prominence of the veins when erect:

> Even now he got a blue veiner every time he held a bar of soap. (Wambaugh, 1975)

To have *blue balls* (of a male) is to suffer from gonorrhoea.

board to copulate with.

Of the male, usually outside marriage, from naval imagery:

> I am sure he is in the fleet: I would he had boarded me (Shakespeare, *Much Ado*)

or, less subtly:

> I tried to board her at Kiva, but the caravanserai was too crowded. (Fraser, 1975—he lusted after a woman in the party)

The 19th-century British *board lodger* was a prostitute who obtained her finery from a pimp, thus staying under his control; or a prostitute working on her own but paying commission to the bawd of the brothel to which she brought men:

> Board lodgers are those who give a portion of what they receive to the mistress of the

brothel in return for their board and lodging. (Mayhew, 1862)

For *board a train* see PULL A TRAIN.

boat people refugees from Vietnam fleeing by sea.

Not those normally living or working on boats but the victims of Hanoi's tyranny, its NEW ECONOMIC ZONES, and the poverty which resulted from its adherence to Marxist economic dogma. In obsolete English use to *boat* was to send convicts to penal settlements abroad, whence to imprison anywhere.

bobby a policeman.

The name comes from a pet form of the Christian name of Sir Robert Peel, who reorganized the London police with his New Metropolitan Police Act of 1828:

> The bobbies over there came across it as a matter of routine. (Bagley, 1977)

It is often forgotten that as Secretary for Ireland Peel had done the same service for Dublin somewhat earlier. Still commonly used of the British police, especially those in tall hats on foot patrol. See also PEELER.

bobtail a dishonourable discharge from the army.

US Army jargon—the bit about 'honorable and faithful service' was clipped off the bottom of the printed certificate of discharge. In obsolete English use a *bobtail* was either a eunuch, from *bob*, to cut off, or a prostitute, from *bob*, to move up and down, and *tail*, what she might be moving.

bodice ripper a novel containing pornographic scenes.

Usually written by women, and featuring an aggressive male attitude to casual copulation:

> Anthony Looch's guide to bodice rippers. (*Daily Telegraph*, 17 December 1994, in a section otherwise devoted to literature)

bodily functions urination and defecation.

The equally important breathing, sweating, digesting, etc. do not count:

> You slept there, bathed, performed your bodily functions. (Sanders, 1973)

bodily wastes urine and faeces.

Discharged in the BODILY FUNCTIONS. Occasional figurative use:

> ... the fan is full of bodily wastes. (M. Thomas, 1987—an allusion to the expression 'the shit (has) hit the fan', which refers to the unpleasant consequences of a particular course of action)

body a corpse.

Short for *dead body*:

> At Worcester must his body be interr'd. (Shakespeare, *King John*)

If you *count bodies* it can be the living or the dead but a *body count* is only of the living.

body bag a flexible container for the transport of a corpse.

As distinct from a *sleeping bag*:

> The KIAs were provided with HRPs which earlier wars had called body bags; the new public relations title translated as 'human remains pouches'. (Simpson, 1991—a *KIA* was 'killed in action')

body odour the smell of stale sweat.

The advertising invention of Lifebuoy Soap, which claimed to correct the condition. Now normally shortened to *BO*:

> SA and BO, perfect rubbish and bosh—one was a beauty or a *jolie laide* and that was that. (N. Mitford, 1949—'SA' was sex appeal)

body of opinion support for a policy advocated by a political party.

The body is seen to be *growing* even while public enthusiasm remains tenuous:

> Jeffrey has been quoted all week as an Unofficial Spokesman, an Informed Source, Feelings in the Party, Sources Close to the Leadership and as a Growing Body of Opinion. (Lynn and Jay, 1989)

body rub masturbation by a prostitute.

One of the services offered, usually to males, in a *massage parlour* by a *body worker*. *Body wax* is not what you get in your ears, or to help with removing hairs on the body, but a turd.

body shaper a corset.

An invention of advertisers to persuade fat women that they are neither fat nor buying a corset. Also as *body briefer*, *hugger*, and *outline*.

boff[1] to copulate with.

Of a male. It probably comes from the slang *boff*, to hit, with the common violent imagery:

> He boffs her or he doesn't boff her. She leaves. (Sanders, 1977, of a man who might have copulated with a woman)

On the other hand, it could be a corruption of *buff*, to rub, and that is surely the etymology of the other sexual meaning, 'to masturbate', used of a male.

boff[2] to fart.

Common British school slang. The etymology is obscure.

bog a lavatory.

A shortened form of *bog-house*, probably from the marshy ground which used to surround it before the advent of the cesspit or the septic tank:

At the court held in October 1753... Edward Clanvill was charged with a 'public nuisance in emptying a bogg house (privy) in the street'. (Tyrrell, 1973)

And in modern use:

... been in the bog a little while... What do you suppose he's doing there? (Theroux, 1973)

In male slang, to *bog* is to defecate.

bogy a policeman.

Probably from the meaning 'a devil', although a *bogy* was really an apparition with special powers of inducing nursery terrors, making you lose your way in the dark, causing you to be chased by phantoms, etc.:

Well, the bloody bogies are cleaning the streets up. There won't be a girl about. (Kersh, 1936, of clearing prostitutes from the London streets prior to a coronation)

Now also of a military foe, often in the spelling *bogey*:

... the target-identification aircraft, which could vector the fighter bombers on to any bogey appearing on their radar screens. (de la Billière, 1992—in this case, the *bogeys* were Iraqi)

Horses were especially sensitive to bogies, and would rear, or *boggle*, when they saw one in their path, as our minds still do in the cliché.

boiled drunk.

The common American culinary imagery:

A crowd that can get boiled without having to lie up with Dr Verringer... (Chandler, 1953)

boiler house an operation for the unscrupulous selling of shares.

From the intense heat applied and the international successor of the *bucket shop* (see at BUCKET[2]). Also as *boiler room* or *shop*:

... an ex-con called Sidney Coe who had time for a boiler room operation in Kansas City. (Sanders, 1990)

The Dutch authorities are finally acting to close down the 'boiler-shop' share-pushing operations based in Amsterdam. (*Daily Telegraph*, August 1986)

boilerplate comprehensive disclaimers and provisions in an agreement.

As used on battleships, with an inference of elaborate and excessive protection for the lawyers rather than the client:

... so that the attorneys for the underwriters could satisfy themselves on matters of title and other boilerplate. (M. Thomas, 1982)

bollocks the testicles.

A venerable usage. The variant *ballocks* suggests derivation from BALLS and the words are synonymous as a vulgar riposte. *Bollocky* of a male means 'naked':

I'm going bollocky. I don't even care. (Theroux, 1989, of a swimmer)

bolt to leave your spouse for another heterosexual partner.

Usually of women, from an unmanageable horse:

He mightn't want to send you off, but he'll be jolly pleased now you've bolted. (Murdoch, 1978)

and of a male:

... was it conceivable that her husband was preparing to bolt? (Murdoch, 1974)

A *bolter* is a person, usually female, who so acts.

bolt the moon see MOONLIGHT FLIT.

bomb[1] an atomic bomb.

As in the slogan *Ban the Bomb*, the taboo being against nuclear fission rather than high explosive —war loses its glory when it threatens to get too close to home:

The French say they will soon have a Bomb. (N. Mitford, 1960)

bomb[2] see BOMBED OUT.

Bombay milk-cart a vehicle used to empty the contents of lavatories.

Like the English milk-cart, it had two wheels and carried its load in a churn:

Waste from the latrines was collected by the sweepers and removed in a conveyance drawn by a pair of oxen and known as the 'Bombay milk-cart'. (Allen, 1975, writing of India)

Bombay oyster a laxative.

This compound of milk and castor oil was given to young British sailors. As with an oyster, you just swallowed it down.

bombed out under the influence of narcotic drugs or drunk.

Or both:

... he was dropping acid and bombed out of his gourd most of the time on pills and booze. (Sanders, 1977)

A *bomber*, *bomb*, or *bombita* (the Spanish diminutive) is usually a marijuana cigarette or a dose of cocaine.

bona-roba *obsolete* a prostitute.

From the fine clothes she wore to attract custom:

She was then a bona-roba. (Shakespeare, *2 Henry IV*, of Jane Nightwork)

bondage sexual activity involving physical constraint or abasement.

Literally, a condition of slavery or of being tied up. Prostitutes' and sexual deviants' jargon.

bonds of life being gradually dissolved dying slowly.

Bath Abbey, from which this example comes, offers many such delightful if morbid evasions, as well as a potted geography of the British Empire:

> The Bonds of Life being gradually dissolved
> She winged her Flight from this World in expectation of a better, the 15th January, 1810.

See also LONG ILLNESS (A).

bone[1] to steal.

Bone may mean 'a finger', which has overtones of thieving as in FINGER-BLIGHT, or there could be an allusion to the ossivorous habits of canines:

> From her grave in Mary-bone
> They've come and boned your Mary.
> (Hood, c. 1830—he certainly worked on his puns, of which this is by no means the feeblest)

Grose gives *boned* as 'taken up by a constable'. The modern American *boning* is enrichment through sharp practice, perhaps from improving the edible weight of meat by removing the bone before sale.

bone[2] the erect penis.

From the rigidity, and as *boner*:

> She was coming on to me outside the men's room. I've got a boner like Babe Ruth's bat. (Bryson, 1991)

Bone-ache, syphilis, probably comes from the general bodily malaise, or it may be a pun.

bone[3] associated with human death.

What is eventually left after burial, along with the teeth, if any. Many obsolete uses such as *bone-house*, a coffin, *bone hugging*, carrying a corpse to the grave, *bone-orchard* or *bone-yard*, a burial ground, etc. You may still meet occasional jocular use, as:

> ...we usually plant one or two in the bone-orchard before we start for home.
> (N. Mitford, 1960, of a party of elderly tourists in Spain)

To *make old bones* is to live long; it is normally used in the negative as a prediction of a short life. The modern American *make bones* is to kill, with special significance in the Mafia, where a candidate must *make his bones*—kill someone—before qualifying for full membership. Some figurative use, meaning 'to receive full acceptance in a group or society':

> The men behind him were old-time spooks who had made their bones on the Berlin Wall when the concrete was not even dry. (Forsyth, 1994)

boner see BONE[2].

bonk to copulate.

The usual violent imagery, from the sense 'to hit'. The *Mail on Sunday* of 9 August 1987 devoted an article to the etymology, which was speculative rather than enlightening. *Bonkers* has long meant 'slightly mad' or, in obsolete British naval slang, 'intoxicated'.

booby a mad person.

Literally, a fool. An American *booby hatch* or *hutch* is an institution for the insane:

> A year later the bride was in the booby hatch. (Sohmer, 1988)
>
> Check the booby hutches... for escapees. (Sanders, 1981)

Boobies, a female's breasts viewed sexually, and its shortened form *boobs*, may be a corruption of the American *bubbies*, all being slang rather than euphemism.

book a sentence in prison.

In America *a book* is a sentence for a year, *the book*, a sentence for life. See also THROW THE BOOK AT, which may explain the derivation. In obsolete use *books* or the *devil's books* were playing cards and to *plant the books* was to cheat by stacking the pack.

book club see CLUB[2].

bookmaker a person who accepts bets for a living.

Book is a 19th-century shortening of *betting book* and this punning standard English use developed soon after, to be followed by other euphemisms for this profitable but once disreputable profession.

boom-boom[1] defecation.

American nursery usage, the image coming perhaps from the firing of ordnance.

boom-boom[2] copulation.

Again from the firing of a gun? That would imply male activity, but it is used of either sex:

> 'No more boom-boom for that mamma-san,' the Marine said, that same tired remark you heard every time the dead turned out to be women. (Herr, 1977)

As you might expect *boom-boom-boom* is repeated copulation:

> You get that madam into bed, boom-boom-boom—why, she's glad to lodge you for a week. (Fraser, 1971)

A *boom-passenger* was not a passive partner on such an occasion, but, in 18th- and 19th-century British usage, a convict who was chained to the boom on deck during his sea voyage to a penal colony.

boondock to court sexually.

Supposedly from the Tagalog *bundok*, a mountain, the isolated place where the car might be parked, thus to America from the Philippines. *Boondagger*, a female homosexual taking the male role, may be a punning corruption of *boondocker*.

boost[1] to steal.

Literally, to give a lift to:

> You were in Fulton Superior Court apologizin' for boosting car radios. (Diehl, 1978)

A *booster* is such a thief, perhaps concealing the proceeds in a *booster bag* or *booster bloomers*.

boost[2] to make or accept a fraudulent bid at auction.

American auctioneers' jargon, again from giving something a lift.

boot (the) summary dismissal from employment.

From the kick to speed the departing servant, which today would land you in court if not in hospital:

> You know they can't sack teachers. You've got to do something really drastic before they give you the boot. (Sharpe, 1976)

To *boot out* is to dismiss summarily.

bootleg smuggled or stolen.

Originally of intoxicants, supposedly from the flat bottles concealed on the legs when transporting supplies to American Indians illegally. Standard English for smuggled intoxicants during Prohibition:

> ...had got his hands on some bootleg liquor and was giving a party. (Theroux, 1978)

Now of anything which has been stolen. Also used as a verb:

> Do you think... that he might come back and bootleg a copy and give it to you? (Colodny and Gettlin, 1991, reporting the cross-examination of Admiral Welander in 1971)

booze to drink spirituous intoxicants.

The word followed the common drinking progression from any liquid to intoxicants:

> There let him bowse and deep carouse. (Burns, 1786)

The date of this poetic use is inconvenient for those who would suggest that the word owes its origins to the 19th-century Philadelphian distiller Mr E. G. Booze. A *boozer* is a habitual drunkard or an inn, and *on the booze* means drinking intoxicants, including beer, to excess.

boracic *British* indigent.

From the rhyming slang *boracic lint,* skint. Usually denoting a temporary embarrassment, when the sufferer may describe himself as *brassic.*

border incident see INCIDENT.

born in... an impolite way of indicating that someone is subject to an imperfection associated with the supposed natal place. Thus *born in a barn* may greet the failure to close a door:

> Henno called him back to ask him to close the door; he asked him if he had been born in a barn. (R. Doyle, 1993)

Born in a mill indicates that the person so designated is not listening or paying attention, rather than being deaf.

Born in the vestry denoted that you were a bastard, because your parents had not been married in a church.

borrow to steal or take.

An army usage if the article has substantial value:

> In the Army it is always considered more excusable to 'win' or 'borrow' things belonging to men from other companies. (Richards, 1936—you must not steal from your mates)

Also civilian use referring to social or trivial consumption, for example a match, *borrowed* to light a cigarette, but never repaid. In obsolete use *borrowed time* was life after 70, the biblical three-score years and ten.

both hands a prison sentence of ten years.

American underworld slang, from the number of fingers.

both oars in the water *American* mentally normal.

Always in the negative, from the uneven progress of a boat propelled with one lateral oar:

> They're not exactly demented, but neither Isaac Kane nor Sylvia Mac has both oars in the water. (Sanders, 1985)

both-way having both heterosexual and homosexual inclinations.

> Maybe he wasn't a fag. One of those both-way people you were always reading about. (Goldman, 1986)

See also SWING BOTH WAYS.

bother to make unwelcome sexual approaches to.

Especially of a male within marriage, from the sense 'to trouble' rather than 'to fight':

> ...grandma whispering hoarsely, 'Leave me alone, will you?'... I only knew he was 'bothering her'. (Cookson, 1969, of a child sharing her grandparents' bedroom)

bottle[1] **(the)** addiction to intoxicants.

Bottles and intemperance have long gone together, especially if the preference is for wine and spirits:

> The bottle was enjoyed by both as a launching pad for the missile of social grace. (Ustinov, 1971)

To *take to the bottle* is to be an alcoholic:

> Mitzi had taken to the bottle, since reality was too bleak for her. (Ustinov, 1966)

The regimen of the baby invites many puns such as *on the bottle,* drinking intoxicants to excess. *Bottled* means 'drunk':

> We none of us were ever quiet when we was bottled. (Cookson, 1967)

bottle[2] an act of urination.

From the shaped glass container used by a recumbent male in hospital or sickbed:

> You don't want the bottle, or anything like that? You're ready to see your visitor? (Price, 1979, of a male hospital patient)

bottle[3] male prostitution.

From the rhyming slang *bottle and glass*, arse. Also as a verb meaning 'to bugger':

> I want to bottle you, mate, Tom says. Kim has never heard the expression but he immediately understands it. (Burroughs, 1984, of an invitation to buggery)

bottle-blonde a woman with dyed hair.

From the bottle which held the dye or bleach. Formerly in America also assumed to be a prostitute. Despite the general acceptance of hair dyed for cosmetic reasons, this usage retains its derogatory overtones. Also in America as *bottle-blond*, using the adjectival masculine even though the subject is feminine.

bottom the buttocks.

Not the soles of your feet, which might be more logical:

> God gave them bottoms to be smacked on. (Bradbury, 1976)

An American *bottom woman* is (illogically) a pimp's favourite prostitute. A *bottomless* woman is naked, usually working as a waitress at one remove from TOPLESS.

bought and sold *obsolete* bankrupt.

From the disposal of the debtor's possessions:

> For Dickon thy master is bought and sold. (Shakespeare, *Richard III*)

bought it killed or severely wounded.

Mainly a wartime usage, perhaps from notionally acquiring the missile which hits you; or it could just be from *buying* your ticket back to your homeland—see BLIGHTY:

> They bought it—all except me. I'd gone for a walk... You know, with a spade. (Manning, 1978—a soldier's mates had been killed while he was defecating)

American servicemen tend to *buy the farm*, from a city dweller's dream of retirement, perhaps:

> Who knows when M. M. will buy the farm? (Deighton, 1982, referring to a fighter pilot in the Second World War)

Occasionally civilians may today buy other appropriate real estate:

> ... if I got things wrong in the next few hours and bought the Elysian fields... (Hall, 1988)

bounce[1] to copulate.

From the motion, especially on a sprung mattress:

> We all bounced about in bed together from time to time and enjoyed it. (Fraser, 1970)

A *bounce*, or *bouncy-bouncy*, is an act of copulation:

> One bounce with that female Russian shotput and you'd bust your truss. (Sharpe, 1977)

bounce[2] to be dishonoured by non-payment.

Of cheques, because they are returned to the person who drew them, like a rubber ball dropped to the ground and caught again.

bounce[3] to dismiss peremptorily from employment.

Again from the rebounding after hitting another surface, such as the sidewalk:

> If the case is cleared, or I get bounced, the two of you go back to your regular duties. (Sanders, 1985)

The *bounce* or *grand bounce* is such dismissal, and also the ending of courtship, usually by the female. A *bouncer* evicts the unruly or unwanted from a nightclub.

bounce[4] to persuade by violence.

Criminal and police jargon, of extortion, forcibly extracting a confession, etc.:

> You push the victim on the floor. When he comes out this time we're going to grab him and bounce him a little. Nothing heavy. (Sanders, 1977)

The word is also used of rushing another party into an ill-considered contract:

> Soviet support for the heavy Cuban involvement in Angola... was achieved... through 'bouncing' the Russians. (*Sunday Telegraph*, November 1983)

bouncers the breasts of an adult woman.

Not rubber balls but from their pendulous tendency when unsupported:

> 'Look at the bouncers on that one.' Hossbach smacked his lips, eyeing the new girl. (Moss, 1987)

bout an act of heterosexual copulation.

The imagery is from wrestling:

> I was sorry to hear that Sir W. Penn's maid Betty was gone away yesterday, for I was in hope to have had a bout with her before she had gone, she being very pretty. (Pepys, 1662, who adds 'I have also a mind to my own wench, but I dare not, for fear she prove honest and refuse and then tell my wife')

bowel movement defecation.

Medical jargon:

> Most constipation is 'imagined'. A daily bowel movement can be a needless fetish. (Hailey, 1979)

See also MOVE YOUR BOWELS.

bowler hat discharge, especially premature discharge, of an officer from the services.

What was once the standard business headgear replaced the uniform cap:

> Command in the desert was regarded as an almost certain prelude to a bowler hat. (Horrocks, 1960, of British armies in North Africa in the Second World War)

Used also of routine retirement and as a verb:

> If Frank had been bowler-hatted and replaced by Bret... (Deighton, 1988—neither Frank nor Bret was a serviceman)

A *golden bowler* is a substantial payment to an officer retired prematurely, usually against his wishes.

box[1] a coffin.

Today people tend to this use when reference is made to their own death, as 'When I'm in my box...'. Formerly to *box* was to place a corpse in a coffin for burial:

> Ol Joe Sharman died. Donald made the coffin and they'd boxed him. (Emerson, 1892)

To *box the Jesuit and get cockroaches* was not to bury a priest, but (of a male) to masturbate, involving a selection of unsavoury puns.

box[2] a shield for the male genitalia.

Mainly sporting use but now extended to riot protection:

> The cricket boxes issued to constables as items in their 'new protective equipment range' are made of nasty plastic with very little room for accommodation. (*Police*, July 1981)

box[3] *American* the vagina.

Viewed sexually as a container for the penis:

> Her box is so big she wouldn't even feel your hand unless you wore a wristwatch. (Wambaugh, 1975)

boy an adult male.

Of any age, including geriatrics:

> Boy. He must be forty-four. (Collins, 1981)

Only white people used the word for a black servant, which is now considered offensive as well as inaccurate.

boy scouts (the) *American* state police.

They enjoy lower status than some of their colleagues and often wear clothes like a Baden-Powell scoutmaster. Citizens' Band slang.

boyfriend a male extramarital sexual partner.

In heterosexual use, he can be any age over puberty:

> ...occasional liaisons which she alluded to by saying... 'He's an old boyfriend of mine.' (Theroux, 1976)

In homosexual use, usually an adult:

> It is not known whether he will take his South African boy friend, ballet dancer [a masculine name], with him. (*Private Eye*, March 1981)

Girlfriend is used in the same way, of a sexual mistress or of a woman's regular homosexual partner:

> What was he so worried about? Maybe he'd got himself a girlfriend. (Kyle, 1975)

The phrases are also used of normal courtship.

boys[1] a lavatory for exclusive male use.

A variant of MEN, not reserved for juveniles:

> I went into 'Boys' and looked around. (Theroux, 1979, referring to a lavatory)

And as with *men's room*, so with *boys' room*:

> You should know we never lock the boys' room. (Sharpe, 1977)

boys[2] any group of men engaged in a nefarious enterprise.

In this use, they are never juveniles. *The boys* may mean members of a criminal gang. The American *boys in the backroom* are those who dictate policy to elected politicians, as distinct from the *backroom boys*, who innovate on behalf of an employer. Also American, the *boys upstairs* control vice without making an appearance in person, from the normal location of managerial offices relative to the workplace of other employees:

> Snyder had appealed to Christiansen for a reduction of his weekly quota. Christiansen... said he'd talk to the boys upstairs. (Weverka, 1973)

The American *boys uptown* are corrupt political bosses anywhere, from the location in New York City of Tammany Hall, which housed the crooked Democrat politicians who used to run the city through those whose election they contrived. In Rhodesian use, the *boys in the bush* were black terrorists supplied from Zambia and Mozambique:

> There are still going to be some boys in the bush dreaming of marching into Salisbury. (*Sunday Telegraph*, December 1970)

brace to kill.

Not, I think, from the putting of arms (Old French *bras*) around, from which the standard English meaning 'to strengthen' comes; perhaps from an American slang meaning 'to waylay'; but most likely from the rigor mortis. All these suggestions are guesses.

> You and your friend go up to brace him. (Sanders, 1973, of a killing)

In criminal use, *bracelets* are handcuffs.

bracer an intoxicant.

Literally, something which stiffens, whence a tonic of any kind. It refers only to a spirituous drink.

Brahms *British* drunk.

From the rhyming slang *Brahms and Liszt*, pissed. See also MOZART, which is rarer.

branch water *American* water which is offered from a bottle.

It is supposed to come from an unpolluted tributary, or branch, of a stream and therefore not to spoil the taste of your whisky with the taint of chlorine. Only bartenders can say how much *branch water* does not come out of the common faucet.

brand X a competitor's product.

Advertising jargon, to avoid being sued for defamation when you have said how inferior it is to your own. As you will have subjected both to selective and unscientific tests and then paid a woman to purport to examine the test results before endorsing your claim, it is sensible to use evasive language. For a manufacturer, *brand X* may be your competitor's product, possibly better than your own, about which you need to know more before writing the specification for development work.

brass a prostitute.

From the rhyming slang *brass nail*, tail. We now meet it only in the cliché *as bold as brass*.

brassière a garment to contain women's breasts.

In French originally a sleeved garment, thus becoming euphemistic there before English adopted it to cover the taboo breasts with a double evasion. The native *bodice*, from *bodies*, was a garment covering the upper part of the body, whence the *bust bodice*, which held the breasts only:

> Others have compared them to Madonna's bust bodice. (A. Waugh, *Daily Telegraph*, 14 December 1994—*them* were two brick cones containing a theatre and *Madonna* is an American singer and actress)

As *bust bodice* became explicit, it was reduced to the initials *BB*:

> Start-rite shoes for Wills and Rory, summer vest for Aunt Dolly, esoteric haberdashery for the Duchy (fearful things like dress preservers), BBs for Clary and Polly. (Howard, 1993—giving a shopping list for someone visiting London in the Second World War)

This usage lapsed when admirers of the actress Brigitte Bardot appropriated those punning initials for her, their *bébé*. We can note with relief that the American *boobytrap* failed to gain popularity.

bread-and-butter letter a routine acknowledgement of hospitality.

A communication indicating breeding rather than imparting information:

> This is obvious from the cool tone of the bread-and-butter letter Charlotte penned to Mrs Smith. (L. Gordon, 1994—Charlotte Brontë was writing to the mother of her publisher George Smith, whose self-styled 'benevolence' in the 1880s involved evicting a family to the workhouse because the wife's screams after constant beatings by her husband upset other tenants. He also inaugurated the *Dictionary of National Biography*.)

break a commandment to copulate outside marriage.

Yes, the one proscribing adultery:

> Look, there is a pretty man; I could be contracted to break a commandment with him. (Pepys, 1665—the speaker was the 'bonny lass' Lady Robinson)

break a lance to copulate.

Of a male, punning on *lance*, to copulate, and the rigid *weapon* which returns to a flaccid state—is *broken*—after that kind of use. The American prostitutes' jargon, *break luck*, means merely to secure the first customer of the day.

break the news to obtain a confession or other information by violence.

The American criminal is forcibly made aware of the extent of his predicament:

> 'Breaking the news'... and numerous other phrases are employed by the police... as euphemisms to express how they compel reluctant prisoners to refresh their memories. (Lavine, 1930)

break the pale *obsolete* to copulate extramaritally.

The *pale*, as in *paling*, was a piece of wood, then a fence, then a fenced-in curtilage, and finally a district under the control of a centre with hostile natives prowling round outside. In standard usage, to *break the pale* was to get where you shouldn't be:

> ... he breaks the pale,
> And feeds from home.
> (Shakespeare, *Comedy of Errors*)

break wind to belch or fart.

See WIND[1]. It is more respectable when used of belching, although Shakespeare indicated farting in his complex vulgar pun:

> A man may break a word with you, sir; and words are but wind;
> Ay, and break it in your face, so he break it not behind.
> (Shakespeare, *Comedy of Errors*)

And in modern use:

> I'll kill the first son of a bitch who even breaks wind. (West, 1979)

The American *break the sound barrier*, to fart or to belch, refers to a sonic boom.

break your elbow *British, obsolete* to give birth to a bastard.

The fracture was sometimes caused by a figurative bed:

> And so she broke her elboe against the bed (Heath, 1650, of a woman who had a child while unmarried)

but if she *broke her elbow at the church*, she was no worse than a poor housekeeper after marriage. To *break your knee*, a direct translation of a French euphemism, was to copulate outside marriage for the first time, of a woman. An unmarried pregnant woman might also be said to *break her leg above the knee*, referring to a ruined horse and to the location of her genitals:

If her foot slip and down fall she,
And break her leg above the knee.
(Fletcher, 1618)

The putative father, if brought to book, was also said to have *broken his leg*.

break your neck to have an urgent desire to urinate.

Normally of a male, with no fear of suicidal tendencies. It originally meant that *break-neck speed* was needed for this activity.

break your shins against Covent Garden rails see COVENT GARDEN.

breathe your last to die.

Perhaps circumlocution and evasion rather than euphemism, as you cannot expect to live more than two or three minutes after the event:

...the quicker that one breathed his last, the better, so I hurried up with my lance...and drove it into his throat. (Fraser, 1969)

breezy drunk.

An American use, from the bonhomie which may be detected in some at a certain stage of drunkenness.

brew to burn.

Of an armoured vehicle, from the habit of British soldiers, who brewed tea over an open fire, often one raised by burning petrol being allowed to soak into sand:

You'll have seen a tank being brewed, Johnny. (Seymour, 1980)

British Second World War tanks were usually inferior to their German opposition and their crews were often burnt to death if the tank was hit by anti-tank weapons.

brewer's goitre frontal obesity in a male.

The thyroid gland is situated elsewhere, but the big belly is caused by persistently drinking too much beer:

—the crenellated face, the brewer's goitre slung under his belt. (Keneally, 1985—in fact the belt is often ineffectively slung under the protuberance)

brick a kilogram of marijuana.

In America it may be packed in the form of a block, the size of a normal building brick.

brick short of a load (a) of low intelligence.

Of the same tendency as FIFTY CARDS IN THE PACK.

Bridewell a prison or police station.

The original London Bridewell was a holy well with supposed medicinal properties, then a hospital for the poor, then a prison:

Crowley went to the 'nearest bridewell' and told the officer of his wife's accusation. (Pearsall, 1969—Crowley was accused of raping his own four-year-old daughter)

Bridport dagger *obsolete* a hangman's rope.

The Dorset town was, and still is, famous for its rope, twine, and net-making, the industry developing from the facility with which flax was once grown in the vicinity. To be *stabbed with a Bridport dagger* meant 'to be killed by hanging'. Perhaps still some literary use.

brig a prison.

Shortened form of *brigantine*, a ship often used as a naval prison:

I'm not saying he'll end up in the brig, but he'll lose all rank. (Higgins, 1976)

Civilian as well as military use.

Brighton pier a male homosexual.

Rhyming slang for 'queer'. An obsolete British use—there were two piers for trippers to the Sussex town.

bring off[1] to cause to achieve a sexual orgasm.

Of either sex, in copulation or masturbation:

He remained in her for what seemed like hours...bringing her off again and again. (M. Thomas, 1980)

bring off[2] to cause the abortion of a foetus.

It is physically removed from the mother:

I was left in the club...like any tiresome little skivvy, but unlike her we were able to arrange to have it brought off. (P. Scott, 1975)

bring out the highlights to dye parts of the hair blonde.

The jargon of barbers and of manufacturers of dyestuffs for women's hair. The pretence is that human hair, which is dead matter, has lively properties which can be brought into play by the application of a lotion.

bring your heart to its final pause to die.

Unless you experience an extra-systole, prior pauses are unlikely:

...and bring his heart to its final pause. (Eliot, 1871)

Bristols the breasts of an adult female.

From the rhyming slang *Bristol City*, titty. Mainly male salacious use:

LAIDBACK, FUNLOVING AUTHOR, 44, is anxious to meet respectable bit of stuff with big bristols and own teeth. (advertisement in *Private Eye*, November 1988—I suppose respectability is a subjective matter)

The compiler of the spelling checklist on my word processor considers this usage noteworthy while the venerable City and County is omitted.

broad a prostitute.

The 15th-century adjectival meaning 'vulgar' survives only when we speak of humour or the accents of country folk. As a shortened form of *broad woman*, it alludes to her moral laxity and not to her girth:

> Give me some pictures where the good guys get the dough and the broads once in a while. (Deighton, 1972)

In obsolete British use *the broads* were not a collection of such females but playing cards, with a suggestion of cheating:

> Will you have a ... touch of the broads with me? (Mayhew, 1851, of an invitation to play cards)

Broad coves, fakers, men, pitchers, etc. were common cheats.

Bronx cheer a fart.

Because the noise is simulated by the mouth to indicate derision, there are some who forget the original meaning, taken from one of the less fashionable parts of New York City.

bronze eye the anus.

In homosexual parlance:

> ... he didn't mind sodomizing a client, but his own bronze eye was closed to all comers. (Fry, 1991)

broomstick match see JUMP THE BESOM.

brother a professional or an associate.

Used to denote a calling thought socially undesirable, if not criminal. Thus a *brother of the bung* was a brewer, a *brother of the gusset* a pimp, a *brother starling* was a man with whom you shared your sexual mistress. In modern British, the accident of birth apart, our *brothers* are fellow socialists, trade unionists, or members of secret societies, where like-mindedness is taken for consanguinity. In America a *brother* may be a fellow black who dislikes a society dominated by whites:

> ... dude called Washington Lee was a brother, not the house nigger on some editorial board. (McInnerney, 1992)

brought to bed see BED[1].

brown dead.

From the rhyming slang *brown bread*.

brown (do a) see BROWN-HATTER.

brown-hatter a male homosexual who indulges in buggery.

From the anal content of the sexual partner:

> Harrison's are a lot of wankers and Slymne's go in for brown-hatting. (Sharpe, 1982)

Some figurative abusive use, too:

> A lot of brown-hatters and word-merchants. (Sharpe, 1974)

Sometimes shortened to *brownie*. To *do a brown* is to bugger a male or a female.

brown-nose to flatter.

Not from exposure to the sun but from the figurative proximity to the anal area of the object of your sycophancy:

> Hungerford—you missed the beginning but this is a course you can't fail so there's no need for brown-nosing. (Goldman, 1984—a pupil had been flattering his tutor)

brown paper bag *American* an unmarked police car.

Nothing on or about it advertises its identity in this Citizens' Band use. To *brown bag* is to take your lunch to work with you and eat it at your desk.

brownie[1] see BROWN-HATTER.

brownie[2] a spirituous intoxicant.

Whisky or brandy, from the colour:

> I had to toddle off to the sherbert cupboard and administer a stiff brownie and water. (*Private Eye*, July 1981)

brownie[3] an amphetamine tablet.

Again from the colour.

brownie points the supposed rewards of currying favour with your superiors.

Earned by juniors in any hierarchical organization who ostentatiously perform, often at the expense of their peers:

> Then you'll find out who slid the blade into Sidney Leonides. And you'll get brownie points for clearing a homicide. (Sanders, 1987)

From Baden-Powell's *Brownies*, which punned on the colour of their uniform and the benevolence of the creatures which are supposed to perform good deeds around the house by night. Promotion in the pack, to the exalted position of sixer or beyond, is obtained by winning points for good works or achievement.

browse to steal and consume food within a store.

From animals' selective feeding habits. In this refinement of supermarket theft, you carry the stolen goods past the checkout in your alimentary canal.

brushfire war a war in which a major power is not directly involved.

It involved figuratively the undergrowth and not the timber:

> The language of the mad foments [violence] ... 'Brushfire wars', 'limited actions', 'clean atom bombs'. (West, 1979)

brute *American, obsolete* a bull.

Of the same prudish tendency as COW BRUTE. For a list of similar euphemisms see BIG ANIMAL.

bucket[1] a place for defecation.

A male usage, especially where a smaller tin for urination is inside communal living quarters and defecation on a larger receptacle takes place outside. Also some figurative use:

> Get off the bucket, I'm serious. (Theroux, 1978)

bucket[2] a prison.

From the rhyming slang *bucket and pail*, jail. In obsolete use, *the bucket* was also peremptory dismissal from employment, probably from the phrase *kick the bucket*, to die (see at KICK[1]). A *bucket shop* is not an ironmonger selling pails but an organization trying to sell securities of little or no value to the public or airline tickets at discounted prices.

bucket[3] to kill by drowning.

A favoured way of disposing of an excess of kittens:

> Hadn't someone better bucket them at once? (N. Mitford, 1960, of newly-born kittens)

buckle to to copulate.

In standard usage, to tackle a task vigorously:

> ... threatening to feed me to the sharks unless she buckled to with him. (Fraser, 1977)

However, the perhaps obsolete English *Bucklebury* is buggery, of which DSUE says 'Ex the Berkshire locality'; for the good name of the residents I suggest that it is from the phonetic similarity.

Buddha stick *American* a marijuana cigarette.

The common association of the mysteries of the east and narcotics.

budget cheap.

Advertising jargon. The implication is that the cost will not exceed the amount which you have allowed for the purpose.

buff the bare skin.

A shortened form of (flenched) *buffalo*, which survives as *in the buff*, naked. To *buff* is also to rub or polish, whence perhaps the 19th-century meaning 'to embrace sexually':

> I wor fit for booath coartin' and buffin'. (Mather, 1862)

bug[1] to conceal an apparatus for eavesdropping in.

From the size, colour, and shape of the device:

> He was ready to give me permission to bug his church pew. (Diehl, 1978)

A *bug fix* is introduced to foil or detect such activity by your opponent:

> It took us months to crack the 'bug fixes' and find our way inside. (Deighton, 1981)

bug[2] a mark indicating the use of union labour in manufacture.

Mainly in the American printing trade, perhaps punning on BUG[1]. The mark is not authorized but condoned by the employer.

bug out to retreat.

This Korean war usage came from the American slang meaning 'to quit rapidly'. Whence *bug-out fever*, cowardice:

> 'Bug-out fever', the urge to withdraw precipitately in the face of the slightest threat from the flank. (Hastings, 1987 of American soldiers in Korea, many of whom had been happier in their previous station in Japan)

bughouse *American* mad.

Perhaps from the insects figuratively in the head:

> It's enough to make a man bughouse when he has to play a part from morning to night. (Conan Doyle, 1917)

The noun denotes an institution for the mentally ill:

> You're bigger bloody fools than anybody outside a bughouse. (Marsh, 1941)

bull[1] to copulate with.

Of a male, from the function for which uncastrated animals are mainly preserved:

> He would guarantee all his female slaves had been bulled by his crew. (Fraser, 1971, writing of the early 19th century—a black slave pregnant by a white man is said to have commanded a premium)

Occasionally also as *bullock*:

> ... giggling about how he'd bullocked the headman's wife on his last leave. (Fraser, 1975)

A *bull* is also a male who regularly seeks casual copulation:

> He is the village bull. The women dare not refuse him. (Manning, 1960, of a priest)

bull[2] egocentric boasting.

Shortened form of *bullshit*:

> You're full of bull this morning. (Steinbeck, 1961)

—but not a shortened form of *bullshitter*, a person who boasts or who acts officiously. In army use *bull* is no more than strict compliance with military standards of drill or tidiness. The initials *BS* apply both to *bullshit* and *bullshitter* in all senses:

> He was a great romancer and wrote the biggest B.S. of them all. (Richards, 1933)
>
> ... sitting around, BS-ing, talking about how law school was coming. (Goldman, 1986)

The uses may come from the 19th-century term *bull-scutter*, defined by Dr Wright as 'the liquid excrementum of a bull after gorging new grass' and figuratively as 'anything worthless or nasty' (EDD).

bull[3] a policeman.

Perhaps from his size and aggressive mien, as it is an American use. Originally only a detective:

Only on rare occasions will the cop... offer any information to the 'bull' or 'dick'. (Lavine, 1930)

Now applied to any uniformed protector of property. A *bullpen* is either a prison:

...ordered them thrown into the bull pen (ibid., of men arrested)

or any common dormitory for males.

bull[4] a female homosexual taking the male role.

I think just a shortened form of *bull-dyke*, meaning the same thing:

So you gave that old bull a key (Theroux, 1976, of an elderly female homosexual)

and

I know the model. Bull dyke. (Sanders, 1977)

bullet (the) peremptory dismissal from employment.

The imagery comes from FIRE and at second remove from the standard English *discharge*. A *bullet* also means 'death by shooting', especially in espionage fiction:

...never knowing whether they're getting a medal or a bullet. (le Carré, 1980)

Fatalistic soldiers say that the bullet which hits them has their name on it:

...he could no more remove the name from his heart than avoid the bullet which had his name on it. (Price, 1978)

bum *American* a prostitute.

Originally in American slang a vagrant or an idle person who hangs around saloons etc.:

It's a well-dressed class of customers. No bums. (Dreiser, 1900)

A *bum* is also the male counterpart of a *bunny* in various sports (see BUNNY[1]). *Bum*, the buttocks, in British English was once a genteel usage although *bum-shop*, a brothel, and *bum-fighting*, copulation, were not. For a dissertation on *bum fodder* see AMMUNITION[1]. In current use, a *bum bandit* is a bugger, *arse bandit* being a rarer form.

bump[1] **(the)** peremptory dismissal from employment.

From the sudden displacement. Less often as *bump off*:

They got bumped off the staff of the hospital. (Chandler, 1939)

In American use *the bump* can either be contriving the dismissal of a senior in a hierarchical system, to obtain that post for yourself, or posting an unwanted senior employee in a union shop from one task to another, each less attractive than the last, until he ends up cleaning the lavatories and hands in his own notice. This process circumvents strict 'first in, first out' labour contracts.

To *go bump* is to be bankrupted.

bump[2] **(the)** pregnancy.

Literally, any swelling in the body, usually caused by a blow. This rather tasteless use is uncommon. A *bumper* is not the putative father but a stripper in an erotic stage show.

bump[3] to copulate.

Probably from the repeated pushing of the bodies against each other:

One could imagine brother and sister bumping like frogs in broad daylight. (Theroux, 1978, of incest)

Occasionally as *bump bones*.

bump[4] a killing.

From striking heavily:

Normal routine in the case of a bump is to stay clear. (Hall, 1969, of a murder)

Bump off, to kill by violence, is commoner:

'He had to take risks.' 'Like bumping chaps off?' (le Carré, 1980)

Sometimes shortened to *bump*:

I don't go round bumping everyone I meet, you know. (Keneally, 1985—in an Australian novel)

An American *bump man* is a professional killer.

bump[5] to cause a pre-booked passenger to travel by a later aircraft.

Airline jargon for the practice of ensuring that staff and those in a position to cause trouble about overbooking are not among those left behind:

17 passengers were 'bumped' in all: although after the desk closed he heard the girl being told to allow for six to eight extra Sudan Airways personnel on the flight. (*Private Eye*, December 1981)

bump off see BUMP[1, 4].

bun a prostitute.

This leads us to the mariner's fetish about rabbits, which must not be mentioned before a voyage if you wish to avoid ill luck. The taboo dates back to the days when fraudulent chandlers supplied cheap rabbit meat, which doesn't keep when salted, for pork, which does. The superstitious had, before a trip, to touch the tail of a hare or, if none were to hand, the pubic hair of a woman, including one who might for a fee allow hers to be touched. Thus the *bun* was a shortened form of *bunny*, itself the diminutive for the rabbit, the hair, and the prostitute. In obsolete British use a *buttered bun* was a prostitute who copulated successively with different men on the same occasion—perhaps punning lewdly on the seminal deposits of previous partners. Grose however says of such a prostitute: She is said to have a buttered bun.

bun in the oven (a) pregnancy.

Second World War service use, punning on the rising of cake mixture and the hidden growth of the foetus:

I rather fancied she had a bun in the oven. (Theroux, 1971, of a pregnant woman)

bun on (have/get a) to be or become drunk.

Perhaps a shortened form of *bundle*, a quantity of anything. Also as *tie a bun on*:

We'll celebrate tonight, if you do. And if you don't, well, then we'll tie a bun on anyway, just to forget it all. (van Druten, 1954)

The Australian *buns on* means 'menstruating', from a slang word for the towel worn.

bun-puncher a person who never drinks intoxicants.

British army usage, where abstention from intoxicants can be as taboo as drunkenness in civilian life:

If a teetotaller he was known as a 'char-wallah', 'bun-puncher' or 'wad-shifter'. (Richards, 1933—of a tea and cake man)

bunch of fives a punch with a clenched fist.

Hardly euphemistic when it refers to the fist alone, but the implication is always that it is to be used aggressively, *given* or *handed out* to the victim.

bundle to copulate before marriage with the consent of your family.

The New England custom of 'bundling', namely the supposedly chaste lying in bed together of young, affectionate, unmarried persons of opposite sexes for the sake of company and the saving of fuel. (Graves, 1941, writing of the 18th century—the partners nearly always married each other)

Similar customs obtained in southern Scotland and elsewhere. The English dialect *bundle with* can still mean 'to marry':

My God! do you expect to bundle with that 'un? (Cookson, 1967, referring a male suitor)

bung[1] *obsolete* a drunkard.

Literally, a stopper for a cask:

Away... you filthy bung. (Shakespeare, *2 Henry IV*)

Bung also meant 'drunk', as you were if someone said you had been to *Bungay fair*, punning on the Suffolk market town.

bung[2] contraception by the use of a pessary.

From the method of vaginal introduction and the function it is expected to perform:

I used to be Bung but now I'm Pill. (Bradbury, 1975)

bung[3] a bribe.

Perhaps from the habit of pushing notes into a pocket, although the OED says 'Origin unknown'. Specifically in the UK denoting illegal cash payments to footballers being sold by one club to another:

Arsenal sack Graham over cash 'bung' for transfer. (*Daily Telegraph*, 22 February 1995)

bung[4] *Australian* dead.

From an Aboriginal word according to the OED, but the concept of something stopped up or finished suggests an alternative derivation, especially as it also means 'financially ruined'.

bung up and bilge free copulating.

British naval use, from the recommended way of storing a cask of rum aboard ship, whence anything in good order. The *bung* is a reference to the filling of an orifice rather than to a particular method of contraception.

bunk flying boasting.

American air force usage. The exploits which redound to your credit are dreamed or otherwise invented in bed.

bunny[1] a homosexual who takes the female role.

It can be a male or female but, if a male, usually a prostitute. The use may come from BUN but I doubt it because *bunny* is a pet name given to someone with timid and feminine characteristics. *Beach, jazz, ski, surf*, etc. *bunnies* are heterosexual girls who consort with males adept at those pastimes—*bums*—with whom they copulate promiscuously.

bunny[2] a towel worn during menstruation.

From its shape and feel? I have seen no literary use.

buoys a lavatory for exclusive male use.

Punning on *boys*, and favoured by American restaurants near the sea, where females are confronted, perhaps perplexed, by *gulls* for girls.

burial of an ass *obsolete* no burial at all.

The hole had to be too big to excavate in rocky terrain:

He shall be buried with the burial of an ass, drawn, and cast forth beyond the gates of Jerusalem. (Jeremiah 22:19)

buried *American* in prison.

Referring to withdrawal from society.

Burke *obsolete* to murder.

From the celebrated Irishman who killed people to replenish his stock of corpses, which he sold for medical dissection until he was hanged in Edinburgh in 1829. See also RESURRECTION MAN for a dissertation on this trade. Care should be taken to avoid confusion with *berk*, a stupid person, which comes via rhyming slang from the *Berkshire* or *Berkeley Hunt*, viewed figuratively and not anatomically.

burn[1] *obsolete* to infect with venereal disease.

From one of the symptoms felt by the male:

> Light wenches will burn. Come not near her. (Shakespeare, *Comedy of Errors*)

Whence *burned*, so infected:

> No heretics burned, but wenches' suitors. (Shakespeare, *King Lear*)

A man who *burned his poker* was so infected and a *burner* was the infection.

burn[2] to kill.

Originally by electrocution, from the singeing of contact points on the corpse. Latterly in America of any violent death, especially by shooting:

> Do you really think Knorr burned Kipper and Stonehouse? (Sanders, 1980, of two such killings)

burn[3] to extort from criminally or to cheat.

Probably derived from the phrase *put the burn on*, to compel, through figurative application of heat or physically by contra-rotating the skin at the wrist:

> I thought he was the one who burned me. (Theroux, 1976, of a cheat)

There are two American narcotics uses—to take money for illicit supplies and fail to deliver; and to give information to the authorities about another's addiction.

burn out *mainly American* to withdraw from narcotics addition.

Especially of an older addict, with obvious imagery.

burn with a low (blue) flame *American* to be very drunk.

The imagery is from a dying fire, about to go out.

burra peg see CHOTA PEG.

burst to have an acute need to urinate.

What you fear will happen to an over-extended bladder and usually in the idiom, *I'm bursting*, with *for a pee* left unsaid.

bury to inter (a corpse).

So long standard English that we assume the thing being buried is a dead body, unless we elaborate by saying *buried alive*. The root meaning was 'to protect, cover' (OED). So too with *burial*, with its assumption of prior death.

bury a quaker to defecate.

In Irish use a *quaker* was a turd. Unless it is from a lack of sympathy for their beliefs among those strongly committed to Rome, the etymology escapes me. A *quaker's burial ground* was a lavatory.

busby a woman's pubic hair.

In standard usage, the tall fur hat of a soldier:

> ...she had him growling in her busby. (Sanders, 1982, describing cunnilingus)

bush[1] the pubic hair.

Of male and female. Partridge says 'Low (mid-c. 19–20) after being a literary euphemism' (DSUE). Still widely used:

> The small, trimmed bush, soft as down. (Sanders, 1982, describing a naked female)

The American *bush patrol* means 'extramarital copulation', punning on the pubic hair, the seclusion sought, and the military exercise so named. In parts of the world where jungles persist, a *bush marriage* is performed without due ceremony:

> ...most of them were bush marriages performed by some joker wearing a coconut mask and a feathered jock-strap. (Sanders, 1977)

In the same areas, a bastard may be said to have been *made in the bush*:

> Yeboa had been 'made in the bush', and was lighter in colour than Anokye. (ibid.)

bush[2] marijuana.

Another term for the slang *grass*. The obsolete British *bush house* was a private house which sold beer or cider on fair days, hanging out a bush to denote it was open:

> Starting from the 'Bush-house' where he had been supping too freely on the fair-ale. (EDD, quoting a source from 1886)

bushwack to ambush.

Literally, to hack a path through woods or to propel a craft upstream by pulling on overhanging foliage:

> ...had bush-wacked a Russian baggage train and were busy looting it. (Fraser, 1973)

business any taboo or criminal act.

It may be defecation and, less often, urination —by humans:

> Frensic finished his business in the lavatory (Sharpe, 1977)

or by animals:

> Clem, a pedigree Labrador, evidently feeling at home, did his business. (Sharpe, 1976)

It may be copulation, usually extramarital and often prefixed with *amorous*, *night's*, etc.:

> ...gave her the business for all I was worth on top of her dressing-stool (Fraser, 1982)

or illicit narcotics aids such as eye-droppers, bent spoons, syringes, etc., perhaps alluding also to the theatrical jargon meaning.

It may denote prostitution; to be *in the business* is to be a prostitute:

> Mine was a large lady, already in the business for some time. (Londres, 1928, in translation)

A *business woman* is prostitutes' jargon for a prostitute.

It may also mean 'killing':

...you'd tried to give the Führer the business. (Price, 1978, of someone who had tried to kill Hitler)

businesslike and friendly totally unproductive.

The phrase is used to describe a meeting of parties with inflexible and opposing views which neither modified during the discussions.

bust the consequence of any taboo course of conduct.

It may refer to financial ruin, a drunkard, a drunken carouse, the unwelcome and violent entry of the police into a gathering, a violent robbery, an arrest, etc. The victim may thereupon be *busted*:

Professor Philip Swallow... was among sixteen people arrested on Saturday... 'I've never been busted before,' he said. (Lodge, 1975)

In America a *bust* is particularly used of police action against illegal narcotics use, but the drug users' jargon *bust a cap* means 'to ingest an illicit narcotic', from the breaking of the seal on the phial.

bust bodice see BRASSIÈRE.

bust your nuts to achieve a sexual orgasm.

The American phrase is used of both sexes, despite the masculinity and function of the *nuts*, the testicles. In figurative use, merely to try hard.

buster *American* a fart.

Presumably because it is the substance which causes you to BREAK WIND.

busy *English* a policeman.

Probably a shortened form of *busybody*:

... don't hang around. The bloody street's alive with busies. (Kersh, 1936)

butch a female homosexual playing the male role.

Perhaps an abbreviation of *butcher*, whence a rough or uncouth male. Also used adjectivally. Rarer use of a male homosexual acting in an ostentatiously masculine way:

He marked them down as very butch guys. (B. Forbes, 1986, of two homosexuals)

In Manx dialect a *butch* was a witch, but without any suggestion of homosexuality.

butt the buttocks.

So long standard usage in American that it perhaps hardly qualifies as a euphemism, even as a shortened form:

Without exception the papers had replaced the offending words with ellipses or dashes, or else had changed them to something lighter —making 'kissing your ass' into 'kissing your butt', for instance. (Bryson, 1994, of reportage of an argument in 1991)

buttered bun see BUN.

butterfly a male homosexual.

From the light and pretty appearance of the diurnal insect:

... if it ever comes out that Dunce's top aide is a butterfly, it's not going to do his candidacy any good. (Sanders, 1984)

buttock a prostitute.

I suppose from a part of the body brought into play. A *buttock ball* was copulation, according to Capt. Grose. A *buttock and twang* was just a prostitute but a *buttock and file* was a prostitute who would rob you. To *buttock* was to copulate extramaritally, whence *buttock-mail*, a fine if you were found out:

Yer buttock-mail, and yer stool of repentance. (W. Scott, 1814)

I think all these British uses are obsolete.

button[1] the clitoris viewed sexually.

Perhaps from the appearance. It is to be found in both obsolete British and modern American slang. The punning *button-hole* was to copulate and a *button-hole factory* was a brothel.

button[2] *American* a professional killer.

Presumably you press him for action when you need it:

Know what a button is, DeLoroza? A shooter. (Diehl, 1978)

Also as a *button man*:

His head was alive and jumping with notions of button men. (M. Thomas, 1980)

button[3] *American* a policeman.

From those on his uniform:

The buttons won't have any time to worry about what's going down on East 55th Street. (Sanders, 1980)

buxom fat.

Of women. There was formerly an implication of comeliness as, for example, in this reference to Dr Spooner's barmaid:

In the daytime she is fair and buxom. (Old saw)

buy a brewery to become an alcoholic.

Or consume its output:

Then the jackaroo married the station and bought a brewery. (Kyle, 1988—in fact, he married the owner of the property and became an alcoholic)

buy it see BOUGHT IT.

buy love to copulate with a prostitute.

Normally of a man but perhaps some homosexual use:

'I don't buy love,' I warned her, 'but how much do you generally get?' 'From one dollar to five.' (Harris, 1925)

Like other songwriters, the Beatles were not averse to punning lyrics.

buyer a person addicted to illegal narcotics.

Addict jargon—he probably also buys food and clothing from time to time:

> The label is drugs... Converse was a heavy buyer. (Ludlum, 1984)

buzz on (a) being mildly intoxicated or under the influence of narcotics.

From the ringing in the ears?

> ... we'd drink, get a little buzz on, and then go into the ocean to swim and sober up. (Theroux, 1973)

buzzed[1] killed.

First World War usage, from the noise of the bullet which hit you. Not heard in the Second World War, despite the destructive power of the *Buzz Bomb* or V1.

buzzed[2] drunk.

Having a *buzz on* (see BUZZ ON (A)):

> He seemed a trifle buzzed when he arrived, blew the ceremony several times, most noticeably when he forgot the business with the ring. (Goldman, 1984, of a drunken priest officiating at a wedding)

by(e) an indication of bastardy.

Literally, ancillary. *By(e)-blow*, *-chap*, *-scape*, etc. meant 'a bastard' and *by(e)-come*, *-begit*, etc., bastardy. Still used despite the fact that bastardy is almost respectable:

> I really was a niece of a one-time Governor and not some by-blow of Lili Chatterjee's family. (P. Scott, 1973)

In obsolete British use *by-courting* was done on the sly without intention of marriage:

> Bitterly did I regret I had done my by courtings so near home. (Crockett, 1896)

A *by-shot* was an elderly unmarried woman, though not because Cupid's aim was poor:

> If she cannot restrain her loquacity, she is in danger of bearing the reproach of a by-shot. (Tarras, 1804—reminding us of the old Scottish taboo in respect of unmarried females)

by yourself mad.

The etymology is unclear. We would not use the old form:

> But monie a day was by himself,
> He was so sairly frighted
> (Burns, 1785)

although we retain the idiom *beside himself with rage*.

C

C anything taboo beginning with the letter C.

If denoting a disease, it is the dread cancer, which is the *big C;* if narcotics, cocaine.

CC pills US army laxatives in the Second World War.

And some civilian use thereafter. The *CC* is not, I think, from cubic centimetre, common carrier, chief clerk, cricket club, closed circuit, or carbon copy.

cabbage[1] to steal.

Cabbages were odd snippets of cloth which were traditionally the perquisite of tailors in employment, who often made sure good material fell into the same category:

> If I cabbage that ring tonight, I shall be all the richer tomorrow. (N & Q, 1882, quoted in EDD)

Still in common British use.

cabbage[2] the female genitals.

From a supposed visual resemblance. Modern male American use, but obsolete in the British Isles.

cactus juice *American* an intoxicant.

Because it is supposed to be able to save the life of a traveller dying of thirst. Usually tequila, especially in states bordering on Mexico.

cadge *obsolete* to pilfer.

Dr Johnson gives '**cadger** A huckster; one who brings butter, eggs and poultry, from the country to market', whence an itinerant vendor who had a reputation for thieving as he moved about:

> A thieving set of magpies—cadgin' 'ere and cadgin' there. (Ward, 1895)

In modern use merely to beg, especially off acquaintances.

cage a prison.

Dysphemism perhaps. The derivation is obvious, but there is a 19th-century reference to the yellow clothes worn by dangerous convicts, coloured like the popular cage-bird, the canary.

Cairo crud see CRUD.

California sunshine lysergic acid diethylamide.

Better known by the initials *LSD.* In obsolete American use a *California widow* was any deserted wife whose husband had left her in the literal or metaphorical hope of striking gold. *California blankets* were newspapers used by the destitute as bedding, as they still are to pad clothes against cold weather. For *California kiss-off* see KISS-OFF[1] (THE).

call (the) death.

Your God needs you elsewhere:

> Hersel' was the first gat the ca'. (Grant, 1884, of a dead woman)

We no longer *call souls* from a flat tombstone on Sunday mornings, when deaths might be announced along with any other gossip:

> Last Sunday fwornuin, after service... the clark caw'd his seale. (Anderson, 1805)

Call it a day means 'to die', from 'to finish work'. *Called,* dead, is usually amplified by the addition of a destination such as *home* or *away*:

> He had been ca'ed awa atween the contract an' the marriage. (Wilson, 1836, of the death of an engaged man)

God is also referred to as a lender requiring satisfaction:

> So he was only sixty-six when God called in the loan. (Lyall, 1985)

Called to higher service is perhaps my favourite euphemism, embodying in one phrase an avoidance of direct reference to death, an implication that the dead person was specifically wanted by a deity, the recognition of meritorious work of a religious nature on earth, and the acknowledgement that heaven is the destination:

> In March 1875, Mr Empson was stricken down with paralysis, and was called to higher service on June 28th the next year. (Tyrrell, 1973)

call-button girl a prostitute.

She usually operates in airport hotels, where there is a constant harvest of transient males:

> Prostitutes, 'call-button girls' as they call themselves, roam from airport to airport, operating from the airport hotels. (Moynahan, 1983)

See also CALL GIRL.

call down to announce publicly that you will not pay your wife's debts.

I give this obsolete British usage to remind us that it was not very long ago that a married woman had no property rights and could only pledge her husband's credit for food and clothing, called by the jargon word *necessities.* The *calling down* was done by the Town Crier and exonerated the husband from meeting any debts contracted by his wife in his name from then on. Failing a Town Crier, notice might be inserted in the local newspaper.

call girl a prostitute.

She used to operate from or in a *call house,* a shabby brothel:

> ... it's no worse than playing the piano in a call house. (Perelman, 1937)

This was changed by the advent of the telephone, by which she may now be summoned:

> A low-church missionary who was discovered as being the business manager of a ring of syphilitic call-girls. (Ustinov, 1971)

Today *call girls* tend to be elegant, expensive, and free from disease, so that their services can safely be bought to bribe businessmen and politicians. The rare *call boy*, apart from summoning performers to the stage on cue, is a male prostitute:

> He made an additional two hundred as a call boy for discriminating gay customers. (Wambaugh, 1981)

call it eight bells let us drink intoxicants.

Noon, or eight bells of the watch, is the earliest time in British naval etiquette when intoxicants can be drunk publicly.

call of nature the need to urinate or defecate.

The visit is demanded by your bodily mechanism:

> I was probably off the road, behind the bush, answering a call of nature. (Follett, 1978)

The imagery appears in several phrases:

> 'When nature calls, heh, heh, heh,' he'd said ... and made his way out into the trees. (M. Thomas, 1980)

call off the bets to die.

If a horse is withdrawn from a race under certain conditions in America, the wagers are invalid.

call out *obsolete* to challenge to a duel.

The resulting contest was staged in the open air:

> If you were not my brother I'd call you out for saying that. (Deighton, 1987—a son had spoken disparagingly of their father)

call the tricks see TRICK.

called see CALL (THE).

caller (a) menstruation.

A usage which shares with *visitor* the imagery of a temporary imposition (see VISITOR (A)).

callisthenics in bed copulation.

Callisthenics is training in graceful movement:

> ... other than callisthenics in bed, and from what rumours I hear, you're getting plenty of that. (Hailey, 1979, describing a libertine)

calorie counter a fat person.

Advertising jargon, subtly suggesting that blame for obesity does not attach to the sufferer:

> ... don't risk offending them by calling them fat. Their ads are addressed to 'weight watchers' and 'calorie counters'. (Jennings, 1965)

camera a police radar set.

I suppose from the resemblance and from its ability to hold a record of the speed indicated for inspection by the motorist. In American Citizens' Band slang a *folding camera* is a speed monitoring device carried in a police patrol car.

camisole *American* a straitjacket.

Medical jargon. Literally, a woman's bodice, which can look something like a straitjacket.

camp[1] homosexual.

The word is used of either sex and, although the progression from using exaggerated gestures to male homosexuality is well documented (see OED), the etymology is obscure. That has not deterred lexicographers from speculation, with Ware suggesting 'probably from the French', without further elucidation, and Partridge urging us to consult the EDD, where Dr Wright gives us a choice from 'an ancient form of the game of football', 'gyrating in the air', 'gossiping', and 'a heap of potatoes or turnips earthed up in order to be kept throughout the winter'. Unfortunately we shall never know which of these caught Partridge's fancy. The use developed in the mid 1940s:

> The Red Shadow is at large. Did you ever see anything quite as camp? (P. Scott, 1975—the dialogue about a male homosexual act in 1946 may have been anachronistic when placed in India)

To *camp it up* is to display or accentuate male homosexual characteristics and, in America, to take part in a male homosexual orgy. To *camp about* may mean no more than to act jokingly:

> ... just words, they weren't meant seriously, I was just camping about. (Bogarde, 1981)

camp[2] see CONCENTRATION CAMP.

camp down with to cohabit and copulate with extramaritally.

Permanence is implied in the arrangement, but not homosexuality:

> Race left Linda with a weeks-old baby and camped down with his House of Commons harpie/secretary. (*Private Eye*, July 1981)

camp follower a prostitute.

From the days when armies had a trail of tradesmen and providers of services moving about with them:

> ... to prevent their men from contracting certain indelicate social infections from —hem hem—female camp-followers of a certain sort. (Fraser, 1975, writing in 19th-century style)

campaign an organized course of action by a pressure group.

Less aggressive than a *front* and occasionally concentrating on important issues, such as the Campaign for Real Ale. More often the title masks just another minority pressure group or supposed grievance.

can[1] *American* a lavatory.

Originally of the bucket which was emptied from time to time but now used of all degrees of plumbing sophistication:

> Snyder had paced the small office and gone to the can a couple of times. (Weverka, 1973)

can[2] *American* to dismiss from employment.

The dismissed person is figuratively put in the *ash-can*, or even in the *can* (see CAN[1]):

> He worked for maybe a month and then he was canned. (Sanders, 1980)

There is also a college jargon meaning 'dismissal of a student for misconduct or academic failure'.

can[3] *American* a prison.

From the meaning 'a container'. Usually of a short-stay location in a confined cell:

> You wanna sit in the can for twenty years? (Weverka, 1973, seeking to emphasize the boredom of close confinement)

can[4] a prostitute. See CANHOUSE.

can on (a) drunkenness.

Probably from the American use of *can* to denote a drinking vessel (WNCD). The phrase antedates the fashion of selling beer in cans. *Canned* means 'drunk' with *half-canned* no less so. A *can* may also be an ounce of marijuana and *can on* and *canned* probably are already used in speech for narcotic stupefaction.

canary an informer to the police.

He 'sings' out criminal information, as in the cliché *singing like a canary*, of a suspect who tells all. Thus the *canary trap*, to catch those who improperly abstract, copy, and then circulate confidential documents, a practice which almost gave us the verb to *pont*, from a celebrated exponent of the art, Clive Ponting:

> 'What about internal security—the project documents?' ... 'You mean canary traps? ... You use the machine to make subtle alterations in each copy of important papers.' (Clancy, 1988)

A *canary* or *canary-bird* was a dangerous convict, because of the yellow clothing he had to wear. In obsolete British use, a female *canary* was either a sexual mistress—'kept' in her figurative 'cage'—or a burglar's assistant:

> Sometimes a woman, called a 'canary', carries the tool, and watches outside. (Mayhew, 1862)

candy *American* illegal narcotics.

Specifically at one time cocaine, and then marijuana or LSD on a sugar cube, but now any narcotic. *Nose-candy* is sniffed:

> C'mon t'daddy little girl. C'mon an' get your nose candy. (Collins, 1981)

The punning *candy man* is a dealer in illegal narcotics.

canhouse *American* a brothel.

Can is rare slang for the buttocks, whence in obsolete use a prostitute:

> The little girls, looking so sweet and demure, knew all the words for canhouses ... and seemed ready to illustrate them with anyone. (Longstreet, 1956)

canned see CAN ON (A).

canned goods *American* a person who has not copulated.

Usually of an adult female: untainted (free from disease) and unopened (with maidenhead intact). Occasionally of a male.

cannibal a person who practises fellatio or cunnilingus.

Usually of a prostitute, *eating* the penis of her American customer.

cannon a pickpocket.

The imagery is from the pool table. You bump into your victim, make him stumble and rob him in the confusion which follows. In underworld slang a handgun may also be called a *cannon*.

canoe to copulate casually.

If a young man took a woman for a trip in such a craft, there was no room for a chaperone, which gave them unwonted seclusion when they went ashore:

> Her Old Man ... had been hearing about me and Daisy canoeing from the first night we'd got together. (L. Armstrong, 1955, meaning copulation and not aquatic sports)

The modern *canoodle*, to fondle sexually, is perhaps a compound of *canoe* and *cuddle*, although you can't do the latter in the former unless you want to swim:

> Helen had fallen from a balcony while canoodling with a Dutch sea captain. (*Private Eye*, May 1981)

canteen medal an exposed trouser fly-button.

A *canteen*, originally a wine cellar, is a place of refreshment for soldiers. The British army did not rate highly medals awarded other than for acts of bravery.

cap *American* to buy, open, or use illicit narcotics.

Drug users' jargon, from the seal on the phial which is broken prior to use, or from a shortened form of *capsule*.

capital involving killing.

Literally, of the head, but now seldom referring to beheading. A *capital crime* is murder, involving a suspect in a *capital charge* before the courts and, in some places, in *capital punishment*, death. A *capital sentences unit* is not a piece of writing using capital letters but the place in America where judicial killing is done.

capital act (the) copulation.

The head would seem to play a minor role in the proceedings, nor is the activity confined to a principal city:

> ...crashed through the top of the bed into the fireplace, and completed the capital act among the warm embers. (Fraser, 1985)

capon a male homosexual.

Literally, a castrated cock. In obsolete standard English it meant 'a eunuch' but the current American use may merely imply that his sexual activity does not lead to procreation.

captain is at home (the) *?obsolete* I am menstruating.

From the British officer's red coat.

card *Irish, obsolete* to punish by laceration.

A 19th-century weapon of the Irish against their landlords:

> Two widows... who... had paid their rents in full, were visited by a party of women with blackened faces and were 'carded'—had sheep's combs drawn through their flesh. (Kee, 1993)

cardiac incident a malfunction of the heart.

Cardiac means 'pertaining to the heart' and an *incident* is a happening. Medical jargon, but every heartbeat might with accuracy be so described. *Cardiac arrest* is a jargon way of saying that the heart has stopped beating.

cardigan *American* a contraceptive sheath.

This use is at two removes from the Crimea, where the pugnacious earl gave his title to an article of clothing.

cardinal is at home (the) *?obsolete* I am menstruating.

From the red biretta and other robes of office.

cards (your) dismissal from employment.

British revenue stamps were affixed weekly to cards, originally to provide various insurance and pension rights but latterly as a tax on employment, paid by both employer and employee. Without your *cards* properly stamped you might not draw state aid when unemployed or for long retain any new employment:

> Get your cards! You take a week's pay and you get out of my place. (Deighton, 1972)

The dismissal is not necessarily peremptory and an employee, wishing to leave employment, might *ask for his cards*.

care *British* the guardianship of children by a local authority.

Often the children, who may be unruly or criminal or have parents unfit to look after them, are confined in an institution. A child subject to such guardianship is said to be *in care*, which is not to say that children living normally at home with their families are uncared for. See also CARING.

careful[1] stingy.

From the old concept that thrift is praiseworthy while meanness is odious:

> He is careful. In the narrow financial sense he always seemed to enjoy receiving hospitality. (Bevins, 1965, reported speech, of Harold Macmillan)

careful[2] using a contraceptive during copulation.

Originally applied to the prevention of impregnation but today as relevant in the context of disease:

> If you can't be good, be careful. If you can't be careful, remember the date. (Old saw)

caress yourself to masturbate.

Of a female, from the proper meaning 'to touch gently':

> She admitted having caressed herself ever since she was ten. (Harris, 1925)

cargo any valuables fortuitously received.

A generic term, especially in the South Pacific, for manufactured articles, food, etc. which strangers may abandon for the natives. A *cargo cult* is the religion in such places of looking skywards for 20th-century manna.

caring the ostentatious display of social conscience.

Originally used by those alert to hypocrisy or self-advertisement in others:

> They will probably become nuns or prison wardresses or join the 'caring' professions. (A. Waugh, *Private Eye*, July 1980)

Now standard English of nurses, home helps, and the like, or *carers* in the jargon, with little thought that those not so employed may not of necessity be careless. *Uncaring* has retained its meaning of cruel, selfish, or insensitive, often in a double negative:

> Ulyatt, who was not a cruel man, or an uncaring one, simply shut his eyes. (Kyle, 1978)

carnal pertaining to copulation.

Literally, it means no more than 'of the flesh'. Legal jargon and standard English in various phrases. A *carnal act* is copulation:

> ...the only time I've completed the carnal act with my nose full of water was in Ranava Ilona's bath. (Fraser, 1977)

Carnal knowledge is extramarital and usually with a taint of wrongdoing:

> 'Know you this woman?' 'Carnally, she says.' (Shakespeare, *Measure for Measure*)

Carnal necessities alludes to the supposed ill-health of those denied copulation:

I have been afflicted for ninety years by the carnal necessities of women. (Sharpe, 1978—the venerable speaker was a libertine)

Carnal relations is copulation, of either sex:

Maitland had had carnal relations with several other women during this period. (Condon, 1966)

A *carnal stump* is an erect penis:

To see some brawny, juicy rump
Well tickled with my carnal stump.
(Rabelais, in translation)

carpet[1] to reprimand.

Unlike the workshop or servants' quarters, the master's room had a carpet on which the defaulter had to stand. Whence *on the carpet*, liable for punishment. Beware the French *sur le tapis*, which merely means 'up for consideration'.

carpet[2] a wig.

Worn by an American male; a variant of RUG:

...snowy-white hair. If it wasn't a carpet, it had enjoyed the attention of an artful coiffeur. (Sanders, 1979)

carpetbagger a plunderer.

Originally, an absconding American banker, who so carried the reserves with him, but in wide use of the Northerners who sought pickings in the defeated South after the Civil War. Modern use is of touts for merchant bankers etc. in rich but primitive societies, whose activities may reflect the cultured fee-grubbing of some of their principals in London, New York, etc.

carry[1] to be pregnant.

Of the same tendency as BEAR[1] and long standard English. A pregnant woman may be said to be *carrying*, without specification of the burden:

She was in the seventh month of pregnancy and carrying big. (Collins, 1981)

To *carry a child* is specific:

Mrs Thrale is big, and fancies that she carries a boy. (Dr Johnson)

carry[2] to have an illegal narcotic on you.

A shortened form of *carry drugs*. Because of the risk of detection in a body search, a rule among drug users says 'Never carry when you can stash'.

carry[3] *American* to be in possession of a handgun.

A shortened form of *carry a gun* but, unlike *carrying* narcotics, this covers both legal and nefarious possession:

'Ahh, I'm carrying', Boone said. 'Someone will spot the heat.' (Sanders, 1977, of a policeman)

carry[4] to drink intoxicants without overt drunkenness.

With or without intermittent urination:

...as gentlemen should, carried their two bottles of an evening. (Strachey, 1918)

carry a card to be a member of the Communist Party.

The use developed in the 1920s when such membership was not flaunted because it might lead to ostracism:

Maurice Dubb who was probably the first academic in Britain to carry a card. (Boyle, 1979)

With the collapse of Communism, the euphemism is no longer needed.

carry a (heavy) load to be drunk.

This common American expression refers to the weight consumed, although not necessarily of beer. See also LOAD[1].

carry a torch for to desire sexually.

The imagery probably comes from the religious processional light:

Maggie Young-Hunt came in today. Out of coffee, so she said. I think she's carrying a torch for me. (Steinbeck, 1961—a male was talking about a woman who was stalking him)

carry off to kill.

An epidemic or sudden illness is said to *carry off* the person who dies. And in the passive:

...if one of the characters did happen to be carried off in the course of nature. (N. Mitford, 1949)

carry on with to copulate with regularly outside marriage.

Of either sex. Perhaps the 19th-century use meant no more than to consort with (a person of the opposite sex):

I carry on with him now and he likes me very much. (Mayhew, 1862—a young woman was speaking of her swain)

In modern use the relationship is explicit and censurable:

...administered a public wigging to Princess Margaret when she was carrying on with that nancy-boy pop singer. (*Private Eye*, April 1981)

carry the banner to be out of work.

Perhaps from the activities of the Salvation Army, whose members may so advertise their faith and always try to help the poor or unfortunate. The American *carry the balloon* has the same meaning, and refers to the *balloon* or bedroll of the itinerant, which is carried above the pack.

carry the can to get undeserved punishment while the culprit goes free.

If the *can* were the receptacle for urine, the derivation might be deduced. The fuller version, *carry the can back*, suggests a less malodorous and burdensome journey, possibly back to the cookhouse with an empty communal food container. The phrases are also used of a guilty person who is singled out for punishment among several miscreants. Also as *take the can (back)*:

Nobody wanted to take the can back. (B. Forbes, 1986, of people trying to avoid blame)

carsey a lavatory.

From the Italian *casa*, a house, or, as defined by Dr Johnson: 'A building unfurnished'. Also as *carsy*, *karsey*, *karzey*, and *karzy*:

'Mens Retiring Room' which he assumed was the carsey. (Follett, 1991)

case[1] a brothel.

Again from the Italian *casa*, a house, and sometimes so spelt:

Four casas, four women, often four Frenchwomen, to the square hectare.
(Londres, 1928, in translation, of Buenos Aires)

Grose reports that a *case vrow* was 'A prostitute attached to a particular bawdy house'. The form *caso* survived into the 1930s at least:

Some people used to call her Caso Maggie. She got her money out of girls. (Kersh, 1936)

A *casita* was a small brothel:

... the representative of the law hurries to the *Casita* and the woman pays at once. (Londres, 1928, in translation)

While all these uses are obsolete, we retain our *casinos*, for public gambling.

case[2] anything that is the subject of a taboo.

In medical use, especially where it might be a breach of confidence to divulge the name of the patient; in funeral jargon, a corpse:

We cremate quite a few cases. (J. Mitford, 1963)

The recidivist is a *hard case*.

cash flow problem an insolvency.

Cash flow is literally the difference between the receipts and payments of a business on a continuing basis. It becomes a *problem* when companies knowingly continue to trade while insolvent, which makes the directors personally liable for the debts then contracted:

Once *that* word gets round we are to have what is euphemistically called a cash flow problem. (Sharpe, 1977)

You may also hear the expression used, perhaps humorously, of temporary personal indigence.

cash in your checks to die.

Equally common in the form *cash in your chips*, both American phrases alluding to the gambling counters for which you get money when you leave the game, there being an assumption that you may be going to your final reckoning. In similar fashion, *checks* or *chips*, may be *passed in*. But if you announce in company that you are going to *cash a check*, you are leaving to urinate.

casket a coffin.

Literally, a small box, usually containing something of value, then a receptacle for human ashes, whence the now standard American English use.

cast to give birth to prematurely.

A standard usage, normally of animals, from the meaning 'to cause to fall':

Just a pair still-born at the hinner een',
Puir dwarfed last anes,
Wee, deid, cast anes.
(Lumsden, 1892, of lambs)

Whence perhaps the two punning phrases which meant 'to give birth to a bastard': *cast a girth* used equestrian imagery and *cast a laggin girth* came from the displacement of the hoop which held the staves of a tub, causing them to spill outwards:

... slipping a foot, casting a leglin-girth or the like. (W. Scott, 1822)

In obsolete British use, if you were *cast for death* you were not playing Julius Caesar but terminally ill, *casting* being divination by magic:

He's cassen her planets, and he's sure she'll dee. (Peacock, 1870)

cast your pellet to defecate.

From *cast*, to let drop:

... the squatting early morning figures of male labourers casting their pellets upon the earth. (P. Scott, 1973, of India)

casting couch (the) extramarital copulation between a female seeking a favour and a male in a position to grant it.

Originally used with reference to actresses seeking roles:

... married a veteran Hollywood stunt man ... saved her from being just another hooker working the casting couches. (Collins, 1981)

Less often outside the theatre:

Young lady, I do not need a casting couch, I can have any woman I want. (*Private Eye*, May 1981, quoting a male journalist)

casual (the) an institution which housed the destitute.

A shortened form of the British *casual ward*, *casual* meaning 'coming at uncertain intervals'. Those who used such accommodation were also known as *casuals*.

cat[1] a prostitute.

Still found in America but obsolete in the British Isles. *Cat-house*, a brothel, is used generally:

'What are those places?' asked Treece. 'Warehouses,' said Jenkins. Treece thought he said whore-houses ... They didn't look like his idea of a cathouse. (Bradbury, 1959)

A *cat* may now also be a male prostitute:

If you want to bugger a male cat, that means you're a queer (Theroux, 1973)

or any male who frequents brothels:

These cats comin' here for a good time. (Collins, 1981, writing about a brothel)

cat[2] *mainly American* the vagina.

Probably from the resemblance of the female pubic hair to the feline fur and see also PUSSY:

> The rest of them were putting cigarettes in their cats and puffing on them. (Theroux, 1975)

catch a rich marriageable adult.

The imagery is from fishing: the *catcher* improves status or security through a marriage. Anyone described as a *good catch* may well have so far eluded the matrimonial net. In Pidgin *catch* also means 'to copulate with', along with 'infinite different meanings' (Jennings, 1965).

catch a cold[1] to contract gonorrhoea.

Second World War British army use, punning on the slang meaning 'to get yourself into trouble'. Perhaps Shakespeare had the same thing in mind when he wrote:

> A maid, and stuft! there's goodly catching of cold. (*Much Ado About Nothing*)

catch a cold[2] to have a trouser zip inadvertently undone.

An oblique warning from one male to another, received by me on the quay at Destin, Florida in November 1987.

catch a cold[3] to suffer a loss.

Normally as a speculator or gambler:

> The 1980s speculative bubble burst and while the rest of the world caught a cold, Japan got pneumonia. (*Daily Telegraph*, 5 December 1994)

You may also in such circumstances *lose your shirt*.

catch a fox see FOXED.

catch a packet to be killed or severely wounded.

Packet is a diminutive of *pack*, and the use meant being struck by something solid, such as a shell, in the First World War. After that war, it meant getting into any kind of serious trouble until the Second World War, when two distinct new meanings developed. Of a regiment or locality, it meant 'to be subjected to a heavy attack from the ground or air':

> The same thing's happening to the 2nd Northants, they've caught a packet too. (Price, 1978, of a badly mauled regiment)

Of a serviceman, it meant 'to catch venereal disease'.

catch fish with a silver hook to pretend to have caught fish which you have bought.

An expression among anglers, where such behaviour is opprobrious. Unlike ANGLE WITH A SILVER HOOK it does not also mean 'to bribe' but see also SHOOT WITH A SILVER GUN.

catch the boat up to have contracted venereal disease.

Naval usage. Jolly (1988) suggests a derivation from the days of pressing, when seamen were not allowed ashore for fear that they would desert. A sick boat would circulate among the fleet and take patients, often with venereal disease, to a naval hospital ashore. In the First World War a seaman so afflicted was obliged on release to travel to regain his ship, and this may be the origin, but it would be an etymology unsatisfactory to those naval purists who are distressed when the layman confuses *ships* and *boats*.

category killer *American* a cut-price store in a shopping complex.

The idea is to attract customers for the cheap article, who may then do other shopping in the same precinct:

> ... unenclosed developments, usually built in a U-shape around a central parking lot and containing at least one *category killer* store—a place like Toys 'R' Us or Circuit City selling a particular type of product in such volume and at such low prices as to deter any nearby competitor. (Bryson, 1994)

cattle[1] negro slaves.

They were herded, exploited, bred, and sold like animals:

> Could be payin' (a right nice price) for the right kind of cattle. (Fraser, 1971, writing of the early 19th-century southern states of America)

cattle[2] to copulate.

From the rhyming slang *cattle truck*, but now only used figuratively:

> I don't give a flying cattle if you give me fifteen thousand pounds a week. (Kersh, 1936)

Cattle, sometimes also defined as prostitutes, seems to be based in that sense solely on John Evelyn's 'Nelly ... concubines and cattell of that sort' (Evelyn, before 1700). The diarist might as easily have written 'people of that sort' and this seems slim evidence on which to assume that *cattell* are *sui generis* with Mistress Gwynne and other concubines.

caught[1] pregnant.

The *catching* takes place when she copulates and is impregnated. Mainly female usage, of unwanted pregnancies:

> If the girl gets caught and pregnancy results ... (Harris, 1925)

caught[2] infected with venereal disease.

Medical practitioners tell me that this is the commonest way in which their diseased and embarrassed young patients introduce the subject of their visit.

caught short having an urgent desire to

urinate or defecate in an inconvenient place.

Of both sexes, from the days when coaches or trains stopped at regular intervals but offered no lavatory accommodation between one stage or station and the next:

Well, this virus carried a gun. I nearly got caught short. (Steinbeck, 1961)

cavalry prostitutes who solicit from motor vehicles.

I have not met this use in America or the British Isles but it is common in the English-speaking Far East. The *infantry* solicit on foot.

cease to be to die.

Hardly euphemistic for an atheist. Of more interest perhaps is:

It ceased to be with Sarah after the manner of women. (Genesis 18:11—she had stopped menstruating)

ceasefire a continuation of fighting.

You meet this when the opponents are operating under different rules, and especially if the politicians on one side wish to pretend that the war is coming to an end:

It is mined by the Viet Cong—even more furiously since the cease-fire (which is, willy-nilly, a painful euphemism). (Theroux, 1975)

Lord Carrington will negotiate no more ceasefires in Bosnia until the warlords there have reached stalemate or exhaustion, he announced yesterday. (*Daily Telegraph*, 24 July 1992)

Cecil the penis.

One of the male forenames adopted—see JOCK, JOHN THOMAS, etc.:

I know all he wants is to dip Cecil in the hot grease. (Sanders, 1981, referring to copulation)

celebrate to drink intoxicants to excess.

It is a social convention that intoxicants should be available on a festive occasion. But when a drunk person is said to have been *celebrating*, there is no suggestion of prior festivities.

celebrity a nonentity employed as an entertainer.

Literally, deserving fame. Jargon of the entertainment industry:

On the fringe of the famous... constantly invaded by idle chatter and envious gossip which inevitably, it seems, surrounds what is euphemistically called today a celebrity. (Bogarde, 1978)

cement to prevent defecation.

Of medicine taken after an attack of diarrhoea, although *concrete* might seem more appropriate:

I'd already got the trots. They're supposed to cement you up. (P. Scott, 1975, of pills)

And in various compound uses, as:

The water came from a communal tap down by the road, so it was cement-sandwich country as far as I was concerned. (Lyall, 1972—he wished to avoid defecating)

In American addict jargon *cement* is a poor quality illegal narcotic.

cement shoes (in) murdered and hidden.

American underworld use, from the practice of setting the lower legs in a concrete block to prevent subsequent floating and discovery of the corpse:

There were more bodies down there at the bottom of the lake with cement shoes than there was garbage. (Weverka, 1973)

See also CONCRETE SHOES (IN).

certain age (a) old.

The precise figure is often uncertain:

They were a certain age; they had humps and braces and wooden legs. (Theroux, 1979)

certain condition (a) see CONDITION[2].

certified compulsorily confined in an institution for the mad.

Under British law two medical doctors and a magistrate had to certify that a lunatic should be involuntarily incarcerated:

He's out of his skull... he's ready for certifying. (Bogarde, 1981)

Certifiable means either mad but still at liberty:

I won't put him in an asylum. He really and truly isn't certifiable (McCarthy, 1963)

or, loosely, irrational by the commission of a single stupid act. See also SECTIONED.

chair (the) death by electrocution.

From the piece of furniture to which the victim is strapped:

We get a lock on the case, you could face the chair. (Mailer, 1965)

In obsolete use, your *chair-days* were your old age:

...in thy reverence and thy chair-days, thus
To die in ruffian battle.
(Shakespeare, *2 Henry VI*)

A *chair* is now also a man or woman who conducts a meeting.

chalkboard a blackboard.

A usage of educationalists obsessed with skin pigmentation.

challenged differing from the norm in a taboo fashion.

Not by a sentry or in a duel:

They are mostly feckless, ill-informed and otherwise unemployable people. One or two are parentally challenged. (*Daily Telegraph*, 20 November 1993, quoting Howard Davies, Director General of the Confederation of British Industry, who was accusing some journalists of being bastards)

This usage is one of the more enduring and endearing to have proliferated, especially among the illiterate. Thus a deaf person can be *aurally challenged* and a blind person *visually*, an idiot can be *cerebrally*, *developmentally*, or *mentally challenged*, an acne sufferer *dermatologically*, a bald man *follicularly*, a dwarf *horizontally*, a lame person *physically*, and so on.

challenging unprofitable.

One of the code words used by chairmen in public statements, ignoring the fact that their firm is *challenged* wherever it faces competition:

> Trading conditions in Continental Europe, however, remain challenging. (Pilkington plc Chairman's Report, June 1994)

chamber a receptacle for urine.

Shortened form of *chamber-pot*, from the room in which it is kept in a small cupboard or under the bed. The punning modern American *chamber of commerce* is a lavatory but in obsolete use was a brothel. The obsolete *chamber-lye* was fomented bottled urine, used in the washing of clothes and in dressing wheat, etc.:

> We leak in the chimney; and your chamber-lye breeds fleas like a loach. (Shakespeare, *1 Henry IV*)

chambering copulation.

The activity normally takes place in an upstairs room:

> Harriet heard more than she wanted of the chambering next door. (Manning, 1978)

champagne trick a rich customer for a prostitute.

Prostitutes' jargon, not the ability to open a bottle without spilling any. *Champagne* for the expense and see also TRICK.

chance conceived out of wedlock.

From the unplanned nature of such impregnation. In many standard English and dialect forms such as *chance-bairn*, *begot*, *born*, *child*, and *come*:

> 'Chance children', as they are called ... are rare among the young women of the costermongers. (Mayhew, 1851)

A *chanceling* was a bastard, both literally and as an insult:

> Offspring of a pair a conncelins. (Bywater, 1853)

change[1] (the) the menopause.

Female usage and shortened form of the standard English *change of life*:

> Too young for the change, I suppose. (J. Trollope, 1992)

change[2] *obsolete* to be an idiot child.

Babies born wise and beautiful grew up stupid and mischievous if the jealous fairies did a switch in the cradle:

> My granny never liked her, said she was 'changed'. (Service, 1887)

Thus a *changeling* was an idiot. This usage demonstrates how those in isolated communities rationalized the results of their incest and other inbreeding.

change[3] to replace by a clean one a soiled napkin on (a baby).

The baby remains unchanged, albeit usually cleaner and sweeter smelling:

> The baby now began to scream. 'I expect he wants changing,' said David. (N. Mitford, 1960)

change someone's voice to injure (a male).

In fact or figuratively by a blow to the testicles, although the vocal modification is seldom permanent:

> Damn, if anybody'd talked that way about Cathy I'd have changed his voice for him. (Clancy, 1987)

change your luck to seek to copulate casually with a black woman.

The use is by a white American male, from switching to black in the game of roulette after failure to win on the red.

channel a vein into which narcotics are illegally injected.

Usually in the crook of the elbow and scars there can indicate such practices.

chant falsely to describe a horse for sale.

Literally, to sing. I include this obsolete British use to illustrate that horse dealers had the same reputation and used the same deceits as salesmen of second-hand cars today. Thus a *chanter* or *horse-chanter* lied about the beast's age, temper, hooves, soundness, teeth, coat, etc.—or you might say about its mileage, roadworthiness, tyres, compression, petrol consumption, bodywork, etc.

chap a male suitor.

Originally, a buyer, and then in colloquial use, any man:

> On the suspicion of an offence, the 'gals' are sure to be beaten cruelly and savagely by their 'chaps'. (Mayhew, 1851)

Chapping, courting by a female, is obsolete. The *old chap* was the devil (see also OLD A' ILL THING):

> Speak truth, then ye needna fear
> Tae meet the auld chap face to face.
> (Thomson, 1881)

chapel of ease a mortuary.

Originally, a place of worship for the convenience of parishioners who live a long way from the parish church. Now funeral jargon:

> From 'undertaker' tout court to 'funeral parlor' to 'funeral home' to 'chapel' has been the linguistic progression. (J. Mitford, 1963)

Also, from the supposition that the body is biding its time against its resurrection, as *chapel of rest*:

> 'James' had already mercifully been removed to the 'Chapel of rest'. (Murdoch, 1978)

In obsolete British use a *chapel of ease* was a lavatory, punning on the ecclesiastical sense and on *ease yourself*, to urinate.

chapter eleven see GO[2].

character conventional.

Real estate jargon. The standard English word comes from a Greek instrument for marking or engraving (SOED), whence anything distinctive. But a *character* bungalow can be expected to offer no more than repetitious banality.

charge[1] an erection of the penis.

DAS suggests derivation of this American use from 'activation through an electric charge and/or the sensation of electric shock'. Many males would find the concept less appealing than that of loading a piece of ordnance.

charge[2] an illegal narcotics injection.

The same imagery, of revitalizing, appears in the phrase *a shot in the arm*, which may be what a troop of Brownies is said to have had when it has received some gift or other, as though the small girls and Brown Owl were about to behave with unnatural vigour after being injected with heroin. *Charged up* means 'under the influence of illegal narcotics'.

charity girl a female who copulates free of charge and restraint.

She was usually from a respectable family and responded in this fashion to the early American chauvinism of the Second World War. A *charity dame* was her mother, acting in the same sense.

Charlie a substitute word for a taboo subject.

It may, according to the context, mean a policeman, a homosexual male, a prostitute, an idiot (or *right Charlie*), the male or female genitalia, menstruation (in the phrase *Charlie's come*) and, for a time, a Vietcong adherent:

> They could sure as shit believe that Charley was shooting them. (Herr, 1977, writing about Vietnam)

Some espionage use, too, referring to the Central Intelligence Agency, from the phonetic alphabet:

> It's just the sort of kit Charlie would have been supplying. (Lyall, 1985)

In the British meaning 'a ponce', it is from the rhyming slang *Charlie Ronce*, although neither he nor his namesake *Joe Ronce* (see JOE[1]) have earned a mention in the DNB. If you are told that *Charlie's dead*, you are receiving an oblique warning that your trouser zip is undone or that your petticoat is showing beneath your hem (see also SNOWING DOWN SOUTH). The British army *Charlie Uncle*, a fool, took two letters of the phonetic alphabet, the *cu* of *cunt* used figuratively. The 19th-century *charlies*, a woman's breasts, is obsolete, along with *dairies* and *bubbies* (*The Slang Dictionary*, 1874).

charms the anatomical parts of a female which excite male lust.

A *charm* was originally the singing of a song, whence a magic incantation, hence *charms*, which work such magic:

> I had a full front-view of all her charms. (Cleland, 1749)

In obsolete use to *charm* was to cure by magic:

> Soom folk says it's hall bosh about charmin' yer cock... Mah feyther took a feather o' his cock to t'old witch an' she charmed un. (*Good Words*, 1869, quoted in EDD—an avian remedy was supplied, not an aphrodisiac)

A *charmer* was a good-intentioned or *white* witch and *charming*, in good health:

> An' how's Coden Rachel?—She's charmin', thankee. (Quiller-Couch, 1890)

charter of labour the feudal conditions of work in Nazi Germany.

Industrial and farm workers were tied to their jobs by a repressive law dated 20 January 1934, which stifled trade union rights and freedom to choose or change your job. Presented as a law to guarantee workers' rights, this 'Law Regulating National Labour' was one of a series of acts taken by the Nazis as soon as they won power, to suppress individual liberty.

charwallah a teetotaller in the army.

Originally, the *wallah*, or man, who brought round the *char*, or tea, for the sale to British troops serving in India, whose expression it was. See also BUN-PUNCHER.

chase to seek to copulate with extramaritally.

Of a male, from following in a predatory way. In America you normally *chase* something specific, such as *hump*, *skirt*, or *tail*:

> ...known to tipple a bit and chase hump. (Mailer, 1965, of a drunken womanizer)

chaser an intoxicant different in kind from that just taken.

It follows the previous libation down the throat. Usually of beer after spirits or vice versa, but sometimes too of a further portion of what you have just drunk.

chattering classes (the) narrow-minded bigots.

A British usage describing the phenomenon of intelligent, educated, and opinionated people who have seldom experienced at first hand any process connected with the creation of wealth:

> And when our cheese lovers discover that they can no longer eat Brie unless it is smuggled in via the Channel Tunnel, even the

chattering classes might at last wake up to just how completely and insanely that Euro-train has gone off the rails. (Christopher Brooker in *Sunday Telegraph*, 5 February 1995)

cheap John *American* a brothel.

Originally, a travelling salesman, whence a low or dilapidated saloon and perhaps punning on JOHN[1] or JOHN[2]. I have not seen this expression in print outside a dictionary.

cheap money a policy of allowing money in circulation to increase faster than the generation of wealth.

Before long it leads to inflation and wasted resources. Politicians seeking Utopia or re-election continue to pursue this chimera despite the problems inherent in repealing the law of supply and demand.

cheat to copulate with someone other than your regular sexual partner.

Of either party, within or outside marriage, from the deception usually involved:

> Eight months married and cheating on me with a piece of merchandise like that. (Chandler, 1943)

cheat the starter to conceive a child before marriage.

Sporting imagery, from starting a race before the signal to go. As with BEAT THE GUN the phrase may refer to any detected pre-marital copulation, even without impregnation.

cheaters cosmetic padding.

From the attempt to deceive. They are worn by American women to improve the appearance of thighs, buttocks, chest, etc.

check out *American* to die.

The imagery is said to have come from medical examination on demobilization, but leaving a hotel or cashing in when you quit gambling are just as likely:

> If you get found, you check out. See you in the morgue. (Chandler, 1953)

check the seat covers look in that car for an attractive woman.

The advice is sent on American Citizens' Band radio from one trucker to another. See also SEAT COVER.

cheesecake an erotic picture of a female.

The word puns on the sweet confection and the smile-inducing *cheese* demanded by photographers. Mainly Second World War use. See also *beefcake* at BEEF.

chef a person who prepares opium for smoking.

American addict use, from the heating, whence too the *lamp habit*, ingesting it. (Note that the *chef* or principal cook may be the only one, just as the *Ober* needs no underlings.)

chemically inconvenienced drunk or under the influence of illegal narcotics.

You are far more likely to *inconvenience* others. You may also be described as *chemically affected* etc.

chère amie a sexual mistress.

> Phryne, the chère amie of a well-known officer in the Guards... (Mayhew, 1862)

The French euphemism is carried into English and even occasionally translated as *dear friend*, retaining the same meaning.

cherry the hymen.

From the colour and appearance?

> ... asking me to look after you was the most risky thing she could do if you wanted to hang on to that cherry of yours. (P. Scott, 1968—a libertine was talking to a virgin)

Whence *cherry-picker*, a libertine who seeks out young females for casual copulation.

cherry-pick fraudulently to select bargains.

Trades done in the morning may show gains or losses by the time written records have to be made, allowing any gains to be credited to favoured parties, and the losses to others:

> Since rules for designating customer accounts are lax, the broker can do blank trades in the morning, then 'cherry-pick' the profitable ones at lunchtime, and allocate them to the intended beneficiary. (*Sunday Telegraph*, 27 March 1994—the beneficiary is said to have been Hillary Clinton)

chew to practise fellatio.

From the action of masticating:

> 'He wanted you to gobble ze goo?' she asked. 'What?' 'Chew on his schlong,' Maddie said impatiently. (Sanders, 1981)

chew a gun *American* to kill yourself.

You put the barrel in your mouth and aim upwards:

> Doing good deeds apparently keeps people from chewing on guns. (Wambaugh, 1981)

chi-chi of mixed white and Indian ancestry.

A derogatory use. It means 'dirty' in Hindi.

chic sale a primitive outdoor lavatory.

From the American humorist, Chic (Charles) Sale, who described in his stage act and book the considerations to be taken into account when building outside lavatories.

chick a prostitute.

The common avian imagery. Occasionally also of a male prostitute, in homosexual use. Also denoting any young female viewed sexually by a male:

> What was the name of the chick with the big behind who sat on my knee in the car? (Bradbury, 1959)

The American *chickie* is always a prostitute:

> Mayhew got himself a little number down at China Beach, little chickie workin' the scivvie houses there. (Herr, 1977)

A *chicken ranch* is a brothel.

chicken a youth or youths attractive to homosexuals.

> Chicken worried him, though. These were the children of eleven, twelve and thirteen. (Fry, 1991)

A *chickenhawk* is a homosexual male who seeks out boys for sexual purposes, punning on the preying sparrowhawk:

> I just happen to like boys... but I don't do chickenhawks. (M. Thomas, 1980)

No CHICKEN does not mean 'heterosexual'. See also CHICK.

child of God a member of the untouchable class in Hindu society.

Dirty work, especially the collection of human excreta, is their monopoly and any other Hindu touching one is defiled:

> She decided he was a Harijan, a child of God, an untouchable. (P. Scott, 1971)

child of sin a bastard.

The *sin* was copulation outside marriage, at least so far as the mother was concerned:

> I have fallen! I am a mother, and my poor dear boy is the child of sin. (Mayhew, 1862)

Also as a *child of grief*:

> She's never been the same since she lost that child of grief. (Macdonald, 1971)

See also LOVE CHILD.

child of Venus a prostitute.

She has been mothered by the Goddess of Love. The phrase is also used of a female who enjoys copulation:

> ... a merry little grig and born child of Venus. (Harris, 1925—a *merry grig* is a lively, jolly person)

chill to kill.

The common cooling imagery:

> A hundred guys could have chilled this little wart. (Chandler, 1939, of a corpse)

Chinese copy a production model stolen from another's design.

Usually exactly copied without acknowledgement, consent, licence, royalty, etc. Used also as a verb:

> ... some big-time outfit'll Chinese-copy his equipment and take his market away by underpricing him. (M. Thomas, 1982)

The practice persists both in China and in Taiwan.

Chinese paper a subordinated debenture.

From its use as a flimsy wall:

> When something happened to break the flow, it came tumbling down on leverage, until it stuck... was buried under the Chinese paper. (Train, 1983)

Paper is financial jargon for a stock, share, bond, debenture, etc. issued by a commercial undertaking without recourse to a third party. Subordinated debentures were one of the devices whereby speculators used a corporation's own assets to finance their takeover.

Chinese parliament a disorganized discussion group.

It would seem the converse of what happens in Beijing, where mute and subservient delegates appear to be harangued interminably by their leaders:

> We call this stage of planning and preparation 'walk through, talk through', and operate a Chinese parliament while we're doing it. (McNab, 1993)

Chinese three-point landing a crash on the runway.

Second World War usage. An elaboration of the First World War *Chinese landing*, which punned on the mythical Oriental, Wun Wing Lo.

Chinese tobacco opium.

From the smoking and the association of opium with China. But *China white* is heroin of good quality, perhaps from the colour:

> ... offered me a whole piece of unstepped-on China white. (Wambaugh, 1981, of pure heroin)

Chinese wall the pretence that price-sensitive information will not be used by an adviser to his own advantage.

Said to have been first used in this context by F. D. Roosevelt in 1927. The pretence allows a banker etc. to retain clients with conflicting interests, or for one part of the organization to make a profit on inside information available to another:

> The next hurdle for the Swiss is the 'inquiry' —the exchange uses the word investigation —by the Securities and Futures Authority, which is trying to decide whether the bank's Chinese walls were breached by its dealings. (*Daily Telegraph*, 19 January 1995—the bank's market-makers had accumulated shares in electricity companies while it was also advising a predator on a possible electricity takeover)

chippy[1] a low-grade prostitute.

Possibly from *chip*, a bit or piece:

> He pays some chippie fifty to gobble his pork. (Diehl, 1978, referring to fellatio)

A *chippie-joint* is an American brothel.

chippy[2] to take illegal narcotics on an irregular basis.

Where the ingestion of illicit narcotics is the norm, non-addiction may be taboo. In such a culture a non-addict may wish to avoid being thought stuffy by not entirely eschewing narcotics, as it were merely chipping at a mass. Lingemann suggests a derivation from *'chippying,* dallying with (a chippy): hence *chippy-user*, one who uses narcotics infrequently'—but that seems unnecessarily complex. In American black slang a *chippy* may also ingest regularly strong illicit narcotics, which adds to our knowledge and confusion, if not to elucidation of the etymology.

chirp to be an informer to the police.

This American underworld slang uses the common *singing* imagery.

chisel to steal or cheat.

The imagery is from the removal of slivers with a sharp instrument. Thus the thefts may be minor and repetitious and the cheating mean:

> Gotham liked to chisel whatever 'float' it could over the weekend. (M. Thomas, 1980, describing the banking practice of stealing the interest on customers' money by being dilatory on Fridays)

On the chisel is so behaving:

> He'd be pretty sore if I was on the chisel. Not that I don't like money. (Chandler, 1953)

To *chisel*, meaning 'to avoid compliance with an unpopular law', is a new American use.

choke your chicken to masturbate.

Of a male, from the likeness of the erect penis to a chicken's plucked neck:

> I went to Chi Town to clean up, but I ended up choking my chicken. (CBSLD—the speaker sought casual copulation in Chicago)

A *chicken-choker*, or masturbator, is said to be 'a friendly term truckers use for each other' (ibid.). I have a single use of the phrase meaning 'to urinate':

> Whenever Walker was about to go and answer a call of nature, he would announce 'Well, I'm gonna choke my chicken'. (de la Billière, 1992—I am assuming the gallant general was not mistaken in his assumption of his colleague's intentions)

chokey prison.

The Hindi *chauki* is a space surrounded by walls, whence a prison in British Indian use and common British slang:

> I've got to cart Voluptia off to chokey. She's been interfering down in the circumcision booths. (Bradbury, 1976)

chop[1] to kill.

Originally standard English, meaning to kill an animal before sportsmen had been given an opportunity of hunting it to death, but now used of killing humans:

> Unless he chopped us both (which seemed far-fetched, pirate and Old Etonian though he was)... (Fraser, 1977)

To *get the chop* is to die, especially in wartime. In newspaper jargon a *chop shot* is a picture of a death:

> You don't get many chop shots these days. (M. Thomas, 1980, referring to a public execution)

chop[2] **(the)** sudden dismissal from employment.

The imagery is from the act of chopping and is used of individual rather then multiple dismissals. To *chop* is so to dismiss:

> Joint editor Allan Segal was chopped last month. (*Private Eye*, May 1981)

For the chop describes a candidate for dismissal:

> Tusker had been for the chop the moment Solly Feibergerstein set eyes on him. (P. Scott, 1977)

chopper[1] a machine gun.

Underworld jargon from army use, referring to the chopping down of the victim; see also CHOP[1]:

> The man with the chopper. (Chandler, 1950, of someone so armed)

A *chopper* may also be a killer by firearm or, in aviation slang, a helicopter.

chopper[2] the penis.

Perhaps from its divisive sexual function in its erect state. (My daughter, whose job included editing etymological puzzles, once erred by passing 'a butcher's chopper' as a crossword clue for *cleaver*, which calls to mind Dr Wright's definition:

> Broach—a butcher's prick. (EDD)

If she had studied Grose she might also have had qualms about *cleaver*—'One that will cleave; used of a forward or wanton woman'. To *cleave* was to copulate:

> How many different men have cleft thee this last six month? (Fowles, 1985, writing in archaic style))

chota peg a spirituous intoxicant.

Chota is 'small' in Hindi, although the measure may not be:

> Better a few too many chota pegs than the possible alternative. (P. Scott, 1968)

A *burra peg* is a large measure of spirits. These British Indian uses are now seldom heard in the British Isles. See also PEG.

chubby fat.

Literally, like the thick, coarse-fleshed fish, whence agreeably plump, especially of babies. Used in advertisements to avoid disturbing mothers who have to select capacious clothes for an obese child.

chuck (the) peremptory dismissal from employment or as a suitor.

You would be wrong to infer that the parting was necessarily forcible.

chuck up to vomit.

Not playing catch and a variant of THROW UP.

chucked drunk.

Probably from the feeling of rotation, as of an object being turned in the chuck of a lathe. It owes nothing to being *chucked out* of an inn for drunkenness. The American *chuck horrors* are either the acute symptoms which may follow withdrawal from narcotics dependence or the claustrophobic symptoms of someone confined or afraid of being confined in a prison cell.

church house *obsolete* an institution for destitutes.

I give this entry to explain the many *Church House Inns* in Devon, which you will find near the church and parsonage:

> Wi' croping church-house grules long fed. (Rock, 1867—the grules were miserly)

church triumphant the dead.

A Christian use, especially of those considered to have faithfully served the *church militant here on earth*, while others of us supposedly languish among the vanquished.

churn to deal unnecessarily in a client's securities in order to inflate commission.

Stock dealing jargon and practice. The imagery is from constantly turning the milk to obtain butter or cream, the principal being milked and the agent getting the cream:

> Your account can be 'churned' even though you haven't signed for discretionary trading. (Chase, 1987)

circular error probability the extent to which ordnance will miss the target.

A Gulf War usage:

> There was something called circular error probability, which simply meant the area where a bomb or missile was likely to fall. (Simpson, 1991)

circular protector *obsolete* a male contraceptive sheath.

Taken literally, it could mean almost anything, from sheep fencing to an envelope for junk mail. Once much used when advertisements were couched in circumlocutory terms.

circus an obscene stage performance.

American jargon among those who promote pornographic shows. The round or oval Roman amphitheatre gave its name to the performances held there and thence to the modern clown and elephant routines.

cissy a male homosexual.

Literally, an effeminate man, probably from *sissy* or *sis*, colloquial abbreviations of *sister*:

> You know how cissies hate pregnant ladies. (N. Mitford, 1949—in fact, I did not know)

Civil Co-operation Bureau a unit established by government to try to perpetuate white political domination.

A South African variant of an older theme:

> After telling the enquiry about the plans of a Civil Co-operation Bureau to hang the dead foetus of a monkey outside Archbishop Desmond Tutu's house, General Rudolph 'Witkop' Badenhorst, Chief of Military Intelligence, complained that he was receiving anonymous calls at home. (*Sunday Telegraph*, March 1990—the practice of designating a classification 2–2 in a university examination as a 'Desmond' after the famous cleric is worthy of notice but no euphemism)

civil rights a political cause seen as a justification for illegal action.

It should mean no more than the freedom under law which any individual is entitled to enjoy. Those who use the phrase are seldom heard to speak of *civil duties*. See also ANIMAL RIGHTS.

civilian impacting the killing or wounding of civilians in error.

A Gulf War neologism:

> Some of the military spoke of 'civilian impacting'. (Simpson, 1991)

claim to steal.

This English use seeks to put a gloss of legality on the taking, usually of items of low value. To *claim responsibility for* arson, murder, etc. is terrorist jargon for admitting to the perpetration of the crime specified. To *claim* is also used journalistically as a device for imputing a lie:

> 'He claims he saw...' was synonymous with 'He is lying when he says he saw...' (D. Francis, 1985, referring to a scurrilous newspaper)

claimant a poor person supported in part or whole by the state.

They *claim* money etc. from public agencies:

> Reductions for Students, OAPs and Claimants. (Theatre Wales poster, October 1981—despite the long standard English use of *claimant* as anybody who makes a claim, a self-sufficient person would have been given short shrift if so bold as to ask for a cheaper ticket)

A *Claimants' Union* is an association of those receiving assistance from the state whose object is to maximize their receipts from the public purse.

clap gonorrhoea.

Not a euphemism except in the 19th-century phrase *come home by Clapham*, to be so infected, Clapham being in those days an outer suburb of London. Pope was clearly under a medical misapprehension when he wrote 'Time, that at last matures a clap to pox'.

claret blood.

Boxing jargon, from the colour, and usually of bleeding from the nose, or *claret-jug*.

class warfare the tension between various citizens in a non-Communist state.

From the days when the manual workers were supposed to be in permanent conflict with the rest. Agitators from the left still have confidence in such activity benefiting mankind or themselves, and may also refer to *class conflict, education, law, literature*, etc.

classic pretentious and costly.

Originally, belonging to the literature of Greek or Latin antiquity, whence 'remaining in vogue over a long period'. Its use in advertising jargon may owe something to *classy*, or *superior*, which did not at one time prevent the British *Classic* cinema chain specializing in pornographic films.

classic proportions obesity.

Of women, because Rubens and other *classic* painters seemed to favour plump women.

clean free from unpleasant or illegal taint.

See DIRTY, of which in many uses this is the opposite. Thus if you are *clean-living*, you refrain from promiscuous copulation:

> VD is dangerous. Clean living is the real safeguard (L. Thomas, 1977)

and *clean* indicates the absence of venereal disease:

> I was a lucky devil to drop on such a lovely clean skirt. (Richards, 1936—the female was disease-free)

A *clean* atom bomb has less fall-out than a *dirty* one:

> The language of the mad... 'Clean atom bombs'. (West, 1979)

A *clean locality* may be free of the enemy:

> ... this village is clean and this village is all Charley. (Theroux, 1975, writing about places in Vietnam)

You are *clean* if you are not illegally carrying narcotics or a handgun:

> 'What's the point if he's clean?' 'If he's carrying something?' (Kyle, 1978—a person was suspected of having a pistol)

A miscreant may escape—*get clean away*—or avoid arrest:

> Drunk as he'd been the night before, he thought he'd made it home clean. (M. Thomas, 1980)

To have *clean hands* is to be uninvolved in bribery.

clean up[1] to bring the proceeds of vice into open circulation.

A variant of LAUNDER using the same imagery:

> The money from this stuff needed cleaning up. (Davidson, 1978)

Or the money may be sent *to the cleaners* for the same purpose:

> Black money tucked away ready to go to the cleaners. (ibid.)

clean up[2] *American* to copulate casually.

Of a male, from the slang meaning 'to have a successful foray':

> I went to Chi Town to clean up. (CBSLD)

cleaners (the) the actuality of being robbed or cheated.

A losing gambler or a dupe by any trickery within or outside the law may be said to *have been, been sent, been taken*, or *gone to the cleaners*, punning on the removal of extraneous matter from clothing in the dry-cleaning process.

cleanliness training *American* teaching children controlled urination and defecation.

Not telling people how to wash their hands or scrub floors. A less common version of *potty training*, which is specific (see POT[3]).

cleanse[1] to free from enemy occupation.

Achieved by bombing or otherwise dominating formerly hostile villages etc.:

> ... paramilitary elements trained and drilled in a special school and sent to 'cleanse' (US word) 'pacified' hamlets. (McCarthy, 1967)

See also ETHNIC CLEANSING.

cleanse[2] to remove the placenta from domestic cattle.

Veterinary jargon:

> I was 'cleansing' a cow (removing the afterbirth). (Herriot, 1981)

cleansing personnel garbage collectors.

The heirs of the old-time dustmen. In practice, during the discharge of their duties they may leave a malodorous mess through careless handling.

clear-out (a) defecation.

Medical jargon and male vulgarism. Of defecation whether or not associated with the taking of a purge.

clear up *American* to desist from regular use of illegal narcotics.

Literally, to tidy up or to redress any ill. Drug users' jargon.

cleavage the visible division between a clothed woman's breasts.

Literally, the action of splitting apart. With the relaxation of taboos concerning the breasts, now standard English.

cleave see CHOPPER[2].

cleft the vagina.

As viewed sexually by a male, from the literal meaning 'a crack or fissure in a rock':

> Oh! let me view the small, dear, tender cleft. (Cleland, 1749, of a vagina)

click[1] to steal.

Literally, to snatch or seize hold of:

> ... wanting to cleik the cunzie (that is hook the siller). (W. Scott, 1814)

In obsolete use, an Englishman might have bought stolen goods at *Clickem Fair* or have been robbed at *Clickem Inn*. A *clicker* was a thief, a body-snatcher, or a pestering touting shopkeeper.

click[2] to conceive a child.

From the onomatopoeic sound of a connection being made:

> I let him into the secret. 'Irene's hoping we've clicked.' (Barber, 1981)

clicket to copulate.

From the French *cliqueter*, to make a clicking noise, used of foxes and sometimes of deer and hares. Occasionally used figuratively of humans.

climax a sexual orgasm.

> A climax was never reached by either of them, but that did not spoil their pleasure. (P. Scott, 1968, referring to copulation)

Literally, the culmination of anything. Hudson attributes the first euphemistic use in the sexual sense to Marie Stopes.

climb to copulate with.

Of a male, from the action of *mounting* the female:

> You mean you are looking to climb some gorgeous chorus girl. (Condon, 1966)

Climb in with refers to no higher an ascent than the female's bed:

> I'd just as soon go to bed with a giant clam as climb in with Eva Wilt. (Sharpe, 1979)

Climb into bed is specific:

> ... sufficient affection and desire for her still to want to climb into bed if I got half a chance. (Fowles, 1977)

climb the ladder *obsolete* to be hanged.

Either from the ascent to the scaffold or because the *ladder* itself was the scaffold:

> When he was upon the ladder he prayed that God would inflict some visible judgement upon his Uncle. (Wallace, 1693)

A female who *climbs the ladder on her back* is one who seeks to turn male advances to advantage and advancement.

climb the wooden hill see WOODEN HILL.

clink a prison.

Originally the jail in Southwark but then used generally, helped no doubt by the onomatopoeic attractions of keys in locks and heavy doors shutting:

> ... the more troublesome firebrands ... were popped neatly into clink. (P. Scott, 1971)

The obsolete Scottish *clink off* was to die, from a meaning 'to depart':

> In God's gude providence she just clinkit off hersell. (Ramsay, 1859)

clip[1] to swindle or rob.

From the shearing of sheep or the venerable practice of cutting the edges off silver coins. Now usually of the picking of pockets. A *clip-artist* is a swindler and a *clip-joint* a night club or similar establishment where customers are overcharged and otherwise cheated.

clip[2] to hit with a bullet.

Literally, to cut or shear, whence to mark as by removing cardboard from a ticket. In America the person *clipped* is usually killed but in the British Isles only slightly wounded, without being incapacitated. The rare *clip the wick*, to kill, comes from snuffing out a candle by cutting it below where it is burning:

> Maitland found out. So they clipped his wick. (Sanders, 1977, of killing, not circumcising)

cloakroom a lavatory.

Coats are commonly stored in or near lavatories:

> To a small boy looking urgently for the cloakroom ... (Jaeger in Morley, 1976)

Very common British usage. The shortened form *cloaks* is used to refer to the storage of outer garments.

clobbered *American* drunk.

From *clobber*, to beat up, with the common imagery of disarray and general malaise which is met at some stage of drunkenness.

clock see PUT THE CLOCK BACK.

cloot the devil.

Literally, one of the divisions of a cloven hoof, a physical characteristic shared by Satan and cattle:

> I hate ye as I hate auld Cloot (Barr, 1861)

and as *clootie*:

> Auld Hornie, Satan, Nick, or Clootie. (Burns, 1785)

I think this Scottish usage is obsolete and doubtfully euphemistic, but I give it, as with GOOD MAN, to illustrate the superstition of our quite recent forebears, who set aside and left untilled *clootie's croft* in the hope that the devil, having his own piece of land, would leave the rest of the farm alone:

> The moss is soft on Clootie's craft. (Henderson, 1856—I suppose all the thanks the farmer got was a seeding harvest of weeds)

close[1] see NEAR[1].

close[2] copulating with extramaritally.

Literally, related to or well acquainted with:

Mr ——and Mrs ——a widow... have been close for two years. (*Daily Telegraph*, 28 March 1994—he was separating from his wife and said he intended to 'continue his relationship' with Mrs ——)

A *close relationship* is specific:

Di was having a close relationship with the muscular Tommy Yeardye. (Monkhouse, 1993—*Di* was the actress Diana Dors)

close an account to kill.

With imagery from banking or story-telling?

We were all hoping you would close his account. (Sohmer, 1988—an FBI agent had been told where a murderer was hiding)

close friend see FRIEND.

close its doors to fail.

Of a bank, though in fact it prudently closes its doors with some care after the close of business every day, in the hope of re-opening them on the morrow:

... if the run persisted, cash reserves would be exhausted and FMA obliged to close its doors. (Hailey, 1975)

close stool a portable lavatory.

Your lion, that holds his poll-axe sitting on a close-stool, will be given to Ajax. (Shakespeare, *Love's Labour's Lost*—see JAKES for the punning *Ajax*)

Originally for use in the CLOSET[1] but now in any sick-room.

close the bedroom door to refuse to copulate with your spouse.

The female usually does the closing:

From the moment he had been a gubernatorial candidate she had closed the bedroom door. (Allbeury, 1980)

I have heard the phrase used figuratively where spouses have continued to share a bedroom but have ceased at the instance of one to copulate with each other.

close your eyes to die.

Or explicitly, *close your eyes for the last time*. If you close another's eyes, that person is already dead and you are arranging the corpse:

I trust that I shall be able to close your eyes in peace. (Hughes, 1987, quoting a letter from a convict in Australia to his parents in England)

closet[1] a lavatory.

Literally, a small or private room. It is normally used with the descriptive prefixes *earth* or *water*, whence the initials *EC* and *WC*.

closet[2] concealing in public your homosexuality.

Again from the small or private room where you act according to your true inclinations. Usually of a male, as a *closet queen* or *queer*, and occasionally of a female:

I often wondered if she was a closet lez. (Sanders, 1977)

In occasional and convoluted use, a *closet homosexual* may be a male who affects homosexuality so as to avert the suspicions of a cuckolded husband:

Dexter Dempster, New York's leading closet homosexual. (M. Thomas, 1980, of a man who so behaved)

clout to steal.

Probably from the meaning 'to hit', whence by transference from the American *hit*, to rob. Normally of minor thefts from shops or cars. In obsolete British use a *clout* was a cloth worn to absorb urine or menstrual discharge:

Their bottles o' pap, an' their mucky bins, an' the clats an' the clouts. (Tennyson, 1885)

The saw *ne'er cast a clout till may be out* refers to clothing and the tree blossom, but millions have sweated unnecessarily through torrid springs and 'keep them on up to the middle of May at least' (Kersh, 1936, referring to winter woollies).

club[1] see IN THE CLUB.

club[2] an agency promoting the sale of a specific product.

There is usually an attempt by the promoters to involve customers—'members'—in a continuing obligation to buy, as with a *Book Club*, but no true association of like-minded people:

Now, alas, the eel and pie shop was a video rental 'club'. (Deighton, 1988)

club[3] a business which contrives to evade regulations.

Usually selling intoxicants without a licence, with customers being enrolled as *members*, whence the British *club licence*, permission to sell intoxicants to a restricted clientele. Also denoting cinemas showing pornographic films of which full public display would be illegal.

clunk a corpse.

Literally, the sound of a blow or a dull person, neither of which elucidates the origin of this American use:

He'll be a clunk before he hits the floor. (Sanders, 1973, of a killing)

cluster the male genitalia.

They are certainly proximate, and even more so in tight trousers:

'The cluster,' he replied, 'is prominent these days.' (Matthew, 1983—a shop assistant was trying to sell tightly-cut trousers)

co-belligerent a former enemy helping the winner in continuing war.

By 1943 COLLABORATOR had become pejorative and so:

> the word 'co-belligerent' was invented to proclaim her new status. (Jennings, 1965, of Italy)

The wise Italians had shown little belligerence before the about-face, and had no reason to show any more after it.

co-operate[1] voluntarily to assist an invader of your country.

Literally, to work with:

> ... people in his area have begun to 'co-operate' with the Americans—the word 'collaborate' is avoided. (McCarthy, 1967)

co-operate[2] to assist another through fear or duress.

Usually of criminals supplying information to the police, who would so describe the arrangement. Thus the former Soviet empire controlled and exploited its subject states economically through what it styled the *Economic Co-operation Council*, or Comecon.

co-ordination the ensuring of submission to totalitarian rule.

Nazi jargon and concept, largely implemented through an *Enabling Act*, which conferred absolute power on Hitler. (This is one of the many lessons the Nazis taught the Communists—have ready and put into effect a law which gives the leader absolute and irreversible power as soon as you seize control.)

co-pilot a person who watches another who has illegally taken LSD.

The hallucinogen can convince the taker that he can fly. Even if he remains earthbound, he may injure or kill himself under its influence.

co-respondent a male accused of having copulated with another's wife.

Legal jargon for the man who has to *respond* jointly with the wife to a husband's petition for divorce on the grounds of adultery:

> Merrick was, in his romantic way, a sort of professional co-respondent. (Bradbury, 1959)

If you wear *co-respondent's shoes* or, less often, other articles of clothing so described, you are dressing in a style affected by men thought to be philanderers. An American *correspondent* can be a woman:

> ... doubled as a paid correspondent in divorce cases—'the Woman Taken in Adultery'. (McCarthy, 1963)

Under English divorce law, the woman whom the wife named as having copulated with her husband was the *party cited*, so that she might defend her good name, if she had one, but without any liability for the payment of damages or costs to the wife.

coast *American* to be under the influence of illegal narcotics.

Freewheeling, usually on heroin.

cobbler a forger.

Criminal or espionage use, for someone who prepares bogus credit cards, passports, etc. To *cobble* is to mend any article in an untidy or rough way, although we expect a neat job from the person who mends our shoes.

cobblers the testicles.

From the rhyming slang *cobblers awls*, balls. DRS says surprisingly 'Accepted as a euphemism and not recognized as reduced rhyming slang'—with which I cannot agree. Mainly figurative use, in that *balls* means 'nonsense'—*a lot of old cobblers*. The obsolete northern English *cobs*, the testicles, probably came from the meaning 'small stones' and was not a shortened form of *cobblers* or a variant of the American 'nuts'.

cock to copulate with.

The treatment of this word by Dr Wright in EDD sets him apart from other lexicographers. Suffice it to say here that the meaning of the noun *cock*, 'the penis', is venerable:

> Pistol's cock is up (Shakespeare, *2 Henry IV* —another of his vulgar puns)

and that to *cock* was to use the penis in copulation. In modern speech we *cock a leg* across, athwart, or over a female, which is not punning on the 'penis' meaning:

> ... to prevent him cocking his lustful leg over my loving Elspeth (Fraser, 1975)

and:

> ... all the more difficult for me to cock a leg athwart Miss Fanny. (Fraser, 1971)

A *cocksman* is a philanderer, from the penis and not the leg movement:

> He didn't think of himself as a cocksman but every now and then ... something would get loose in his system. (M. Thomas, 1982)

In black American use, a *cock* is also the vagina. This is also a meaning of *cockpit* generally, punning on the scene of avian contest and the repository for the penis:

> ... the rose-lipt overture presenting the cockpit so fair (Cleland, 1749)

and the Second World War vulgarism 'She was only a pilot's daughter but she kept her cockpit clean'.

I think Partridge is wrong when he says of 'by cock, they are to blame' (Shakespeare, *Hamlet*) that '*Cock* is a euphemism of God'. Because everyone knows that it is vulgar to refer openly to the penis and *cock* is the commonest word for it, prudery has led to many evasions, which I examine at ROOSTER.

cock-eyed drunk.

Literally, askew, which is what your clothing might be or your gait. The obsolete *cocked*, drunk, may have come from a pistol prepared for firing with, as normal in drunkenness, the half equalling the whole:

> Half cock'd and canty, hyem we gat. (Wilson, 1843—*canty* means 'cheerful')

cock the leg to urinate.

Normally of a dog, but not of a bitch:

> The poodle... shivered and cocked its leg nervously against the front door. (Bogarde, 1978)

Sometimes used humorously by and of male humans.

cock the little finger to be a drunkard.

From the manner in which some hold a cup:

> Some say she cocks her wee finger... In short that she's gien to the drink. (Barr, 1861)

The phrase is also used of addiction to alcohol short of dipsomania.

cockchafer[1] a treadmill.

From the way the flesh was rubbed raw by the coarse cloth used in prison garments and punning on the beetle, *Melolontha vulgaris*, known also as the *May-bug*:

> He 'expiated', as it is called, this offence by three months' exercise on the 'cockchafer' (treadmill). (Mayhew, 1851)

See also EVERLASTING STAIRCASE.

cockchafer[2] a prostitute.

From the soreness which might result.

cocked see COCK-EYED.

cocktail[1] an alcoholic mixed drink.

I like to think the derivation is from the Aztec *xoc-tl*, after the maiden Hochitl who introduced the king to a concoction devised by her father, thereby winning his heart and immortality. Derivation based on a feather used for stirring, or the French *coquetel*, a mixed intoxicant, is much less interesting, whatever the probabilities. A late contender came to my notice when I read the admirable *The Story of English* (McCrum et al., 1986), which tentatively suggests derivation from the Krio *koktel*, a scorpion. Commoner in America than the British Isles, where it tends to mean a mixture based on a spirit and no more:

> They had started having cocktails every night (McCarthy, 1963—which would have included beer or straight spirits)

and you find *cocktail bars*, *hours*, *lounges*, etc. everywhere. In obsolete uses a *cocktail* was a flaming tankard of ale in Yorkshire; a six-oared boat used by smugglers in Kent; and in the British Isles generally, a prostitute, involving an obvious pun:

> Such a coxcomb as that, such a cocktail. (Thackeray)

A *cockatrice* was also a prostitute, as well as being the fabulous serpent which killed by its glare.

cocktail[2] a cigarette of tobacco and other illicit narcotics.

From the mixed components. In American slang it is also a tobacco cigarette used to enable you to smoke in its entirety the butt of a marijuana cigarette.

coco mad.

A Second World War usage, perhaps after a famous clown called *Coco*. Sometimes as *cocoa*:

> I mean for a minute he sounded perfectly normal, or is he really cocoa? (M. Fraser, 1992)

coffee grinder *American* a prostitute.

From the rotary motion of the pelvis rather than the pigmentation of the skin. The term also means 'a belly dancer' or 'stripper', and the three callings are not mutually exclusive. The 19th-century British *coffee shop* seems also to have been 'a watercloset, or house of office' (*The Slang Dictionary*, 1874).

coffin varnish *American* inferior whisky.

This phrase puns on the look, the taste, and the container in which you may precipitately find yourself if you drink much of it.

cohabit to copulate with extramaritally on a regular basis.

Literally, merely to live with in the same abode, as do parents and children:

> My staff are all highly trained in the Swedish technique and strictly forbidden to cohabit with the customers. (B. Forbes, 1986, of a bawd)

In fact *cohabitors* normally share a residence as well as a bed.

coil (the) a mechanical contraceptive device used by a woman.

From its shape:

> Zoya thumbed a dish of birth control pills. Polish pills. She refused to use the coil. (M. C. Smith, 1981)

A woman who describes herself as *coil* indicates that she copulates regularly, seeks to avoid conception, and uses this device.

coition copulation.

It started by meaning 'mutual attraction', as of planets. The usage is now confined to etymologists, purists, and the writers of verse:

> When Titian was mixing rose madder
> His model sat poised on a ladder.
> Her position, to Titian,
> Suggested coition...
> (Vulgar limerick)

cojones *mainly American* the testicles.

From the Spanish, where it is also a euphemism 'often used to indicate machismo. Offensive.' (*Dictionary of Cautionary Words and Phrases*, 1989 —this fascinating work, prepared by the US newspaper industry's 'Multicultural Management Program', is a delight for the connoisseur of newspeak.)

coke cocaine used illegally.

Probably no more than a shortened form with no pun on the beverage:

> Out of the apartment houses came cokies and coke peddlers, people who look like nothing in particular. (Chandler, 1943)

Coked is under the influence of narcotics:

> ... 'coked' or 'bopped up' gunmen. (Lavine, 1930)

A *coke-hound* is a cocaine or heroin addict:

> He's a coke-hound and he talks in his sleep. (Chandler, 1939)

cold[1] dead.

Usually but not exclusively of hot-blooded creatures, and sometimes of unconsciousness —*knocked out cold*—rather than death, despite the retention of body heat. A source of many phrases alluding to death in morbid 19th-century humour, some of which have survived or been reinvented, for example *cold meat*, a corpse:

> If you bother with us, I will make meat of you —see!—cold meat. (Harris, 1925)

A *cold-meat party* is a funeral; a *cold-box*, a coffin; a *cold-cart*, a hearse; a *cold cook*, an arranger of funerals; and *cold storage*, the grave.

cold[2] not easily susceptible to sexual excitement.

The opposite of HOT[1] but also of someone who fails to be sexually aroused on a specific occasion:

> I have often been asked why on my African travels I was so cold in regard to the native women. (Harris, 1925—what strange inquisitors he must have had)

As in Shakespeare's *cold chastity*, more of women than of men.

cold[3] an excuse for an ailment which is taboo or concealed.

Thus Second World War servicemen who contracted gonorrhoea might say that they had *a cold* and:

> Andropov spent half his 15 months in power seriously ill, supposedly suffering from a 'cold' but in fact lying in a Kremlin hospital hitched up to a dialysis machine with kidney failure and diabetes. (*Sunday Telegraph*, 27 March 1994)

See also CATCH A COLD[1, 2] and DIPLOMATIC COLD.

cold feet cowardice or fear.

There is a physical justification for this standard English use—we have all shivered with fear:

> I think I must have the merest touch of claustrophobia—or cold feet as they would call it in the mess. (Price, 1978, of a tank commander who admitted fear)

cold turkey the effect of sudden deprivation of illegal narcotics.

From the resemblance of the sufferer to a bird after it has been plucked. Less often of withdrawal from alcoholism:

> You can't suddenly sign the pledge, go cold turkey. (B. Forbes, 1986)

In American use other birds may replace the *turkey* in this phrase.

cold-water man a person who drinks no intoxicants.

A perhaps obsolete Scottish use, in a society where abstinence is taboo:

> 'Dae ye drink?' He's a cauld-water man. (Gordon, 1894)

collaborator a traitor.

He works disloyally for the conqueror, not loyally with another like-minded person:

> The English so often have these unknown French friends... Collaborators one and all. (N. Mitford, 1960)

The proportion of the population of France which decided to undergo the enormous risks associated with opposing the German occupiers or the Vichy government was small. Liberation at the hands of the Anglo-Saxons was, and perhaps still is, difficult for the French to come to terms with.

A *collaborationist* is always a traitor:

> I told him I was not a collaborationist, that I was a doctor. (Fowles, 1977, of a Greek in the Second World War)

To *collaborate* is so to act:

> They express their willingness to collaborate. (Goebbels, 1945, in translation, of foreign workers under German control—they had little option at the time)

In Communist East Germany a *Mitarbeiter* was a police spy.

collapsible container a contraceptive sheath.

American police jargon which transfers the collapsing to the contraceptive:

> In any police report when you refer to a collapsible container, it's a rubber. (Wambaugh, 1981)

collar[1] to steal.

The thing stolen may be any small object, probably from putting a collar on a dog to secure it.

collar[2] an arrest.

From the act of grabbing a suspect by his collar and perhaps so leading him away:

> But the evidence is of such a nature that it doesn't justify a collar—an arrest. (Sanders, 1973)

To *collar* is to arrest. An American *accommodation collar* is an arrest made to fill a quota and so show a superior that you are working well. By transference, a *collar* may also be an American policeman.

collateral damage killing or wounding civilians by mistake.

From the sense 'something running alongside':

> What an odd term, he thought, *Collateral Damage*. What an off-hand way of condemning people whom fate had selected to be in the wrong place. (Clancy, 1989)

collect to accept a bribe.

Usually of taking bribes on a regular basis:

> Woe to the cop who collects anything... and doesn't 'see the sergeant'. (Lavine, 1930)

This American use is of the same tendency as the obsolete British *collector*, a highway thief, who asked you to stand and deliver.

colleen the erect penis.

From the rhyming slang *colleen bawn* for 'horn', referring to the heroine of 'Lily of Killarney'. I suspect this British use is now obsolete.

colonial old.

American real estate jargon of buildings which were not always there before the 1780s. *Ante-bellum*, referring in the same sense to the 19th-century civil war, is more likely to be authentic.

colony a distant territory ruled by expatriates.

Literally, a place to which people emigrate in order to live, but most British, French, and Italian *colonies* were populated by their indigenous population under the military, economic, and political control of London, Paris, or Rome. The British *Crown Colony* of Hong Kong has over 98 per cent of Chinese residents and under 2 per cent of others, including British.

colour-tinted dyed.

Of hair, where it is more than circumlocution because the process is more drastic than the variation of a shade:

> For dry, damaged, bleached or colour tinted hair. (instruction on bottle, 1980)

Also as *colour-corrected*:

> She 'mutates' or 'adapts' or 'colour-corrects' her hair. (Jennings, 1965)

coloured[1] not of exclusively white ancestry.

> There are already white tables so why not have a table for the coloured fellers. (Theroux, 1973, but not of foresters)

No skin pigmentation is colourless and those with pale skins undergo considerable expense, risk, and discomfort to darken them by exposure to ultra-violet rays, under sun-lamps or distant skies.

This common evasion for a while replaced *black*, except in South Africa, where it denotes those of mixed ancestry. Those looking for new euphemisms, which in this area of language is just about everyone, may refer to a *person of the coloured persuasion*:

> I am not a black. I am a person of the colored persuasion. (Sanders, 1977, although it would seem not to have been a matter of choice or argument)

Person of colour is another variant.

coloured[2] (of hair) dyed.

No hair is colourless:

> He could see the spark of rouge on her cheeks, the perfect part in her colored hair. (Turow, 1990—*part* means 'parting')

colt[1] *obsolete* to impregnate a woman.

From the servicing of a mare in order to produce a colt, perhaps punning on the meaning 'to cheat':

> She hath been colted by him. (Shakespeare, *Cymbeline*)

colt[2] a fine extracted from a recruit by old employees.

The money was then spent on communal intoxicants in a process known as *shoeing the colt*. I give this obsolete western English example of a practice that was widespread in the days when apprenticeships were hard to secure but ensured a good livelihood once completed.

Columbian gold high-quality marijuana.

From the colour and the profits. It is grown near Baranquilla in Columbia, where the millionaire traders are called *marijuaneros*:

> Pot-smokers the world over recognize the taste of its product, known as Columbian Gold. (Theroux, 1979)

combat fatigue *American* the inability to continue fighting.

Not just weariness from broken nights, too much moving about, and poor food, but the result of prolonged exposure to danger:

> He is suffering from what I suppose you call combat fatigue, and is subject to fits of depression and hallucination. (Shaw, 1946)

In wartime it is not always easy to distinguish between those suffering from a genuine psychological illness and cowards affecting the same symptoms. See also BATTLE FATIGUE and SHELL SHOCK.

combat ineffective dead or badly wounded.

Not a gun which doesn't shoot straight:

> If he became combat ineffective, a subtle way of saying wounded or killed. (Coyle, 1987)

come to achieve a sexual orgasm.

Of both sexes. There are similar euphemisms in French and German. Much use both spoken and written:

> 'I don't know why I let you come this evening,' says Flora. 'You haven't let me come,' says Howard. (Bradbury, 1975)

The alternative *come off* is used more of male than of female orgasms. *Come* is also the fluid secreted by the male and the female respectively during copulation:

'It's Bernard Shaw's semen.'...'You mean it's come?' 'Yes.' (Bradbury, 1976)

There are three obsolete phrases meaning 'to copulate' which are probably not punning on the achievement of an orgasm: *come about*, a female use of a male copulating with her, stems from the meaning 'to visit for a purpose'; *come over* puns on the attitude assumed by the male and his authority over his mistress:

To have no man come over me (Shakespeare, *Much Ado*)

and *come together* is a standard English use, indicating physical proximity rather than simultaneous orgasms:

When as his mother Mary was espoused to Joseph, before they came together, she was found with child by the Holy Ghost. (Matthew 1:18)

The obsolete *come aloft* was to have an erection of the penis:

I cannot come aloft to an old woman. (Dryden, 1668)

come across[1] to do something unwillingly under coercion.

Or to accede to a suggestion. In American use, of extortion or bribery:

...ask why he had to pay when the other bird didn't come across. (Lavine, 1930)

come across[2] to copulate casually.

Again from the meaning 'to accede to a suggestion'. Of a female, usually on a single occasion.

come across[3] to defect.

Espionage jargon, from the actual or figurative passage of a frontier or line of battle:

He's defected. He came across and that's that. (Seymour, 1980)

come again to resume your living physical state after death.

An eagerly awaited expectation by some devout people, despite the manifest problems of space etc.:

He shall come again in His glory, to judge both the quick and the dead. (*Book of Common Prayer*, 1662)

Come back in the same sense is obsolete.

come aloft see COME.

come around see COME ON[1].

come down to cease to be under the influence of illicit narcotics.

From the feeling of levitation:

Floating. When she came down it was pretty grim. (Bogarde, 1981)

Also referring to the unpleasantness and ill-temper which follow drug-taking. *Come off*, for such cessation, is now rare.

come home by Clapham see CLAP.

come home feet first to be killed.

Corpses are usually carried that way round:

Whoever came home feet first, it wasn't going to be him. (Fraser, 1977)

come in at the window to be a bastard.

The newcomer is figuratively introduced into the household by any aperture other than the front door. Following the window in popularity were the side door, the back door, the wicket, or the hatch:

In at the window or else o'er the hatch
...I am I, howe'er I was begot.
(Shakespeare, *King John*)

A *come o' will* was also a bastard—the will of God rather than the mother's:

Little curlie Godfrey—that's the eldest, the come o' will. (W. Scott, 1815)

come into the public domain to cease being a secret.

Of embarrassing or scandalous information which politicians or public employees wish to conceal:

Naturally we are, all of us, in the Service, concerned that advice one has given could be misunderstood if it were to come into the public domain. (Lynn and Jay, 1986)

come off[1] see COME.

come off[2] see COME DOWN.

come on[1] to menstruate.

Obvious derivation and wide female use:

Have you come on badly or something? (P. Scott, 1968—one woman was asking another about menstruation)

The American *come around* is equally logically used in the same way.

come on[2] an invitation to make a sexual approach.

Usually by a female:

'Did she touch the young guy?...Stroke his hair. Put her hand on his arm. Anything like that?' 'You mean was she coming on?' (Sanders, 1981)

Less often of a male approach:

He's come on to me, you know. His own son's wife. (M. Thomas, 1985)

come-on a deceptive inducement to enter into a long-term commitment.

Advertising jargon for an offer intended to tempt or trap the unwary:

The electricity bill, a come-on for *Time/Life* books... (Allbeury, 1980, listing the contents of mail)

come out to announce your availability as a sexual partner.

This was social jargon up to the 1950s for the mainly London parade of rich marriageable girls before supposedly eligible bachelors:

> Girls had to come out, I knew. (N. Mitford, 1949, of that experience)

Now widely used of homosexual males revealing their tastes publicly:

> The Bishops' group also says that a homosexual who has 'come out' should offer his resignation to his bishop. (*Daily Telegraph*, October 1979)

Often in the fuller form, *come out of the closet* (see CLOSET[2]):

> Lord Mountbatten was definitely gay himself though he never had the courage to come out of the closet. (*Private Eye*, May 1981)

come over see COME.

come through to act under duress.

From the meaning 'to reach a goal'. It is used of paying a bribe, or responding in the desired manner to such a payment; or of making any payment or giving information under duress:

> They'll snatch your wife or take you out in the woods and give you the works. And you'll have to come through. (Chandler, 1939)

come to to copulate with.

Particularly in marriage where the spouses occupy separate beds:

> I have come very seldom to you in the last few years. (Bogarde, 1981, of a husband who seldom copulated with his wife)

come to a sticky end[1] to fail disastrously but deservedly.

The fate of an insect on fly paper. It may describe an untimely death of a dissolute or criminal person, the incarceration of a rogue, an unwanted pregnancy of a flighty girl, a woman's unhappy marriage especially outside her social class, or any other unpleasant upshot.

come to a sticky end[2] to masturbate.

A male vulgarism, from the ejaculation, with perhaps a hint of the fate supposedly awaiting those who thus indulge.

come to see to court.

Literally, to visit, but a man who *comes to see your sister* is unlikely to content himself merely with a visual inspection. See also SEE[1].

come to your resting place to die.

Not reaching your overnight hotel but the common imagery of likening death to resting while you await resurrection:

> He drove me direct to this bungalow and then to the resting place which she had come to just the day before. (P. Scott, 1973)

You may also *come to the end of the road*:

> She came to the end of the road only five months after we had laid father to rest. (Tyrrell, 1973)

Some might prefer the obsolete Shetland Island *come to yourself*, with its overtones of Buddhism:

> I faer dis ane 'ill come to himsel'. (*Shetland News*, 1890, quoted in EDD)

come to your time to give birth to a child.

Short for *come to your time for birth*. Standard English but now perhaps obsolete.

come together see COME.

come up with the rations to be awarded as a matter of routine and for no particular merit.

Especially with reference to British wartime campaign medals or those awarded other than for a specific act of gallantry:

> 'Bit of decoration. Congratulations.' 'Came up with the rations.' He took the ribbon. But if he joked he was pleased in his soul. (Lee, 1937)

Also civil awards to public servants by virtue of their tenure of an office for a period without any overtly scandalous incident.

come your mutton to masturbate.

Of a male, from the slang *mutton*, the penis.

Comecon see CO-OPERATE[2].

comfort break a pause in a meeting which allows the participants to urinate.

> But it was in one of the comfort breaks from the negotiations with the NUJ ... that I realised my arguments had outstayed their welcome. (J. Cole, 1995)

comfort station *American* a lavatory for public use.

Of the same tendency as COMFORTABLE[1]:

> Ari habitually terminated the beach section of his run by the comfort station coyly labelled 'Boys' and 'Girls'. (L. Thomas, 1979)

In an American bus, where it lies behind the back axle, it can be anything but comfortable.

comfort women prostitutes.

The lot of many Korean, Chinese, and Dutch females in territories conquered by the Japanese:

> ... the forced recruitment of 'comfort women' by the Japanese army in the Second World War. (*Daily Telegraph*, 7 March 1994)

comfortable[1] having urinated.

A genteel usage, often in the phrase *make yourself comfortable*, sometimes shortened to *comfy*:

> She went in to make herself comfy;
> Sat on the seat and could not get her bum free.
> (Vulgar song—'Three Old Ladies Locked in a Lavatory')

comfortable[2] *American* drunk.

From the feeling of well-being induced at some stage.

comfortable[3] not in mortal danger.

Hospital jargon, although a patient so described would seldom admit to being 'free from pain and trouble' (OED):

> 'Well, how is he?' 'He is said to be comfortable.' If so it must be the first occasion for weeks. (E. Waugh, 1955, of someone admitted to hospital)

We can all sympathize with Mr Steve Wickwar, 27, who:

> sustained severe cuts after being attacked by a two-year-old male leopard... His condition at Northampton general hospital was said to be comfortable. (*Daily Telegraph*, April 1982—but why not 'a leopard, 2'?)

coming of peace a military defeat.

The words used by Hirohito in his broadcast of 15 August 1945, when he announced Japan's capitulation in such evasive and formal language that some of the military misunderstood his message and carried on fighting. He also asserted, with what seemed at the time to be considerable understatement, that the war 'had turned out not necessarily to Japan's advantage' (Keegan, 1989).

commerce copulation.

Literally, exchange or dealings between people, but long used of copulation, especially if it is outside marriage. Explicitly as *sexual commerce*. *Sinful commerce* implies copulation with a prostitute:

> Jenny the tavern-girl was not alone in this world of sinful commerce. (Monsarrat, 1978, writing in archaic style)

There is now also some homosexual use.

A *commercial gentleman* was a salesman or representative whose job entailed travel and absence from home, the mobility in the pre-motor age earning him and his job title a reputation for lechery. To avoid such stigma, the title changed over the years through *traveller*, *agent*, and *representative* until they mostly came to call themselves *managers*.

commercial sex worker a prostitute.

Not a salaried nurse running a VD clinic or someone employed to categorize chicks:

> A St John Ambulance worker... tells me that she is only allowed to describe those women engaged in the oldest profession in the world as CSWs—short for Commercial Sex Workers. (*Daily Telegraph*, 5 January 1994)

commission a bribe.

Not the warrant to do something for another but a reward in percentage terms for doing it. Commercial jargon, especially of dealings with corrupt or poor countries where you may have to bribe officials to get orders but it is wise to throw a gloss of legality over the process:

> As for bribes... this is a capitalist society, General. We prefer to talk about commissions and introducer's fees. (W. Smith, 1979)

commission agent a person who accepts bets for a living.

He is neither an agent of those who place bets with him nor is he rewarded by the receipt of commission. The now old-fashioned English use seeks to mitigate the opprobrium which attaches to his work.

commit to consign to an institution for the insane.

Literally, to give in charge, and clearly a shortened form of a longer phrase:

> Polly, you ought to commit your father. (McCarthy, 1963—he was mad but at liberty)

commit a nuisance to urinate publicly somewhere other than in a urinal.

Usually of a male, where *commit* means 'to perform' and *nuisance* is legal jargon for an offensive act:

> These are the same naughty young men who 'Commit a Nuisance'... Or it could be some old rustic twelve-pinter who is past caring. (Blythe, 1969)

We now see the phrase on old signs, mainly in the negative, where we are abjured to *commit no nuisance*.

commit misconduct to copulate extramaritally.

Of either sex in 19th-century speech:

> ... moments of passion reduced to 'committing misconduct'. (Pearsall, 1969, of Victorian journalism)

committed dogmatic as to political or social views.

Literally, devoted, but you are often left to guess the object of the devotion:

> Committed to what? Abortion, Marxism or promiscuity? It's bound to be one of the three. (Sharpe, 1976)

The word is used by bigots and enthusiasts of themselves but pejoratively of them by others.

Committee of State Security a Soviet instrument of internal repression.

Or *Komitet* in Russian, whence *KGB* (*Komitet Gosudarstvennoĭ Bezopasnosti*), sometimes shortened to *Committee*:

> 'The Committee's involved,' Suchko went on, using the standard euphemism for the KGB. (Moss, 1985)

The Cuban *Committee for the Protection of the Revolution* is a system involving neighbours spying on each other, in the manner of the Nazi *Blockwächter* organization.

commode a portable lavatory.

Originally, a woman's tall headdress, whence a tall chest of drawers, which many of these lavatories were designed to look like:

> An ice-box built in a Marie-Antoinette commode. (Ustinov, 1971—he was shocked by the vulgarity)

commodious too large.

Literally, convenient, but in real estate jargon, where we might have expected elegant spaciousness, all we find is a place too big to heat or keep in repair.

common customer *obsolete* a prostitute.

Although *supplier* might seem more appropriate:

> I think thee now some common customer. (Shakespeare, *All's Well*)

Also as *common jack, maid, sewer,* and *tart,* or *commoner o' th' camp* in obsolete use.

common ground a policy favoured by civil servants.

Bureaucratic jargon, especially if a policy is being persisted in contrary to what the government wants or thinks is operative:

> In order to guide the Minister towards the common ground, key words should be inserted with a proposal to make it attractive. (Lynn and Jay, 1981)

common house[1] *obsolete* a brothel.

Common perhaps from the sharing:

> Do nothing but use their abuses in common houses. (Shakespeare, *Measure for Measure*)

See also HOUSE[1].

common house[2] *obsolete* a lavatory.

Again from the sharing. A feature of old British terraced cottages. See also HOUSE[2].

communicable disease a venereal disease.

For medical practitioners the phrase has two meanings: it denotes both a disease like meningitis, a case of which must at once be reported to the authorities; and a disease which may be transmitted from one sufferer to another. In lay use it refers normally only to a disease which is sexually transmitted.

communication see COMPREHENSIVE.

community affairs tension between black and white people.

Social service jargon which has come into common use, seeking to avoid reference to skin pigmentation. Used in various phrases associated with the distrust, jealousy, fear, and antagonism which may be present when black and white people move into each other's territory. Thus a *community affairs correspondent* exists because of and writes about discord between black and white people. *Community relations* has much the same use.

community alienation lawlessness.

Social service jargon to avoid blaming anti-social behaviour—it does not mean that the place is full of foreigners:

> The village now exhibits the signs of this community alienation with its smashed telephone kiosks, litter and graffiti painted on its mellow walls. (*Thatch*, March 1982)

In America the vandals might be sent to a *Community Treatment Centre,* a prison, not a doctor's surgery.

companion a person with whom you cohabit extramaritally.

Of either sex; see also CONSTANT COMPANION. Originally, in this sense, an employee who attended to and lived with another, and then of homosexual relationships:

> I'm thinking of getting a new companion. There's a little actress on the train who would suit me. (G. Greene, 1932—the speaker was a female homosexual)

Now the equivalent of *spouse* where a male and female live and copulate together although not married to each other:

> Princess Stephanie of Monaco and her companion, Daniel Ducruet, her former bodyguard, pose with their second child, Pauline. (*Daily Telegraph*, 7 June 1994)

See also PARTNER.

companion spaces adjoining lockers in a permanent store for corpses.

American funeral jargon, especially when trying to sell graves to the living:

> Cosy 'companion spaces' for occupancy by husband and wife. (J. Mitford, 1963)

company[1] a person with whom you copulate outside marriage.

Literally, companionship. In American use, often of a transient relationship:

> And your wife on the outside, looking around for company. (Sanders, 1977)

See also KEEP COMPANY WITH and STEADY COMPANY.

company[2] **(the)** the main US organization for espionage and foreign subversion.

A pun on the initial letters of *Central Intelligence Agency* and the Spanish *Cía.*, an abbreviation for *compañía*, company:

> Your outrageous statement that we intend to commit bodily harm tarnishes our friends in the Company. (Ludlum, 1979, of the organization)

completion a sexual orgasm.

Usually of the female, whether final or not:

> In thanks, he summoned up a patient rigidity which brought her to six vast, grunting completions before she subsided into sleep. (M. Thomas, 1980)

complications the swelling of an adult's testicles during mumps.

This symptom, additional in some cases to a swelling of glands in the neck, is very painful and may lead to infertility:

> Measles without complications at nine and mumps when he was still too young for complications at ten. (Price, 1972)

complimentary included in the price.

Often the usage seeks to mask an inconvenient substitution for a discontinued service, such as a paper strip to clean your shoes in place of a night porter. An irritating concept but 'we will shortly take your beverage order. The wine in your basket is complimentary' (Republic Airlines Flight RC 207, Greenville/New York, May 1981) was an exception.

compound with *obsolete* to copulate with.

Literally, to mingle with:

> My father compounded with my mother under the dragon's tail. (Shakespeare, *King Lear*)

comprehensive offering non-selective entry.

Of British secondary schools, which are not necessarily characterized by comprehension:

> 46 per cent of children now leaving Mrs Williams' 'comprehensive' secondary school system [are] unable to read or write. (A. Waugh, *Private Eye*, July 1981)

Comprehension is British educational jargon for the ability to read, especially on the part of those who have not acquired that skill in their years of schooling. In the case of such people, *communication* means 'writing' and *numeracy* 'arithmetic':

> MID GLAMORGAN ADULT LITERACY/NUMERACY SERVICE For help with: READING SPELLING ARITHMETIC. (Advertisement in *Rhymney Valley Express*, noted in *Private Eye*, October 1981)

compromise[1] to involve in extramarital copulation.

Of a man it formerly meant 'to be caught acting dishonourably'. Now usually of a female:

> He began to fiddle with his clothes ... is he going to do it here, in public, to compromise me? (Bradbury, 1959)

compromise[2] to kill.

The language of espionage and those who fantasize about it:

> He was killed—and he *was* killed—because whatever that woman told him was so conclusive he had to be compromised hours later. (Ludlum, 1984)

compromise[3] to expose or endanger.

Again espionage jargon, which is near to the standard English use, although that relates to destroying a person's reputation rather than blowing his cover. Less often of general exposure to danger:

> I'd learned from an unidentifiable source that Flight 306 to Bangkok was compromised. (Hall, 1988—it was about to be blown up in mid-air)

comrade a fellow Communist.

From the sharing of a *camera*, or bed-chamber, by Spanish soldiers. Grose gives it as 'cambrade', of the same derivation, 'generally used to signify companion'.

con to trick.

Not relating to the path of a ship or 'set in a notebook, learned, and conned by rote', but a shortened form of *confidence trick*:

> Most of the people you meet will be out to con you. (Sanders, 1980)

Confidence in this derogatory sense was first used in 1866 of the advisers of the Confederate President Davis. A *con-man* or *con-artist* is a person who tries to obtain things by deceit:

> Don't pull that con artist crap with me, pal. I've seen you working this street for three days. (Weverka, 1973)

concentration camp a place for arbitrary imprisonment of political opponents and others.

They were originally the areas in which civilians were concentrated by the Spanish in Cuba and the British in South Africa to prevent the feeding and hiding of men engaged in fighting against them. The Nazis adopted the tactic and the terminology (*Konzentrationslager*); their prisons, which started as places for extortion, ransom, and humiliation, become depots for slave labour, starvation, and extermination:

> There are not only prisons now, there are concentration camps. (Manning, 1962, of the Second World War)

Also abbreviated to *camp*:

> ... three-fifths of them had disappeared into camps that used the new scientific methods ... They had an official name —Vernichtungslager, extermination camp. (Keneally, 1982, of Polish Jews in the Second World War)

concentration problem (a) idleness.

Educational jargon among those for whom there are no lazy or stupid children:

> You clearly have a concentration problem, 'are an idle bitch', and I was wondering ... (Amis, 1978)

concern political dogmatism.

Literally, care or interest:

> ... the Claimants' Union, a focus of responsibility and concern ... (Bradbury, 1975)

In the same sense *concerned* means 'dogmatic':

> This kind of *decent*, modest radicalism ... was a perpetuation of the concerned student politics ... (Bradbury, 1965)

Much pejorative use; see also COMMITTED. The obsolete British *concerned* merely meant 'drunk', and was probably a shortened form of *concerned in liquor*:

> He never call'd me worse than sweetheart, drunk or sober,
> Not that I knew his Reverence was ever concern'd.
> (Swift, 1723)

concert party the concerted buying of shares in a company in different names.

Stock Exchange jargon for an attempt to build up a key or powerful holding without putting on notice the Board of the Company or the Stock Exchange.

concessional free or subsidized.

The use tries to mask the granting of a privilege or charity both to individuals (*concessional* fares for old people) and to countries (*concessional* exports of food etc.).

concoct to falsify.

Originally, to form from different ingredients and then to invent:

> I never knew anybody—anybody—concoct his expenses like you. (L. Thomas, 1989)

concrete shoes (in) *mainly American* murdered and hidden.

> ... it's tough to play golf in concrete shoes. *Comprende*? (M. Thomas, 1980)

A more accurate version of *cement shoes* (see CEMENT SHOES (IN)). Also as *concrete boots*. The victim may sometimes be said to wear a *concrete overcoat*:

> Aiden ... has a three-day plan; repay two grand he owes his pornographer boss or else find himself trying on a concrete overcoat. (*Empire*, August 1993)

condition[1] an illness.

Literally, any prevailing circumstance, but in matters of health any *condition* is bad, be it of the heart, bladder, liver, or whatever:

> Throughout the aircraft, the old, then those with pre-existing medical conditions, began to die. (Block, 1979)

condition[2] a pregnancy.

This usage is not reserved for unwanted or difficult pregnancies and merely avoids direct reference to the taboo:

> Naturally, Melinda did not mention her condition. (Boyle, 1979—Melinda Marling was pregnant by the spy Donald McLean before their marriage)

The *condition* may be *delicate* or *interesting*, these adjectives being descriptive of the woman's situation or state of health. It is often referred to obliquely as a *certain condition*:

> He said that a young woman who was obviously in 'a certain condition', but not having a ring ... (Lodge, 1975)

conditioner a mild acid.

Soap, including soft soap or shampoo, is alkaline, and to restore pH to neutral after washing anything, including hair, a corresponding acid may be introduced. Vinegar was the normal substance so used until the cosmetic industry came up with its more expensive *conditioner*. A further twist in this story of exploitation and folly is when a soap is sold with a pH of 7 claiming that it combines two properties in a single (more expensive) product.

confederation a pressure group.

Literally, an alliance or union of states for joint action. The *Confederation of British Industry* exists to further the interests of employers, managers, and, to a much lesser extent, shareholders; the *British Confederation for the Advancement of State Education* strives not to improve the quality of teaching in state schools but to eradicate private schools.

conference a period in which you wish to avoid callers.

The excuse of businessmen and their secretaries, the evader being *in conference*:

> Ah'm afraid Miss Brimley is in conference. Can someone else answer your query? (le Carré, 1962)

The 'formal ... meeting ... for consultation and discussion' (SOED) is where we go when we are *at a conference*. For medical practitioners or academics, a *conference* is a holiday which they enjoy with their peers in a congenial place at the expense of a third party.

confinement the period of childbirth.

Literally, no more than the fact of being shut up, as in prison:

> The women continue working down to the day of their confinement. (Mayhew, 1851, of childbirth)

To be *confined*, which might be taken as being unable to leave a sick room, is to be giving birth to a child, this usage having superseded the 19th-century meaning 'constipated'. We have also lost the obsolete *bond* of childbirth, which used the same restricting imagery.

conflict a war.

Literally, a strong disagreement or a single battle. But it sounded better than *war*, especially when the Korean *conflict* burst upon us in 1950, so soon after the Second World War.

confrontation a war.

Literally, a meeting face-to-face. Indonesia's 1963 attack on Malaysia was so described but limited to local armed raids and subversion. By the time of Vietnam, it meant that kind of war. Now also used by terrorists of their violence in waging war indiscriminately against society:

> Well for one thing we haven't ruled out the possibility of confrontation. (Theroux, 1976 —terrorists were discussing tactics)

confused drunk.

It certainly can take you that way:

> I gather our son was very confused *that* night; which is a mother's way of saying he was plastered. (Ludlum, 1979)

congress copulation.

> I had heard precisely how that acrobatic quartet achieved congress. (Fowles, 1977 —four people were copulating)

Literally, a coming together, as in the *Congress System* established by the victors over Napoleon's France or the Indian Congress Party.

Normally with the prefix *sexual*:

> Eight days later in the little summer-house sexual congress takes place. (Boyd, 1987)

Occasionally as *male*:

> She's been repeatedly raped, both by male congress and by instruments. (West, 1979)

conjugal rights copulation with an unwilling wife.

Legal jargon for marital copulation. Etymologically, *conjugal* means 'yoked together' and in some societies the placing of a symbolic double yoke formed part of the marriage ceremony. A woman seldom seeks these *rights* from her husband, except in satire:

> Wilt had enough troubles with his own virility without having Eva demand that her conjugal rights be supplemented oralwise. (Sharpe, 1976)

A woman who applies to the Court for *restoration of conjugal rights* seeks pecuniary rather than physical gratification.

conk (out) to die.

From the unplanned stoppage of an engine and the end of the movement:

> Jassy and Victoria will scream with laughter when I finally do conk out. (N. Mitford, 1949)

The American *conk* is rarer:

> ... the paintings would automatically increase in value once Maitland had conked. (Sanders, 1977—Maitland was an artist)

connect[1] to copulate.

The imagery is from joining or fastening together:

> ... two beautifully engraved figures of man and woman who were connecting at every tick of the clock. (Richards, 1936)

Connection normally denotes extramarital copulation:

> Of course there had been no connection (Harris, 1925, writing for once of an absence of copulation)

but can also refer to bestiality:

> ... others were homosexual, others who sought connection with animals (an ill-documented area even in sexology). (Pearsall, 1969)

The plural, referring to systematic conduct, is rare:

> Privates in the Blues ... often formed very reprehensible connections with women of property, tradesmen's wives, and even ladies. (Mayhew, 1862)

To *have connection with* is to copulate extramaritally:

> ... the gentleman had connection with me. (Harris, 1925—obviously no more a gentleman than Mayhew's 'ladies' (above) were ladies)

connect[2] to find a source of illegal narcotics.

The reference is probably to joining the chain of distribution, or *connection*, which in the jargon runs from manufacturer to retailer.

connection see CONNECT[1, 2].

connections people susceptible to bribery.

Literally, people to whom you are related or whom you know well. They may be influential, whence the implication that their influence will be improperly used or may be obtained through bribery:

> ... the redoubtable lady was able first to defraud the public and then to evade the consequences because she had 'connections'. (Shirer, 1984—she used bribery)

connubial pleasures copulation.

Although *connubial* literally means 'to do with marriage', the *pleasures* can be taken by either party within or without that institution:

> She never married, but it didn't prevent her from enjoying connubial pleasures. (Ludlum, 1979)

conquer a bed *obsolete* to copulate extramaritally with its usual occupant.

Of a male; see also BED[2]:

> When you have conquer'd my yet maiden bed. (Shakespeare, *All's Well*)

consenting adults male homosexuals aged over 18 who engage in a physical sexual act with each other.

British legal jargon:

> Two consenting adults had been ejected from the Gents. (Sharpe, 1975—they were not 'consenting' to the ejection)

The former British stringent but largely unenforceable laws against overt expressions and even private acts of male homosexuality have been amended to permit almost any private behaviour so long as minors are not involved. The same laws never applied to female homosexuality, reputedly because Queen Victoria thought there was no such thing.

console to copulate with extramaritally.
Literally, to alleviate sorrow. It is used of either sex, especially when a regular sexual partner is absent:

> Another girl of similar type, who had briefly consoled him in France. (Boyle, 1979, of the spy Kim Philby)

Consolation is such copulation:

> Men whose wives were sent out of harm's way were quick to find consolation. (Manning, 1977, writing of the Second World War)

consort with to copulate with extramaritally.

> Some of them consorted with—with the worst type of native woman. (Fraser, 1975)

A *consort* is someone who keeps you company.

constant companion a person with whom you are seen publicly and regularly to copulate extramaritally.
Journalistic jargon, when a public figure might sue for defamation if the relationship were spelt out:

> Miss Kristina Olsen, his close friend and constant companion. (Allbeury, 1976, of such an arrangement)

Constant companions refers to both parties to such an arrangement. See also COMPANION.

constructed brought to submission by force or intimidation.
The jargon of Vietnam:

> A 'constructed' hamlet meant not a newly built one but a former Viet Cong hamlet that had been worked over politically. (McCarthy, 1967)

It was *reconstructed* if it had changed hands three times, with intervening Vietcong control.

consultancy the humouring of a dismissed employee by giving him money.
A business practice of continuing to pay an unwanted senior employee, to speed his departure, avoid recrimination, ease his tax position, save his face, protect trade secrets, etc. A *consultant* is so treated.

consultant[1] see CONSULTANCY.

consultant[2] a dispenser of bribes.
Literally, one who is consulted, whence a specialist, especially in medicine, who sells advice or services. In the export of capital goods, the *consultant* may be the person who accepts a fee in return for acting as the conduit for the payment of bribes which would be risky and illegal if paid directly by the vendor. See also THIRD PARTY PAYMENT.

consulting Mrs Jones urinating.
An American excuse, explaining the absence of a third party. I have no clue as to the etymology.

consummate (a relationship) to copulate.
Originally, to accomplish to the full.

> I have had to learn [self-control]. She has refused to consummate our relationship. (Townsend, 1982, of unmarried sweethearts)

Consummation is one of the essential ingredients of a Christian marriage, in default of which a Court will on request grant an annulment.
To *consummate your desires* implies extramarital activity and less discrimination in the choice of a partner. It is usually used of men, but:

> ... there is a house in Regent Street, I am told, where ladies, both married and unmarried, go in order to ... consummate their libidinous desires. (Mayhew, 1862)

consumption pulmonary tuberculosis.
Prior to penicillin, this was the dread disease which wasted away, or *consumed*, the sufferer:

> The girl had died also since then. Consumption devoured her. (Keneally, 1979, writing of the 19th century—the young woman had not been eaten by a lion)

Also known as the WHITE PLAGUE, both phrases were replaced latterly by *TB*, for tubercule bacillus.

contact with copulation with.
From the touching, but in this use the male is the only one to do it, despite the mutuality of the encounter:

> ... he would need ... to augment his size and permanence by food, booze, contact with a woman. (Keneally, 1982)

contagious and disgraceful disease a venereal disease.
Legal jargon in the English law of slander. If you wrongly imputed it in another, the plaintiff had no need to prove special damage. Women had further protection which was withheld from men: the Slander of Women Act of 1891 made an imputation of unchastity by a woman actionable without proof of special damage, but a man could not sue successfully without such proof.

content kept involuntarily under heavy sedation.
Medical jargon:

> ... the few violent cases we have are kept pretty, uh, content. (Sanders, 1979, of an institution for the insane)

content your desire to copulate.
Normally referring to casual copulation by either sex:

> It was the Doctor who undertook to content her desire. (Harris, 1925)

continent see INCONTINENT.

continuations trousers.

They *continued* a Victorian male's waistcoat in a direction too delicate to mention. See also UNMENTIONABLES[1] for more of these quaint obsolete uses.

contour a fat shape.

Literally, the outline of any figure, but jargon for those who sell clothes, exercise equipment, or dietary foods to the obese. To *reduce your contour* is to slim or become thinner.

contract[1] a bribe.

Used of American payments to politicians, the police, etc. in an attempt to put a gloss of legality on the transaction.

contract[2] the hiring of a killer.

Underworld jargon for a murder treated as a commercial transaction:

> There's a contract out on Billson and he's still alive. (Bagley, 1977—the implication is that he should be dead)

To *put out a contract* is to arrange a specific murder. The *contract* may also be the potential victim:

> I want you to know that you could become a contract. (Deighton, 1981)

contribution[1] a quantity of urine.

Medical coyness when asking a patient to provide urine for analysis:

> 'The usual contribution, please,' she said motioning towards the lavatory door. (Sanders, 1981)

contribution[2] a bribe.

Paid, usually from corporate funds, to secure either favours for an individual (e.g. a knighthood in the British Isles) or privileged treatment for a corporation from government.

control unit *American* a cell for solitary confinement.

It is, I suppose, one way of *controlling* a violent prisoner.

controlled economy a strict socialist state.

The jargon of Marxism, where an attempt is made to regulate production and individual economic activity through a bureaucracy.

controlled substance a narcotic.

So called because its legal manufacture and distribution are regulated and supervised:

> ...there was no evidence that he was dealing in what the government laughingly calls a 'controlled substance'. (Sanders, 1987)

controversial[1] politically damaging.

Bureaucratic Whitehall jargon, mainly to be used when deflecting a politician from a course which a civil servant does not like:

> 'Controversial' only means 'this will lose you votes'. 'Courageous' means 'this will lose you the election'. (Lynn and Jay, 1981)

controversial[2] disreputable and untrustworthy.

Journalistic jargon, especially of company directors etc.:

> ...[the] moving spirit of the now defunct stockbroker——is working on a diamond company float on the Moscow stock exchange [and is] now sharing a Knightsbridge office with the controversial financier——. (*Sunday Telegraph*, 17 December 1994)

convalescent home see REST HOME.

convalescing see EXTREMELY ILL.

convenience[1] a lavatory for public use.

Literally, anything which accommodates. Often specifically described as *public*, *men's*, or *ladies'*, or merely in the plural:

> ...another tin outhouse with a sign saying Conveniences. (Theroux, 1983, of a lavatory on a camp site)

convenience[2] see FOR YOUR CONVENIENCE.

convenient[1] *obsolete* a prostitute.

She restricted her clientele to one regular customer:

> Dorimant's Convenient, Madam Loveit. (Etherege, 1676)

convenient[2] tiny.

Real estate jargon, usually describing a lazy person's garden which is manifestly inconvenient for drying washing, privacy, lighting bonfires, growing produce, etc.

conventional not involving nuclear or germ warfare.

There is something bizarre in the notion that any weapons for killing or maiming are sanctioned by general agreement or established by social custom.

conversation see CRIMINAL CONVERSATION.

convey *obsolete* to steal.

From the carrying off of the article taken. A *conveyor* was a thief:

> ...conveyors are you all,
> That rise thus nimbly by a true king's fall.
> (Shakespeare, *Richard II*)

Conveyance was theft, and *conveyancing* swindling, this usage long pre-dating the exorbitant fees charged by lawyers for the simple process of transferring title to real estate.

convince to compel by force.

Underworld and police slang:

> He knew exactly what methods Willi Kleiber would use to 'convince' Colonel Pitman to open the safe. (Deighton, 1981)

convivial habitually drunken.

Usually written of a prominent male:

> ...obituaries are simply eulogies of the great and the good, any of whose peccadillos (unusual sexual tastes, drunkenness and so on) are tactfully powdered over with euphemism ('flamboyant', 'convivial' etc.). (Lewis Jones, *Daily Telegraph*, 1 December 1994, explaining why as Obituaries Editor he selected franker contributions)

cook[1] to kill.

I think not from culinary imagery, even as a shortened form of *cook your goose*, to cause to fail. Perhaps from execution by electricity:

> Those fucking sketches could cook him if we found the girl. (Sanders, 1977)

Also of stock in a hot drought-ridden country:

> A drought...would cook half the stock in the country. (Boldrewood, 1890)

cook[2] fraudulently to alter.

As in *cooking the books*, preparing accounts dishonestly, from the culinary art of re-arranging ingredients to make a more acceptable dish. Sometimes too of records:

> It is better not to use the verb 'cook' in connection with either books or minutes. (Lynn and Jay, 1989)

The phrase was first used of George Hudson, the 19th-century British 'Railway King', after he had falsified accounts to pay dividends out of capital. Among his other attainments, he had a hand in building a fifth of the British railway network, devised the 'clearing' system for shared services, was elected to parliament, and was sent to prison for debt.

cook[3] an opium user.

From the act of preparing opium over a flame. Now in America anyone illegally preparing a narcotic for injection through the application of heat.

cooked drunk.

The common American culinary imagery.

cookie a promiscuous female.

Supposedly warm, sweet, and fresh, and usually found in an American bar:

> ...you might come clean about that blonde cookie you've parked on big-hearted Mrs Swallow. Rumour has it that she's pregnant. (Lodge, 1975)

A *new cookie* is a younger female for whom a man has abandoned his wife.

cookie pusher a male employee who curries favour with his boss.

From a man handing round the cakes at a function largely attended by women and owing nothing to the obsolete *cookie*, cocaine:

> ...do you see that furry-headed little cookie-pusher Brittan is having the fountains in Trafalgar Square drained for New Year's Eve? (*Private Eye*, December 1983, of the British Home Secretary)

Also used generally of male homosexuals.

cool[1] dead.

The common imagery from the loss of body heat:

> ...if the old lady hadn't been cool for a month even, the will certainly wouldn't have been proven. (Lyall, 1969)

A *cool one* is a corpse:

> Mr Yow would not have brought me here if he'd known there was a cool one in the car. (T. Harris, 1988)

In rare American use the verb to *cool* is to kill.

cool[2] not carrying illegal narcotics when searched.

The reverse of *hot* (see HOT[2]) and perhaps owing something to the meaning 'poised and unruffled'.

cooler[1] a prison.

Common imagery for the place where you are sent to cool down:

> We could be put in the cooler for these. (Theroux, 1973)

cooler[2] an intoxicant which is diluted and served in a large glass.

Normally with ice in it. It is intended to cool you down.

coop *American* a prison.

In this case, for humans and not for hens or rabbits:

> 'No convictions, but prints on file.' 'Been in the coop.' (Chandler, 1958)

cop[1] to steal.

Literally, to catch or seize:

> I was taken by two pals to an orchard to cop some fruit. (Horsley, 1887)

cop[2] a policeman.

Usually said to be a shortened form of COPPER but perhaps also because he *cops* or seizes you:

> The fuzz—that's what they call them now, not cops any more. (Ustinov, 1971)

A *cop shop* or *house* is a police station:

> I have to go to the cop house just about now. (Chandler, 1958)

I think *cop*, prison, is obsolete:

> I saw a policeman taking two men to cop. (EDD)

cop[3] *American* to obtain illegal narcotics.

Through buying, stealing, or howsoever.

cop a cherry to copulate with a female virgin.

Perhaps obsolete. See CHERRY.

cop a packet to be killed or severely wounded.

A variant of CATCH A PACKET in all its senses, including, for a Second World War serviceman, the contraction of venereal disease.

cop an elephant's to become intoxicated.

From the rhyming slang *elephant's trunk*, drunk. See also ELEPHANT'S and JUMBO.

cop it to be killed or die.

Again from the catching:

> I was really rather lucky. A lot of my mates copped it. (Manning, 1977)

Occasionally as *cop out*.

cop out[1] see COP IT.

cop out[2] *American* to plead guilty to a minor offence among several.

Part of the process of plea-bargaining. The prosecution drops more serious charges if the accused admits to lesser offences.

cop the bullet to be dismissed from employment.

As distinct from *copping a bullet*, being killed or wounded. See also BULLET (THE).

cop the drop to accept a bribe.

From the passing of money into an upturned palm or from DROP[5].

copper a policeman.

It is always said to be from the metal buttons on their 19th-century uniforms, but see COP[2] for an attractive alternative etymology:

> An' up comes a bleedn' rozzer an' lumbers me. Wot a life! Coppers. (Kersh, 1936)

copulate to fuck.

Originally, to link together, whence to become joined together, from which the common imagery of fucking. As it is explicit in standard English and less jarring than *fuck*, I use the word, along with *copulation*, throughout this dictionary.

cordial[1] *?obsolete* an intoxicant.

Originally, any food or drink which comforted the person who ingested it:

> ...make invitation the one to the other for pipes and foreign cordials. (Blackmore, 1869)

In modern use, the drink so described, for example *lime juice*, is calculated to please but not intoxicate.

cordial[2] cold and unfriendly.

Diplomatic jargon which indicates the opposite of the correct meaning 'warm and friendly'.

cordless massager a masturbating machine for women.

It is shaped like a penis and used as a vibrating dildo. A more explicit, and also euphemistic, name for it is VIBRATOR.

corked (of a person) drunk.

Precisely what a *corked* bottle of wine should not be. The American imagery is unclear.

corn[1] low-quality whisky.

From the raw material, and often home-made. In many compounds, such as *corn juice, corn mule, corn waters*:

> Various sorts of distilled spirits, particularly one named Cornwaters. (Hibbert, 1822)

Corned, drunk, may also come from drinking too much *corn*, although a more likely etymology is from the old meaning 'pickled', particularly as the use has recurred at various times since the 18th century in America, England, and Scotland.

corn[2] copulation with a woman.

From the food a horse likes best and regularly. Usually of extramarital copulation. The obsolete *cornification*, lust, comes from the Latin *cornus*, a horn.

corn-fed *American* plump.

Of a woman, especially one below the age of thirty, from the fattening of livestock on an augmented diet.

corned see CORN[1].

corner[1] to establish a monopoly in a product.

Probably from driving cattle into the corner of a yard rather than from storing something in a hidden place. To *corner* in the jargon of British criminals is to sell shoddy goods at above their worth by persuading greedy buyers that they are stolen or in short supply.

corner[2] a urinal.

Male use, from the facility to urinate in an open space so long as the penis is concealed:

> Oh, I'm so sorry, I was looking for a corner. (Olivier, 1982, quoting Churchill who entered a theatre dressing-room in 1951)

corner[3] the penis.

Seen sexually, as in one phrase, *get your corner in*, to copulate:

> ...if he did get his corner into a nice mine wife... (Keneally, 1979, of the Chairman of a 19th-century American miners' Benevolent Fund)

cornhole *American* to bugger.

Cornhole, as the anus, seems to have failed to attract the notice of most lexicographers, including those responsible for the 2nd Edition of the OED and its 1993 Additions:

…you think I'm gonna want the whole world watching him cornhole me. (Goldman, 1986)

coronary inefficiency a weak heart.

Medical jargon which borders on mere circumlocution or pomposity:

A coronary inefficiency had made it necessary for Robert Winthrop to use a wheelchair. (Ludlum, 1979)

corporal and four see MOUNT A CORPORAL AND FOUR.

corporate entertainment bribery.

Those who are invited to enjoy such HOSPITALITY might be less relaxed if the cost were merely handed over in the form of notes:

The boxes [at Covent Garden Opera House] are largely used for corporate entertainment, that is to say buttering up clients. (H. Porter, *Daily Telegraph*, 22 October 1994—he also noted *business entertainment*, *freebie*, *conference*, *jolly*, *jaunt*, *concessionary fare*, *facility trip*, *sale preview*, and *hospitality*, as 'distant cousins' or 'blood relations' to the bribe)

corporate recovery the management or winding-up of insolvent companies.

British accountants' jargon which seeks to draw attention to the often slim hopes of revival rather than the probability of demise:

This compares with the 75 p.c. growth in the insolvency side—which the firm delicately calls 'corporate recovery'. (*Daily Telegraph*, 28 November 1990: the firm was the accountants, Peat, Marwick, McLintock)

correct unswervingly accepting or in line with current party policy.

In standard use it means 'adhering to high moral standards'. Nazi and Communist jargon:

To ensure that political affairs would be handled correctly in an emergency. (Goebbels, 1945, in translation)

From the correct point of view there are no contradictions. (M. C. Smith, 1981, of Russian policy)

correction[1] a serious fall.

Stock Exchange jargon for tumbling prices, which seeks to imply that they had previously risen too high:

…there were sufficient signs on the horizon to indicate that some major correction—for which read 'collapse'—is called for. (M. Thomas, 1982)

correction[2] see HOUSE OF CORRECTION.

correctional of or pertaining to prison.

From the theory that convicts are there to be taught better ways. Thus the American *correctional facility* is a prison, a *correctional officer* is a jailer, etc.

corrective training political imprisonment.

Communist jargon and practice:

Those who said that…underwent corrective training that proved fatal in most cases. (Amis, 1980)

The activities were carried out in 'Corrective Labour Camps', controlled by the *Glavnoe Upravlenie Ispravitel'no-Trudovykh Lagerei* (from which was formed the acronym *Gulag*).

correspondent see CO-RESPONDENT.

corrupt to copulate with outside marriage.

Literally, to spoil. In literary use it is the male who does this kind of spoiling:

Angelo had never the purpose to corrupt her. (Shakespeare, *Measure for Measure*)

The word is now rightly used of any conduct which leads another astray.

cosa nostra the Mafia.

'Our thing' to the gangsters from Sicily and their adherents. Unhappily they too often share it with others.

Costa Blanca a term of vulgar abuse.

Rhyming slang for 'wanker' (see WANK).

costume wedding the marriage of a pregnant bride.

Her physical indications rather than remorse at the pre-marital loss of virginity may inhibit the wearing of the traditional white gown. Cheshire.

cottaging seeking a male homosexual partner in a public urinal.

Perhaps from *cottage*, a slang name for a urinal:

The Tea Room Trade they call it in America; in England, Cottaging. (Fry, 1991)

cotton *American* a female's pubic hair.

From the fluffy appearance of the bush. It need not be white.

cottonwood *American, obsolete* pertaining to death by hanging.

From the prevalence and convenience of the tree in those areas of America which needed ad hoc gallows. Thus if you *decorated a cottonwood tree* or *looked through the cottonwood leaves*, you were killed by hanging.

cough[1] (of a criminal) to give information to the police.

A common variant of the *singing* theme, which can include making a confession of your own guilt:

I could go up to Grosvenor Square and cough it all. (Theroux, 1976, of a threat so to give information)

cough[2] to die.

From the laboured breathing of the terminally ill:

> All a matter of luck, whether one man stands his ground and wants to take people with him when he coughs. (Seymour, 1977, of terrorists)

cough medicine a spirituous intoxicant.

Usually whisky in male humorous speech. And as *cough syrup*. In criminal slang *cough syrup* may also be a bribe to prevent an informer talking —see COUGH[1].

counsellor one who intrudes on those suffering misfortune.

Literally, anyone offering advice, whence the practitioners so named:

> 'I wish you'd take my advice and see a Counsellor.' 'Everyone wants me to see a shrink!' she burst out. (Sanders, 1981)

This usage seeks to avoid an implication of mental instability, as most attorneys in America also call themselves *counsellors*. *Counselling* has become a growth industry in the 1990s, with busybodies and do-gooders joining the ranks of supposedly trained professionals to force their services upon the victims of any unusual event:

> There are only three recession professions. One is garden design. The others are counselling and consultancy... two of these activities involve people who are not sure what they ought to be doing telling other people who are not sure what they ought to be doing what they ought to be doing. (Victoria Glendinning in *Daily Telegraph*, 27 January 1994—fortunately garden designers do nobody any harm)

Thus the wreck of a ship off Shetland in January 1993 led to suggestions that all schoolchildren in the islands might need *counselling* about the incident, while neighbouring Orkney provided its citizens with the services of three professional HIV *counsellors* even though no inhabitant had been affected with the virus.

count (the) death.

Boxing imagery. The *long count*, though rarer, shows greater knowledge of the sport. To *put out for the count*, again from boxing, is to make unconscious and usually not to kill. To *count the daisies* is to be dead, the sums being done from the roots upwards. (See also POPPING UP THE DAISIES, PUSH UP THE DAISIES, and UNDER THE DAISIES.)

counter-force capability the ability to hit military targets with nuclear weapons.

The language of the Pentagon; in a saner world it might mean how well you can push a laden wheelbarrow up a slope. Also as *counter-force attack*:

> The Russians, then, had an advantage in a potential 'counterforce' attack—the sort aimed at weapons rather than people. (Clancy, 1988)

Counter-value capability is the ability to hit civilian targets only. For all these activities you use a *counter-force weapon*, or nuclear bomb.

counter-revolution any opposition to a totalitarian state.

Mainly the jargon of the Communists to describe things such as owning a typewriter or listening to foreign broadcasts, because the only authentic revolution was the one which brought them to power.

counterattack an unprovoked aggression.

There is no longer the prior requisite of an attack to counter:

> Thus did the Nazi dictator and his cohorts in Berlin see the German 'counterattack' on Poland become a European war. (Shirer, 1984, describing the invasion of Poland in 1939)

counterinsurgency waging war in another country.

The *insurgents* are the native inhabitants who want to establish their own administration in place of one imposed by those who use this phrase, such as the French in Algeria, the Americans in Vietnam, and the British in various places:

> Kennedy men revealed the need for brand-new tactics with brand-new names: counter-insurgency, special warfare. (McCarthy, 1967)

country not reconstituted.

The language of the coffee shop(pe):

> Your choice of three crisp slices of bacon... served with one large country egg. (American Holiday Inn menu, May 1981—in fact the short-order cook 'chose' the slices for me. The same menu offered an omelet with my 'favorite' filling, making two rash assumptions —that I favour one particular filling and that they would be able to serve it if I decided to order it.)

country-club girls prostitutes operating out of town.

When the law closed the New Orleans brothels in 1917 as part of the war effort, many of the prostitutes moved to country clubs out of town:

> The country-club girls are ruining my business. (Longstreet, 1956, referring to a city operator)

country in transition a poor and backward country.

The phrase fails to specify in which direction they are moving:

> ... those thrilling economies known to the IMF as Less Developed Countries or (a new euphemism) countries in transition. (*Daily Telegraph*, 15 September 1994—the IMF is the International Monetary Fund)

country Joe *American* a local policeman.
Joe is the name often given to someone considered simple. Citizens' Band slang.

couple[1] to copulate with extramaritally.
A rare transitive use:

> Thou hast coupled this Hindoo slut. (Fraser, 1975, writing in archaic style)

Its standard meaning is 'to marry' of humans, 'to copulate' of animals. *Couple with* is more common:

> Only ten minutes ago she had been coupling with me on the bed. (Fraser, 1969)

In obsolete use a *coupling house* was a brothel.

couple[2] a woman's breasts.
Viewed sexually, usually by a male but:

> Reminded her of a girl at prep school who was voted best couple in the yearbook. (McInerney, 1992, of a girl with big breasts)

courses *?obsolete* menstruation.
From the meaning 'a period of time':

> I had my courses, my flowers. (Fowles, 1985, writing in archaic style of a woman denying that she had been pregnant)

courtesan a prostitute.
In the 15th century it referred to someone at court. The derivation is more likely to be from the Italian *cortigiana*, despite the morals and opportunities of Tudor courtiers:

> He regularly visited a famous courtesan in the Srinagar bazaar, and enjoyed other favours too. (Masters, 1976)

Standard English.

courtesy included in the price.
From the meaning 'given freely'; but the *courtesy coach* takes you to an inaccessible hotel which you would not have used without it. See also COMPLIMENTARY.

cousin Cis a drunken carouse.
Rhyming slang for 'piss' which, in the expression *piss-up*, has the same meaning. DRS says *sis*, I think wrongly.

cousin John see JOHN[1].

Covent Garden *obsolete* involved in or ancillary to prostitution.
The London district, with the neighbouring Drury Lane, was a 17th-century centre of prostitution (see also DRURY LANE AGUE). As *Covent* is a corruption of convent, there were many ecclesiastical puns and witticisms. Thus a *Covent Garden abbess* kept a brothel, or *garden house*, which contained *Covent Garden goddesses*. They often infected their customers with *Covent Garden ague* or *garden gout*, venereal disease, and they were then said to have *broken their shins against Covent Garden rails*.

cover[1] to copulate with.
Standard English of stallions, from the mounting of the mare, and other mammals, and occasionally of humans:

> He'll ask you why you did it. 'Because your overseer's covering 'em,' you'll say, using a lady-like term. (Fraser, 1971—the overseer was copulating with the black slaves)

In Coverdale and the Geneva Bible, to *cover your feet* was to urinate, being a literal translation from the Hebrew.

cover[2] to dye.
Referring to men's hair. A 1983 advertisement for dye described it as 'covering grey hair'.

cover story a lie.
Espionage jargon but also used by errant husbands and others. The story tries to *cover up* the truth, whence a *cover-up*, an attempt wrongly to conceal by deceit.

covert act any illegal behaviour.
Espionage jargon. *Covert* means 'hidden' or 'secret':

> 'Do you mean acts of sabotage?' 'Er... could I just say covert acts?' (Lyall, 1985)

cow brute *American, obsolete* a bull.
From the olden days, when a bull was too overtly sexual to talk about. For a list of similar euphemisms see BIG ANIMAL.

crabs *American* syphilis.
In standard usage, short for *crab-lice*, which favour pubic hair, and therefore medically inexact.

crack[1] to rob.
By forcible entry of a building or specifically by *cracking* a safe. A *cracksman* specializes in the latter art.

crack[2] to hit or kill.
Not necessarily from a blow that damages the skull:

> I figure you cracked him in anger. (Turow, 1987, referring to a murder)

crack a bottle to drink intoxicants.
Perhaps the very impatient may break the neck to get at the contents more quickly, but the phrase is used of any wine drinking.

crack a Jane *American* to copulate extramaritally with a female virgin.
From CRACK[1] or from *cracking* a problem? A Briton speaks of *cracking a doll* or *a Judy*. The obsolete British phrases for the same achievement, to *crack a pitcher* or *pipkin*, showed more imagination; both the pieces of pottery would remain serviceable after the *cracking*, but not as desirable as those without blemish.

crack your whip to copulate extramaritally.

Of a male, punning, I suppose, on the mastery of an animal trainer and the *whip*, or penis:

She was crazy for me to get her that guy who wrote about cracking his whip all the time. (Sharpe, 1977)

cracked mad.

The imagery is from having a flaw, whence the synonym *crackpot*:

Now the necessary removal of Bayldon was threatened by a hijack engineered by some crackpot group. (B. Forbes, 1989)

In various other phrases such as *crack-brained* and the common slang *crackers*.

cracked in the ring see RING[1].

crackling a female viewed sexually by a male.

Literally, the crisp and tasty outside of roast pork. She is usually described as a *bit* or *piece of crackling*.

cradle-snatcher an older person marrying one much younger.

In everyday English, someone who steals a baby (not its bed). The term is used of either sex, with disapproval if the older person is a woman or a much older man. To *rob a cradle* is so to act. See also BABY-SNATCHER.

cram to copulate.

Of a male and of the same tendency as STUFF[2]. The usage survives in the American vulgar injunction *cram it*.

cramps (the) menstruation.

General female usage, from one of the symptoms. Also as *stomach cramps*:

... the stomach cramps ... happen quite regularly in the first week of every month when a certain software salesman is in town. (J. Trollope, 1992, of a woman staying away from work)

crap to defecate.

Not a euphemism, as that was its original meaning in Old Dutch. Later dictionary definitions, such as 'dregs', were themselves euphemistic. *The Slang Dictionary* of 1894 gives '*Crapping case*, or KEN, the water-closet. Generally called CRAPPING-CASTLE'. Today's *crapper* is more often the lavatory than the person using it:

He couldn't even take a little crap without two of his men checking out the crapper first. (Collins, 1981)

These uses long antedate the 19th-century English ironfounder called Thomas Crapper, who invented the syphon system used in modern flush lavatories and proudly cast his name on the elevated water tanks he manufactured. The relaxation of the bowel at death led to the association of crapping and hanging:

The hangman was Jack Ketch ... the crap merchant, the crapping cull, the switcher, the cramper, the sheriff's journeyman, the gaggler, the topping-cove, the roper or scragger. (Hughes, 1987, describing 19th-century criminal slang)

crash to return to normality after taking an illegal narcotic.

To descend from a *high* (see HIGH[2]):

Brodie had said ... 'I'm crashing.' And she had gone to the mantelshelf ... and taken out a vial of powder. (Theroux, 1976)

Also used by those who take such drugs to mean 'to fall asleep', following the ingestion of a narcotic.

crawl[1] to copulate with extramaritally.

Of a male in occasional spoken use—I suppose it is one way of approaching the encounter. To *crawl in with* someone is specific, of male or female, being no more than a shortened form of *crawl into bed with*.

crawl[2] to curry favour.

By flattering a superior or other subservient behaviour. A *crawler* so acts.

cream to ejaculate semen.

From the colour and texture. The noun *cream* is also a slang term for semen:

At the sight of his bride
When he got her inside,
He creamed all over the bedding.
(*Playboy's Book of Limericks*)

Thus the phrase *cream your jeans* refers to male sexual excitement. *Cream* is also used of the vaginal discharge of a sexually aroused female, whence the common *cream for*, to desire a male sexually:

'Honey,' he said, 'You're still creaming for me.' (Mailer, 1965)

And a *creamer* is a young female seeking extramarital copulation:

Plenty of young creamers ready to spread their pussies. (Sanders, 1982)

cream crackered exhausted.

Rhyming slang for the vulgarism *knackered* (see KNACKERS). Used on BBC Radio on 21 September 1987 by a reporter describing his exhaustion after a period of competitive rowing.

crease to kill by violence.

Mainly in America, I suppose from the collapse of the victim. In British use it also means to hit with a bullet without severely wounding.

creative disputatious or dishonest.

Thus for churchmen, *creative conflict* is a bitter doctrinal argument. For a businessman *creative bookkeeping* is false accounting:

They give you a lot of crap about 'creative freedom' but all they're really talking about is 'creative bookkeeping'. (B. Forbes, 1989)

Creative freedom, for artists or academics, can mean anything you want it to other than doing something original. *Creative tension* was how the relationship between the chairman and the chief executive of British Airways was described in the aftermath of the airline's exposure for unethical behaviour against a competitor:

> He denied that... relations were strained between Lord King and Sir Colin but conceded there had been 'creative tension' between two able executives. (*Daily Telegraph*, 16 January 1993)

creature (the) spirituous intoxicating liquor.

Literally, something created, and perhaps only a shortened form of *creature comforts*:

> When he chanced to have taken an overdose of the creature. (W. Scott, 1815)

The use survives in Irish English, often spelled *cratur*, *crathur*, or *crater*:

> Never a drop of the crater passed down Chancy Emm's lips. (Mayhew, 1862)

The obsolete *creature of sale* was a prostitute:

> The house you dwell in proclaims you to be a creature of sale. (Shakespeare, *Pericles*)

credibility gap the extent to which you are thought to be lying.

Or, which is more honourable, reluctant to come to terms with unpalatable truth. The phrase comes from US strategic analysis in the 1950s and was used by Gerald Ford in this sense in 1966 when questioning President Johnson's statements about the extent of American involvement in Vietnam.

> We do not recognize them, helmeted, in a bomber aiming cans of napalm at a thatched village. We have a credibility gap. (McCarthy, 1967, referring to American pilots in Vietnam)

A *serious credibility gap* means that everyone thinks you are a liar.

creep joint *American* a brothel.

Originally, an illegal gambling operation without a liquor licence which moved from place to place to avoid detection by the police:

> Wieland says you were in Sampaloc. In a creep-joint. (Boyd, 1993—he had been visiting a brothel)

The British *brothel-creeper* is a rubber-soled suede shoe worn by young men.

Cressida a prostitute.

She was the lady who gave Troilus a hard time and Chaucer, Shakespeare, and others a plot:

> The girl was a born Cressida, a 'daughter of the game'. (Manning, 1960)

crib *American* a brothel.

From the American dialect term meaning 'a house', and usually implying meanness:

> Miserable naked girls in the twenty-five and fifty-cent cribs. (Longstreet, 1956)

A *crib man* is a thief who specializes in robbing private homes.

crime against nature sodomy or bestiality.

As proscribed in the laws of many American states:

> Most states have laws against fornication and even masturbation lying somewhere on their books... One of the most popular phrases is 'crime against nature'... but almost never do they specify what a *crime against nature* is. (Bryson, 1994—as he concludes, it could be anything from walking on grass to chopping down trees)

criminal assault the rape of a female.

In fact any force offered against another intentionally is a crime, whether or not sexually inspired:

> ... leading a criminal assault by several Indians on an English girl. (P. Scott, 1975, of a rape)

The woman may be said to have been *criminally used*:

> She was dragged from her bicycle into the derelict site... where she was criminally used. (P. Scott, 1971, of the same event)

criminal connection *obsolete* extramarital copulation.

The *connection* is as in CONNECT[1], but it was never a crime in the British Isles to copulate if the female were old enough and consented:

> These women seldom or never allow drunken men to have criminal connection with them. (Mayhew, 1862, of prostitutes)

criminal conversation *obsolete* adultery.

Usually of the woman and abbreviated to *crim. con.* in legal jargon:

> In 1837, Mrs Charlotte Trevanien née Brereton, of Cornwall, was accused of having criminal conversation with a man (Pearsall, 1969)

and:

> It consists of a number of cases, crim-con and divorce. (ibid., of *Annals of Fashionable Gallantry* of 1885)

Conversation, meaning 'copulation', and now fortunately obsolete, must always have been somewhat misleading:

> His conversation with Shore's wife. (Shakespeare, *Richard III*)

See also INTERCOURSE.

criminal operation an illegal abortion.

Not a planned robbery or cutting a hostage's finger off.

crinkly old, or an old person.

As wrinkled as a WRINKLY:

> ... there was no sign of the yachting-capped assholes and bejewelled crinkly women who must make them prosper in the summer. (Bryson, 1991, of shops on Capri out of season)

critical power excursion a nuclear meltdown.

Jargon in an industry which insists it won't happen.

croak[1] to die.

From the death rattle of a dying person unable to clear mucus from his throat:

> They go mouching along as if they were croaking. (Mayhew, 1851)

Less often *croak* means 'to kill':

> ...the guy who had guts enough to croak 'Tough Tony'. (Lavine, 1930)

To *croak yourself* is to commit suicide.

croak[2] *British, obsolete* to whinge.

A common usage from the 17th century to the Second World War, and considered bad form among the officer class:

> ...they were civilians and, like all civilians, spent their time either in pettifogging or in 'croaking'. (Farrell, 1973, writing of mid 19th-century British India)

A *croaker* was one who so acted, but might also have been a doctor, either from his gloomy prognoses or from the results of his professional shortcomings.

crock a drunkard.

Literally, a bottle. The American *crocked*, intoxicated, puns on the meaning 'injured':

> In New York they prefer to arrive crocked... sorry, smashed...and sober up during the interview. (B. Forbes, 1972)

crook the elbow *Scottish* to drink intoxicants.

A variant of *bend the elbow* (see BEND). A person who is reputed to *crook his elbow* is a drunkard.

cross to copulate with.

Male usage, from the attitude assumed on the female:

> They found in the grass
> The marks of her ass
> And the knees of the men who had crossed her.
> (*Playboy's Book of Limericks*)

The obsolete English *cross girl* was not an angry young woman but a cheating prostitute who broke her contract, reaching accord and taking payment but not giving satisfaction, a *double-crosser* in fact.

cross bar hotel *American* a prison.

Prisons are described as *hotels* in various underworld euphemisms. In this punning use, the bar must cross and secure the gate.

cross-dress to be a transvestite.

Usually of male homosexuals playing the female role:

> She had never accepted his desire to cross-dress, regarding him as 'perverted' and 'disgusting'. (*Listener*, 12 July 1984)

cross the floor *British* to change political allegiance.

The seating arrangements in the House of Commons have the opponents facing each other across the floor of the House. If you change parties, you sit the other side. Sir Hartley Shawcross, known to be increasingly disenchanted with the Labour Party, of which he was a member, acquired the nickname 'Sir Shortly Floorcross'.

cross the Styx to die.

In classical mythology, you were ferried to the other side of the Styx by Charon, so long as your relatives had remembered to put the fare in your mouth when they buried you. A dead Christian may figuratively *cross the River Jordan* which is toll-free.

cross your palm to bribe.

The derivation is probably from the request of a gypsy to have her palm *crossed with silver*, after which she will tell you your fortune. Divination falls within the sphere of influence of the devil, whose powers can be negated only by the use of the Christian cross. The gypsy keeps the silver.

crower *American, obsolete* a cock.

Another variant from the days when it was thought indelicate to talk about bulls, cocks, stallions, and asses.

crown jewels see JEWELS.

crud human shit.

Originally, curdled milk, although the American army use may have other origins (see Jennings, 1965):

> ...'crud' has come into popular use as a euphemism for 'shit'. (ibid.)

Much figurative use, as a noun or the adjective *cruddy*:

> This Reape apparently was a cruddy character. (Sanders, 1980)

Cairo crud is one of the many alliterative geographical terms for diarrhoea.

cruise[1] to seek a sexual partner at random.

Usually of a male, seeking someone of either sex according to his predilection, on foot or in a car, on the street, or at a party:

> I don't want to cruise any more. I'm afraid I won't be able to get it up. (Sanders, 1982)

A *cruise* is such a foray:

> A spell behind bars for a sexual misdemeanour and recent cruises around New York's gay clubs. (*Private Eye*, May 1981)

In obsolete London use a *cruiser* was a prostitute who solicited custom from a hansom cab.

cruise[2] to be under the influence of illegally taken narcotics.

The imagery appears to be from flying:

> Directors didn't seem to drink much. A little champagne or white wine. Although at least

six of them were cruising at five thousand feet
on something else. (Wambaugh, 1981)

cruise[3] a missile that is programmed to hit a distant target.

Precisely targeted and very different from a sea trip calling at different ports:

Apart from spotting the missiles guarding the Bunker he had filmed the first pictures of a cruise fired in anger. (Simpson, 1991)

crumpet a female or females viewed sexually by a male.

Literally, a cake made of flour and yeast:

Never short of crumpet. That's one thing about this job. (Deighton, 1972, of opportunities for extramarital copulation)

Usually as a *bit* or *piece of crumpet*. A distinguished lexicographer assures me that a female may now also so describe a male, or males generally.

crush a sexual attraction towards another person.

From the wish to embrace the object of your affections? American *crushes* are heterosexual in the main while British schoolgirls in single-sex schools have *crushes* homosexually, usually on an older female (see also PASH):

These are schoolgirl dreams. And why pick on me for your 'crush'? (Murdoch, 1977)

crypt *American* a drawer in which dead bodies are stacked.

The crypts facing the corridor are called 'Mausoleum crypts'... Crypts facing outside... are now called 'garden crypts'... 'It's all part of the trend towards outdoor living,' explained the counsellor. (J. Mitford, 1963)

Funeral jargon. Literally and etymologically a hidden place, whence the cellar of a church used for burial.

crystal concentrated cocaine.

Also known as *crack*:

She was into crystal like it was gonna be banned tomorrow. (M. Smith, 1993—the lady was not a collector of cut glass)

Cuban heels thick soles and heels to enhance height.

As worn in the Caribbean and by the vain:

The prosecution had alleged that the bantam-weight Basham, who stands only a fraction over five feet three (without his Cuban heels) had committed assault. (*Daily Telegraph*, 11 March 1995)

cube a lump of sugar with LSD on it.

From *cube sugar*. An addict is a *cube head* and LSD *cube juice*.

cuckoo[1] a male profligate.

He does not necessarily cuckold another, although the word comes from *cuckoo*, the bird which makes use of another's nest:

The cuckoo then on every tree
Mocks married men.
(Shakespeare, *Love's Labour's Lost*)

This obsolete English use is no great loss to the language but we could still use the colourful phrase, to *cuckold the parson*, which was not to copulate with his wife but with your betrothed before the parson married you.

cuckoo[2] insane.

The cuckoo has the reputation for being a silly bird:

Old defectors, old spies, they get a bit cuckoo. (le Carré, 1980)

cuff to arrest.

From placing handcuffs on the victim rather than hitting him about the ear:

I figure if you move fast, you should be able to cuff him tomorrow. (Sanders, 1977, of a prospective arrest)

cull to kill.

Originally, to select, whence to select for rejection, as deer, seals, etc. The standard English use is never of the killing of humans.

cult dated and of limited appeal.

As in *cult movie*. An editor told me that this work might turn out to be a *cult book*—but not too soon, I hope.

cultural having characteristics differing from the norm.

Originally, relating to good taste, manners, etc. but:

'Cultural'... is the sociologists' jargon for saying as Lewis Carroll once put it 'the word means what I choose it to mean'. (Shankland, 1980)

Thus any citizen, especially one of foreign origin or ancestry, said to be *culturally* deprived or dispossessed has not mislaid a painting or had a book stolen but is poor and perhaps forlorn, and may be thought to suffer from *cultural bias*:

Eventually the pen-and-paper tests were dropped altogether because they were 'culturally biased'. (*Sunday Telegraph*, 20 November 1994, of unsuccessful attempts over a period of ten years by the New York Police Department to introduce examinations which did not result in a higher proportion of white policemen being eligible for promotion to sergeant than black)

Mao's *Cultural Revolution* was the encouragement of anarchy by an unbalanced tyrant to preserve his autocracy.

cumshaw a gratuity.

At one time used in the Far East:

> The expression was originally 'come ashore money', a sailor's tip to the launch boatman. (Jennings, 1965)

Still British naval slang for anything of value obtained without payment.

cunning man *obsolete* a wizard.

Cunning meant 'knowing' and as most wizards are in league with the devil, you had to talk nicely about them:

> A 'cunning man' was long resident in Bodmin, to whom the people from all parts went to be relieved of spells. (R. Hunt, 1865)

cup a protective shield for the male genitalia.

A sporting use, from its shape, and more genteel than BOX[2].

cup too many see IN YOUR CUPS.

cupcake a male homosexual.

Not, I think, from CUP, nor am I sure why the inoffensive confection should be so translated:

> 'Odd? Queer? Gay?' Audley raised an eyebrow. 'A cupcake? I heard that word recently.' (Price, 1982)

Cupid's measles syphilis.

The symptoms are not, at one stage, dissimilar:

> ... it was on this leave that he contracted his umpteenth case of Cupid's measles. (M. Fraser, 1992, of a soldier in the Second World War)

Also in America as *Cupid's itch*. Cupid, the son of Venus and Mercury and a god of love in his own right, gave us *cupidity* and, under his Greek alias *Eros*, *eroticism*. Numerous 19th-century vulgarities for the vagina such as *Cupid's arbour*, *cave*, *cloister*, *corner*, *cupboard*, and *hotel* have happily gone out of use, but we still describe lips of a particular shape as a *Cupid's bow*, and name the blue and white flower *Catananche caerulea*, *Cupid's dart*.

cure (the) treatment for a taboo condition.

From the regimen formerly available at spas to alleviate the pangs of rheumatism etc. Now for alcoholism, addiction to narcotics, or obesity:

> I haven't seen him for—well, since May. He was going to try the cure again. (Steinbeck, 1961, of an alcoholic)

To have *taken the cure* implies abstinence on the part of a former addict.

curious homosexual.

From the meaning 'unusual':

> He was my tutor. Surely you don't imagine I go to curious parties with Pinkrose. (Manning, 1965)

currency adjustment a devaluation.

Political use, seeking to disguise the failure of the policy which led to the necessity. No politician hesitates to call a revaluation by its correct name. See also ADJUSTMENT[1].

curse (the) menstruation.

A shortened form of *the curse of Eve*, who thus burdened all females:

> You've probably got the curse or something. (Bogarde, 1978)

Very common.

curtains death.

From the end of a play, the darkening of a room for a wake, or the screening of an execution. Also some figurative use:

> To have given Nixon knowledge of even the smallest part of that particular Haig connection would have meant curtains for Haig as Nixon's Chief of Staff. (Colodny and Gettlin, 1991)

A common use in America despite the fact that rooms are normally screened by drapes.

custody suite *British* a prison cell.

Usually in a police station:

> The police claimed that they had been instructed to refer to custody suites ... (*Daily Telegraph*, April 1986, and not just for minors)

custom and practice a vested interest.

A ratchet device whereby British trade unions sought to retain every advantage secured in past negotiation or confrontation without regard to technological progress or what an employer might see as commercial necessity:

> We have the custom and practice of the shop floor behind us. (Price, 1978—a trade union official was opposing a proposal for change)

cut[1] to render (a male) sexually impotent.

Of humans by vasectomy; of domestic animals by castration:

> The bull calves are cut. (Marshall, 1818)

It was also used as a euphemism meaning 'cut off the penis' in the Bobbitt case in America.

cut[2] to dilute in order to cheat customers.

Mainly of intoxicants and illegal narcotics, from the practice of dividing and adulterating:

> The real thing. Pharmaceutical coke. Not the cut street stuff. (Robbins, 1981)

cut[3] drunk.

From the dialect meaning 'tacking' or 'weaving'? Often as *half-cut*:

> On many a night we left the canteen half cut. (Richards, 1936)

cut[4] an illegal commission.

Common criminal and commercial use, again from the dividing. Whence the verb to *cut*, to take such a share:

Crap games were played in the corridor with the keeper 'cutting' the game. (Lavine, 1930)

cut[5] a reduction in the size of the increase desired or expected by the recipient.

Political use, of spending by government agencies etc. where decades of routine increases and consequent profligacy have led to the expectation that anything in the public sector must keep growing and the wages of those employed rise, whatever the state of the economy as a whole.

cut[6] to kill.

Not necessarily with a knife:

You Americans—you are so strange. You 'put a man down', or you 'cut him' or you 'burn him', or you 'put him away' or 'take him for a ride'. But you will never say you killed him. Why is that? (Sanders, 1970—the questioner had clearly not studied the language of taboo)

Occasionally also as *cut down on*:

They want me to cut down on him... I am to burn this man. (ibid.—the speaker was an assassin, not a worker in a crematorium)

cut a cheese to fart.

From the rich and not always pleasant smell which may escape. In Somerset you may *cut a leg* in the same sense. The more general use is merely to *cut one*:

... none of them would say anything if he cut one. (McInerney, 1992)

Grose gives 'Cheeser. A strong smelling fart'.

cut down on see CUT[6].

cut off dead.

Always of premature or untimely death, with imagery from the gathering of a flower in bloom:

... whose headstones record an early death, a cutting-off before the prime. (P. Scott, 1968)

cut out to deprive (someone) of something valuable.

Said formerly by sailors, from singling out a ship in the opposing fleet for concentrated attack and capture. The term is also used of displacing a female's partner, especially on the dance floor.

cut the mustard to copulate.

Of a male; *cut* means 'to share in' but why *mustard*, unless from the German *Senf*, meaning 'mustard', with a slang sense 'pizazz'?

You can't cut the mustard but how about watching? (Theroux, 1973)

Some figurative use, meaning 'meet the required standard':

None of this bailing out firms that can't cut the competitive mustard. (M. Thomas, 1982)

Also shortened to *cut it*:

'Are you married?' 'Divorced.' 'Ha! Couldn't cut it, huh?' (Collins, 1981)

cut the painter to die.

Like a boat cast loose on the water and mainly used of old seamen. *Cut adrift*, of the same tendency, is probably obsolete. *Cut your cable* should indicate suicide but in fact is used of natural death in old age.

Cyprian *obsolete, literary* a prostitute.

Aphrodite, the Greek Venus, was associated with Cyprus:

The Burlington Arcade, which is a well-known resort of Cyprians of the better sort. (Mayhew, 1862)

Adjectivally, it may refer to any copulation:

... a fulgently glamorous singer starring locally was absorbing all my Cyprian energies. (Monkhouse, 1993)

The *Cyprian sceptre* was the erect penis.

Cythera *obsolete, literary* extramarital copulation.

From the Cretan centre of the cult of Aphrodite, the Goddess of Love:

... nor indeed were we long before we finished our trip to Cythera. (Cleland, 1749, meaning they had copulated)

Cytheromania means 'nymphomania'.

D anything taboo beginning with the letter D.

Usually *damn, damned, damnable*, and the like, which used to carry more weight than they do today:

> And at last he flung out in his violent way, and said, with a D, 'Then do as you like.' (C. Dickens, 1861)

The *big D* is death:

> The systematic encroachment of the big D. (le Carré, 1980)

D and C the abortion of a foetus.

The medical abbreviation for *dilatation and curettage*, otherwise known as a 'scrape', a common operation for women in middle age but, of the young, sometimes used as a pretext for an abortion:

> ... going in for a d and c and coming out foetus-free but permanently stained. (P. Scott, 1975, of a young female)

DCM a notice of dismissal from employment.

The initial letters of 'Don't come Monday' pun on the Distinguished Conduct Medal. Mainly American—in British railway use, it denoted suspension for one day only.

dabble to use illegal narcotics irregularly.

From the proper sense 'to give casual or intermittent attention to'. The obsolete British *dab* was a bawd:

> Their scrutinizing pow'r severe
> Discerns a vestal from a dirty dab.
> (Pindar, 1816)

and *dab it up* meant 'to copulate', of a male, from the meaning 'to thrust'.

dad *American* a mild oath.

DAS says it is a euphemism for *God*.

daddy a man who keeps a much younger woman as a sexual mistress.

Perhaps a shortened form of SUGAR DADDY.

dairies see CHARLIE.

daisy[1] *obsolete* a grave.

From the flowers on the turf. However the association of *daisies* with death continues. If you *count* or *push up daisies*, you are dead.

daisy[2] a male homosexual taking the female role.

From the common female name, whence the punning American *daisy chain*, a male homosexual orgy with heads threaded through stems.

daisy chain[1] see DAISY[2].

daisy chain[2] *American* a body of investors concertedly inflating the price of a quoted security.

Except insofar as it may be a conspiracy to defraud others, perhaps no more than one of many devices whereby dealers maintain movement in stock markets, institutional investors generate short term profits, and the general public gets cheated.

dally to copulate with extramaritally.

Originally, to talk idly, but men do it more than women:

> On the night of the divorce he was out with Australian harpie ——with whom he dallied for a year or two. (*Private Eye*, April 1981)

Dalliance is so behaving:

> What time the gifted lady took
> Away from pencil, pen, and book,
> She spent in amorous dalliance
> (They do those things so well in France.)
> (Parker, 1944, of George Sand)

damaged[1] drunk.

Mainly American use, from the temporary incapacitation.

damaged[2] having copulated before marriage.

Of a woman who under the former convention would have become less desirable as a bride, and hence was described as *damaged goods*.

damp down to increase saleable weight by adding water.

A usage mainly in the coal trade, where you purport to lay dust before bagging and weighing the coal.

dance[1] to be killed by hanging.

From the gyration and kicking of the victim:

> Spring's passage out was going to be at the end of a rope, and unless I shifted I'd be dancing alongside him. (Fraser, 1982)

You might also be said to *dance on air, at the end of a rope, off, upon nothing, the Tyburn jig*, etc.:

> Matthew would be dancing on air by next sun-down. (Monsarrat, 1978, writing in archaic style)

The *dance-hall* was the condemned cell and the *dancing master* the hangman. To *dance a two-step in another world* is to be dead, but not necessarily by hanging:

> ... no good keeping souvenirs of that sort when any moment we may be dancing a two-step in another world. (Richards, 1933, writing of First World War trench life)

dance[2] to be involuntarily under another's control.

You have to move as another tells you, and not necessarily because a gunman is shooting at your feet. Much figurative use.

dance at *obsolete* to court heterosexually.

Not, I think, from the activities of Salome. This expression suggests greater decorum than our present methods:

> I should have no opinion of you, Biddy, if he danced at you with your consent. (C. Dickens, 1861)

dance barefoot *obsolete* to marry before an unmarried elder sister.

Of a girl and perhaps from the absence of a dowry:

> I must dance barefoot on her wedding day. (Shakespeare, *Taming of the Shrew*)

The economic pressure on those unwed females who were not allowed to work to keep themselves made it socially desirable that sisters should marry in descending age order. If in Yorkshire you remained a bachelor while your brother married, you might be said to *dance in the half-peck*, the derivation of which has eluded me.

dance-hall hostess a prostitute.

From her place of work; see also HOSTESS:

> A night-club or dance-hall hostess... are the modern equivalents of the old-time disorderly house and of the street walker. (Lavine, 1930)

dance the mattress quadrille *obsolete* to copulate.

The common *bed* imagery:

> I could have had the buxom piece dancing the mattress quadrille within the hour. (Fraser, 1975, writing in archaic style)

The obsolete London *dance a Haymarket hornpipe* was something of an in-joke meaning 'to copulate with a prostitute', as they frequented the Haymarket; it also includes two vulgar puns:

> Perhaps we'll dance another Haymarket hornpipe before long. (ibid., of copulation)

danger signal is up (the) I am menstruating.

From the redness of the blood and the advice to stay clear.

dangerous to women adept at persuading a female to copulate casually.

You do not normally so describe a rapist:

> 'Is Morny dangerous to women?' 'Don't be Victorian, old top. Women don't call it danger.' (Chandler, 1943)

Lady Caroline Lamb implied as much of Lord Byron when she confided to her journal that he was 'Mad, bad, and dangerous to know'.

Darby and Joan an elderly married couple living together.

From the characters in Woodfall's 18th-century ballad, who grew old together:

> Darby and Joaning it into the sunset. (Bogarde, 1981)

Also used of a pair of homosexuals in the British Indian army:

> The attitude of other soldiers towards the 'Darby and Joans' of the regiment was generally good-natured. (Allen, 1975)

dark[1] closed.

Theatrical jargon, from the absence of advertising, footlights, etc. when a play has flopped or a theatre management failed:

> The theatre is now 'dark'—only the bars and a buffet are open to earn money. (*Sunday Telegraph*, November 1981)

dark[2] (of people) black.

Sometimes as *dark-skinned* or *dark-complected*:

> I tried to tell him a dark-complected man is nothing in this country without an education to stand on. (Macdonald, 1952)

Dark meat is a black American woman (or women collectively) viewed sexually by an American white male:

> Bill, you better try some dark meat and change your luck. (Sanders, 1982, of a man unable to copulate with his white wife; see also CHANGE YOUR LUCK)

A *darky*, a person who is not white, is offensive:

> Was it something about not taking on the darkies as conductors? (le Carré, 1983)

dark man the devil.

A rare variant of the BLACK GENTLEMAN, the *darkness* coming from his evil ways and from the soot which adhered to him as he made his way down the chimney:

> A drunk of really a noble class that brought you no nearer to the dark man. (Hardy, 1874)

dark meat[1] flesh of poultry other than the breast.

The breast was the *white meat* and twin evasions, which remain in common use, allowed our prudish ancestors to skate round two taboo parts of the body, breasts and legs. The use of *drumstick* also avoids mention of legs, but it predates the 19th-century American prudery which spawned *lower limb* and *trotter*. See also ROOSTER.

dark meat[2] see DARK[2].

dark moon *obsolete* a wife's secret savings.

A 19th-century English expression, when a wife was allowed no possessions and hid any savings away to provide for future disaster without telling her husband:

> The farmer was delighted at the discovery of this wife's dark moon. (*Notes & Queries*, 1867)

darn a mild oath.

A shortened form of the obsolete *tarnation* which 'was a blend of, "damnation" and "by the 'tarnal" (eternal)' (Jennings, 1965). Still widely used for *damn*, which itself is now mild.

dash[1] to adulterate an intoxicant.

Literally, to mix or dilute, as in a culinary recipe:

> This beer's dashed an' 'er aulus do dash it. (EDD—late 19th century)

Obsolete, but the practice in the British Isles of watering beer is not.

dash[2] a mild oath.

From the literary convention of replacing a taboo word such as *damn* with a dash. I disagree with Partridge who says: 'Euphemistic only when used as an evasive for *damn*' (DSUE). The obsolete British *dasher* was not a lady who swore or sprinted, but a prostitute, perhaps punning on *cutting a dash* and the time she accorded to each customer.

date a heterosexual companion.

From specifying the meeting time and a reminder of the days when courtship was the subject of mild taboo:

> ... theories as to the girl's possible date. (Davidson, 1978—they were speculating about her companion and not her age)

On a *blind date* you take pot luck. In America, a *date* may also be a prostitute:

> ... pictures and other materials about the women ... were given to Bailley's DNC contact, so that prospective clients could choose among possible dates. (Colodny and Gettlin, 1991, of facilities for obtaining prostitutes which were available to Democratic visitors to Washington)

Also used as a verb:

> If the Smiths hadn't been there I would have dated her myself. (Theroux, 1978)

dateless *obsolete* senile.

Not leading a celibate life but unable to recall the passage of time:

> We were like to be turned out on t' wide world, and poor mother dateless. (Gaskell, 1863)

daughter of the game a prostitute.

The *game* is prostitution:

> The girl was a born Cressida, a 'daughter of the game'. (Manning, 1960)

Cressida was the lady who, by preferring Diomed to Troilus, gave Chaucer, Shakespeare, and others a plot.

Davy Jones' locker a grave at sea.

Grose says 'David Jones. The devil, the spirit of the sea' and the first literary use was by Smollett in 1751. Derivation from the biblical Jonah is sometimes suggested. The *locker* was his seaman's chest:

> All hands are snug enough in Davy Jones's locker. (Chamier, 1837—they had died at sea)

Davy Jones' natural children were pirates, *natural children* being bastards.

dawn raid the sudden accumulation of a block of shares.

City jargon, from the military surprise attack. A device by or on behalf of a bidder who would have to disclose a gradual accumulation of shares, which would also tend to increase their price.

day of protest a politically motivated strike.

Mainly trade union jargon, for such an effort to influence opinion and coerce government by extra-parliamentary action and wide television publicity. Also, despite the cessation of work, *day of action*; see also INDUSTRIAL ACTION.

dead meat[1] a human corpse.

Criminal jargon beloved by writers of detective fiction. To *make dead meat of* is to kill a human being.

dead meat[2] an older prostitute who still lives by prostitution.

As distinct from *fresh meat*, which means a young prostitute new to the business, or occasionally a sexually lax girl. *Fresh and sweet* is used of a prostitute newly released from jail.

dead soldier an empty bottle of wine or spirits.

Perhaps from the military appearance of a line of bottles:

> Or I'd take it to him if he had a dead soldier. (Sanders, 1980, of a bottle of brandy)

dead to recklessly ignoring.

In Victorian days, usually referring to sexual behaviour, when a person might be *dead to honour* or *propriety*:

> I cannot suppose that he is altogether Dead to propriety, though how long such Restraint will continue I cannot say. (Fraser, 1977—writing in archaic style of how a kidnapper would behave to his female victim)

Dead to the world means 'asleep'.

deadhead *American* a successful scrounger.

Because you can't include him when you count the takings. Of a non-paying spectator at a ball game, a fare-evader on a train, etc. Another American use is of a cadger at a bar who never buys others a drink.

deal to supply illegal narcotics in return for payment.

The American language of commerce is used to conceal criminality:

> 'A little grass now and then. Not from her.' 'But she deals?' (Sanders, 1977, of marijuana)

deal from the bottom of the deck to lie or cheat.

The imagery is from card-sharping:

> For all we know, he could be dealing from the bottom of the deck, just to make more money. (Forsyth, 1994)

dear friend see CHÈRE AMIE.

dear John the ending by a woman of an engagement or marriage.

In the Second World War, the letter of dismissal which so many men serving abroad received started formally rather than by using the warmer appellations which indicate closer affection:

> The colonel confesses he should have got out on receipt of his first 'Dear John' letter, particularly as this coincided with the break-up of his regiment. (*Daily Telegraph*, January 1984)

Dear John is now used to denote any method of communicating such a decision.

death benefit a contribution by the state towards funeral costs.

An unhappy juxtaposition of concepts, except where we talk of a *happy release*; but we have no qualms about using the standard English *beneficiaries* for heirs. See also BENEFIT.

debauch to copulate with extramaritally.

Literally, to corrupt, and today's *debauches* refer only to drunkenness:

> Men so disorder'd, so debauch'd and bold,
> That this our court, infected with their manners,
> Shows like a riotous inn.
> (Shakespeare, *King Lear*—but not of drunkenness)

I cannot decide what Dr Johnson meant when he wrote:

> A man may debauch his friend's wife genteely.

debrief to interrogate aggressively.

In standard usage, no more than to obtain information from someone about a mission for which he has previously been briefed:

> At first I thought the interrogation—or debriefing as it was delicately termed... (Deighton, 1989)

debt of honour money lost at gambling and not paid.

Under English law gambling debts are not recoverable through the courts, but a defaulter may lose his good name and his credit with his bookmaker—or get his teeth pushed in.

decadent not conforming to accepted tastes.

Literally, in a state of decline from past standards. Much used by autocrats of anything of which they disapprove, from homosexuality to artistic style:

> Shetland had accepted eight 'decadent' surrealist paintings that Göring had confiscated. (Deighton, 1978)

decant to urinate.

> 'Just going to decant' (and the awful phrases they come up with). (Barnes, 1991)

Literally, to pour a liquid from one container to another.

deceive (your regular sexual partner) to copulate with another.

> Harper nodded and made a private vow that he would not deceive his wife. (Theroux, 1980, which does not mean that he would pretend only to drink one whisky a day)

Of either sex. Literally, to mislead as to the truth.

decent wearing clothes which hide any suggestion of nakedness.

You do not have to be fully clothed to be *decent* but your attire must not suggest immodesty:

> ...since I could see she was clothed—'decent', as girls used to say. (Styron, 1976—and they still do)

deck *American* a packet of illicit narcotics.

Usually heroin, from being wrapped in paper like a pack of cards. To *deck up* is to pack heroin for retail sale.

decks awash *American* drunk.

Applied not only to sailors. See also HALF-SEAS OVER.

decline the fatal onset of pulmonary tuberculosis.

A euphemism which became specific before being replaced with CONSUMPTION:

> She said one of his suitors was like to die of decline. (Hughes, 1856)

Obsolete as a euphemism for TB since the Second World War, but used today of failing mental or physical health.

decontaminate *American* to embalm.

The majority of corpses are no more *contaminated* than is a leg of mutton, a flitch of bacon, or a side of beef (see also HYGIENIC TREATMENT):

> The incentive to select quality merchandise would be materially lessened if the body of the deceased were not decontaminated and made presentable. (J. Mitford, 1963—the survivors will spend more if the corpse is spruced up)

It is also possible to *decontaminate* a file or tape by removing or excising any matter which might prove embarrassing on later investigation.

decoy an unmarked police car.

Literally, a lure. You may be encouraged to keep pace with it and then be booked for speeding. American Citizens' Band use.

deed (the) copulation.

Often extramarital, and always so if 'dirty' or 'vile':

> ... one that will do the deed
> Though Argus were her eunuch.
> (Shakespeare, *Love's Labour's Lost*, of Rosaline)
> 'You may not remember but ...' 'Oh hell, we haven't done the deed, have we?' (Fry, 1994—a man had failed to recognize a woman to whom he was talking)

deep freeze a prison.

The common American cooling imagery:

> If the cops didn't grab him and toss him in the deep freeze ... (Chandler, 1958)

deep interrogation see INTERROGATION (WITH PREJUDICE).

deep-six to kill or destroy.

The original meaning was merely 'to dispose of', not from the statutory depth in feet of a grave but from the lowest mark on a naval heaving line in fathoms, below which everything vanished:

> You can deepsix that crap. Eighty years old and still fucking. That I don't need. (Sharpe, 1977)

Whence to destroy:

> How do you propose we deep-six that Straton. (Block, 1979—they were trying to make an airliner crash)

And so to killing:

> Barney would have expected his friend to deep-six it out of the window. (M. Thomas, 1980, referring to a suicide)

deep sleep an extended period of inactivity by a spy.

To obtain a position of acceptance or trust in the society to be attacked, the SLEEPER is a long time awakening:

> Hard-picked subjects, with good career prospects, psychologically equipped for deep-sleeping ... (Keneally, 1985)

defecate to shit.

The original meaning was 'to purify or cleanse'. Thus William Harvey could write in the 17th century:

> The blood is not sufficiently defecated or clarified, but remains muddy. (Harvey, 1628)

Now *defecate* and *defecation* are used in medical and polite standard English for 'shit'.

defence aggression.

This is one of those opposites, like HEALTH and LIFE INSURANCE. A *Ministry of Defence* is no less warlike than a *Ministry of War*. *Active (air) defence* is military jargon for air raids and *defence budgets, debates, votes, procurement, strategy*, etc. are all to do with fighting. The British *D Notice* which is never called a *defence notice* is an instruction to newspapers etc. to suppress news, ostensibly on the grounds that state security is involved.

defend your virtue to refuse to copulate outside marriage.

Usually of a female and indeed:

> A male defending his virtue is always a farcical figure. (McCarthy, 1963)

The phrase is also used for the rejection of homosexual approaches.

defensive victory the postponement of defeat.

Used to mask the reality of military disaster:

> On the Cowland front a complete defensive victory was scored yesterday. (Goebbels, 1945, in translation)

defile to copulate with extramaritally.

Literally, to make filthy. Of a male, with a presumption of female reluctance or resistance:

> Children who only hours ago had been virgins, defiled by men they had never seen before. (Ludlum, 1979)

The male who thus copulates is a *defiler*:

> ... thou bright defiler
> Of Hymen's purest bed.
> (Shakespeare, *Timon of Athens*)

To *defile a bed* does not imply involuntary urination:

> My bed he hath defiled. (Shakespeare, *All's Well*—meaning 'he has copulated with me')

Defilement is what the female undergoes:

> The law prohibits the defilement of girls under the age of 15. (*Daily Telegraph*, 3 June 1994, quoting an Irish judge sentencing the man who had impregnated a girl whose attempt to obtain an abortion in England led to a re-examination of the Irish constitution)

deflower to copulate with (a female virgin).

OED gives a 14th-century quotation from Wyclif in this sense as the first use and Shakespeare speaks of 'A deflower'd maid' (*Measure for Measure*). The imagery is clearly from plucking a bloom. Also used to refer to loss of virginity other than by copulation:

> His female admirers had a model of it made in pure gold and organized a ceremony in which several virgins deflowered themselves on this object. (Manning, 1977)

Defloration is such copulation:

> ... the usual sanguinary symptoms of defloration. (Cleland, 1749)

degrade to reduce by killing or wounding.

> ... an air assault guaranteed to 'degrade' by 50 per cent the strength of the Iraqi forces arranged north of the border. (Forsyth, 1994)

Literally, to reduce a substance in strength or purity.

degraded *?obsolete* having been copulated with extramaritally.

Of a woman. Literally, lowered in rank, which she may be if she is found out:

> 'Do you suppose she has been... degraded?' says he, in a hushed voice. (Fraser, 1971)

dehired *American* dismissed from employment.

This expression would be more accurately used of equipment which is no longer needed after being on hire.

Delhi belly diarrhoea.

An alliterative use not confined to India or its capital:

> Kind of a bowel thing. Up all night. Cramps. Delhi belly. Food goes right through you. (Theroux, 1975)

delicate suffering from pulmonary tuberculosis.

One of the 19th-century euphemisms for the common disease:

> The brother died young, he was delicate. (Flanagan, 1988, writing in 19th-century style)

For the modern *delicate condition*, pregnancy, see CONDITION[2].

deliver to drop on an enemy.

Especially of bombs or ordnance. The imagery makes the transaction more remote and guilt-free for the attackers. In this jargon, a *delivery vehicle* is not a milk float but a missile which carries a bomb.

demands of nature urination and defecation.

You might think gravity came first, followed by breathing:

> ...walking with the sense of purpose proper to a man about to attend to the demands of nature. (Masters, 1976)

demanning the dismissal of employees.

Not an operation to change masculinity. It is usually expected that the reduced workforce will produce as much as did the larger number:

> It is imperative the process of demanning continues. (*Daily Telegraph*, 8 March 1994—a chairman was announcing the dismissal of 2,000 employees)

demi-mondaine a prostitute.

Married people who 'went to the world' in the French Second Empire were the *monde* and women on the fringes of that society unaccompanied by men were the *demi-monde*. The obsolete English *demi-rep*, a shortened form of *demi-reputation*, meant the same thing.

democratic/democracy have always meant different things to different people and virtually never 'rule by the people'. The Athenian state, styled the *Cradle of Democracy*, relied heavily on disenfranchised and slave labour. In modern times the *German Democratic Republic* was Russia's totalitarian satellite. The terms are widely used to describe autocracy and tyranny:

> 'Vietnam's Democratic One-man Rule'—the Procrustean subject was Diem. A democratic 'dictator' or a 'democratic' dictator? (McCarthy, 1967)

It is not only the etymologist who is wary of any political or other institution which incorporates the word *democratic* in its title; we have learned to look for the flaw in any concept or argument said to be based on principles of *democracy*.

demographic strain too many people.

Demography is the study of population statistics but this phrase does not mean your eyes ache from reading too many censuses. It is taboo to suggest that poor countries face starvation because ignorant people breed too fast and medical science allows too many to survive.

demonstration an attempt to coerce or influence by force.

Literally, a showing, illustration, or proof. The type of vociferous and sometimes violent rally so styled is a useful device for protestors and agitators, because any action or failure to react by Government assists their cause:

> He never took part in demonstrations or marched in May Day parades. (McCarthy, 1963)

The short form *demo* has no non-euphemistic meaning.

demonstrator a car used by a motor trader and then sold as new.

Motor trade jargon. The implication is that it has only been used for giving test drives to potential customers. It is the system through which the trader, his senior staff, and his family obtain the constant use of new vehicles without having to meet the cost of purchase, depreciation, or repair.

demote maximally to kill one of your associates.

Espionage jargon. The career as a spy of the person concerned certainly comes to an abrupt close:

> Jonathan smiled at the cryptic jargon... in which 'demote maximally' meant purge by killing. (Trevanian, 1972)

A *maximum demote* is such a killing:

> The assassinations are called 'sanctions' if the target is someone outside the CIA, and 'maximum demotes' if the target is one of their own men. (Trevanian, 1973)

deniable (of a lie or deceitful action) hard to expose.

Usually of a statement made off the record or an act taken by a third party on behalf of another:

> ...the small country could inflict wounds itself, or even more safely, sponsor others to

do so—'deniably'—in order to move its larger opponent in the desired direction. (Clancy, 1987)

Whence *deniability*:

...nothing more than an exercise in keeping its own nose clean—not being seen to be involved. Deniability was the polite word for it. (Mason, 1993)

denied area a place in which it is unwise or forbidden to operate.

Espionage jargon:

His field was offensive intelligence in what the Secret Intelligence Service euphemistically described as Denied Areas. (M. Smith, 1993)

depart this life to die.

The implication is that you will arrive in another state of existence:

Things went on smoothly for a dozen years, when the old Frenchman departed this life. (Mayhew, 1851)

Take your departal is obsolete:

When my father took his departal to a better world. (Galt, 1823)

The *departed* are the dead:

Ground given over to the accommodation of the departed (Amis, 1978)

and the *dear departed* either a dead individual or the dead generally. *Departure* is death:

This unsound mode of transport would have been her only criticism of William's orchestration of her departure. (Archer, 1979, of a funeral)

An *unauthorized departure* in prison jargon is the escape of a prisoner.

departmental view the policy favoured by the senior civil servant.

The jargon of the civil servant based in London. A *ministerial view* is what the government thinks, but if that differs from the *departmental view*, the latter will in the medium term prevail.

dependency[1] a subject territory.

British imperial use for those parts of the globe which were not dominions, colonies, or protectorates but ruled from London.

dependency[2] an addiction to narcotics or alcohol.

The victim *depends* on regular ingestion:

It is estimated that at least two million women have dependencies—addiction would be a better word—on prescriptive drugs. (Sanders, 1981)

depleted poor.

Literally, emptied or reduced in quantity:

Clara twice a week drove her Seville to the city's depleted neighbourhoods for the morning. (Turow, 1990—she went slumming)

deposit a turd.

Usually in the phrase *make a deposit*, to defecate:

Never read when you eat, guys, but always read when you go make a deposit. (Theroux, 1993, and not of visiting a bank)

deprived poor.

Literally, having lost something which you once had, which is not so for most paupers, unless they were names at Lloyds:

Deprived Families on Increase. (headline in *Daily Telegraph*, 4 October 1983—the article claimed that there were more poor families, not that poor families were having more children or any of the other constructions that might be put upon the four words)

A 1965 Jules Feiffer cartoon shows the progression from 'poor' to 'needy' to 'deprived' etc. (Pei, 1969)

derailed mad.

The common transport imagery:

Was her father derailed, off his trolley, losing hold. (Turow, 1990)

derrière the buttocks.

The French too have behinds and use the same euphemism, although without our salacious overtones:

...there were mischievous triple-rilled *derrières*. (E. S. Turner, 1952, writing of American advertising of tight skirts)

deselect to dismiss (an incumbent).

Referring to an elected representative who loses favour with the caucus responsible for the selection of candidates.

designer stubble unshaven facial hair.

Neither bearded nor cleanshaven but cultivating the appearance of a vagrant who may be unable to shave regularly:

He sported dark glasses, his usual 'designer stubble' and wore a single-breasted pinstripe suit. (*Daily Telegraph*, 22 June 1994, of a musician said to be worth £70 million who was seeking to avoid the contract under which he was remunerated because he found its terms inequitable)

designs on (have) to wish to copulate with.

Not just wearing a patterned dress or having plans:

...they contain no mention of his having had designs on the local girls. (Bence-Jones, 1987, of the dissolute Earl of Leitrim who was murdered in Donegal in 1878)

destabilize to overthrow.

Literally, to create instability. Espionage jargon for attempts to create conditions under which a new government more to your liking might assume power.

destroy to kill a domestic animal.

The meaning 'to kill' has been standard English since the Middle Ages but there has developed a jargon use, referring to sick, old, or unwanted pets:

> If he makes another mess... I'll have him destroyed. (N. Mitford, 1945, of a dog)

destruction *obsolete* extramarital copulation with a female.

Especially where there was no prospect of subsequent marriage:

> I gather from it [a remark] that you are one of those who go through life seeking the destruction of servants. (Bence-Jones, 1987—a young member of the Kildare Street Club in pre-1914 Dublin had drawn the attention of an older man to a pretty servant girl cleaning the windows of a house across the street)

Discovery or impregnation would destroy the girl's reputation and cause her to lose her job.

detainee a political prisoner.

Each of us becomes a *detainee* when our train is held up at the signals. *Detain* is used in this sense in societies where such imprisonment is against public sentiment and tradition:

> ... they were stoned and scourged and imprisoned—or 'detained', as the authorities called it. (Seymour, 1977)

developing poor and relatively uncivilized.

Of states:

> ... countries which have successively and with increasing euphemism been termed backward, under-developed, less-developed and developing. (FDMT)

In more industrial societies, *developing* may express aspiration rather than reality, as with the former British *Development Areas*, those parts of the country which had fallen economically behind the rest of the country.

developmental associated with ignorance or lack of ability.

Educational jargon. A *developmental class* is for the unruly or stupid and a *developmental course* is what used to be called cramming.

device any object that is the subject of a taboo.

Literally, a mechanical contrivance. Of armament where, for a short while, *nuclear device* was thought to sound more acceptable than *atomic bomb*; also of contraception:

> The pharmaceuticals don't agree with me. I had to go to a doctor and get a device. (Keneally, 1985)

devil disease (the) AIDS.

From its effect if not its cause:

> The Devil Disease has driven the flying-fuck brigade underground. (Blacker, 1992—the imagery takes some working out)

devil's mark (the) congenital idiocy.

In British rural use:

> That's where your village idiots come from. They call it the Devil's Mark, I call it incest. (le Carré, 1962)

God and the devil seem to have caught the blame equally for the results of inbreeding and incest—see GOD'S CHILD.

diaphragm a female contraceptive device.

Literally, any dividing membrane. It is worn internally and usually fitted medically:

> Having her fitted for a diaphragm by one more of Larry's associates. (Styron, 1976)

dick[1] the penis.

Probably rhyming slang for PRICK, but the penis is often given common male names—see, for example, JOHN THOMAS, JOCK, and WILLY:

> What she had said about things like his dick. (Amis, 1978, of the penis)

The American *dicked* means 'buggered':

> ... six bad [years] in San Quentin gettin' dicked by the residents. (Collins, 1981, of a male prisoner)

Dick around is a variant of *fuck around* for promiscuous male copulation, and is also used figuratively:

> ... dicking around with his cows and windmills. (M. Thomas, 1982, of a painter)

Dick's hatband, male homosexuality, referred to the effete Richard Cromwell and the crown he was unfit to wear in succession to his mighty father, Oliver:

> Hollo, thinks I, he ain't one of the Dick's hatband brigade, surely. (Fraser, 1977, writing in 19th-century style)

To *wear Dick's hatband* was to be known as a male homosexual.

dick[2] a policeman.

Usually a detective:

> One of the more ambitious would go to the Detective Bureau and become a dick. (Lavine, 1930)

Although *Dick* is a shortened form of *Richard*, and this use is said to be from the rhyming slang *Richard the Third*, turd (see RICHARD), I have no real idea of the derivation. An American *dickless Tracy* is a policewoman, punning on her femininity (see DICK[1]) and the cartoon character.

dickens the devil.

On the common expression *what the dickens* Partridge says 'In origin a euphemistic evasion for *devil*' (DSUE), but the connection is not obvious.

dicky unwell.

From the rhyming slang *Uncle Dick*, sick. Widely used to refer to our own mild indispositions; with reference to others, it usually signifies a chronic state of ill-health, such as a *dicky heart*:

> ... sent me home. Said I had a dicky heart. (Theroux, 1974, of a former colonial resident)

dictatorship of the proletariat a self-perpetuating oligarchy.

Marx used *dictatorship* and *rule* interchangeably, which gave Lenin, Mao, and others the high authority for their systems which precision of speech might have denied them.

diddle[1] to urinate.

Literally, to jerk from side to side, which a male may do to his penis when he has urinated; but *Dicky Diddle* is also rhyming slang for 'piddle'. Mainly nursery use.

diddle[2] to masturbate.

Of both sexes, again from the jerking movement:

> ... she caught Leslie, then three, diddling herself and forced her to wear hand-splints. (Styron, 1976)

Diddle is also used of copulation, but I have not traced any literary example.

diddle[3] to cheat.

Of uncertain origin and probably not euphemistic. In obsolete use it also meant 'to kill'.

die to achieve a sexual orgasm.

Of a male:

> I will live in thy heart, die in thy lap, and be buried in thy eyes (Shakespeare, *Much Ado*)

or a female:

> These lovers cry—Oh! Oh! they die. (Shakespeare, *Troilus and Cressida*)

And much subsequent poetic use but now obsolete. To *die in a horse's nightcap* or *die in your shoes* was to be killed by hanging. In obsolete Kentish use, to *die queer* was to kill yourself.

diet of worms a corpse.

Modern science tells us that the process of corporal dissolution is fungal, with worms obtaining little sustenance. Happily Marvell knew better:

> ... then worms shall try
> That long preserved virginity,
> And your quaint honour turn to dust,
> And into ashes all my lust.
> (Marvel, c. 1670)

The *Diet*, or assembly, was held in the Rhineland city of Worms in 1521 and is remembered by generations of schoolchildren for the pun in English rather than for Luther's courage in attending. In the same sense *worm-food* is a corpse:

> You have to be faster, or you are worm-food. (Seymour, 1977)

and we all eventually become *food for worms*, unless cremated:

> But it was William who became food for worms. (Macdonald, 1976)

dietary difficulties the barring of Jews from the German Imperial navy.

German anti-semitism was not a Nazi invention:

> Jews unwilling to give up their faith and be baptized were barred from the Imperial Navy; the official excuse was 'dietary difficulties'. (Massie, 1992)

differently affected by a taboo condition.

In a cringing series of phrases such as *differently abled*, crippled or of low intelligence; *differently advantaged*, poor; *differently weighted*, obese:

> ... it can only be a matter of time before the differently-weighted push for job quotas in the fire departments and the police. (*Sunday Telegraph*, 6 March 1994)

difficult particularly objectionable.

You may say this about other people's children, but it is wise to keep out of earshot of their parents if you do so.

dikey see DYKE.

ding-a-ling the flaccid penis.

From the pendent position of a bell clapper. Mainly nursery use:

> The quads have been reporting progress on papa's dingaling daily. (Sharpe, 1979—papa had damaged his penis on rose thorns)

Wider American than British use, which explains but does not excuse the naïvety of the BBC in permitting the repeated broadcast of a lyric in which a male singer invited his auditors to 'play' with his *ding-a-ling*. Also in America as *ding*, *ding-dong*, or *dong*, with some adult use:

> His dong was never as all-fired important to Wally as yours is to you. (Hailey, 1979)

dip[1] to steal.

Literally, to put into liquid, which necessarily involves a downward movement, and so specially of picking pockets:

> Dipping; lifting money out of a mug's pocket. (Kersh, 1936)

A *dip* or *dipper* so acts:

> Twenty years of muggers and dips, safe men and junkies. (Mailer, 1965—but don't rely on that kind of *safe man* unless you wish him to break open a safe for you)

In airline jargon, to *dip* is to steal alcohol (provided for passengers who have paid for it in the ticket price) and sell it to 'tourist' or 'economy' passengers:

> During take-off ... the steward busily funnels liquor from the first-class bottles into miniatures. This is called 'dipping'. (Moynahan, 1983)

In obsolete British derogatory use, a *dipper* was an Anabaptist, from the total submersion of initiates during christening.

dip[2] a drunkard.

Perhaps just a shortened form of *dipsomaniac* but the American *dip your beak* or *dip your bill* means 'to drink intoxicants to excess'.

dip your wick to copulate.

Common male punning usage, from the immersion of a wick for lighting in oil, and see also WICK:

> Worms who had had an exhausting time dipping his wick, as he called it, all over Wimbledon. (Bogarde, 1978)

Usually of chance male extramarital copulation.

diplomatic cold a bogus excuse for non-attendance.

Politer, it is thought, than an unvarnished refusal. First contracted by Mr Gladstone but the indisposition may be less specific, as a *diplomatic illness* suffered by someone who wishes to keep out of the public eye for a while:

> This was interpreted by some as a 'diplomatic' illness, allowing him to dissociate himself from the campaign if it went disastrously wrong. (*Daily Telegraph*, 31 December 1994, of Yeltsin's absence due to a supposed minor ailment at the time of the Russian invasion of Chechnya)

direct action unlawful violence or trespass.

Usually in support of a political cause, a labour dispute, or a legal activity, such as hunting, to which objection is taken by a minority:

> 'I mean direct action,' said Araba, ignoring Brodie. 'In a word Susannah—violence.' (Theroux, 1976)

direct mail unsolicited enquiries sent by post.

It seeks an order, a subscription, a donation, political support, etc. but the delivery is no more or less direct than the rest of your mail, most of which you actually want to read. The American *junk mail* is accurate but not euphemistic.

dirty pertaining to anything harmful or damaging which may be the subject of a taboo.

A *dirty* atomic bomb is going to go on killing more life for a longer period in a nastier way than a *clean* one. A narcotics addict is *dirty* not from failure to wash but from being caught carrying illicit narcotics. A *dirty joke* usually involves copulation or homosexuality. The *dirty deed* is extramarital copulation by a male:

> ... my mind leaped to the conclusion that he meant he had taken her from me, and done the dirty deed on her. (Fraser, 1977)

A *dirty old man* seeks a sexual arrangement with a much younger woman, or, if he is homosexual, with much younger males. A *dirty weekend* may be fine and sunny but it is passed in overnight clandestine copulation:

> They've simply gone for a dirty weekend at the Spread Eagle. (Matthew, 1978)

See also CLEAN.

Also as a verb. For example, to *dirty your pants* is not to splash mud on them but to urinate or defecate in them.

Dirt is used as a noun in these and similar uses and has a criminal jargon meaning, 'information which is damaging to another'.

disabled crippled or mentally ill.

Literally, rendered incapable, but so long standard English that we forget that there is normally no suggestion that the condition could be wilfully brought about. Now also of mental illness and imbecility:

> The passage of the Americans with Disabilities Act in 1990 extended the same legal protections against discrimination ... to an estimated 43 million disabled Americans. (*Chicago Tribune*, 20 May 1991)

Whence *disability*, of those so categorized:

> Since the term 'disability' can include a former addiction to cocaine, marijuana etc., this means that an employer cannot enquire into past use of drugs, even for jobs such as airline pilots. (*Sunday Telegraph*, 6 March 1994—what a web we liberals weave for ourselves by the misuse of language!)

disadvantaged poor.

Sociological jargon which would literally indicate that an *advantage* had been taken away at some time, whereas those so described were generally not born into a wealthy home or subjected to better education or training:

> I do want to help him—because he's black and probably grew up disadvantaged. (Theroux, 1982)

A 1965 Jules Feiffer cartoon shows the progression from 'poor' to 'needy' to 'deprived' to 'disadvantaged'. (Pei, 1969)

disappear[1] to be murdered.

The implication is that the body is unlikely to be found:

> ... then he, Danny Lehman, might disappear for a minimum of thirty years, or he might disappear, period. (Erdman, 1987—the alternatives were imprisonment or death)

disappear[2] to urinate.

Mainly female use. Women do not in fact vanish after telling you that they are going to *disappear*, but they pay a fleeting visit to a lavatory.

discharge to ejaculate semen.

As in Pistol's gun:

> I will discharge upon her, Sir John, with two bullets (Shakespeare, *2 Henry IV*—the *bullets* were the testicles)

or today:

> The executioner was the first to break away, for he had discharged between Erik's golden thighs. (Genet, in 1969 translation)

Discharge in the sense of leaving employment is a standard English use, from the meaning 'to free' or 'to rid', but see FIRE.

disciple of a person addicted to or participating in (the activities of someone associated with something taboo).

Thus a *disciple of Bacchus* is a drunkard; a *disciple of Oscar Wilde* is a male homosexual:

> When I asked if you were a disciple of Oscar Wilde I meant it not only in the sense of literature. (Burgess, 1980)

discipline see DOMINANCE.

discomfort agony.

The supposedly comforting language of dentistry. If your dentist, drill at the ready, informs you that you may feel a little *discomfort*, it is time to grip the arms of the chair and think pure thoughts.

discouraged drunk.

This American use is odd because alcohol is supposed to make you brave.

disease of love a venereal infection.

For many *love* is synonymous with copulation, which may result in such disease:

> ... advertisements for doctors who cured 'all the diseases of love'. (Manning, 1977)

disgrace to copulate with outside marriage.

But only if the news gets about, I suppose:

> So don't talk about people *making little* of other people, or of him *disgracing* me. (Binchy, 1985—the speaker was pregnant and unmarried)

dish a sexually attractive woman.

A male use, with the common culinary imagery.

dishonoured copulated with outside marriage.

Of a woman who has lost her HONOUR:

> ... he could think of a number of ways for a dishonoured woman to spend the rest of her life. (Farrell, 1973)

The concept and use are now obsolete.

disinfection mass killing.

From the Nazi pretence that Jews, Gypsies, and others killed by gassing were being put into a confined space for the purpose of eliminating lice etc.:

> The underground chambers were named 'disinfection cellars', the above-ground chambers 'bath-houses'. (Keneally, 1982, of Auschwitz)

disinformation see INFORMATION.

disinvestment the disposal of shares etc. as a political gesture.

Not just a normal sale for economic reasons. Those who advocate such a course do so for political or emotional reasons, because they are opposed to a foreign government, vivisection, etc. The rare alternative *divestiture* literally means 'dispossession' and has clerical overtones, because that is what can happen to naughty parsons.

dismal trade the arranging of funerals for payment.

> There is no reason to believe the big-volume concerns will demonstrate a more tender regard for the pocket-books of their customers than has traditionally been the case in the Dismal Trade. (J. Mitford, 1963)

Dismal means 'dreary' and was used to denote the devil or a funeral mute (OED). A *dismal trader* arranges funerals but *dismals*, mourning clothes, is obsolete.

disorderly house a brothel.

Originally 19th-century legal jargon and still used in legal parlance, even of the most tidy and well-conducted brothel:

> If the neighbours chose to complain before a magistrate of a disorderly house... (Mayhew, 1862)

disparate impact *American* a difference in intelligence or ability.

Sociological jargon for the result of any examination or test where one group consistently achieves better results than another:

> Wherever there is 'disparate impact'—one race getting more marks than another—the Government assumes bias in the methodology of testing. (*Sunday Telegraph*, 20 November 1994, also reporting the failure of over 10 years of the NYPD to evolve tests which enabled whites and blacks to achieve equal results)

dispatch to kill.

Despatch was the older spelling, and the only one recognized by Dr Johnson. From the meaning 'to send', it has long been used of the killing of both animals and humans:

> ... we are peremptory to dispatch
> This viperous traitor.
> (Shakespeare, *Coriolanus*)

Today it implies efficient and unspectacular killing:

> If custody was out of the question, employ all feasible measures for dispatch. (Ludlum, 1979)

dispense with someone's assistance to dismiss someone from employment.

Usually peremptorily and with dishonour if not worse, referring to a senior official etc. although used humorously of lesser beings:

The Führer will dispense with his assistance. (Goebbels, 1945, in translation, of such a dismissal)

See also HELP[1], which has common imagery.

disport amorously to copulate.

Literally, no more than frolicking with sexual overtones:

Same old rut. A Richmond resident tells me that it is once again that time of the year when the deer in Richmond Park are disporting themselves amorously. Notices in the park are models of tact. They read demurely: 'Warning, Excessive Deer Activity'. (*Daily Telegraph*, October 1987—who was being warned, we must ask ourselves)

disposal a killing other than by process of law.

Espionage and criminal jargon, from the need to get rid of the body:

Disposals are not in our line of country. (Allbeury, 1981, of such a proposed killing)

Thus *disposal facilities*, the ability to conceal the corpse:

'Have you got disposal facilities if it's necessary?' 'There's hundreds of acres of woods.' (ibid.)

dispute a strike.

Shortened form of British *industrial dispute*, a variant of INDUSTRIAL ACTION. Used twice in three minutes by BBC Radio 4 on 15 June 1983:

A dispute among Southern Region guards has led to the cancellation of trains

—they were not arguing about politics or soccer among themselves—and:

A dispute among camera and technical staff has prevented the televising of sporting events.

In both cases the people concerned were striking without notice in breach of their contracts of employment and in defiance of their trade union. Purists will note and savour the use of *among*.

dissolution[1] death.

Literally, the splitting up into constituent parts, as the corpse into bones, or the body from the soul:

A fetch ... comes to assure ... a happy longevity or immediate dissolution. (Banim, 1825)

dissolution[2] a persistent course of licentious behaviour.

Of casual copulation, homosexuality publicly flaunted, heavy gambling, drunkenness, or the use of illicit narcotics. In each case normal restraint is *dissolved*. He who so acts is *dissolute*.

distinguished aged over forty.

Of male politicians, entertainers, and other public figures. Those who also have white hair may be *very distinguished*.

distressed mentally ill.

Medical and sociological jargon. It literally means 'sorely troubled' but today you call those people distraught.

distribution the payment of a bribe.

Usually where there are several recipients, or where the organizer of a corrupt deal hands on bribes to others, which may then be called a *secondary distribution*:

I also want acknowledgement from every recipient in the 'secondary distribution', as you so nicely put it. (Erdman, 1981, referring to such a transaction)

disturbed[1] naughty or ill-disciplined.

Sociological jargon which does not imply that the miscreants have been interrupted in their activities:

Boys and girls who steal or vandalize, or wet the bed, or are found by their teachers or doctors disturbed ... (Bradbury, 1976)

disturbed[2] mad.

Medical use—literally it means showing no more than unease:

He had stopped looking for the hospital ... 'Are you disturbed?' went on the lunatic. (Amis, 1978—the lunatic was using the language officially used of himself)

ditch to crash an aircraft in water.

From the drain dug to receive water, whence the standard English meaning to discard in such a drain, or elsewhere, any unwanted object. Originally a Second World War punning use but now of any aircraft so crashing.

dive[1] *obsolete* to pick pockets.

From the movement of the hand:

In using your nimbles, in diving in pockets. (Ben Jonson)

Grose notes *diver* as a pickpocket but the use is now perhaps obsolete.

dive[2] a place for the sale and drinking of intoxicants.

Often low-class and from the American use of cellars, where the rent was lower. In the same sense Grose gives *diver* as 'one who lives in a cellar'.

dive[3] to pretend to have been knocked down.

Used of a boxer who goes of his own volition to the canvas. Also as a noun:

Some gamblers tried to scare him into a dive. (Chandler, 1939—they wanted a boxer to throw a fight)

Also used of a soccer player who tries to win an undeserved free kick.

divergence homosexuality.

Moving away from the norm:

> Miles's divergence had been one of his most valuable assets. (Trevanian, 1972—Miles was a homosexual)

diversity *American* the presence of black and white employees.

Literally, the condition of being different or varied.

> The company selected the black candidate because only two of its 82 managers were from ethnic minorities and the board was feeling the pressure of federal rules demanding 'diversity' in the workplace. (*Daily Telegraph*, 9 April 1995—the unsuccessful white candidate was awarded $425,000 damages against the company and the judgement was confirmed by the Supreme Court; see also REVERSE DISCRIMINATION)

divert to steal.

Usually of embezzlement:

> Like the wharfingers, the lock-keepers had ample opportunities to 'divert' a certain amount of cargo. (Burton, 1989—the goods were not merely misdirected)

Whence *diverted*, embezzled:

> ... a large proportion of the profits had been, shall we say, diverted. (Erdman, 1987)

do[1] to copulate with.

Mainly male usage, from his supposed initiative:

> Doing a filthy pleasure is, and short. (Ben Jonson)

> 'Where you might meet anyone and do anything.' 'Or meet anything and do anyone.' (Bradbury, 1975)

Both sexes *do it*—see IT[2]:

> Always wanted to do it outside, you know, ever since I read *Sons and Lovers*. (ibid.)

To *do a perpendicular* is to copulate while both parties are standing. *Do what comes naturally* is of either sex:

> Their pimps would come round and collect, do what comes naturally, and cut out. (L. Armstrong, 1955)

Do yourself is to masturbate, usually of a female as there are so many alternative phrases for male masturbation:

> The thought of him inside her, made her squirm; for an instant she considered doing herself. (M. Thomas, 1980)

Also *do it with yourself*, again mainly of females:

> 'Have you ever done it with yourself?' Dottie shook her head violently. (McCarthy, 1963)

do[2] to kill.

Usually expanded to *do for* or *do in*:

> Some of our chaps say that they had done their prisoners in whilst taking them back. (Richards, 1933)

To *do yourself in* is to commit suicide:

> He has written a letter to my parents. I might as well do myself in. (Townsend, 1982)

In underworld slang to *do* or *do over* also means 'to maim'.

do[3] to cheat or rob.

Perhaps only a shortened form of *do the dirty* etc. Also as *do over*.

do[4] a battle.

In standard usage, a party or function. Usually of a less successful and bloody encounter, such as the British *Arnhem do*.

do[5] to charge with an offence.

Police use:

> She's been done twice for drunk in charge. (Allbeury, 1976)

And a person charged, especially with a motoring offence, will refer to having been *done*.

do a brown see BROWN-HATTER.

do a bunk to urinate.

There are numerous slang and dialect phrases meaning 'to urinate' or 'to defecate' which employ the verb to *do*. I have listed many, SLASH for example, under the noun because slashes etc. are *had*, *done*, or *gone for*, and the noun itself imparts the meaning 'urination'. *Do a bunk* comes from the meaning 'to depart quickly'. The obsolete British *do a dike* denoted urination or defecation, from the trench used in primitive arrangements. The punning *do a job* is to defecate, as in BIG JOBS, but mainly in adult use. To *do a rural* is to defecate out of doors, and may be obsolete. To *do a shift* uses the same imagery as *do a bunk*.

do a number (of a criminal) to give information to the police.

A variant of SING:

> Look, if Keiser's doing a number, I've arranged for you to get fifty to knock him off. (Maas, 1986)

do-gooder a self-righteous person who forces his concerns on others.

Nearly always used derogatively:

> ... hated to ... make the other policeman think he was a do-gooder. (Wambaugh, 1975)

Do-gooding is so acting:

> What were her do-gooding parents but pious cheats? (Theroux, 1976)

do-lally-tap mad.

From the transit camp at Deolali near Bombay where time-expired British soldiers were sent to await repatriation. The heat and the boredom were accentuated by the vagaries of intercontinental transport in the days of sail. If you arrived at the camp in the wrong season, you could be stuck there for six months, which would be additional to your contracted service:

In India he had had a touch of the sun, which we old soldiers called 'Deolalic Tap' (Richards, 1933)

but the 'Old Soldier' also uses another spelling:

Oh, he's got the Doo-lally tap. (Richards, 1936)

In the Second World War sometimes shortened to *tap*:

I was sure by now that this was your natural wild man, and not permanently tap. (M. Fraser, 1992)

do over to steal from.

It may also mean 'to beat up or damage':

Sometimes I'd go with a friend to France for the weekend, expeditions that were financed by him doing over his aunty's gas meter. (McNab, 1993)

do the right thing to marry the woman you have impregnated.

The initial extramarital copulation was the wrong thing:

He Did The Right Thing, by a girl who had only six months to live. (Lyall, 1982)

do your duty by to impregnate (your wife).

Referring to an obligation to produce an heir, when it may also be used of a wife having a son by her husband, with a side glance perhaps at the ordinance in the marriage service concerning the procreation of children:

I regard it as my duty to have an heir. If my husband refuses to do his duty by me I shall find someone who will. (Sharpe, 1975)

dock to copulate with (a female).

This expression was at one time confined to copulation with a virgin, using the imagery of pruning. This is a convenient place to note a common characteristic of etymologists—they welcome any chance to disagree with each other. Farmer and Henley trace this meaning of *dock* to the Romany *dukker*. Partridge in DHS and DSUE looks to 'the standard English *dock*, to curtail' which, in his judgement, 'is obviously operative'. Grose makes no suggestion as to derivation but reports 'Docked smack smooth; one who has suffered an amputation of his penis for a venereal complaint'. EDD correctly reports that *dock* meant 'to undress', as in 'mun dock this gownd off'. OED does not give this use but improves our knowledge by deriving the *dock* in which a prisoner stands in court from the Dutch word for a rabbit hutch. Having discovered that euphemisms may have varied parentage, even being reborn after long disuse, my contribution to the debate is to draw attention to a marine *dock*, a long, moist, narrow space into which a ship moves and may fit snugly. I am sorry that we shall never know what Alfred Holt, the erudite author of *Phrase and Word Origins*, thought.

doctor to change through deception.

By adulterating intoxicants; administering drugs to racehorses to change their performance; falsely adjusting accounts; or castrating tom-cats:

One doctors a cat or a company's accounts (Howard, 1978)

and, as you would expect, of Watergate:

They've doctored the tapes. (Colodny and Gettlin, 1991)

All from the noun denoting a medical practitioner, who need not be, but usually is, a *doctor of medicine*.

dodgy indicating some characteristic that is taboo or of doubtful legality.

Thus for a sailor a *dodgy deacon* is a homosexual priest; for a truck driver, a *dodgy night* is one spent at home but entered on his time sheet as being passed with his vehicle:

If you check your overtime sheets, or the appropriate lay-bys, you will find out... what the drivers' jargon 'dodgy nights' means. (Holder, 1992)

doe a woman who goes to mixed parties on her own.

This American use involves the feminine of *stag*, denoting males only. However much we legislate to the contrary, males remain suspicious of females who show that they need no male protection. In 17th-century England a *doe* was a prostitute, and in later use any single woman at Oxford University.

dole a payment by the state to the involuntarily unemployed.

Originally, a portion, whence a gift made regularly to the poor, as *dole-bread* or *dole-money*—and at funerals *dole-meats*. One of a succession of euphemisms—see RELIEF[1].

doll[1] a sexually attractive female.

Dr Johnson reminds us that *Doll* was a contraction of *Dorothy* as well as being 'A little girl's puppet or baby'. A female so described may be beautiful but slow-witted, although *real doll* implies beauty and brains.

doll[2] a narcotic in pill form.

Usually a barbiturate or amphetamine. The punning title of Jacqueline Susann's novel *Valley of the Dolls* started or sanctified this usage.

dollar shop a store which will not sell in the local currency.

A feature of former Communist regimes, where luxuries were reserved for tourists and party officials. The currency was not necessarily the dollar—almost any non-Communist currency or plastic would suffice.

dolls see GUYS.

dolly a prostitute.

Perhaps from DOLL[1] but also owing something to her smart dress—*dolled-up*:

It seemed rather steep of my father to keep his dolly at home with my wife there. (Fraser, 1969, writing in 19th-century style)

The 19th-century *dolly-common* or *dolly mop* was a part-time prostitute:

Maid-servants, all of whom are amateurs, as opposed to professionals, more commonly known as 'Dollymops'. (Mayhew, 1862)

domestic *?obsolete* a servant in the home.

A shortened form of *domestic help*; see also HELP[1]:

We used to call them servants. Now we call them domestic help. (Chandler, 1953)

domestic afflictions *?British* menstruation.

It could mean myriad other things which cause unhappiness in the home.

dominance a sexual perversion in which a woman inflicts pain on a man.

Literally, authority or control over another. A male pervert is said to be *into dominance* and may go to a specialist prostitute for *discipline* or *dominance training*.

Don Juan a male philanderer.

The successful practitioner in seduction inspired the music of Mozart and the words of Molière, Byron, and Shaw, to mention but a few. Whence *Don Juanism*, such behaviour:

Etlin has great courage and charm, yet his Donjuanism somehow detracts from his authority. (Read, 1986)

done for subjected to a major misfortune.

By death—see DO[2]:

'They're both done for'... George lay spread-eagled at my feet (Fraser, 1971)

or serious wounding, especially in battle; or bankruptcy; or defeat.

dong see DING-A-LING.

don't-name-'ems *obsolete* trousers.

A 19th-century example of the great trouser taboo—see also UNMENTIONABLES[1].

doorstep to abandon a baby.

You left it on the step of a prosperous house and, if brave, rang the bell before you made off:

When it became obvious... from the hour of my conception, that my parents intended to doorstep me... (N. Mitford, 1945)

This use, now probably obsolete, reminds us of the former stigma of bastardy affecting both the mother and the child and the lack of facilities for unmarried mothers and their children.

dope a narcotic.

Originally, a thick liquid, from the Dutch *doop*, sauce. Whence prepared opium, which looks like that at one stage:

A younger sister whom she loved... had taken to dope. (Harris, 1925)

To *dope* is to give illicit narcotics to horses or greyhounds to alter form, whence *the dope*, inside information about a race—which beast has been drugged? *Dope*, a simple person, comes from his opiate mien and behaviour.

dose a venereal infection.

Usually gonorrhoea in a male. Literally, an amount of medicine; the use may allude to the attempted cure:

And if I give that man a dose, that's my pleasure and he just gettin' what he's payin' for. (Simon, 1979, of a prostitute)

doss-house *American* a brothel.

Doss means 'sleep' and a *doss-house* is normally a sleeping place for the destitute.

dotty eccentric or mentally ill.

Originally, of unsteady gait, whence 'feeble' and so 'feeble-minded'. It is used of eccentricity as much as of madness:

There might be a basis of truth, but I felt she was pretty dotty. (Manning, 1965)

double[1] an alcoholic drink containing two standard portions.

Scotty has really cashed in... and was ordering doubles like a man possessed. (*Private Eye*, May 1981)

The volume of the standard portion seems to vary in relation to the excise duty of the country concerned, so that one man's *single* is another man's *double*.

double[2] a spy working contemporaneously for two opposing parties.

A shortened form of *double agent*:

How else does anyone play a double? (le Carré, 1980, referring to such a spy)

In similar vein to *double* is to pimp for two prostitutes:

In the meantime I'm going to 'double' you. (Londres, 1928, in translation—the speaker was going to take a second woman)

double depth *American* (of corpses) buried one above the other.

Funeral jargon, describing a way of saving cemetery space:

The companions will repose one above the other in a single grave space, dug 'double depth', to use the trade expression. (J. Mitford, 1963)

double entry dishonest.

A double-entry man. Hong Kong's full of them. Twisters. (Theroux, 1982)

Double-entry bookkeeping is a self-balancing method of keeping simple accounts. This use puns on keeping two sets of accounts, of which one is intended to deceive.

double-dipper a person in illegal receipt of a second source of income or bribery.

Not someone taking a classic sauna, where he would pass from hot to cold, but someone who is in receipt of a second, illegal, source of income:

> Keegan was an academy graduate who had put in his thirty and retired to become a double-dipper. (Clancy, 1986)

double-gaited having both homosexual and heterosexual tastes.

DAS says 'An extension of the horse-pacing term' and I have no better suggestion:

> '... homosexuality isn't the handle it once was'... 'Pascoe's wife didn't know he was double-gaited.' (Bagley, 1982)

double-header copulation by a male with two females in each other's presence.

Prostitutes' jargon, from the use of two locomotives to pull a train and perhaps punning on *head*, the glans penis:

> ... she wasn't interested in the hundred-dollar bag of bones who Juicy Lucy said was coming back at eight o'clock for a doubleheader. (Wambaugh, 1981)

double in stud *American* to copulate with two people in each other's presence.

Of either sex, despite the maleness of STUD:

> ... maybe there were some who doubled in stud. (Longstreet, 1956)

double time *American* copulation outside marriage.

From the increased payment for overtime working and perhaps also alluding to TWO-TIME:

> Your wife is standing right beside you and you are practically accusing her of a little double time. (Chandler, 1953)

douceur a bribe.

Literally, a gratuity, in French and English:

> I bet he's had some little douceur slipped into his hand. (Manning, 1965)

I prefer the 19th-century spelling:

> Nobody is allowed to take dowzers. (EDD, from 1885)

dove an appeaser or pacifist.

From the symbol of peace and the opposite of HAWK. The use is not necessarily pejorative and became hackneyed in the Kennedy and Johnson presidencies.

down among the dead men in a state of deep drunkenness.

The *dead men* are the skittles which have been knocked over in the western English inn game, whence too the rarer *in the down-pins*.

down below the genitalia.

Of either sex, despite the location of that part of the body above the legs. In nursery use a child or geriatric may be asked if he has dried himself *down below*. And sexually, where the logic is even less certain as the parties would probably be horizontal:

> We take it in turns to stroke and massage each other anywhere but what you used to call down below. (Amis, 1978)

Also as *down there*:

> The first time she touched him 'down there' she thought she would die of mortification. (Forsyth, 1994)

down for good dead.

The imagery is from boxing where it means 'knocked out unconscious'.

down on see GO DOWN ON.

down the line[1] in or to prison.

This may have come from the direction a First World War prisoner went after capture, although I suspect the usage antedates that conflict. *Down* also refers to the cells, which may be situated in the cellar of the building where the court sits, and the prisoner who is *sent down* by the magistrate, and *taken down* by the tipstaff or his equivalent, may in fact descend no lower than from the dock to the courtroom floor on his way to prison.

down the line[2] in the seedier parts of a city.

American use for the location of brothels, gambling joints, etc. It may owe something to the linear arrangement of most American towns with a railroad as the axis, and the gravitation of such establishments to the periphery. The obsolete British *house in the suburbs* (see at HOUSE[1]) employs the same idea.

Down's syndrome a congenital disorder due to a chromosome deficiency.

The physical characteristics of this disorder were formerly thought to be reminiscent of Mongolians and thus it was known as *mongolism*. We now seek to hide both the stigma and this implicit racial sneer by naming it after the English physician John Langdon Down (1828–96):

> People they spoke to about mongolism —Down's Syndrome, as Angela insisted on referring to it. (Lodge, 1980)

downs depressant narcotics.

The reverse of *ups*:

> ... took his pills by the fistful, downs from the left pocket of his tiger suit and ups from his right. (Herr, 1977)

In America also as *downers*:

> He hoped there might be some downers left... where his girlfriend left a small cache. (Wambaugh, 1975)

downsize to dismiss employees.

The volume you wish to reduce is that of the payroll:

> It was an unhappy time. We had to downsize the company substantially and we had quite a serious divergence of opinion between the management and the workforce. (*Sunday Express*, 12 February 1995—which was not that surprising, as the managers tend to be the ones who keep their jobs)

See also RIGHT-SIZING.

downstairs[1] *obsolete* the house servants.

From their normal location at one time in a semi-basement of a town house. Whence the British television serial *Upstairs, Downstairs*.

downstairs[2] the genitalia.

A genteel use by male and female. A nurse may ask a patient if he has a problem *downstairs*.

downward adjustment a devaluation or an economic depression.

Economic or political jargon to avoid the panic which the truth might bring and imply that human agency still controlled events:

> ...the worst America has to endure is a 'downward adjustment of the economy'. (Jennings, 1965, noting the euphemism)

doxy a prostitute.

Originally, a sweetheart, from the Dutch *docke*, a doll:

> A party taken on a cruise by wealthy degenerates, who had sold their doxies at various places in the Caribbean. (Fraser, 1971, writing in 19th-century style)

Obsolete except for literary use.

drag[1] the clothing of the other sex worn by a homosexual.

From the theatrical use, referring to a male actor (not necessarily a homosexual) in female clothes, the long train being *dragged* on the floor. A homosexual so attired is said to be *in drag*:

> A cop tried to intervene and was promptly felled by someone in drag. (Sharpe, 1977)

A *drag* is also an American male homosexual party.

drag[2] an illicit narcotic in cigarette form.

In tobacco smoking, a *drag* is a single puff at a cigarette, sometimes one being smoked by another. The imagery in each case is from the inhalation.

drain off to urinate.

Usually of a male, with obvious imagery:

> Weak bladders, old men... Might as well drain off himself. (Grayson, 1975)

dram a drink of a spirituous intoxicant.

You used to buy spirits from apothecaries, who used their own measurements, in this case one eighth of a fluid ounce (originally the weight of a drachma), corrupted to *dram*:

> 'Come over for a dram,' he urged them. (Boyle, 1979)

The Scottish *wee dram* is larger than the standard size.

draw a bead on to shoot at.

The *bead* is the foresight of the old-style rifle, rather than the bullet. Also used figuratively of anyone whose actions you watch closely:

> I am going to draw a bead on this gentleman. I am preparing an operation to liquidate him. (Goebbels, 1945, in translation—he was particularly upset by the way in which the inhabitants of his home town had welcomed the Anglo-American invaders)

draw a blank to be very drunk.

An American use, perhaps now obsolete, punning on the loss of awareness and an unsuccessful entry in a lottery:

> For after the funeral I drew a near-blank, as they said in those days about drunkenness in its most amnesic mode. (Styron, 1976)

draw the enemy into a trap to retreat involuntarily.

Military use when you want to disguise your predicament in order to keep up morale:

> Of course the officers knew, but they were telling us we were drawing the enemy into a trap. (Richards, 1933, of a First World War retreat)

draw the king's picture to counterfeit banknotes.

Or the queen's, or the president's, as the case may be, from forging the likeness.

draw the long bow to boast or exaggerate.

The longer the bow, the further its potential range:

> ...draw the long bow better now than ever. (Byron, 1824, of boasting)

In the same sense you might also *pull* it:

> You will say, 'Ah, here's Flashy pulling the long bow,' but I'm not. (Fraser, 1973, writing in 19th-century style)

See also SHOOT AMONG THE DOVES.

draw too much water to outrank other officers or officials and require their obedience.

Naval use and imagery of someone who makes a decision which seems questionable but must be obeyed. Also some bureaucratic use, where the same conventions apply.

dream dust powdered illicit narcotics.

Especially heroin. A *dream stick* was opium. *Dreams* are associated with the illegal ingestion of narcotics in many addict phrases.

dress for sale a prostitute.

Apart from the fact that most women wear dresses, this American Citizens' Band usage is doubly misleading. The *dress* is not what's on offer and the transaction contemplated is of hire or licence, rather than sale. The obsolete London *dress lodger* was a prostitute clothed in suitable style by a pimp, working from a brothel called a *dress-house*:

> The dress-lodger probably lives some distance from the immoral house by whose owner she is employed. (Mayhew, 1862)

Today an American pimp who decks out a prostitute is said to buy her *bonds* or *threads*.

dress on the left to be a male homosexual.

Reputedly from the words used in the enquiry by a bespoke tailor who asks his customer which side his penis normally rests, so that the trousers can be cut accordingly:

> I wondered if the senator was attempting to discover whether I was 'dressing on the left' (as the London master tailors put it). (Behr, 1978)

See LEFT-HANDED[2] for the association with *left*.

drill[1] to kill by shooting.

The imagery is of boring holes:

> I could drill you and get away with it. (Chandler, 1958—the speaker was not an army sergeant)

drill[2] to organize and train civilians in an illegal militia.

From instilling the first rudiments of military training:

> ...the Ulster Volunteer force went on drilling —and not with dummy weapons. (Foster, 1988)

drink[1] an intoxicant or to drink intoxicants.

The commonest euphemism for anything to do with intoxicants. *Drunk* is so well-established as standard English for being intoxicated that I use it as a definition, despite its euphemistic origins. Thus if a friend offers you a drink, you do not expect water. It is found in many phrases, for example *drink taken*, slightly drunk, and *in drink*, drunk (see also DRINK[3] (THE)). To *like a drink* is to be addicted to alcohol. To *drink too much* is to take intoxicants to excess, on a single occasion or habitually:

> He sometimes drank too much. (Harris, 1925, of an alcoholic)

Given to the drink, so addicted, is perhaps obsolete:

> Some say she cocks her wee finger. In short that she's gien to the drink. (Barr, 1861)

To have a *drink* or *drinking* problem is to be an alcoholic, not to have a restriction of the throat:

> ...her father had had a drinking problem. (Theroux, 1982)

To *drink at Freeman's Quay* was to cadge intoxicants from others. A *non-drinker* takes no intoxicants and a *drinker* too many:

> Aunt Estelle was no 'drinker' and her 'wildness' was the merest good spirits. (Murdoch, 1978)

A *drunk* may also be a habitual drunkard or a carouse:

> He also had some glorious drunks with the men he had met. (Richards, 1933)

drink[2] a bribe or a tip.

To give an excuse for the handing over of cash, but less common than the corresponding *pourboire*:

> 'Has any money changed hands?' 'I dare say Jimmy was offered a 'drink' of some sort.' (Read, 1979)

Also used non-euphemistically for a tip.

drink[3] **(the)** the sea.

Used by airmen when forced to put down on water, or *in the drink*.

drink milk *Indian* (of a baby) to drown.

The Parsees set a higher value on male children and used to drown females in milk:

> ...if it were a daughter, Bapaiji swore she would make it drink milk; all good women, so she contended, hated their sex. (Desai, 1988)

drive a ball through to kill by shooting.

Using the same imagery as DRILL[1]:

> Supposing, he asked, landlords refused to give any reduction of rent: what were they do to? 'Drive a ball through them.' (Kee, 1993—the advice was tendered by a man in the crowd attending one of Parnell's meetings)

droit de seigneur copulation by a male employer with a female employee.

It means literally the right of the lord (of the manor), but other dominant males, especially in the entertainment industry, have claimed similar privileges:

> The droit de seigneur died with the Hollywood czars. (Deighton, 1972)

The rights of a Norman lord of the manor were reputed, fictitiously in most cases, to include that of copulating with each virgin in his territory. It is entirely predictable that this particular manorial right, or perhaps duty, survives in our memory and languages when other more important but commonplace rights have been forgotten. The feudal system functioned primarily on the lord's ability to demand free labour from tenants or villeins, which they euphemistically called *boonwork*, *love-boonwork*, or *bederipe* (reaping by request)—just as the Tudor and Stuart monarchs called forced loans *benevolences*, until the 1689 Bill of Rights brought that practice to an end.

drop[1] to kill.

By shooting, after which the victim falls. Standard English of animals and underworld slang of humans. In the days of the Chicago mobs—those of the 1920s—*drop down the shute* meant 'to murder', from sliding garbage down a shute in an apartment block:

> If he's alive, put him on ice until tonight. Then drop him down the shute. (Weverka, 1973)

> But they got so close there was no way they were going to avoid us, so we dropped them. (McNab, 1993—'they' were four Iraqi soldiers)

drop[2] a quantity of intoxicant.

Usually of spirits and seeking to imply moderate consumption:

> The rum came up with the rations and was handed over to the Company-Sergeant-Major. If he liked his little drop, he took his little drop. (Richards, 1933)

Occasionally as a *drop of blood*:

> 'Give me a drop of blood, will you?' The bourbon tasted like linseed oil. (Mailer, 1965—perhaps the opinion was that of a habitual Scotch drinker)

With a drop on means 'intoxicated':

> Two of our chaps with a drop on shot all the bottles and glasses in a cafe. (Richards, 1933)

A *drop taken* also means 'intoxicated':

> My father was always giving out about it when he had a drop taken. (Flanagan, 1979)

drop[3] to die.

Usually suddenly, of natural causes. Perhaps only a shortened form of *drop dead*:

> Louie's out mowing that lawn and he drops... Like that. The ticker. (Sanders, 1977)

The *long* or *last drop* is death by hanging, occasionally *tout court* as *drop*:

> Unlike the festive hangings of earlier times, the drop was performed in a church stillness. (Keneally, 1982)

To *drop off* is also to die of natural causes, from either the fate of an old gate which *drops off the hooks* or from an avian demise:

> The soo took the fever, the kye drappit off. (A. Armstrong, 1890)

Thus the fuller form:

> It's the dropping off the perches... Soon we shall all have gone. (N. Mitford, 1949)

To *drop in your tracks* is to die suddenly (from racing imagery), but not necessarily of natural causes:

> ...if Kramer had not been so inconsiderate as to drop in his tracks. There was nothing like death for spawning myths. (Francis, 1978)

Drop your leaf refers to deciduous trees rather than gate-leg tables, and is used of natural death in old age. To *drop off* is also to fall asleep when you should be keeping awake.

drop[4] to give birth (to).

Usually of quadrupeds, but to *drop a bundle* is used of women, meaning 'to have an induced abortion':

> Ask the girls who dropped their bundles... (W. Smith, 1979, of such abortions)

drop[5] a place where stolen goods are left.

Espionage and criminal use, and sometimes also of the person who does the dropping.

drop[6] to fail to select.

Of a team player when another has been preferred. See also DESELECT.

drop acid see ACID.

drop anchor fraudulently to cause a horse to run slowly in a race.

The imagery is naval and the practice associated with crooked gambling.

drop beads *American* to identify yourself esoterically to another homosexual.

By speech or body language. The wearing of beads by a male may imply effeminacy. If the string breaks, they distribute themselves over a wide area.

drop-dead list a list of names of people to be dismissed from employment.

The offensive American expression *drop dead* expresses rejection.

drop off to die.

As illustrated in DROP[3]. It is more commonly used of falling asleep or napping.

drop the boom on to discriminate against.

From the defensive obstruction to navigation. Used to refer to the withdrawal of credit facilities, exclusion from confidence, or dismissal from employment:

> [He] still worried that Harold would drop the boom on him. (McInerney, 1992—he was afraid of losing his job)

drop the hook on *American* to arrest.

The imagery has to be from fishing:

> The buttons in the prowl car were about ready to drop the hook on him. (Chandler, 1953)

drop your arse to have diarrhoea.

Not merely to lower yourself into a chair:

> A guard appeared each time and dragged me down to the toilet, then stood over me while I dropped my arse. (McNab, 1993)

drop your drawers to copulate casually.

Of a female, with obvious imagery:

> ... those pressed, permanented country-club types ... would drop their drawers for a New York Jew. (M. Thomas, 1980—*permanented* means having had their hair 'permanently waved')

A British female is more likely, if so inclined, to *drop her pants*, the equivalent of the American *underpants*. The obsolete *drop your flag* was to surrender, of a ship, not a female.

drop your guts to fart.

Not the result of a hernia. Also as *drop your lunch*.

droppings the excreta of animals.

Standard English in this sense since at least the 16th century:

> There were steaming piles of elephant droppings bang in the middle of the road. (Allen, 1975)

Animals might also *drop the crotte*, from French *crotte*, dung:

> Buller splayed out and dropped his crotte on the edge of the path. (G. Greene, 1978—the English word *crottels* for 'horse dung' is obsolete)

Humans may literally *drop their wax* or *a log*, the latter also being achieved figuratively:

> 'He find a rat outside the guy's fat head,' Willie said. 'I almost dropped a log.' (Theroux, 1993—Willie was amazed at what he had seen)

drown your sorrows to drink intoxicants to excess.

Alcohol in sufficient volume is supposed to bring solace to the unfortunate:

> If I didn't know you better I'd have said you'd been drowning your sorrows. (Amis, 1978)

The obsolete Scottish *drown the miller* meant either 'to be made bankrupt' or 'to add too much water to whisky', both being derived from the proverb 'o'er muckle water drowned the miller'.

drug an illegal stimulant, narcotic, or hallucinogen.

Literally, a medicinal compound affecting the body or mind but, as with DRINK[1], now also standard English for what is taboo. Thus *drug abuse* is not taking too many aspirins, a *drug dealer* or *pusher* is not a retail pharmacist, and a *drug habit* is not the regular ingestion of beta blockers for hypertension.

drumstick see DARK MEAT[1].

drunk see DRINK[1].

Drury Lane ague *obsolete* venereal disease.

Drury Lane, adjoining COVENT GARDEN, was a notorious brothel area of pre-20th-century London. A *Drury Lane vestal* was a prostitute.

dry[1] prohibiting or not offering the sale of intoxicants.

It does not mean that nothing is available to quench your thirst, but you will find no *wets* there (see WET[2]).

dry[2] wanting an alcoholic drink.

Usually of beer with a pretence of temporary dehydration:

> You dry, lad? S'm I, begod! mouth like an ash pit. (Cookson, 1967)

dry[3] to forget your lines.

Theatrical jargon, a shortened form of *dry up*, for something which should not happen to a professional actor:

> I delivered the previous lines right on cue. But after the Yorick speech I let them think I'd dried. (Deighton, 1972)

dry bob copulation without ejaculation.

Partridge suggests 'ex dry bob, a blow that leaves the skin intact' (DSUE) but I suspect that is too intellectual an etymology if we are familiar with the normal derivation of these vulgarities. Consider rather *dry*, the absence of semen, and *bob*, from the association with women's hair, both punning on the English schoolboy who eschews rowing in favour of cricket. A *dry run*, copulation during which the male wears a contraceptive sheath, is a triple pun, on the absence of a seminal free discharge, on the feeling to the woman, and on the meaning 'a practice or rehearsal'. The obsolete British *dry pox* was syphilis:

> The disease communicated by the Malays, Lascars, and the Orientals generally ... goes by the name of the Dry ——. (Mayhew, 1862 —he isn't always so squeamish)

dry out to desist from drinking alcohol after a period of excess.

Not what you do in front of the fire after a walk in the rain, nor even derived from DRY[1]:

> I have been at a health farm in the depths of Suffolk, slimming and drying out before the summer holiday. (A. Waugh, *Daily Telegraph*, 13 August 1994—the satirist was not seriously suggesting that he had suffered from undue intemperance)

duck a urine bottle for males.

American hospital jargon, from its shape. (A *duck*, a zero, is generally thought to come from the shape of an egg, just as *love* in tennis comes from *l'oeuf*, although some still say it comes from playing for *love*, for nothing.)

duff[1] see FLUFF YOUR DUFF.

duff[2] the buttocks.

Presumably from the suety pudding or pastry. This American usage appears to be unconnected with the slang expression *duff up*, to belabour.

duff[3] *American* a male homosexual.

Because he masturbates—see FLUFF YOUR DUFF—or because he is considered by those who use the expression to be something worthless?

dull to kill.

With imagery from making dark rather than stupidity:

He dulled them, turned, quietly left the room. (Goldman, 1986, of a double murder)

dummy[1] a stupid person.

Don't look so worried, it's not that difficult. I picked it up in no time, and I'm a dummy. (Blair, 1990)

Literally, a representation of the human form, from the original meaning 'a dumb person'. It may denote someone who is momentarily unthinking or distracted or it may refer to the mentally ill:

So don't get the idea all of Ellerbee's patients are dummies. (Sanders, 1985—Ellerbee was a psychiatrist)

dummy[2] the penis.

From the shape and usually in the phrase, *flog your dummy*, to masturbate.

dump[1] to defecate.

An obvious and rather distasteful male usage:

Everything hinged on that first dump of the day. (Theroux, 1971, referring to defecation)

Also some figurative use:

But maybe you also recall how your Service dumped all over us on that one? (Lyall, 1985)

dump[2] *American* to kill.

Literally, to discard, which someone has to do with the corpse:

Now we've got to go back and check everyone ...to find out where they were when Dolly was dumped. (Sanders, 1986—Dolly had been murdered)

duration the time occupied by the Second World War.

Shortened form of *duration of the war*; common British usage at the time when the outcome was uncertain and there was a taboo about predicting the future:

...you'd never get back to England. You'd be stuck there for the duration. (Barber, 1981)

dust[1] illicit narcotics in powdered form.

From the visual similarity; found in various forms such as ANGEL DUST and DREAM DUST:

He pays off with the dust, and it's party time every Saturday night. (Sanders, 1990, of someone who remunerated helpers with narcotics)

dust[2] to kill.

Probably wiping off rather than turning to dust, with the common blackboard imagery:

The question is...did she hate him enough to dust him. (Sanders, 1985)

dustman a corpse.

An obsolete punning British usage, *dust* being what the body may eventually become. The *dustbin* was a grave.

Dutch appears in many offensive and often euphemistic expressions dating from the 17th-century antagonism between England and the Low Countries. Thus anything described as *Dutch* is likely to be bogus or inferior, as in the entries that follow.

In Dutch means 'in trouble':

Got me in proper Dutch, you did. (B. Forbes, 1986—a nurse had been exposed to criticism by a patient's action)

The exception is the *Dutch cap*, a contraceptive device worn by women internally, named for its shape and perhaps its country of origin. It was a male commentator on a hockey match between the ladies of Holland and England who described the goalkeeper of the visiting side as 'very experienced—she has thirty-two Dutch caps'.

Dutch (do the) to kill yourself.

You're not going to do the dutch, are you?... Commit suicide? (Sanders, 1980)

Dutch act (the) suicide.

Dutch auction an auction in which the auctioneer drops the price until a buyer makes a bid, being the reverse of a normal auction in which bidders raise the price until only one remains in the auction.

Dutch bargain an unfair and unprofitable deal.

Dutch cheer a spirituous intoxicant—the Dutch are supposedly gloomy otherwise.

Dutch comfort an assumption that things cannot get worse.

Dutch concert music played out of tune or a drunken carouse.

Dutch consolation an assurance that although things are bad, they could have been worse.

Dutch courage bravery induced by intoxicants, implying that a Dutchman is a coward when he is sober.

Dutch feast an occasion when the host becomes intoxicated while his guests are still sober.

Dutch fuck lighting one cigarette from

another, perhaps because the action is soon over and costs nothing.

> ...then lit his cigarette from mine...'That's a Dutch fuck, old chum.' (Barnes, 1991)

Dutch headache a hangover.

Dutch reckoning an inflated bill without details.

'...as brought at spunging or bawdy houses' (Grose—the tradition dies hard).

Dutch roll combined roll and yaw in an aircraft—it behaves like a drunken sailor.

This phrase, noted by Moynahan in 1983 as modern airline pilots' jargon, shows how unforgiving of their historic enemies English-speakers can be.

Dutch treat entertainment or a meal to which you are invited, but where you have to pay for yourself.

> She and Callahan enjoyed the better restaurants in town, and never ate at the same place twice. It was always a Dutch treat. (Grisham, 1992)

It is also called *going Dutch*:

> 'Here,' Ardis Peacock said half-heartedly, 'let's go Dutch.' 'No way...I asked you to lunch.' (Sanders, 1980)

Dutch uncle someone who reproves you sharply or gives you solemn advice, unlike the geniality of real uncles.

> I talked to him like a Dutch uncle. It doesn't seem to have done him any good. (Baron, 1948)

Dutch widow a prostitute.

Dutch wife a bolster, the sole bedmate of many white bachelors serving in the Far East.

> ...he clutched tightly the bolster—sweat-absorbing bedfellow of sleepers in the East—known as a Dutch wife. (Burgess, 1959)

duty defecation.

Probably from the requirement placed daily on children:

> Many an unwary person has been knocked off his toes by a charging porker before the completion of his duties. (Simon, 1979, of open-air defecation in India)

dyke a female homosexual.

In a homosexual partnership she plays the male role. The use is perhaps not euphemistic, as I can trace no etymology:

> 'Good God, what was she, a dyke?' asked the President. 'No, a woman in her middle forties, unmarried, sentimental.' (Ustinov, 1966)

The adjectival form is *dikey*:

> Dikey lady, white hair, jolly. (le Carré, 1989)

dynamite a marijuana cigarette.

There are two other American drug users' meanings: a mixture of marijuana and heroin; and cocaine and heroin taken together by any method.

E

EC see EARTH CLOSET.

ESN see EDUCATIONALLY SUBNORMAL.

ear a microphone used in secret surveillance.

The jargon of espionage or of spy fiction:

> If they think you've got something to hide, they'll plant another ear. (Francis, 1978)

early bath see TAKE AN EARLY BATH.

early retirement dismissal from employment.

Not retirement to bed before 10 o'clock:

> Paul Bergmosen, in charge of purchasing, who was given 'early retirement' in 1977. (Lacey, 1986—the gift was probably unwanted, but the device allows employers to charge redundancy costs against pension funds)

The expression is more accurately used to describe the choice made by people who voluntarily give up work before reaching the normal retirement age of 65.

Also as *early release*, which might suggest that those dismissed should be grateful for being granted their freedom from working:

> *Early Release Schemes* The group expects to reduce the number of employees by about 15,000 during each of the next two financial years. (British Telecom Report, 1993)

earn to steal.

A military use, seeking to show entitlement—see also LIBERATE[2]. In the Ottoman Empire, to *earn a passport* was to be lent by the Sultan from his harem to a minister, whom you were required to kill, by means other than assiduous concubinage:

> Her task accomplished, she was re-integrated into the Royal household and rewarded for her services. In the argot of the Seraglio, this was known as 'earning a passport'. (Blanch, 1954)

earnest *obsolete* homosexual.

Of a male:

> To be 'earnest' was also slang for being homosexual in Victorian London. (*Sunday Telegraph*, 12 February 1995)

The possible pun was perhaps another reason for Wilde's choosing *The Importance of Being Earnest* as the title for a play.

earth to inter (a corpse).

Not to fit with an anti-static device, as are living workers in an electronics factory.

> There was a multitude fit for a city procession saw her earthed. (O'Donogue, 1988)

earth closet a non-flush lavatory.

First used to denote those arrangements where earth was used as an absorbent, and then for any *closet* or lavatory which does not use a water flush or chemical dispersion. Commonly abbreviated to *EC*.

earthy vulgar.

A venerable usage:

> Certainly we know that he enjoyed an earthy story. (Bryson, 1994, of President Lincoln)

ease springs to urinate.

Of a male, from the military action of *easing springs*, in which the rifle bolt is rapidly moved up and down the breech, which has a tenuous similarity to the stroking of the penis to prevent a drip of urine. Jolly (1988) suggests that a sailor who excuses himself from the company so that he can urinate may pretend to be seeing to the *springs* or mooring lines of a ship, which may need easing according to the current or tide. To *ease your bladder* applies to either sex:

> One man I knew used to swear that he only eased his [bladder] once a month. (Richards, 1936, discussing service in India)

To *ease yourself* is to urinate or defecate and to *ease your bowels* is to defecate:

> I had dismounted to ... try to ease my wind-gripped bowels. (Fraser, 1973)

A *house of ease* is a lavatory, as was a CHAPEL OF EASE.

East to your death.

From the direction in which Polish and other Jews were sent by the Nazis to the death camps:

> 'Where has Herr Hirschmann gone?' I was able to ask. 'The Germans sent him east.' (Keneally, 1985, of Second World War Belorussia—it meant he was dead)

East African activities extramarital copulation.

A *Private Eye* refinement of the in-joke *Ugandan affairs* (see UGANDA):

> I am distressed to see the old French word 'romance' used as a code name for East African activities. (A. Waugh, *Private Eye*, December 1980)

East Village a less fashionable area of New York.

Used by realtors and others to exploit the cachet of *The Village*:

> Property speculators tried to call the East Side of [10th Street] 'the East Village' but there were not many takers. (Deighton, 1981)

London too has its SOUTH CHELSEA.

Eastern substances illegal narcotics.

From the association of China with opium or from the other sources of much modern illicit supply:

The smell of exotic Eastern substances grown on the premises that wafts gently across the square. (*Private Eye*, May 1981, of cannabis)

easy terms hire purchase.

The use is so widespread that we no longer address our minds to the reality that everything involved becomes more expensive and difficult, except finding the initial down-payment.

easy way out (the) suicide.

Not an open door. The use implies a lack of courage:

...they've told me it's cancer and I'm taking the easy way out. (James, 1972, from a suicide note)

easy woman a female with no reservations about casual copulation.

Easy in the sense 'compliant'. They are not necessarily prostitutes:

Whether we worked in a Massage Parlour or were rich ... we were still the same to you. Easy women. (Bogarde, 1978)

See also LADY OF EASY VIRTUE.

eat[1] to engage in fellatio or cunnilingus.

Usually specifying what is being figuratively consumed—pussy, meat, pork, or other slang names for the vagina and penis:

Wouldn't you like to eat my pussy? (Robbins, 1981, said by a woman suggesting cunnilingus, not the sacrifice of a pet)

Occasionally as *eat out*:

She used to give hand jobs. She let Moochie eat her out. (Theroux, 1989)

See also EAT FLESH.

eat[2] to ingest illicit methedrine.

American drug users' slang.

eat a gun to commit suicide with a firearm.

The only sure way of killing yourself is to load it, point it upwards through your mouth, and then pull the trigger:

...his back against the filthy tiled wall, and he was trying to eat his gun. (Sanders, 1977, of an attempted suicide)

eat flesh to copulate with a woman.

A venerable pun:

Suffering flesh to be eaten in thy house, contrary to the law. (Shakespeare, *2 Henry IV*)

See also EAT[1].

eat for two to be pregnant.

The theory, unjustified in affluent families, is that a woman needs double rations during pregnancy:

'Do you ever remember me on a diet, Edie?' 'No, I can eat for two.' 'You don't mean ...?' 'Edie!' she laughed. (Deighton, 1972)

eat-in kitchen there is no separate dining-room.

American real estate jargon for a small house or apartment:

Eat-in kitchen, lovely porch overlooks private yard. (*Chicago Tribune*, 30 July 1991)

eat porridge to be in prison.

From the British prison diet:

The best offer you're going to get, mate, is to eat your porridge here for a respectable time. (C. Thomas, 1993, of someone accused of a crime abroad)

See also PORRIDGE.

eat stale dog *American* to take a deserved reprimand.

I think this is analogous to *eat dirt*, with *dog* being a shortened form of *dog shit*:

I can eat some stale dog and get by. (Chandler, 1939, of someone detected in wrongdoing)

eat the Bible *American* to perjure yourself.

False evidence given on oath, with the swearing usually on a bible, constitutes the crime of perjury:

...told the lieutenant not to count on me to 'eat the Bible'. (Lavine, 1930)

eating disorder anorexia nervosa or bulimia.

Not spilling egg down your shirt, but the inability to eat a normal diet as a consequence of psychological or emotional disturbance.

ebony elite wealthy or well-known black people.

In Britain they are often talented sportsmen:

Fashanu, a leading role model for young blacks ... one of Britain's 'ebony elite'. (*Sunday Telegraph*, 5 December 1993)

eccentric mad.

Literally, the word means 'not moving on a centrally placed axis', whence, of human behaviour, whimsical or unusual:

The poor man is crazy; the rich man is eccentric. (Old saw quoted in Sanders, 1977)

economic storm a slump.

Perhaps no more than circumlocution, but stockbrokers, economists, etc. go to great lengths to avoid any of the terms which might recall the events which happened between 1929 and 1935.

economical with the truth lying.

Said by the Secretary of the British Cabinet, Sir Robert Armstrong, in an Australian legal action in 1986. His presence, and the action, were the result of the determination of Mrs Thatcher to prevent the publication of confidential, inaccurate, and largely inconsequential allegations, in which she went against firm and wise advice. As a result she enriched the author, his lawyer, and the language. Also as *economical with the actualité*:

> Mr Clark admitted that he had been economical with the *actualité*. (*Sunday Telegraph*, 20 March 1994—Clark was a British minister who had approved the sale of lathes, described by the British media as *arms*, to Iraq after the Iran–Iraq war. Subsequently customs and excise officials, allegedly anxious to score off the Foreign Office, launched a vindictive and ill-advised prosecution against the directors of the Iraqi-owned British company involved, leading to its collapse, the loss of several hundred jobs, and a public inquiry.)

economically abused poor.

Those for whom inadequacy, misfortune, ill-health, or fecklessness lead to comparative poverty may also be described as *economically exploited* or *marginalized*.

economically inactive unemployed.

The actions of each of us impinge on the economy, whether or not we create wealth:

> Both men claimed there had been an unlawful interference with their rights as EU citizens when they became 'economically inactive'. (*Daily Telegraph*, 21 March 1995—a Portuguese man with three dependants and an Italian who had migrated to England, where they had both been supported at public expense without working, were claiming damages because they had been told to return to their countries of origin)

economy cheap.

Literally, the avoidance of waste, but does that mean airline travellers not in inferior seats are feckless? In supermarket jargon, *economy* is meant to mean 'cheap' but in fact means 'large' (see LARGE[2]).

écouteur a person who obtains aural gratification from the sexual activity of others.

Literally, the French word means 'a person who listens', but never a simple radio fan nor even an eavesdropper:

> The shrieking bed springs were no accident. The manager's wife was an écouteuse. (Condon, 1966—you will no doubt have already observed that I follow the literary convention of using the masculine case where either sex is imported)

ecstasy an illegal stimulant.

Easier to pronounce than methylene dioxy-methamphetamine.

edged slightly drunk.

The obsolete Suffolk dialect use was probably not the direct parent of the modern American, but both must have derived from being on the *edge of drunkenness* or some such phrase:

> When he was nicely edged he was a pretty good sort. (Chandler, 1934)

edie *obsolete* a prostitute.

From the woman's name. It denotes the cheaper type of London prostitute:

> The Edies of the East End, Piccadilly and the railway stations... (Gosling and Warner, 1960)

educable see EDUCATIONALLY SUBNORMAL.

education welfare manager a truancy officer.

It is no longer acceptable to recognize that children play truant through their own naughtiness or the inability of schools to interest or discipline them:

> The case was adjourned while the disease was investigated, despite objections from the local education welfare manager, as truancy officials are now called. (*Daily Telegraph*, 25 May 1994—the *disease* which made it impossible for the child to attend school was the newly identified 'School Phobia Syndrome')

educationally subnormal moronic.

The jargon of educationalists, often abbreviated to *ESN*, and in the process of being superseded by the dominant series of euphemisms coined on CHALLENGED. In fact the lower half of any grouping is subnormal, but every child who attends ordinary classes must be described as *normal*, however far below the norm his achievement may be. Similarly the American *educable* is often used to mean 'extremely stupid', with just a possibility of learning something from going to school.

effeminate (of a male) homosexual.

Literally, having the characteristics of a woman:

> She wondered for a moment if he might be what people called effeminate. (Follett, 1978)

effing see F.

effluent a noxious discharge.

Of sewage or industrial waste, but literally it means anything which flows out, including a trout stream. *Sewage* itself started life in this sense as a euphemism, from its original meaning 'draining off water'.

effusion *obsolete* an ejaculation of semen.

> The mere effusion of thy proper loins. (Shakespeare, *Measure for Measure*)

Literally, a pouring forth, which is why the word is often used of wine.

egg a bomb.

Laid by Second World War airmen, the technology, the word, and the imagery divorcing the perpetrator from the carnage.

elbow bender see BEND.

electric cure death by judicial electrocution.

Of those sentenced to the *chair* (see CHAIR (THE)) but not of those subjected to shock therapy to cure depression.

electric methods torture.

A refinement of Nazism:

> Bienecke used the 'electric methods' pioneered by the SD in France—not the sort of scientific advance to crow about. (Keneally, 1985, of Second World War German behaviour in Russia)

See also APPROPRIATE TECHNOLOGY.

electrical surveillance see SURVEILLANCE.

electronic counter-measures spying.

The phrase is used of both optical and electronic aids. The euphemism is in the *counter-measures* because, far from *countering* anything, you are invading another's privacy. If your operation remains undetected, you achieve *electronic penetration*.

elephant and castle the anus.

Rhyming slang for 'arsehole', from the area of south London named after a public house which stood at the start of the old London to Brighton road.

elephant's drunk.

From the rhyming slang *elephant's trunk*. To COP AN ELEPHANT'S is to become drunk. See also JUMBO.

elevate to make drunk.

Perhaps from the feeling of power or levitation experienced at some stage of inebriation. *Elevated* and *elevation* mean respectively 'drunk' and 'drunkenness'.

eliminate to kill.

A variant of LIQUIDATE with the same origin and the same connotations, usually of political or espionage killings:

> We will just have to eliminate him. No time. No publicity. (G. Greene, 1978)

An *elimination* is such a killing:

> Elimination is rather a new line for us. More in the KGB line or the CIA's. (ibid.)

embalmed very drunk.

The derivation is not from *barm*, froth, nor even from the punning *embalming fluid*, cheap whisky, but almost certainly from the lifeless condition of the subject.

embraces copulation.

Literally, clasping in the arms affectionately, with familial or sexual fervour:

> ...solicited the gratification of their taste for variety in my embraces. (Cleland, 1749)

A woman who *shares her embraces* copulates regularly with two or more males contemporaneously. *Illicit embraces* means 'extramarital copulation':

> Harold and Noreen must have been surprised again in their illicit embraces. (McCarthy, 1963)

You occasionally see *embrace* in the singular:

> When a girl's lips grow hot, her sex is hot first and she is ready to give herself and ripe for the embrace. (Harris, 1925)

embroider to exaggerate or invent.

The derivation is from ornamental needlework; the word is used to avoid direct accusation of lying. Raleigh accused the ancient Greek historians of 'embroidering and intermixing' fact and fiction although, as a historian himself, he should have known the way most history is written.

emergency a war.

Used by those who think the opposition is unworthy of them, as for example the British in the Malayan civil war against Chinese Communists; or those with a guilty conscience, such as the Dublin government in the Second World War:

> Not only must the war be referred to as 'the emergency' but nothing could be printed which could conceivably offend either side. (Fleming, 1965, describing the censorship imposed by de Valera, which was from time to time wittily circumvented by the Irish Times. He was to be the only head of state to present his formal condolences on Hitler's death.)

emergent poor and uncivilized.

The use is mainly of former colonial territories in Africa, some of which, far from *emerging*, are retreating into greater poverty and the tribal divisions which their colonial masters had tried to break down:

> To avoid embarrassing its trading partners in emergent Africa, South African officials and trade organizations will not disclose the destination of its £800m. annual food exports. (*Daily Telegraph*, October 1981)

Emerging is used in the same sense:

> Except for King Paul of Greece ... they came from the emerging nations. (Manchester, 1968, of Mali, Yemen, Nigeria, etc.)

emotional drunk.

Intoxication makes some people excitable or sentimental:

> Tired and emotional after a long flight from Australia ... (*Private Eye*, September 1981, of a drunken person)

employ *obsolete* to copulate with.

The man is the employer:

> Your tale must be, how he employ'd my mother. (Shakespeare, *King John*)

employment unemployment.
This is one of those evasive opposites like DEFENCE, HEALTH, etc. Thus a *Department of Employment* is concerned with finding jobs or providing for the unemployed.

empty your bladder to urinate.
Perhaps circumlocution rather than euphemism, except that it could as well refer to letting air out of an old-fashioned football:

> Go to the bathroom, empty your bladder. (McCarthy, 1963)

To *empty yourself* is to defecate:

> It was the period when some men ate, or read, or wrote home, or dozed, or just went to the lavatory and emptied themselves. (Forsyth, 1994)

enceinte pregnant.
It means 'surrounded'—and is also euphemistic—in French and so, when we use it, we are doubly evasive:

> The idea that Kay might be enceinte had stolen more than once into her quiet thoughts. (McCarthy, 1963)

end to kill.
The common scepticism about reincarnation:

> This sword hath ended him. (Shakespeare, *1 Henry IV*)

The *end of the road* may describe any situation after which there will be no further developments, including death:

> Cheeky servants and cunning poachers ceased to annoy the Rev. Francis in 1811, for that year he came to the end of the road. (Tyrrell, 1973)

There are other similar phrases, such as the American police *end of watch*:

> Knuckles Garrity went End-of-Watch forever in the old police station parking lot. (Wambaugh, 1975)

Endlösung see FINAL SOLUTION.

endowed see WELL ENDOWED.

energy release *American* an accidental escape of radioactive material.
An atomic power station should only release energy which is converted into electricity. This nuclear jargon seeks to play down the health hazards following the accidental release of radioactive materials.

enforcer a criminal who terrorizes under orders.
He may threaten or maim, to cow or punish illegally on behalf of a gang leader, bookmaker, etc.:

> She was a freelance enforcer, renowned for her skill in getting any job done quickly. (Collins, 1981, of one such)

A British municipal council's *enforcement officer* threatens you and seeks to punish you within the law if you fail to comply with any of the myriad regulations which now affect industrial life.

engine the penis.
Viewed sexually and a variant of MACHINE or TOOL:

> ...too much desirability can freeze a man's engine. (Keneally, 1985)

engineer a catch-all title to enhance status.
More American than British. Not illogically, it takes an engineer to drive a railway engine on one side of the ocean, a task performed by an engine-driver on the other. Found in many compounds, some accurate but many pretentious. See, for example, EXTERMINATING ENGINEER.

English denoting or pertaining to sexual deviation.
A usage not found in the British Isles but common in America to indicate bondage or masochism, which may be advertised as *English arts*, *assistance*, *discipline*, *guidance*, *lessons*, *treatment*, etc., none of which have anything to do with elocution or any other kind of instruction in the most elegant and versatile of languages. In the 19th century the *English vice* was not a piece of mechanical equipment secured to a bench but a predilection said to have developed from the experience of boys and their masters at single-sex boarding schools:

> The popularity of flagellation—known on the continent as the 'English vice'—created a large corpus of literature. (Pearsall, 1969)

This did not however deter the French from using the *capote anglaise* in heterosexual encounters.

English disease[1] male homosexuality.
Again, not a common usage in its country of origin:

> We call this thing a disease and sometimes the English disease. (Burgess, 1980—a New Yorker was talking about male homosexuality)

English disease[2] a propensity to strike.
Used both in England and of the British by foreigners whose knowledge of the other constituents of the United Kingdom is limited. It has become less common since the laws relating to trade unions were modified. See also SPANISH PRACTICES.

enhance to alter or increase in a surreptitious way.
Thus dye may *enhance* the latent blondeness of hair; an *enhanced radiation weapon* is a neutron bomb, not a sun lamp; *enhanced contouring* is cosmetic padding:

> ...her bra comes with 'built-in emphasis' or 'enhanced contouring'. (Jennings, 1965)

Revenue enhancement is the raising of taxes by politicians who prefer that option to frugality in state expenditure.

enjoy to copulate with.

Usually of the male, from the days when the pleasure was supposed to be his alone:

> You shall, if you will, enjoy Ford's wife. (Shakespeare, *Merry Wives of Windsor*)

Denoting copulation in general:

> He felt entitled to enjoy a woman (Follett, 1978)

but a man may also *enjoy favours* (see FAVOURS):

> He regularly visited a famous courtesan in the Srinagar bazaar and enjoyed other favours too (Masters, 1976)

or *enjoy hospitality*:

> The scandal mags said Kennedy, quote, Enjoyed her hospitality, unquote (Sanders, 1977)

but you would be old-fashioned if you *enjoyed her person*:

> ... prostituted for some time to old men, who paid a high price for the enjoyment of her person. (Mayhew, 1862)

To *enjoy yourself* is to masturbate:

> I was not the only European officer in the jungle who enjoyed himself secretly on occasion. (Barber, 1981)

An *enjoyed* female is one who has copulated, whether or not her partner found it pleasurable:

> After Mrs Mayhew when I was seventeen, no mature woman who had been enjoyed attracted me physically. (Harris, 1925)

enjoy a drink to be a drunkard.

You may also be said to *enjoy* a *glass, drop, jug, nip, the bottle*, etc.

enjoy Her Majesty's hospitality to be in prison.

In jail you do not have to pay for your keep. The phrase is suitably adjusted for kings and presidents.

enlist the aid of science to undergo cosmetic surgery.

The 'scientist' removes wrinkles, causes superfluous hair to vanish, or implants it where it is scarce, etc.:

> A few years ago when my hair began to recede I enlisted the aid of science. (Murdoch, 1978)

entanglement an embarrassing or clandestine association.

Of an ill-advised sexual relationship, an unpublished business relationship, or something which politicians would rather remained secret:

> Mr Hurd sought to extricate Lady Thatcher and other ministers from responsibility for the 'temporary and incorrect entanglement' of arms and aid in a protocol signed by Lord Younger. (*Daily Telegraph*, 3 March 1994—the British government had agreed to give money to the Malayan government to fund a civil engineering project of doubtful economic worth, while at the same time the Malayans bought arms from Britain)

enter to copulate with.

Of the male:

> She let out her breath in a long quavering moan as he entered her. (Masters, 1976)

The entry of a tongue into the mouth of another is called *French kissing*.

enter the next world to die.

In various forms, indicating devout belief or scepticism:

> It was better to enter the next world with a full belly. (Richards, 1933)
>
> ... within a month or so I shall have entered the great 'Perhaps', as Danton I think called 'the undiscovered country'. (Harris, 1925)

entertain[1] *American* to beat a prisoner violently.

The 'hospitality' is shown to coerce those in police custody:

> The more they protested the more they were *entertained*. (Lavine, 1930)

entertain[2] to copulate with outside marriage.

Another way, I suppose, of keeping a man interested or amused:

> She had 'entertained' him before and each time he had nearly ripped her in half. (Collins, 1981)

entertain[3] to bribe.

Commercial use, denoting excessive prodigality, such as the British practice of paying for surgeons and their wives to attend 'conferences' in holiday playgrounds, so long as they continue to insist on the sole use of your product in their operating theatre. *Entertainment* is such bribery.

equipment a word used to cover up and evade in any matter that is the subject of a taboo.

Sexually it may refer to the breasts or vagina of a female, the genitalia of a male; when used of illegal narcotics it denotes a supply or the means of introducing it into the body; in the airline industry it is something which goes wrong when your flight is delayed because of *unserviceable equipment*—nobody must suggest there is something wrong with the flying machine.

equity retreat a stock market fall.

Almost any phrase will do so long as it does not suggest another crash. The American *equity equivalent contingent participation* is the process by which a lending bank seeks to get round the law that, if it lends money, it cannot obtain a share of the profits arising from an increase in the price at which the borrower's stock is traded:

Our interest wouldn't be in stock, of course Glass-Steagall rules that out. It'd be what they call 'an equity equivalent contingent participation'. (M. Thomas, 1987)

erase to kill.

A version of *rub out*, which uses the same imagery:

I'd have hired a drunken lorry driver and had her erased on a zebra crossing. (Sharpe, 1977)

erection an enlargement of the penis due to sexual excitement.

Literally, the condition of being upright. Standard English of both buildings and penises:

...his toilet closet choc full of Japanese erection lozenges and love elixirs. (Ustinov, 1971)

Whence *erect*, having such an enlargement:

He had woken erect himself. (P. Scott, 1975—he had not been sleeping standing up)

err to copulate outside marriage.

Literally, to stray or wander, whence to sin generally and then specifically of copulation:

That every woe a tear can claim,
Except an erring sister's shame.
(Farrell, 1973, quoting from a 19th-century verse)

Whence *errant*, so engaged:

...serving legal papers on reluctant defendants and following errant wives (Deighton, 1981, of a private detective)

and the crossword clue:

Where to find errant pairs. (*errant* also signalling an anagram of pairs)

escort a paid heterosexual partner.

Originally, a body of armed men, whence one person accompanying another. Usually in this sense a female who also, on payment of a further fee, reveals herself as a prostitute:

One was a persistent 'escort' of Arabs. (*Private Eye*, July 1981, of a young woman—the inverted commas indicate prostitution)

An *escort agency* provides such people:

But escort agency meant hookers for hire. (Theroux, 1982)

essence semen.

From the meanings 'the essential being' and 'what is left after distillation':

I want to drink your essence and I will. (Harris, 1925, referring to semen)

essentials the male reproductive organs.

The brain, heart, or liver assume less importance:

...once your essentials are properly trapped in the mangle there's nothing to do but holler. (Fraser, 1985)

estate a grave.

The last property you occupy, in funeral jargon:

...the section of the cemetery, where your family estate is located. (J. Mitford, 1963)

eternal life death.

A monumental favourite, although those who order the inscription are seldom anxious personally to put their faith to the test. The obsolete *eternity box*, a coffin, represented a more pragmatic approach.

ethical investment a policy of purporting to buy only those stocks which do not offend the prejudices of dogmatists.

The morals of those doing the investment or of the managers of the firms in which they invest are not taken into account:

The latest craze to be imported from America is for 'ethical investment'. Almost every week there seems to be a new unit trust launched which promises to invest your money only in 'socially screened' firms. (*Daily Telegraph*, 25 September 1987)

ethnic not of exclusively white ancestry.

The commonest euphemism of the 1980s and 90s, with *ethnic minority* coming to mean black or Hispanic people according to where in America or the British Isles you lived:

The car had been stolen the previous night from outside a block of high-rise apartments in Brixton chosen because of its ethnic inhabitants. (B. Forbes, 1986—Brixton has become a London ghetto)

The word *ethnic* means 'pertaining to nations not Christian or Jewish' (OED), from which anyone who wasn't a Christian or a Jew. As the practice of these religions was largely confined to Europe or those of European descent elsewhere, *ethnic* came to mean those not of that skin pigmentation.

ethnic cleansing the elimination of inhabitants of a race or religion different from your own.

Practised first in Bosnia and Croatia by the Serbs after the break-up of Yugoslavia, but then imitated by their enemies. The favoured methods were murder, starvation, expulsion, forcible confinement, and rape:

But the Baijo-Lukans sponsored some of the most savage ethnic cleansing of the Bosnian conflict. (*Sunday Telegraph*, 17 January 1993)

ethnic loading making appointments for reasons other than those of suitability or qualification.

A way of achieving a QUOTA:

America's problem is that its 'intellectual elite' is now chosen by a system of positive discrimination and ethnic loading. (A. Waugh, *Daily Telegraph*, 10 April 1995)

Eumenides the Furies.

The Greek word means *kindly ones*, which the Greeks called them because they might get even more angry if their real attributes were mentioned. Similarly they called the stormy and fearsome Black Sea the *Euxine*, the hospitable. Not so long ago we tried to appease the devil and the fairies in the same way—see GOOD MAN, GOOD FOLK—and some Christian prayers to an all-powerful and avenging God make strange reading.

evacuate to defecate.

Medical jargon and a shortened form of *evacuate your bowel*. Whence *evacuation*, defecation:

> ... supported the dysentery cases as they trembled and shuddered during their burning evacuations. (Boyd, 1982)

evacuation[1] see EVACUATE.

evacuation[2] mass murder.

The Nazi *Aussiedlung*, under which the victims were taken from their homes for forced labour or murder, usually in the EAST.

evasion a lie.

More than an avoidance of the truth:

> I should say she indulged in certain evasions. (Styron, 1976, of a liar)

Eve a female.

Especially viewed sexually outside marriage:

> ... a local 'Eve-teasing' problem. The sexual harassment of women in public places, sometimes quite open, was a problem all over India. (Naipaul, 1990)

You may also see *Eve* as an indication of sex on a lavatory door, with the corresponding ADAM. For *Eve's custom-house* see ADAM'S ARSENAL.

evening of your days old age.

The common comparison of a lifetime to a single day:

> ... his mother came to reside with him for the evening of her days. (Tyrrell, 1973)

Whence the *eventide home*, an institution for geriatrics.

everlasting life death.

Another trustful monumental favourite. The words are sometimes inverted in the verse of *Hymns Ancient and Modern* and in the poetry of Cranmer, lately rejected by the Church of England.

everlasting staircase a treadmill.

> The convicts' names for [the treadmill] were expressive: the everlasting staircase, or, because the stiff prison clothes scraped their groin raw after a few hours on it, the cockchafer. (Hughes, 1987)

A screw might be turned in the device by a jailer to make the operation more arduous, which enriched the language and gave Henry James a title. See also COCKCHAFER[1].

ewe mutton an older woman who affects the style of a younger.

Derogatory female use of another. I include it to give the obsolete British meaning 'an elderly prostitute'.

exceptional failing to achieve or differing from the norm.

Of a stupid child in educational jargon, an imbecile or a paraplegic or lame person in adult use. Literally, it can and does also mean 'exceptionally good'.

excess *American* to dismiss from employment.

Used when the employer wants to cut costs by getting rid of *excess* labour:

> Workers are never laid off; they're 'redundant,' 'excessed,' 'transitioned,' or offered 'voluntary severance'. (*Wall Street Journal*, 13 April 1990, quoted in *English Today*, April 1991)

To *excess* something is also to charge extra for it, referring for example to overweight luggage on an air flight.

exchange flesh *obsolete* to copulate.

As distinct from swapping titbits of meat. This may be no more than Shakespeare's vivid imagery:

> She would not exchange flesh with one that loved her. (*Winter's Tale*)

exchange of views a disagreement between dogmatically opposed parties.

Mainly diplomatic use. Qualifications such as *cordial* or *helpful* do not indicate greater amity nor is an *exchange of ideas* more propitious.

exchange this life for a better to die.

A perhaps obsolete monumental message:

> After a long illness which she bore without a murmur exchanged this life for a better on the 23rd day of March, 1815. (Monument in Bath Abbey)

execute to murder.

Literally, to carry out any task, whence to effect the sentence of a court, especially a death sentence. It became standard English for beheading and today terrorists have adopted the word to try to cloak their killings with legality:

> 'The execution of hostages will begin then.' 'Execution.' She was using the jargon of legality. (W. Smith, 1979, of a terrorist)

In the same way *executive action* is CIA jargon for an unlawful killing.

executive measure a political murder.

Yet another of the Second World War Nazi evasions:

'Lohse, I recommend that your office initiate an executive measure aimed at Oberführer Willi Ganz'... *Executivmassnahme*, a classic 'soft word' whose intent can be convincingly denied long after the corpses are counted. (Keneally, 1985)

exemplary punishment death by hanging.

Not being made an example by having to stand in the corner for a few minutes:

Few people want to take direct responsibility for hanging; understandably they prefer abstractions—'course of justice', 'debt to society', 'exemplary punishment'—to the concrete fact of a terrified stranger choking and pissing at the end of a rope. (Hughes, 1987)

exercise your marital rights see MARITAL RIGHTS.

exhibit yourself to show your penis to a stranger in a public place.

A persistent form of male sexual gratification, the erect penis being displayed to women or children:

...a wealthy old man charged...with exhibiting himself to toddlers. (Sanders, 1973)

To *make an exhibition of yourself* is merely to behave stupidly.

exhibition a pornographic display.

Literally, a showing of anything, from fine art to kindergarten basketwork:

The card had half a dozen choices on it: blue movies; girls; boys; exhibition; massage. (Theroux, 1973)

The Spanish *exhibición* is an American use for a trip into Mexico to see pornographic items which would be barred north of the border.

expectant pregnant.

A shortened form of *expectant mother*, who is said to be *expecting*. We take for granted that what she *expects* is the birth of a baby to herself, and not a birthday present or an increase in salary.

expedient demise an unlawful killing by a government agency.

The pretence is that the death was natural—a *demise*—but timely:

You had to give orders for the expedient demise of two men. (Deighton, 1981—he called the book *XPD*)

expended killed.

This use treats soldiers like ammunition:

'And what do you mean about me being expended'...'He has wanted to kill you.' (L. Thomas, 1978—this is a rare non-military example)

Expendable is the number of soldiers you can afford to have killed or wounded in a battle:

'You're what they call "expendable".' Clark nodded with sad honesty. (ibid.)

expenses an additional tax-free income.

In standard usage, the amount incurred by an employee in the course of his work and reimbursed by the employer. Too often the disbursement has not been made as claimed, or at all. *Expense account* living is the profligacy which results either from payment in respect of fraudulent claims or from an employee's extravagance when he can charge what he spends to his employer's account.

experienced having copulated.

Of either sex:

Stephanie too was 'experienced'. Whatever had it been like for her with all those men? (Murdoch, 1977)

Whereas, in most disciplines, to gain *experience* you must practise often and become adept, in this activity a single essay may be enough.

Of an American motor vehicle, it means no more than that it is not new.

expert a person who makes a living by professing knowledge.

Others often find a claim of omniscience spurious:

The directorate at ARCOS was topheavy with so-called 'experts'. (Boyle, 1979)

An *expert witness* is a person with impressive qualifications who is paid to support on oath the case of a litigant.

expire to die.

To breathe out, but for the last time:

As to other euphemisms—of words which connote death...'expire' for 'die'. (J. Mitford, 1963)

Expire in an obsolete doubly euphemistic use meant to reach a sexual orgasm—see DIE:

When both press on, both murmur, both expire. (Dryden)

expletive deleted an obscenity.

Part of the linguistic debt to Richard Nixon, and perhaps also to Rose Mary Woods who transcribed the tapes:

Suddenly hearing that his words were being overheard by newsmen, Thompson ended with a grin and the words 'expletive deleted'. (Hackett, 1978)

The transcriptions also used *adjective deleted* and *characterization deleted*, neither of which has passed into the language:

I will never forget when I heard about this (adjective deleted) forced entry and bugging. (Colodny and Gettlin, 1991, quoting Nixon from a tape of 18 February 1973)

What in (characterization deleted) did (Segretti) do? (ibid.—Nixon again on the same tape)

expose yourself to show your penis to a stranger in a public place.

A British variant of EXHIBIT YOURSELF but found in America also:

> He... had rung the doorbell and introduced himself to Stacie, then had exposed himself. (Condon, 1966)

Exposure and the legal jargon *indecent exposure* are such a display.

extended dull.

This American educational jargon seems to be based on the premise that simple work which those so described find difficult *extends* their faculties, as indeed it may.

exterminating engineer a controller of pests or vermin.

I give this American sample to illustrate the popular pastime of upgrading our job description to gratify our self-esteem, and that of our spouses. Logically, this particular ENGINEER might be in the process of personal dissolution, and even if we accept that he is *exterminating* something else, the choice is large. The British *rodent operator* is no less pretentious and illogical —might he not provide performing shrews for a circus? Mencken (*The American Language*) gives many examples of this restyling of jobs, some of which are ephemeral, some circumlocutory, and a few euphemistic.

extra-curricular activity extramarital copulation.

Literally, anything at school, college, etc. which is done in addition to the prescribed course of study. *Extra-curricular sex* is more specific:

> An opportunity for extra-curricular sex occurred, and he hadn't fought it. (Hailey, 1979)

extra-mural sexual activity outside marriage.

Literally, outside a normal course of study. The verbal use is rare:

> Besides she's always liked to extra-mural a bit. (Bradbury, 1983, of a promiscuous wife)

extramarital excursion a single case of copulation outside marriage.

It might be, but is not, a skittles tour with the lads or a day at the seaside with the Mothers' Union:

> ... similar situations—in reverse—when *he* returned from extra-marital excursions. (Hailey, 1979)

extras bought sexual gratification.

The service provided is usually masturbation or copulation in a brothel which calls itself a *massage parlour*:

> Mr Bircher admitted giving the service with 'extras' on request, consisting of acts of masturbation by him and his wife. Basic massage was £15. Erotic massage cost £20. (*Daily Telegraph*, January 1984, of a Cornish MASSAGE PARLOUR)

extreme physical duress torture.

Perhaps circumlocution rather than euphemism:

> Tell him he will be interrogated under extreme physical duress. (Hall, 1979)

extreme prejudice see TERMINATE[1].

extremely ill under sentence of death.

The coded public language of the Chinese Communist autocrats:

> ... if a high official is said to have a cold he's likely to be fired, if he's 'convalescing' he has been exiled and if he is 'extremely ill' he is about to be murdered. (Theroux, 1988)

extremely sensitive source an illegal interception of messages.

Not praising the quality of the equipment used:

> ... being careful not to mention the phrase wiretapping, but using instead the standard cover language, 'extremely sensitive source'. (Colodny and Gettlin, 1991, of the Watergate fiasco)

eye the anus.

Male homosexual use, either *tout court* or in a compound, as BRONZE EYE. An *eyeball palace* is an American male homosexual bar.

eye-candy *American* a nubile young woman.

Good-looking and by implication sexually promiscuous:

> I have this gorgeous stick of eye-candy (LA-speak for glamourpuss) that absolutely *nobody* knows about. You want her number? (*Daily Telegraph*, 6 December 1994)

eye in the sky a police helicopter.

This American Citizens' Band expression is used to warn other truckers of possible speed assessment by police working from that vantage point.

eye-opener an intoxicant or stimulant taken on waking.

Punning on the meaning 'a surprise':

> A morning eye-opener (brandy, Scotch or whatever) would also be provided. (Sanders, 1980)

The usage seems to have originated in France during the First World War, especially but surprisingly among airmen. Now generally used by people addicted to alcohol or drugs who need topping up before they can face another day.

F

F fuck.

Nearly always for the verb. Also as *the F word*:

The 'f' word was broadcast on Radio 4 yesterday. (*Daily Telegraph*, 12 January 1995)

The common *effing* is only used figuratively:

It wasn't a case of where's my effing breakfast. (Allen, 1975)

FA see FANNY ADAMS.

facile sexually complaisant.

Of women and in one of its senses a synonym of *easy* (see at EASY WOMAN):

... he soon made the acquaintance of Mme de Warens, a woman of facile morals. (Boyd, 1987)

facility[1] a lavatory.

Literally, anything which makes a performance easier:

A small outdoor facility and the forest. (Poyer, 1978, describing a chalet on the edge of a village)

Often seen in the plural, despite there being only one:

... containing a washbasin, a folding table and two seats, one of which contained what the timetable coyly called 'facilities'. (Francis, 1988, of a compartment in a railway carriage)

facility[2] an agreement to lend money by a bank.

Banking jargon of the limit to which you may borrow. I suppose it makes life easier, for a while.

fact-finding mission a holiday with salary and expenses paid.

The participants, who may well be politicians, tend to seek out the facts in distant and agreeable places:

But it was hard to suppress the thought that the final touch was provided by a be-suited Commons Select Committee junket (sorry: *fact-finding mission*) to France. (*Daily Telegraph*, 18 April 1995)

Also as *fact-finding tour*.

facts (of life) the human process of reproduction.

Thus breathing, eating, and growing old are not the *facts of life*, while conception, pregnancy, menstruation, birth, etc. are:

I sometimes think your children are right and you don't know the facts of life. (N. Mitford, 1949)

Sometimes shortened to *the facts*:

Linda's presentation of the facts had been so gruesome that... their future chances of a sane and happy sex life [were] much reduced. (N. Mitford, 1945)

A *fact of life* is an unpalatable truth.

fade away to die.

Especially of former soldiers:

Frank wrote to me regularly until he faded away, in 1961. (Graves, in an introduction to a reprint of Richards, 1933)

To *fade* is underworld slang for 'to kill':

'You fade him?' 'Not me. I just found him as he was.' (Lyall, 1965, of a corpse)

fag a male homosexual.

Probably from the fact that male cigarette—or *fag*—smokers were in the 1920s thought effeminate by pipe and cigar smokers:

An eager young fag, very pert in urchin cut and ear-rings, had accosted him. (Davidson, 1978)

I think the longer form, *faggot*, comes from *fag* in the way that *pooftah* comes from *pouff*, and this despite the attractions or otherwise of *faggot*, a wretched old woman, from which came the obsolete British use, to copulate with a prostitute.

faggot see FAG.

fail to win to lose.

The excuse of the pusillanimous Unionist General McClellan in the American Civil War:

McClellan insisted that he had not lost; he had merely '*failed to win*' only because overpowered by superior numbers. (G. C. Ward, 1990—in fact the numbers opposing him were inferior, but better led)

fair[1] poor.

A classification denoting scholastic performance or quality of goods and services which is just above the lowest rating or outright rejection. It should mean favourable, or at least halfway between good and bad.

fair[2] unfair.

One of the opposites so loved by politicians. Thus the British term for a rent controlled below the open market rate was a *fair* rent:

Their regulated rent (euphemistically called a 'fair rent' by law) would buy dinner for one at a local restaurant. (*Private Eye*, July 1981)

See also DEFENCE, HEALTH, and LIFE INSURANCE.

fair-haired boy *American* someone unfairly favoured.

He may be dark-haired or bald, but he is being helped to political office or other promotion beyond his deserts:

Alexandrov's too old to go after the post himself... Gerasimov's his fair-haired boy. (Clancy, 1988)

fair lady *obsolete* a sexual mistress.

But not necessarily a blonde or free from prejudice.

fair trader a smuggler.

Paying no excise duty, he charged his customers less:

> I am what is called a fair trader—in other words a smuggler. (Pae, 1884)

fairy a male homosexual.

Usually denoting a male taking the female role, but used collectively without such connotation:

> A mob of howling fairies, frenzied because the best part went to younger stars who didn't lisp. (Theroux, 1976)

The imagery is from the modern Christmas pantomime concept of fairies, who used to be nasty creatures. A *fairy lady* is a female homosexual taking the female role.

faithful not copulating with anyone other than your regular sexual partner.

Usually of such behaviour within marriage. Literally, true to your word or belief, but in this sense true to merely one of the marriage vows:

> He loved his beautiful wife and, so far as I know, was faithful to her. (Murdoch, 1978)

fall[1] to copulate outside marriage.

The imagery is from *falling from grace*:

> It is their husbands' faults,
> If wives do fall.
> (Shakespeare, *Othello*)

Less often as a noun:

> The Queen was convinced that what she called 'Bertie's fall' was at least in part responsible for Prince Albert's death. (Massie, 1992—Bertie had fallen in, with, on, and for Nellie Clifton, who had been introduced to his bed by fellow officers in camp in Ireland. He was thereafter reluctant to eschew the delights revealed to him.)

In modern slang, if you *fall for* someone, you become besotted with them.

fall[2] to become pregnant.

A common modern use, which may have come from FALL[1]. Of pregnancy within or outside marriage:

> Annabel Birley has fallen again and delivered another (legitimate) Goldsmith into the world. (A. Waugh, *Private Eye*, 1980)

To *fall for a child* or to *fall in the family way* both suggest pregnancy outside marriage:

> The girl fell in the family way and was sent out of the house. (Mayhew, 1862)

Fall wrong to, of unmarried pregnancy only, is obsolete:

> There was a lass... who fell wrong to a farmer's son where she had been serving, and he wouldn't marry her. (Saxon, 1878)

To *fall pregnant* is specific:

> ...one of the Emalia girls fell pregnant, pregnancy being, of course, an immediate ticket to Auschwitz. (Keneally, 1982)

fall[3] to die.

On military service, if not in battle, from being hit by a bullet etc.:

> John Cornford had fallen the day after his coming of age. (Boyle, 1979)

In Hitler's case, it covered suicide but the intention was to show he died fighting:

> Adolf Hitler fell in his command post in the Reich Chancellery. (official German announcement of Hitler's death, 1 May 1945, in translation)

Both soldiers and civilians *fall asleep*:

> ...fell asleep in Jesus... of enteric fever in Mesopotamia. (memorial in West Monkton church to First World War soldier)

You may *fall off the hooks* or *the perch* as easily as you may *drop*—see DROP[3]:

> If the excitement of sharing a bedroom with a shapely lass should cause Fred to fall off the perch... (*Sunday Express*, March 1980)

The American *fall out* probably employs military imagery. The *fallen* are those who have died on military service, unless they be *fallen women* (see FALLEN WOMAN). There was an obsolete southern English use of *fall* which conversely meant to be born, of animals:

> The calf is lately fell. (Ellis, 1750)

Fall about, to be ready to give birth, is still occasionally used of humans.

fall[4] to be sentenced to prison.

I suppose from the reversal of fortune:

> I want you to follow my instructions when the case is tried, and if I fall I will find no fault with you. (Moore, 1893)

A *fall* is a prison sentence. An American *fall*, apart from being the British autumn, is an arrest and if you think that is likely, you may keep your *fall money* by you, to pay for a lawyer, put up bail, bribe the police, etc.

fall among thieves to admit to intoxication.

Males use this biblical reference to explain to their wives what they suggest is untypical behaviour attributable to the wiles of others. Less often they may admit to *falling among friends*:

> ...'the Fleetsh all lit up' commentary by Cdr Tommy Woodrouffe, who had lately fallen among friends. (*Daily Telegraph*, June 1990, in the obituary of the officer who had arranged the lighting for the Spithead Coronation Review of 1937 which is remembered, if at all, for the Commander's inebriated radio commentary)

Mainly humorous use.

fall asleep see FALL[3].

fall off the back of a lorry to be stolen.

Now specifically of stolen goods sold below market value in public houses etc.:

> You wouldn't believe what I paid for them. Fell off the back of a lorry. (Theroux, 1976

—and even then it would be stealing by finding)

Valuables also *fall off the back of wagons, trucks,* and even, in the Second World War, *half-tracks*:

Scotch, which is said to have fallen off the back of an American half-track. (Price, 1978, referring to the Second World War)

fall off the hooks see FALL[3].

fall off the roof to start menstruating.

Commonly in American female use, and as *I fell off,* I started menstruating, but the etymology escapes me, unless it implies a wound from falling.

fall out to die. See FALL[3].

fall out of bed to fail commercially.

This American phrase indicates failure after some carelessness:

But if Seaco fell out of bed, or the bond market cracked. (M. Thomas, 1982, referring to a failing corporation)

fallen (the) see FALL[3].

fallen woman a female who has been caught in extramarital copulation.

From FALL[1] and normally, but not necessarily, of a prostitute:

Let's face it dear, we are nothing but two fallen women. (N. Mitford, 1949)

You must remember to choose your words with care when a lady trips over her skis or her shoelaces.

falling evil epilepsy.

Evil in England, *sickness* in Scotland, but Webster gives neither. The *falling* is one of the symptoms:

To cure the falling sickness wi' pills o' pouthered puddocks. (Service, 1887—*puddock* does not here have its normal meaning, 'a kite or buzzard', but is a corruption of *paddock*, a frog or toad)

fallout radioactive matter introduced into the atmosphere by human agency.

Now standard English and no longer used of seemingly innocuous substances such as volcanic ash, the most likely proximate cause of the next Ice Age.

false engaging in copulation outside marriage.

The opposite of TRUE:

False to his bed. (Shakespeare, *Cymbeline*)

Of either sex.

false market improper rigging of share prices.

Stock exchange jargon, especially where someone buys or sells using advance knowledge. If, however, the information reaches brokers or professional investors first, to the detriment of the general public, the phenomenon is called *normal market intelligence*.

falsies pads concealed under clothing for females.

Mainly used of padding breasts and thighs to make them more alluring, but today seldom used of either. The padding of men's jackets, to make the wearer look broad-shouldered, is almost universal, but not the subject of euphemism or derogatory comment.

familiar with copulating with extramaritally.

The adjective *familiar* originally meant 'relating to your family', whence it was used of someone with whom you associated freely. It may refer to either sex:

The intimation is that you have been indecorously familiar with his sister. (Jennings, 1965—was he bad-mannered in his copulation?)

Over-familiar usually describes a male's rejected sexual approach to a female.

family[1] not pornographic.

Not as modern as we might think—Bowdler called his emasculation of the Bard *The Family Shakespeare*. Used of entertainment, with the assumption that an individual can see without being corrupted what the family cannot. A *family show* is one in which the vulgarity may be muted.

family[2] the Mafia.

It ain't gonna be easy now, keeping the Feds and the Family from tumblin' on to me. (Diehl, 1978)

Again nothing is new—in 18th-century England a *family* was an association of thieves.

family jewels see JEWELS.

family planning contraception.

This standard English use in fact denotes the reverse of planning a family for most people most of the time, especially for the unmarried. In many compounds, such as *family planning requisites*, contraceptives.

family way see IN THE FAMILY WAY.

fan club[1] a group or person who clandestinely follows someone.

From those who unthinkingly adulate and follow an entertainer:

...he chose a zigzag route to throw off any fan club. (Lyall, 1982, of someone who suspected that he was being followed)

fan club[2] people copying the actions of another.

Stock market jargon, where the following may be of a successful manager or of an investor, especially if there is a suspicion that the investor enjoys inside knowledge:

> While there is a distinction between a legal 'fan club' and an illegal market support operation, the black and white turns to grey when the 'fans' were selling Guinness 'short'. (*Private Eye*, August 1989)

fancy to desire sexually.

Either sex may *fancy* the other:

> You can't do it to an ordinary woman just because you fancied her at school. (Murdoch, 1978)

A usage dating from the 18th century, and predating the 19th-century meaning of the noun *fancy*, 'a girl's suitor':

> Crokey and lawn tennis for't young misses and their fancies. (*Weekly Telegraph*, 1894, quoted in EDD)

In ante-bellum America a southern *fancy* was a black female slave with some white blood:

> These yellow wenches ... being graceful delicate creatures of the kind they called 'fancy pieces', for use as domestic slaves. (Fraser, 1971, writing of the early 19th century—they were also used in brothels)

A *fancy man* is someone with whom a woman regularly copulates outside marriage:

> I can only remember two of them that had regular fancy-men. (Richards, 1936, of soldiers' wives)

Today a *fancy piece* or *bit* is synonymous with *fancy woman*, a man's sexual mistress:

> They supposed that Donald must be keeping 'a fancy woman' in New York. (Boyle, 1979—in fact Maclean was keeping a rendezvous with his Russian spymaster)

A *fancy seat cover* in American Citizens' Band use is a sexually attractive woman spied in a car.

fanny the vagina or buttocks.

Asexually of the buttocks, it can refer to male or female, as in the expression *sitting on your fanny*. Sexually, it can refer to the vagina:

> She'd have your fanny for a dishcloth (Sharpe, 1977)

or to a woman who looks ready for copulation:

> Great fanny, the wife of the KGB Captain. (Seymour, 1982)

Although derivation from a shortened form of *fantail* has its advocates, it probably comes from Cleland's *Memoirs of a Woman of Pleasure*, which relates the adventures of Frances (Fanny) Hill. The descriptions of her life as a prostitute in 18th-century London are almost Shakespearean in the fertility of their sexual imagery. Cleland made his heroine live happily ever after, and that alone is worthy of note. He would rejoice to know that the Sybil Brand Institute, a women's prison on rising ground in Los Angeles, is popularly known as *Fanny Hill*.

Fanny Adams nothing.

She was murdered in 1810 but her memory was kept alive in naval slang for tinned meat. The meaning comes from the shared initials with *fuck all*. (I have known two ladies called *Fanny Adams*, one who married a Mr Adams and the other with careless parents.) Also as *sweet Fanny Adams*, and often abbreviated to *FA*, or *sweet FA*.

far from staunch cowardly.

> I would inevitably learn later, that some Americans had been far from staunch. (Hastings, 1987, quoting the British General Mansergh on the Korean campaign)

An example of the euphemistic use of understatement.

Farmer Giles haemorrhoids.

Rhyming slang for 'piles'.

fast ready to copulate casually.

Mainly of women, from the meaning 'high-living':

> Anglo-Indians (regarded as 'fast') swinging their bums. (Theroux, 1973)

fast buck (a) money obtained unscrupulously.

The dollars come quickly and easily, and not necessarily dishonestly. Perhaps punning on the stag, which is fleet of foot, but perhaps not. The expression is also used where the unit of currency is other than the dollar.

fat cat a person who prospers, especially through the exploitation of a senior position.

Usually of politicians, professional men, and civil servants. The ingredients are success, sleekness, and self-satisfaction—actual purring is not expected:

> There's a fat cat called ... who used to be in very big with Heath and who now floats round the City. (*Private Eye*, November 1980, referring to a British politician)

fate worse than death unsought extramarital copulation by a woman.

A pre-Second World War use, with some reference to the convention that women should be virgins when they married:

> So being rattled stupid by Solomon would be no fate worse than death to her. (Fraser, 1977)

Still used humorously.

father of lies the devil.

Dysphemism rather than euphemism, from Satan's being credited with the invention of lying:

> Terry Reeves believed this fantastical personage to be the Father of Lies himself. (Graves, 1941, writing in 18th-century style)

fatigue mental illness.

In medical jargon, *mental fatigue* has in places replaced NERVOUS BREAKDOWN. The First World

War SHELL SHOCK became BATTLE FATIGUE in the Second World War.

fatigued drunk.

A rarer version of TIRED[2].

favour to copulate with extramaritally.

A form of Dr Johnson's *regarding with kindness*, I suppose, without some of the overtones of FAVOURS:

> He thanks our transport lady whom Mr Muspole claims to have favoured in the snooker room. (le Carré, 1986—he did not give her an easy break)

favours copulation outside marriage with consent.

Distributed, given, granted, sold, etc. by whichever sexual partner is in the subservient role:

> The small luxuries of life that plenty of women were prepared to exchange their favours for (G. Greene, 1978)

and of males:

> A fondness amounting to sexual mania for the favours of young men. (Sharpe, 1977)

To *force favours from* is to rape:

> But even so he forced his favours from her. (Keneally, 1987—or should it have been 'her favours'?)

feather-bed to grant excessive indulgence towards.

From the warmth of bedding so stuffed. Holt, who is usually right (or, as with Fowler, I think he is because I always agree with him), suggests in *Phrase and Word Origins* that the derivation is from the Rock Island Railroad; when train crews complained of hard bunks, they were asked sarcastically if they wanted feather beds. Today it is also trade union jargon for forcing an employer to pay unnecessary labour, which may enable all to work less hard. 'Feather-bed' Evans was a British post-Second World War Labour minister who used the term to refer to farmers, winning both immortality and dismissal from office—if you are going to be outspoken in politics, you must also be wrong. In obsolete British use a *feather-bed soldier* was one who went whoring a lot.

feather your nest to provide for yourself at the expense of others.

Now standard English, from avian imagery. You can either do it by dishonesty:

> Mr Badman had well feathered his nest with other men's goods and money (Bunyan, 1680)

or through unprincipled self-enrichment:

> They have planned Germany's subjugation with an eye to feathering their own nest. (Goebbels, 1945, in translation, of the British)

Formerly, a man who married a rich widow was also so described.

featuring having in the cast.

A *feature player* is one with a leading role. In the entertainment business, from which this usage comes, your billing may be more important than the money you receive or the quality of your performance. Few laymen appreciate the niceties of distinctions among *starring, co-starring, guest-starring, also starring, introducing, featuring*, etc.

feed to suckle.

You avoid mentioning the taboo breasts:

> Louisa was feeding her second baby in Scotland. (N. Mitford, 1945)

Not to feed a baby does not mean that you condemn it to death by starvation.

feed a dog to urinate.

A male expression, and he may also say he is going to feed a *horse, parrot*, or even a *goldfish*.

feed a slug to kill by shooting.

The *slug* is a bullet (see SLUG[1]):

> ...rubbing his greasy hair, and then feeding him a slug while he was still purring. (Chandler, 1943)

Feed a pill is rarer:

> I want to make certain that both you and your friend feed Danny Boy the pills. (Sanders, 1973—two people were to be implicated not in medical care but in a shooting)

feed from home to copulate casually.

Perhaps just another Shakespearean image, but it deserves its place:

> ...he breaks the pale,
> And feeds from home.
> (*Comedy of Errors*)

feed the bears to receive a ticket for a traffic offence.

American Citizens' Band use, the *bears* being the police, who are *fed* by a fine, which may get as far as the local municipality and may not.

feed the fishes to be seasick.

Old humorous use, but never funny to the victim. You do not actually have to vomit over the rail.

feed the meter illegally to extend a period of parking.

To prevent people hogging parking space, you should move on after the parking period which you have paid for has finished.

feed your nose to inhale illicit narcotics through the nose.

Usually cocaine or heroin:

> A woman like that... has got to be on. I'd be willing to bet she's feeding her nose. (Sanders, 1977)

feed your pussy to copulate.
Occasional female use; see also PUSSY.

feel to excite sexually with the fingers.
Either sex may *feel* the other, or themselves:
> Blank reached inside his coat pocket to feel himself. (Sanders, 1981)

A male who persuades a female to allow this activity is said to *cop a feel*:
> I... with my beloved Maria did not even try to cop a feel. (Styron, 1976)

To *feel up* is only done by a male to a female, for obvious reasons:
> He had probably been in the kitchen feeling Ella up (Follett, 1979)

and a *feel-up* is what he does:
> How is this genital whatname different from a feel-up? (Amis, 1978)

feel a draft *American* to sense prejudice.
The *draft* is felt by blacks in the company of whites, with imagery from the invisible but uncomfortable household phenomenon (the British *draught*).

feel no pain to be drunk.
From the numbing effect of the intoxicant rather than unconsciousness:
> 'But they wasn't drunk.' 'Feeling no pain?' 'Not even that.' (Sanders, 1981, suggesting mild inebriation)

feel the need to want to defecate or urinate.
A genteel usage which could mean almost anything:
> If she goes off to the bathroom when she feels the need, it's surely a good sign. (Francis, 1981, of someone in a state of shock)

feet first dead.
This is the way corpses tend to be carried:
> Cut up rough and you'll go out feet first. (Deighton, 1981)

fell design a male attempt at casual copulation.
Fell means 'cruel' or 'clever', this derivation being, I suppose, from the former:
> 'Are you a virgin?' he said suddenly, stopping right in the middle of his fell design. (McCarthy, 1963)

Now only humorous use.

fellow commoner an empty bottle.
Originally, an 18th-century student at Cambridge or Oxford universities who was wealthy and thus supposedly empty-headed as he did not need to work. Still heard in some British academic circles.

fellow traveller a Communist sympathizer or apologist.
Trotsky's *poputchnik* and Lenin's USEFUL FOOL:
> I knew you had some Communist friends... They thought you were a sentimental fellow-traveller, just as we did. (G. Greene, 1978)

female see MALE.

female domination a male fetish involving obtaining sexual gratification from being assaulted or tied up by a female, who is usually a prostitute.
> 'Big item in the FD market.' 'The what?' George asked. 'Female domination. Whips and bonds.' (Lyall, 1982)

Not a realistic name for marriage.

female oriented homosexual.
Of women and American, or else it would be *female orientated*. *Female identified* means the same thing.

female physiology menstruation.
Physiology is literally the functioning of the body; several bodily functions are exclusive to women:
> I held her lightly, protectively, then murmured in her ear, 'Beastly female physiology.' (Fowles, 1977, of a male wishing to copulate with a menstruating female)

female pills medication to abort a foetus.
It was not until 1950 that a British Code of Standards was introduced banning misleading or dishonest medical advertising and:
> the use in any advertisements for medicines or treatments for women of any phrases implying that the product could be effective in inducing miscarriage—for instance 'Female Pills', 'Not to be used in cases of pregnancy', and 'Never known to fail'. (E. S. Turner, 1952)

feminine complaint an illness which affects only adult females.
Not that her husband is out drinking again:
> 'Probably a feminine complaint,' Scaduto's wife said. When I squinted she said, 'Plumbing.' (Theroux, 1982)

feminine gender the vagina.
Oddly, in languages where it is declined, it is usually male, as for example *con* or *cunnus*:
> She went in to adjust her suspender.
> It got caught up in her feminine gender.
> (Old vulgar song)

feminine hygiene see PERSONAL HYGIENE.

femme a homosexual female playing the female role.
Originally an American usage, and occasionally of a male homosexual playing the female role.

femme fatale a woman considered by men to be irresistibly attractive.

She only kills figuratively:

> I suppose such corny little manifestations of endearment were what she thought appropriate to her role as a femme fatale. (Deighton, 1985)

A *femme du monde* is much the same as a *woman of the world* (see at WOMAN OF THE TOWN).

fence a dealer in stolen property.

He provides a screen between the thief and the eventual buyer. To *fence* is so to act:

> He used to take things home and 'fence' them. (Mayhew, 1862, referring to stolen goods)

ferret to look for clandestine listening devices.

A word beloved of the spy writers, with its imagery of going down another's burrow, of chasing larger prey to the surface, and of operating under the control of a remote and powerful master.

A *ferret* is a person who looks for such equipment or occasionally an agent who intends to create a reaction from the opposition:

> A shadow executive for the Bureau is a ferret and they'd put me down the hole. (Hall, 1979, of such a spy)

fertilizer the excreta of cattle.

It should mean anything which adds fertility to the soil, including compost and seaweed:

> Today's 'fertilizer' was 'manure' yesterday and 'meadow dressing' the day before. (Jennings, 1965)

We differentiate factory-made chemical products as *artificial fertilizers*.

fetch[1] *obsolete* to steal.

Of animals rather than humans:

> The fox fetched the last duck I had. (EDD)

fetch[2] *obsolete* a ghost.

Its appearance presaged imminent death —*fetching you away*—or long life. If the viewer did not die of fright, the alternative role was necessary, if the phantom were not to be discredited.

fiddle to steal by cheating.

From playing the stringed instrument, as certain untrustworthy itinerants once did, whence acting irresponsibly and specially applied to manipulation of accounts. A *fiddle* is any trick, even within the law, whereby someone may be cheated or overcharged. The obsolete Scottish *fiddle* was a child abandoned by gypsies, who also favour the instrument.

fidelity copulation only with a regular sexual partner.

It means faithfulness, in all its senses:

> ...expecting complete fidelity from Christine. (Green, 1979, of Rachman and Keeler—she must not copulate with anyone else)

field associate *American* a police officer charged with detecting police corruption.

An unpopular and taboo task for which the name changes from time to time.

fiend the devil.

This dysphemism comes from a Germanic word meaning 'to hate', whence 'enemy'. It is a rare example of the devil getting his nominal due. *Fiend* is also used of anyone obsessed with any activity, and especially if addicted to illegal narcotics, as *dope-fiend*.

fifth column traitors within your ranks.

General Mola, investing Madrid in 1936 with four columns of soldiers, said that he had a fifth column already in the city, meaning covert Nationalist supporters. The use today usually implies treachery:

> Their supporters here would know about it, and would be making preparations to join in, as a fifth column. (Masters, 1976)

fifty cards in the pack *American* of low intelligence.

You need fifty-two cards, except for tarot.

fifty up masturbation

By a male, who may count the strokes:

> ...hence the old tombola call: *'Five-oh, under the blanket—fifty'*. (Jolly, 1988)

fight in armour *obsolete* to copulate in a contraceptive sheath.

Boswell used both the pun and the appliance, and had cause for regret when he omitted to do so (Boswell, c. 1792).

file a waste-paper basket.

Perhaps jocular use by those who do not like storing documents and tend to make a fetish of their prejudice or idleness. An American would speak of *file thirteen* or *file seventeen*:

> They won't give them time off, or they'll put the application in 'File 13'—the waste-paper basket. (McNab, 1993)

fill a pannier *British, obsolete* to impregnate.

From the hooped framework or *pannier* which distended a skirt and concealed the swelling of pregnancy.

fill full of holes *mainly American* to kill by shooting.

You may also, if so unfortunate, be *filled with lead* or *daylight*.

fill in to maim or torture.

The origin of the slang meaning 'to beat up' appears uncertain, except that it seems to have been British naval slang:

> Then I realized that though the people sounded more in control, if they filled me in

they'd do it more professionally. (McNab, 1993, referring to his capture and subsequent torture by Iraqis)

fill in the blank spaces in our history to cease lying.

Referring to the Russian tendency to allow some of the truth about the excesses of Stalin and Communism to be publicly revealed:

> In the past two weeks, the pace of the debate over 'filling in the blank spaces of our history', to borrow Mr Gorbachev's own euphemism, has quickened. (*Daily Telegraph*, September 1987)

filler a cheaper substitute surreptitiously introduced to increase apparent weight or volume.

The stratagem increases the profit on sale by weight or volume. Small coals and slack were used as a *filler* in bags for large, but not today, when all are equally dear. But china-clay, inert, odourless, colourless, and tasteless, goes into many things from 'cream' chocolates to latex thread. In the same sense a *filler* is a trivial news item used to occupy space in a newspaper:

> We used to produce fillers, which is what the papers use to cement the real news to the adverts: 'Sacked stripper organizes strike.' You know the sort of thing. (Deighton, 1972)

fillet to steal.

Literally, to remove flesh from the bone:

> We did think some spare parts might be filleted; but luckily nothing's gone. (*Sunday Telegraph*, October 1981—a manager was talking about a factory in which employees had trespassed)

filly a young woman viewed sexually by a male.

Literally, a female horse less than four years old:

> We pre-war soldiers always made enquiries as to what sort of a place it was for booze and fillies. (Richards, 1933)

To *slip a filly* was to have an abortion; see SLIP[1].

filth (the) the police.

The jargon of criminals; dysphemism rather than euphemism:

> He's a big wheel in the filth, Mr Nolan. Y'know... assistant chief constable and all that. (Wainwright, 1979)

filthy relating to any taboo sexual act.

Referring to masturbation, attempted male fondling of an unwilling female, or even attempted rape:

> ... that sailor tried to be filthy. (L. Thomas, 1977, of such an attempt)

The obsolete *filth*, a prostitute, came too from the dirt:

> Wisdom and goodness to the vile seem vile: Filths savour but themselves. (Shakespeare, *King Lear*)

Filth and *filthy* are also used of talk or humour involving particularly copulation or homosexuality.

final solution the killing of all Jews.

The Nazi *Endlösung*, which emerged in 1938 when the hopes of deporting German Jews to colonize Madagascar vanished. The policy was ruthlessly and systematically applied from 1941 to 1945.

financial assistance *American* state aid for the poor.

True as far as it goes, but it could as well be a loan, gift, or subsidy to the rich:

> 'You're on welfare?' 'Financial assistance,' she said haughtily. (Sanders, 1985)

financial engineering false accounting.

Not controlling a metal-working shop through the use of figures:

> ... as we have seen elsewhere, financial engineering cannot conceal the truth indefinitely. (*Daily Telegraph*, 16 November 1990)

financial products forms of moneylending.

Moneylenders like to use the word PRODUCT because it suggests their activities are beneficial:

> ... proliferation of new instruments and 'financial products'. Reshapings and reclothings of lending and borrowing packaged to the advantage of a now totally institutionalized market. (M. Thomas, 1987)

financial services money-lending.

The language of those who offer credit or small loans at high rates of interest to the relatively poor. It should mean accounting, banking, or money-changing.

find[1] to steal.

From the pretence that the goods have been lost or abandoned, which is even barer when you *find something before someone has lost it*. The imagery is as old as stealing, with the obsolete Scottish phrase *found a thing where the Highlander found the tongs*:

> Spoken when boys have pick'd something and pretend they found it. (Kelly, 1721—to Lowlanders the Highland Scots were remorseless thieves)

A 19th-century *finder* was a thief:

> The 'finders' and 'stealers' of dogs were the most especial subject of a parliamentary enquiry. (Mayhew, 1851)

find[2] to fabricate evidence.

The language of the Nixon White House—you *find* it after you have manufactured it:

There was no prior evidence of such a relationship between Rutherford and Anderson, and Stewart refused to try to 'find' one. (Colodny and Gettlin, 1991—Nixon was wrongly convinced that there was a homosexual relationship between the journalist and the naval yeoman; Stewart was an investigator.)

finger[1] to inform on or point out in a criminal context.

The person who *fingers*, or betrays, another criminal to the police may do his pointing only figuratively:

Snyder had hoped to pick up a few hundred bucks by fingering Hooker to Amon Lorrimer. (Weverka, 1973, referring to the pointing out of a small-time crook to another gang boss)

To *put the finger on* is also to betray. To *finger a job* is to draw the attention of criminals to an opportunity for crime:

I figure he knew them, and they knew him. Maybe he fingered the job. (Sanders, 1977)

Finger-man has three meanings: he may identify either a victim to other criminals or a suspect to the police; or he may inspect the site of a potential crime; or he may be a killer, with his finger on the trigger:

... the finger man loiters ahead undetected till the target blunders into him. (le Carré, 1980)

An American *finger-mob* commits crimes without police intervention in return for information about other, more serious, criminals.

finger[2] to masturbate.

Of a woman:

... her other hand fingering, all five fingers fingering like a team of maggots at her open heat. (Mailer, 1965, of a woman masturbating)

A *finger-artist* is a female homosexual.

finger-blight the reduction of a crop due to stealing.

A usage with apples in the cider country, *blight* being a natural phenomenon, like the stealing of fruit from trees. *Scrumping*, or picking up windfalls, was permitted, whence *scrumpy*, the rough cider made from them.

fingers get close to the thumb favouritism is shown to relatives or friends.

From the clenched hand:

Yes, sir, the fingers have got pretty close to the thumb. (Egerton, 1884, of nepotism)

finish[1] to kill.

Of humans and animals. If they have been previously wounded or are sick, you *finish them off*.

finish[2] to achieve a sexual orgasm.

Very common use, referring to either sex. To *finish yourself* is to masturbate to ejaculation or orgasm.

finish off[1] see FINISH[1].

finish off[2] to wipe dry your genitalia.

Children or geriatrics are told to *finish yourself off* after someone has helped dry them, which avoids having to refer to taboo organs.

fire to dismiss peremptorily from employment.

Punning on *discharge*, which is standard English, meaning 'to dismiss from employment':

'Working?' 'Nope, I got fired.' (Theroux, 1976)

Although the dismissal is usually peremptory, I do not think there is any derivation from *fire*, to shoot a projectile from a gun, where the ignition of powder operates the *fire-arm*.

fire a shot *American* to ejaculate during copulation.

Of obvious imagery. To *fire blanks* (of a male) is to copulate without impregnation due to sexual impotence or other cause, and is now also used by men who have had a vasectomy. To *fire up* is to copulate with a woman extramaritally on a single occasion, punning on the meaning 'to excite' (from stoking a steam boiler) and the ejaculation.

fire has gone out the engine has stopped.

Second World War navy fliers' usage, with imagery from steam-driven ships. It described a crisis, as the single-engine aircraft were often a long way from land or a carrier.

fireman[1] a motorist exceeding the speed limit.

American use from the corny question, traditionally asked by a traffic policeman, 'Where's the fire?'

fireman[2] a person to whom unpleasant duties are delegated.

With obvious imagery:

Since starting at the Pentagon, Buhardt had been a fireman helping ... stave off or limit the fallout from a variety of scandalous episodes. (Colodny and Gettlin, 1991)

See also VISITING FIREMAN[2].

fireman[3] a Jewish policeman in a camp for Jews.

The Nazis so styled the Jews they enlisted in death camps to enforce their rules as the lowest rung of authority.

firewater whisky.

As well as burning your throat and guts, it is flammable:

Would I be consultant in exchange for a generous consignment of firewater? (*Private Eye*, September 1981)

Fired up means drunk, probably from starting an engine and not necessarily after drinking whisky.

firk *obsolete* to copulate with.

From the meaning to beat, using the common violent imagery:

> I'll fer him, and firk him. (Shakespeare, *Henry V*, with only beating in mind)

Partridge suggests 'partly a euphemistic pronunciation of *fuck*', of which the present participle is often articulated thus.

firm (the) a clandestine, illegal, or bogus organization.

A 19th-century usage of the Fenians in Ireland and adopted by espionage writers:

> Ever since he joined the firm as a young recruit... (G. Greene, 1978, of a spy)

Also used by the Windsor family:

> ...masses of photographs of 'The Firm', as they somewhat affectedly style themselves. (A. Clark, 1993, reporting on the contents of a royal palace)

first people the indigenous Canadian Indians.

A usage which does not ruffle the feathers of those so described although it is less acceptable to the proponents of Adam and Eve:

> The constitution is filled with modish catchphrases of the late 20th century, affirmative action, first people (natives), collective human rights, and the equality of female and male persons. (*Daily Telegraph*, 23 October 1992, of a proposed Canadian innovation which the electorate roundly voted down)

first strike unannounced aggression.

A use referring to an attack before war has been declared; otherwise it would not be a euphemism. A *first-strike capability* is an ability to attack the enemy with nuclear bombs without prior warning. See also SECOND STRIKE.

first world rich.

The language of those for whom talk of poverty or backwardness is taboo:

> And then 50,000 First World citizens—Brits, Americans, French, German, Spanish, Swedish, Danish—name it. (Forsyth, 1994)

The 'second world' was inhabited by Russia and its vassal states. See also THIRD WORLD.

fish[1] a heterosexual woman.

The word is used among American female homosexuals, for whom heterosexuality in women is taboo, and the imagery is from her being caught by a male. A *fishwife* is the wife of a male homosexual. The American term *fish*, a prostitute, is obsolete.

fish[2] a prostitute's customer.

To be caught and cheated:

> You may sit and drink if you wish. I shall tell the girls that you are not a fish. (Trevanian, 1973—he was in a brothel)

fish[3] a torpedo.

Second World War jargon, when an EGG was a bomb and a less dangerous mission was a MILK RUN:

> We had a fish coming at our ship at about 265 degrees. (N. Lewis, 1989—it had been fired by a German E-boat)

fish hook see HOOK[1].

fish market see FISHMONGER'S DAUGHTER.

fishing expedition[1] a foreign trip to seek a husband.

Single British girls were sent to Malta or India, where they might meet naval or army officers on extended tours in societies with a dearth of unmarried white females. The annual excursion was called the *fishing fleet*:

> ...girls who had come out from England...as members of the 'fishing fleet' to find a husband. (Farrell, 1973)

fishing expedition[2] an attempt to obtain gratuitous information.

Not knowing what you may catch, as cross-examining counsel, detective, journalist, or spy:

> ...things that an investigative fishing expedition into the break-in could uncover and exploit politically. (Colodny and Gettlin, 1991, quoting a White House tape of June, 1973)

fishmonger's daughter a prostitute.

As with the obsolete *fish market*, a brothel, the allusion is to the vaginal smell:

> Excellent well; you are a fishmonger. (Shakespeare, *Hamlet*, says Hamlet to Polonius, implying not that he was a pimp but that his daughter, Ophelia, was a prostitute. Polonius misses the point (along with many distinguished lexicographers), only to take another behind the arras in the Third Act.)

fishwife see FISH[1].

fishy homosexual.

Of a male, from the meaning 'queer' or 'irregular':

> ...her only husband had been as fishy as Dick's hatband. (Fraser, 1975)

fistful a prison sentence of five years.

Prison slang, a variant of FIVE FINGERS and HANDFUL[1].

fit up to incriminate falsely.

Another way to *frame* (see FRAME[1]):

> ...some of the criminals changed their stories and admitted PC Cooley had been 'fitted up'. (*Daily Telegraph*, March 1990—they were not alleging that the constable had been issued with uniform or riot gear)

five-fingered discount *American* a reduction due to stealing.

Referring to goods stolen and then sold below their market value. See also FALL OFF THE BACK OF A LORRY.

five-fingered widow male masturbation.

British army use for those long absent from the company of white women:

> The red light districts... were strictly out of bounds... Many turned, as a last resort, to the 'five-fingered widow'. (Allen, 1975, of service in India)

five fingers a prison sentence of five years.

A variant of FISTFUL and HANDFUL[1].

five or seven drunk.

This pre-Second World War London use comes from the standard court sentence on conviction of being *drunk and disorderly* or *drunk and incapable*—five shillings fine or seven days in jail.

fix[1] to make an illegal arrangement.

In standard usage, to mend or adjust. It is an omnibus word for various types of criminal behaviour, including bribery:

> To a Metropolitan policeman fix could only mean nothing other than a bribe. (Deighton, 1978)

Also of damaging a rival:

> ... specifically named in several of the White House tapes whom Nixon planned to 'fix' after he had been reelected (Colodny and Gettlin, 1991)

and especially of a gambling coup:

> There's eight or nine races on the card and... the fix can be in any time somebody says so. (Chandler, 1953)

fix[2] to castrate.

Of domestic animals in America, the arrangement or mending being less to their liking than that of their owners.

fix[3] an injection of illicit narcotics.

Usually heroin:

> Frank, had you had a fix? (Davidson, 1978, asking about narcotics use, not navigational verification)

A *fix* is the quantity needed for one injection and to *fix* is to supply with illegal narcotics.

fix up *American* to hire a prostitute for another's use.

The practice arises from the wish to influence visiting buyers etc. and this usage comes from the general sense 'to arrange something for someone'. In American business practice, the operation of a flexible price structure, as everywhere else in the world, is forbidden by Federal Law, and the law, though widely ignored, is often invoked when a disappointed competitor peaches to Washington. Thus you may be heavily fined or imprisoned for giving a good customer who pays quickly a discount, but not for *fixing him up*—with a prostitute. Whatever the effect on normal business, it is all good for prostitution.

fixer an arranger of embarrassing, dubious, or illegal business.

In commercial jargon, the *fixer* is the person who bribes where needed to secure a contract, isolating his principals from overt illegality by charging them excess commission to finance the bribery. The word also has the various meanings of FIX, except perhaps as a castrator of animals. A *fixer* is also a person who deflects unwelcome publicity:

> He's a fixer, a smoother-out. (Price, 1970)

fizzer an accusation of a military offence.

A British army pun on *charge*, now also met in civilian use.

> 'I'll put you on a fizzer!' Vince shouted as he went out and took over from Stan on the Minimi. (McNab, 1993—a *Minimi* is a type of automatic gun)

flack out *American* to die.

Literally, to lose consciousness, due to drunkenness or from lack of oxygen while skin-diving.

flag is up (my) I am menstruating.

Punning on the redness of the danger flag, the towel, and the blood. Female use, also in a number of related expressions such as *flag of defiance*, *fly the (red) flag*, and *baker flying* (see at BAKER). The obsolete British *flag of distress* was no more than a tactful warning in company that a man's shirt-tail was hanging out.

flake[1] *American* an eccentric or strange person.

Of uncertain derivation, although there are plenty of theories to choose from. Dr Johnson gives 'Any thing that appears loosely held together'. The imagery of disintegration is common of mental illness:

> 'What a character she is,' he said. 'A real flake.' (Sanders, 1986)

flake[2] *American* cocaine.

Perhaps a shortened form of something, or from chipping it from the mass—see CHIPPY[2].

flamboyant homosexual.

Literally, colourful or showy, which is how some male homosexuals are thought to comport themselves:

> ... obituaries are simply eulogies of the great and the good, any of whose peccadillos (unusual sexual tastes, drunkenness and so on) are tactfully powdered over with euphemism ('flamboyant', 'convivial' etc.). (Lewis Jones in *Daily Telegraph*, 1 December

1994—as the editor responsible for obituaries, he eschewed such evasion)

flap a military crisis.

Second World War usage, from the waving of arms or wings in agitation:

I didn't know till I got there that there's a flap on. (Manning, 1977)

flapper *obsolete* a young woman who flouts convention.

In northern English dialect, a young prostitute; in western England, a 19th-century under-petticoat; in OED a 'young wild duck or partridge'; and in the 1920s, a participant in the *flapper era*:

I was sure I would have enjoyed being a rich Canton flapper with a peacock called Bluey too. (Irvine, 1986, of a white expatriate)

flash[1] to display your penis in public to a stranger.

You open your raincoat *in a flash* and thereby become a *flasher*:

These men were rapists or Peeping Toms or flashers or child molesters. (Sanders, 1973)

An American police officer who *flashes tin* does no more than identify himself, the *tin* being his badge of office:

Chief, should Jason Two flash his tin or work undercover? (Sanders, 1977)

The obsolete *flash girl*, *flash woman*, or *flash-tail* was a prostitute who was especially ostentatious:

...keeping a cold eye on the more obvious thieves and flash-tails (Fraser, 1977, writing in 19th-century style)

and Grose tells us that thieves and prostitutes congregated in a *flash panney*.

flash[2] to vomit.

American use, usually after drunkenness.

flat on your back copulating.

Of a woman, from her probable posture:

...if I can't charm this one flat on her back, I've lost my way with women. (Fraser, 1971)

The American *flat-backer* in black speech is a prostitute.

flatfoot a policeman.

From the pounding the beat, which is now no longer necessary to earn the appellation. In America shortened to *flat*.

flawed drunk.

Perhaps a pun on *floored* and on having an imperfection, as drunkards tend not to remain in pristine condition. Grose says *flawd*.

fledgling nation a poor country.

Said to have been coined by Eleanor Roosevelt, with avian imagery, when colonies were becoming countries and had to make their own way in the world. Now overtaken by DEVELOPING.

fleece to defraud.

By robbery or overcharging, from the shearing of sheep:

...all the petty cutthroat ways and means with which she used to fleece us. (Cleland, 1749, of a cheating bawd)

It is still in use today:

I knew I was being fleeced. (Theroux, 1983, of an excessive hotel bill)

flesh your will *obsolete* to copulate.

Of a male: it is hard to say whether Shakespeare invented the imagery and exactly what vulgar pun he had in mind:

This night he fleshes his will in the spoil of her honour. (*All's Well*)

fleshpot a brothel.

Originally, a vessel in which meat was cooked, whence a source of luxury and debauchery offering a variety of vicious attractions:

...found the 'fleshpots' of Nairobi to be 'insidious and most likely to corrupt'. (Allen, 1975)

fleshy part of the thigh the buttocks.

It was here that a military bulletin said Lord Methuen had been wounded in the Boer War. Apart from late 19th-century modesty, to be wounded in the buttocks might imply that you were not facing the enemy.

flexibility the abandonment of principles in pursuit of any object.

The object for bankers is profit; for soldiers, promotion; for politicians, power:

Conservative MPs, impatient for the pre-election bribery to start, call for 'flexibility'. (*Financial Times*, December 1981)

Such conduct is described as being *flexible*:

Pym is preparing...a swift twitch of the rug from under the few remaining loyalist sheepshaggers. This is called being flexible. (*Private Eye*, May 1982—as Foreign Secretary Pym opposed the bellicose stand taken by Mrs Thatcher, but Argentine impetuosity and stupidity prevented him making any move to divest the Falkland Islands of British patronage)

flight see FLY[2].

fling (a) copulation outside a regular relationship.

From the meaning 'indulgence in any unaccustomed excess':

I had my fling with the Tanglin wife whom I reported as being 'ever so nice'. (Theroux, 1973)

flip your lid to go mad.

The *lid* is slang for the head:

> ... you suddenly decide to answer questions today? And from the Press? You must have flipped your lid. (Lynn and Jay, 1989)

Of temporary rather than permanent derangement.

flit[1] **(do a)** to leave accommodation without paying rent due.

Shortened form of *do a moonlight flit* (see MOONLIGHT FLIT):

> The family on the corner, two years in arrears on the rent, were doing another flit, all their furniture ... stacked up on creaking barrows. (Bradbury, 1976)

In obsolete Scottish and northern English use, to *flit* was to die:

> She canna flit in peace until she sees you. (W. Scott, 1816)

flit[2] a male homosexual who usually plays the female role.

> He assured me that he had a luscious ass ... Flits have always been attracted to me. (McCarthy, 1963)

He affects female mannerisms by *flitting about*. To be *flitty* or *flit* is to act in this way.

float paper to issue cheques unsecured by bank deposits.

Pre-computer delays in the banking clearance system enabled a shrewd operator to generate credit balances during the four or five days it took to clear inter-bank cheques, by which time a fresh deposit in the paying bank would cover the initially uncovered cheque:

> He could probably stall [bankers] for the necessary twenty-four hours. It wouldn't be the first time Lorrimer had floated paper for a day or two. (Weverka, 1973)

floater *American* an undesirable.

Either literally, as a corpse in the water:

> Floaters ... are another matter; a person who has been in the Bay for a week or more ... (J. Mitford, 1963)

or figuratively, as a vagrant or itinerant worker.

floating *American* drunk or under the influence of narcotics.

From the feeling of levitation or mental detachment.

flog off to masturbate.

Of a male, using the common *beating* imagery. He may also *flog the bishop*, his *dummy*, his *mutton*, *beef*, *donkey*, etc.:

> ... dragged off to jail everytime he ... flogged his dummy on the porch. (Wambaugh, 1975)

flop to copulate casually.

Of a woman, implying that she will readily drop to a prone position:

> Lois flops at the drop of a hat. (Chandler, 1943, of one such)

floral tribute a wreath presented at a funeral.

Tribute as protection money or rent paid on a regular basis has evolved in standard English into the meaning 'a gesture of respect or praise on a single occasion'. Brides prefer to carry bouquets.

flower[1] *obsolete* the virginity of a woman.

What you lose when you are deflowered:

> Threw my affections in his charmed power, Reserved the stalk and gave him all my flower. (Shakespeare, *Lover's Complaint*)

flower[2] *American* a male homosexual.

In this use, he is said to take the female role, but I have not traced a literary example.

flowers the menstrual flow.

Normally expanded to *monthly flowers*, from flowing rather than flowering, it might seem:

> I had my courses, my flowers. (Fowles, 1985, of a woman denying pregnancy)

flowery a prison cell.

From the rhyming slang *flowery dell*; sometimes used to refer to the institution as a whole .

flowery language swearing.

Embellishments unnecessarily added to normal speech. In the 19th century *flowery language* was blasphemy.

fluff your duff to masturbate.

Of a male, probably from the suety dish, with the same imagery as PULL THE PUDDING:

> What are you doing here in the dark—fluffing your duff? (Sanders, 1982)

flush down the drain to dismiss peremptorily from employment.

The imagery is from the lavatory:

> If I bounce him and ask Thorsen to get me another man, he'll flush Boone down the drain. (Sanders, 1977)

fluter *American* a male homosexual.

Probably punning on American *flute*, the penis, and the act of fellatio.

flutter[1] a wager.

A 17th-century use which is still current, from the excitement of gambling. Normally referring to a small bet placed by someone who is not a habitual gambler, but heavy punters also so describe their bets, to minimize the extent of their addiction.

flutter[2] to copulate extramaritally.

Of either sex, and again from the excitement. The obsolete English *flutter a skirt* meant 'to be a prostitute', from her method of advertising her wares.

flutterer a machine to assist in lie detection.

From the variations recorded of the subject's pulse, temperature, sweat, etc.:

> What we used to call a lie detector, sir. A polygraph, known in the business as a flutterer. (le Carré, 1989)

flux menstruation.

Literally, the condition of flowing or, as with solder, causing to flow:

> Even her body's flux, which she could feel in a gentle, almost controlled flow, wasn't the inconvenient and disagreeable monthly discharge... (James, 1980—although that is what it was)

fly[1] *American* in plain clothes.

Of a policeman, from the meaning 'knowing' or 'cunning'. It is especially used in the expressions *fly ball, bob, bull, cop, dick*, etc., referring to a detective assigned to duty away from his normal precinct to avoid criminal recognition.

fly[2] to be under the influence of illegal narcotics.

Usually the sense of levitation from smoking marijuana:

> This is top-grade grass, the real stuff. We'll fly. (Sanders, 1982)

Flight or *flying* is the state induced.

fly a kite[1] to write or cash an unbacked cheque.

See KITE.

fly a kite[2] to write a begging letter.

This was a considerable industry and art in 19th-century London, made possible by the advent of the penny post. See also JUNK MAIL.

fly a kite[3] to smuggle a package into or out of prison.

The trade inwards is usually narcotics, and outwards, letters.

fly-by-night[1] an absconding debtor.

Not from the witch, on or off her broomstick, or the ominous bird, also known as the whistler or gobbleratch, whose nocturnal flight presaged imminent death, nor even the unfortunate tourist on the cheapest package holiday, but the tenant with unpaid rent, who took all his goods with him to prevent distraint by the landlord. Now standard English for anyone who is financially unreliable. See also MOONLIGHT FLIT.

fly-by-night[2] drunk.

Rhyming slang for 'tight', with perhaps a sideswipe at the unreliability of drunkards.

fly one wing low to be drunk.

RAF slang from the Second World War; an aircraft with a wing down was likely to crash:

> ... half the officers in the club house were flying one wing low already. (Deighton, 1982)

fly the blue pigeon to steal lead.

Usually from the roof of a church, where it is known as *bluey*:

> And there's the bluey ... the lead from the pipes and the roofs like of churches. (L. Thomas, 1981)

Whence the pun on the colour, as to *pigeon* is to cheat or rob, normally from an innocent dupe; a shortened form of *pluck the pigeon*. The sailor who *flies the blue pigeon* swings the lead, literally not metaphorically. I heard the phrase used at Musselburgh in October 1981, when a thief, arrested and bailed one morning, was back on the roof again ripping off lead the same afternoon.

fly the red flag see FLAG IS UP (MY).

flyblow a bastard.

Punning on the deposit of eggs left in meat by flies, and the taint:

> She is still a bairn. And the flyblow of the system. (Cookson, 1969—her autobiography was largely about her own bastardy)

flying handicap diarrhoea.

This English phrase puns on the celerity needed, the disability, and a typical name for a horse race.

flying low inadvertently having a trouser zip undone.

Used as an oblique warning from one male to another. See also SNOWING DOWN SOUTH (IT'S).

flying picket *British* a crowd from afar trying to stop others working.

They travelled by road rather than by air, and sought to intimidate by a concentration of forces.

flying squad a police detachment organized for rapid movement.

There have been earthbound *flying squadrons* from the 17th century. The London version which was formed in the 1920s is also known in rhyming slang as the *Sweeney*, from the demon barber *Sweeney Todd*.

fog to kill.

Presumably from the disappearance of the victim. The American *fog away*, to kill by shooting, may also allude to the smoke from the gun.

foggy drunk.

Your eyes may be and your memory becomes. Also as *fogged*.

foil *American* a small packet of illicit narcotics.

From a normal way of packing tablets.

foin *obsolete* to copulate.

Literally, to make a thrust with a sharp weapon:

> When wilt thou leave fighting o' days and foining o' nights. (Shakespeare, *2 Henry IV*)

fold to fail in business.

Either personally or corporately, from the collapse of a structure. In America it sometimes also means 'to die'.

folding camera see CAMERA.

follower *obsolete* a male who is courting a female.

From the servant who *followed* his master, it acquired the specific meaning of a man who courted a domestic servant girl:

> No, sir, missus don't permit no followers. (Mayhew, 1862)

Then in upper-class use—those who had servants—of courting any girl:

> If she had no followers at all they would say she's a Lesbian. (N. Mitford, 1960)

To *follow* was to court:

> He followed his wife ten year afore they were wed. (*Leeds Mercury*, 1893, quoted in EDD)

fondle to caress sexually.

> ... she had learned to slide her hand into his slitted pocket and fondle him. (Sanders, 1973)

Literally, to handle (something or someone) fondly.

fool (about) with yourself to masturbate.

From the inconsequential action of *fooling (about)*:

> Honey ... you don't care if I fool with myself a little. (M. Thomas, 1982, of a woman wishing to masturbate)

foot *obsolete* to levy an imposition on new employees to buy intoxicants.

Perhaps a shortened form of *foot the bill*:

> When he wor lowse on his prentis-ship his shopmates fooited him. (Treddlehoyle, 1875)

A *footing* was such a levy:

> I paid five shillin' for footin when I started. (Pinnock, 1895)

This is a sample entry; many euphemisms were used to refer to the common British 19th-century practice by established workers of taking money from a new employee or apprentice.

footpad see HIGHWAYMAN.

for the high jump in deep trouble.

Not participating in an athletics meeting. It originally meant sentenced to death by hanging, whence to die by any means. Now used figuratively of peremptory dismissal or other disciplinary action:

> Satchthorpe and Frimston are for the high jump ... the Chief Constable's ... practically said as much. (Grayson, 1975)

for your convenience provided ostensibly as a special service (although it is one for which you have already paid indirectly).

This slightly objectionable pretence of giving you something extra or special seems to have spread from America:

> The notice said they were sanitized under infra-red and ultra-violet light for Koolman's protection and convenience, but I suppose anybody would get the same kind of towel. (Deighton, 1972)

However, all is forgiven when you meet an unconscious pun:

> For your convenience—Sanitor tissue seat covers. (a lavatory in Fall River, Massachusetts, May 1981)

for your (own) comfort and safety to prevent a stampede.

The wording in the standard injunction from airline cabin staff telling passengers to stay in their seats with their safety belts fastened while the aircraft is making its final approach to the ramp. The staff themselves show no signs of discomfort or danger as they move freely through the cabin at this time.

forage to steal.

Originally, it was a noun meaning 'food for cattle'. Such food was traditionally stolen for their horses by armies on the march. Whence the act of looking for any food to steal and the verb to *forage*, mainly used by soldiers:

> 'Where the devil did you come by this?' 'Foraged, sir.' (Fraser, 1969, of something stolen)

force-put job the marriage of a pregnant woman.

Force-put is a matter of necessity in Devon dialect. Still heard in the South Hams despite the wide availability of contraception.

force your ardour upon to copulate with.

The male does the *forcing* on a supposedly reluctant female, usually outside a regular relationship:

> This was the evening when the conquerors of the Afrika Korps were to force their pent-up ardour on the ladies of Alexandria. (Manning, 1977—she meant 'the conquerors who were in the Afrika Korps' and not the British 8th Army, who eventually conquered it)

Also as *force your attentions on*:

> Willie tried to force his attentions on her. (Kee, 1993—Willie O'Shea was trying to copulate with his wife Katie at a time when she was bearing children by C. S. Parnell)

To *force favours from* indicates a greater degree of female reluctance.

forced labour imprisonment under severe conditions.

Not euphemistic in Nazi Germany during the Second World War, despite the brutality and starvation, because it was just that. In Communist Siberia, the economic function in *forced labour camps* was subordinated to the penal.

forget yourself to be guilty of a solecism.

As by swearing where swearing is out of place; by making a sexual approach to a woman who has not signalled that she would welcome it; by urinating involuntarily, especially of a geriatric; etc. The use does not really indicate amnesia.

fork to copulate with.

Of a male, punning on the 'pronging' and the place where the legs join the trunk. Referring to the latter, Shakespeare used the *face between her forks* for a woman's frontal crotch:

> Behold yond simpering dame,
> Whose face between her forks presages snow.
> (Shakespeare, *King Lear*)

The obsolete *wear a fork* was to be a cuckold, from the proverbial horns, and the *forked plague* was cuckoldry:

> This forked plague is fated to us. (Shakespeare, *Othello*)

form a criminal record.

Police jargon, probably from horse-racing, although there is usually a special form on which these things are recorded:

> With regard to a police record, Artie Johnson is the only one with any form. (Davidson, 1978)

forspeak *Scottish, obsolete* to call up evil spirits.

This is the first recorded meaning, but it must have meant 'to deny' originally. *Forspoken* means 'bewitched'. It also meant 'to speak ill of':

> We hae forespoke the Brownie. They say, if ye speak o' the deil, he'll appear. (Hogg, 1866)

forty-four a prostitute.

An unusual example of American rhyming slang, for 'whore'.

forward drunk.

An obsolete English use which may refer to the truculence associated with drunkenness or to making progress in that direction:

> Twer querish tack—beer and reubub weind an' bacca juice a-mixed, but I knowed we could get furrud on't. (Buckman, 1870—a mixture of tobacco juice, beer, and rhubarb wine is queerish tack indeed)

The modern American *forwards*, amphetamine pills, might have the same derivation, if we could be sure what that was. The obsolete northern English *forward at the knees* meant aged, from the way old people walk.

foul to defecate in an unaccustomed place.

Usually of dogs on carpets or sidewalks but occasionally also of humans:

> Who had fouled his home? (Boyd, 1982, of a house in which troops had defecated everywhere)

To *foul yourself* is to defecate into your clothing:

> They fouled themselves where they lay. (Fraser, 1971)

It is also used to describe vomiting over your clothes, but not of falling into a patch of mud.

foul ane the devil.

Another common Scottish dysphemism; not all the shortened forms are obsolete:

> Our deacon wadna ca' a chair
> The foul ane durst him na-say.
> (Fergusson, 1773)

Also as the *foul thief*:

> Seek the foul thief onie place. (Burns, 1785)

Shortened to *foul* in imprecations such as *foul skelp ye*, the devil take you, and *foul may care*, devil-may-care.

foul desire a wish to copulate.

Foul meaning disgusting; it seems that linguistically only males are thus taken:

> If foul desire had not conducted you.
> (Shakespeare, *Titus Andronicus*)

You may also have *foul lusts* for or *foul designs* on a female who is not your wife. If you proceed to *have your foul way with her*, you copulate with her, although not necessarily against her inclination. Still used humorously.

foul play murder.

British police jargon, and not merely the way professional footballers play football:

> 'He was shot.' 'Foul play... Isn't that what you British call it?' (Deighton, 1978)

foundation garment a corset.

The imagery comes from building, although *buttress* might seem more appropriate:

> ... she may be half-perishing in the clutch of a 'foundation garment'. (Jennings, 1965, noting the euphemism)

four-letter word an obscenity.

Jennings (1965) demonstrates that there are only eight among the catalogue of obscenities which contain four letters, and then, as always, enlightens and entertains the reader with an analysis of their use. However, the most hackneyed obscenities do have only four letters. You may apply one of them to an unpleasant person—a *four-letter man*.

four sheets in the wind see SHEET IN THE WIND.

fourth a lavatory.

The use seems to have originated in Cambridge University, possibly from the lowest category of degree then awarded on graduation. The three 'Estates of the Realm' were the peers, the bishops, and the commons. 19th-century society delighted in nominating a fourth, the press being the favourite:

> Just to make sure that the food and drink were equally up to the expectations of the fourth estate. (Deighton, 1982, of journalists)

Carlyle says Burke first suggested the press, but Macaulay has a better claim.

foxed drunk.

Literally, deceived, and so a variant of the obsolete 'deceived in liquor', which implies that it was not your fault:

> ... poured drink into himself until he was completely foxed. (Fraser, 1970)

As usual, the half is the same as the whole:

> Here I was, half-foxed and croaking to myself in a draughty shack. (Fraser, 1971)

Both uses are now dated. *Catch a fox*, to be drunk, is obsolete.

foxy *American* eager for copulation.

> Over forty and feeling foxy. (On a woman's apron, JFK Airport, 1979)

fractured drunk.

Another American example of the *broken* imagery.

frag to kill.

Shortened form of *fragmentation device*, which is a long-winded way of saying 'hand grenade'. See also DEVICE:

> Molly Turner was important to me and you fragged her. (Sanders, 1984—she had been murdered)

A Vietnam innovation, where the grenade was used as a way of killing over-keen or unpopular officers, often white, by GIs, often black.

fragile suffering from sub-acute alcoholic poisoning.

Usually an admission of a HANGOVER by the sufferer himself.

frail[1] suffering from sub-acute alcoholic poisoning.

A variant of *fragile* above.

frail[2] a female viewed sexually by a male.

Probably from, and as obsolete as, the *weaker sex* concept, because a *frail*, in male usage, might denote any female:

> In persuading frails to divulge what they know... (Lavine, 1930)

A *frail job* is a single act of casual copulation by a male. The frailty of a *frail sister*, a prostitute, was moral rather than physical, her manner of life seeming to call for considerable stamina.

frame[1] to incriminate falsely.

Like mounting a picture so that you can see it better:

> I take it you don't want your daughter-in-law framed. (Chandler, 1943, referring to such incrimination)

The result is a *frame-up*:

> It's a frame-up as sure as ever I saw one. (Deighton, 1981)

frame[2] a male who is attractive sexually to male homosexuals.

Not necessarily a homosexual himself, although dress and posture may send such a signal to other homosexuals, in which case he takes the female role. Perhaps from the slang *frame*, a body.

frank[1] *obsolete* copulating promiscuously.

Dr Johnson gives 'licentious', from the early meaning 'liberal' or 'generous':

> Chaste to her Husband, frank to all beside.
> A teeming Mistress but a barren Bride.
> (Pope, 1735)

frank[2] unfriendly and without consensus.

Of political talks between fundamentally opposed parties:

> Mr Mugabe had agreed on the need for urgent and 'frank' talks. (*Daily Telegraph*, December 1980)

Full and frank in a communiqué tells you that the parties failed to agree on anything.

fraternal associated with politically.

Much loved by parties of the left (see also BROTHER). Thus for the Communists *fraternal assistance* included anything from subsidizing the *Daily Worker* to invading one of your neighbours:

> But the decisions to say 'counter-revolution' instead of 'uprising', 'people' instead of 'party', 'fraternal assistance' instead of 'invasion' are choices of the highest solemnity. For Communism exists by casting spells—change the language and the world itself will change. (*Sunday Telegraph*, March 1989)

fraternize to copulate with civilians in militarily occupied countries.

Strictly, to be friendly with the natives, but the *frater* was of less interest than his sister:

> Relics of the Great Fraternization Period, you remember. (Bogarde, 1981, referring to the many post-Second World War German bastards fathered by occupying soldiers)

Mainly denoting such activities in Austria and Germany after the Second World War.

fratricide inflicting casualties by mistake on your own troops.

Killing your brothers-in-arms:

> ...it is very difficult to avoid blue-on-blue, or fratricide, as the Americans call it. (de la Billière, 1992)

See also FRAG and FRIENDLY FIRE.

freak a devotee of a taboo or unconventional activity.

Literally, an irrational event or a monster. An *acid freak* uses LSD and a *surf freak* is addicted to the rollers. A *freak* (*tout court*) is a male homosexual:

> They wanted to go down to Greenwich Village to see the freaks. (Sanders, 1981—they were not hoping to visit a circus with monsters on show)

A *freak trick* in prostitutes' jargon is a man who ill-treats her or demands abnormal sexual activity. To *freak* is to take an illegal hallucinogen and to *freak out* is to be under its influence.

free[1] unmarried.

The use deprecates the ties imposed by wedlock. Less often as *in freedom*:

> I've been alone all this time, I've stayed in freedom because of you. (Murdoch, 1978, explaining a reason for not having married)

A person described as being *free of fumbler's hall* was supposed to be unable to impregnate his wife.

free[2] inducing you to buy something which you may not want.

You are encouraged to buy something else at a price which includes the cost of the *free* item. Advertising jargon which persuades the gullible to subscribe to a magazine, buy a packet of cornflakes, take out insurance, etc.

free a man for duty to appoint a female to a military non-combatant role.

Perhaps not a euphemism even for those who believe that women are physically capable of sustaining the role of a fighting soldier. The usage is not popular among feminists.

free from infection not suffering from venereal disease.

In the army, a soldier can have measles and a heavy cold, but still be so described. Usually abbreviatied to *FFI*.

free love unrestricted copulation outside marriage.

The use implies an absence of concealment and disregard of convention. It is used for either sex:

> Dismal free love at a summer camp. (G. Greene, 1932)

free of your hips willing to copulate casually.

Of a woman, whose hips may play some part in the action:

> Free of her lips, free of her hips. (Old proverb)

free relationship licence within a partnership to copulate with third parties.

Usually denoting a heterosexual marriage, with the implication perhaps that a normal union in which the partners copulate only with each other involves sexual servility:

> Our marriage had broken up over my jealousy. Esther wanted a free relationship. (McCarthy, 1963—Esther fancied other men)

free samples copulation permitted by a woman prior to marriage.

From a taster or trial quantity offered by a trader. The euphemism was used when there was a prospect of engagement, and of betrothed couples.

free trade *obsolete* smuggling.

It was a way of avoiding excise duty:

> My father let me have a horse from the stable and a ling-tow over my shoulder to go out to the free trade among the Manxmen. (Crockett, 1894)

In commercial use, *free trade* is commerce between international partners without tariff restrictions, although it seldom operates freely due to non-tariff barriers and other factors.

free world the countries not under Communist control.

Any other tyranny can be included, whatever the imperfections of its political arrangements:

> The Western countries call themselves collectively the 'Free World'. (Jennings, 1965)

freebie see FREELOADER.

freedom fighters terrorists.

> We are not murderers...we are freedom fighters against international imperialism. (Sharpe, 1979)

Even when opposing an autocratic regime, they normally seek to replace it with autocracy.

freelance (of a woman) to copulate regularly with different men.

A complex pun on *lance*, to copulate (though normally of the male), being *free* from involvement with a pimp or not even demanding payment, and the *freelance* who works for more than one employer. Such a woman is called (among other things) a *freelance* or *freelancer*. Occasionally of males in the same sense.

freeloader a thief or systematic cadger.

Less often one who simply steals:

> Though gas meters were considered more difficult to tamper with, this had not deterred some ambitious free loaders. (Hailey, 1979)

Normally of those who gormandize at receptions etc.:

Only 400 of the most abject freeloaders bothered to turn up. (*Private Eye*, March 1980)

An occasion when *freeloaders* have the chance to indulge their gluttony is called a *freebie*. To do so is to *freeload*.

Freemans cadged cigarettes.

Army use, referring to a fictional brand supposedly smoked by habitual cadgers. See also *drink at Freeman's Quay* at DRINK[1]. The obsolete British *freeman of Bucks* was a cuckold, punning on the county and the horns.

freeze[1] an attempt to contain public expenditure by reducing wages.

In an inflationary economy, absence of an increase effectively reduces pay. Also used to refer to an attempt to restrain the inexorable increases in public expenditure:

There aren't any music teaching jobs, said Michael, they've all been cut back in the freeze. (Lodge, 1980—Michael was complaining that those requiring such training might be asked to pay for it themselves rather than getting it provided free by the community)

freeze[2] refusal by a female to copulate with her normal partner.

A practice as old as time, especially as a weapon to discipline or to secure assent. First noted in Australia but now common.

freeze off to kill.

The common chilling imagery:

Frisky Lavon got froze off tonight. (Chandler, 1939)

freeze on to to steal.

From the adhesive quality of ice. Referring to minor speculation and stealing by finding.

freeze out to eliminate minority shareholders unfairly.

Commercial jargon, from the meaning 'to exclude arbitrarily'. British law provides scant protection for minorities in an unquoted company and almost none if their interest amounts to less than 25 per cent. Thus the majority can ignore them and do, or neglect to do, things which may oblige the minority shareholders to accept their fate or sell out on unfavourable terms.

freezer a prison.

The frequent American *ice* imagery:

You didn't spend three days in the freezer just because you're a sweetheart. (Chandler, 1953)

French[1] see FRENCH WAY (THE).

French[2] an excuse for swearing.

You pretend the taboo word is foreign:

...not when some poor fucker...you'll excuse the French, Mr Carter... (Seymour, 1980)

French ache syphilis.

We all name bad things after our enemies—treachery for the Romans was *Punic faith* and for the Carthaginians *Roman faith*. Shakespeare refers to the baldness caused by this supposed import:

Some of your French crowns have no hair at all. (*Midsummer Night's Dream*)

In former times you might, if so unlucky, contract *French disease, fever, gout, measles*, or *pox*, and so become *Frenchified*, syphilitic.

French article brandy.

A euphemism of smuggling, which passed into general use and is now obsolete. Also as *French cream, elixir,* and *lace*. A *Frenchman* was a single bottle of brandy. I'm not sure why in Ireland *French cream* was whiskey:

Might he have the pleasure of helping her to a little more of that delicious French cream. (Kennedy, 1867, of whiskey)

French blue an illegal narcotic.

Usually amphetamine, from its colour and the country of manufacture.

French leave unauthorized absence.

Usually of a soldier, implying a propensity in French troops for desertion. Some civilian and figurative use:

We could still, if we wished, take 'French leave' of Vietnam. (McCarthy, 1967)

French letter a contraceptive sheath worn by a male.

The term may come from their being packed in small envelopes, coupled with the supposed Gallic penchant for frequent copulation, or from the use of the product in France and its quality of *letting* or preventing, as in the phrase *without let or hindrance*:

...keep in their bags not even small change, only a powder-puff, a lipstick, a mirror, perhaps some French letters. (G. Greene, 1932)

Abbreviated to *FL*:

Preyed on his mind, all those FLs did. (Sharpe, 1974)

Frenchie is rarer:

You can't feel a thing with a Frenchie. You get more thrill with a pill. (Sharpe, 1976)

Also as *French tickler*

...you were screwing matron with a French tickler. (Sharpe, 1982)

Froggie is a naval corruption, from the derogatory name for a nation noted for the delicacy of frogs' legs. The 19th-century *French renovating pills* were taken in the hope of ending a pregnancy that earlier resort to a *French letter* might have averted (see Rawson, 1981).

French way (the) cunnilingus or fellatio.

From supposed Gallic tastes. The Victorians called it the *French vice*:

> ... sodomy, incest, buggery, or the fashionable 'French' vice clinically called *cunnilictio* and its corresponding variation, *fellatio*. (Pearsall, 1969, writing of the 19th century)

Sometimes shortened as a verb to *French*:

> Only fooled around with him a little. I wasn't Frenching him. (Wambaugh, 1975)

A *French kiss* is a kiss with the mouth open.

fresh drunk.

Perhaps a shortened form of *fresh in drink*, with *fresh* meaning 'lively', and so not far gone:

> He wa' to say drunk—on'y fresh a bit. (Pinnock, 1895)

Inns used to stay open all day in English market towns and drunken farmers used to return home *market-fresh*:

> ... was already 'market-fresh' when we started back. (*Cornhill Magazine*, 1896, quoted in EDD of a drunken farmer)

More logically the 19th-century Scottish *fresh* referred to sobriety before drinking intoxicants:

> There is our great Udaller is weel enough when he is fresh. (W. Scott, 1822)

fresh meat see DEAD MEAT[2], MEAT.

freshen a drink to serve more alcohol.

Formerly denoting the addition of more soda water to a partly drained glass, but now you add whisky etc.:

> 'Let me freshen your drink.' Delaney said. He went over to the liquor cabinet, came back with new drinks for both of them. (Sanders, 1973)

freshen up to urinate.

The standard American invitation to an arriving traveller, who may also change his shirt, defecate, and take a shower:

> Why don't you just freshen up and then stroll on down the path, first right, to my lodge? (M. Thomas, 1980—the standard reply is 'Thank you, I'd be glad to wash my hands.')

fricasseed drunk.

The common cooking imagery. Of note also as an example of the American English conjugation of a French word.

fried drunk.

American culinary imagery again.

friend an extramarital sexual partner.

Heterosexual:

> You got a friend that don't work and a husband that works, you're all set. (Chandler, 1943)

or homosexual:

> I have a very nice friend. It's against the law of course. (G. Greene, 1932)

See also LADY FRIEND, WOMAN FRIEND, and MAN[1]. A *rich friend* is always the male in such a relationship. *Friends*, *close friends*, or *just good friends* are such partners:

> ... he mustn't say *good friends*, that was always taken as a euphemism for extreme intimacy (Price, 1974)

and:

> She managed to let me know ... that Dylan Thomas had once been a 'close friend'. (Fowles, 1977)

friend has come (my) *?obsolete* I am menstruating.

Punning on the arrival for a limited period and perhaps the relief at not being pregnant. Females also used to have *a (little) friend to stay*.

friendly[1] subject to colonial rule.

Churchill understood better than Roosevelt what Stalin meant at Yalta when he said he wanted to have *friendly* governments in eastern Europe after the Second World War.

friendly[2] lacking in accord or sympathy.

The language of diplomacy of discussions between mutually suspicious or antagonistic parties. The discussions may also have been described as *frank* (see FRANK[2]), in which case the parties probably came near to throwing punches at each other.

friendly fire being bombed or shelled by your own side.

The use seeks to play down one of the hazards of war:

> ... strafed and bombed by American planes. (Afterwards the ghastly error was described in military double-talk as 'friendly fire'). (Hailey, 1979)

An older version of BLUE-ON-BLUE or FRATRICIDE.

frig to copulate.

From *frig*, to rub, despite the etymological attractions of the Old Cornish *freg*, a married woman, and of *Frigga*, Odin's wife, the aptly-named Norse goddess of married love commemorated in the word *Friday*:

> I kept on frigging her with my man-root. (Harris, 1925)

It is also used of male masturbation, also from the rubbing:

> ... under a haystack in the country we gave ourselves to a bout of frigging. (ibid.)

You now only hear it in the expletive *frigging* for *fucking*.

frightener a person paid illegally to intimidate.

Usually for the collection of a usurious debt or to prevent the giving of evidence:

> 'Why are you bothering?' he asked. 'I don't like frighteners.' (Francis, 1988—the speaker was being so threatened)

fringe insubstantial or fraudulent.

Close to the edge of propriety or convention in the arts, as in *fringe theatre*, or of honesty in commerce:

The Bank of England's least favourite 'fringe' banker. (*Private Eye*, March 1981)

fringes extra payment in kind.

Short for *fringe benefits*, which some employees receive in addition to the emoluments they reveal to the Revenue or their stockholders.

frippet a sexually available young woman.

Probably from *frippery*, clothes, going back to *frip*, a scrap of cloth or something worthless:

I'll take my Bible oath *you've* got your little bit of frippet tucked away nice and convenient. (Amis, 1990)

Now standard English.

frog a policeman.

From his manner of walking or from jumping on delinquents? It is childish and insulting to call a Frenchman a *frog*.

froggie see FRENCH LETTER.

front[1] an organization hiding its real objectives so as to appeal to the gullible and well-meaning.

The method was first devised by Munzenberg, the Communist propagandist, and has been widely used by Communists and others ever since:

The World Committee for the Relief of Victims of German Fascism set the pattern for all future camouflaged 'front' organizations. (Boyle, 1979)

Today:

We have Fronts for this and Fronts for that ... One should always ask what ... is behind a Front. (Francis, 1978)

front[2] a seemingly honest person or business shielding an illegal operation.

Criminal jargon and practice, also used of espionage:

... invested in a wide range of new enterprises one of which was a 'front' for the Gehlen organization. (Allbeury, 1976)

To *front* for someone or something is so to act:

Where's the front money coming from? ... I think he's fronting for someone. (Deighton, 1981)

front door (the) copulation.

As distinct from the *back door*, the anus or buggery. The *front door* or *parlour* might also be the vagina, viewed sexually by a male:

You'll be able to hand out radical deliverance to both of them now. One at the front door, and one at the back. (Bradbury, 1975—a profligate lecturer was being addressed by his wife on the break-up of his marriage)

To *use the back door* is to bugger a female:

So if you're courting a girl from a bourgeois family and she's willing, she'll probably tell you to use the back door. (Moss, 1987, not of a female suggesting the use of the tradesmen's entrance but of a Brazilian girl expected to remain a virgin until married)

front line opposing.

Used to describe the black states lying to the north of and critical of South Africa when it was under white control, implying a state of constant belligerence rather than the actuality of economic dependence:

We would only upset a lot of front-line African states if we got involved. (Lynn and Jay, 1989, of an African political issue)

front loading the avoidance of risk through excessive down payment.

Commercial jargon for a common practice in international sale to an unreliable customer or a poor country. The surplus provides some hedge against cancellation or the inability of the buyer to complete the deal. It is facilitated by the eagerness of agents and those whom they are bribing to touch cash as soon as possible. Also as *front money* or *money up front*.

front office a police station.

From the location of main offices in American factories etc. at the front of the building.

front-running dealing illegally as an insider.

A dealer in a commodity market may anticipate price movements through prior knowledge of orders about to be placed on behalf of third parties:

The alleged offences include 'front-running', in which dealers hurriedly execute trades for themselves, knowing prices will move in their favour by client orders they have just received. (*Daily Telegraph*, August 1989, of Chicago commodity dealers)

frontier guards troops used for an invasion without a declaration of war.

The Chinese Communists so described their armies which invaded India in the 1960s and Vietnam in the 1980s:

Then a reference to 'Chinese Frontier Guards' alerted me. (Naipaul, 1964—he had thought he was listening to an Indian broadcast about the invasion until the euphemism told him the source was Chinese)

... Chinese soldiers fighting in Vietnam but marked as 'Frontier Guards in South China'. (Theroux, 1988)

fruit[1] a male homosexual.

Which came first, the RAISIN or the *fruit*? I suspect *raisin*, from the French meaning 'lipstick', but this conjecture is unsupported:

Pastor was screwing that Mexican fruit. (Deighton, 1972)

The punning *fruit picker* is a male who occasionally seeks a homosexual partner. In Far Eastern use, a *fruit fly* is a prostitute:

> ... sailing two fruit-flies as scrub-women (greasy overalls covering silk cheongsams). (Theroux, 1973, of smuggling prostitutes out to a freighter)

fruit[2] see FRUITCAKE.

fruit machine a mechanical gambling device.

From the symbols on the rotating discs in the early versions:

> As army-surplus dealer, a scrap-metal merchant, a fruit-machine importer—or a property man. (Green, 1979)

The alternative name, *one-armed bandit* from the actuating lever, is more fitting.

fruit salad[1] a mixture of illegal narcotics.

Supplies are pooled and each participant in the *fruit salad party* is meant to take a bit of everything. A *fruit salad* is also what you end up with if you take one pill of each narcotic from the family medicine store.

fruit salad[2] medal ribbons.

A derisive British use, from the multicoloured cloth used for campaign ribbons, especially those worn by the US military.

fruitcake a mentally abnormal person.

From the cliché *as nutty as a fruitcake*:

> God knows they've got their share of armed fruitcakes. (Lyall, 1985)

Occasionally shortened to *fruit*, it can refer to anything from mild eccentricity to lunacy.

fry *American* to kill or be killed.

Of judicial killing through electrification:

> Frying some druggie-pirate-rapist-murderers would surely appeal to the citizens of the sovereign state of Alabama. (Clancy, 1989)

and less often of other killing:

> If I don't get 'em off in them, they'll fry. (Marmur, 1955, of soldiers surrounded on a hostile beach, subject to attack by the enemy and not sunstroke)

fuddled drunk.

Literally, confused, and descriptive of the drunken state in which you think you can drive but cannot find the keys of the car.

fudge[1] to deceive by making wrong entries.

Especially of falsification of accounts, being a corruption of the standard English *fuddle*, to confuse:

> Perhaps he had been fudging his tax returns. (Chandler, 1958)

To *fudge an issue* is a correct use, albeit something of a cliché.

fudge[2] *American* to masturbate a person of the opposite sex.

I suspect this use comes from the meaning 'devise as a substitute' (WNCD).

fulfilment copulation.

Literally, the accomplishment of anything:

> In the corners couples embraced and fondled, stopping just short of actual fulfilment. (Bradbury, 1959)

full drunk.

It survives in the Scottish *fou*:

> The cup that cheers, but maksna fou (Tester, 1865—he meant tea)

and in various American and Australian clichés of which the commonest is *full as a tick* and the least common *full as a boat*.

full and frank see FRANK[2].

full-figured fat.

Of women rather than men. To be other than *full-figured* is not to be short of an anatomical appendage. Dressmakers advertise their wares for women with the *fuller figure* and:

> Arabs and Turks are said to appreciate the fuller figure in a woman. (A. Waugh, *Daily Telegraph*, 11 July 1994)

Full-bodied is also occasionally used in the same sense.

full in the belly pregnant.

Not merely having eaten a hearty meal. In various forms:

> He had run away from a girl with a full belly and a father with a loaded musket. (Monsarrat, 1978)

full treatment (the) copulation.

The language of those heterosexual brothels which operate under the cover of massage parlours etc.:

> Is it just your neck that's giving you trouble, or do you require the full treatment? (Matthew, 1978)

fumble to copulate with.

Literally, to use your hands awkwardly, whence to caress:

> The dish you was trying to fumble up the hall. (Chandler, 1958, and not of a waiter)

A *fumble* is a single act of extramarital copulation by a male:

> I must have carried twenty females to the barges (and none of them worth even a quick fumble). (Fraser, 1975)

An 18th-century *fumbler* was a sexually impotent male, who could do no more than caress a woman, whence *free of fumbler's hall* (see at FREE[1]). In American English, if you *fumble for a check*, you do not copulate for payment but try to get out of paying for a shared meal.

fun extramarital copulation.

Originally *fun* meant 'a hoax or trick', whence amusement:

> Country gentleman, 45, wealthy, tall, educated, is looking for an attractive young mistress. For fun. (advertisement in *Private Eye*, April 1980)

Fun and games is general sexual promiscuity:

> She was a bit of an all-rounder. Both sexes. General fun and games. (Davidson, 1978)

Fun-loving means 'promiscuous' (used of either sex), or addicted to alcohol, of a male:

> The *Washington Post* had described him as 'fun-loving', which was journalese for a hearty preference for alcohol or sex. (M. Thomas, 1980)

Fun stick, the penis or 'prick' in punning rhyming slang, is happily rare. A *fun-house* is a brothel:

> I'm exaggerating, but it *was* splendidly furnished, with more mirrors than a fun house. (Sanders, 1986)

funeral director an arranger of funerals.

Pretentious rather than euphemistic perhaps, although I hesitate before disagreeing with Hugh Rawson, who gives *funeral service practitioner*, *sanitarian*, and the delightful *embalming surgeon*, among other synonyms. I would also find myself in conflict with the guru of American funerary jargon:

> Certain American euphemisms—'funeral director' instead of 'undertaker'. (J. Mitford, 1963)

funny[1] unwell.

From the meaning 'strange' or 'unusual', usually as to *feel funny* or have a *funny tummy* etc.

funny[2] drunk.

We hesitate to blame our indisposition on our intemperance.

funny[3] homosexual.

Of a male, from the meaning 'odd':

> And you said last night he was 'that kind'... funny, kinky. (Bogarde, 1981)

funny[4] mad.

Again from the oddness and usually in the phrases *funny farm*, *house*, or *place*, an institution for the insane:

> Wasn't that the first picture of Pound to appear after he was let out of the funny farm? (Theroux, 1978)
>
> ... if Harold were really worried about joining his mother in the funny place, he should see a psychiatrist. (Wambaugh, 1975)

funny money cash that cannot be spent openly.

Of counterfeit notes or the proceeds of vice:

> As quick as he finds out that's funny money he'll put the finger on you. (Weverka, 1973)

furlough *American* involuntary dismissal from employment.

Literally, paid leave of absence, whence suspension from duty without pay and then dismissal. Used in America in the 1980s for the dismissal of airline pilots and cabin staff.

furry thing a rabbit.

British seamen must not mention rabbits before putting to sea, under an old taboo based on the substitution by fraudulent chandlers of rabbit meat, which did not keep, for salt pork, which did. In obsolete British use a *furry tail* was a worker who refused to join a trade union, a *rat* to some.

fuzz the police.

Probably a shortened form of *fuzzy bear*; see also BEAR[2]:

> The fuzz—that's what they call them now, not cops any more. (Ustinov, 1971)

fuzzed drunk.

Literally, blurred, which things tend to become at a certain stage. Also in American use as *fuzzy* or *fuzzled*.

G

G anything taboo beginning with the letter *G*.

It may be a mild expletive, usually spelt *gee*, a shortened form of *jeez*, from *Jesus*; or a verb meaning 'to rob', perhaps from the American *gyp*; or the leader of a *gang* of convicts; or a *gallon* of whisky. The commonest, and non-euphemistic, use of *G* is for a *grand*—$1,000. *Geed up*, crippled, may have arrived in America from the old Scottish dialect word *jee*, crooked:

> On a sair jee'd moss-grown stane. (Mucklebackit, 1885)

The American *G-man* is a special kind of Federal policeman, with *G* for 'government'. A motion picture graded *G* can be shown generally.

gage[1] cheap and unpalatable whisky.

The American container so called holds a quart. Whence *gaged*, drunk, but not necessarily on whisky.

gage[2] a marijuana cigarette.

Is derivation from the token of defiance, the measurement of the quantity, or merely a shortened form of *engage*? None is convincing. To *gage* is also to ingest illicit narcotics:

> The people who didn't smoke or gage, get razored in barrel-houses... (Longstreet, 1956)

gain to steal.

In the 15th century *gain* was booty. Today the verb means 'to acquire something of small value without payment or detection'.

gallant a woman's extramarital sexual partner.

In standard use, an adjective meaning 'chivalrous', which may be in contrast to her boring or boorish husband:

> Elspeth would be back in the saddle with one of her gallants by now. (Fraser, 1971, of a profligate wife)

Today *over-gallant* describes a male's unwanted sexual attention to a female:

> Sammy was... How shall I put it? I think the kindest way would be 'over-gallant'. (Boyd, 1982)

In obsolete use to *gallant to* a woman meant to copulate with her extramaritally:

> Is it the case you had been gallant to her before marriage? (Galt, 1826)

Gallantry was extramarital copulation by either sex:

> She was not without a charge of gallantry. (Hutchinson, of an adulteress in the mid-17th century)

gallop to copulate with.

Of males, using the common equine imagery:

> ... beaky, sharp-eyed old harridans whom I wouldn't have galloped for a pension. (Fraser, 1971)

A *gallop* is an act of copulation or the woman with whom you do it:

> She was a fine, rousing gallop, all sleek hard flesh. (ibid., of a woman)

To *gallop your maggot* was to masturbate, of a male.

game[1] wild animals killed primarily for human amusement.

Standard English for those hunted in the wild, birds conserved so that they can be shot, and certain large fish. *Big game* describes large quadrupeds, mainly in Africa, where they have not yet been hunted to extinction.

game[2] **(the)** prostitution.

The same imagery as *sport* (see SPORT (THE)) but a serious business if it is your livelihood:

> I'm old at the game (Harris, 1925, quoting an old prostitute)

and for Boswell the *noble game* was copulation, with an actress whom he paid. A working prostitute is said to be *on the game*:

> Every girl in Bayswater bangs to him if she wants to stay on the game. (Turner, 1968)

In the game means no more than being a prostitute already:

> They don't take only women who are in the game already. They get hold of innocent women. (Londres, 1928, in translation, of pimps)

In obsolete use a *game pullet* was a young prostitute and a *gamester* any prostitute:

> She's impudent, my lord, and was a common gamester to the camp. (Shakespeare, *All's Well*)

Game, of any female, indicates a supposed willingness to copulate casually. The jocular *national indoor game* is copulation. At one time, if you were detected in adultery in Scotland, the church demanded a *game fee*:

> Niest ye maun pay down the game fee,
> An' nae mair we sal trouble thee.
> (Liddle, 1821)

gamester[1] see GAME[2].

gamester[2] a gambler.

Gaming has meant 'gambling' since the 16th century, because people have long wagered on the outcome of games:

> The credit of a race-horse, a gamester, and a whore, lasteth but a short time. (Torriano, 1642—surely he meant 'or a whore')

In the ABC used by myself as a child and by my children:

> G was a gamester who had but ill luck (and we saw no impropriety in 'U was an usher who loved little boys')

gander-mooner *obsolete* a husband copulating outside marriage.

The month after the birth of a child was known as the *Gander Month* or *Gander Moon*, from 'the month during which the goose is sitting when the gander looks lost and wanders vacantly about' (EDD). For this period the English father and husband was supposedly given licence to copulate with someone other than his unavailable wife. Both the use and the giving of licence are obsolete.

gang

This has many of the euphemistic uses of GO, some obsolete and some Scottish.

gang-bang see BANG[1].

gap the vagina.

From the opening between the thighs. Some medical use.

The obsolete Lincolnshire *gap-maker* was a poacher, who might break through a hedge to avoid detection.

garb of Eden nakedness.

Without even a fig-leaf:

> ... usually clothed in her 'garb of Eden' —starkers. (Theroux, 1992)

garçonnière an apartment used by a married man for copulation with his sexual mistress.

Literally, the French word means 'a bachelor flat':

> ... the distinguished brick-faced town house in which he kept his garçonnière. (M. Thomas, 1982, of such a place)

garden[1] the vagina viewed sexually by a male.

If he is so vulgar as to use the expression, he sees it as a place for cultivation, or pleasure, or both. In rarer use as *Garden of Eden* or *Pleasure-garden*.

garden[2] to sow mines in water from the air.

It is important to adhere to a pattern and this Second World War imagery is from planting bulbs etc. The use comforted the airmen by avoiding explicit lethal terms and also reflected the comparative safety of such an operation.

garden crypt *American* a drawer facing outward in a store for corpses.

Funeral jargon, to persuade living customers to pay more so that the dead may enjoy the ambience:

> Crypts facing outside ... are now called 'garden crypts' ... 'It's all part of the trend towards outdoor living,' explained the counsellor. (J. Mitford, 1963)

garden gout see COVENT GARDEN.

garden house see COVENT GARDEN.

garden of remembrance the curtilage of a crematorium.

Usually a few seats, some roses, a path, and a lawn, all of which are soon forgotten:

> There is something comfortlessly empty about a 'garden of remembrance' after the loquacious populated feeling of a graveyard. (Murdoch, 1978)

A *Garden of Honor* is the part of an American cemetery in which you can pay to put up a plate naming a dead soldier.

gardening leave suspension from office on full pay.

Often a preliminary to EARLY RETIREMENT for a senior manager or civil servant, who may well live in a town house without a curtilage, or employ someone else to tend the lawns and rosebeds:

> ... given £228,000 in redundancy payments after just nine weeks on an American posting and eight months 'gardening leave'. (*Daily Telegraph*, 5 February 1994)

gargle to drink intoxicants.

Literally, to suspend a liquid in the throat for medical purposes:

> 'Let's ... gargle.' He poured drinks. (Chandler, 1939)

Gargle as a noun is usually whisky.

gas deliberately to kill or injure by poison gas.

Of troops in the First World War; civilians in modern Iraq; the chronically unfit, gypsies, and Jews under the Nazis; and convicted murderers in some American states:

> He's not around any more to be asked. They gassed him. (Chandler, 1953)

Occasionally as *get the gas pipe*:

> You may go down the toilet there, Victor, but I get the gas pipe. (Diehl, 1978—Victor was liable merely to imprisonment)

gash a woman viewed as an object of casual copulation.

Although *gash* is a vulgarism for the vagina, this use probably comes from the meaning 'something acquired for nothing' or 'surplus to another's needs':

> Maybe there's some of that Swedish gash hanging around. (Sanders, 1977—men were looking for women to pick up)

gassed intoxicated.

Possibly from the way you may feel or act, or from the slang American *gas*, satisfying, or from taking the slang *gas*, an inferior crude spirituous drink. A *gas-house* is a saloon which sells beer.

gastric flu diarrhoea.

Gastric refers to the stomach and *flu* or influenza is a viral disease which most of us confuse with a common cold, to obtain sympathy or excuse absence from work, but they don't add up to diarrhoea unless you wish to disguise the precise nature of your affliction out of prudery or delicacy.

gate[1] to confine to college as a punishment.

Originally used with reference to those colleges in Cambridge and Oxford which had formidable barriers to prevent unobserved access and formidable porters in the gatehouse.

gate[2] **(the)** peremptory dismissal from employment.

The exit for the last time from the factory, which you are *given* in America and *shown* in the British Isles:

> Amtrak board facing the gate. (*New York Post*, September 1981, of their threatened dismissal en bloc)

Like many of the 'dismissal' euphemisms, it can also mean the summary unilateral ending of courtship, almost always by the woman.

gathered to your fathers dead.

You do not necessarily have to be buried in a family vault. Less often we may be *gathered to our ancestors* but not, as yet, to *our mothers*. Legislation apart, this last is unlikely because the use is based on deep tribal concepts, and the human is rare among mammals in requiring the mature female rather than the male to leave the tribe. For those interested, this phenomenon is also reflected in most primitive conceptions of incest—the taboos are much stronger against marrying paternal relatives than maternal.

You may also be *gathered to God:*

> Jane's father Patrick had been gathered to God some six summers... (Fry, 1994)

gay[1] enjoying or doing something that is the subject of a taboo.

Literally, happy or cheerful, a standard English use now vanished under the pressure of GAY[2]. In the 19th century to *gay it* was to go with prostitutes, a *gay house* was a brothel, and the *gay life* was that of a prostitute who might also be called a *gay lady* or *girl*:

> I went through all the changes of a gay lady's life. (Mayhew, 1862, quoting an old prostitute)

Until the 1960s, *gay* meant 'intoxicated' or under the influence of narcotics:

> It wasn't a very serious crime—getting three amorous Kanaka girls gay on... gin. (Alter, 1960)

gay[2] homosexual.

Now virtually standard English, from GAY[1]:

> Investigations were proceeding with a gay club. (Davidson, 1978)

Gay Liberation Front has the distinction of combining three euphemisms—see LIBERATE[3] and FRONT[1]:

> Now if this Piper was a gay liberationist Jew-baiter with a nigger boyfriend from Cuba called O'Hara. (Sharpe, 1977)

Gay deceivers are not worn to confuse homosexuals but are pads worn in the clothing to enhance a bust.

gazelles are in the garden something is not quite as it should be.

Said when someone wants to tell you your nose is dripping, your trouser zip is undone, or as the case may be.

gear anything that is the subject of secrecy or taboo.

Literally, equipment. In obsolete Scottish use it meant 'smuggled spirits':

> There were... two kinds of the lads who brought over the dutiless gear from Holland. (Crockett, 1894)

In modern sexual use, the male or female genitalia. In the jargon of narcotics users, the *gear* is the apparatus you use to introduce the substance into your body. A burglar's *gear* is his specialized tools. *Gear*, homosexual, may refer also to the indicative clothing worn, and in any case it is rhyming slang for 'queer'.

geared up intoxicated.

American use, perhaps from GEAR, or from a feeling of acceleration imparted by the alcohol. I have no record of its use with reference to narcotics.

gears have slipped mind is deranged.

Motoring imagery—you still move but ineffectively:

> Her gears have slipped. Not a lot, but some. (Sanders, 1982)

Similarly they may not mesh:

> It was just that the things she said and did were slightly askew. Her gears weren't quite meshing. (Sanders, 1986)

geezer an intoxicant or illegal narcotic.

Perhaps from the Scottish *guiser*, a drunkard, which came from *guised in liquor*—or perhaps not, since it is also a term of derision (a *silly geezer*); or, from the Icelandic, something which gushes, which drunkards do when they are sick. (Whence the old gas-fired *geyser*, which used to gush hot water into the bath, if it did not explode first.) *Geezed up* means 'drunk' or 'under the influence of narcotics'.

gender-bending the deliberate rejection of the characteristics of a sex.

No longer a pupil's struggles with Latin grammar. Usually denoting flagrant homosexuality or bisexuality in dress, etc. A *gender-bender* is a homosexual.

gender norming accepting different standards from women.

A phenomenon of public sector employment, where the expense is borne by the taxpayer and there is thought to be political advantage:

> ...uncongenial to most ordinary soldiers, whose prospects for promotion are already limited by 'gender norming'—the deliberate skewing of test results to make sure that more women pass. (*Sunday Telegraph*, 11 April 1993)

The British forces (and taxpayers) subsequently had to pay vast sums in compensation to pregnant women who left the service because the authorities had been so stupid as to think that motherhood and a military career were incompatible.

general discharge dishonourable dismissal from the US forces.

Normal people get an *honorable discharge*.

genital sensate focusing digital foreplay to copulation.

Medical jargon which borders on circumlocution:

> He would have been at it anyway without having ever heard of genital sensate focusing. (Amis, 1978, of such activity)

gentle art (the) copulation.

As Voltaire and others have pointed out, it does not especially bring to mind tameness or moderation:

> ...a fine, fat little rump she was...but no great practitioner of the gentle art. (Fraser, 1985)

gentle people fairies.

Fairies, prior to the Christmas pantomime, were nasty people and to assuage them, you attributed to them the major quality in which they were defective. In Ireland they were also known as the *gentry* but I am not sure whether this was because the local gentry were worthy of respect, or because the gentry would be Anglo-Irish, and as nasty as the fairies themselves:

> Biddy was known, too, to have the powers of seeing the 'gentry', beings who creep out from every mousehole and from behind every rafter the minute a family has gone to sleep. (Lawless, 1892)

Hawthorns were called *gentle thorns* despite their pricks, because the fairies put spells on them.

gentleman was once used of any occupation or state that was the subject of vilification or taboo. Thus in obsolete British use he was without work, in the days when that also meant without money, a grim joke on the wealthy who did not need to work:

> He is a gentleman now, without seeking the shelter of the workhouse. (O'Reilly, 1880)

If a tramp, he was a *gentleman at large*. The *gentlemen* were smugglers:

> If the gentlemen come along don't you look out o' window. (Egerton, 1884)

A *gentleman of fortune* was a pirate and a *gentleman of the road* a highway thief, who might also be called a *gentleman's master*, because gentlemen stood and delivered to him. A *gentleman's gentleman* was a personal servant or valet. Then there were numerous oblique descriptions of work or taste, whereby you might be described as a *gentleman of the cloth*, a tailor (not a parson), *gentleman of the quill*, a clerk, *gentleman of the scalpel*, a surgeon, *gentleman of the back door*, a bugger, etc.

gentleman commoner *?obsolete* an empty bottle of intoxicants.

In the 18th century certain wealthy students at Cambridge and Oxford were so called because they paid higher fees, enjoyed various privileges, and were thought by some to be empty-headed.

gentleman cow a bull.

As we note elsewhere, polite 19th-century Americans talked about bulls and cocks, if at all, with such ingenious evasion that we are left wondering how they described a *cock and bull story*. A bull might also be called a *gentleman ox*. For a list of similar euphemisms see BIG ANIMAL.

gentleman friend an older man with whom a woman regularly copulates extramaritally.

Mainly female use. He does not have to be 'of gentle birth' or behave chivalrously towards her or anyone else.

gentlemen a lavatory exclusively for male use.

Less often in a compound by the addition of *convenience* etc. than is the case with LADIES, but often shortened to *gents*:

> I always thought wearing a kilt was a pretty daft idea, but they do save time in the Gents. (*Private Eye*, August 1980)

gentry see GENTLE PEOPLE.

geography the location of a lavatory.

In genteel usage to a visitor, to avoid the need for exploration:

> Let me show you the GEOGRAPHY of the house. (Ross, 1956)

George a turd.

British rhyming slang from *George the Third*, whence also defecation, as in the expression *a morning George*. In dated American slang *George* meant 'pleasurable', whence, briefly, to entice into copulation.

Georgian old.

British estate agents' jargon for a house usually in poor repair. As the first four Georges ruled from 1714 to 1830, and the last two from 1910 to 1952, the net is spread quite wide.

German distorted to fit Nazi dogma.

Everyone writes history from a nationalist viewpoint and excess of hindsight, although the

Nazis and Communists brought to the task added ruthlessness and cynicism. The Nazis set themselves two further targets—moulding scientific facts to fit their often bizarre theories, and excising any Jewish contribution to learning. The resulting *German* chemistry, mathematics, etc. would have been ludicrous had not the tragedy been so awful. *Germanization*, the adoption and adaptation by the Nazis of anything foreign, led also to ridiculous and horrifying behaviour, such as the placing of fair-haired children from occupied countries in Nazi families to be brought up as Germans:

> It is believed they were the rejected ones from the Germanization program. (Styron, 1976, of mass killings of such children)

German Democratic Republic see DEMOCRATIC, DEMOCRACY.

Gestapo see SECRET (STATE) POLICE.

get a bullet to be killed violently.

Usually by being shot, but also by any other method.

> He will probably finally learn to understand the true nature of bolshevism only when he gets his bullet. (Goebbels, 1945, in translation, referring to the Finnish Prime Minister Paasikivi)

get a leg over to copulate.

This expression and its variants *get your leg across* and the obsolete *get your leg dressed* denote copulation by the male. Many of the 'copulation' phrases starting with *get* are dealt with elsewhere and to avoid repetition and tedium this entry ignores them. For example, *get in the saddle* may be located at SADDLE UP WITH. The expressions afforded a separate entry are not conveniently noted elsewhere.

get a marked tray *?obsolete* to have contracted venereal disease.

American hospital jargon. To avoid cross-infection, the crockery was not used by other patients.

get along to grow old.

I suppose a shortened form of *get along the road of life* or *in years*:

> He be gettin' along, and we can't expect him to be as nimble. (Hayden, 1902)

Get on is standard English for reaching old age:

> ...socialists conceding the excellence, which they could afford to do since there was only one of him and he was getting on. (N. Mitford, 1949)

Up along in English dialect means 'elderly'.

get away *Scottish, obsolete* to die.

The soul departs from this tiresome place:

> The Laird, puir body, has gotten awa. (Thom, 1878)

If a soldier *gets it*, he is killed:

> Richards got it in Danang. (Theroux, 1973, of a death in Vietnam)

A dying Christian may *get the call* from a waiting deity.

get down to business to copulate.

Of either sex, usually extramaritally:

> ...going to bed with Lola in a great creaking four-poster which swayed and squealed when we got down to business. (Fraser, 1970)

get fitted to wear a contraceptive device.

Short for *get fitted with the loop* etc. and usually of a female for the first time:

> ...asking them if they would like to come in and, as he puts it, get fitted. (Bradbury, 1976, of young women)

get in her pants to copulate with a female casually.

It does not imply that you are a transvestite:

> He'd tell a woman *anything* to get in her pants. (Sanders, 1977)

get into bed with to copulate extramaritally on a single occasion.

> ...to get voluntarily into bed with a wanted murderess. (Sharpe, 1979)

get into her bloomers *American* to copulate with a female casually.

A version of GET IN HER PANTS:

> ...those motel units where you're planning to get into my bloomers. (Sanders, 1982)

get it in/on/off/up to copulate.

Of a male; see also IT[2]:

> I know a pillar of the community who gets it off with alligators. (Sanders, 1982)
>
> He was too drunk to get it up even with the help of a crane, darling, but he paid his ten dollars. (Archer, 1979)

Get it on is also used of male homosexuality:

> ...an amusing set of photographs of one man getting it on with a couple of sailors. (M. Thomas, 1980)

get laid to copulate or have copulated.

Usually extramaritally:

> A place where even the most diffident foreigner can get laid. (Theroux, 1975)

This expression has totally supplanted the English dialect *get laid*, to secure rest and quiet. A lady would be misunderstood today if she said

> I couldn't git myself laid for the noise he mead. (EDD)

get off[1] to achieve an orgasm.

Of either sex:

> At my age, just getting off takes my breath away. (M. Thomas, 1980—not referring to subsequent disengagement from the female)

get off[2] to marry.

Of a woman who might be anxious to find a husband. The phrase is also used intransitively when a mother wants to see marriageable daughters suitably married:

> You'd think she'd want to get her off all the quicker. (N. Mitford, 1949, of such a mother)

get off[3] to take narcotics illegally.

From the feeling of floating. To *get on* is to start taking illicit narcotics regularly.

get off with to pair sexually with another.

Of either sex, usually of a person met socially on a single occasion.

get on see GET ALONG, GET OFF[3].

get on your bike to be dismissed from employment.

> They'll still keep him on. There's no talk at all of telling him to get on his bike. (*Private Eye*, July 1980)

From the days when the majority cycled to and from work. In the same period an unemployed person would also use a bicycle in an effort to find work, as suggested by the British minister Norman Tebbitt in 1981, describing how his father, when unemployed in the 1930s, 'got on his bike and looked for work'. The expression *on your bike!*, meaning 'get moving!', has come also to mean 'make an effort to find employment for yourself', with some attributive use:

> There is ... an alternative last resort ... and that is the 'on yer bike' solution, going abroad to find new fortune. (*Spectator*, 19 December 1991)

get round to copulate with (a female) extramaritally.

From the meaning 'to cajole'.

get the shaft *American* to receive rough or unfair treatment.

With many variations as to the tactics or weapons which may be used:

> The executives continue to raise their pay and their perks while the workers got the shaft. (*New York Times*, 17 March 1992)

get the shorts to be insolvent.

From the shortage of funds:

> Suddenly he's got the shorts ... he can't come up with the scratch and he's hurting. (Sanders, 1977)

get the upshoot to receive vaginally the male ejaculations.

Another of Shakespeare's lewd puns:

> 'Then will she get the upshoot by cleaving the pin.' 'Come, come, you talk greasily; your lips grow foul.' (*Love's Labour's Lost*)

get there to copulate with a woman for the first time extramaritally.

> Never seen her before tonight in my life. Bet I get there, though. (Bradbury, 1959)

get through to copulate extramaritally.

Of a male, punning on making a connection.

get up to copulate (of a male).

Common use since Sir Walter Raleigh's 'getting up one of the mayds of honour' (Aubrey, 1696) and probably long before that.

get with child *obsolete* to impregnate (a female).

Within or outside marriage:

> At that time he got his wife with child. (Shakespeare, *All's Well*)

It should mean, literally, acquiring a step-child on marriage.

get your end away to copulate with a female.

get your end in to copulate with a female on a single occasion.

> We could both get our end in there. (Keneally, 1985—two men were discussing copulating with a woman)

get your greens to copulate regularly.

Of either sex and often within marriage, probably from the supposed benefits of a diet regularly containing brassica:

> She's not getting what I believe is vulgarly called her greens. (G. Greene, 1967)

get your hook into to copulate with (a woman).

It uses angling imagery:

> 'I'd like to get my hook into *her*,' Davis said. (G. Greene, 1978)

get your muttons *obsolete* to copulate.

Of either sex; see also MUTTON:

> They couple like stoats, by the way, but only with men of proved bravery ... you have to be blood-thirsty in order to get your muttons. (Fraser, 1977)

get your nuts off to copulate.

Usually of the male, implying ejaculation.

get your rocks off to copulate.

Of either sex, despite the maleness of *rocks*, the testicles:

> ... thanks for coming over, we got our rocks off. (M. Thomas, 1980)

get your share to copulate frequently.

Usually of a male, and one who has more than one partner:

'Everyone talks about what a stud he was'... 'He was getting more than his share even then.' (M. Thomas, 1980)

get your ticket punched to hold an office which will assist your career.

From the evidence that you have made a specific journey. Also as *have your ticket punched*:

He had come to Washington to have his ticket punched, that is, to hold down a Pentagon desk assignment, a pre-requisite in the modern Navy for being awarded the rank of admiral. (Colodny and Gettlin, 1991)

get your way with to copulate casually.

Of a male and again implying female reluctance.

get your will of to copulate.

Of a male; rather dated and implies reluctant submission by the female:

When he had got his wills o'her. (Kinloch, 1827)

ghost[1] a fictitious employee.

A name on a payroll, either invented or of someone who does not work, so that another can steal the wages:

As for the ghosts, some African governments have them on their payrolls. The Congo has just laid off 5,000 of them—fictitious employees... created to allow people to obtain five or ten salaries each month. (*Daily Telegraph*, 26 March 1994)

ghost[2] a writer whose work is published under another's name.

Used by public figures without literary expertise:

Ghost! Good God! The greatest political story of the century, and they're looking for a 'ghost'. (A. Clark, 1993, of Mrs Thatcher's memoirs)

ghost does not walk *?obsolete* the cast will not be paid.

Theatrical jargon, the *ghost* being the cashier. From the Hamlet of pre-union days, when only Marcellus spoke of striking, with his partisan.

gift of your body *obsolete* extramarital copulation by a female.

Usually of a virgin, and anyway *loan* would be more appropriate:

He would not, but by gift of my chaste body
To his concupiscible intemperate lust.
(Shakespeare, *Measure for Measure*)

Both the use and the concept are dated.

giggle stick[1] the penis viewed sexually.

Rhyming slang for 'prick' and punning on the stick for agitating a glass of sparkling wine, usually champagne or *giggle water*.

giggle stick[2] *American* an illegal marijuana cigarette.

From one of its effects.

gild to tell a lie about.

Literally, to cover thinly with gold, and perhaps alluding to that misquoted cliché *gild the lily*—Shakespeare actually wrote 'To gild refined gold, to paint the lily' (*King John*). Normally people *gild* the facts, the truth, etc.:

'He lied to me about the security clearance.' 'It's a bad word to use in law. I'd agree he gilded the proposition.' (West, 1979)

ginger a male homosexual.

From the uncommon rhyming slang *ginger beer*, queer. Sometimes written in full:

I can usually detect anything that's ginger beer. (B. Forbes, 1986, of Donald Maclean)

gippy tummy diarrhoea.

A corruption of *Egyptian tummy*, suffered by white visitors rather than the local inhabitants, who may also catch it anywhere in the world, it seems:

She knew she was in for a further attack of 'Gyppy tummy'. (Manning, 1977—she was about to get diarrhoea again)

Both spellings are common, as with *gipsy* and *gypsy*.

girl[1] a prostitute.

I suppose from the meaning, 'a sweetheart':

They turn the young Jewesses... into what are generically known as 'girls'. (Londres, 1928, in translation)

Often expanded to more explicit phrases such as *girl of the streets*:

The veritable girl of the streets is too 'vicious'. (ibid.)

To *girl* is to seek after prostitutes, and a *girler* is a womanizer:

I hear this Frank Sinatra's a fearful girler. (Theroux, 1978)

Girlie often indicates that the women are being exploited for pornography or prostitution, as in *girlie bars, clubs, houses, parlours*, etc., which are brothels:

...a front for the girlie house Billie ran upstairs (Weverka, 1973)

and:

...direct traffic up to Billie's girlie parlor. (ibid.)

A *girlie magazine* is an erotic publication and a *girlie show* holds itself out as being a lewd live performance by females.

girl[2] any female of less than 50 years.

Literally, a female child or servant. The use seeks to imply that the ageing process has been retarded:

...she was only a slip of a girl—what was she now—twenty-seven or eight. (Collins, 1981)

give to copulate.

Rare on its own, of male or female:

Maybe Bill gives at the office. (Sanders, 1982—not of charitable donations but of a man who did not copulate with his wife)

give a little to copulate occasionally.

Of a female, usually married or engaged to be married:

She still give you a little? (Wambaugh, 1975, of an ex-wife)

give access to your body to copulate.

Of a female extramaritally, perhaps for payment:

She decided to ... give all soldiers who wished to take advantage of her free access to her body. (Richards, 1936)

give green stockings see GREEN GOWN.

give head to practise fellatio or cunnilingus.

The old bastard had his son-in-law giving him head in the back seat. (Diehl, 1978)

give it to to copulate with.

Of a male, despite the convention that the female is the generous party:

You been giving it to her, have you? (Allbeury, 1976, referring to copulation)

give out to copulate.

Of a female. Not the opposite of giving in:

A guy buys gifts for his wife because he knows she won't give out if he don't. (Sanders, 1970)

give someone the air to dismiss someone from employment.

Presumably after working in an office or factory. The many variants of this are dealt with under the appropriate noun; see BAG[2] (THE), BOOT (THE), BULLET (THE), etc.

give someone the good news to kill someone.

Whatever can the bad news be like?

As the boy shouted Mark gave him the good news. His body disintegrated in front of my eyes. (McNab, 1993)

give the ferret a run to copulate.

Of the male, with obvious imagery.

give the time to to copulate with.

I was personally acquainted with at least two girls he gave the time to. (Salinger, 1951)

Of either sex. The etymology is obscure.

give time to other commitments to be summarily dismissed from employment.

A superficially face-saving form of words for senior employees:

[He] is 'giving time to his other commitments', according to the board. (*Daily Telegraph*, 8 October 1993—he was dismissed after less than seven months in office after 'disappointing figures')

Others similarly dismissed profess to wish to *give time to other interests*.

give to God to commit (a child) to a priestly or monastic life.

Most such donors were Irish:

Every good Catholic family, he says, gives someone to God. (Burgess, 1980)

give up the ghost to die.

The *ghost* is the spirit which you surrender to heaven, or as the case may be, when your body dies:

Man dieth, and wasteth away; yea, man giveth up the ghost. (Job 14:10)

There are many obsolete dialect expressions indicating that the dead will make no further demand on terrestrial resources, such as the Lancashire *give up the spoon*—you will eat no more:

Johnny gan up his spoon one day beawt havin' any mooar warnin' nor other folk. (Brierley, 1865, of a sudden death)

give up your treasure to copulate.

Of a female, for the first time and extramaritally:

The summer solstice, when maids had given up their treasure to fructify the crops. (McCarthy, 1963)

give way to copulate.

Of a female, usually after male urging and extramaritally.

give your all to copulate extramaritally.

Of a female, but not a testatrix:

Magill wasn't the first time I've given my Little All for my job. (Lyall, 1985)

give your body to copulate.

Of a female casually:

I loved a man, gave him my heart and, God help me, gave him my body. (Higgins, 1976—it sounds almost like helping a transplant surgeon)

give yourself (to) to copulate (with).

Of a female, with implied extramarital *giving* by a virgin and the implication that her virginity, once donated, cannot be returned:

She was resolved not to give herself completely (Harris, 1925)

or, naming the donee:

In small families the servants often give themselves to the sons. (Mayhew, 1862)

Also some homosexual use:

... despite his decision to give himself to me, he was postponing the moment of going to bed. (Genet, 1969, in translation, of male homosexuals)

given new responsibilities demoted.
Of managers whose performance is not quite bad enough to justify dismissing them:

> ... the two existing top managers ... have been given new responsibilities. (*Daily Telegraph*, 2 September 1994 under the headline 'Simpson shakes up Lucas')

given rig *Scottish, obsolete* an uncultivated piece of ground left to placate the devil.
Another version of the practice noted elsewhere of trying to mollify the devil by leaving him some land of his own:

> 'The Gi'en Rig', which was set apart or given to the Diel, to obtain his good will. (Gordon, 1880)

given to the drink see DRINK[1].

glands *American* taboo parts of the body.
An evasion when used to refer to the female breasts; misleading when the male testicles are intended.

Glasgow kiss a head butt.
Parts of the conurbation have an unenviable reputation for violence:

> This is a Glasgow kiss, I said, and butted him in the face. (Barnes, 1991)

glass *obsolete* an intoxicant.
Intoxicants in general:

> The glass marched pretty quick. (Cleland, 1749)

but now usually spirits:

> He, too, was happy to drink a glass. (Kyle, 1978)

except in Ireland, where a *glass* is half a pint of beer or stout. A *glass of something* is a spirituous intoxicant and a *glass too many* implies drunkenness.

glass ceiling the level above which certain categories of people are unlikely to be promoted.
It is there but cannot be seen. Mainly used by women in a hierarchical structure:

> 'Don't whinge about glass ceilings', is Prue Leith's advice to budding business women. (*Daily Telegraph*, 3 April 1995)

Also used on BBC News, 10 April 1995, with reference to a politician accused of impropriety, implying that he would not achieve further promotion unless he cleared his name. See also OBLIGATORY and STATUTORY.

glass house an army prison.
British army use, from the glass roof of the one in Aldershot, and perhaps too the figurative *heat* applied to the inmates.

glassy-eyed drunk.
From the vacant and watery gaze rather than the container which carried the damaging fluids.

glazed drunk.
An American version of GLASSY-EYED, using the same imagery.

glean to steal.
Literally, to pick up ears of corn left by reapers. Now a rare British use for pilfering small articles. At one time it also was applied to capturing stragglers from a losing army after a battle.

globes the breasts of an adult female.
Of obvious derivation and doubtful taste:

> There was a white and sunny skin,
> Revealing currents warm within,
> The Graceful peak where beauty sits,
> The swelling globes, the pouting teats.
> (Pearsall, 1969, quoting a verse from 1860)

glove money *obsolete* a bribe.
By ancient custom, you gave gloves to anyone who had done you a favour or who might be persuaded to do you one, concealing a bribe inside. Sir Thomas More, when Lord Chancellor of England, kept the gloves which Mrs Croaker gave him but returned the hidden £40. We should not then be surprised that he was later beatified.

glow *obsolete* to sweat.
Women and horses were said to *glow* from the visual effect of moisture on the skin. Sweat was the subject of taboo because of the smell from that secreted in the armpits and crotch, which in males has a biologically sexual function as with other mammals, although manufacturers of deodorants would have us believe otherwise.

glow on (a) a state of mild drunkenness.
From the sweating which you experience:

> I didn't feel like getting a glow on. Either I would get really stiff or stay sober. (Chandler, 1953)

glue *American* to steal.
Implying pilferage, the object sticking to you.

go[1] to die.
From the departure of the soul, or body, or both:

> ... he said 'I think I'm going, Peters.' He didn't speak again. (Manning, 1977)

Also found in innumerable phrases alluding to the manner of going, the destination, etc., a selection of which is given below. In many of them, the obsolete or dialect *gang* is found as well as *go*.

go[2] to become bankrupt.
As *go* (*tout court*) in the phrases *on the go* or *they've gone*, but usually in phrasal verbs.
The obsolete London *go due north* was to the debtor prison in White Cross Street, on the northern side of the city.

Go to staves or *fall at the staves* came from the collapse of a barrel when the hoops are removed. I have not traced the origin of the obsolete English *go up Johnson's* (or *Jackson's) end*, but many still GO FOR A BURTON, which used to mean no more than visiting a pub to drink beer brewed there by Bass.

Go smash is barely euphemistic:

> If a shopkeeper conducted his affairs upon such a principle we would go smash. (Flanagan, 1988)

The American *go Chapter Eleven* is to continue trading while insolvent, under court protection:

> The Lelands had first approached him in the summer of 1921, six months before they were driven to file Chapter Eleven. (Lacey, 1986)

The American Bankruptcy Reform Act of 1978 includes other 'Chapters', which denote differing degrees of insolvency, *Chapter Thirteen* indicating a more dire situation than *Chapter Eleven*.

See also GO WEST.

go[3] to urinate or defecate.

Short for *go to the lavatory*:

> ... especially Lally who was longing to 'go' as much as we were. (Bogarde, 1978)

As stated above, in many phrases *go* and *gang* were once more interchangeable than they are today; indeed, in obsolete British use a *gang* was a lavatory, probably from *gang* or *going*, a drain, while *geing* or *goung* was human excrement:

> No man shall bury any dung or goung within the liberties of this city. (Stowe, 1633, referring to London)

There are hundreds of *go* phrases for urination and defecation, of which, to avoid tedium and save paper, I give a selection only in the entries that follow.

go about your business to defecate.

It refers to a common use of BUSINESS:

> They should go about their private business one hundred yards from the ordinary encampment. (Harris, 1925)

go abroad *British, obsolete* to accept a challenge to a duel.

In the 19th century duelling was illegal in the British Isles but not in France; not to accept a challenge implied cowardice:

> I have called frequently today and I find that you are not going abroad. (Kee, 1993, quoting a letter from O'Shea to Parnell, whom he had challenged to a duel. Parnell wisely ignored it.)

go again to reappear after death.

> ... hes Vauther went agen, in shape of a gurt voul theng. (*Exmoor Courtship*, 1746, quoted in EDD)

go all the way to copulate after a series of sexual familiarities.

Teenage use.

go aloft to die.

For sailors, punning on the rigging and heaven.

go (any) further (of a woman) to allow a man to copulate with you.

It denotes actual or attempted female prohibition in fondling short of copulation:

> ... though I wouldn't let him 'go any further' as we used to say, I did like that kiss more than I've liked anything for years. (Read, 1979)

go away to die.

Not common and can cause confusion:

> Not since my wife, Miriam ... went away. (Diehl, 1978—she could merely have left him)

go case to work as a prostitute.

Probably from CASE[1] rather than the client you find yourself, despite its prevalence among women working in night clubs etc.:

> I was green. It took me a week to realize that I was the only girl in the club not 'going case'. (Irvine, 1986—she was not suggesting that she and others were pregnant)

go corbie *Scottish, obsolete* to die.

The *corbie-messenger* returned late or not at all:

> Hadna Pyotshaw grippit ma airm he was a gone corbie. (Gordon, 1885—a *corbie* is a crow)

go down[1] *obsolete* to be hanged.

In odd contrast to GO UP[1]:

> The lasses and lads stood on the walls, crying, 'Hughie the Graeme, thou'se ne're gae down.' (W. Scott, 1803)

go down[2] to be sent to prison.

It probably refers to the descent from the dock to the cells, but see also DOWN THE LINE[1]:

> I often heard talk about criminals ... If they got you, then you went down. (Simon, 1979)

go down[3] to crash.

Not just ceasing to be airborne:

> A plastic card in the seat-pocket in front of me read: *In case of an Emergency* ... 'Forget that, muffin. If we go down, we're history.' (Theroux, 1993)

A naval pilot who crashes into the sea may be said to have *gone in*. See also HEAVY LANDING.

go down on orally to stimulate the genitalia of.

Heterosexual and homosexual use, with the active partner *down on* the other:

> 'When I'm up, Barbara's down,' says Howard ... 'When you're up who, Barbara's down on whom?' asks Flora. (Bradbury, 1975)

go down the nick to die or be killed.

Usually of animals:

> Looks like they're all goin' to go down t'nick. (Herriot, 1981, of a herd of cattle)

This particular *nick* is not a police station.

go Dutch see DUTCH TREAT.

go for a Burton to die.

Alluding to the British habit of slipping out for a beer brewed in that Milwaukee of the Midlands. It also means the failure of any enterprise.

go for a walk with a spade to defecate in the open air, burying the turds.

> I'd gone for a walk... You know, with a spade. (Manning, 1978)

Mainly military use.

go forth in your cerements *obsolete* to die.

Suggesting that the corpse might have been wrapped in waxed wrappings. The expression is used by a character called Stringer in Powell's *Dance to the Music of Time* novels.

go home to die or come to an end.

This should imply a return to heaven, but it is also used of inanimate objects which are worn out. The heavenly savour is lost in phrases such as *go home feet first* or *go home in a box*, referring to the return of a corpse for burial.

go into to copulate with.

Of a male. It leaves little to the imagination. When Baroness Burdett-Coutts, a friend of Queen Victoria, married a man forty years her junior, the *Pink 'Un* published the following announcement:

> AN ARITHMETICAL PROBLEM: How many times does twenty-seven go into sixty-eight and what is there over? (quoted by Harris, 1925)

go into the ground to die and be buried.

Not to take the subway:

> ... he is not sick, that he doesn't have to go into the ground with her. (T. Harris, 1988—his wife was dying)

go into the streets to become a prostitute.

From the open soliciting:

> While my boy lived, I couldn't go into the streets to save my life or his own. (Mayhew, 1862, of a prostitute)

go native (of a politician) to accept bureaucratic attitudes and aspirations.

Whitehall jargon of British politicians, from the behaviour of a white person who when abroad adopts the indigenous life-style:

> When a Minister is so house-trained that he automatically sees everything from the Civil Service point-of-view, this is known in Westminster as the Minister having 'gone native'. (Lynn and Jay, 1981)

go off[1] to die, especially of pulmonary tuberculosis.

From the gradual consuming nature of the disease:

> But he went off and was laid by soon after. (EDD)

Obsolete since the Second World War.

go off[2] to achieve a sexual orgasm.

Of both sexes:

> There was an old whore in Montrose
> Who'd go off any time that she chose.
> (*Playboy's Book of Limericks*)

go off[3] to start acting in a cowardly manner.

In standard usage, to putrefy or deteriorate. Military jargon of soldiers who behave in this way after being subjected to prolonged exposure to danger:

> But no shell had hit near Sergeant Porter. He had just gone off for no reason. (H. Brown, 1944—Porter had not deserted or moved away but had ceased to perform)

go off the hooks to die.

From the image of a falling gate. A rarer version of *drop off the hooks* (see DROP[3]).

go on[1] to die.

Common with reference to both humans and animals, giving both an implied option of a further stage of existence.

go on[2] to copulate with

Of the male, from the normal posture.

go on the box *obsolete* to be ill.

Long before television, the *box* was a sick club, from the weekly collection of subscriptions which were put in a box. If you were ill for a long time and ran out of benefit, or if you recovered and went back to work, you went *off the box*. To *go on the club* in the same sense was specific, being a shortened form of *sick club*.

go on the coal to urinate, of blacksmiths.

Urine serves both to damp the slack and to ammoniate it, making it better in the forge. By tradition, only a British working smith is allowed to *go on the coal*, the urine of others perhaps lacking certain desirable properties.

go out to die.

Used in the First World War, perhaps from the still current meaning, 'to lose consciousness'. The modern cliché *go out like a light* is to become unconscious or to go to sleep quickly.

go over[1] to die.

Spiritualist and to a lesser extent Christian use, from the soul's passage to the *other side* (see OTHER SIDE (THE)) or another imprecisely defined destination.

go over[2] to defect.

Short for *go over to the other side* and until 1991 usually to or from Communist allegiance. For the devout Anglican, to *go over* was to join the Roman Catholic Church, a fearsome treachery in the days when people worried about these things:

> Evangelical of course. No, I was glad that Wilfred didn't go over. (James, 1975, of an Anglican clergyman)

Go over the wall is used in the same sense, punning on the meaning 'to escape from prison' and the infamous wall which once imprisoned the East Germans under Russian domination:

> He didn't go over the wall until he had to. (Allbeury, 1981, of Kim Philby)

To *go over the hill* is to escape from prison or to desert from the army:

> I guess he figured you'd gone over the hill. (Deighton, 1982, of an army absentee—it also means to have passed your peak)

Sailors *go over the side* if they desert their ship, or go ashore without a pass.

go over the heap to urinate, of British colliery workers.

The *heap* is the adjacent spoil tip.

go over the top[1] see OVER THE TOP.

go over the top[2] to foul deliberately in soccer.

The players are meant to kick the ball, not each other. In a tackle, with the ball between the opponents, he who *goes over the top* of it seeks to hurt or disable his adversary.

go places to urinate or defecate.

From leaving company in order to do so:

> What am I to do? I can't follow them when they go places. (Manning, 1977)

go right *obsolete* to die and go to heaven.

> I knowed 'e went right, far a says t'I, a says 'I'a sin a angel'. (EDD)

go round a corner to urinate, of a man, in the open air.

It probably owes nothing to CORNER[2], a urinal.

go round land *obsolete* to die.

A contradiction, when you consider that in burial the land goes round the corpse:

> He went round land at las', an' was found dead in his bed. (Quiller-Couch, 1893)

go short to copulate less often than you might wish.

Literally, to lack a quantity of anything. This is a male usage, though men may use it of a female if her denial might be construed as a defect in themselves:

> When I say she sometimes bored, don't think I mean she's goin' short... I'm 'bout wore out pilin' inter that li'l darlin'. (Fraser, 1971—a man was speaking of his wife)

go-slow *British* a deliberate failure to complete the work allocated.

A bargaining tactic which may in the short term cause an employer loss without corresponding hardship to his employees.
See also SLOWDOWN and WORK TO RULE.

go the length to copulate.

Extramaritally, not from lying down but indicating the extent of the mutual attraction:

> I had satisfied myself that his enthusiasm for Elspeth wasn't likely to go the length. (Fraser, 1977, writing in 19th-century style)

go the other way to be a homosexual.

Other than heterosexual. See also SWING BOTH WAYS:

> 'Well, you think I'd ever go the other way?' 'No... Not you, the old Davenport cocksman.' (Sohmer, 1988)

go the whole way to copulate after preliminary fondling.

> If it had gone the whole way and the man had aroused her senses, the poor child was in a fix. (McCarthy, 1963)

go the wrong way to die.

After an illness, usually of animals—to recover is the *right way*:

> ... a chronic state of diarrhoea under which the animal wastes away and dies. This is what is perfectly understood as going the wrong way. (EDD, from western England)

go through to copulate (with).

> ... nudging each other in the ribs saying '*I wouldn't mind going through her on a Saturday night*'. (Lodge, 1988)

Of the male, but in the sense 'to experience' rather than 'to transfix', I would suggest.

go to a better place to die.

Only one among dozens of destinations, of which many denote a heavenly reward—*glory, life eternal, rest*, etc.:

> I expect he's gone to his rest long since, poor man (James, 1972)

and others suggest a continuation of a favoured occupation on earth:

> Now Sam's gone to the great massage parlor in the sky. (Sanders, 1977)

go to bed with to copulate with.

Of either sex, with the activity figuratively taking place anywhere:

> Years ago she had gone to bed with him for a few weeks. (Amis, 1978—you might suppose they were a pair of invalids)

It is also used of homosexual activity:

> 'The idea of going to bed with Donald,' he spluttered. (Boyle, 1979—the splutterer was Guy Burgess)

go to Cannes to urinate or defecate.

A pun on *can*, a lavatory, but I imagine you have to write it down before anyone sees the attempt at humour.

go to Denmark to have a sex-change operation.

The pioneer work on this advance in surgery and the human condition was carried out in Denmark, a country long in the forefront of sexual licence and experimentation. This American phrase is also used of such surgery wherever undertaken.

go to grass *obsolete* to die.

Implying burial in the green English churchyard, some going there 'with their teeth upwards'.

go to ground *British, obsolete* to defecate in the open.

> 'Going to ground' is a phrase well known to the surgeons in the Birmingham hospitals. (EDD)

Perhaps from the normal meaning 'to hide'.

go to heaven in a string to be hanged.

The fate of 16th-century English Roman Catholics:

> Then may he boldly take his swing,
> and go to Heaven in a string.
> (T. Ward, 1708, quoted in ODEP)

go to it to copulate.

Of either sex. Sufficiently obsolete in 1940 for Herbert Morrison's wartime slogan in the British Isles to be taken by most as an exhortation to work harder.

> The fitchew nor the soiled horse goes to't
> With a more riotous appetite.
> (Shakespeare, *King Lear*)

go to Paul's for a wife *obsolete* to seek a prostitute for copulation.

Prostitutes used to frequent the fashionable walks around London's St Paul's cathedral, whence Falstaff's allusion:

> I bought him in Paul's ... an I could get me but a wife in the stews. (Shakespeare, *2 Henry IV*)

go to the bathroom to urinate or defecate.

Other phrases using common euphemisms for lavatory are found under their respective nouns. Such phrases can also be used where no lavatory is involved:

> ... he went to the toilet down a bit of hosepipe through Missis Kilmartin's car window. (R. Doyle, 1993, referring to urination by a naughty child)

go to the Bay *obsolete* to be transported to Australia as a prisoner.

The bay was Botany Bay in New South Wales:

> 35 per cent are known to have been charged with as many as four earlier offences before they 'napped a winder' or 'went to the Bay'. (Hughes, 1987, rebutting the myth that the original involuntary white settlers in Australia were not hardened criminals)

go to the wall (of a business) to fail or (of a person) to die.

The origin of this phrase is obscure.

go to your reward to die.

I suspect most of us hope there will be no posthumous accountability:

> But it's a glory to know he has gone to his reward. (Sanders, 1980)

go to yourself to die.

A Shetland Islands use; I give it for its overtones of Buddhism.

go under (of a business) to fail or (of a person) to die.

Not specifically referring to people who drowned:

> He had once said to Victoria that he did not cling to life (as she did) and that, if he had a severe illness, he would go under. (Pearsall, 1969, of Prince Albert, who died of typhoid caught at Windsor castle, although his widow preferred to think it was in part from mortification at the perceived frailty of his son and heir, of which more in FALL[1])

go up[1] *obsolete* to be killed by hanging.

Especially in 19th-century America. Occasionally also used to refer to a natural death:

> You'd better give it up if you don't want to go up. (Cookson, 1969—the person so addressed was being poisoned by working with lead paint)

Go up the gate referred to the entry to the churchyard, and not to the portal over which St Peter watches. *Go up the chimney* referred to the burning of the corpses of their victims by the Nazis:

> The air stinks now but it might get better. A lot of Jews going up the chimney. (Styron, 1976)

go up[2] to be under the influence of illegal narcotics.

From the feeling of levitation. Other phrases relating to illegal narcotics and starting with *go* are dealt with, if at all, under the following word.

go up the river to be sentenced to jail.

From the location of American penal institutions in New York, New Orleans, and elsewhere:

> The long-term prisoners waiting to go up the river. (L. Armstrong, 1955, of New Orleans)

go upstairs to urinate or defecate.

From the location of the lavatory where there is only one in the house, and for the lavatory reserved for female guests where there is more than one:

> 'Do you want to go upstairs, Emma?' she asked... 'I'll come too,' said Louis... 'You can't come where she's going.' (Bradbury, 1959)

go west (of a person) to die; (of a company) to become bankrupt.

Representing a long tradition based on the setting of the sun. The ancient Egyptians called their dead the *westerners*.

go with to copulate extramaritally with.

More often used of women than of men.

> She hurt him terribly when she went with other men. (Green, 1979—Christine Keeler hurt Rachman)

go wrong to copulate with a man casually.

Doubly blameworthy if you were also impregnated:

> 'When I was sixteen,' she said, 'I went wrong.' (Mayhew, 1862)

goat a male who habitually copulates casually.

From the Grecian god Pan and the general reputation of billy-goats. *Play the goat* is so to act, although *play the giddy goat* is merely to act stupidly. In obsolete British use a *goat-house* was a brothel and a *goat-milker* a prostitute.

gobble to practise fellatio (on).

Also in various phrases using meat imagery:

> If he pays some chippie fifty to gobble his pork... (Diehl, 1978)

Gobble a pecker leaves less to the imagination:

> I had her gobbling my pecker behind the lifeboats. (M. Thomas, 1980)

Also used as a noun:

> ...the search for a half-decent English gobble has been my Holy Grail. (Blacker, 1992—the suggestion was that native females were unwilling or incompetent participators in fellatio)

In an American prison, a *gobbler* is a male homosexual.

God's child an idiot.

Another obsolete British dialect phrase where the results of interbreeding were attributed to divine rather than parental agency:

> Such as him were called 'God's Children'. (O'Reilly, 1880, of an idiot)

God's (own) medicine a narcotic.

Of opium in the 19th century, when it was freely available in nostrums for infant and adult use. Now of morphine, mainly used illegally and abbreviated to *gom*.

God's waiting room a resident institution for geriatrics.

Making a charitable assumption about posthumous selection:

> In a private nursing home—one of those places they call God's waiting room. (B. Forbes, 1986)

gold see SILVER.

gold-digger[1] a woman who consorts with a man because he is rich.

She may be working a single vein or be staking more than one claim.

gold-digger[2] *obsolete* a person employed to remove human excreta.

From the colour and of the same imagery as HONEY. He might also have been called a *gold-finder* and the excrement *gold-dust*. The use survives in the American *goldbrick*, a turd, which is used more figuratively than literally:

> Tarrant was the biggest goldbrick on the base. (Deighton, 1982)

gold water urine.

Presumably from the colour; see also WATER:

> Often the names imposed on them had been unattractive to begin with, as with Geldwasser ('gold water'), a venerable euphemism for urine. (Bryson, 1994, who also reminds us that Schmuel Gelbfisz called himself Samuel Goldfish for some years before deciding that Goldwyn—a combination of Goldfish and his partner Selwyn—suited him better)

golden excessive.

Financial jargon, of payments to managers etc. First seen in the *golden handshake*, where a larger sum than is contractually due is paid to a dismissed employee to secure silence, preserve secrets, avoid litigation, and set a precedent against your own departure:

> They have something called a 'Golden Handshake'. If they want to get rid of a foreigner they offer him a chunk of money as compensation for the loss of his career. (Theroux, 1977)

Whence other compounds. A *golden hello* can be a capital payment, share options, or other inducements to recruit an experienced employee, whose loyalty may then be bought by the use of *golden handcuffs*:

> It gives employees an equity-type stake in the bank as well as acting as a form of 'golden handcuffs'. (*Daily Telegraph*, 18 March 1994)

If you are concerned about the security of your own job, you seek a *golden parachute* to ease your descent from high office:

> But when a person fell from a position of influence, there was no safety net, no golden parachute. (Sohmer, 1988)

The less common punning *golden retriever* is the bribe paid to a former employee to come back to work for you.

golden ager *American* a geriatric.

Not someone who lived in a mythical *golden age*, but one supposed to be enjoying the *golden years* (see GOLDEN YEARS (THE)).

golden bowler see BOWLER HAT.

golden shower see SHOWERS[1].

golden triangle the Siamese opium poppy district.

From the geographical shape and the rewards of those who engage in the business:

> I was up in the north, where they grow poppies for opium and heroin. So-called 'golden triangle'. (Theroux, 1975)

Other dictionaries tell me that *gold dust* is cocaine from any source, reflecting what it costs to buy illegally.

golden years (the) old age.

Referring perhaps to ripened corn or your *golden wedding* rather than the incipient setting of the sun—and certainly not to the prosperity of living on a pension:

> They are addressed as 'senior citizens' and congratulated on their attainment of the 'golden years'. (Jennings, 1965, of old people)

goldilocks a policewoman.

An extension of the American BEAR[2] sequence. She may be brunette or red-headed.

golly a mild oath.

Perhaps the commonest corruption of *God*. This 19th-century use antedates by some forty years Florence Upton's Golliwogg books (of which I still have six of my father's). Other corruptions include *goles, golles, gollin, golls, gull, goom, gomz, gom, goms, gommy*, and *gum*, as in *by gum*, etc.

gone[1] dead. See GO[1].

gone[2] pregnant.

Usually indicating the period since conception:

> What's he going to do about our Doreen who is six months gone? (Tidy, in *Private Eye*, March 1981)

The rare and perhaps obsolete British *gone after the girls* meant infected with a venereal disease, the *girls* being prostitutes (see GIRL[1]).

gone[3] drunk or under the influence of narcotics.

Rational behaviour and comprehension have departed:

> She was so 'gone' by the time I finished clearing up... (Bogarde, 1981—she had ingested a narcotic)

gone about besotted with.

From the symptoms of infatuation:

> Mr Hawkins was *fearfully* gone about Francie Fitzpatrick—oh, the tender looks he cast at her. (Somerville and Ross, 1894)

More often as *gone on*, less as *gone over*.

gone walkabout been stolen.

From the practices of the Australian aborigines. Of anything from minor thefts to complex frauds:

> ... the whole of the money put in for the development of the DeLorean Car had disappeared—or 'gone walkabout'. (Cork, 1988)

goner a person who is about to die or recently dead.

> I thought she was a goner, I'm afraid. You've never seen anybody so pale. (Fry, 1994)

Pronounced, and sometimes spelt, *gonner*:

> Better say your prayers. If we crash, you're a gonner. (Manning, 1962)

gong *British* a medal or award.

The allocation of awards to others is seldom welcome—see COME UP WITH THE RATIONS—and those vainglorious persons who conduct themselves so as to secure such RECOGNITION are stigmatized as *gong-hunters*:

> He was fixing his miniature medals to his jacket. It was a rather meagre display of gongs. (Deighton, 1994)

good unwilling to copulate outside marriage.

Usually of a female:

> If you can't be good, be careful. (Old saw)

But a *good-natured* woman, as recorded in 19th-century Somerset, was anything but *good*, her promiscuity coming from her willingness to please men in that regard:

> Her was always one o' the good natur'd sort. (EDD)

good folk the fairies.

These malevolent creatures, like the devil, had to be placated by flattery:

> The guidfolk are not the best of archers, since the triangular flints with which the shafts of their arrows are barbed do not always take effect. (Hibbert, 1822)

For the Scots they were also *good neighbours*:

> If ye ca's guid neighbours, guid neighbours we will be;
> But if ye ca's fairies, we'll fare you o'er the sea. (Ayrshire Ballad, 1847)

In Ireland they could be the *good people*:

> ... so young that you were in girls' skirts lest you be carried away by the good people.

(Flanagan, 1988—until the 19th century, Irish fairies tended to steal baby boys but were apparently deceived by *cross-dressing*; see CROSS-DRESS)

good friends see FRIEND.

good man the devil.

As with the GOOD FOLK, you dare not suggest otherwise:

The Goodman will catch you in his net. (Henderson, 1856)

Farmers placated him by allocating to him an uncultivated piece of land which was known as the *goodman's craft*, *field*, *rig*, or *taft*:

Bonny's the sod o' the Goodman's taft. (ibid.)

good time a single act of extramarital copulation.

A fairly conventional introduction by a prostitute:

I'll try to give you a good time (Harris, 1925)

and a *good-time girl* is a prostitute. Men occasionally offer women a *good time*, meaning 'copulation':

The man was offering her a drink and a good time in Spanish. (Theroux, 1979)

good voyage the use of a warship for commercial transport.

British naval officers until the beginning of the 20th century used their ships for the transport of civilian cargo, especially to remote destinations or where there was risk of interception. Naturally some succumbed to the temptation of putting trade before the flag:

The practice variously known as freight or 'good voyages' was in Pepys's eyes the most pernicious of all. (Ollard, 1974)

goodbye peremptory dismissal from employment.

It is the employer who decides when to say the farewells:

...since released, not surprisingly, to pursue 'other business interests', the banking euphemism for goodbye. (*Private Eye*, April 1988)

goods (the) something illicit or harmful in your possession.

Used of stolen property or illegal narcotics; and figuratively of any information of a damaging or shameful nature, which can be used in extortion or as evidence of guilt:

But what if a twist exactly like her was a suspect, and you had to get the goods on her? (Sanders, 1980)

goof a habitual user of illegal narcotics.

Literally, a stupid person, whence many meanings to do with unsophistication and incompetence. A *goofball* may be an addict:

Clearest of all was that solitary *hoo* of the goofball in the crowd (Theroux, 1978)

or it may be an illicit narcotic:

Goofballs are one of the barbiturates laced with benzadrine. (Chandler, 1953)

Goofed is under the influence of illegal narcotics.

goof up *American* to kill.

By any method. For a while *goof* became something of an omnibus word, but this use is now probably obsolete.

goolies the testicles.

We can, I think, reject derivation from *gool*, an outlet for water and Partridge's suggestion that it comes from *gully*, a game of marbles. The Hindi *goli* meant 'a ball' and this is another word brought into the language by British Indian troops:

Then when he's off guard you give it to him in the goolies. (Sharpe, 1974)

goon squad members of a police or military unit capable of acting violently and ruthlessly.

Either Jericho has been taken and has told the goon squad everything, or he's up to something. (Forsyth, 1994—Jericho was an informant in Iraq)

A *goon*, although not noted by Dr Wright, was in dialect speech a simpleton, from which it came to denote a German guard in a prisoner-of-war camp.

goose[1] a prostitute.

The common avian imagery. If she were a *Winchester goose*, she had syphilis, from the insalubrious church-owned property in south London where the poorer prostitutes lived:

...but that my fear is this
Some galled goose of Winchester would hiss. (Shakespeare, *Troilus and Cressida*)

Compton Mackenzie reported that a *goose girl* was a female homosexual, although nothing seems more innocuous than that Danish porcelain figure.

goose[2] to pinch the buttocks of.

A male sexual approach as delicate as a nip from the bird's beak—or as indelicate:

Leroy goosed the girl from behind, causing an alarmed but happy squeak to emerge from her lips. (Collins, 1981)

Women occasionally *goose* men:

They chivvied each other and laughed a lot. Once she goosed him. (Sanders, 1982)

DAS says that a finger has to be poked into another's anus which, even with the intervention of clothing, brutalizes the less offensive activity.

goose[3] an act of copulation.

Occasional use from the rhyming slang *goose and duck*, fuck. Not used figuratively.

gooseberry[1] the devil.

Usually as *old gooseberry*:

> Th' match ther wur betwixt a tailior and owd gooseberry. (Axon, 1870)

The use survives in *play gooseberry*, to play the devil with a courting couple by keeping them company when they would far rather be left alone.

gooseberry[2] *American* to steal clothes from a washing line.

From the noun meaning, 'a line with clothes on it', from which stealing is as easy as taking berries off a bush. Thus a *gooseberry lay* is any crime easily carried out.

gooseberry bush see PARSLEY BED.

Gordon Bennet(t) a mild oath.

> 'Gordon Bennett!' Jack said, and in spite of myself I laughed out loud at the exclamation. (Theroux, 1982)

You can use it for *God*. From the American press proprietor who sponsored H. M. Stanley in his African travels and balloon races, not the London stipendiary magistrate who came into prominence some decades later.

governess *obsolete* a female bawd.

The 19th-century brothel she ran brought back memories of the schoolroom:

> The most professional of the 'governesses' who ran brothels for flagellants was Mrs. Theresa Berkley of 28 Charlotte Street. (Pearsall, 1969)

gow an erotic or salacious magazine cover.

Literally, a narcotic, from the Cantonese word meaning 'sap', and now used in that sense mainly of illegal marijuana. I suppose the cover is intended to stupefy a browser into buying, especially as that kind of American publication is packaged in such a way as to prevent a prospective purchaser discovering the repetitious banality of the inside pages.

grab[1] to steal.

The common imagery which links seizing and theft:

> 'How are you going to get the money?' I asked. 'Grab it. Steal it,' he said. (L. Thomas, 1977)

grab[2] to accept a tenancy after an eviction.

During the agrarian disturbances and disputes over the occupation of farms in 19th-century Ireland, to accept a tenancy after the previous occupant had been evicted was deplored by almost everyone but the landlord:

> But Mick Tobin, now... he was prepared to grab. (Flanagan, 1988—Tobin was later killed by his ejected predecessor)

A *grabber* was one who so acted.

gracious old and expensive.

Of dwellings, in the language of estate agents. Other advertisers use *gracious* when they tell us that, if we buy their product, we can lead a life free of menial labour, giving sumptuous dinner parties to exciting people without having to shell the peas.

graft bribery.

In standard usage, hard work, from the original meaning 'grave-digging'. The euphemism, however, may have come from the meaning 'to insert a bud into other living stock to induce more profitable growth'. The obsolete English *graft*, to cuckold, came from the figurative *grafting*, implanting, the symbolic horns on the victim's head.

grand bounce see BOUNCE[3].

grande horizontale see HORIZONTAL[1].

grandmother to stay menstruation.

Another of the *visitor* images, referring to an inconvenient but limited disruption of the normal tenor of your ways.

grandstand play accentuation of difficulty in order to win praise.

From the location of the spectators, but not used only of athletes:

> ... kept details to yourself. A real grandstand play. (Diehl, 1978, of a policeman who tried to solve a case on his own)

In American pejorative use, to *grandstand* is so to act:

> I relied on you to grandstand enough to let her get wise to you. (Chandler, 1958)

grape (the) wine.

Standard English since the 17th century. The punning but perhaps obsolete *grape-shot* means 'drunk' but a *whiff of grapeshot* is not the bouquet —see WHIFF.

grass[1] to inform against.

Probably from the rhyming slang *grass in the park*, copper's nark:

> 'Favours. Grassing.' Blamires said, 'I've nobody to grass on.' (Kyle, 1975—the police were trying to get information from a suspect about a third party)

A *grass* is an informer:

> There's a copper in that boy, you mark my words. He's a natural grass. (le Carré, 1986)

In obsolete English use to *grass* was also to kill, from the felling or the burying.

grass[2] marijuana.

Shortened form of *grass-weed*; perhaps the commonest drug users' term for this:

> Frank was restive about the marijuana. 'You surely wouldn't make trouble about a scrap of grass.' (Davidson, 1978)

Occasionally as *green grass*:

> We are smoking too, man, you know? Grass. Green grass. You know what I mean? (Simon, 1979)

grass[3] the female pubic hair.

American use among both males and females.

grass bibi/bidi see BIBI.

grass-widow a woman of marriageable age divorced or separated from her husband.

Literally, in modern use, a woman whose husband is absent for a long time because of his job. *Grass* might appear to be a corruption of *grace* but the derivation is in fact from the flora of the Indian hill-station to which a wife resorted while her husband sweated it out on the plains. In obsolete British use there were three specific meanings: a woman who copulated extramaritally while her husband was away (and many acquired that reputation, whether deservedly or not); a discarded sexual mistress; and a woman who became pregnant before marriage:

> Grass widows with their fatlings put to lie in and nurse here. (Hunt, 1896)

grasshopper a park policeman.

American Citizens' Band use, perhaps from his tendency to appear unexpectedly.

gratify see GRATITUDE.

gratify your passion to copulate with a woman.

A mainly 19th-century usage, in the days when such passions were supposedly confined to males:

> He cannot afford to employ professional women to gratify his passions. (Mayhew, 1862)

Women may still *gratify a man's desires*, if so inclined, but *gratify your amorous works* is obsolete:

> She did gratify his amorous works. (Shakespeare, *Othello*)

Gratification was extramarital copulation by a male:

> ... since the Roman Church regarded such errors as venal ... I had much gratification at little expense. (Graves, 1940, describing a soldier's service in Ireland)

gratitude a bribe or payment for illicit services.

To express thanks in words is not enough:

> Gratitude was a pay-off. Gratitude was drink and diamonds. (Keneally, 1982)

In obsolete Scottish use to *gratify* meant 'to tip' or 'to bribe':

> People were still obliged to gratify the keepers for any access they had to visit or minister to their friends (Wodrow, 1721, of prisoners)

and *gratification* was a tip or a bribe.

grave (the) death.

Standard English figurative use:

> There will be sleeping enough in the grave. (Franklin, 1757)

In obsolete Scottish use, the *gravestone gentry* were the dead:

> My bed is owre amang yon gravestane gentry. (A. Murdoch, 1873)

gravy[1] *American* an intoxicant.

A variant of *sauce*, which is now much more common.

gravy[2] a supplemental benefit received gratuitously.

The pleasant but unnecessary complement to the meat. To *ride the gravy train* is to receive such benefits on a regular basis:

> The gravy train has not stopped entirely for Grub Street hacks. (*Private Eye*, July 1981)

graze to steal and eat food in a supermarket.

Like cattle in a pasture, you eat what you take between the rows and pass the checkout desk with empty hands and a full stomach.

The obsolete British *graze on the plain* or *the common* was to be dismissed from employment, presumably as a house servant, or merely to be evicted from your house:

> He turnde hir out at durs, to grase on the playne. (Heywood, 1546)

grease[1] to bribe.

The usage predates OIL but the concept of making something run easily is the same:

> With gold and grotes they grease my hands,
> In stede of ryght that wrong may stand.
> (Skelton, 1533)

and today:

> He lacked the financial resources with which Oskar greased the system. (Keneally, 1982)

As well as *greasing hands*, we *grease palms*, *paws*, etc. *Grease* is bribery, *a grease* a bribe, but beware of the obsolete British *grease the wheel*, which was to copulate, punning on the lubricity.

grease[2] *American* to kill.

Perhaps from converting the body into a fatty substance, or merely a corruption of CREASE:

> If ... he makes any threatening movement —anything at all—grease him. (Sanders, 1973)

greased drunk.

The common American culinary imagery. Things may indeed seem to run more smoothly for a time:

You come over early and we can get greased before the mob arrives. (Sanders, 1982—the speaker was hosting a party)

great pregnant.

Shortened form of *great with child*:

O silly lassie, what wilt thou do,
If thou grow great they'll heez thee high.
(Herd, 1771—society would reward her not with a weekly stipend and visits from sympathetic officials but with death by hanging)

great and the good (the) people comprising or approved by the British political establishment.

An often derogatory description of the rich or powerful clique who seek to parcel out awards, rewards, and positions among themselves:

Maynard, astute businessman ... Maynard, supporter of charity ... Maynard, the great and good. (Francis, 1985, describing a rogue conspiring to be knighted)

great certainty death.

Certain that you will die one day but uncertain for most of us what happens next:

'The Great Certainty looms,' said Mr Flawse. (Sharpe, 1978)

Also as the *great out, leveller*, or *perhaps*:

I thought this is the end, China, and you're going to find the Great Perhaps. (M. Fraser, 1992)

'Now for the Great Secret', remarked the dying King Charles II, shortly before or after expressing concern about 'poor Nellie's' future.

great leap forward reckless industrialization of a peasant country.

The Chinese Communist government in 1958 so titled their plan to increase the productivity of agriculture and industry by a compound annual rate of 25 per cent. By 1961 the vast country faced starvation and economic ruin.

great majority the dead.

A shortened form of the *great majority of souls*, who are supposed to be in heaven or limbo:

Life is the desert, life the solitude; Death joins us to the great majority. (Young, 1721)

Greek Calends (the) never.

The Romans were meant to settle their taxes and other accounts on the calends of each month, but the Greek calendar had no calends:

The emergence of chaos in Germany ... would put off the pacification of Europe to the Greek Kalends. (Goebbels, 1945, in translation)

Greek way (the) buggery.

From the supposed practice of the ancients, whence too the obsolete verb *Greek*, to bugger. If you think our language is not even-handed between antagonists see TURK.

green fruit *obsolete* a female virgin.

In many varieties of fruit, greenness denotes unripeness:

In any event the virgins were still being bought, and the men were still enjoying what was described as 'green fruit'. (Pearsall, 1969)

green goods *American* counterfeit banknotes.

Goods for the stolen element (see GOODS (THE)) and *green* for the colour of US currency. A *green goods man* is either a forger or someone who passes forged banknotes:

He was just in here looking for a green-goods man. (Weverka, 1973, of a passer of bills)

green gown *obsolete* an indication of female copulation.

Stained by being pressed on the grass:

Then some greene gownes are by the lassies worne
In chastest plaies.
(Sidney, 1586)

Much clothing was soiled on the eve of May Day, when convention allowed the lads and lasses to spend all night in the woods, ostensibly gathering flowers. A woman *with a green gown* has copulated before marriage:

... she had had the salutation 'with a greene gowne' ... as if the priest had been at our backs, to have married us. (G. A. Greene, 1599, quoted in ODEP)

The *green sickness* was 'The disease of maids occasioned by celibacy' (Grose) and to *give green stockings* was to commit the solecism of getting married before your elder sister. All these useful phrases are obsolete.

green grass see GRASS².

green stamp collector a policeman with radar.

Green from the colour of bills, *collector* for the keenness of some American police in levying fines, and *stamp* from the tokens given to shoppers which they may later exchange for goods.

greenmailer a corporate raider who seeks to get paid to go away.

The green of the US dollar replaces the black of BLACKMAIL:

... the first place to which takeover artists and greenmailers and LBO peddlars came for cash and complicity. (M. Thomas, 1987, of a bank —an *LBO* is a leveraged buy-out)

greymail *American* a threat to tell state secrets if prosecuted.

See also BLACKMAIL and GREENMAILER—the variations of shade reflecting the different degrees of legality perhaps:

He would also use a 'CIA defense', so called greymail tactics that had been successfully practised by other defendants involving national security. (Maas, 1986)

grief therapy charging money for activities purporting to alleviate bereavement.

American funeral jargon for persuading the survivors to spend a lot making the corpse look like a healthy person and otherwise seeing it off in style:

> 'Grief therapy', the official name bestowed by the undertakers on this new aspect of their work. (J. Mitford, 1963)

Grief therapist thus becomes a new name for an arranger of funerals.

grim reaper see REAPER (THE).

grind to copulate.

Probably from the rotary pelvic pressing motion:

> ... a young person of Harwich,
> Tried to grind his betrothed in a carriage.
> (*Playboy's Book of Limericks*)

A *grind* is either an act of copulation or a female participant, and then always referred to in flattering terms—where do all the *bad grinds* go? *Grind* is also used less often of masturbation. A *grinding-house* or *grind-mill* is a brothel:

> It was a business in the grind-mills...
> (Longstreet, 1956, of New Orleans brothels)

A *grinding employer* was neither a miller nor one who copulated with his female staff, but a man who expected a lot of work for a little pay:

> ... grinding, or being compelled to do the same or a greater amount of work for less pay. (Mayhew, 1851)

The obsolete British *grind the wind* was to be punished on a treadmill:

> The prisoners style the occupation 'grinding the wind'. (Mayhew, 1862)

groceries bombs.

Second World War Royal Air Force jargon.

groceries sundries intoxicants.

Bought from the grocer and thus described on the delivery note so that the nature and extent of the purchases could be hidden from the servants and, very often, from the unknowing husband. Obsolete since the Second World War.

grog on board drunkenness.

Grog was originally rum, from the nickname of the British Admiral Vernon (1684–1769) who instituted a ration for sailors and wore a grogram coat. It may now mean 'any spirituous intoxicant'. *Grogged* means 'drunk' and a *grog-hound* is a drunkard.

groin the genitalia.

Literally, the place where the abdomen meets the thigh. In sporting jargon, he who is said to have received a nasty knock in the *groin* has suffered a more telling and painful blow. Non-sporting use is less common:

> He was grabbed by a sensitive portion of his lower groin. (Lavine, 1930)

To *rub groins together* is to copulate:

> ... they should get to know one another better ... by rubbing their groins together. (*Sun*, March 1981)

grope to fondle another person sexually.

Literally, to use the hands for feeling anything. Usually of a male whose activity may be inexpert and unwanted:

> You mean fornicating in the sauna or in a mop closet or underwater groping is okay? (Sanders, 1973)

Thus a *groper* is an unattractive male suitor, replacing the two more realistic obsolete meanings 'a blind person' or 'a midwife'.

gross height excursion a dangerous and unplanned loss of aircraft height.

Civil aviation jargon in an environment where nothing must be acknowledged as dangerous or unplanned:

> ... a nose dive is never called a nose dive. It is a 'gross height excursion'. (Moynahan, 1983)

gross indecency see INDECENCY.

ground-sweat *obsolete* death.

Originally, in obsolete use, the grave, whence the old proverb: 'A ground-sweat cures all disorders'. A *ground lair* was a family burial place and *ground-mail*, a burial fee:

> 'Reasonable charges?' said the sexton; 'ou, there's ground-mail—and bell-siller—and the kist—and my day's work.' (W. Scott, 1819—a *kist* is a coffin)

I include these old British phrases because they represent a line of euphemism which has lapsed and because, as the quotation shows, nothing changes over the years when it comes to fixing charges in the funeral business.

grounded forbidden to fly an aircraft.

Of a pilot who is unwell or subject to disciplinary action. Those of us who are not pilots are in this sense permanently grounded. A pilot who is *grounded for good* is dead.

Some figurative use of a child or teenager who is punished by being confined to its home.

group sex a sexual orgy.

It could mean no more than an outing to the Mothers' Union, if the parson stays away, they being all females. It indicates a disregard for pair bonds and normal sexual practices:

> If God had meant us to have group sex, I guess he'd have given us all more organs. (Bradbury, 1976)

grow your greens to urinate outdoors.

British male usage, combining the ridiculous with the notion of urine as a fertilizer.

growler-rushing *American* drinking intoxicants at home.

A *growler* is a large pitcher which was taken to a bar to be filled with cool beer for consumption at home, in the days before general domestic refrigeration. You had to *rush* to stop it warming up on the way. Still usually of beer but occasionally also of other intoxicants:

> Meanwhile my jug is getting low. How about rushing the growler for me? (Sanders, 1980)

growth a carcinoma.

Literally, something which has grown and, even of human tissue, not necessarily malignant. A common usage to avoid direct reference to the dread cancer.

grumble an act of copulation.

From the rhyming slang *grumble and grunt*, cunt. Less often as *gasp and grunt*, despite the imagery that that suggests.

grummet the vagina viewed sexually.

Literally, a metal eyelet. Some American use and also of a single act of extramarital copulation.

grunt to defecate.

American nursery use, perhaps from the initial training in controlled defecation, where a grunt may accompany the muscular effort.

grunter a pig.

Used among fishermen to avoid speaking of something taboo, the word *pig* being sure to lead to a bad trip:

> When Kate referred to a pig she said grunter. (Cookson, 1969—Kate was married to a mariner)

Probably from the sickness caused by rotten salted pork. See also FURRY THING.

guardhouse lawyer *American* an opinionated know-all and trouble-maker.

Variant of BARRACK-ROOM LAWYER. Guard duty involves long periods of boredom, providing the right environment for bores and agitators, as in other activities where people are paid to sit around waiting for something to do. Originally an army use but now general.

Guatemala (go to) to take illegal narcotics.

The country is a favoured source of marijuana:

> 'Hey, where are we going?' Hood said, 'Guatemala.' Murf smiled. He understood the euphemism. (Theroux, 1976—although the scene is set in London, Hood was an American)

guest a prisoner.

Usually the jocular *guest* of Uncle Sam, Her Majesty, etc.:

> To book a prisoner—I beg your pardon, 'guest'. (Lavine, 1930)

The obsolete Scottish *guest* was only a ghost, an unwelcome visitor or a linguistic corruption:

> Brownies, fays and fairies,
> And witches, guests.
> (Liddle, 1821)

guest-artist a paid performer making a single appearance.

Entertainment jargon. Such a performer fills in the time allotted and helps to break the tedium.

guidance towards change a compromise under pressure.

Under such compulsion a priest can afford to abandon principle:

> The Holy Spirit could guide the church towards that change. (BBC 2 September 1986—a Cardinal was conceding that the shortage of bachelors in Latin America might lead to the waiving of the requirement that Roman Catholic clergy be celibate)

guidelines the discouragement of wage increases.

More tentative and less pretentious than other euphemisms called in aid by ministers seeking to prevent inflationary wage increases in a market economy—see FREEZE[1], PAUSE[2], RESTRAINT[1], etc. The *guidelines* were on one occasion at least illuminated by a *guiding light*:

> ...the Government expected that a 2.5 per cent 'guiding light' would be observed. (Crossman, 1981, of a Conservative White Paper issued in February 1962. The expectation was not fulfilled.)

guinea-hen *obsolete* a prostitute.

A pun on her fee and the common avian imagery:

> Ere I would say, I would drown myself for the love of a guinea-hen, I would change my humanity with a baboon. (Shakespeare, *Othello*)

gulag see CORRECTIVE TRAINING.

gull *American* a prostitute who frequents naval bases.

The common avian imagery, with a suitable choice of bird.

gulls see BUOYS.

gumshoe *American* an investigator in plain-clothes.

From his ability to walk quietly on rubber soles. As a policeman:

> Don't you call me 'sister' you cheap gumshoe. (Chandler, 1958)

or as a private detective:

> The president's private eye...had become for all intents and purposes the exclusive gumshoe of White House counsel John Dean. (Colodny and Gettlin, 1991)

gun[1] a criminal who carries a handgun.

Criminal jargon for someone who is also ready to use it:

> Especially if they're killers—guns for hire. (Bagley, 1977)

But an American *gun* can also be an unarmed professional thief, from the Yiddish *gonif*.

To *gun down* is to kill or wound humans, singly or collectively. Animals are always *shot*.

A *son of a gun* was a bastard, conceived on a long voyage and of doubtful paternity, although the phrase had long ceased to give offence to those so addressed until it passed from common use.

gun[2] a hypodermic syringe.

Used illegally for a *shot* of narcotics.

gun[3] the penis.

Heard in America but obsolete in Britain. The association of ideas is not new—see PISTOL.

gun down see GUN[1].

gunner's daughter a flogging.

Literally, the barrel of the gun over which the victim was strapped, thus *kissing* or *marrying* her:

> I was made to kiss the wench that never speaks but when she scolds, and that's the gunner's daughter. (W. Scott, 1824)

guys a public lavatory for exclusive male use.

With its counterpart *dolls*, perhaps from the musical play. Not common and relatively inoffensive, as these things go.

gyppy tummy see GIPPY TUMMY.

gypsy's warning no warning at all.

Only used in this sense in America. Elsewhere it implies the foretelling of misfortune from whatever source. Female gypsies have long been credited with the power to see into the future, so long as their palm has first been crossed with silver. This precaution is to negate the power of the devil, fortune-telling coming normally within his sphere of influence, although the silver is never returned. In obsolete Irish use, a *gypsy's warning* was gin, perhaps from what happened to those who became overfond of it.

H anything taboo beginning with the letter H.

Usually *hell* in the expression *What the H?* In addict use, *heroin*:

> Daddy is fillin' the gun full of beauteeeful H. Soon you will be ridin' a wave. (Collins, 1981)

H & C is a mixture of heroin and cocaine, punning on the plumbing abbreviation for *hot & cold*.

habit an addiction to narcotics.

Not used of the equally addictive alcohol or tobacco, nor of power. Your preference may be indicated by a modifier, for example *nose habit*, and the degree of addiction in a phrase assessing the cost:

> ... $50-a-day habit. (Lingemann, 1969)

To *kick the habit* is to stop taking illegal narcotics.

habit of your youth masturbation.

Probably of males only:

> Does Mrs. Hagerty oblige or are we reduced to the habits of our youth. (le Carré, 1989—Mrs Hagerty was the housekeeper)

had it dead or beyond repair.

Of man, beast, or worn-out machinery:

> You've had it. You're snuffed. You're wiped out. (Theroux, 1976)

hail Columbia *American* an expression of annoyance.

Hail perhaps from hell, and *Columbia* is the USA:

> I got Hail Columbia from Father for that escapade. (Sullivan, 1953)

To *raise Hail Columbia* is to cause a fuss.

hair of the dog a morning drink of an intoxicant.

Usually after too many the previous evening. Shortened form of the *hair of the dog that bit you*, which is supposed to provide some protection against rabies:

> Do you feel like swilling the hair of the dog with me? (Francis, 1978)

In American use also as the *horn of the ox*:

> ... three guys bellying up to the bar in an adjoining room, starting their day with a horn of the ox that gored them. (Sanders, 1979)

hair stylist a barber.

It avoids mentioning anything so menial as washing, cutting, or dressing another's hair. You will find *hair sculpture* even more pretentious and expensive.

haircut a financial collapse.

All the locks are shorn:

> The total of the Golden Grove haircut was less than $200 million in capital and reserves. (M. Thomas, 1982)

hairpiece a wig.

Literally, no more than a piece of hair, on or off the scalp:

> He patted his hairpiece lovingly. (Moss, 1987)

half a can a quantity of beer.

The American volume is indicated by the container:

> 'Bring me half a can.' A half-can meant a nickel's worth of beer. A whole can meant a dime's worth. (Longstreet, 1955)

The British *half a pint* or *pint* similarly indicate quantities of beer:

> Pints were for men and... only boys drank halves. (Sharpe, 1975—today women seem to as well)

The obsolete British *half-pint* as a verb was to drink beer:

> Two miners 'half-pinting' in the public house. (Hunt, 1865)

half and half drunk.

Usually less than half sober:

> 'Were you drunk at the time?' 'Well, I'll tell you what it is, gentlemen, I was half-an-half.' (*Evesham Journal*, 1879, quoted in EDD)

Half and half and *top and bottom* also describe a mixture of mild and bitter beers in the same glass:

> He would not play except for a pint of half-and-half. (Mayhew, 1862)

In American prostitutes' jargon *half and half* is oral followed by vaginal sex (DAS).

half-baked bread *obsolete* a person of mixed Indian and white ancestry.

British Indian deprecatory use when there were strong taboos among the British expatriates about interbreeding:

> They used to call us *Kutcha butcha*, that is to say, half-baked bread. (Allen, 1975, of such a person)

half-canned drunk.

In intoxication, the half is seldom less than the whole and the attempt to understate the condition deceives nobody. Also as *half-cooked*, *-corned*, *-cut*, *-foxed*, *-gone*, *in the bag*, *on*, etc.

half-deck a lunatic.

From the partly open craft which is less seaworthy than one fully decked:

> But all those people on Doctor Diana's list sound like half-decks. (Sanders, 1985, of a psychiatrist's patients)

half-inch to steal.

Rhyming slang for 'pinch', mainly of pilfering:

> You used to 'arf inch suckers orf the barrers. (Kersh, 1936, referring to stealing oranges from street traders)

half-seas over drunk.

All the other states of drunkenness preceded by *half* indicate a condition of intoxication no less than the whole. In this case there is no *seas-over* to be halved. It is used either of total drunkenness:

> I'm half-seas o'er to death (Dryden, 1692)

or of a milder state:

> It was no longer the custom to get drunk, but to get half-seas over was still fairly usual (Harris, 1925)

and as *half-sea*:

> Hoarse elder John sat at his knee,
> In proper trim—more than half-sea.
> (Spence, 1898—but we must beware of forced rhymes)

hammer[1] to declare a defaulter.

London Stock Exchange jargon, from the hammering to obtain silence in which to make the announcement on the once noisy and crowded floor.

hammer[2] a philanderer.

> I used to be a great hammer, you know...Not any more...You wouldn't believe the really small kids who've tried to bring it back to life. (Amis, 1988—see IT[2] for further elucidation)

The common violent imagery, with no thought to Thomas Cromwell, the *malleus monachorum*, who proved more adept at harrying monks than picking a wife for his sovereign.

hammered *American* drunk.

Your head sometimes feels like it, especially after poor red wine.

hampton the penis.

From the rhyming slang *Hampton Wick*, a district to the west of London, for 'prick':

> No worse off physically than for a couple of smart tweaks of the hampton. (Amis, 1978)

Unusually, both words in the phrase are commonly used in the slang meaning but whereas *hampton* is only met literally, WICK is also used figuratively.

hand an employee.

A venerable usage, especially in compounds such as *deckhand* for an unskilled seaman, *farmhand*, *cowhand*, etc. The modern American general usage seeks to play down any suggestion of servitude on the part of the employee. To *give a hand* is still to render voluntary assistance and an *old China hand* does not work in a crockery store but is credited after residence in the Far East with understanding the intricacies of oriental affairs. See also HELP[1] and OBLIGE[2].

hand-fast *Scottish, obsolete* to cohabit and copulate extramaritally.

I include this entry to show that what seems new in social behaviour may not be so:

> It was not until more than twenty years after the Reformation that the custom of 'hand-fasting', which had come down from old Celtic times, fell into disrepute, and consequent disuse. By this term was understood cohabitation for a year, the couple being then free to separate, unless they agreed to make the union permanent. (Andrews, 1899)

The custom survived until the 19th century in Selkirk and Dumfriesshire, and now seems once again to enjoy universal approbation, except among those of us for whom the change in convention came too late.

hand in your dinner pail to die.

From the common imagery of making no further demand on resources:

> Uncle Wilberforce having at last handed in his dinner-pail...[he] had come into possession of a large income. (Wodehouse, 1930—the *he* was a legatee)

hand job the masturbation of a male by someone else, especially a prostitute.

> He declined her offer of a compensating handjob. (M. Thomas, 1980, of someone who had hoped to copulate)

Hand relief usually denotes self-masturbation.

hand trouble *mainly American* unwelcome male attempts to fondle a woman sexually.

She, not he, has trouble with his hands:

> Bonnie had encountered men with hand trouble. (Hynd, 1949)

handbook *American* a place away from a racetrack where bets are placed illegally.

Horseracing usage, from the recording of the wagers. Sometimes the person who accepts the bets is also called a *handbook*.

handful[1] a five-year prison sentence.

Criminal jargon. Literally, what you can hold in your hand. See also FISTFUL, FIVE FINGERS.

handful[2] a badly behaved person.

Either a precocious child or a spouse whose prior habits suggest that matrimony will demand excessive forbearance on the part of the partner.

handicap a physical or mental defect.

Literally, a disadvantage imposed on a contestant to make an equal contest. The American euphemism of 40 years ago is now standard English, as evidenced by:

> We fight shy of abbreviations and euphemisms. They rejoice in them. The blind and maimed are called 'handicapped', the

destitute 'underprivileged'. (E. Waugh, 1956, comparing American and British speech)

Today idiots and lunatics are *mentally handicapped*, lame people are *physically handicapped*, the blind are *visually handicapped* (even without a white stick), the deaf are *aurally handicapped*, etc. See also CHALLENGED.

handle[1] to embrace a woman sexually.

Literally, to touch or hold with the hands:

A did in some sort, indeed, handle women. (Shakespeare, *Henry V*)

The obsolete southern English dialect use was not euphemistic:

In love making, where the swain may not have the flow of language, he may sometimes attempt to put his arm around the girl's waist; this is called 'handlin' on her'. (EDD—as ever Dr Wright uses *lovemaking* for courtship)

If you were said to *handle a woman* today, you would be a pimp.

handle[2] power over another to coerce or extort.

From the leverage:

In this permissive age homosexuality isn't the handle it once was. (Bagley, 1982)

handle the truth roughly to lie.

Political use where a direct accusation of lying contravenes convention.

handout[1] a payment by the state to the poor.

Originally food and clothing, but now cash payments. Much pejorative use among those in work, whether or not they pay taxes.

handout[2] *American* a bribe.

Police jargon:

Six weeks' suspension and six weeks at reduced pay for taking a handout. (Diehl, 1978, of a policeman)

handout[3] a written statement to the press containing inaccurate, incomplete, or misleading information.

In standard usage, a summary in written form intended to record or amplify verbal information:

The question which has not been raised in the Press here, force-fed as it is on NASA hand-outs. (*Private Eye*, July 1983)

handshake a payment on leaving a job.

Usually on being dismissed and a shortened form of *golden handshake* (see at GOLDEN):

Had he agreed to suppress his feelings for five months—thereby collecting a full pension and a brigadier's handshake over £8,000... (M. Clark, 1991—of a First World War British brigadier in the Indian Army whose pay-off would have exceeded £8,000)

handyman special a derelict building.

American real estate jargon:

* HANDYMAN SPECIAL * Huge house w/lots of potential. (*Chicago Tribune*, 30 July 1991—lots of cockroaches, dry rot, expense, and other problems too, no doubt)

hang to kill by breaking the neck through suspension.

Now the standard English meaning, but it formerly meant 'death by crucifixion':

'No, Grace, we don't hang them any more.' 'Not even murderers?' 'Specially not them.' (N. Mitford, 1960)

The past participle is *hanged*—except in colloquial speech, only paintings are *hung*:

'You'll probably be hung.' 'Not now I won't, get ten years.' (Murdoch, 1977, deliberately reflecting vulgar speech)

A *hang-fair* was an execution by hanging in public:

The innkeeper supposed her some harum-skarum young woman who had come to attend the 'hang-fair' next day. (Hardy, 1888)

A *hanging judge* readily sentenced convicted people to death:

He'd got one or two unlikely convictions out of them. A hanging judge, some people said. (Christie, 1939)

However, the obsolete English *hang in the bellropes* was merely to be jilted or to delay a wedding after the calling of banns, from denying the campanologists their fun and reward:

...the 'deserted one' is said to be hung in the bell-ropes. (N&Q, 1867, quoted in EDD)

hang a few on to drink intoxicants to excess.

Of an American male alone, or in company with other males:

He had only hung a few on and was, for him, slightly sober. (Longstreet, 1956)

Less often as *hang one on*, which means many more drinks than one. The obsolete British *hang it on with* was to copulate regularly with your sexual mistress.

hang a red light on *American* to drive out of business.

The imagery is from a closed road—for once *red light* does not signal a brothel:

I have enough influence around this town to hang a red light on you. (Chandler, 1958)

hang in the bellropes see HANG.

hang on the bough *Scottish, obsolete* to remain unmarried.

Of a woman. See also WITHOUT A HEAD for a dissertation on the economic problems of 19th-century unmarried females. This use takes its imagery from unplucked and wasted fruit:

Ye impident woman! It's easy seen why ye were left hangin' on the bough. (Keith, 1896)

hang out the besom see BESOM.

hang out to dry *American* to be exposed publicly to protect others.

From exposure on the washing line:

Mitchell and Dean gave him assurances that he wouldn't be hung out to dry. (Colodny and Gettlin, 1991)

Whence a *hang-out*, such an exposure:

Is it too late to go the hang-out road? (ibid. —Nixon was asking if his accusers might be bought off by sacrificing one White House witness)

hang up your hat[1] *obsolete* to marry a woman wealthier than yourself.

Of a man whose bride provided the matrimonial home:

Snelling 'hung his hat up'—that is the local phrase—at the abode of Ephraim Shorthouse, whose daughter Cecilia had grown to a marriageable age. (D. Murray, 1890)

More generally, of any fortunate or fortune-hunting bridegroom. In similar fashion, a man might *hang up his ladle*, being nourished in future at his wife's charge.

hang up your hat[2] to die.

When this expression was common, the motor car had not discouraged the population from wearing hats out-of-doors, as shown in old photographs. Equestrians might also *hang up* their harness or tackle, and all could *hang up* their boots, an imagery which is now used of those who give up participation in a sport:

I'd always thought of thirty-five as approximately hanging-up-the-boots time. (Francis, 1985, referring to steeplechasing)

hangover symptoms of prior sub-acute alcoholic poisoning.

From the *hanging over* of the ill effects until the next day:

'How's the hangover?' From the sound of it, on the mend. The hair of the dog had bitten. (Francis, 1978)

Hung over is so afflicted:

He put down the receiver with all the gentleness of the badly hungover. (ibid.)

Occasionally shortened to *hung*:

'Sweating out sour booze?' 'You look hung yourself.' (Mailer, 1965)

hanky-panky extramarital copulation.

Originally, trickery. Favoured by mothers when telling daughters what not to get up to if spending an evening with a male.

Hanoi Hilton a North Vietnamese prisoner-of-war camp.

Of the same tendency as POTSDAM:

... two other general officers had been excused a stay in the Hanoi Hilton because of him. (Clancy, 1989, of someone who rescued American fliers shot down in Vietnam)

Hansen's disease leprosy.

The Norwegian physician was not himself a sufferer, so far as I am aware, dying at the age of 71:

The widely feared affliction now known as Hansen's disease. (*National Geographic Magazine* 1979—it will be widely feared whatever you call it)

happen to to cause to die.

Things *happen* to us every moment of our lives, but this particular *happening* we prefer not to spell out. Whence the phrase *If anything happens to me...* by a geriatric anxious about the welfare of a pet etc.

happenings *American* illicit narcotics.

From the effects of ingestion and perhaps too after the *happenings* of Allan Kaprow, a performance artist working in New York in the 1960s. There is some use too referring to any illegal or questionable events:

If you only have a list of *happenings*—is that the latest Whitehall word? (Lyall, 1985)

happy slightly drunk.

The symptoms for some, at some stage of drunkenness, include a sense of well-being. *Happy hour* is the period—not necessarily restricted to sixty minutes—during which bars sell intoxicants at reduced prices and give away food to encourage people to drop in on their way home from work; in theory to relax and relieve the tensions of the day, but then you may run into worse trouble when you arrive home drunk, broke, and late:

I bought two more: it was, after all, Happy Hour. (Theroux, 1979, of beer)

happy dust cocaine.

Used by American addicts:

... that happy dust gonna take you a real great snow ride. (Collins, 1981)

happy event the birth of a child.

Of planned and unplanned, wanted and unwanted pregnancies alike.

happy hour see HAPPY.

happy pill a poison which kills immediately.

Espionage jargon:

A happy pill... When you don't want to talk, but you think you may, then you crunch it... and you don't talk ever again. (Price, 1987)

happy release the death of a terminally ill patient.

We use the phrase of others, although they may feel otherwise. Occasionally in the same sense you may hear *happy despatch*, a translation of the Japanese *hara-kiri* but not importing suicide. The *happy hunting grounds* are the post-mortem destination of American Indians, and perhaps too of the scriptwriters who churn out the Westerns.

happy sock *American* male masturbation.

The derivation is unclear:

> Apart from these civic obligations, for most it was back to a creaking cot and the 'happy sock'. (Forsyth, 1994—the *civic obligations* were, for soldiers in the Gulf War, to drive past where females were sunbathing)

hard an extreme version of anything or anyone taboo or shameful.

Thus *hard core* is specific pornography, as distinct from *soft*, which is suggestive only; *hard drink, liquor, stuff*, etc. are spirituous intoxicants as opposed to *soft*, non-alcoholic, drinks:

> If I don't soon have a drop of hard I'm for it. (Cookson, 1967, of a man wanting whisky)

A *hard drinker* habitually drinks to excess. To *harden* a drink is to add spirits to it:

> I carried it to the kitchen and hardened it up from the bottle. (Chandler, 1943, of a drink)

A *hard drug* is an addictive and dangerous illegal narcotic such as heroin, while a *soft drug* such as cannabis is considered less immediately dangerous; a *hard case* can be a confirmed criminal, an alcoholic, or someone addicted to almost any taboo activity.

hard of hearing deaf.

Deafness, unlike blindness, when so described is not understood to be an absolute condition, the sense being only partially impaired except where otherwise stated, as *stone deaf*:

> 'I'm hard of hearing, you know,' she said. 'Practically deaf.' (Sanders, 1980)

Similar euphemistic forms exist in French and German.

hard-on an erection of the penis.

Of obvious derivation:

> ... getting a hard-on listening to a beautiful woman screwing another guy. (Diehl, 1978)

To *have a hard-on for* may be merely to lust after a woman:

> And this Piper guy had a hard-on for old women. (Sharpe, 1977, although normally the male's desires are more focused)

Occasional figurative use as an insult:

> 'Jesus,' she says, groaning, 'what a hard-on you are.' (Sanders, 1977—the groan was out of frustration, not desire)

hard room a prison cell.

As distinct from those with soft furnishings:

> ... defacing the walls of some of the subterranean 'hard-rooms'—a polite departmental euphemism for prison cells. (Deighton, 1985)

hard up poor.

Usually of a temporary shortage of funds and perhaps a shortened form of *hard up against it*.

hardware[1] whisky.

A 19th-century American use revived during the Prohibition years, perhaps punning on *hard* liquor but also seeking to conceal the nature of the purchase as in GROCERIES SUNDRIES.

hardware[2] any modern armaments.

Military jargon for things made of metal such as tanks, bombs, planes, guns, and missiles:

> 'You're talking about hardware.' ... 'We don't buy machine guns at the local ironmongers.' (Theroux, 1976)

hardware[3] medals and awards.

> He wanted to have on his chest a bauble to compare with all the glittering hardware they'd accumulated in a lifetime of soldiering. (Deighton, 1994)

A deprecatory usage. See also GONG for a further disquisition.

harlot a prostitute.

The word was used of a disreputable person of either sex until *varlets* became male and *harlots* female in the 17th century. Shakespeare writes of 'the harlot king' (*Winter's Tale*) but also 'Portia is Brutus' harlot, not his wife' (*Julius Caesar*). It is a corruption of *harlet*, a small hiring, although Harlotte, the mother of the bastard William of Normandy, has had her share of unfair attributions down the ages.

harness bull *American* a uniformed policeman.

The *harness* is the uniform and *bull* a policeman.

harpic mentally unbalanced.

From the brand name of a substance claimed to clean a lavatory bowl 'right round the bend':

> God, he must be harpic. (M. Fraser, 1992—using Second World War slang)

Harry the devil.

Usually as the *old Harry*, the *Lord Harry*, or the *living Harry*:

> By the livin' hairey, if I could win ower tae them. (Wardrop, 1881)

Still used in exclamations and when we *play old Harry* about something which upsets us.

hash marijuana.

A punning shortened form of *hashish*. A *hash-head* is an addict.

hat and cap gonorrhoea.

American rhyming slang for 'clap'.

hatch the birth of a child.

Emergence from an egg is clearly less taboo than the method of mammalian delivery:

> The female mind... takes an interest in the 'Hatch, Match and Despatch' of its fellow creatures. (Payn, 1878)

hatchet man someone entrusted with a job requiring ruthlessness.

Not he who chops the kindling in the back yard. He may be a professional killer:

> ... regular hatchetmen on their payrolls who were available to kill at a moment's notice (Lavine, 1930)

or an employee entrusted with a difficult or dubious transaction:

> 1981 is not exactly turning out to be a vintage year for ——, Sir James Goldsmith's hatchet man (*Private Eye*, April 1981)

or a manager entrusted with the peremptory dismissal of employees. Occasionally shortened to *hatchet*:

> If he's dead, he's worth five grand to you and five to the hatchet, so go and fix it. (Francis, 1988)

A *hatchet job* is the result of his activities, literally or figuratively:

> This series is going to be very sympathetic to the police... I'm not out to do a hatchet job. (Sanders, 1973)

haul a quantity of stolen property.

Literally, the thing or amount gained from a profitable transaction, with much figurative use. For the police, to *haul in* is to arrest. In obsolete British use, to *haul ashore* was to retire from employment, from the beaching of an old boat.

haul your ashes to copulate with a female.

Haul has a meaning 'to harm another physically', which is a common image where copulation by males is concerned; and *ashes* are *hauled* or pulled out from the small door at the foot of the furnace, which is red and glowing within. But I don't find either of these clues to the etymology convincing. A male may *get his ashes hauled*:

> I pop in a red, get a little shot, you get your ashes hauled. Same dif. (Diehl, 1978, or in translation 'I like self-induced narcosis, you prefer promiscuous copulation')

Whence *hauled*, having copulated, usually extramaritally.

haute cuisine small portions of expensive food.

Literally, high-quality cooking:

> When I'm away I live in hotels, where I get junk tricked out as haute cuisine... (Follett, 1979)

Havana rider a passenger who seeks to take control of an airliner in flight.

American airline use, from the preferred destination of many such pirates:

> Research in America has come up with a fair picture of the 'Havana riders', as airline staff call them. (Moynahan, 1983)

have to copulate with.

The word is used of either sex. Of an individual act:

> I was so impatient I had her without getting out of my chair (Fraser, 1969)

or of copulation generally:

> You must have had a lot of men?... Have you enjoyed it? (Amis, 1978)

Most of the uses of *have* referring to sex in the entries that follow are part of daily talk and literature to the extent that we forget their intrinsic stupidity. Only hermaphrodites do not *have sex*; and we *have something to do with* every person of the opposite sex whom we meet in our daily lives. See also HAVE RELATIONS.

have a man/woman to copulate extramaritally.

With the action being more important than the choice of partner.

have a screw loose to be mentally ill or eccentric.

The imagery of falling apart. The condition may range from mild and temporary forgetfulness to serious mental illness:

> The first time I met [him] I thought he was mad. And I don't mean mad as in zany or whacky, I mean mad as in screw loose or tonto. (Barber, 1991)

Or it can mean merely 'to behave rashly':

> It was me having a screw loose and going out looking for trouble. (*New York Times*, 14 June 1992)

See also SCREWY.

have a slate loose to be mentally ill.

The imperfect covering of the roof of a house is transferred to the head. In obsolete use a *slate off* was an idiot:

> He left aw 'at he hed to his slayatt hoff of a nevvy. (EDD—the beneficiary was his nephew)

have at to copulate with.

Of a male, from the meaning 'to attack':

> I woke up and had at her again. (Fraser, 1970)

have it to copulate.

Of either sex, this is often expanded to *have it away* or *have it off*:

> The true test of love is when you can watch your wife having it off with someone else and still love her. (Sharpe, 1976)

Have it off is also used of homosexual activity:

> Khaliq will insist on having it off with the other ranks. (M. Thomas, 1980)

have relations to copulate.

Genteel usage:

You perhaps ought to have relations once to make sure of a happy adjustment. (McCarthy, 1963)

Occasionally, and perhaps obsolete, as *have been in relation with*:

This is of course Weguelin, so she must have been in relation with both [O'Shea and Parnell]. (Kee, 1993—Weguelin had been a resident in the O'Shea household before Parnell)

To *have sexual relations* is explicit:

I'm not 'very highly sexed'. I can live perfectly well without sexual relations. (Murdoch, 1978)

have sex to copulate.

Within or outside marriage. It is also used of homosexual activity.

have something to do with to copulate with.

Usually of a male:

The euphemistic modern to have (something) to do with a woman. (Partridge, 1947)

have the painters in to be menstruating.

Common female usage, with reference to the staining and colour, the protective sheeting, the temporary dislocation, and the inconvenience.

have your end away to copulate.

Of a male, the *end* being the penis rather than the objective:

He has been having his end away. (P. Scott, 1977—he has been copulating)

Also as *have your end off*, which does not imply unwanted surgery.

have your way with to copulate with.

Of a male extramaritally, with a suggestion of duress:

Piper prowled the dark streets in search of innocent victims and had his way with them. (Sharpe, 1977)

To *have your filthy way* denotes no lesser degree of male cleanliness.

have your will of to copulate with.

Again usually of the male, especially if the will is *wicked*:

...sweeping her off at his saddlebow and having his wicked will of her. (Fraser, 1982)

hawk a person who advocates aggression as a way of defence.

The idea comes from Calhoun's *War Hawks*, a political party of 1812, and was revived during the confrontation after the Second World War between the USA and Russia. The obsolete English *hawk* was only a policeman, watching and seizing his victims.

hawk your mutton to be a prostitute.

To *hawk* is to offer for general sale. See also MUTTON. In rarer form, she might *hawk her pearly*, the oyster being a bivalve to which the vagina is coarsely likened:

I told her to hawk her pearly somewhere else. (Sharpe, 1976)

The obsolete English to *hawk your meat* was to display an immodest amount of bosom as an allurement to males.

hay marijuana.

An American use: older, perhaps, than the now fashionable *grass*.

hazard of the town the contraction of venereal disease.

The town implied enjoyment of its debauched pleasures:

...in his fears of the hazard of the town, he had been some time looking out for a girl to take into keeping. (Cleland, 1749)

The phrase is obsolete but the hazards persist.

hazy drunk.

Things may seem a little misty to the drunkard, and his memory defective.

he-biddy *American* a cock.

Another example of 19th-century prudery.

he-cow *American, ?obsolete* a bull.

More prudery. Also as *he-thing*. For a list of similar euphemisms see BIG ANIMAL.

head[1] to kill by beheading.

As in the modern use, where we *head* gooseberries etc., taking the top off:

Has not heading and publickly affixing the head been thought sufficient for the most atrocious state crimes? (Maidment, 1868)

A *heading* was such an execution, carried out by a *heading-man*, often upon a *heading-hill*, for the convenience of onlookers. Many seem not to have been deterred from treason by the severity of the penalty.

head[2] a lavatory on a ship.

Originally in a warship, but now general:

There was a small head off the little cabin. (Sanders, 1977)

Usually in the plural:

He heard the liquid pour in the bowl of the heads. (W. Smith, 1979)

Occasionally used in affectation by yachtsmen ashore.

head[3] the incidence of sub-acute alcoholic poisoning.

Shortened form of *headache*, which you are likely to get if you have been drunk.

head[4] see GIVE HEAD.

head[5] a narcotics addict.

Perhaps from the effect on alertness. Usually in combination, for example *snow-head*, a person who is addicted to cocaine. A *head-kit* is the apparatus used for illegal administration of narcotics.

head-case an idiot.

A *case* of weakness in the head, referring to a single aberration or general deficiency:

> His teachers at school didnae think he was very bright. They thought he was a head case. (Theroux, 1983)

head-count reduction the dismissal of numbers of employees.

Not a diminution in the frequency of counting them. Industrial jargon where a policy decision is made to reduce the payroll either systematically over time or by a single major reduction.

head job (a) fellatio.

Prostitutes' jargon, from the *head*, glans penis:

> ... receiving a listless headjob from an ageing black prostitute. (Wambaugh, 1975)

A *head chick* is a prostitute who offers such a service.

head-shrinker a psychiatrist.

Punning on the cerebral centre of their investigation and the practices of primitive tribes apropos their enemies:

> One day I may need some headshrinking work done. (Ustinov, 1971)

Also shortened to *shrink*:

> ... ending up on some shrink's couch twice a week. (Hailey, 1979)

The evasion is needed because consulting a psychiatrist, though a status symbol for some, is a shameful matter for many people.

headache[1] see HEAD[3].

headache[2] a signal of unwillingness to copulate.

Used by women, the signal is usually given to a husband or other regular partner, but:

> You were glad you found out about the headache before you invested too much time and money and hope in her. (Chandler, 1953)

headache-wine *obsolete* a narcotic made from the petals of English poppies.

In English/Irish use, common corn poppies were called *headaches* and unmarried girls had a fetish against touching them, perhaps because the drowsiness and feeling of goodwill induced may have made them easier to seduce:

> Corn-poppies, that in crimson dwell,
> Call'd head-aches from their sickly smell.
> (Clare, 1827)

(I include this entry to show that the narcotics problem is not new, and to provide an excuse for my single quotation from John Clare.)

headhunter[1] a recruiting agent who entices employees to change jobs.

Punning on seeking out a person with special talents and the neighbourhood sport of Papuans. The object is to obtain for the new employer, who pays a large fee, the knowledge, contacts, skills, and experience which the employee has acquired serving a competitor.

headhunter[2] a police internal disciplinary inspector.

American police jargon. He is looking for dishonest policemen who will have no future in the police service if discovered:

> Headhunters made rank consistently better than other investigators. (Wambaugh, 1975)

health illness.

Here, as with DEFENCE or LIFE INSURANCE, you avoid mentioning the taboo by stating the converse. Thus the medicine industry refers to its commercial activity as *health care*, selling its products to the sick; the British *National Health Service* provides for the ill and dying, without, however, neglecting the well-being and security of its administrative staff; and we refer constantly to *health clinics*, *health insurance*, etc.

hearing-impaired partly deaf.

To *impair* indicates a weakening in quality or strength. Those so described may have suffered an unchanged degree of deafness since birth.

heart condition a malfunction of the heart.

Medical jargon, in which all *conditions* are bad:

> He had suffered from a heart condition for several years. (*Daily Telegraph*, November 1980)

Sometimes shortened to *heart*, as in *having a heart*, but who doesn't?

hearts *British* penurious.

From the rhyming slang *hearts of oak*, broke, and punning on an affluent building society of the same name. Sometimes in the full form:

> It left me 'earts-of-oak. (Kersh, 1936)

heat[1] action causing alarm or anxiety.

From the rise in body temperature when we are in danger. Of illegal coercion:

> It's life or death, nothing in between. This is immediate heat. (M. Smith, 1993, of a blackmail threat)

It is also used of concerted police activity against specific crime; of military attacks on rebels or terrorists; and of a government enquiry into a bureaucratic scandal or administrative practice.

heat[2] a handgun.

From the warmth of the barrel and perhaps punning on *firing*:

> 'Ahh, I'm carrying,' Boon said. 'Someone will spot the heat.' (Sanders, 1977)

Also as *heater*:

> 'All right, Dad. Shed the heater.'... He put his enormous frontier Colt on the floor. (Chandler, 1939)

To *pack heat* is to carry a gun.

heaven dust cocaine.

Used by addicts. *Heaven and hell* is phenocyclidine, which involves painful withdrawal symptoms. *Heavenly blue* or *blue heaven* are other narcotics used illegally, perhaps punning on the colour of the tablet.

heavy date an important tryst for courtship.

Heavy means 'important', and see also DATE:

> Thought you had a heavy date tonight, Molly? (Deighton, 1981)

Heavy necking or *heavy petting* indicates sexual activity just short of copulation. A *heavy involvement* indicates an extramarital arrangement which includes copulation, often where marriage cannot be contracted. The obsolete Scottish *heavy* was a shortened form of *heavy of foot*, meaning 'in a late stage of pregnancy':

> James cam to me ae morning when she was heavy o' fit. (Service, 1887)

heavy landing an aircraft crash on the runway.

Aviation jargon for a mishap in which nobody is injured:

> ...a DC 10 of the big American carrier Overseas National careered off the runway at Istanbul after a heavy landing. (Moynahan, 1983)

The phrase avoids taboo words such as *mistake* or *crash*.

heel-tap a small volume of alcohol left in a glass.

> Seize the bottle and push it about; Don't fill on a heel-tap, it is not decorous. (A. Boswell, 1803)

The *tap* was the sole of a shoe in cobblers' jargon, although Holt identifies it as 'the layers of the leather heel' (*Phrase and Word Origins*, 1936) —and he is usually right. Whichever the case, it lay at the bottom of the shoe. Today the admonition *no heel-taps* means that the glass must be drained before replenishment.

heeled[1] *American* drunk.

Probably from being tilted over.

heeled[2] carrying a gun.

> I noticed Collins's hand stray under his jacket, and wished I'd thought to come heeled myself. (Fraser, 1982)

Literally, armed or equipped. *Well heeled* means 'wealthy'.

heels foremost dead.

From the normal direction in which corpses are carried.

height of connubial bliss copulation.

Scaled within marriage, although the cliché plumbs the depth of banality.

heightened interrogation torture.

As practised by the Nazis:

> Down in the cellar the Gestapo was licensed to practise what the Ministry of Justice called 'heightened interrogation'. (R. Harris, 1992)

heist *mainly American* to steal.

A variant of HOIST[1]. It refers normally to taking a truckload of goods or to an armed robbery. A *heist* is such a robbery:

> 'This is a heist!' Frisky yelled. 'Out of there and line up.' (Chandler, 1939)

helmet a police officer in uniform.

The derogatory jargon of those who are permitted to wear plain clothes:

> They had a taste for lapel pins... All things which said 'I am not a helmet'. (*Daily Telegraph Magazine*, August 1990)

help[1] a domestic servant.

Shortened form of the American *hired help* and implying voluntary assistance rather than servitude:

> I don't want my help to know or guess. (Harris, 1925—he claimed to be copulating with an American housewife)

Help is also used of any American employee.

help[2] the services of a ghost-writer.

Publishing jargon:

> The odd thing about this kind of collaboration is that the celebrity... in seeking 'help' with a novel, inevitably appears dimmer than if she had never done a book at all. (*Daily Telegraph*, 9 September 1994)

There are also codewords for detecting who actually wrote the book:

> Naomi awards Caroline 'my very special thanks' in small type on the flyleaf. In publishing circles, this is the accepted way of acknowledging a ghost-writer. (ibid.)

help the police (with their inquiries) *British* to be in custody and presumed guilty of an offence for which you have not been charged.

The purpose of this phrase is not to pre-judge guilt, to avoid the possibility of a subsequent conviction being quashed:

> When someone is helping the police with their inquiries into a murder it may not be proven that he is a murderer but the suggestion is there. (Sharpe, 1976)

To *assist the police* means the same thing. In American police jargon and practice, *helping with inquiries* means something quite different:

> 'He is helping us with our inquiries.' 'What a pompous phrase for torture.' (Theroux, 1977)

A transatlantic visitor, reading British newspaper reports, might be led to judge the suspects so described a very accommodating and gentlemanly crowd.

help yourself to steal.

Literally, not to await service by another. Now used of unpremeditated pilfering, especially where the goods are unguarded.

hemispheres *obsolete* the breasts of an adult female.

Viewed sexually or poetically, although not necessarily half the size of GLOBES.

hemp[1] *obsolete* pertaining to death by hanging.

From the *hemp-string*, noose:

> In a' probability he wad form a bonnie tossil at the end of a hemp string. (Willock, 1886 —a *tossil* was a tangle)

The *hempen fever*, *hemp strung*, or *hemp quinsy* were death by hanging:

> The hemp quinsy, as the lags call hanging. (Keneally, 1987)

A *hempen widow* had so lost her husband, who might be known as a *hempshire gentleman*, punning on the English county of Hampshire.

hemp[2] marijuana.

A shortened form of *Indian hemp*:

> Reefers, grefa, musta, the hemp. (Longstreet, 1956, listing illegal narcotics in New Orleans)

hen-silver *northern English, obsolete* extortion at the church door before a wedding.

In a refined form, firearms were used:

> Formerly a gun was fired over the house of a newly married couple, to secure a plentiful issue of the marriage (probably to dispel the evil spirits that bring bad luck). The firing party had a present given them... and this was termed hen-silver. (*Penrith Observer*, September 1896)

Money collected generally for intoxicants at a wedding, or *hen-drinking*, was called *hen-brass*. This usage shows that marriage was an occasion for drunkenness, extravagance, and refined extortion for our ancestors also. See also BELL MONEY.

herb (the) marijuana.

An American variant of the commoner *grass*.

hereafter (the) death.

Religious use, anticipating the life to come, or as the case may be:

> The contents of that box were all that held off the Hereafter. (Francis, 1978)

hermaphrodite a male homosexual.

The nymph Salmacis was the result of the fusion of the bodies of her parents, Hermes and Aphrodite, whence the fusion of names also. Literally, a creature combining the features of both sexes, or neither, but venerable, if rare, use of homosexuality.

hick *American, obsolete* a corpse.

In standard usage, an unsophisticated country dweller. It is said that this use came from the country dweller's availability to the anatomical dissection industry, if he ventured alone into town, an understandable concern about the risk of resurrection of an incomplete body giving rise to a cadaver shortage on both sides of the Atlantic. The 19th-century *hic jacet* was a tombstone, punning on the coat and the Latin *here lies*:

> By the cold Hic Jacets of the dead. (Tennyson, 1859)

hide the weenie *American* to copulate or bugger.

From the nursery word for penis:

> 'So, he was acey-deecy.' 'He was a devout catholic. A daily communicant.'... 'Listen, Doc. Lots of old altar boys play hide-the-weenie when they shouldn't.' (Sohmer, 1988)

high[1] *British, obsolete* infected with venereal disease.

This imagery was a reference to rotten meat.

high[2] drunk or under the influence of narcotics.

From the feeling of elevation, but you do not use it for those lapsed into torpidity or unconsciousness. Of drunkenness:

> We'd had some people in for cocktails, and we all got quite high (McCarthy, 1963)

and of illegal narcotics:

> The user smokes them in big puffs getting high. (Longstreet, 1956)

Whence a few clichés, such as *high as a kite*. An adverse reaction to the ingestion of narcotics is a *low*. To *live on the high-fly* in 19th-century England was to rely on the proceeds of begging letters, a career made possible by the introduction of the penny postage.

high forehead (a) baldness.

We comment on this only in other men:

> 'And the receding hairline?' 'Receding what?' ... Godfrey swung round. 'High forehead,' he said. (Lynn and Jay, 1986, of a politician being groomed for a television broadcast)

high in the belly *British, obsolete* in an advanced state of pregnancy.

Pregnancy was the subject of many old wives' tales and to be *carrying high* meant different things in various parts of the country, few of them according with medical science.

high jinks extramarital copulation.

Originally, a carouse which you had if you played the game involving dice, the repetition of verses, and forfeits; the more you drank, the more mistakes you made, the more forfeits you paid, the more you drank, and so on. Daughters today are still advised by anxious mothers not to get up to any *high jinks*.

high jump see FOR THE HIGH JUMP.

high-yellow girl see YELLOW[2].

highball a measure of whisky with ice etc. in a tall glass.

High from the glass and was the *ball* a bowl?

> Aren't you coming up with me to have a highball for the road? (Ustinov, 1971)

For American railroad engineers it meant 'a clear track', which has no bearing, I think, on the etymology.

highgrade to steal.

From the useful American phrase meaning 'to take the easiest pickings', of timber from a forest, ore from a mine, etc. Thus a *highgrader* is a discriminating and selective thief.

highjack see HIJACK.

highwayman a thief on the highway.

Not just any wayfarer. So described, he was usually on horseback, when he was a *high pad*, as distinct from a *footpad*, who robbed on the *pad*, path, on foot. Such robbery was called the *high law*, and the thief the *high lawyer*.

hijack to take illegal possession of (a vehicle).

Doubtfully euphemistic and now standard English. Originally American Prohibition use, when it became easier to steal from smugglers than to smuggle on your own account. The command to put up your hands was a laconic *High, Jack*. A *highjacker* (or *hijacker*) so operated:

> Highjackers stopped cargoes at interurban boulevards. (Longstreet, 1956, describing the Prohibition days in America)

While lorries and their cargoes are still thus stolen, a post-1970 refinement has been the seizure of commercial aircraft in flight to extort political or financial concessions:

> A man armed with grenades hijacked a Russian jetliner yesterday and took the plane on a three-country odyssey. (*Star-Ledger*, 21 February 1993)

Also much figurative use:

> But the environmentalists are the main interest group to have figured out that science can be hijacked for ideological purposes. (*American Spectator*, February 1994)

hike[1] to forge an increase on a negotiable bill.

From the meaning 'to hoist' or 'raise'. In commercial jargon, a *price hike* is an unwarranted increase in a selling price usually due to a shortage:

> I...expect that allowing for the effect of the oil price hike the inflation figures will begin to improve well before Christmas. (*Guardian*, 25 September 1990)

hike[2] to dismiss peremptorily from employment.

Hooking the employee out of his job and, in the past, sending him on a long walk:

> Another minute an' he'll hyke me aff. (Proudlock, 1896, of such dismissal)

See also TAKE A HIKE.

hillside men *obsolete* armed outlaws.

Or 'freedom fighters', although when the expression was in use in 19th-century Ireland, most of the population, as today, were opposed to force as a solution to political differences. Some were merely landless peasants:

> He was no bog-trotter...but ranged on the side of the moonlighters and the hillside men. (Flanagan, 1988—see MOONLIGHT[2] for *moonlighters*)

hindside the buttocks.

BACKSIDE is more common. Also figurative use:

> Although Richard had a tendency to look after his bureaucratic hindside, Barcella knew him and trusted him. (Maas, 1986)

hips a female's body used for copulation.

As in the old adage 'Free of her lips, free of her hips'.

historic old.

A usage of estate agents which sometimes traps them in tautology:

> Historic Saxon barn. (*Sunday Telegraph*, May 1981, implying construction before 1066)

hit[1] drunk.

From the rhyming slang *hit and miss*, piss, for a carouse—see ON THE PISS:

> Sorry about my breath—I've been out on the hit and miss. (*Daily Telegraph Magazine*, August 1990)

hit[2] to kill.

Usually of assassination, which is also called a *hit*:

> This is some kind of Mafia hit? (Diehl, 1978, describing such a killing)

A *hit man* is a professional killer:

> You've narrowed the field down to a couple thousand hitmen. (ibid.)

Criminals apart, you are also *hit* when you are struck by a bullet etc. To *hit*, from the obsolete rhyming slang *hit and miss*, meant 'to kiss' and earlier still, 'to copulate', with the common violent imagery:

> She'll find a white that shall her blackness hit. (Shakespeare, *Othello*)

hit[3] to steal.

There are two distinct American meanings; street robbery, which may have developed from the hobo jargon *hit*, to beg with threats; and a planned major theft, in which a bank or warehouse may be *hit*.

hit[4] to administer an illegal narcotic.

To another or yourself, and usually in cigarette form:

> I want another hit before you bring him in. I want to be really up for what I have to do. (Robbins, 1981)

To *hit the pipe* is so to smoke opium or marijuana.

hit on to attempt or achieve copulation with a female.

In the case of an attempt, perhaps no more than trying to *make a hit*, a good impression:

> ... people start sending drinks over to me, like fifty at a time. Then they're all hitting on me. (Theroux, 1990)

The usual violent imagery emerges where there is copulation:

> Did you hit Sonny because he was a Russian or because he was hitting on me? (de Mille, 1988—Sonny and the speaker had been copulating with each other)

hit the bottle to drink intoxicants to excess.

The commonest alternative is *hit the hooch*. Of a single debauch:

> We were kind of hitting the bottle a little. I guess we were pretty noisy (Chandler, 1943)

or of sustained excesses:

> ... hitting the hooch like you birds been. (ibid.)

Less often as *hit it*:

> ... poor old Carlisle, who between you and me had been hitting it a bit of late. (*Private Eye*, September 1981)

hit the bricks[1] *American* to go on strike.

From the sidewalk where the strikers might congregate after leaving a factory, or merely from the act of walking out.

hit the bricks[2] *American* to escape or desert.

Of a prisoner, again from reaching the sidewalk. To *hit the hump* also means 'to escape' or 'to desert from the armed forces', from the figurative hill over which the fugitive disappears.

hit the sack with to copulate with extramaritally.

Of either sex and not interchangeable with *sleep with*, because that euphemism covers marital copulation too:

> ... blame a Colonel for hitting the sack with a hooker. (Ustinov, 1971)

hit the silk see SILK (THE).

hoary-eyed drunk.

From the look of frost over them. A cockney editor could have advised DAS not to give a separate entry for London's 'oryide'.

hobby-horse a prostitute.

The article in Morris dancing which became a children's toy. This obsolete use puns too on *hobby*, a wanton, and the common equine copulation imagery:

> My wife's a hobby-horse. (Shakespeare, *Winter's Tale*)

hochle to copulate extramaritally and openly.

Literally, to sprawl about. Dr Wright defines this perhaps obsolete Scottish word as 'To tumble lewdly with women in open day'. (EDD—a fine example of his Cleland-like definitions where he has to define any impropriety. Do not be misled into thinking that there was once an 'open day' for tumbling lewdly with women.)

hoist[1] to steal.

The common lifting concept. In 19th-century England specifically denoting pilferage from retail shops, whence *on the hoist*, so engaged. Today it applies to any theft. A *hoister* is a thief.

hoist[2] to drink intoxicants.

From lifting the vessel to the mouth, in phrases such as *hoist one* or *a few*:

> The pub was full of hollering men ... Murf said, 'I think I should split.' 'Forget it. Let's hoist a few.' (Theroux, 1976)

hoist[3] *American* peremptorily to dismiss from employment.

Perhaps from the sudden *lifting* out of a job.

hoist[4] *obsolete* to kill by hanging.

Normally an impromptu event rather than death on the gallows.

hoist your skirt to copulate casually.

Of a female, and implying an element of spontaneity:

> Every girl in the reseau would hoist her skirt for *you*. (Allbeury, 1978)

hold to possess narcotics illegally.

Whether for your own use or resale:

> Never hold when you can stash. (Addict proverb)

hold-door trade (the) *obsolete* prostitution.

From the prostitutes who seek customers when lounging against a partly open door:

> Brethren and sisters of the hold-door trade. (Shakespeare, *Troilus and Cressida*)

The phrase is obsolete but the practice persists.

hold-up a robbery.

Literally, a delay of any kind, and I suppose a considerate thief may *hold up* his hand to stop you, before taking your valuables:

> You'll hold me up now, I suppose! (Chandler, 1939)

Formerly of stage coaches but now of any robbery, especially where violence is threatened.

hold your liquor to drink a lot of alcohol without appearing drunk.

Without vomiting either, but intermediate urination does not disqualify you:

> He can't drive, he can't cook, he can't hold his liquor. (Theroux, 1978)

Strictly speaking, each of us with a charged glass in his hand holds his liquor.

hole[1] to kill.

From the entry of the bullet or the excavation for the grave—I suspect the latter:

> Keep yourself from being holed as they holed Muster Bingham the other day. (Trollope, 1885)

The modern idiom *a hole in the head* is not your mouth but death from a bullet.

hole[2] the vagina viewed sexually by a male.

A common vulgarism:

> When a girl gets hot, her hole gets bigger. (Theroux, 1989)

Whence in compounds the *hole of contentment* or the punning *holy of holies*:

> I want to see the Holy of Holies, the shrine of my idolatry. (Harris, 1925)

A woman who copulates promiscuously may be lewdly described as a *good bit of hole*, despite any manifest moral or physical shortcomings.

hole[3] see BLACK HOLE.

holiday a term in prison.

A jocular explanation of absence:

> Not since I took that little state-financed holiday. (Lyall, 1969, referring to a prison sentence)

holiday ownership a compounded annual rent paid in advance.

As TIME-SHARING became discredited another expression had to be coined to ensure that the gullible would continue to part with their money:

> So you must agree that buying a holiday ownership apartment at the Lanzarote Beach Club will actually SAVE YOU £20,000. (*Daily Telegraph*, August 1989—see SAVE to understand better this proposition)

hollow legs the ability to drink a lot of liquor without getting drunk.

The volume has to be stored somewhere, it seems:

> Born with hollow legs! I watched with fascination while the gold liquid disappeared like beer. (Francis, 1978)

This common cliché is also used of a glutton.

holy of holies[1] see HOLE[2].

holy of holies[2] a lavatory.

There are several puns here, in addition to the allusion to the quiet and secret place. The Latin *sanctum sanctorum* is rarer and loses all in translation.

holy wars the expansion into the Middle East in the Middle Ages by western adventurers.

We know them better as the Crusades. The prime cause was probably not religious—although that was the excuse—but the over-population of western Europe prior to the fortuitous onset of the Black Death. After culling humanity for over a century, the problem was starting to recur when the Age of Discovery revealed softer victims in the Americas, Africa, and the East.

holy week the period of menstruation.

Not the week before Easter but punning on the proscription of copulation, the duration of the disability, and perhaps on *hole*, the vagina (see HOLE[2]).

home an institution.

For orphans, chronic invalids, young criminals, geriatrics, etc. We play down the formal and alien nature of such places by stealing one of the most basic and emotive of our familiar concepts. But compare a *rest home*, such an institution, and a *home of rest*, a morgue. The builder or estate agent who tries to sell you a *home* when he is offering a house seeks to appeal to the same concepts.

home economics cooking and housekeeping.

It would not do to be seen teaching girls a sexually stereotyped subject.

home equity loan a second mortgage.

Usually to fund consumer extravagance and on onerous terms:

> 'Home equity loan' sounded ever so much more palatable than 'second mortgage', palatable to the extent of seventy-five billion dollars already on the banking system's books. (M. Thomas, 1987)

homelands nominally independent areas of South Africa into which black families were transported.

Political jargon of the whites, as part of the former policy of SEPARATE DEVELOPMENT:

> South Africa's ethnic homelands are crumbling from internal corruption and bankruptcy and outside pressures by President

F. W. de Klerk and the African National Congress. (*Sunday Telegraph*, March 1990)

homely plain.

'Unaffected naturalness' (WNCD) becomes 'plainness' in a woman, and even downright 'ugliness':

It was the homeliest members of your class who became teachers. (McCarthy, 1963)

homework[1] a woman with whom a man copulates outside marriage.

In standard usage, a task undertaken additional to and outside your normal duty or curriculum, such as a school exercise which has to be done in the evening at home. In this use, the activity can hardly be classed as work and it is never done at home. A secondary American meaning is sexual excitement prior to copulation, perhaps punning on the preparation aspect of homework.

homework[2] preparation before a discussion.

In the cliché *doing your homework*, referring either to research done for you by others or to a quick glance at the papers before a meeting starts. The use seeks to imply virtuous zeal and efficiency.

homo a homosexual.

A shortened form, and used usually of the male, despite being derived from the Greek word meaning 'same' rather than the Latin for 'man':

I'll never understand women. Sometimes I think these goddamned homos have got something. (Deighton, 1982, the usage implying that all *homos* are male)

The Russian spies recruited in the 1930s at Cambridge were called the *homintern* because of their homosexuality:

... bound to him intellectually, emotionally, and sometimes physically as active members of what has since been aptly nicknamed the 'Homintern'. (Boyle, 1979)

honest not copulating extramaritally.

Not necessarily truthful or trustworthy in other respects:

I do not think but Desdemona's honest. (Shakespeare, *Othello*)

And a man may still *make an honest woman of* someone by marrying her after impregnating her or some other sexual familiarity with her.

honey *American* human excrement.

Referring no doubt to the colour and texture rather than the smell or sweetness. In the army you might fill, and then empty, a *honey bucket*:

'I emptied the honeybucket!' shouted an American voice. (L. Thomas, 1981)

The aggregate is carried away in a *honey wagon* or, in the navy, a *honey barge*. In airline jargon, a *honey cart* empties the lavatories after each long flight:

... the sanitary servicing vehicle ('honey cart' to the crews). (Moynahan, 1983)

A *honey-dipper* is not a bee but a lavatory cleaner:

The V.C. get work inside all camps as shoeshine boys and laundresses and honey-dippers. (Herr, 1977, referring to Vietnam—the V.C. were the Vietcong)

honey man *American* a person with whom a female regularly copulates extramaritally.

From giving 'sweeteners' rather than getting sweetness.

honey trap an attempt to seduce.

The use of sweet, normally female, wiles to compromise a target or obtain information:

Mossad had twice used him for honeytrap operations. (Forsyth, 1994)

Also as a verb:

... the arrest of a Marine Embassy official who had been 'honey-trapped' by a woman working for the KGB. (Pincher, 1987—he meant an embassy official who was a Marine)

honeydew melons see MELONS.

Hong Kong dog diarrhoea.

Only of the affliction if contracted in Hong Kong. Because it *dogs* the sufferer? I doubt it.

honk to feel the genitals of a male.

Prostitutes' and police jargon denoting a sexual approach, usually in a public place, where the action resembles squeezing a bulb horn:

Sabrina ... gave his genitals a squeeze ... He knew he had been 'honked' as the vice cops called it. (Wambaugh, 1975)

honked drunk.

Probably from the American *honk*, to vomit, but it may be used of drunkenness even if you don't throw up.

honour the avoidance of extramarital copulation.

Of female integrity in this respect, if no other:

You sitting there with your legs crossed and a hole in the head and me trying to explain how I shot you to defend my honour. (Chandler, 1958)

Of a male it meant 'the avoidance of being cuckolded':

You're a sneekin' varmint to take advantage of a man's hospitality to try and steal his honour. (Fraser, 1971, writing in 19th-century style—Flashman had copulated with the wife, as usual)

honours a system whereby politicians reward or bribe supporters and discourage dissidents.

As in the British *honours system*, *list*, etc., which should imply the recognition of outstanding merit or service. Although nominally under royal patronage, a government routinely selects participants on populist or political grounds. For civil servants *awards* are given or withheld as a disciplinary measure:

> They certainly wouldn't bother [to ingratiate themselves with royalty] if they knew how the Honours system actually operates. (A. Clark, 1993, of fawning businessmen at a meeting attended by the Prince of Wales)

hoof (the) *British* peremptory dismissal from employment.

A variant of KICK[2] (THE).

hook[1] to steal.

Normally by using bent metal on a pole through a window, such as the modern East African *fish hook*, but without the Kenyan refinement of razor blades let into the pole to deter you from grasping it as you see your valuables vanishing through the shutters:

> I guessed he had hooked it from the Miskito Indian on the Rio Sico, after his showerbath. (Theroux, 1981, of pilfered soap)

In obsolete British use a *hooker* was such a thief, working *on the hook*.

hook[2] a threat used to influence conduct.

The imagery is from angling:

> He had a hook of some sort into her. (Chandler, 1958, referring to such coercion)

hook[3] an enticement leading to trickery.

Again, angling imagery:

> 'Let's hear what the guy has to say.' The hook was in. (Weverka, 1973)

hook[4] a hypodermic needle.

From the usual shape of a needle used in surgery. The word is also sometimes used of the illegal narcotic, punning on your addiction to it.

hooked under an addictive compulsion.

Angling imagery again. The addiction may be to golf, surfing, an author, or anything else, especially narcotics:

> The kid never did get hooked on the hard stuff. (Sanders, 1977)

hooker[1] a thief. See HOOK[1].

hooker[2] a prostitute.

The derivation is from catching a customer:

> Some nights we go about and don't hook a soul. (Mayhew, 1862, of a prostitute and not a Salvationist)

General Hooker's exploits in Washington brothels came later and the abundance of prostitutes in Corlears Hook or the Caesar's Hook districts of New York helped to keep the usage alive:

> Even the hookers had done no more than cast an eye. (Mailer, 1965, of prostitutes)

Hook-shop, a low brothel, is obsolete.

hooky human excrement.

Perhaps from the shape it sometimes takes. It is used for *shit* in the literal, allusive, figurative, and expletive senses of that overworked word.

hoosegow a prison.

From the Spanish *juzgado*, a court, and from being judged in court and sent to prison:

> In that case, stew in a French hoosegow for the rest of your natural. (Sharpe, 1982)

hoovering abortion of a foetus.

Specifically by vacuum aspiration under medical supervision:

> I already had two hooverings when I wasn't sure. (McInerney, 1992)

hop a narcotic.

Originally, opium and I think from the twisting vine rather than a corruption of some Chinese word, as is sometimes suggested:

> They take him over to the hospital ward and shoot him full of hop. (Chandler, 1943)

A *hophead* is an addict:

> Frank wasn't just a deviant and not just a hop-head. (Davidson, 1978)

A *hop joint* is where you can buy and ingest illegal narcotics and become *hopped*, under their influence:

> 'Coked' or 'hopped up' gunmen... (Lavine, 1930)

hop into bed to copulate casually.

Usually on a first or single occasion and not propelling yourself on one leg only:

> 'How about hopping into bed?' 'At half-past four on a Sunday afternoon?' (Francis, 1978)

The American *whore-hopping* is not brothel leapfrog but copulation with several prostitutes in succession:

> Red-necks who had come down for the beer-drinking and the whore-hopping. (Theroux, 1979)

The obsolete British *hop-pole marriage* was living together unwed, or a marriage hastily contracted to avoid the bastardy of a child, in which case the arrangements might be signified by the couple actually jumping together over a stick or hop-pole.

hop off to die.

This and *hop the twig* employ avian imagery to denote departure. After passing into disuse, the phrase was taken up by Second World War fliers and is still used:

> It's not often multi-millionaires hop their twig. (Bagley, 1982—and even they do it only once)

hopper *American* a lavatory.

Literally, an inverted cone through which solids are discharged into a container, which explains the imagery:

Mom on the hopper with her knees pressed together. (Theroux, 1973)

hopping-Giles *British, obsolete* a lame person.

St Giles was the patron saint of cripples when they were accepted as a common and unremarkable feature of society. It is only comparatively recently that the word *cripple* has become taboo, although we are slightly less squeamish about the adjectival *crippled*. It is questionable whether the array of euphemisms we choose to deploy is kinder than the nickname *Hopkins*, to which not long ago any lame person would reply.

horizontal[1] pertaining to copulation.

From the normal attitude of the parties. A *horizontal life* was living as a prostitute:

> ... through this horizontal life I have risen from being a homeless waif to become a famous lady. (L. Thomas, 1977)

A *horizontal conquest* is an act of casual copulation, the victor usually being the male, but not always:

> ... diamonds and rubies ... and the other battle honours of her horizontal conquests. (Ustinov, 1966)

The *horizontal position* means 'copulation':

> Propinquity—that's what leads to the horizontal position. (Barber, 1981)

Horizontal aerobics involve both parties:

> When their horizontal aerobics are concluded, they lie awhile, insensate and numb (Sanders, 1987, referring to copulation)

as does *horizontal jogging*

> ... women didn't go in for all this casual, take-it-or-leave-it horizontal jogging that seems to lie at the very root of our society today. (Matthew, 1983)

A *grande horizontale* was a famous prostitute:

> Lola Montez, the grande horizontale, began her whoring in Simla. (Theroux, 1975)

horizontal[2] drunk.

A rare British usage, from actually or figuratively lying on the ground.

horn[1] the erect penis.

Common enough in the 16th century for Shakespeare's punning vulgarism:

> I can find out no rhyme to 'lady' but 'baby' —an innocent rhyme; for 'scorn', 'horn', a hard rhyme. (*Much Ado*)

To *have a horn* is to be thus sexually aroused. A *horn-emporium*, for one author at least, is a bookshop which sells erotic literature for males:

> Scrutinising the neighbourhood for a new, more convenient horn-emporium, was a pressing need. (Amis, 1988)

The *horn of fidelity* was a drinking cup, doubtless with sexual overtones, which Morgan le Fay sent King Arthur to enable him to test the chastity of the ladies of his court. Legend sadly records that only four of the hundred examinees managed to *drink cleane*, thus preserving the liquid and their honour.

horn[2] to cuckold.

The traditional *horns* of cuckoldry, with their male sexual insinuations, were figuratively placed on the head of the deceived husband:

> ... by those that do their neighbours horn (Colvil, 1796)

and today:

> ... evidence of Julie and Ronnie putting horns on the head of [her husband]. (Sanders, 1979)

To *wind the horn* or *blow* it was to acknowledge that you had been cuckolded, by a *horn-maker*:

> Virtue is no horn-maker. (Shakespeare, *As You Like It*)

He who was *horned* was cuckolded:

> Our horn'd master (waes me for him)
> Believes that sly boots does adore him.
> (Morison, 1790—but not sly enough to deceive the servants too)

This field of literary ingenuity, of which I here only explore the headlands, is no longer cultivated. Perhaps we have ceased to care about cuckoldry.

horn of the ox see HAIR OF THE DOG.

horner a sniffer of cocaine.

With reference to the nose, which becomes sore and runny as the narcotic damages the mucous membranes, along with much else.

horny[1] the devil.

He has them on his head. Usually *old* and in various spellings—*hoorny*, *horney*, and, in Scotland and Ireland *hornie*:

> Should Hornie, as in ancient days,
> 'Mang sons o' God present him.
> (Burns, 1785)

To many of the 19th-century Irish population, *horny* or *hornie* was a policeman.

horny[2] anxious for copulation.

Surprisingly—see HORN[1]—used of either sex. Of a male:

> Even if they did put bromide in the tea he still felt horny every morning and woke up with an erection like a tent pole (Bogarde, 1978)

and of females:

> The stewardesses were plain and presumably horny. (M. Thomas, 1980)

horse[1] gonorrhoea.

From the rare rhyming slang *horse and trap*, clap.

horse[2] to defecate.

Again from the rhyming slang *horse and trap*, this time for 'crap'.

horse[3] *American* a corrupt prison warder.

He carries loads of contraband in and out of prison.

horse[4] heroin.

Probably a corruption of *heroin*, despite the attractions, etymologically speaking, of *riding* under its influence. Widely used, as in Deighton's punning title for a novel, *Horse under Water*. *Horsed* means 'under the influence of narcotics'.

horse apples the turds dropped by a horse.

Especially in a street, where in days gone by they might pile up like apples on a fruiterer's shelf:

> ... 'horse apples', 'cowpats', 'prairie chips', 'muck', 'dung', etc. (Jennings, 1965, listing words commonly used as euphemisms for animal shit; but he missed *manure*, from *main-d'œuvre*, manual labour)

horse collar an expression of disgust.

An American may select some part of the animal or its accoutrement when it is not appropriate to say *horse shit*. The obsolete British *horse-leech* was an extortioner, after the blood-sucking worm of the class Hirudinea; he was usually a male but, if female, demanded payment for sexual services or for silence about past sexual encounters.

hose[1] to cheat.

From the spraying with water or bullets rather than the stocking on the leg:

> I know about Marcus Wheatley ... who hosed someone on a dope deal. (Turow, 1987 —Marcus was later murdered for his folly)

In common American slang to *hose* someone is to flatter him.

hose[2] *American* the penis.

Of obvious imagery, although the usage refers to its sexual rather than its urinary function, whence the ancillary slang meanings 'copulation' and, of a male, 'to copulate'.

hospice an asylum for the incurable or dying.

Originally, a resting place for travellers, especially pilgrims, and often run by members of a religious order. The current use emerged first in Dublin at the end of the 19th century.

hospital[1] an institution for the insane.

An American use originally, which glosses over the nature of the affliction both for the patient and for his family:

> American insane asylums are now simple *hospitals*. (Mencken, 1940)

Much of the taboo associated with madness came from the fact that a majority of the lunatics were at one time alcoholics or syphilitic.

hospital[2] an unauthorized prison.

Jargon of the American Central Intelligence Agency for a place where suspects may be held illegally.

hospital job a contract which can be loaded with excess charges.

In normal manufacturing use, a job you can work on when you have nothing else to do, delivery not being urgent. The dishonesty starts when you start loading such a contract with waiting time and scrap because your customer, usually the government, is too inefficient to detect malpractice.

hospitality free intoxicants.

In standard usage, the provision of a welcome and entertainment to a visitor:

> The landlord ... was happy to stay open as long as Seddon Arms wanted a drink. Maxim was beginning to guess at the scale of 'hospitality' which the arms business could afford. (Lyall, 1980)

Whence the television *hospitality room*, in which the tongues of amateurs are loosened and to which the staff repair for free supplies:

> In the hospitality room George Foster stood with his clip-board in one hand. (Allbeury, 1982)

hostess a prostitute.

In one of those places where arrangements for casual copulation are made as an adjunct to the provision of food, drink, accommodation, etc.:

> Once a hostess, always a hostess. You always were a bit of a whore. (Kersh, 1936)

This may explain why *air hostesses* prefer to be called *cabin flight attendants*.

hot[1] eager for copulation.

> I have never in my life seen so many ladies so hot in such a small place. (Green, 1979, of lustful females)

From the increased bodily temperature and flushing caused by sexual excitement; also used of other emotions, such as anger, which give rise to the same symptoms. *The hots* indicates lust for a specific person:

> Now he's got the hots for this young chick. (Sanders, 1973)

hot[2] obtained or retained illegally.

Usually of stolen goods, which may figuratively *burn* you if you touch them:

> Boudreau sold cheap liquor and handed fixes downtown and sometimes sold hot goods. (Weverka, 1973)

You dispose of stolen goods in a *hot market*:

> Not rich enough for the hot market. (Price, 1979, of stolen property)

hot[3] infected with venereal disease.

Normally of a male, from the burning sensation when urinating, but perhaps also from the risk of infecting another.

hot[4] radioactive.

Nuclear jargon, probably taken from the *hot spot* on a bearing, where heat indicates danger or trouble.

hot[5] *American* drunk.

Alcohol can make you perspire.

hot back (a) *obsolete* lust.

I am uncertain why the back becomes heated:

> When gods have hot backs, what shall poor men do? (Shakespeare, *Merry Wives of Windsor*)

hot for lusting after.

Of either sex.

hot-house *obsolete* a brothel.

It puns on the horticultural structure:

> She professes a hot-house, which, I think, is a very ill house too. (Shakespeare, *Measure for Measure*)

hot pants an urge to copulate.

Normally of a female:

> If she ever got hot pants, it wasn't for her husband (Chandler, 1953)

and occasionally of a male:

> I've still got hot pants for her, if you want to call that love. (McCarthy, 1963)

hot-pillow motel a place which lets rooms for extramarital copulation.

The pillows have little time to cool down:

> That notorious hot-pillow hotel on the far side of San Jorge. God knows, Stone had never been fastidious about where he'd take his girls for a quickie. (Deighton, 1972)

Likewise the *sheets*:

> The hotel was noted for its hot-sheet business. (M. Thomas, 1980)

hot place (the) hell.

Where the fires for ever burn. Now rare even among evangelicals.

hot seat[1] **(the)** an uncomfortable position of authority.

Usually in circumstances where something has gone wrong and there is nobody else on whom you can throw the blame. A figurative use, punning on the electric chair (see HOT SEAT[2] (THE)). In modern commercial use *hot-seating* is the sharing of desks or computer equipment by office shift workers.

hot seat[2] **(the)** the electric chair used for execution.

> The killers who end up in the gas chamber or the hot seat. (Chandler, 1953)

Also as the *the hot squat*.

hot shot *American* a fatal dose of illegal narcotics.

The impurities of illegal narcotics, often adulterated along the distribution chain, constitute an additional addict risk.

hot stuff a person who copulates enthusiastically on a casual basis.

Usually without payment or much encouragement.

hot-tailing extramarital copulation with more than one member of the opposite sex on a regular basis.

Punning on a rapid sequence of events. Whence a verbal use:

> She's going to be hot-tailing it with every... (Price, 1982—a man with a broken spine was speaking of his wife)

hot time (a) frequent and animated copulation.

Within or outside marriage:

> Reggie gives me a hot time in bed. (overheard at a ladies' bridge table in 1948)

hot tongue open-mouthed kissing.

It may also describe a sexually aroused female so behaving.

hot-wire to steal (a vehicle) by bypassing the ignition switch.

Punning perhaps on HOT[2] and the modification of the circuitry:

> ...faking a left to land a right, hot-wiring a car, and finding a place to dispose of the body. (Theroux, 1976)

hourly hotel a place which lets rooms for casual copulation.

Day or night, you are charged for a short period of occupation, often with a prostitute:

> ...bustin' the massage parlours, movie pits, hourly hotels. (Diehl, 1978)

house[1] a brothel.

Literally, a dwelling or any other building given over to a special purpose, such as a theatre or debating chamber. The use *tout court* for a brothel is obsolete:

> Some of the girls about here live in houses. (Mayhew, 1862, but not chastely with their parents)

There is usually amplification, as in the phrases listed below, most of which are now obsolete.

A *house of accommodation* was a place which let rooms to prostitutes:

> They enter houses of accommodation, which they prefer to going with them to their lodgings. (Mayhew, 1862)

A *house of assignation* was rather like an HOURLY HOTEL:

> ...keepers of houses of assignation, where the last-mentioned class might carry on their amours with secrecy. (Mayhew, 1862, of 'ladies of intrigue')

A *house of ill-fame* was a brothel, which might at the same time have a reputation for excellence:

> There is in Exeter Street, Strand, a very old established and notorious house of ill-fame. (Mayhew, 1862)

A *house of ill-repute* was also a brothel, again often with a good reputation for its prostitutes and decor, if not for decorum:

> A girl who had been forced into a house of ill-repute... (Lavine, 1930)

A *house in the suburbs* was a brothel in the 16th century, when they tended to be located on the edge of a town. A *house of profession* was also a prostitute's workplace:

> I am as well acquainted here as I was in our house of profession. (Shakespeare, *Measure for Measure*)

A brothel might be described as a *house of resort*:

> Shall all our houses of resort in the suburbs be pull'd down? (ibid.)

or a *house of sale*, which was not an auction room:

> I saw him enter such a house of sale —Videlicet, a brothel. (Shakespeare, *Hamlet*)

You might also find *houses of civil reception, evil repute, pleasure, sin, tolerance*, etc. or, inverted, a *scalding house*, where you might contract a venereal disease, a *common house*, or an *ill-famed house*:

> Lord Euston was said to have gone to an ill-famed house (Harris, 1925)

or an *introducing house*:

> His eager beaver interest in an 'introducing house' in St George's Road, near Lupus Street, was particularly resented by his colleagues as it catered almost exclusively to Members of Parliament. (Pearsall, 1969—*he* was Gladstone, whose obsession with female prostitution would cause greater comment today than it did when he was the Grand Old Man)

house[2] a lavatory.

Again the building given over to a particular purpose. Although Dr Johnson defines lavatories as *houses*, he does not so define a *house*. In various compounds, all of them obsolete, for example *house of commons, of ease, of lords*, and *of office*:

> I had like to have shit in a skimmer that day over the house of office. (Pepys, 1660)

house[3] *obsolete* an institution for the homeless.

A shortened form of the dread *workhouse*:

> Many old people... have to enter the 'house', as it is nick-named, like humble suppliants. (Gordon, 1885)

A workhouse was also known as a *house of industry*:

> The House of Industry for the reception of the poor of eleven of our fourteen parishes. (Peshall, 1773)

house[4] intended to avert criticism for prejudice.

Referring to the appointment of a nonentity to appease militant lobbyists:

> ... dude called Washington Lee was a brother, not the house nigger on some editorial board. (McInerney, 1992)

See also OBLIGATORY, STATUTORY, and TOKENISM.

house-cleaning[1] *American* a reorganization which leads to dismissals.

Denoting the kind of infrequent investigation in the police or bureaucracy when inefficiency or corruption have reached levels which can no longer be ignored. The imagery is from the annual major assault made domestically by some people in the spring on curtains, carpets, etc.

house-cleaning[2] the destruction of embarrassing records.

Again, the domestic imagery:

> In the afternoon hours of August 8, Ford staff members learned of frantic housecleaning under way at the White House. (Colodny and Gettlin, 1991, of the aftermath to Watergate)

house man *American* a security guard.

Police jargon:

> I'm the house man here. Spill it. (Chandler, 1939)

house of correction a prison.

So named in the hope that there will be no recidivism:

> Lyburn ... is unlike any other house of correction in the world. (Ustinov, 1971)

The modern American *correctional facility* uses the same imagery. *House of detention* is explicit:

> Incarceration in the House of Detention means loss of wages and a job. (Lavine, 1930)

house-proud obsessed with domestic cleanliness and tidiness.

This tedious affliction, normally of childless wives, may have little to do with pride in the family residence.

house-trained no longer given to involuntary urination or defecation.

Usually of pet animals and sometimes too of young children. Thus of any person induced to perform a subservient or unusual pattern of behaviour, such as a working husband who also undertakes an undue portion of the domestic chores or a politician who goes along with bureaucratic practices and abuses:

> The Civil Service phrase for making a new Minister see things their way is 'house-training'. (Lynn and Jay, 1981)

housekeeper a resident sexual mistress.

Most ladies who follow the occupation of keeping house for a bachelor or widower lead sexual lives of impeccable propriety, and who are we to censure those few who are willing to change their status and testamentary expectations?

Several housekeepers ... chosen for their willingness to endure the bed and board of old Mr. Flawse. (Sharpe, 1978)

how's your father casual copulation.

A male expression, perhaps from an opening conversational gambit. Also used of unmarried pregnancy:

The girl was in the club, knocked up, a bun in the oven—'ow's yer father. (Lyall, 1982)

Frequently as 'a bit of how's your father'.

hulk a prison.

Originally, a ship, and then the hull of a ship which was no longer seaworthy but good enough for the confinement of convicts. Often in the plural:

From his 'unhappy position' in York Castle, awaiting transfer to the hulks ... (Hughes, 1987, of a 19th-century convict)

human difference a faculty below the norm.

Not used to refer to all the infinite variety of the specimens of *Homo sapiens*, nor even to those with acute eyesight or hearing:

... many people in the deaf community define their deafness not as a disability, but merely as a 'human difference'. (*Chicago Tribune*, 20 May 1991)

human intelligence the use of spies.

Not the 'sapiens' of *Homo sapiens* but the acquisition of *intelligence*, information, by human rather than other agencies such as interception of radio signals, satellite photography, etc.

human relations copulation.

Our family in the narrow sense comprises our *human relations*, while in the broader view we have *human relations* with everyone we come into contact with:

She had no idea of elementary human relations. (Fraser, 1969, meaning she knew nothing about how babies are conceived)

See also HAVE RELATIONS.

human resources personnel.

This usage, often shortened to *HR*, originated in America but is unhappily no longer confined to those shores. It has the worthy intention of dignifying the relationship between employer and employee. Thus personnel officers, directors, and so on are now *HR managers*, *directors of human resources*, etc. Pretension rather than euphemism perhaps, but certainly a pervasive evasion.

human rights individual licence beyond that permitted by existing institutions.

The phrase comes from the 1948 United Nations' *Universal Declaration of Human Rights*, a concept to which no exception can be taken by those who consider mankind to be paramount on earth. In practice *human rights* sometimes provides a slogan for those who wish to overturn an established form of social living acceptable to a majority, using violence if necessary.

human waste sewage.

Not discarded crisp packets or bottles, amputated limbs or corpses, the unemployed or those without fulfilling lives. Mainly the jargon of civil engineers, to distinguish it from other forms of garbage and from surface water.

hummingbird *American* an instrument of death by electrocution.

The 'chair' gives a noise when the current is switched on.

hump to copulate with.

Possibly from carrying a load. Grose says 'once a fashionable word for copulation' and the fashion has returned:

His trouble was seducin'. Story is he humped the faculty wives in alphabetical order. (Bradbury, 1965—perhaps using their maiden names)

In the 19th century a male might *hump the mutton*, punning on porterage:

She completed her undressing while we were positively humping the mutton all the way to the couch. (Fraser, 1977, writing in 19th-century style)

hump it *obsolete* to die.

Probably from the departure of a porter after taking up his load.

hung see HANGOVER.

hung like claiming the fabled sexual prowess of.

Of the male in various clichés. Thus *hung like a bull*, *stallion*, or *horse* implies large genitalia:

I hear he's hung like a horse. (Sanders, 1986, of an Italian male)

Hung like a rabbit suggests a penchant for frequent copulation.

hunt to seek a homosexual partner.

Of a male, and often in public urinals:

Gilbert's given up 'hunting', he says all he ever wanted was love and he's got mine. (Murdoch, 1978)

hunt the brass rail *American, ?obsolete* to frequent bars selling intoxicants.

There used to be a brass rail in many bars which you could rest a foot upon:

Virgins, reporters, house-wives, kept-wenches, customer's men hunt the brass rail. (Longstreet, 1956)

hunt the fox down the red lane *British, ?obsolete* to become drunk.

The *red lane* is the throat:

I am sorry, kind sir, that your glass is no fuller ...

So merrily hunt the fox down the red lane. (Dixon, 1846)

husband[1] a pimp.

Of his relationship to the senior of the women in his stable:

... to denounce a woman to her 'husband' if the creature makes advances to you. (Londres, 1928, in translation)

In obsolete English use a *husband* was also a prostitute's regular customer:

I know very many sailors—six, eight, ten, oh! more than that. These are my husbands. I am not married, of course not. (Mayhew, 1862)

husband[2] a homosexual who takes the male role.

Male or female, cohabiting sexually with another homosexual:

The 'husband' he tripped with a heel behind her ankle. (Sanders, 1982, describing a man's fight with a pair of female homosexuals)

hush *obsolete* to kill.

From the ensuing silence.

hush money a bribe to ensure silence.

Made to an extortioner, blackmailer, or former employee who may be tempted to talk out of turn. Less often as *hush payments*:

People objected to the bald language, the discussions of hush payments and stonewalling. (Colodny and Gettlin, 1991, of the Nixon tapes)

hush-shop *obsolete* an unlicensed inn.

After licences became mandatory in the 19th century, many establishments continued selling intoxicants without proper authority:

'Hush' signifying that the company frequenting such places were expected to conduct themselves as orderly as possible that no alarm might be given to parties in authority. (Brierley, 1865)

hussy a woman who habitually copulates casually.

A corruption of *housewife*, from the days when only the marital bargain granted the right, or duty, of copulation. Soldiers still call their small sewing pack, or housewife, a *hussif.*

hustle[1] to steal or swindle.

Originally, to push, or knock about, whence to obtain by such activity. The words *hustle*, a theft, and *hustler*, a thief, are now mainly used of American addicts seeking cash by any means so as to buy a supply of illegal narcotics.

hustle[2] to engage in prostitution.

From the vigorous importuning in a public place:

I hustled at a dead run until the streets were empty and the bars closed. (Theroux, 1973)

In very common use a *hustler* is a prostitute:

I don't think she's an out-and-out hustler. (Allbeury, 1976)

The 19th-century British meaning 'to copulate' has lapsed.

hustle[3] to sell dearly articles of low value.

Usually accompanied by skilful banter or other pressures, such as a suggestion that, because no duty is payable, you are being offered a bargain:

Duty-free baubles were interminably hustled by stewardesses. (Deighton, 1988)

hygiene see PERSONAL HYGIENE.

hygienic free from venereal disease.

But not necessarily clean or healthy in other respects:

But there were a few men in formal evening dress with stiff collars, looking for company that was certified hygienic. (Moss, 1987—they were seeking disease-free prostitutes)

See also CLEAN.

hygienic treatment *American* the temporary preservation of a corpse.

Funeral jargon, which ignores the fact that newly dead meat is aseptic. We are conditioned to the sides of bacon or the pheasants hanging in the butcher's shop, but we regard with distaste the corpses of those formerly near and dear to us:

Although some funeral directors boldly speak of 'embalming', the majority consider it preferable to describe the treatment by some other term as ... 'Hygienic Treatment'. (J. Mitford, 1963)

See also DECONTAMINATE.

hymenal sweets copulation by a male.

A pedantic and probably obsolete phrase, alluding to the hymenal membrane, although not necessarily to the female's first essay in copulation. If you do find anyone using this expression, he is likely also to refer to the vagina, if at all, as a *hypogastric cranny*, from the lower part of the abdomen or *hypogastrium* and not rhyming slang for the coarse 'fanny'.

I

I am listening I have already made up my mind.

Used by anyone who is not interested in the proposal being put to him, *listening* not importing considering or paying attention.

I hear what you say I do not agree with you.

A convenient form of words, especially for bureaucrats, as it obviates the need to enter into discussion or argument.

I must have notice of that question I am not going to answer you.

This response is best used in an interview broadcast live when you wish to hide known facts as well as ignorance. Radio and television are too ephemeral for there to be a risk of your bluff being called.

ice1 *American* a bribe.

Not from the slang meaning 'diamonds', but from the stickiness of frozen water. However DAS suggests derivation from the initials of *Incidental Campaign Expenses*, which is at least ingenious.

ice2 *American* to kill.

Probably from the permanent lowering of body temperature rather than the ice in pre-refrigeration morgues:

> I heard what The Bat did to you for icing High Ball Mary. (Diehl, 1978)

To *put someone on ice* is also to kill:

> Somebody put this Domino on ice about four hours ago ... it wasn't no amateur hit. (ibid.)

ice3 a narcotic or stimulant.

Formerly only cocaine, from the numbing sensation perhaps. See also ICE CREAM:

> I'll just be snorting some ice around the USA. (M. Smith, 1993)

Now also methylamphetamine, or SPEED, which is smoked rather than sniffed.

ice box1 *American* a prison.

Originally, a cell used for solitary confinement, where you were sent to cool down:

> A prisoner sent to the 'ice-box' or solitary ... (Lavine, 1930)

but later any place of confinement:

> He has so far stayed out of the icebox. (Chandler, 1953, of a criminal who had escaped prison)

On ice usually means 'solitary confinement', which may be spent in the *icehouse*:

> ... three days in the icehouse. (ibid.)

The *iceman* who used to deliver ice to your home, in the days before refrigerators, had a reputation for copulating with housewives, whence O'Neill's punning *The Iceman Cometh*.

ice box2 a mortuary.

The ice was to stop the corpses decomposing prematurely. This American use has survived refrigeration:

> He's got seven stiffs down there in the icebox. (Diehl, 1978)

ice cream *American* illicit narcotics in crystalline form.

From the appearance and the vanilla colour. The *ice cream man* supplies addicts, but an *ice creamer* with the *ice cream habit* is only moderately addicted.

ice queen a reserved and chaste young woman.

Male use, from her supposed frigidity; not a champion skater:

> Her nervousness got her the reputation of an ice queen and she was not often asked out. (Follett, 1991)

ideal for modernization dilapidated.

In this estate agents' newspeak, *ideal* means 'only fit for':

> Stone-built semi-detached cottage. Ideal for modernization. (*Western Daily Press*, May 1981)

identification the ability to pay.

In an American hotel, a passport will not suffice. The desk staff will want cash or the imprint of a credit card before they hand over the key to your room. Whence the greeting, 'May I see your identification, please?'

idiosyncrasy homosexuality.

> She seemed quite comfortable in the company of Anthony Blunt, even after his 'idiosyncrasy' was known. (*Daily Telegraph*, 25 March 1995—'she' was the Queen, for whom Blunt was employed as Surveyor of the Queen's Pictures)

Literally, any tendency or unusual preference.

ill1 menstrual.

Common female usage and imagery:

> 'When were you ill last?' 'About a fortnight ago,' she replied. (Harris, 1925—as ever, he was concerned about impregnating one of his conquests)

Mrs Pepys was *ill of those*:

> Thence home and my wife ill of those upon the maid's bed. (Pepys, 1669)

ill2 suffering from venereal disease.

Usually of a prostitute and in prostitutes' jargon:

> The poor girl may not even have known she was ill [syphilitic]. (Harris, 1925)

ill[3] in the custody of an espionage agency.

The natural state of one in a CIA HOSPITAL[2]. The Russians said that Taraki, the Afghan President shot in 1979, died *of a serious illness*, which means the KGB caused him to be killed.

ill[4] drunk.

Of the same tendency as MIGRAINE:

> 'Roddy felt ill,' I said... 'Ill,' said Jerry. 'Drunk, you mean.' (Deighton, 1988)

ill-adjusted mentally sick or compulsively destructive.

But no amount of tinkering with the mechanism will put things right:

> We aren't here to provide a haven for the ill-adjusted. (Bradbury, 1959)

ill-famed house see HOUSE[1].

ill man the devil.

A reference not to his health but to his character. He might take you away to the *ill place* or the *ill bit*, hell:

> The devil... took him awa' to the ill bit. (A. Armstrong, 1890)

Whence the obsolete British *ill* or *ill-wished*, bewitched:

> ... the child had been ill-wished and... would never be better until 'the spell was taken off her'. (R. Hunt, 1865)

The malady could be cured by a visit to the conjuror, or white witch.

ill-used having copulated outside marriage.

Of a female, supposedly against her will and regardless of the tenderness which the male may have displayed:

> I cannot believe that she will be... ill-used, in any way, if you follow me. (Fraser, 1977, writing in 19th-century style)

Now probably obsolete.

illegal operation an induced abortion.

In the days before such became legal:

> What about you, doctor—and your little professional mistake? Illegal operation, was it? (Christie, 1939)

illegitimate born a bastard.

> A yearly average of 1,141 illegitimate children thrown back on their wretched mothers. (Mayhew, 1862)

Of interest because this is a dog-Latin word coined in an age which worried a lot about paternity; the meaning 'unlawful' seems to have developed later.

illicit pertaining to extramarital copulation.

Literally, unlawful, although English common law, perhaps wisely, saw no criminality in adultery, leaving jurisdiction to the Church. Usually in phrases such as *illicit embraces, connection, commerce, intercourse*, etc. See also CRIMINAL CONVERSATION.

illuminated drunk.

Another version of *lit up* or LIT.

imaginative journalism sensationalist invention.

It is unwise to call a journalist a liar because he will have many more chances of damaging you than you will of hurting him:

> ... a piece of imaginative journalism was being perpetrated by one of its own reporters. (*Private Eye*, June 1981)

imbibe to drink intoxicants.

Usually to excess, if someone says you *imbibe*. It really means to drink anything.

immaculate in fair decorative order.

No used residence is ever 'spotlessly clean or neat; perfectly tidy, in perfect condition' (SOED). This is real estate puffing for a house which you can move straight into, if you can stand the wallpaper.

immediate need on death.

American funeral jargon, referring to those whom they consider improvident because they have not paid their burial expenses several years in advance:

> The American Cemetery reports a discussion on 'Immediate-need selling'. (J. Mitford, 1963)

immigrant a black person living in Britain.

Any white, such as a person of mixed Greek and German ancestry born in Argentina, can make his home with due consent in the British Isles and not be classed in popular speech as an *immigrant*. Conversely:

> Most 'immigrants' have been here for many years, and two of every five of them were born in the United Kingdom. (Howard, 1977, referring to black people)

immoral associated with prostitution.

Literally, contrary to virtue, but confined to sexual misbehaviour in various legal jargon phrases. Thus *immoral earnings* are what a prostitute gets paid and a pimp takes from her:

> It would mean my arrest on a charge of living off immoral earnings. (Theroux, 1973)

Immoral girls are prostitutes:

> Though they'd twice given him the boat fare home he had spent it on drink and probably on immoral girls. (Bradbury, 1976)

An *immoral house* was a brothel:

> The dress-lodger probably lives some distance from the immoral house (Mayhew, 1862)

and today a building used for *immoral purposes* is either a brothel or where prostitutes take their customers:

...full of brothels, almost every house being used for an immoral purpose. (ibid.)

The American Mann Act, known as the *Immorality Act*, makes it unlawful to transport a female across a state line with intent to 'induce, entice or compel her to give herself up to the practice of prostitution, or to give herself up to debauchery, or any other immoral purpose'. It is happily a defence to plead that the seduction came as an afterthought.

impaired hearing deafness.

In standard usage, to impair is to damage or weaken:

...the deaf shall be described as 'people with impaired hearing'. (*Daily Telegraph*, 1 October 1990, quoting a memorandum issued by the Derbyshire County Council's Equal Opportunities and Race Relations Department, a body with a staff of 36 and an annual budget of £630,000; the Police in the same county were characterized by inefficiency due in part to chronic underfunding)

See also MOBILITY IMPAIRED.

impale to copulate with.

Of a male. Originally, to surround with a fence, whence to thrust a stake into a body:

Before she could turn round I had impaled her, and was subsiding into a chair with her on my lap. (Fraser, 1971)

implement the penis.

In its sexual role:

...he was such a big man and could nigh-on dig post holes with that there implement of his. (Keneally, 1979, of a libertine)

implemented with torture.

Literally, no more than being put into effect:

I can go through implemented interrogation inside Lubyanka and come out sane. (Hall, 1988—but not undamaged, it would seem)

importune to offer sexual services for money.

Literally, to beseech earnestly, but in legal jargon especially of prostitutes who approach potential customers in a public place, since the 19th century.

impotent sexually infertile.

Literally, powerless in any respect, but used in this sense of either sex, and also of males who cannot achieve an erection of the penis:

...advertisements for doctors who cured 'all the diseases of love' and promised the impotent 'horse-like vigour'. (Manning, 1977)

improper involving extramarital sexual behaviour.

Either heterosexual or homosexual, being in either case a deviation from propriety in this limited area of behaviour. Thus the obsolete *improper house* was not badly constructed, but a brothel:

Neither are the magistracy or the police allowed to enter improper or disorderly houses, unless to suppress disturbances. (Mayhew, 1862—he seems to discount the possibility of their being customers)

An *improper suggestion* is a homosexual approach or an invitation by a member of either sex to copulate casually:

...one of the tarts plucked at Kavanagh's sleeve and made an improper suggestion. (Fraser, 1975)

improvement[1] forcible depopulation.

The Scottish Highland clearances replaced people with sheep to increase income and give the clan chiefs more spending money:

The necessity for reducing the population in order to introduce valuable improvements. (Prebble, 1963, quoting Sir George Stewart Mackenzie of Coul, who goes on to advise his evicted clansmen to 'find happiness as the servants of servants')

Many of the dispossessed found their way to America.

improvement[2] a reduction in quality or service.

Any statement that a change in procedures is to *improve service to customers* should be viewed with as much scepticism as an assertion by a civil servant which is prefaced by *of course*. It usually conceals the intention of improving profitability by selling less for the same money:

Improvement means deterioration. (Hutber's Law—Patrick Hutber was City Editor of the *Sunday Telegraph*)

improver *British* an underpaid worker.

A device under which supposed prospects were traded against a living wage. Obsolete since the Second World War.

in[1] imprisoned.

Criminal jargon; a shortened form of *in prison*:

She was in the first time for robbing a public. (Mayhew, 1851)

in[2] copulating with.

Of a male:

Climbing into bed with...Lady Fleur, when that noble lord was not only in it but in her. (Sharpe, 1978)

in a certain condition see CONDITION[2].

in Abraham's bosom dead.

Where Dives reputedly saw Lazarus, although it seems poor recompense for a lifetime of penury and abuse:

The sons of Edward sleep in Abraham's bosom. (Shakespeare, *Richard III*)

in bed see BED[2].

in calf pregnant.

It is male rather than female practice to describe pregnancy by the standard English terms for farm and other domestic animals. Farmers tend to use *in calf*. *In foal* is more general:

> She had just discovered ... that she was in foal for the ninth time. (Fraser, 1975, of Queen Victoria)

In kindle, in standard usage with reference to hares or rabbits, is occasionally used of women. *In pig* is quite common:

> 'I'm in pig, what d'you think of that?' 'A most hideous expression, Linda dear.' (N. Mitford, 1945)

In pup is widely used, not only of bitches or by dog-lovers. *In pod* is from the fruit of a leguminous vine, although a *pod* was also a protuberant stomach:

> I've 'ad seven girls i' pod and wor going wi' a married woman. (Bradbury, 1976)

I have never seen the common British *in for it* in print.

in care see CARE.

in Carey Street bankrupt.

From the location of the Bankruptcy Court in London. You might have thought that *in the Crown Office* meant the same thing, and would have been wrong—it meant 'drunk', punning on the head and a government building in London.

in circulation ready to copulate.

There are two distinct uses referring to females; of a normal sexual partner, to indicate that she is not menstruating or incapacitated by parturition; and of a female earning her living by prostitution:

> ... cannot conceive that a grown-up girl can earn her living in any other way. At twelve she is in secret circulation. (Londres, 1928, in translation)

in conference see CONFERENCE.

in-depth study industrial espionage.

A pretentious and often tautological phrase in ordinary usage for an investigation or survey. Business jargon for the illegal acquisition of information, drawings, etc. which you pay a third party to obtain in an attempt to avoid overt criminality.

in drink see DRINK[1].

in flagrante delicto in the act of extramarital copulation.

Legal jargon for those observed copulating outside marriage. Often shortened to *in flagrante*. Occasionally, the French equivalent is used:

> In the old days you at least knew this death *en flagrant délit* meant hell-fire for ever. (Read, 1979)

The phrase is also used to describe detection in other kinds of wrongdoing, or being caught *red-handed*, the victim's blood being figuratively upon you.

in foal see IN CALF.

in for it see IN CALF.

in freedom see FREE[1].

in full fling *obsolete* engaging in regular extramarital copulation with one partner.

A *fling* is a temporary bout of uncharacteristic hedonism:

> It seems she's in full fling with Valhubert. (N. Mitford, 1960)

in heat see ON HEAT[1].

in heaven dead.

Mainly tombstone usage among Christians, or monumental masons, along with *in the arms of Jesus, his Maker, the Lord*, etc. These uses soften or mask the reality of death, especially to children who may be told that a deceased relative is *in heaven* etc. rather than in the graveyard.

in kindle see IN CALF.

in left field *American* mad or eccentric.

A baseball term, with perhaps a hint of the normal sinister connection:

> Sometimes they make sense and sometimes they're way out in left field. (Sanders, 1985)

in liquor see LIQUOR.

in name only without copulation.

Of marriage, especially where the parties have continued to live together:

> My husband was ... in name only. (Ludlum, 1979)

in need of supervision criminal.

Police and prison jargon of convicted children and juveniles who have to be locked away. It might suggest that well-behaved children do not need the supervision of adults.

in pig see IN CALF.

in pod see IN CALF.

in protection see PROTECTION.

in pup see IN CALF.

in purdah menstruating.

Not always segregated, as in some Hindu and Muslim societies:

> Do we know how long she's going to stay in purdah? (B. Forbes, 1990—a woman had started to menstruate)

in rut copulating.

Literally, in the state of sexual excitement of a stag during the appropriate season:

I could hear Deborah in rut, burning rubber and a wild boar. (Mailer, 1965)

in season able to conceive.

Of mammals other than humans, which are also said then to be *in use*:

That bitch director marched through the offices like a gorilla in season. (Ludlum, 1979)

Unlike other mammals, a woman conceives when not bleeding after ovulation and the phrase has the wider meaning of 'available for copulation':

The point of women being in season all the time with only brief interruptions... (Amis, 1978)

in the altogether naked.

From the biblical passage:

Thou wast altogether born in sins (John 9:34)

with the association of nudity; or perhaps merely a shortened form of *altogether without clothes*.

in the arms of Jesus see IN HEAVEN.

in the arms of Morpheus asleep.

The expression is used by and of those who should have kept awake:

At this hour when it is so very hot he is usually to be found 'in the arms of Morpheus' which means, I understand, that he is asleeping. (Farrell, 1973)

Morpheus, the god of dreams, was the son of Hypnos, the god of sleep, but those unversed in Greek mythology are liable to confuse euphemistically the two deities.

in the bag[1] taken as a prisoner of war.

Sporting imagery, from what the hunter shoots and so carries:

Tell him if he tries to stick it out, he'll only end in the bag. (Manning, 1977, of the Second World War)

in the bag[2] *American* drunk.

The slang phrase indicates accomplishment or confidence of success, from the fate of hunted game, which may not take us much further:

He had a shotgun next to the chair, and he was half in the bag from booze. (Clancy, 1989)

See also IN THE TANK.

in the barrel *American* about to be dismissed from employment.

Or *fired*, which makes it twice removed from the standard English *discharged* (see also FIRE).

in the black see BLACK MARKET.

in the box copulating.

Of a male, from BOX[3], the vagina. When there was more piety and less overt obsession with copulation, a *good man in the box* was a rousing preacher, the *box* being the pulpit.

in the business see BUSINESS.

in the cart in serious difficulty.

You could only ride in the cart in medieval Europe if you were a female, a child, or an old, sick, or wounded male. Thus a fit adult male only found himself *in the cart* on his way to execution, where the method of conveyance both degraded the victim and facilitated the hanging, as the cart could be driven off once the noose was secured.

in the club pregnant.

Shortened form for *in the pudding club*, which is itself a shortened form of the punning *plum(p) pudding club*:

Chaps having it off get taken aback when young women are put in the club. (Davidson, 1978)

Whence *join the club*, to become pregnant.

in the family way pregnant.

Probably an alteration of *in the way of having a family* although only used of a mother. There is often a suggestion that the pregnancy may be unwanted:

But she's not so fucking happy when she's in the family way. (Manning, 1977)

in the glue in difficulty.

Unable to move freely:

What about you, are *you* in the glue? (T. Harris, 1988)

Also as many other figurative vulgarisms such as *in the shit*, *in the nightsoil*, etc.

in the hay copulating.

Literally, in bed, from the stuffing of a mattress with hay; or making use of a haystack:

Tell me friend, what's she like in the hay? (Fraser, 1971)

in the mood ready to copulate.

Female usage, especially in the negative when she wishes to avoid copulation with her regular partner:

'I'm not in the mood tonight,' Saroya told Robin. (*Daily Mirror*, February 1980)

in the rats suffering from delirium tremens.

Army usage. Pink elephants, snakes, and rats are the reputed visitors to the delusions of those so afflicted:

Seeing the pool of sacred snakes... sent him 'in the rats'. (Richards, 1936)

in the raw see IN THE SKIN.

in the red owing or losing money.

From the days when bankers and others used red ink for debit balances and black for credits.

in the ring engaged professionally in cheating or thieving.

A *ring* is a cartel, from people meeting in a circle (see also RING[2]). In modern use of fraudulent antique dealers who combine to buy cheaply at auction and of manufacturers exploiting a joint monopoly. In the 19th century it was used for stealing:

> These parties are connected with the thieves, and are what is termed 'in the ring', that is, in the ring of thieves. (Mayhew, 1862)

in the sack copulating.

Literally, in a bed, and usually referring to extramarital copulation:

> A medical examiner took a smear. The German girl has been in the sack tonight. (Mailer, 1965)

See also INTO THE SACK.

in the saddle copulating.

Of either sex, using the common equine imagery:

> Elspeth would be back in the saddle with one of her gallants by now. (Fraser, 1971)

in the skin naked.

Particularly of nudity in public and breach of convention:

> She must sunbathe in the skin. (L. Thomas, 1979—for the absence of strap marks)

The common *in the buff* comes from *buff* as a shortened form of *buffalo*, whence *hide*, whence *leather*, whence again *skin*. *In the raw* also refers to nakedness where you might be reasonably expected to be wearing clothes:

> I know what you were doing in the middle of the bay in the raw. (Sharpe, 1977)

in the sun(shine) see SUN HAS BEEN HOT TODAY.

in the tank *American* drunk.

From the *drunk tank*, the cell into which inebriates are placed to sober up:

> Spermwhale was almost in the tank, a fifth of bourbon or Scotch in the huge red hand. (Wambaugh, 1975)

in the trade earning a living from prostitution.

The phrase covers anyone, pimp, bawd, or prostitute, who makes money out of casual copulation. The British *in trade*, on the other hand, is —or was—used derogatively by landed or professional people, of those who manufacture or distribute goods, being an interesting combination of jealousy and snobbery.

in trouble pregnant or criminal.

A young woman so described may be pregnant and unmarried; a person of either sex so described may have been accused or convicted of an offence. See also TROUBLE.

in your cups drunk.

You need only one cup, if it is large enough or refilled sufficiently often:

> ...in his cups, could do an admirable soft shoe clog. (Sanders, 1973)

Also as a *cup too many*, again without necessarily changing your drinking vessel. In obsolete use a *cup-man* was a drunkard.

inamorata a sexual mistress.

> As a member of the Souls and for twenty years the *inamorata* of the painter, Edmund Burne-Jones... (S. Hastings, 1994)

From the Italian *innamorata*, literally no more than a female with whom someone is in love. *Inamorato* (Italian *innamorato*), used of a male, although given in SOED, is rare.

incapable drunk.

From the British legal offence *drunk and incapable*, of a drunkard who had lost physical control, as against *drunk and disorderly*, of a rowdy drunkard:

> She was so drunk. Incapable—isn't that the word they use? (Theroux, 1976)

incident war.

Literally, a single occurrence. In a *border incident*, opposing soldiers start shooting at each other:

> The Phantom dived past, pulling up sharply, with another thunderclap of sound. 'We're going to end up with another incident,' Kit said grimly. (Masters, 1976)

Such incidents have no fixed duration:

> ...the 'China incident', the cruel war which had now been raging for four years against the Kuomintang government. (Keegan, 1989)

inclusive language changing the literary convention that the male may also import the female.

The purist may find it more intrusive than inclusive:

> It is a matter of 'gender', or 'inclusive language' as the feminists call it. (*Sunday Telegraph*, 9 May 1993)

(In the 2nd edition of my book, *Thinking About Management*, we thought it prudent to insert the following caveat: 'In accordance with literary convention and to avoid the inelegant use of language, the male gender may also import the female.' I am not sure whether this was enough to assuage the susceptibilities of the businessperson, businesswoman, or businessman who chanced to read it or to spare her, his, or their feelings when I was writing about the chair, chairwoman, or chairman, etc.)

income protection arranging your affairs to avoid tax.

Although legal, it is looked upon with disfavour, especially by those who have no opportunity to do it themselves:

> Tax avoidance, or as Mr Treyer preferred to call it, Income Protection. (Sharpe, 1978)

income support *British* money paid by the state to poor people.

It is paid to those with low incomes and those with no income at all. A recent coinage in an area where the language needs constantly to be updated to mask any suggestion of charity:

> ... she was only £10 a week better off than when she was on income support (as national assistance is now called). (A. Waugh in *Daily Telegraph*, 8 October 1994, of a woman earning £20,500 a year)

incompatible with diplomatic status spying.

Activities, behaviour, etc. are *incompatible with diplomatic status* when a diplomat is caught spying in a host country and declared *persona non grata*.

inconstancy regular extramarital copulation.

Literally, a propensity for change:

> Inconstancy was so much the rule among the British residents in Cairo, the place, she thought, was like a bureau of sexual exchange. (Manning, 1978)

incontinent[1] copulating extramaritally.

Of either sex. Literally, lacking self-restraint. If you are *continent* you copulate only within marriage, if at all:

> He had rekindled her ... she had never been particularly continent. (le Carré, 1980)

Continence or *continency* is such behaviour:

> In her chamber, making a sermon of continency to her.
> (Shakespeare, *Taming of the Shrew*)

incontinent[2] urinating or defecating involuntarily.

Again from the literal meaning, 'without interval', and the opposite in this sense too of *continent*. Medical jargon for a common manifestation of extreme age or sickness, but not used of babies before they have learned the requisite control:

> The geriatric ward, where ... he found himself surrounded by the senile and incontinent. (G. Greene, 1978)

Incontinence is the state of being so afflicted:

> ... embarrassed at the incontinence which had overtaken him. (M. Thomas, 1980)

incontinent ordnance *American* mishits.

Figuratively defecating without control at the wrong time in the wrong place:

> Bombs dropped outside the target area are 'incontinent ordnance'. (Commager, 1972)

inconvenienced with permanently impaired facilities.

A mainly American use, as in *The National Inconvenienced Sportsmen's League* (quoted in Rawson, 1981), in which you could only take part if you were confined to a wheelchair, blind, one-armed, or otherwise impaired, and, presumably, male. Today to be *aurally inconvenienced* is to be deaf, to be *visually inconvenienced*, blind.

incurable bone-ache *obsolete* syphilis.

Not rheumatism or arthritis. Until Fleming's discovery of penicillin, the condition might be arrested but not cured, and mental institutions were full of patients suffering from neurosyphilis, or general paralysis of the insane:

> Now the rotten diseases of the south ... incurable bone-ache. (Shakespeare, *Troilus and Cressida*)

incursion an unprovoked attack.

> The White House, describing the invasion (or, as it preferred, 'incursion') of Grenada ... (McCrum et al., 1986)

Literally, a running into, but used in this military sense since the 15th century.

indecency an illegal male sexual act.

Literally, unseemliness of any kind, and used of both homosexual and heterosexual behaviour. *Gross indecency* is buggery or bestiality:

> ... he was arrested by members of the Metropolitan vice squad for an act of gross indecency in Hyde Park. (B. Forbes, 1986)

An *indecent offence* is usually an illegal homosexual act:

> Accused by fellow officers of an indecent offence with a local youth ... (*Private Eye*, July 1980)

An *indecent assault* is nearly always against a woman, with the man seeking sexual gratification, by force if need be, and pinching her buttocks may qualify, if she objects. For *indecent exposure* see EXPOSE YOURSELF. In the 19th century *indecency* also covered extramarital copulation:

> Numbers sleep on the kichen floor, all huddled together, men and women (when indecencies are common enough). (Mayhew, 1851)

indescribables *obsolete* trousers.

From the vintage years of 19th-century prudery. See also UNMENTIONABLES[1].

indiscretion *obsolete* a bastard.

It was the mother who was supposed to have been indiscreet, rather than the father.

indiscretions[1] *obsolete* involuntary bodily discharges.

Vomiting, urine, and faeces:

> Nurse's vails, *obs.*, a nurse's clothes when penetrated by nepial indiscretions. (EDD —*nepial* means 'childish')

indiscretions[2] repeated acts of adultery.

In standard usage, acts taken without caring about the embarrassment or distress they may cause:

> The Princess of Wales, who normally overlooked her husband's indiscretions... (Massie, 1992—of Alexandra, not Diana)

indisposed[1] menstruating.

Literally, unwell, with common allusion to sickness:

> *Flag* 3. A sanitary pad or towel. Hence, *the flag (or danger signal) is up*: she is 'indisposed'. (DSUE)

indisposed[2] drunk.

Again from feeling unwell later, and sometimes used to excuse an absence:

> When a rich man gets drunk, he is indisposed. (Sanders, 1977)

Hudson points out that public performers are never *ill* when they fail to appear, they are *indisposed* (DDE).

individual behavior adjustment unit *American* a cell for solitary confinement.

Circumlocution combined with evasion: it could be anything from a dose of medicine to a turnstile.

indoctrination camp a political prison.

These were the special prisons to which the Chinese Communists sent those who had worked closely with Russians between 1947 and 1959, from which contact they had to be decontaminated.

indulge to drink intoxicants.

Literally, to humour or gratify, and used normally of those who say they won't or don't:

> 'Drinks, Chester,' she said. 'The usual for the Reverend and me. Mr Bigg isn't indulging.' (Sanders, 1980)

If you *overindulge*, you get drunk.

industrial action industrial inaction.

Trade union jargon in the British Isles for a strike, which has almost become standard English. Not used in the plural of more than one strike:

> Khafiq's flight was delayed, successively by industrial actions involving baggage handlers at Heathrow and air controllers in France. (M. Thomas, 1980—an American misunderstanding of British English usage)

Industrial relations is not the interplay between supplier and customer but the dialogue, or lack of it, between employer and employee.

industrial logic greed.

A bid to take over another corporation is often justified on this nebulous ground. The true motives are usually megalomania, the elimination of competition, the need of professional advisers to earn fees by stimulating business, or a chance of asset-stripping, especially where the victim has been careless about its use of working capital.

industrializing country a poor and relatively uncivilized state.

A recent coinage based on aspiration rather than realism:

> The term 'developing nation' was to be superseded by 'industrialising country'. (*Daily Telegraph*, 12 May 1993, quoting a directive issued by Leeds Metropolitan University—let us hope it does not profess to teach the English language)

inexpressibles see UNMENTIONABLES[1].

infidelity clandestine extramarital copulation.

Literally, an absence of faith, whence acting dishonestly in any respect:

> In conducting these amours they perpetrate infidelity with impunity. (Mayhew, 1862)

Infidelities are a consistent pattern of such conduct with different partners:

> Mavis had seized the opportunity to catalogue his latest infidelities. (Sharpe, 1979)

informal acting illegally or without required permission.

Literally, casual or easygoing, which is not one of the properties of a receiver of stolen property, or *informal dealer*:

> No action would be taken against 'informal' dealers who came forward, and nor would the money be confiscated. (Davidson, 1978)

A street market in the British Isles which is not shut down despite its lack of official licence is called an *informal market*.

information lies and a selection or suppression of the truth.

A usage of governments and public bodies. In the Second World War there was a British *Ministry of Information* which suppressed, edited, distorted, or invented 'news', although without the verve or artistry of Goebbels in Germany. In peacetime, the Foreign Office continued the good work:

> Indeed he chose Sir John Rennie, a career diplomat and one-time head of the Foreign Office's Information Research Unit, responsible for what had once been termed psychological warfare. (N. West, 1982)

Disinformation is the publication of rumours and lies to confuse and mislead.

informer a private individual who reports the activity of another surreptitiously to authority.

Dr Johnson gives 'One who discovers offenders to the magistrate', but now also used of police spies:

> I was aware of the likelihood that he was an informer, planted by those who wished me ill. (Cheng, 1984)

initiation the first act of copulation.

In standard usage, becoming a member of a club, etc., usually with due ceremony. Of either sex. *Initiation into womanhood* is specific, as well as being ridiculous when applied to an adult:

> She thought vaguely about the morning and her 'initiation into womanhood'. (Boyd, 1982, of a bride on her honeymoon)

initiative a belated reaction.

Literally, a first step. It is often used of someone trying to head off a disaster, such as a *wage initiative* taken by government which is trying to check a sequence of inflationary wage settlements; or to describe bureaucratic machinations:

> ... there was a top-level conspiracy—no, wrong word ... *initiative* ... a top-level initiative among the Joint Chiefs. (Block, 1979)

inner city slum.

Used of derelict housing, abandoned shops, etc. which remain when those who can afford to have escaped to the suburbs to avoid noise, smell, and mugging.

inoperative lying.

Literally, no more than invalid or not functioning:

> ... the press office that had been damaged by being forced many times to retract earlier statements about Watergate as 'inoperative'. (Colodny and Gettlin, 1991)

inquisition torture.

It went a lot further than mere questioning when 16th-century Spanish priests got their hands on heretics:

> ... the priests who worked for the Inquisition three hundred years ago, and who could prove from the Bible that God *wanted* people racked and tortured. (Keneally, 1979)

insatiable having a wish for frequent copulation.

Literally, not capable of satisfaction in any particular respect. Used of either sex, within or outside marriage:

> Her mother had warned her that men were insatiable, especially in heating climates. (P. Scott, 1977)

inside in prison.

Mainly criminal use:

> ... an unfortunate habit to be inside, those who treat H.M.'s prisons as hotels. (Ustinov, 1971)

inside track an unfair or illegal advantage.

19th-century oval racetracks were operated without staggered starts and the animal on the inside had less far to run than the competition.

insider a person using confidential information to his own advantage.

In standard usage, any person with such knowledge, usually of a financial deal, whether or not he abuses the confidentiality:

> As an insider, I'll get my arse in a sling if I wheel and deal. (Sanders, 1977)

Insider dealing is dishonestly using the information.

instant bestseller see BESTSELLER.

institutionalize to confine involuntarily.

Especially if you are mentally ill:

> Nathan is *insane*, Sophie! He's got to be ... *institutionalized*. (Styron, 1976)

instrument the penis.

Viewed sexually and with common imagery, mainly in female use:

> ... he could just touch the cloven inlet with the tip of his instrument (Cleland, 1749)

and Maupassant boasted:

> I can make my instrument stand whenever I please. (Harris, 1925)

integrated casting *American* giving blacks unsuitable roles.

Of occasions where the producer has to discriminate against white actors regardless of the plot, to achieve a quota of blacks.

intelligence spying.

The ability to comprehend has been thus debased since the 16th century.

intemperance regular drunkenness.

The converse of *temperance*, moderation, which is standard English for refusal to drink any intoxicants at all:

> ... had, through intemperance, been reduced to utter want. (Mayhew, 1851)

intentions a resolve by a male to marry a specific female.

As against intending to continue to enjoy the pleasures and rewards of courtship. Although a girl's father may no longer dare ask a young man what his *intentions* are, for fear of being told the truth, the concept survives in the rather dated *my intended*, the person whom you have arranged to marry, English having no convenient equivalent for *fiancé(e)*.

intercourse copulation.

Literally, any verbal or other exchange between people:

> For justifying himself, he wrote a full account of the intercourse he had with the Nun and her complices. (Burnet, before 1714—of Sir (later St) Thomas More)

Now standard English as a shortened form of *sexual intercourse*, which itself could apply to no more than holding hands:

> Have you ever had intercourse, Dorothy? (McCarthy, 1963—Dorothy knew that her inquisitor referred only to copulation)

The euphemistic use of *intercourse* echoes that of the now obsolete *conversation* (see CRIMINAL CONVERSATION).

interesting condition see CONDITION[2].

interfere with to assault sexually.

Journalistic and forensic jargon for extramarital male physical acts against boys and non-consenting females:

> They are quite alive and nobody has interfered with them, not yet. (N. Mitford, 1960, of boys who had absconded from boarding school)

intermediate not heterosexual.

And possibly not homosexual either:

> Membership of the intermediate sex was an excellent excuse for contracting out of society and any sexual embroilment. (Pearsall, 1969, writing of Victorian London)

intermission a period of television advertisements.

Temporary cessation has become constant interruption on American television.

internal affairs *American* the investigation by policemen of allegations against the police.

Most police forces are reluctant to wash dirty linen in public, or at all, and complaints against them, often maliciously inspired, are the subject of taboos:

> In Internal Affairs in his sneakers and sweatshirts, investigating complaints against his fellow officers. (Diehl, 1978)

The Russian *Ministry for Internal Affairs* was the fearsome MVD.

internal security the repression of anti-government action.

In a totalitarian state, if the *internal security* is good the tyrant dies in office and in bed.

international bestseller see BESTSELLER.

interrogation (with prejudice) torture.

The Communist KGB used the same imagery as the CIA—see TERMINATE[1]:

> 'Interrogation with prejudice' left Viskov crippled and his wife mute (a suicide attempt with lye). (M. C. Smith, 1981—they had been imprisoned in Russia)

The British *deep interrogation* in Northern Ireland, which involved prolonged discomfort for the victims, was later held to be 'in contravention of human rights'.

intervention a military invasion.

Literally, placing yourself between two other parties. The continued use of *intervention* by the BBC when speaking of the Russian occupation of Afghanistan in 1979 and 1980 caused offence not only to etymologists, even if the events occasioned only muted protests at the time from the usually vociferous western liberals.

intimacy copulation.

The SOED says 'Intimate friendship or acquaintance; close familiarity ... **b** *euphem.* Sexual intercourse.' Used of both marital and extramarital copulation:

> A social escort who ... would amateurishly offer 'intimacy', as they called it. (Theroux, 1973)

So too *intimate*, copulating with:

> You also need a bath and a change. Especially if you propose to be intimate with anyone other than myself. (Bradbury, 1975)

intimate part the human genitalia.

Or the breasts of a female:

> ... glimpsing an occasional movement of white skin which ... might, for all one could tell, belong to an intimate part. (Farrell, 1973)

See also PRIVATE PARTS.

intimate person the penis.

A refinement of *your person* (see PERSON (YOUR)):

> ... the idea that any of them had ... decorated his intimate person with a doughnut was absurd. (Blacker, 1992)

into the sack copulating.

From the slang meaning 'into bed', but there is less immediacy than with IN THE SACK, just as *into bed* implies more time for contemplation of what is to come than *in bed*:

> 'Would you get into the sack with a phallic symbol?' 'I go to bed with you, don't I?' she said lightly. (Theroux, 1976)

intrauterine device a female contraceptive worn internally.

As with atomic bombs, *device* always indicates a desire to avoid a direct statement.

intrigue (an) extramarital copulation.

In this sense, an *intrigue* is a plot, whence something done surreptitiously. Usually in the plural:

> ... only stipulating for the preservation of secrecy in their intrigues. (Mayhew, 1862, referring to extramarital copulation)

introducer's fee a bribe.

> As for bribes ... this is a capitalist society, General. We prefer to talk about commissions and introducer's fees. (W. Smith, 1979)

Literally, a sum paid to a third party who brings the principals together. Also called an *introduction fee*.

introducing house a brothel for daytime use.

Prostitutes went there to meet customers:

> Introducing houses, where the women do not reside, but merely use the house as a place of resort in the daytime. (Mayhew, 1862)

See also HOUSE[1].

intruder an armed invader.

More sinister than merely arriving without an invitation:

> ... so many intruders from across the Pakistan border killed. (Naipaul, 1990)

invade to copulate with.

The male *invades* the female, on however temporary a basis. Partridge says: 'A literary euphemism' (DSUE) and the OED agrees with him, but only in the sense 'to make an attack upon (a person, etc.)'.

invalid coach *American* a hearse.

An invalid description, even if it takes its cargo to a SLUMBER ROOM.

inventory adjustment a loss caused from prior overvaluation of goods.

The simplest way for managers to inflate profits is by overvaluing stocks or failing to write down those which are slow-moving, damaged, or unsaleable. Nemesis may be postponed if trade picks up, but too often the reality emerges:

> Company officials blame losses on share investments and 'inventory adjustments'. (*Daily Telegraph*, 19 February 1993)

inventory leakage stealing.

Not an imperfectly corked bottle in the stores. Used of regular stock losses due to pilferage by customers and employees in large stores.

invert a male homosexual.

Figuratively turned upside-down, if you are heterosexual:

> 'We don't call anyone a queer, homo, pouf, nancy or faggot.' 'What in hell do you call them then?' ... 'Inverts.' (Bogarde, 1978)

Whence *inverted*, homosexual and *inversion*, homosexuality.

investigate to create, exaggerate, exploit, or distort (a scandal).

But you describe it as enquiry:

> 'What d'you mean—smear?' 'Have it your way—investigate, if you prefer. Just so you keep on digging until something starts to smell. Choose your own euphemism.' (Price, 1979)

Whence, *investigative journalism, reporting*, etc.:

> 'I do investigative reporting when I think it's needed.' 'Yeah, investigative, meaning one-eyed, slanted.' (Hailey, 1979)

investor a gambler.

A usage by promoters of football pools and other lotteries to delude subscribers into the belief that they are not wasting their money.

invigorating cold.

Of water for swimming, weather for walking, etc. Anyone who says that participation in the activity to which they are committed would be *invigorating* wants you to suffer with them.

involuntary conversion a crash.

True, as far as it goes, which is not far enough. American legal jargon for an aircraft crash.

involved actively and uncritically supporting an extreme policy.

Mainly sociological jargon:

> Charming girl, very committed, very involved. You must have read about her campaign ... (Theroux, 1976)

Although *involved* should mean 'complex', the people so described are often simple and unthinking. *Involvement* is such devotion to extremism.

involved with enjoying a sexual relationship with.

Usually not of a transitory nature but:

> Khan cites the case of one off-duty flight attendant who became 'involved' with two passengers and a crewman on a single flight. She had taken anti-depressants and consumed several glasses of champagne. (*Daily Telegraph*, 18 April 1995—she copulated with all three in succession)

Irish[1] Irish whiskey.

Irish[2] illogical or defective.

The prefix appears in many offensive and sometimes euphemistic expressions dating from the time when Irish people were deemed to be backward in both Old and New England.

Irish fever typhus.

The disease was endemic in 19th-century Dublin tenement slums, many of which, prior to the forced Union with England, Scotland, and Wales in 1801, had been the elegant town houses of a capital vying in many respects with London:

> Irish slums were graphically illustrated in the *Builder*; typhus was known as the 'Irish fever'. (R. F. Foster, 1988)

Irish hoist a kick in the pants.

An American usage, despite which it seems that the Irish were generally the recipients.

Irish horse *obsolete* an inedible gobbet of meat.

The British navy called salted beef *salt horse* and Irishness was attributed to the toughest portion.

Irish hurricane a calm sea.

A usage of the British navy.

Irish pennant a loose end.

In both literal and figurative senses, based on an American patronizing view of the Irish:

Always loose ends. You know what they call them in the Navy? Irish pennants. (Sanders, 1985)

Irish promotion a reduction in wages.

The English seem to need constant reassurance about the supposed lower status and intelligence of the Irish. An *Irish(man's) rise* means the same thing.

Irish toothache a pregnancy.

Adverting perhaps to the supposed confusion of the Irish, in English eyes, and the dental troubles of undernourished pregnant women. In the male, it means 'an erection of the penis' and although this condition is not unconnected with pregnancy, the exact process of derivation escapes me.

Irish(man's) rise see IRISH PROMOTION.

iron[1] a handgun.

He punched Malvern with the muzzle of the gun ... 'Keep your iron next to your own belly.' (Chandler, 1939)

The metal is inexactly specified, but a *steel* has long been a sword or a bayonet.

iron[2] a male homosexual.

From the rare rhyming slang *iron hoof*, POUFF.

Iron Curtain the European frontiers of the states of the former Soviet Union.

Churchill never acknowledged his debt to Goebbels for the imagery, nor Goebbels to Schwerin von Krosigk:

As soon as the Soviets have occupied a country, they let fall an iron curtain. (Goebbels, 1945, in translation)

We forget too that the Russian *Literary Gazette* used the same expression in 1930 for what was seen as a western policy to isolate Russia from the rest of Europe.

iron out to kill.

I suspect not from the American *iron*, a gun, but from the flattening of the victim. Occasionally too as *iron off*.

irregular see REGULAR[1].

irregular situation extramarital cohabitation and copulation.

If the male is not a priest, one of the partners is usually married to a third party:

No, no, I mean it's you who've had the bad time and the irregular situation. (Murdoch, 1974—a man was talking to his sexual mistress)

Specifically in the Roman Catholic Church to denote such a relationship entered into by a priest despite his vow of celibacy.

irregularity dishonesty or fraud.

These 'irregularities' had allegedly taken the form of loans she had not repayed (sic). (*Private Eye*, April 1981)

Literally, anything which deviates from the norm.

it[1] the sexual attractiveness of females as perceived by males.

From the 1930s prudery about sex:

'It is not beauty that makes every head (except one) turn on the beach to look at her.' 'It's IT, my boy,' said the Major. (Christie, 1940)

it[2] copulation.

A usage without any previous reference to the subject matter:

I would have asked you anyway ... you see, I like it with you. (Bradbury, 1975, of an invitation to copulate)

it[3] the male or female genitals.

Again, the subject matter has not been previously introduced:

Whereas in Jake's youth he had gawped at a girl with her upper clothing disarranged to reveal a, to him, rare glimpse of 'them', he is now horrified to find himself staring much lower down at a sharply focused full-colour close-up of 'it'. (Muir, 1990, of K. Amis's *Jake's Thing*)

it's a big firm my depredations will pass unnoticed.

Originally army use, to excuse waste or pilfering, and now used in the same senses by those working for a public employer.

itch to feel the desire to copulate.

Usually of a woman, from the supposed aphrodisiac properties of cantharides which, by inducing vaginal itch, is said to stimulate sexual desire:

A tailor might scratch her where'er she did itch. (Shakespeare, *Tempest*, with another of his obscure sexual puns)

Occasionally of men in the same sense:

I was beginning to itch for her considerably. (Fraser, 1969)

Itchy feet is the propensity, especially of women, to leave a regular sexual partner for another, without any reference to fungal or other infection—unless you are a prisoner, when it denotes an intention to try to escape. See also SEVEN YEAR ITCH.

item (an) a continuing sexual partnership between two people.

Perhaps merely an *item* of news or gossip:

The country's new wire-cutter-in-chief and defender of medieval rights of way says it was *The Daily Telegraph* that revealed a few years ago that she and ——, Chairman of the Ramblers' Association at the time, were 'an item'. (*Daily Telegraph*, 15 April 1995)

J

J *American* a marijuana cigarette.

From the *J* in *Mary Jane*—see MARY[2]—and explicitly as *J stick* or *J smoke*.

J. Arthur *British* masturbation.

From the rhyming slang *J. Arthur Rank*, wank:

> ...having to slip into the bog at the office and give yourself a quick J. Arthur into this little bottle. (Matthew, 1983)

Rank, the British miller and lay reader, rather improbably found himself during the Second World War chairing a company dominating the British film business, acquired from the financially versatile Oscar Deutsch, from whom we inherit the Odeon.

jab a vein to inject an illicit narcotic.

Used by addicts:

> ...smoke marijuana or opium, or sniff snow or jab a vein. (Longstreet, 1956)

Occasionally as *jab off*.

jack[1] the penis.

One of the male names often used, whence the American *jack off*, to masturbate (of a male):

> The schmuck hasn't done anything but indict homos and jack-off artists for two years. (Diehl, 1978)

Also as *jack into the mattress*:

> If the day has gone well I'll disappear upstairs for a round of light celebratory masturbation —what Roman would no doubt call 'jacking into the mattress'. (Fry, 1994)

Occasionally *jack* can also be semen. The obsolete *jack of both sides* was a male homosexual, perhaps indicating bisexuality:

> A Godly and necessary Admonition concerning Neutres, such as deserve the grosse name of Jack of both sydes. (Title of Broadsheet, 1562, quoted in ODEP)

A *jack in the orchard* was an act of copulation and *jack in the box* was syphilis, being punning rhyming slang for 'pox'.

jack[2] a policeman.

Most *Johns* are also *Jacks* in familiar speech—see JOHN[4]:

> ...a uniformed cop was using a small walkie-talkie... Another jack was sitting and writing in a notebook. (Lyall, 1972)

Also in American Citizens' Band use as *jack rabbit*.

jack it in to die.

From the meaning 'to give up an attempt or enterprise'. Occasionally as *jack it*.

jack off see JACK[1].

jacket *American* a criminal record.

From the file cover:

> ...you don't think people like that have jackets, do you? (Sanders, 1985, referring to people working in learned professions)

jag house a brothel.

A *jag* was a load, and used to denote drunkenness just as *load* is today:

> A man with a 'fairish jag on' would be one with rather more intoxicants than he could 'carry streck'. (*Yorkshire Weekly Post*, 1899, quoted in EDD)

Thus a *jag house* was an inn where you could get drunk, whence a brothel (today one which tends to cater mainly for male homosexuals).

jagged *American* drunk.

From visiting a *jag house* or merely feeling rough?

jail bait a sexually mature female below the legal age of copulation.

Laws change but human physiology doesn't. Twelve-year-old British girls were considered ready for marriage in the 1920s, as they are elsewhere in the world today. The American *jail bait* or Citizens' Band *San Quentin jail bait* is used for a girl who tempts men to risk imprisonment by copulating with her illegally:

> Two chickies, delicious little morsels of jail bait. (Collins, 1981)

Bait is also a youth attractive to male homosexuals.

jakes a lavatory.

Just as you visit the *john* in modern America, in the past you visited *Jake's place*. Dr Johnson's examples from Shakespeare, Swift, and Dryden are all lavatorial, although he defines the word as 'a house or office', giving two euphemisms for one:

> I will... daub the walls of a jakes with him. (Shakespeare, *King Lear*)

Cleland in 1749 uses the singular '...breath like a jake's' but his punctuation is sometimes imperfect. Wits, especially in the 19th century, also used *Ajax*, punning on the King of Salamis.

jam to copulate.

From the pressing tightly together:

> 'He had a good grip on her and she closed her eyes and they did it.' 'Did what?' he said hoarsely. 'Jammed.' (Theroux, 1978)

jam tart see TART.

jane[1] a prostitute.

In England the derivation was from the rhyming slang *Jane Shore*, whore. Jane Shore was the sexual mistress of King Edward IV:

> Louis Quatorze kept about him, in scores, What the Noblesse, in courtesy, term'd his Jane Shores. (Barham, 1840)

I suspect America has eschewed the monarchic derivation in favour of that from the Hungarian *jany*, a girl:

> He happened to bring a couple of beautiful janes along. (Condon, 1966)

jane[2] a lavatory for the exclusive use of women.

A feminine, or feminist, JOHN[1].

Janet and John an over-simplified summary.

British civil service usage for an aide-memoire or summary prepared for ministers:

> For the most senior grade in the Service it would mean a rise of £26,000 per annum. It hardly seems necessary to mention that in the Janet and John. (Lynn and Jay, 1989, referring to a conversation between two senior civil servants intending to conceal their avarice)

Janet and John are an archetypal middle-class boy and girl in illustrated reading books that used to be widely used for young children.

jar a drink of an intoxicant.

Usually beer, from the container:

> 'Have you been drinking?' 'A jar or two,' I admitted. 'But nothing noticeable.' (Lyall, 1975)

If you *enjoy a jar*, the implication is that you are a drunkard.

Jasper *American* a female homosexual.

Possibly from the meaning 'variegated', with common imagery, or a variant of JOHN[3].

jawbone credit.

You talk the seller into parting with the goods without paying him. Usually in the phrase *on jawbone*:

> Many ranchers did all their Williams Lake business on jawbone, paying once a year when they sold their crops of beef. (St Pierre, 1983)

jazz *American* to copulate.

From the frenzy of the tempo or the ecstatic abandonment? And which came first? To *jazz yourself* is to masturbate:

> ... thought it the apex of bliss
> To jazz herself silly.
> (*Playboy's Book of Limericks*)

jelly roll copulation.

In American black use it has a variety of meanings, examined, along with its African derivation, in *The Story of English* (McCrum et al., 1986).

jerk off[1] to masturbate yourself.

Of a male, from the movement of the hand:

> He's jerking off thirty times a day, that fuckin' guy, and they's all set to give him a medical. (Herr, 1977)

A male may also *jerk* his *maggot* or *turkey*. *Jerk-off* is also used figuratively as an insult to a male:

> It is impossible, even for a flinty-hearted jerk-off such as your narrator, not to be won over. (Bryson, 1989, after visiting Williamsburg)

But to call him a *jerk* would be more usual, if equally disrespectful:

> Look, you think this is some penny-ante organization I'm running, you stupid jerk. (Poyer, 1978)

To *jerk around* is to frustrate or annoy:

> I've got the feeling someone is jerking me around, and I don't like it. (Sanders, 1980)

The obsolete British *jerker* was a prostitute, using the same imagery.

jerk off[2] illegally to inject heroin slowly.

You allow the narcotic to mingle with blood in the phial so that eventually you inject a mixture.

jerry a pot for urine.

Dr Wright says it is a shortened form of 'Jeremiah, a chamber utensil' (EDD) but for the more ambitious it might have been a shortened form of *Jeroboam*, a bowl or bottle which contains 10 to 12 quarts. No connection probably with the obsolete British *Jericho*, a lavatory, which was merely one of those unlikely places to which people said they were going. The German soldier, or *Jerry*, wore a helmet of much the same shape but that too is probably only a shortened form of *German*.

jet-lag sub-acute alcoholic poisoning.

In standard usage, disruption of the biological clock through time change. On long flights many people drink too much alcohol free or at what seems to be a bargain price, to which you can add tiredness, dehydration from high-altitude flying, lack of exercise, and excitement:

> I am still under the weather due to jet lag et al. (*Private Eye*, March 1981—he had a massive hangover)

jewels the male genitalia.

American rather than British use, from their pendulate proclivity:

> If I'd given him a bright, 'Good morning, Sam!' he'd have kicked me in the jewels. (Sanders, 1979)

Sometimes as *crown jewels* or *family jewels*:

> ... draw up the knees to protect the family jewels. (ibid.)

Jews' lightning see LIGHTNING.

Jezebel a prostitute.

She was the naughty wife of Ahab in the Old Testament:

> 'But that's ...' She was about to say 'a mortal sin' but desisted. 'It makes me a Jezebel, doesn't it?' (Read, 1986)

Long ago you qualified as a *Jezebel* merely by wearing make-up. Although it was a stand-by for the vituperative oratory of John Knox and Titus Oates, Dr Johnson missed it; but as he intermingles his *I*s and *J*s as a single letter, we can overlook this rare lapse.

jig-a-jig copulation.

From the movement and mainly Far Eastern use:

> 'Dated her,' I said. 'You mean a little boom-boom.' 'Jig-jig,' he said. 'But it comes to the same thing.' (Theroux, 1978)

Jiggy-jig is the next most common form:

> ...the familar cry of 'jiggy-jig, Sahib.' Very small boys did the soliciting for these native girls. (Richards, 1936, of India)

Then come *jig, jiggle, zig-zig*, etc.

jiggle to masturbate.

Of a male on his own, from the meaning 'to move back and forth':

> 'Nothing of the sort, he lay there jiggling like.' (I guessed what she meant... frigging himself.) (Harris, 1925, of Carlyle's behaviour on his wedding night. Evidently Mrs Carlyle had more to put up with than the celebrated cup of tea, or less.)

Jim Crow the unfair treatment of black people by whites.

> It was my first experience with Jim Crow. I was just five, and I had never ridden on a street car before. (L. Armstrong, 1955)

In early usage in America, any poor man. The character came from a song in the Negro minstrel show written by Tom Price (1808–1860). Occasionally as *Jane Crow* for such behaviour to black women in America.

Jimmy an act of urination.

From the rhyming slang *Jimmy Riddle*, piddle. The punning *Jerry riddle* is obsolete. *Jimmy* (or occasionally *Edgar*) *Brits* is diarrhoea.

jive to copulate extramaritally.

Literally, to dance to swing or jazz music (or, as a noun, the music itself) whence, mainly in black American use, to copulate as a sequel to this. Those who should know say that the etymology is unknown and those who originated the usage may well have been unfamiliar with the Scottish dialect meanings relating to legs.

job an act that is the subject of a taboo.

Especially an act of defecation, mainly in nursery use, and also as BIG JOBS; also denoting robbery, especially if planned in advance or with forcible entry, as in the film title *The Italian Job*; or denoting copulation, as in ON THE JOB.

job action job inaction.

The American equivalent of the British *industrial action*, the procedure through which employees in concert seek to bring pressure to bear on their employer by failing to do their work properly, or at all, while retaining an entitlement to be paid.

job turning *American* reducing the responsibility and pay associated with an appointment.

What happens when a manager is obliged to appoint a less qualified person to comply with regulations or achieve a QUOTA, especially when an employer is obliged by law to put a woman in a job for which he thinks a man more suitable and of which the previous incumbent was a male. A feminist will tell you that this is merely another example of males wanting to retain domination over females. A manager will tell you that the change in status reflects the fitness for the post of the new incumbent and he has to find a way round a stupid law to keep his business running. In varying degrees, both are right.

jock the penis.

Vulgar on its own:

> He washes his jock in public and he's shy? (Sharpe, 1977)

But almost standard English in *jock-strap*, the support worn over the male genitals:

> ...some joker wearing a coconut mask and a feathered jock-strap. (Sanders, 1977)

In obsolete use to *jock* was to copulate with a woman and *jockum*, a shortened form of *jockum gage*, a pot for urine.

jocker a male homosexual.

> Roxie hustles the guys who want a queen, and the kid goes after the ones who want a jocker. (Wambaugh, 1972)

From JOCK. Sometimes too as *jockey*.

joe[1] a ponce.

From the rhyming slang *Joe Ronce*; Ronce's exploits have not merited an entry in the DNB (see also *Charlie Ronce* at CHARLIE). Another use of *joe* is from the rhyming slang *Joe Hunt*, cunt, being used pejoratively and not anatomically. In obsolete American use, a *joe* was a lavatory.

joe[2] a spy.

Espionage jargon, for one of your own spies whom you may prefer not to identify:

> A joe in the parlance is a living source, and a live source in sane English is a spy. (le Carré, 1989)

john[1] a lavatory.

From the *cousin John* men or women said they had to visit as they absented themselves:

> Running back and forth, practically living in the john. (Theroux, 1975)

john[2] *American* a woman's regular sexual partner to whom she is not married.

He need not necessarily be married to a third party.

john[3] a male homosexual playing the male role.

He often lives with another homosexual. There is probably a direct derivation from the obsolete

English *John and Joan*, a male homosexual in the days when all homosexual males were thought also to be heterosexual, and there were no homosexual females.

john[4] *American* a policeman.

So the Johns came for him. (Chandler, 1939)

A shortened form of *John Law*, which is also used to refer to the police collectively:

I'd have no trouble with John Law. (Sanders, 1982)

john[5] *American* a potential customer for a prostitute.

Prostitutes' jargon:

Our hustlers sat on their steps and called to the 'johns' as they passed by. (L. Armstrong, 1955)

John Barleycorn see BARLEY CAP.

John Law see JOHN[4]

John Thomas the penis.

The common use of masculine names and without necessarily sexual implications:

John Thomas doesn't even have a chance to lift his head. (G. Greene, 1978)

I never cease to wonder at the thoughtlessness of a Mr and Mrs Standing of my former acquaintance, who gave their son these Christian names. *John Willie* is rarer:

What I call your penis and what you prefer to regard as your John Willie. (Sharpe, 1978)

Whence perhaps WILLY. The American *Johnnie's out of jail* is an oblique warning of an undone trouser zip.

johnny a contraceptive sheath.

From FRENCH LETTER via the slang *frenchie* and *Johnny Frenchman* and so to the shortened form *Johnny*:

Millroy was unrolling a small tight ring of rubber... 'A rubber Johnny,' Millroy said. (Theroux, 1993)

Also as *Johnnie*.

join[1] *?obsolete* to copulate.

Lovers passed the virulent lice to each other when they joined, fast and secret in some hidden corner. (Keneally, 1982)

Of the same tendency as the common *couple*.

join[2] to be as dead as.

The imagery is of coming together again in some physical or spiritual existence rather than in the grave. Thus you may *join* your deceased spouse or a selection of dead relatives:

He was about to join his ancestors (Sharpe, 1978)

or you may *join* the GREAT MAJORITY:

...he was really doing no more than joining that majority (Price, 1985—he was dying, not joining a political party)

and, if both optimistic and devout, you may aspire to *join your Maker*.

join hands with *obsolete* to marry.

No game of ring-a-roses when marriage too was the subject of reticence and evasion:

...the day before his Highness the Prince of Wales was to join hands with the Princess of Saxe-Gotha. (Fowles, 1985, writing in archaic style)

join the club see IN THE CLUB.

joiner *American* a person who seeks popularity or business by attaching himself to associations etc. in which he has no special interest.

Literally a skilled carpenter. Pejorative use:

He appeared to be a genial greeter and joiner, an intellectual lightweight. (Sanders, 1977)

joint[1] a marijuana cigarette.

Two or three people can get high on one joint (marijuana cigarette). (Longstreet, 1956—he would not need the brackets today)

Formerly, the equipment of an opium user; I see no clear connection with *joint*, the place where you take part in any communal activity, including the illegal ingestion of narcotics.

joint[2] *American* the penis.

...drawings of a man's joint, a woman's cooze. (Sanders, 1982)

I imagine from the meaning 'a piece of meat cut ready for cooking'. To *unlimber your joint* is to urinate (of a male):

...graffiti...where males unlimbered their joints. (Styron, 1976)

joint[3] to incapacitate by shooting.

Another type of butchery:

According to Belfast's grisly argot, he was 'jointed'—shot through both elbows, both knees, and both ankles. (*Sunday Telegraph*, January 1990, reporting on a victim of IRA retribution)

jolly[1] drunk.

An old variant of MERRY:

They're not up all night at balls and parties, and they don't get jolly in the small hours. (Pearsall, 1969, quoting 19th-century music-hall patter)

The obsolete English *jolly* (of a bitch) meant 'able to conceive':

Nine days jolly, Nine weeks in belly, Nine days blind, That's a dog's kind. (Old proverb)

jolly[2] an unnecessary treat paid for by a third party.

Common business use to describe unnecessary conferences held in congenial locations, where the expenses of those attending are paid by their employers or customers.

jolly[3] an act of casual copulation.

Homosexual or heterosexual:

> ... found the names of Thomas J. Kealy and Constance Underwood, and what they had been paying for their jollies. (Sanders, 1984 —the names were in a prostitute's notebook)

jolt anything taboo which gives you a shock or impetus.

For illegal narcotics users, an injection of heroin; for criminals, a term in prison; for drinkers, any intoxicating drink, usually whisky:

> I think maybe I'll get a jolt too (Sanders, 1982)

or specifically:

> I went out to the kitchenette and poured a stiff jolt of whisky. (Chandler, 1939)

Jordan a pot for urine.

> They will allow us ne'er a jordan, and then we leak in the chimney. (Shakespeare, *1 Henry IV*)

Dr Johnson spells it *jorden* and suggests derivation from Greek. ODEE says: 'Early forms with u do not support the conjecture of deriv. from the River Jordan' but I am less sanguine where pre-1850 spelling is concerned and the synonym *chamber* has 19 different ways of being spelt in EDD, which makes a redundant *u* small beer. *Jordeloo* was the warning in Edinburgh after 10pm that urine was about to be thrown into the street from an upper room, a compound of *jordan* and *below* or *l'eau*, just as *gardyloo—gardez l'eau* —heralded the descent of other liquids.

joy[1] extramarital copulation.

For male or female, as in *mutual joy*:

> ... the woman ... seeking mutual joys courts him to run the complete race of love. (Lucretius, in translation)

A *joy girl* or *sister* is a prostitute:

> The gambling casino on the lake, and the fifty-dollar joy girls ... (Chandler, 1953)

She may work in a *joy house* or brothel:

> I ain't been in a joy house in twenty years. (Chandler, 1940)

The punning *joy ride* is a single act of copulation:

> I feel no fatigue, indeed, I feel the better for our joy ride. (Harris, 1925—he never wasted time by taking a girl for a spin in the park)

Joy stick, the penis, is punning if tasteless rhyming slang for 'prick' and the aircraft control column. The *joy bag* is not the scrotum but a contraceptive sheath.

joy[2] the sensation sought by the use of narcotics.

Used attributively in many compounds such as *joy popper*, an occasional user; *joy powder*, morphine; *joy flakes*, cocaine; *joy rider*, a person who takes narcotics on a single occasion; *joy smoke*, marijuana; *joy stick*, an opium pipe.

joy ride[1] see JOY[1].

joy ride[2] to take and drive away a motor vehicle without consent.

Under the former British larceny rules, this was not a criminal offence unless there was proof of intention 'permanently to deprive the owner thereof', and it was usually impossible to establish a theft of fuel. Now a crime in its own right.

Judy a prostitute.

Either from the common girl's name which became a name for common girls or from *Judith*, the beautiful Jewess who is said to have tricked Nebuchadnezzar's general Holofernes in order to save the town of Bethulia, the general first losing his head figuratively and then literally:

> When monks like Negga were shooting down their officers or bribing potential Judiths to seduce their Holofernes ... (Mockler, 1984)

In obsolete British use a *Judy* was a sexual mistress:

> He went tul his wife at Wortley, an his judy went to Rotherham. (*Dewsbury Olm*, 1866, quoted in EDD)

jug[1] prison.

Although probably from the Scottish *joug*, a pillory, as 'He set an old woman in the jougs' (W. Scott, 1814), this 19th-century development may also owe something to the confining nature of a stone vessel.

To *jug* is to imprison:

> He is arrested. He is jugged. (Manning, 1960)

jug[2] *American* a container for intoxicants.

> 'It is necessary to drink alcohol?' 'Well, I'm going to have a pull at the jug.' (Benet, 1943)

Jugged is sometimes used to mean 'drunk'. It probably comes from the pot rather than the common culinary imagery. See also JAR.

juggle *obsolete* to copulate.

If Shakespeare was running true to form, punning on the play with balls:

> She and the Dauphin have been juggling. (*1 Henry VI*)

jugs a woman's breasts.

Probably from their shape and purpose of producing and containing milk:

> Blue eyes. Peaches-and-cream complexion. Big jugs. (Sanders, 1970)

Grose tells us that a *double jug* was a man's backside.

juice[1] **(the)** intoxicants.

The common modern use probably came from the literal meaning, 'liquid of fruit', rather than from the Scottish *juice of the bear*, whisky. *Juniper juice* was gin but *the juice* means 'any intoxicant':

> The cops will probably want you so stay off the juice. (Deighton, 1972)

To *juice* is to drink spirits:

> ... would gather after a long day in the IO shop to juice a little. (Herr, 1977—in fact they

gathered in the IO shop after a long day elsewhere)

Whence too *juiced*, drunk; *juice head*, a drunkard; *juice joint*, a bar.

juice[2] payment made or demanded illegally.

What comes in America if you SQUEEZE. Denoting extortionate rates of interest, made by a *juice dealer*, loan shark, and collected by a hoodlum or *juice man*; the proceeds of organized vice; and bribes:

> The bookie was a big operator and sent his juice money directly to City Hall. (Weverka, 1973—he actually sent it 'direct' without an intermediary but not necessarily promptly)

juiced up desiring immediate copulation.

Of a female, from the vaginal secretion:

> ...he knew how to get a girl juiced up better than anyone she'd ever known. (M. Thomas, 1982, describing a philanderer)

Whence *juicy*, being sexually aroused:

> They will claim that only the other day they saw a man whose bottom reminded them a little of Mel Gibson and that they got really quite juicy thinking about it. (Fry, 1994)

juju a marijuana cigarette.

Probably an American shortened form of *marijuana*, but referring to the black African connection:

> I knew a guy once who smokes jujus. (Chandler, 1940)

jumbo drunk.

From the rhyming slang *jumbo's trunk*. In supermarket language *jumbo* means 'big'.

jump[1] to rob.

From the pouncing. An English 18th-century use since revived in America:

> Instead of 'jumping' those stores for an average of forty dollars... (Lavine, 1930)

jump[2] a single act of copulation.

> You've never had a quick jump in the hay in your life. (Steinbeck, 1961)

Normally a male usage, but he does not have literally to leap on to his partner. To *jump* is to copulate and a *junior jumper* is a youthful rapist.

jump ship to desert.

It should be the opposite of keelhauling. Standard English of seafarers and occasionally used of others:

> Moscow Centre officers who were thinking of jumping ship... (le Carré, 1980)

A prisoner on remand who fails to appear in court may be said to *jump bail* and someone absconding without paying may *jump a bill* or *check*.

jump the besom to cohabit and copulate without being married.

Sometimes after banns had been called, but more often not, a *besom* being also a slang name for a prostitute. Similarly a couple might *jump a broomstick*:

> Besides, I ain't married proper. No more than if I jumped a broomstick. (Cornwell, 1993, reporting 19th-century speech)

Whence the *broomstick match* or common law marriage:

> I never had a wife but I have had two or three broomstick matches. (Mayhew, 1851)

In Kent, when families from the London slums congregated to harvest the hops manually, the piece of wood which was ritually jumped was a *hop-pole*, and the ensuing union a *hop-pole marriage*. A couple might also be said to *leap the broom*.

jump the last hurdle to die.

With steeplechasing in mind.

junk illegal narcotics.

Originally, old rope, whence hemp, whence narcotics generally. A *junkie* is an addict:

> A cheap junkie's arms and legs are covered with unhealed scabs. (Longstreet, 1956)

A *junker* in this world is not a Prussian aristocrat but a peddler in narcotics, as is a *junkman*:

> I just retired a junkman. (Diehl, 1978, of killing one such)

Junked up is under the influence of narcotics:

> Will you go out now, before he gets junked up for the evening? (Chandler, 1939)

junk mail unsolicited advertising matter sent by post.

Not letters conveyed by a Chinese vessel nor packages containing narcotics. The recipient may stem the flow by readdressing the envelope and returning it to the sender without affixing a stamp.

junket an occasion for bribery.

The dessert of flavoured milk curdled by rennet has largely given way to an entertainment provided to obtain advantage at another's expense:

> ...lurking in the background of every junket there is likely to be a provenance or motive that is not especially palatable. (H. Porter, *Daily Telegraph*, 8 October 1994)

Whence *junketing*, accepting such a bribe or inducement:

> Junketing is part of modern life and almost every large scale business concern and most individuals that make up the political and media establishment are familiar with it. (ibid.)

just good friends see FRIEND.

justify *obsolete* to kill by court order.

It meant in Scotland 'to bring to justice', whence either 'to acquit' or 'to execute', which must have been the source of some confusion:

> Our great grand uncle that was justified at Dumbarton. (W. Scott, 1817—we can only learn uncle's fate by reading on)

KGB see COMMITTEE OF STATE SECURITY.

kangaroo court an ad hoc investigation in which the issues are prejudged.

Prison and trade union usage and practice for summarily disciplining those who fail to comply with unenforceable instructions. The offender has figuratively to *jump to it*, like the marsupial. A prison *kangaroo club* is a body of long-serving inmates:

> He was president of the Kangaroo Club and would hold court to instruct them in their duties. (Lavine, 1930, describing the initiation of new inmates)

karsey/karzey/karzy see CARSEY.

kayo to kill.

From the boxing *KO*, or *knock-out*:

> ... this stiff got kayoed around the end of October. (Diehl, 1978)

keel over to die.

> He told me he might keel over at any time. (A. Waugh, *Private Eye*, August 1980)

From the capsizing of a boat.

keelhauled *?obsolete* drunk.

It was Dutch practice to drag defaulters under the keel of a boat for punishment and we still use the word *keelhauling* of a verbal reprimand. If you were dead drunk, you might look and later feel like a victim of *keelhauling*:

> They wad fuddle an' drink till they were keel-haul'd. (W. Anderson, 1867)

keep to maintain (a sexual mistress).

This 16th-century euphemism implies both provision for her upkeep and keeping her sexual activities for yourself:

> One officer offered to keep me if I would come and live with him. (Mayhew, 1862)

A *kept woman* is not one who receives housekeeping money from her husband:

> Most kept women have several lovers ... and in ninety-nine cases out of a hundred escape detection. (ibid.)

Kept mistress is explicit:

> It is a mistake to suppose that kept mistresses are without friends and without society. (ibid.)

Kept wench is rarer:

> Virgins, reporters, housewives, kept wenches. (Longstreet, 1956—but which was the oddity in that class?)

The male was a *keeper*, which we now reserve for a custodian of animals in a zoo:

> ... amongst the kept mistresses ... I hardly knew one that did not perfectly detest her keeper. (Cleland, 1749)

keep company with to copulate with outside marriage.

Literally, to accompany, whence, in standard English, to court:

> Their sweethearts or husbands have been keepin' company with some one else. (Emerson, 1890)

See also COMPANY[1].

keep sheep by moonlight to be killed by hanging.

You watch over them from the gallows. I include this obsolete English usage as an excuse for quoting:

> ... that shepherded the moonlit sheep a hundred years ago. (Housman, 1896)

keep up with the Jones's to live beyond your means or extravagantly because you measure your standard of living against that of your neighbours.

The *Jones's* are your mythical neighbours who always seem to be able to afford the new curtains you have coveted or the garden tractor you have been collecting brochures about.

keep your legs crossed to refuse to copulate.

Of a female, in relation to extramarital copulation:

> I don't think she keeps her legs crossed all the time. (Price, 1972)

Less often as *keep your legs together*:

> ... had kept her legs tightly together. (Price, 1975, of a woman who had not copulated)

kept see KEEP.

kerb crawling looking for a prostitute.

Usually by a man who drives slowly in part of a town frequented by prostitutes. He may also, occasionally, be a pedestrian, and the phrase is used too of prostitutes soliciting male pedestrians while travelling slowly in a car.

key the penis.

From the manner of its entry into a lock, whence *keyhole*, the vagina viewed sexually. A *key party* is a sexual orgy in which females pair with males after the supposedly chance selection of a key from those thrown into a central pile.

Khyber the anus.

From the rhyming slang *Khyber Pass*, arse. You hear it most, but not very often at that, in the vulgar riposte *up your Khyber*, whence a punning British title *Carry on up the Khyber* for a film which contrived to live down to it in vulgarity and banality.

kick[1] to die.

Probably from the involuntary spasms of a killed animal. Usually as *kick in, it, off*, or *up*:

> Thou'se no kick up, till thou's right aul. (Picken, 1813)

The common *kick the bucket*—from which *kick it* probably comes—is supposed by OED and others to derive from the *bucket*, or beam, from which you tied a Norfolk pig to facilitate slitting its throat and which it then kicked in its death throes. Lexicographers are plagiarists, and that kind of derivation tends to have special appeal for them. I suspect *kick the bucket* merely comes from a preferred method of those who kill themselves, or others, by stringing the victim up to a beam, and then kicking away the upturned bucket on which he was standing:

> It all went. So he kicked the bucket, literally. (Sanders, 1977, of a suicide)

The obsolete *kick the wind* was to be killed by hanging, from the death movement and the subsequent tenure of the gallows by the corpse. The rare *kick your heels* puns on idleness:

> In a few moments most of them would be kicking their heels in a different world from this one. (Richards, 1933)

kick[2] **(the)** peremptory dismissal from employment.

Usually affecting a single employee, who *gets the kick*, but the violence is only figurative.

kick stick a marijuana cigarette.

From the *kick*, or thrill, it gives. Smoked alone or at a *kick party*. To *kick the gong around*, to smoke narcotics illegally, puns on the excitement and the symbolic oriental gong.

kick the habit see HABIT.

kick the tyres to take superficial account of deep-seated problems.

Business jargon, from the actions of inexpert buyers of used cars. Some figurative use:

> ...a simplistic agrarian vision which the war-weary nation had bought without kicking the tires. (M. Thomas, 1980)

kickback a clandestine illegal payment.

The derivation must be from the vicious habits of starting handles in the days before cars had electric starters. Of hidden commissions, percentages, or the proceeds of vice or bribes:

> It's the job if I get a kickback. (Chandler, 1939)

kid an adult.

A child since the 16th century, before which it was only the young of a goat. Untypically missed by Dr Johnson and his team. As with MIDDLE-AGED, used to minimize age:

> He was still a kid, no more than thirty, thirty-two. (M. Thomas, 1980)

In obsolete English use to *kid* also meant 'to impregnate', or 'to give birth', of both women and goats.

kill a snake *Australian* to urinate.

Not the usual penis as a serpent imagery, but from going into the bush.

kilo connection *American* a dealer in illegal narcotics.

In this business, a kilo is a large amount, and see also CONNECT[2]. It is at this stage that the narcotics are usually adulterated before being passed down the distribution chain.

kind copulating without payment.

From the meaning 'friendly or considerate' and of a female outside wedlock—within marriage males tend to think they copulate as of right. Of a male, it means 'exercising tenderness or restraint in copulation':

> 'Your highness,' he said at last, 'will you be kind to our treasure'... It's a polite way of suggesting you don't make too much of a beast of yourself on the honeymoon. (Fraser, 1970)

kindness bribery.

Again an extension of the meaning 'consideration':

> ...what hath passed between us of kindness, to hold his tongue. (Pepys, 1668—he was worried that the person who had bribed him might talk about it)

king Lear a male homosexual.

British rhyming slang for 'queer' with perhaps a passing thought to the monarch's madness.

king over the water a Stuart pretender in exile.

Possibly used of Charles II and James II during their 17th-century absences from the throne and certainly much in vogue after the Hanoverian kings took over after Queen Anne died in 1714:

> He so far compromised his loyalty, as to announce merely 'The King', as his first toast ...Our guest...added, 'Over the water'. (W. Scott, 1824)

You normally passed your wine glass over your glass of water without venturing verbal amplification. In retrospect, it all seems a little childish, but loyalty to the Stuarts also implied adherence to Roman Catholicism, which in turn involved civil disabilities if not persecution.

kingdom come death.

> ...Piper being blown to Kingdom Come in the company of Mrs Hutchmeyer. (Sharpe, 1977)

Despite our generally unsatisfactory experience with theocracies, we do not demur at the plea *thy kingdom come* in the Lord's Prayer.

kinky displaying bizarre sexual tastes.

A *kink* is a bend, as in a hosepipe, and *kinky* implies a number of deviations. Formerly only of male homosexuality:

> And you said last night he was 'that kind'... funny, kinky. (Bogarde, 1981)

Now used of any perverted extravagance.

kiss to copulate with.

This dates from the era when you only kissed within the family. If you got that far with someone of the opposite sex to whom you were not related, there was no stopping further progress. Whence the euphemistic definition of Dr Wright: 'Obs. To lie with a woman' (EDD). If you *kissed St Giles' cup*, you were killed by hanging, from the practice of offering the victim a cup of water of St Giles in the Fields on his final journey from Newgate to Tyburn. To *kiss the cap* was to drink intoxicants to excess, *cap* meaning 'cup', and a *kiss the cap* was a drunkard. To *kiss the counter* or *kiss the clink* was to go to prison. To *kiss the ground* was to die:

> I will not yield,
> To kiss the ground before young Malcolm's feet.
> (Shakespeare, *Macbeth*—although here it could mean to pay homage)

All these British uses are obsolete.

kiss-and-tell involving the sale of personal memoirs of extramarital copulation.

By the female, usually to the gutter press, which pays her *kiss cash* for *kissing and telling*. Also denoting salacious material included to boost sales of autobiographies etc.

kiss off[1] *American* to die.

From the gesture of parting.

kiss off[2] to engage in fellatio.

From the use of the mouth. Occasionally too, without much logic, of buggery.

kiss-off[3] **(the)** summary dismissal from employment.

Again, from the gesture of parting. On the American west coast, you may call it a *New York kiss-off*, in New England a *California kiss-off*, which shows we still impute bad habits or behaviour to our rivals. Some figurative use, as when you bring a conversation to a polite close:

> 'Yes. Sure. Fine,' Delaney said heavily, feeling this was just a polite kiss-off. (Sanders, 1973)

Kit has come *British* I am menstruating.

Kit is a shortened form of *Charles* and see also CHARLIE.

kitchen-sinking making excessive provision.

From the cliché *everything bar the kitchen sink*. It is the practice of those taking control of a business which has been doing badly to unveil losses and create reserves so that no blame for past poor performance can be attributed to them. They usually also ensure that some of what is described as loss can at a later time be written back as a profit to the company's and their credit:

> There will be an element of 'kitchen sinking' in these numbers. (*Sunday Telegraph*, 4 April 1993—excessive reserves had been provided)

kite to issue (a negotiable instrument that is not covered by the drawer).

> 'Just don't start kiting checks,' Delaney warned. (Sanders, 1985)

Shortened form of the 19th-century *fly a kite*, which had the same meaning, from launching something without support. The device was widely used to obtain credit through the banking system while a cheque was being cleared, but the advent of computers has stopped it being done with safety on a regular basis. A *kiteman* still tries, however.

kitty the vagina viewed sexually.

A variant of PUSSY, whence the double-punning saw:

> A thrifty tom-cat puts something in the kitty every day.

In obsolete English use a *kitty* was a prison or lock-up, from which we probably get the central pool in a game of cards.

knackers the testicles.

A *knack* was a toy or small object, made by a *knacker*, whence a saddler, who bought old or dead beasts for their hides, whence the modern knacker who disposes of dead cattle. The use may come from the meaning 'small objects'—see also ROUND OBJECTS—but Dr Wright is persuasive when he gives 'Two flat pieces of wood or bone' especially when he adds 'Of unequal length' (EDD). Partridge suggests 'Prob. ex dial knacker, a castanet or other "striker"' (DSUE) and the imagery from the small Spanish chestnut is attractive if unconvincing. To be *knackered*, usually of a male, is to be exhausted, ready figuratively for the *knacker's yard* or merely winded by a painful blow. The possibility of a vulgar derivation has brought into common use the rhyming slang CREAM CRACKERED.

knee-trembler a prostitute who copulates while standing up.

Of obvious derivation. Both parties are said to be *knee-trembling*, if so engaged.

kneecap to maim by shooting the knees.

A disciplinary measure of the IRA in Northern Ireland. See also JOINT[3].

knees up copulating.

Of a female, from a position sometimes adopted:

> ...he's had more hot dinners in my house than I've had nights with my knees up. (Lyall, 1972)

A *knees-up* is no more than a party or informal dance.

knight a person associated with any illegal, taboo, or despised occupation.

A source of much British wit. A *knight of Hornsey* was a cuckold, punning on the London borough and the horn of cuckoldry; a *knight of the road* was a mounted thief; my favourite is the *knight of the Golden Fleece*, for a lawyer.

knob the penis.

A male vulgarity, perhaps using the same imagery as KNOCKER.

knobs a woman's breasts.

A less common vulgarism than KNOCKERS but using the same imagery:

> ...and who do I see in a tight sweater, with knobs like this? (Theroux, 1989)

knock see KNOCKING-SHOP, KNOCK OFF[2].

knock around habitually to beat.

> I gather he likes to knock her around a bit. (le Carré, 1989, of a habitual wife-beater)

Usually it is the male member of the household who offers habitual violence to other members, but occasionally too of a mother hitting children.

knock it back to drink intoxicants to excess.

Once, or many times, perhaps from angling the glass as you drink:

> ...he'd begun to knock it back at half-past ten in the morning. (P. Scott, 1977)

knock off[1] to kill.

As a bird from a branch, but in American use it may also apply to humans:

> So you wouldn't knock him off...but you might throw a scare into him. (Chandler, 1939)

Knock on the head comes from the slaughtering of cattle for meat, but is also used of killing humans by any method. To *knock down* is to kill animals by shooting:

> She knocked down squirrels with exquisite faces. (Mailer, 1965)

knock off[2] to copulate with.

The male *knocks off* the female, usually in a casual relationship in which he is careful not also to *knock her up* (see KNOCK UP). *Knock* is this activity:

> Throw her away and she'll always come back for another weekend of cheap knock. (Fowles, 1977)

See also KNOCKING-SHOP.

knock off[3] to steal.

Of minor thefts, from the concept of dislodging something from a counter or barrow.

knock off[4] to drink (an intoxicant).

Usually beer, and specifying in pints the amount consumed. See also KNOCK IT BACK.

knock-out a fraudulent auction.

There is usually a conspiracy between the auctioneer and some of the bidders. Auctioneers' jargon which puns on *knock down*, to register a sale by the fall of the hammer, and the boxing term to *knock out*, to render unconscious.

knock over to kill.

By shooting, from hunting jargon:

> I heard...he had been knocked over in the last month of the war...the rumour proved false...he is alive and kicking. (Richards, 1933)

knock up to impregnate (a female).

Usually when the result is unwanted pregnancy in an unmarried woman:

> ...they told me that seven of the girls were knocked up—well, pregnant. (N. Mitford, 1960)

This use has now virtually displaced the former meanings, 'to wake by knocking' or, as *knocked up*, 'exhausted'.

knocker the penis.

From the shape of a door knocker and perhaps punning on its sexual function—see KNOCK OFF[2]:

> Susie was a perfect fool for any chap with a big knocker. (Fraser, 1982)

knockers a woman's breasts.

Again from the shape of a door knocker and its movement in a vertical plane when activated:

> I could see a roomful of libidinous Japanese with their mouths open, transfixed by a wobbling pair of Russian knockers. (Theroux, 1988)

knocking-shop a brothel.

From the obsolete *knock*, to copulate:

> At the fifth knocking-shop, I struck pure gold. (Fraser, 1971—the gold was figurative—he found a bawd to hide him)

Formerly too as *knocking-house* or *knocking-joint*. Occasionally as *knocker's shop*:

> ...in twenty minutes they had organized a taxi to a 'knocker's shop'. (M. Clark, 1991, where they 'were taking a look at the tarts')

knot *obsolete* to copulate.

From the meaning 'to unite':

> ...a cistern for foul toads
> To knot and gender in.
> (Shakespeare, *Othello*)

I am not sure about:

> ...young people knotting together, and crying out 'Porridge'. (Pepys, 1662)

know to copulate with.

It was a euphemism in Hebrew, Greek, and Latin, which explains why the translators for King James I (of England) found it so useful:

And he knew her not till she had brought forth her first-born son. (Matthew 1:25, of Joseph and Mary)

We still use *knowledge* for copulation in the phrase *carnal knowledge* (see CARNAL).

know the score see SCORE[1].

known to the police having a criminal record or been suspected of crime.

With or without convictions:

Hamilton was a frightening man, known to the police. (Monkhouse, 1993—he had been arrested three times in connection with crimes of violence but had never been convicted)

The police also know many reputable citizens, magistrates, judges, etc.

knuckle sandwich *American* a punch in the face.

Not meat from near the joint of a pig placed between slices of bread:

First the velvet glove, then the knuckle sandwich. (Sanders, 1977)

L

labour[1] childbirth.

Literally, physical toil, but so long standard English that we do not think about it.

labour[2] unemployment.

The former British Ministry of Labour existed to try to place the involuntarily unemployed in work, operating through a series of *Labour Exchanges*, which also gave state aid to the needy. Thus to be *on the labour* was both to be without work and in receipt of unemployment benefit. See also DEFENCE, HEALTH, and LIFE INSURANCE.

labour education arbitrary imprisonment on political grounds.

A Chinese prison regime of work and harangues. A Chinese woman who wished to marry a French diplomat was in 1981 accused of 'illegally living together with a foreigner' and sentenced to two years' *re-education through labour*. (*Daily Telegraph*, November 1981)

lack of moral fibre cowardice.

Mainly British Second World War military use, often as *LMF*:

> ...stamped on the record of failed officers. *Lack of moral fibre*. If Second-Lieutenant Audley suffered from LMF... (Price, 1978)

lad an exclusive male sexual partner outside marriage.

Used mainly in northern England:

> But when I was nineteen he sought me out and he became my lad. (Cookson, 1969)

So too with LASS.

ladies a lavatory exclusively for female use.

Usually adjacent to GENTLEMEN. Also as *ladies' convenience, room, toilet*, etc.:

> I tapped a kidney in the ladies room. (Theroux, 1978)

ladies' college/ladies' fever see LADY.

ladies' man see LADY-KILLER.

lads a lavatory exclusively for male use.

You may see this in a trendy restaurant, which will then have a corresponding *lassies*, for females. Unhappily this usage is not confined to Scotland.

lady a prostitute.

As in the oldest of jokes:

> 'Who was that lady I saw you with last night?'
> 'That was no lady; that was my wife.'

In obsolete use a *ladies' college* was a brothel, where you might contract *ladies' fever*, syphilis. See also the phrases below.

lady bear see BEAR[2].

lady boarder *American* a prostitute who worked in a brothel.

> ...played for the lady-boarders and their friends. (Longstreet, 1956)

lady dog a bitch.

The very fastidious want to avoid any hint of confusing the inoffensive quadruped with the spiteful and domineering biped.

lady friend a female with whom a male regularly copulates extramaritally.

She does not have to be a woman of breeding or distinction, but the use implies slightly more acceptability than WOMAN FRIEND:

> It's my lady friend. I've reason to suspect that she's getting a bit on the side. (James, 1972)

lady in waiting[1] a concubine in the Japanese court.

> A dozen concubines, euphemistically termed ladies-in-waiting, nightly awaited the drop of the imperial handkerchief at their feet to follow him into his quarters. (Behr, 1989 of the Emperor Meiji, Hirohito's dissolute father)

lady in waiting[2] a pregnant woman.

Mainly humorous use, punning on the Court official. The obsolete English *lady in the straw* was used for a woman in process of being delivered of a baby.

lady-killer a male profligate.

Not Crippen or Christie. The standard English *ladies' man* implies no more than a certain foppishness and ability to charm women.

lady of a certain description a prostitute.

> There are two kinds of person who supply the police with all the information they want; one, that of unmarried ladies of a certain description... (James, 1816)

lady of easy virtue a prostitute.

With *easy* meaning 'compliant'.

lady of intrigue *obsolete* a woman who sought to conceal extramarital copulation.

> By ladies of intrigue we must understand married women who have connection with other men than their husbands and unmarried women who gratify their passions secretly. (Mayhew, 1862)

lady of no virtue a prostitute.

Although no less adept at the job than other prostitutes:

> So when he visited ladies of no virtue, it might be for purposes of fornication ... (Masters, 1976)

lady of pleasure a prostitute.

> Here was my Lord Brouncker's lady of pleasure. (Pepys, 1665)

lady of the night a prostitute.

But no more nocturnal than her colleagues:

> The lady of the night studied Abel carefully. (Archer, 1979)

lady of the stage a prostitute and actress.

> We call them ladies of the stage. They prefer that. Most of them have been in front of the footlights at one time or another. (Innes, 1991)

See also ACTRESS.

ladybird *obsolete* a prostitute.

Apart from the insect, it meant 'sweetheart':

> What, lamb! What, ladybird! God forbid. (Shakespeare, *Romeo and Juliet*)

laid see LAY.

laid out see LAY OUT.

laid to rest dead.

> She came to the end of the road only five months after we had laid Father to rest, so they were not parted long. (Tyrrell, 1973)

A monumental favourite. The obsolete *laid in the lockers* meant 'having died at sea', from the storage of the corpse for subsequent burial on land. But if you died at sea beyond the Thames estuary town of Gravesend, your corpse would be disposed of at sea.

lame duck[1] the holder of an office who has failed to secure re-election.

His successor will have been elected on a different platform and until the handover he lacks effective power. I suspect Peter Pindar was merely being rude when he described Pitt as 'A duck confounded lame Not unattended waddling' because that may well precede the 19th-century use, 'a stockjobber who speculates beyond his capital, and cannot pay his losses. Upon retiring from the Exchange he is said to "waddle out of the alley"' (*The Slang Dictionary*, 1874).

lame duck[2] a failing business.

Especially in a declining industry where overmanning, change of trade, or chronic lack of investment make unaided survival unlikely. The politically inexperienced minister John Davies used the phrase of British firms seeking state aid, and was thereafter accused of a lack of sympathy towards the manufacturing industry.

lamp habit see CHEF.

lance to copulate with.

Literally, to pierce, with the common male thrusting imagery:

> She would fall in a faint,
> And only revive when lanced freely.
> (*Playboy's Book of Limericks*)

land of Nod sleep.

A pun on Cain's travels when he 'dwelt in the land of Nod' (Genesis 4:16). Used in the past to refer to sleep generally:

> There's queer things chanced since ye hae been in the land of Nod (W. Scott, 1818)

but now only nursery use for coaxing children to bed in the frightening dark.

landscaped tidied up.

In standard usage, made to look scenically attractive. Real estate agents and builders use it of any housing development where most of the rubble has been removed from the site or covered with an inch or two of earth. In either case the topsoil will have disappeared.

language swear words.

A shortened form of *foul language*:

> I'll have no man usin' language i' my house. (D. Murray, 1886—he was not a Trappist abbot)

language arts *American* the ability to speak coherently.

Sociological and educational jargon. You must never imply that any child comes from an environment or has a defect which prevents it speaking normally.

lard the books dishonestly to increase a claim for repayment.

You enrich the mix by adding too much fat:

> The housekeeper at Twin Beeches regularly larded her books with non-existent bills. (Deighton, 1972)

large[1] pregnant.

Occasional female use:

> It was when I was large with our Lizbeth. (EDD)

large[2] small.

Well, smaller than *family* or *jumbo* in supermarketspeak:

> The smallest tube of toothpaste you can buy is the 'large size'. (Jennings, 1965)

See also ECONOMY.

larger obese.

In the jargon of the clothes trade, without stating the norm against which it is measured. It may also refer to females who are taller than most:

...a brand aimed at 'larger' women. (*Daily Telegraph*, 15 September 1994)

See also AMPLE, FULL-FIGURED, and PETITE.

lass an exclusive female sexual partner outside marriage.

A usage confined to the young but not to Scottish. See also LAD.

lassies see LADS.

last call death.

In various combinations, and sometimes referring to the dead person's job. Thus the *last curtain* falls on actors who may take their *last bow*, while cowboys head for the *last round-up*. To pay our *last debt* does not imply comment on the solvency of our estate, and the *last trump* is not for card players but for all those who hear the call to the seat of judgement. The *last end* and *last resting place* are specific, at least until the resurrection, as is the *last voyage* or *last journey*:

Just before the armistice George made his last journey to Banbury; a month later everyone in the village knew he was near the last journey of all. (Tyrrell, 1973)

Such a trip is not taken on the *last rattler*, which is not for the commuter who will take the train no more, but from the noise of congested breathing which may precede death. If you are *at your last*, death is near:

...which he sent me when he was at his last. (Mayhew, 1862)

last favour copulation by a female outside marriage.

Granted after prior familiarities. Also as *last intimacies*:

A man... has a secret horror of an innocent young woman allowing the last intimacies to a man whom she does not passionately love. (Pearsall, 1969, quoting Patmore, c. 1890)

For the eminent diarist, the *last thing*:

I had my full liberty of towsing her and doing what I would but the last thing. (Pepys, 1663 —to *towse* was to pull or shake about, whence *towser*, a dog used in bear-baiting and then any mastiff)

last shame (the) a term of imprisonment.

An obsolete British usage, from a time when more stigma attached to criminality.

last waltz *American* the walk to death by execution.

A waltz traditionally ends the ball.

latchkey (of a child) arriving home to an empty house because neither parent is then available, and specifically the mother is absent at work.

With implications of parental neglect:

'In a world of latchkey children,' he said, 'children whose only companion is the television set...' (M. Thomas, 1985)

late[1] newly dead.

Venerable enough to have been used by Caxton in 1490 but still often confused with unpunctuality. The obsolete *latter end* meant 'death'.

late[2] failing to menstruate when expected.

With fears of unwanted pregnancy:

He thought of her telling him she was late, had never been late before, and was he going to walk out on her. (Seymour, 1980)

late developer a poor scholar.

Used by parents who have hope rather than by teachers who have experience:

She was a late developer and a bit of a slow-coach. (Murdoch, 1977)

late disturbances a recent war.

Late means 'former':

The year of 1688 brought to England the worst turmoil since the 'late disturbances', as Mr Pepys had once described a brutal civil war and a royal beheading. (Monsarrat, 1978)

In the same way *late unpleasantness* was used by the southern states after the American Civil War and by the protagonists after the First World War. Another version after the Second World War is *late nastiness*:

...it was a great mercy we couldn't fight tanks in the dark in the late nastiness. (Price, 1987 —by *fight* he meant 'fight with' rather than 'against', night sights not having been invented)

latrine a lavatory.

Like *lavatory* itself, derived from the Latin *lavare*, to wash. Usually denoting primitive and communal structures, as in the army:

Latrines... often consisted of no more than a small mud hut with an open door. (Allen, 1975)

latter end[1] the buttocks.

Of the same tendency as BOTTOM; also in the form *latter part*.

latter end[2] see LATE[1].

laughing academy an institution for lunatics.

Not a school for comedians, but from inappropriate laughter as a symptom of insanity:

The way you're going in to bat to get the old man back in the laughing academy... (Wambaugh, 1975)

launder to bring (tainted cash) into open circulation in apparent legality.

You are *washing* (see WASH[3]) money which has been stolen, is the proceeds of vice, or represents undeclared income:

> ... accused of 'laundering' some of the marked banknotes used to pay the Schild ransom. (*Daily Telegraph*, July 1980)

Also when public funds are secretly diverted from the purpose for which they were voted:

> Cash from various Ministries is 'laundered' and diverted to the secret service. (*Daily Mirror*, February 1980)

A *laundry* is a bank or seemingly legal trading concern through which such money passes.

lavabo a lavatory.

'I will wash', from the Latin Vulgate version of Psalm 26:6 (*Lavabo inter innocentes manus meas*), and still used interchangeably with *lavatory*, but not very often:

> They follow me even to the lavabo. (Theroux, 1975)

lavatory a place set apart for urination and defecation.

Originally, a vessel for washing in, and then the place where you went to wash:

> Remember that our 'lavatory' is really a euphemism. (E. Waugh, 1956, but I use it *passim* to define others)

lavender related to male homosexuality.

Scent is made from the plant and the use of perfume by males used to indicate homosexuality. An American *lavender convention* is a meeting of male homosexuals, or *lavender boys*.

lay to copulate with.

The male usually *lays* the female, from his superior attitude perhaps, or from assisting her to a prone position:

> Laying me's part of your terms of service? (Bradbury, 1975)

Shakespeare used *lay down*:

> The sly whoresons
> Have got a speeding trick to lay down ladies. (*Henry VIII*)

A male may *lay a leg across*, *on*, or *over* a female:

> Whar was a' his noble equals when he bute to lay a leg un my poor lassie? (Graham, 1883)

Shakespeare also uses *lay it* for a woman's restoring a man's penis from an erect to a flaccid state by an act of copulation:

> To raise a spirit in his mistress' circle
> Of some strange nature, letting it there stand
> Till she had laid it and conjured it down. (*Romeo and Juliet*)

Laid, meaning 'copulated with', usually refers to the less experienced party, whether male or female; see GET LAID.

A *lay* is a woman who copulates extramaritally, usually with male laudatory adjectival embellishment—the *bad lays* are not talked about, if there are any. A *lay* is also an act of extramarital copulation:

> He smiled to himself, watching her thinking about the high cost of a free lay. (Weverka, 1973)

(The obsolete English to *lay a child* was not to be a paedophile but to attempt to cure it of rickets by taking it to a smithy where three smiths of the same name worked and there subjecting it to a number of experiences which are detailed in the EDD, none of which we would view with confidence today.)

lay down your life to be killed in wartime.

There are overtones of voluntary sacrifice:

> David Haden-Guest ... also laid down his life. (Boyle, 1979)

You also die if you *lay down your burden*, *knife and fork*, etc. The Scottish *lay down the clay* is obsolete, the *clay* being the human body:

> I'll soon lay down the clay, yet ere I go away
> I'd like to see the brig across to Torry. (Ogg, 1873)

lay hands on to beat.

Someone who expresses a wish to *lay hands on* you is seldom a faith-healer or a bishop seeking to achieve your confirmation. Occasionally it means 'to kill', especially in cases of suicide, with the variant *lay (violent) hands on yourself*.

lay in see LIE IN.

lay off to dismiss from employment.

Formerly for a short period only, until business picked up, but now of permanent dismissal:

> I didn't know my old man had been laid off. (Theroux, 1977, of lost employment)

lay out to prepare (a corpse) for burial.

You straighten the limbs before the onset of rigor mortis might make it hard to accommodate the body in a standard coffin. *Laid out*, meaning 'drunk', probably comes from the slang meaning 'knocked unconscious', although some drunkards look like cadavers.

lay paper to pass worthless cheques.

Perhaps from the old-style paper-chase. In America it also means 'put forged banknotes into circulation'.

lay pipes *American* to seek votes through bribery.

From the engagement of unemployed labourers on public works in return for their votes. But to *lay some pipe* is a male vulgarism for copulating.

lead associated with shooting.

From the composition of the bullet. The victim might have a *bellyful of lead*; be *filled with lead*; be loaded with *lead ballast*:

> You won't float long if I put lead ballast into you. (Fraser, 1970)

He may also *wear lead buttons*:

> Talk to me like that ... and you are liable to be wearing lead buttons on your vest (Chandler, 1943)

or have *leaden fever* or *lead poisoning*; or be given a *lead pill*:

> Hey, reb! Her's a lead pill for your sickness. (Cornwell, 1993)

lead apes in hell *obsolete* to die without having copulated.

> I must dance barefoot on her wedding day,
> And, for your love to her, lead apes in hell.
> (Shakespeare, *Taming of the Shrew*)

Of a woman, who has been denied the opportunity, or has refused, to accommodate sexually the figurative monkeys on earth, perhaps alluding to simian vigour.

lead in your pencil sexual potency.

Almost always an attribute of the male, with the *pencil* being the penis in nursery use (see PENCIL[1]), and likening the ejaculation to the core of graphite (not lead) which is used for writing:

> Wally shook some drops of Angostura into the gin. 'That'll put lead in your pencil,' I said. (Theroux, 1973)

leak[1] an act of urination.

Of obvious derivation. *Leaks* may be *had*, *done*, *needed*, *sprung*, *taken*, etc. by either sex, in mildly vulgar use:

> ... shuffling through the house in carpet slippers to take a leak. (Theroux, 1978)

To *leak* is to urinate:

> ... we were allowed out for twenty minutes drinking and leaking. (Lyall, 1972)

leak[2] to release (information) furtively.

Either of a politician who wishes to sound out public opinion about future policy; or, more commonly, the unauthorized disclosure of confidential information by an employee for financial or political purposes:

> Until fingered by his ex-wife in 1984, former Navy officer John Walker leaked secrets to the Soviets for nearly 20 years. (*Life*, Autumn 1989)

Also some intransitive use:

> Someone in that group was leaking. Newspapers were getting rumours. (Sanders, 1973)

A *leak* can either be such confidential information, or the source from which it escapes. *Leaky* describes a body prone to such disclosures. *Leakage* is now obsolete:

> We discussed leakages. Lady S. said that the surest way of making people repeat things was to say 'Don't quote me'. (Colville, 1967)

leaky[1] menstruating.

> As leaky as an unstanch'd wench.
> (Shakespeare, *Tempest*)

Of obvious derivation. Also used of a person prone to involuntary urination.

leaky[2] see LEAK[2].

lean on to put pressure on (a person) so as to extract a benefit.

The benefit may be silence from a witness, money from a victim, etc. and is secured by actual or threatened violence:

> I know his victims. I know who he leaned on. (Theroux, 1976)

leap on to copulate with.

A variation by males on the verb to *jump* (see JUMP[2]):

> You can't take a vow of celibacy ... You'll end up leaping on somebody and then feeling guilty. (Murdoch, 1985)

In obsolete use as *leap into*:

> I should quickly leap into a wife (Shakespeare, *Henry V*)

and *leap into bed with* is specific. Also of farm animals:

> His bulls leap at 5s a cow. (Marshall, 1811 —five shillings was the stud fee)

A *leaping house* or *leaping academy* was a brothel:

> Dials the signs of leaping houses. (Shakespeare, *1 Henry IV*)

> ... teaching 'em Latin in the environs of a leaping-academy. (Fraser, 1982, writing in 19th-century style)

leap in the dark (a) death by hanging.

From the sack or blindfold to cover the victim's eyes. Hobbes is reported to have so described his own imminent death.

leap the besom to cohabit and copulate without being married.

A variant of JUMP THE BESOM, with the same variants of *broom* and *broomstick*. It could imply a common law marriage where a priest might not be available:

> Leaping a broomstick was the deep country way of wedlock (Cornwell, 1993)

or simply that the couple were so cohabiting.

learn on the pillow to acquire proficiency in a foreign language from a sexual mistress who is a native speaker.

Those who use the expression want to draw attention to the extramarital copulation rather than to any linguistic achievement.

learning difficulties low intelligence.

Not just a lack of concentration:

> ... the mentally handicapped [shall be described] as 'people with learning difficulties'. (*Daily Telegraph*, 1 October 1990)

The term was defined in the British Warnock Report of 1978.

leather to beat.

Long standard English, from the use of a leather belt. In American use, the wearing of leather jackets became a symbol of homosexuality, and a *leather* or *leather-queen* is a male homosexual, often with a penchant for sadism.

leave to desert (a spouse).

When we use this word, we ignore the fact that married couples part company daily, to come together again in the evening:

> He shocked Victorian society even more by leaving her. (Howard, 1978)

leave before the gospel to withdraw from the vagina before ejaculation.

From attending church but forgoing the Mass. Especially Roman Catholic use and practice, mechanical and chemical forms of contraception being eschewed.

leave of absence suspension from employment during investigation of a supposed offence.

Literally, no more than a vacation, but used by an employer to avoid a defamation charge until the offence is proved:

> 'But not canned; just on leave of absence.' 'Without pay,' I said bitterly. (Sanders, 1986—an employee had been accused of theft)

leave shoes under the bed to copulate casually.

Not merely staying in a hotel on business:

> Haven't been leaving your shoes under a strange bed. (Sanders, 1979)

leave the building to die.

The *building* is the body to which the soul is attached while you are living:

> I could quietly die—or as Papa said, 'leave the building'. (Theroux, 1978)

If you affect clichés, you are more likely to *leave the land of the living*:

> Let us cut him off from the land of the living, that his name be no more remembered. (Jeremiah)

Those who *leave the minority* rely on the assumption that the majority are already dead. The American *left town* means 'dead'. Occasionally as *leave* (*tout court*):

> 'I think,' the maid replied, 'Mr Ford will be leaving us.' (Lacey, 1986—Henry Ford was dying)

leave the class to urinate.

Please may I leave the class? echoed through all our schooldays, unless it were *Please may I leave the room?*

leave your pillow unpressed *obsolete* not to copulate within marriage.

Of a male, from sleeping in other, presumably pillowless, beds:

> Have I my pillow left unprest in Rome,
> Forborne the getting of a lawful race?
> (Shakespeare, *Antony and Cleopatra*)

Lebensborn/Lebensraum see LIVING SPACE.

led astray having voluntarily done something of which you profess later regret or shame.

Men most use it as an excuse when they come home drunk, women for eating fattening food, and both for extramarital copulation:

> She had been led astray before I met her... and was a common prostitute. (Mayhew, 1862)

left supporting Marxist theory and practice.

Short for *left-wing*, from the seating plan of the French Estates General, where the Third Estate sat on the king's left. In the 1930s to be *left* was to support Stalinist Russia, as in the *Left Book Club*.

In obsolete British use a *left twin* was not the one with extreme political convictions, nor even the left-handed one, but the survivor and much in demand medically, having 'the power of curing the thrush' (Henderson, 1879).

left-field crazy or unconventional.

> ... a touch fundamentalist, but not too left-field to scare away sensible money. (Barnes, 1989)

The imagery is from baseball, referring to the area less favoured by right-handed hitters.

left-footer a male homosexual.

> I can pass meself off as a left-footer. (Fraser, 1983—he could ape homosexuality)

He may not dig or play football and, if he does, may favour the right foot. In the British navy and other closed societies, a *left-footer* might also be a Roman Catholic, again differing from the norm. See also LEFT-HANDED[2].

left-handed[1] bastard.

From the bar sinister on the coat of arms (a sign of bastardy). A *left-handed alliance* was cohabitation and copulation by an unmarried couple. If there had been a ceremony, the man would have taken the woman's left hand, instead of her right, she becoming a *left-handed wife*. See also DRESS ON THE LEFT.

left-handed[2] *American* homosexual.

From the perhaps sinister deviation from the normal.

left-wing see LEFT.

leg-over an act of copulation.

Usually outside marriage:

> He is on the terrace *tout nu*. She cannot resist him. *Voila*! It is a leg-over. (Mayle, 1993)

A *leg-over situation* is thought to describe an unmarried couple known to copulate regularly. See also GET A LEG OVER.

leg-sliding extramarital copulation.

By either sex, from a movement involved:

Everyone's allowed a bit of leg-sliding these days. (le Carré, 1980)

legal resident a spy accredited as a diplomat.

...he should never have been appointed to the vital position of legal resident in the USA. (Deighton, 1981, of a Russian spy with diplomatic status)

As distinct from an *illegal resident*, a spy who works in another country under cover.

legless drunk.

From your inability to walk steadily, or at all:

Bagley getting legless on Southern Comfort... (*Private Eye*, June 1981)

lend to give.

If you ask someone to *lend* you a match, you do not contemplate repaying him, and in any event 'He that lends, gives' (17th-century proverb, meaning that you won't be repaid, or see your adjustable spanner again). In 1941 the US Congress agreed to lease arms to Britain under the fiction that the *loan* would be repaid after the Second World War, a measure which was helpful to the penurious combatants, the American armament manufacturers, and the US economy. Truman wisely and abruptly ended *Lend-Lease* in 1945.

length a term of imprisonment.

A rare version of the common STRETCH.

lesbian a female homosexual.

It was commonly rumoured that Tanya was a Lesbian. (Bradbury, 1959)

Originally you referred to the poetess herself —see *Sappho* at SAPPHIC—rather than to her island home. Shortened to *les* or *lez*:

She-would-not-screw. I often wondered if she was a closet lez (Sanders, 1977)

and corrupted to *lizzie*:

To get into Mortimer's outfit you have to be a lizzie or a drunk or an Irishwoman. (Manning, 1978, writing of the British First Aid Nursing Yeomanry, shortened to FANY and the members being generally known to Second World War soldiers as *fannies*, to some more intimately than to others)

Lesbianism is female homosexuality:

I practised Lesbianism, which was certainly sterile. (Harris, 1925)

Lesbic is an outdated adjectival form:

...this perverse intertwining of two figures in lesbic passion. (ibid.)

less academic stupid or unteachable.

A group of expressions prefixed *less* glosses over the deficiencies of certain children, about which educationalists choose to speak allusively. Such a child may also be *less able, gifted, talented,* etc. This entry and those which follow are no more than samples of a common euphemistic use of *less* in phrases covering a range of taboo subjects.

less attractive repulsive.

Those who insist that Cinderella's slipper should be of fur, in a correct translation from the Dutch, are likely also to refer to her two *less attractive* sisters.

less developed poor.

Of a country which may combine a seat in the United Nations with poverty and maladministration. *Lesser developed* indicates no worse a state:

...loans to lesser developed countries such as Zaire and Jamaica. (M. Thomas, 1980)

less enjoyable boring.

Of books, plays, etc. especially when the speaker wishes to cover himself against passing a judgement which might indicate a lack of taste or discernment. Normally the full comparison is not made, so that you never learn with what it compares unfavourably.

less prepared of inferior attainment.

The usage seeks to gloss over deficiencies or difficulties arising from heredity, environment, education, etc.:

...the selection of the best-qualified black applicants...in preference to less gifted or less prepared blacks. (Pei, 1969)

let go to dismiss from employment.

The employer seems to imply that in dismissing a worker, he does him a favour:

It wore the sheriff down after a while and he let George go. (Chandler, 1943, of a dismissed deputy)

Very common.

let in to permit (a male) to copulate with you.

From the physical penetration rather than the prior entry into an apartment:

I still thought it good policy not to let him in yet a while. I answered then only to his importunities in sighs and moans. (Cleland, 1749)

let me I will persist without interruption.

In the forms *let me finish, let me say at once,* etc. when politicians are being interviewed and are determined to deliver a prepared statement and avoid answering embarrassing questions.

let off to fart.

A shortened form of *let off wind* rather than from the firing of a gun:

'He keeps letting off,' she repeated in a whisper ... 'I think it's because he's scared.' (L. Thomas, 1986, of farting)

The rarer *let fly* implies a more violent, noisier, release.

let out to dismiss from employment.

Jay Allen, the most brilliant among us younger men, would soon be let out. (Shirer, 1984, of a journalist about to be dismissed)

An American version of LET GO (above) but woe betide the employer who forcibly confines his employees after hours, or indeed at any time.

letter box a place where a spy may leave a message.

But not for the postman to collect. The jargon of spies or those who write spy stories:

The rulebook said Kleiber should be provided with a 'drop' and 'letter box'. (Deighton, 1981)

leverage borrowing.

American rather than British use but in both societies a *leveraged* bid or purchase is one in which a predator borrows excessively in the expectation of repaying his loan from the victim's assets:

Anyway, this investment banker specializes in 'leveraged buyouts'; it's the new thing in Wall Street fashion. (M. Thomas, 1987, who is worth reading for an informed account of how the combination of dishonest corporate officers and greedy bankers robs investors)

levy *?obsolete* male masturbation.

Rhyming slang for WANK on the name of a firm of London caterers, *Levy and Franks*.

liaison a relationship involving repeated clandestine extramarital copulation with the same person.

Originally, the culinary thickening of a sauce, whence a close relationship:

... striking up occasional liaisons which she alluded to by saying ... 'He's an old boyfriend of mine'. (Theroux, 1976)

liaison officer a controller of spies.

Espionage jargon. Semenov of the KGB was so described, his career spanning Montevideo in 1944, where he met Agee, and service in Cuba with the 'General Directorate of Intelligence'.

libation a drink of an intoxicant.

'... this may be a good time for a drink. Do you concur, Senator?' 'A small libation would not be inappropriate,' he said in a wry manner. (Sanders, 1984)

Literally, the ceremonial offering of a drink.

liberal intolerant.

Literally, free-thinking and objective, but appropriated by some so sure of their own rightness that they brook no contradiction:

He is, on the surface, the perfect 24-carat knee-jerk liberal sap. (*Daily Telegraph*, February 1980, of Edward Kennedy)

In pejorative use by extremists, *liberal* may mean supine or indecisive:

I'm beset by a good liberal ambivalence. (Bradbury, 1976)

In obsolete use a *liberal* woman was willing to copulate casually:

It's sign she hath been liberal and free. (Shakespeare, *1 Henry VI*)

liberate[1] to conquer.

Literally, to free:

Egypt would be liberated and Rommel and his men would keep their assignation with the ladies of Alexandria. (Manning, 1977)

liberate[2] to steal.

Originally a Second World War army use when freeing, occupying, and looting tended to go hand in hand:

It's a gold watch—a *liberated* gold watch. (Price, 1978, of the Second World War)

Now in general use:

'Are you going to be warm enough in that jacket?' 'I'm all right, I liberated it from a second-hand shop.' (Theroux, 1976)

liberate[3] to permit or encourage to flout social convention.

Again the concept of setting free:

... her immersion in the pop scene, and how it had liberated her. (Bradbury, 1976, of a woman so behaving)

Liberation is such flouting of convention or denial of manners, of minority dissident groups, etc. In prostitutes' jargon a woman working without a pimp is also said to be *liberated*, but that must be distinguished from WOMEN'S LIBERATION.

lick of the tarbrush see TARBRUSH (THE).

lick the dust to die.

Usually after being killed, from the attitude of a corpse in a dry terrain:

His enemies shall lick the dust. (Psalm 72:9)

See also BITE THE DUST.

lid *American* an ounce of marijuana.

It is the quantity which fits into the lid of a tobacco tin and makes about 40 cigarettes:

Tommy smoked a couple of lids a week. (Wambaugh, 1981)

lie in to await the imminent birth of a baby.

Greek, Latin, and Teutonic roots of *lie* all mean 'bed' where, in the language of euphemism, you only give birth or copulate:

Within ten days she'll be lying in. (Graves, 1940, of a pregnant woman)

Formerly to *lay in* was synonymous:

> When the gal is in the family way, the lads mostly sends them to the workhouse to lay in. (Mayhew, 1851)

A *lying-in* is a childbirth, formerly attended by the *lying-in wife*, the midwife:

> As well as can be expected. That's the answer of a lying-in wife. (Wilson, 1836)

lie with to copulate with.

It has long been assumed that the adult male and female cannot *lie* in each other's company without copulating, within or outside marriage:

> To tell thee plain, I aim to lie with thee. (Shakespeare, *3 Henry VI*)

Lie on might be more accurate but is less used:

> Lie with her! lie on her! (Shakespeare, *Othello*)

To *lie together* implies extramarital copulation:

> Foreign students were positively encouraged to lie together, he said sardonically, so that they didn't go out and pursue the natives. (Francis, 1978, of Moscow University)

The obsolete British *lie backwards and let out your forerooms* meant to be a prostitute, being a rather complex vulgar pun on the female posture in acts of copulation and on letting the more desirable rooms in a house while you remain in the back quarters.

life (the) any taboo way of earning your keep or existing.

Prostitutes' jargon for prostitution, thieves' for stealing, and also used by drug users for addiction to narcotics, especially when they alternate between scheming to get money to buy narcotics and periods under their influence.

life everlasting see EVERLASTING LIFE.

life insurance contracting for a sum to be paid on death.

Living is a more saleable commodity than dying. The *life* is the person whose death triggers the payment, *life cover* is such assurance, and a *life office* is a company which trades in such contracts.

life of infamy prostitution.

How the righteous profess to see it:

> ... she may have been a servant out of a place ... and betaken herself here to a life of infamy. (Mayhew, 1862—*here* was a brothel)

We still talk of a *life of shame* but the concept has tended to go out of fashion.

life preserver a cosh.

It is not intended to preserve the victim:

> Macarthur was hit with a life preserver... on the back of the head. (Christie, 1939—it killed him)

lifestyle sexual orientation.

> This lifestyle, choice—whatever it was called —remained beyond him. Not the acts, but the very philosophy. (Turow, 1990, of homosexuality)

A common usage in questioning potential blood donors about their sexual habits so as to screen out any likely to have HIV or AIDS.

lift[1] to steal.

Usually of pilfering, from the casual taking:

> Billy can lift your jock strap, and you wouldn't feel a thing. (Weverka, 1973)

Specifically too of plagiarism in 20th-century America, of picking pockets in 19th-century England, and of digging up corpses from graves in 18th-century Scotland:

> Resurrectionists... who were as ready to lay their murdering hands on the living, as to lift the dead. (Whitehead, 1876)

A *lifter* is a thief, usually by picking pockets. *Shoplifter*, a thief from a store, has been in use since the 17th century and the verb to *shoplift* since the 19th:

> I know it's bloody for them, but thousands of people shoplift. (Francis, 1981)

lift[2] an ingestion of illegal narcotics.

From the feeling sometimes induced:

> 'Want a lift?' 'I can use something,' Janette said. She took a small vial from the bag. (Robbins, 1981)

lift[3] to arrest.

Mainly police jargon. Also as a noun:

> The lift and then the interrogation, the interrogation and then the imprisonment. (Seymour, 1982)

lift[4] a thick sole and heel to enhance height.

Only remarkable and taboo when worn by a male:

> Beware Greeks wearing lifts. (*Financial Times*, September 1988, quoting a quip about the presidential candidate Dukakis, who was said to be so shod, the motto being after—long after—Virgil's *timeo Daneos et dona ferentes*, I fear the Greeks even when they are bearing gifts)

lift a leg[1] to copulate.

Of a male, from getting himself into a convenient attitude:

> I'll ne'er lift a lawless leg
> Again upon her.
> (Burns, 1785)

See also GET A LEG OVER.

lift a leg[2] to urinate.

In standard usage, of a dog, from the action, and occasionally of a male human:

> She opened the front door, and watched him go over to the hedge where he lifted a leg. (Ustinov, 1966, of a dog)

The obsolete English *lift a gam* was to fart; a *gam* was a leg, and also a school of whales, but their propensity for *blowing* does not seem to have contributed to the derivation.

lift the books *obsolete* to withdraw from regular service in a church.

In the days when church membership was a social necessity, apart from any moral benefits:

> He saved a public scandal by lifting his books ... resigning his membership. (Johnston, 1891, of a churchgoer)

lift your elbow to drink intoxicants.

On a single occasion, from the conveyance of the glass to your mouth, or more often of a drunkard. In the same sense you might also *lift your arm*, *little finger*, or *wrist*:

> Liquors a bit, don't you know; lifts his little finger. (Peacock, 1890)

lift your hand to to hit.

A man who has never *lifted his hand to* his wife has not been delinquent in waving greeting to her—indeed, the contrary is likely to be the case.

light[1] sexually promiscuous.

Of no moral weight and *light wenches* or *ladies* were not successful dieters or nurses emulating Florence Nightingale, but prostitutes:

> Light wenches will burn. Come not near her. (Shakespeare, *Comedy of Errors*—they were not condemned to the stake but would give you venereal disease)

> I wouldn't have thought that many of the light ladies of Calcutta had the opportunity to bestow their favours on the Japanese. (M. Fraser, 1992)

light[2] **(a)** (in a request to light a cigarette) a male homosexual recognition signal.

The universal password, usually in the question *Can you give me a light?*:

> ... it was not granted to me to live a moment of happiness, because a sailor's face in front of me went blank when I asked him for a light. (Genet, in 1969 translation, of a homosexual)

light[3] not obviously black.

The language of American segregation:

> I couldn't go in there with her. Even if I was light enough to pass, like her. (Macdonald, 1952)

light-fingered thieving.

From the propensity to lift small objects:

> ... Rose and Crown public house, resorted to by all classes of light-fingered gentry. (Mayhew, 1862)

An old superstition has passed into oblivion:

> The baby's nails must not be cut till he is a year old, for fear he should grow up a thief, or ... 'light-fingered'. (Henderson, 1879)

The obsolete Scottish *light-footed* or *light-heeled* meant 'promiscuous', of a woman; and the modern American *light-footed* means 'homosexual', of a male. In obsolete English a *light-skirts* was a prostitute.

light housekeeping *American* cohabitation and regular copulation between an unmarried couple.

I suppose a development of LIGHT[1] and a pun on the avocation of the coastguard.

light in the head of low intelligence.

Not a turnip on Hallowe'en:

> The kid's a little light in the head. His brother takes care of him. (Sanders, 1970)

light ladies see LIGHT[1].

light on his toes homosexual.

Of a male, from the mincing manner of walking adopted by some:

> Your assistant in the theatre, sir, your dresser, he's a bit light on his toes as well, isn't he? (Monkhouse, 1993—a policeman was enquiring about sexual activity which appeared to involve homosexuality, bestiality, pornography, and prostitution)

light the lamp to copulate.

Probably punning on the prostitute's trademark of a red lamp, and the sexual stimulation:

> She confided to me that she had lit the lamp four hundred and two times, in one week, in her Casita. (Londres, 1928, in translation, of a prostitute)

A man may use the phrase of casual copulation on a single occasion, usually with a woman who is not a prostitute.

lightning a low-quality spirituous intoxicant.

From the effect when it strikes you. Usually denoting whiskey in America and gin in England. The American *Jews' lightning* is arson, from the phenomenon of business premises being burnt down after being struck by lightning from a cloudless sky when trade is bad, but not so bad that you have neglected paying your fire insurance premiums.

like a drink see DRINK[1].

like that see THAT WAY[1].

lily *American* a male homosexual.

I suppose from the woman's name and the pale colouring, although the flower is an emblem of chastity and innocence.

limb[1] a leg.

A classic of 19th-century American prudery. Even your dining table had *limbs*.

limb[2] a policeman.

Shortened form of *limb of the law*, which used to mean 'a sheriff's officer':

> Be't priest, or laird, or limb o'law. (Nicholson, 1814)

There is no etymological connection with the obsolete *limbo*, a prison, from the place where unbaptized infants dwell along with those who predeceased Christ and various others:

> I have some of them in limbo patrum. (Shakespeare, *Henry VIII*)

limit (the) extramarital copulation.

Usually after a series of lesser sexual encounters with the same person.

limited stupid or incompetent.

Of children, educational jargon to avoid precision as to idleness or stupidity; of adults, lacking in ability or intelligence. One of the sillier euphemisms, as we are all confined within limits, of memory, knowledge, experience, common sense, and physical power.

limited action a war.

The stronger participant so describes it when he wants his adversary to get no help, and to keep his own domestic population inviolate and in ignorance:

> The black magic of violence; and even then the language of the mad foments it...
> 'Bushfire wars', 'limited actions'. (West, 1979)

A *limited covert war* is one financed by the US which Congress should not hear about:

> Para-military action of any type, Tyler argued, was war, and he had gingerly coined the euphemism 'limited covert warfare'. (Woodward, 1987)

limp *American* very drunk.

From the inert posture of the drunkard.

limpwrist a male homosexual.

> He looked like a peroxided limpwrist. (Wambaugh, 1983)

From the action of masturbating. Adjectivally as *limp-wristed*:

> His limp-wristed nancy-boy of a son. (*Private Eye*, January 1980)

line to copulate with.

> Winter garments must be lined,
> So must slender Rosalind.
> (Shakespeare, *As You Like It*)

Literally, of a wolf or a dog and obsolete of humans. *Lined*, pregnant, is also obsolete:

> ...she got lined by a big black buck. (Graves, 1941, writing in archaic style of a pregnant woman)

line your pocket wrongfully to enrich yourself.

The money provides the *lining*:

> ...adept in the field of corruption and lining his own pocket. (Goebbels, 1945, in translation)

In America you may, if occasion arises, *line your vest*:

> I think he'd been lining his vest. (Moss, 1987, not of a tailor but of an official suspected of peculation)

An older form was *line your coat*:

> And throwing but shows of service on their lords,
> Do well thrive by them, and when they have lin'd their coats,
> Do themselves homage.
> (Shakespeare, *Othello*)

link prices to arrange an illegal cartel.

Manufacturers either divide markets on a geographical basis or agree to quote the same prices in each market.

linked with copulating with extramaritally.

A favoured journalistic evasion:

> Since the breakdown of her marriage, the Duchess had been linked with a Texan oil executive...and...her financial adviser. (*Daily Telegraph*, 14 December 1994)

liquid consisting of, serving, or containing intoxicants.

Literally, anything from water to sulphuric acid. Usually in compounds; a *liquid refreshment* is an intoxicant. A *liquid restaurant* serves intoxicants as well as food:

> ...indebted...to the owner of a 'liquid' restaurant. (Lavine, 1930, writing during the Prohibition)

A *liquid lunch* is the drinking of excessive intoxicants at midday, with little or no food:

> Following our liquid lunch, he agreed to totter round the greens with me (*Private Eye*, August 1981—they went off to play golf)

and a *liquid supper* the same thing later in the day:

> Barley and his friends had enjoyed a liquid supper under plastic muskets. (le Carré, 1989—the muskets were the bogus decor of a London pub)

liquidate to kill other than by process of law.

Originally, to clear away, whence the implication of ruthless efficiency:

> The silent liquidation of many friends in the Soviet Union without a single bleat of protest from the freedom-loving West. (Boyle, 1979)

In legal jargon, a *liquidator* also kills off failed companies.

liquidity crisis an inability to pay your debts as they fall due.

Commercial jargon for an insolvency which has not yet been declared, and less often used of personal penury. *Liquid* funds are those at your immediate disposal, unlike *fixed* assets in the form of machinery, buildings, work-in-progress, and other investments often incapable of immediate realization.

liquor a spirituous intoxicant.

Originally, any liquid. In various spellings:

Lecker makes her drunk as David's sow. (*Gentleman's Magazine*, 1742, quoted in EDD)

Some said it was the likker. (Longstreet, 1956)

Liquored means 'partly drunk' and *full of liquor* or *in liquor*, 'drunk':

He was in liquor when he made his first appearance. (Monsarrat, 1978)

lit drunk or under the influence of narcotics.

From the generally exhilarated state rather than the redness of the nose:

An old con like me don't make good prints —not even when he's lit. (Chandler, 1939)

Also as *lit up*.

little bears *American* the local police.

Citizens' Band usage. Most American towns have a local police force who can be zealous about imposing on-the-spot fines for traffic violations, with the money going towards local taxes, if it gets that far. See also BEAR[2].

little bit *American* a prostitute.

Citizens' Band usage. The adjective does not define the lady's girth or height:

There's always a little bit at that truck 'em up stop. (CBSLD)

See also BIT.

little boys' room a lavatory for exclusive male use.

Fairly common adult male usage, despite its cloying imagery. *Little girls' room* for females is equally nauseous but happily less common:

She slid out of her chair, 'Just goin' to the little girls room, hon'. (Collins, 1981)

little folk see LITTLE PEOPLE.

little friend see LITTLE VISITOR.

little gentleman in black velvet a mole, cause of the death of King William III.

The king, hated by the Jacobites, was riding a horse which stumbled on a molehill. He fell off, broke his collar bone and died from complications which ensued. So the Jacobites toasted the mole:

The little gentleman in black velvet who did such service in 1702. (W. Scott, 1814)

They also toasted the KING OVER THE WATER and *limp*—Louis, James, Mary, Prince of Wales —which was hardly a rousing call to arms even without its modern overtones.

little girls' room see LITTLE BOYS' ROOM.

little house a lavatory.

It was often a small detached shed. In obsolete use a *petty house*.

little jobs see BIG JOBS.

little Mary the abdomen.

This 19th-century euphemism can still be heard, reminding us that the abdomen was the subject of taboo because it housed organs with sexual, child-bearing, and defecatory functions. As trousers and legs came under the same types of taboo, we may wonder how so reticent and sheltered a society managed so effectively to reproduce itself.

little people the fairies.

You had to speak allusively of malevolent people who could do you so much harm. Also as the *little folk* and see also WEE FOLK.

little sister see LITTLE VISITOR.

little something an intoxicant.

Denoting the substance on its own, or added to a non-alcoholic drink with which you have already been served.

little stranger an unborn child.

Nursery usage to avoid truthfulness about pregnancy.

little visitor menstruation.

In female use interchangeable with *visitor* (see VISITOR (A)), the affliction being equally severe in either case. Also as *little friend* or *sister*.

little woman a sexual mistress.

In ponderous male humour, literally a wife:

I think we can take it that there's a 'little woman' in the case. (James, 1962)

live as man and wife to cohabit and copulate without being married to each other.

From the supposed regular copulation of married couples:

Irene and I lived together as man and wife. (L. Armstrong, 1955, referring to his sexual mistress)

Indeed a married couple who do not copulate with each other may be said not to *live as man and wife* regardless of their sharing a house. To *live in sin* means the same thing, the *sin* being the mortal sin of adultery, and was until recently often used of a couple who so conducted themselves before marrying each other:

But the first year we lived in sin. (Sanders, 1973, of such a couple)

But then aren't you living in mortal sin? (N. Mitford, 1945)

The obsolete northern English *live tally* was more picturesque, a *tally* being a corresponding piece which exactly fits the other:

Aw'd advise thi t' live tally if theaw con mak it reet wi some owd damsel. (Brierley, 1854 —most men would prefer some young damsel, I suspect)

To *live together* always implies extramarital copulation, but not necessarily cohabitation:

If parties is married, they ought to bend to each other; and won't, for sartain, if they're only living together. (Mayhew, 1851)

Live with is perhaps the commonest use, of either sex:

You lived with women. You lived with that old actress. (Murdoch, 1978)

Live together and *live with* are also used of homosexual arrangements. To *live on* or *live off* another may mean, for a man, 'to be a pimp', and for a woman, 'a sexual mistress':

In this life I have known, loved, lived for, lived on, lived off... many men. (L. Thomas, 1977, of a prostitute)

For *living by trade* see TRADE (THE). A *live-in* girlfriend etc. is a sexual mistress with whom the male also cohabits:

...attending Hollywood high society affairs as his live-in girlfriend rather than as his wife. (*Daily Telegraph*, September 1981)

live source a spy.

One of yours rather than one of theirs, to be differentiated from other *sources* such as telephone tapping, radio interception, satellite photography, etc.:

A live source in sane English is a spy. (le Carré, 1989)

lived-in untidy.

Of another's house, sometimes with an implication of dirtiness. A *lived-in* face indicates premature and incipient decay through excesses.

livener an intoxicant taken early in the day by a drunkard.

Either by a person drunk on a single occasion, or by a habitual drunkard:

Your Lordship has heard of people having 'liveners' in the morning. (*Birmingham Daily Post*, 1897, quoted in EDD)

The sufferer is supposed to be stimulated by the fresh infusion.

living space the territory that Nazi Germany intended to take from Poland and Russia.

Nazi jargon for part of the policy of aggression 'to obtain by the German sword sod for the German plough'. The German word *Lebensraum* is now as notorious:

Lebensraum which should have meant 'living-room' but actually signified the occupation of Europe and as much of Russia as Hitler had been able to lay his hands on. (Sharpe, 1979)

Even more sinister was the lesser known Nazi policy of *Lebensborn*, under which fair Polish or Czech children were taken from home and placed with German families to be raised as Germans and thus augment the Teutonic stock.

lizzie see LESBIAN.

load[1] the quantity of intoxicants which has made someone drunk.

You usually *have* or *carry a load*, or *have* or *get a load on*:

Sure, I seen him drunk. Lots of times. He's have a load on. (Sanders, 1977)

Loaded is drunk:

I'm not loaded, as they haven't told me when the bars around here open up. (Ustinov, 1971)

Also of being under the influence of illicit narcotics.

load[2] the genitalia of a male.

American male homosexual use:

The long-haired youth entered, came close to Firenza's side, pressed his nylon-sheathed load against the doctor's arm. (Sanders, 1977)

loaded[1] see LOAD[1].

loaded[2] fraudulently increased.

The account is made heavier with fictitious or inflated entries.

loaded[3] laced with intoxicants.

Of any non-alcoholic drink:

We sipped our loaded coffee. (Chandler, 1939)

loan a gift.

See LEND. In obsolete use it meant 'a gift or grant from a superior' (OED) before it came to imply an obligation to return or make restitution. *Loan-soup*, fresh milk given to passers-by, came from the *loan*, a cowshed.

local *British* an inn.

Shortened form for *local pub* etc., being the nearby resort of the person, usually male, who habitually drinks intoxicants and uses the evasion. Standard English.

local bear a policeman attached to a small force.

As distinct from a state trooper (see also BEAR[2]). In American Citizens' Band also as *local boy*, *yokel*, etc.

lock out (of an employer) to refuse to make work available to (his employees).

Thieves and others may literally be *locked out* of the workplace each evening, or as the case may be. Industrial jargon even where some of the employees may accede to demands from the employer and continue to work while others oppose them:

We have been given two days... to carry on production or we should be locked out. (Allbeury, 1982)

loco mad.

From the Spanish; the *loco weed* is American slang for *marijuana*:

> The average square would say those animals were all loco. (L. Armstrong, 1955)

log-roller a person who gives selfish and insincere support.

The first *log-rollers* were the neighbours who helped each other manhandle heavy tree branches for winter domestic burning, from which came *log-rolling* with the meaning 'political support in return for another favour':

> The members [of Congress] ... make a compact by which each aids the other. This is called log-rolling. (Bryce, 1888)

The modern American use covers any insincere commendation, as in literature; and any reward for sycophancy:

> If either were appointed ... it would be a piece of disgraceful log-rolling. (Manning, 1965 —both candidates had flattered the patron)

loins the male genitalia and their reproductive role.

Literally, the region of your body between your ribs and your hips:

> A tongueless man may pass through his loins his unsung music. (Kersh, 1936)

Much biblical and some literary use.

lone parent a parent living alone with dependent offspring.

> The main reason given by divorced lone-parents for marital breakdown was infidelity. (Bath Report, June 1991)

Literally, *lone* means no more than 'unaccompanied' or 'solitary'. Less common than the SINGLE PARENT and see also ONE-PARENT FAMILY for a further dissertation.

long-arm inspection *American* a medical inspection of the penis.

DAS says the inspection is 'of the erect penis' and see also SHORT-ARM INSPECTION.

long home death.

Really the grave:

> Horn sent her off to her long hame to lie. (Burns, 1785)

Those who die also go on their *long journey*:

> I expect this is our last time around, Dick, but I hope to take a few of them on the long journey with us. (Richards, 1933, of going back into the First World War trenches after leave)

The *long day* is the Christian day of judgement, when a considerable catalogue of offences comes up for hearing; whence the admonition:

> Between you and the lang day be it. (Pegge, 1803)

A *long walk off a short pier* is murder by drowning:

> ... such topics as hanging, cyanide, and a long walk off a short pier. (Sanders, 1979)

long illness (a) cancer.

The language of the obituary notice, often adding with what fortitude the ordeal was borne. The American *a short illness* may indicate suicide.

long in the tooth old.

> ... he wanted to link up with some nice little bit less long in the tooth. (Christie, 1939)

From ageing a horse by the recession of its gums.

long pig human flesh.

From the similarity in taste:

> The Fijian's chief table luxury was human flesh, euphemistically called by him 'long pig'. (Theroux, 1992—the modern Pacific preference is for the proprietary sweetened processed pork, Spam, which has a similar taste)

long-term friend a permanent sexual partner.

Or as permanent as these relationships ever are. Either heterosexual:

> ... a house in Aylesbury where he lives with a long-term woman friend (*Daily Telegraph*, April 1990)

or, more often, homosexual and a pointer in the obituaries of those who have died from AIDS.

longer-living *American* geriatric.

From the moment we are born, we are longer-living than those younger than ourselves.

loo a lavatory.

> She sat in the loo on the pink tufted candlewick of the seat cover. (Bradbury, 1976)

Probably a corruption of *l'eau*, although this theory does not find favour with the OED.

look after your other interests to be peremptorily dismissed from employment.

A standard excuse, in various forms, when a senior employee leaves unexpectedly and in a hurry. If his other interests are *expanding*, his departure was the more precipitate. Also as *pursue other interests*:

> He suddenly needs more time to pursue that old favourite 'my other expanding interests'. (*Private Eye*, April 1987, of a dismissed chairman)

look at the garden to urinate out of doors.

Males say that they are going to do it, usually on a dark evening. They may also *look at the compost heap, crops, flowers, lawn, roses, vegetable garden,* etc.

look in a cup to foretell the future.

For some, the tea-leaves reveal all:

> I'm just broucht a si o'tea wi' me, an' I wis just wantin' you to luik in a cup fir me. (Stewart, 1892)

The practice is now rare, and not just because of the introduction of tea bags.

looking glass a pot for urine.

This Irish use is now obsolete, which is just as well. I draw your attention to the obvious joke involving the traveller and the waitress in EDD Vol. III p. 635.

lookism a preference for comeliness.

A usage of the POLITICALLY CORRECT as exemplified in a handout issued by Smith College, Massachusetts, in 1990. Their speech is as ugly as those whom they seek to protect:

> ... 'lookism', which oppresses the ugly by supposing 'a standard for beauty/attractiveness'. (*Daily Telegraph*, 23 February 1991)

loop[1] to kill by hanging.

From the noose:

> Like moussie thrappl't in a fa',
> Or loon that's loopit by the law.
> (Ainslie, 1892—the mouse was throttled and a loon is a person of low rank, whence, if a woman, a prostitute)

loop[2] a contraceptive used internally by a female.

From its shape. To be *on the loop* is to be using such a method.

looped *American* drunk.

From *looping the loop*, acting like a fool, or the inability to walk straight? We can only guess.

loopy see UP THE LOOP.

loose[1] willing to copulate extramaritally.

Used of women rather than men from the 16th century, from the relaxation of normal, tighter, standards:

> There were 8,600 prostitutes known to the police, but this was far from... the number of loose women in the metropolis. (Mayhew, 1862)

On the loose was living as a prostitute:

> When I lived with S. he allowed me £10 a week, but when I went on the loose I did not get so much. (ibid.)

The obsolete English *loose in the hilts* punned on a dagger unfit for use:

> A sister damned: she's loose i' the hilts;
> Grown a notorious strumpet. (Webster, 1623, quoted in ODEP)

A *loose house* was a brothel:

> You'd think she had started a loose house dead centre of the village. (Cookson, 1967)

A *loose fish* is a male profligate who has escaped the matrimonial net.

loose[2] suffering from diarrhoea.

Originally diarrhoea was the *loose disease*. Of both humans and animals. To *loosen the bowels* is to cause to defecate:

> It was fit to loosen the bowels of a bronze statue. (Fraser, 1975)

Looseness is mild diarrhoea.

loose cannon a person whose unpredictable conduct may cause difficulties or embarrassment.

Usually of an official, politician, or policeman exceeding authority:

> Mr Clinton's foreign policy team... view Mr Carter as a loose cannon. (*Daily Telegraph*, 19 September 1994)

From the gun on a naval vessel which, if not properly secured, would fire in another direction to the aimed broadside.

loose in the attic mad.

Attic is slang for 'head':

> He's a goddam loony bird. He's just uh... a little loose in the attic. (Diehl, 1978)

You may also be *loose in the head* or any of the other slang words for 'head'.

lord Harry see HARRY.

Lord has him (the) he is dead.

A Christian use, in expectation of joining Jesus in heaven. The Lord may also *send for you*:

> A woman like me doesn't part with pearls and diamonds until the good Lord sends for her. (Sharpe, 1977)

Lord of the flies the devil.

Beelzebub, *fly-lord* in Hebrew, was the Prince of the Flies in Syrian mythology.

lose[1] fraudulently to destroy.

Material so destroyed might include embarrassing files, documents, and tapes.

lose[2] to dismiss from employment.

The essence of loss is that it should be involuntary:

> That'll be fewer breakdowns, less overtime to make up for breakdowns, and of course, I'll be able to lose several men. (Lodge, 1988—an employer was explaining why automated machinery should be installed)

lose[3] to be bereaved of.

> Hendrix, like... John Lennon, lost his mother at an early age. (Murray, 1989)

In specific phrases it means 'to die'. The deceased might have *lost the vital signs* or *lost the wind*. If a British soldier, he might have *lost the number of the mess*, because he would eat there no more. See also LOSS[1].

lose your cherry to copulate for the first time.

Of a female, the *cherry* being the maidenhead:

In thirty years you can get born, grow up, go to college, get married, lose your cherry, have a couple of kids. (Diehl, 1978)

The obsolete Scottish *lose your snood* meant the same thing, the silken snood being worn as a symbol of virginity:

A' body kens it's lang syne you tynd your snood (Hamilton, 1897—'tyne' means 'lose')

and *snooded folk* were unmarried females. To *lose your innocence* is to copulate for the first time, of either sex:

In less than no time I had lost my senses and also my innocence. (Richards, 1936, of himself)

To *lose your virtue* is to copulate outside marriage, of a woman:

Every woman who yields to her passions and loses her virtue is a prostitute. (Mayhew, 1862)

To *lose your character* is either to be guilty of a crime (of either sex), or to copulate extramaritally (of a woman):

I might not lose, with my character, the prospect of getting a good husband. (Cleland, 1749)

To *lose your reputation* is to be known either as a prostitute or as a woman who has copulated extramaritally:

We cannot go there. The night watchman will see us. You will lose your reputation. (Bradbury, 1976)

In this context *lost*, of females, does not imply being unaware of their bearings, but engaged in prostitution:

They weren't by any means all lost women when they came. (Londres, 1928, in translation, of prostitutes in Argentina)

lose your grip to deteriorate mentally.

Not to fall from a bar or entrust your luggage to an airline. Less often as to *lose your hold*:

Was her father derailed, off his trolley, losing hold? (Turow, 1990)

lose your lunch to vomit.

Usually when drunk or on a boat. You may also *lose your breakfast, dinner, doughnuts*, or whatever else you may have eaten.

lose your shirt to be ruined.

Figuratively having nothing left to wear. An American may *lose* his *vest* from a predator who may also take his *pants* off him.

loss[1] a bereavement.

You do not necessarily miss the person who has died:

But she told her other gentlemen she could feel he had had a *loss*. (le Carré, 1980)

loss[2] see NIGHT LOSS.

loss of separation flying dangerously close.

Air traffic control jargon:

... the Tristar then flew within a few miles of an Aer Lingus Boeing 747 heading for Shannon and had a 'loss of separation' (flew closer than the legal safety limit) with two other planes. (*Daily Telegraph*, 21 August 1991).

lost[1] see LOSE YOUR CHERRY.

lost[2] killed.

Usually through violence:

My... my wife and son, sir... lost in the uprising... murdered. (Fraser, 1975)

Lost at sea means 'drowned'.

lot a battle in which there were many casualities.

A First World War usage which sought to play down the horror of the carnage:

I was in the last lot, sir. In Flanders. (Kyle, 1988)

lothario a male who constantly makes sexual proposals to women.

He pointed out the office lothario and the office seductress. (Sanders, 1981)

After the character in a play of 1630, *The Cruel Brother*, by Davenant.

lotion an intoxicating drink.

Originally, the action of washing, whence any liquid applied externally to the body:

I suggested to our noble friend that a lotion might not come amiss. (*Private Eye*, March 1980—he was offering an intoxicant)

love see ABODE OF LOVE, BUY LOVE, FREE LOVE, MAKE LOVE TO.

love affair a relationship which involves copulation outside marriage.

A debasement of the original meaning 'a courtship between unmarried persons'. Now used too of a single act of extramarital copulation:

Do you want me to drop in for a short love affair? (Murdoch, 1978)

As with MAKE LOVE TO, also used of homosexual activity.

love child a bastard.

This use predates the modern assumption that copulation and love go together and might even suggest that children born in wedlock are unwanted:

... little to dispute save the paternity of 'love children'. (Bartram, 1897)

In many similar forms, such as *love-bairn, love bird, love begotten, lover child*, etc.

love-juice the vaginal excretion of a sexually excited female.

I felt her warm love-juice gush. (Harris, 1925)

love nest the place where a sexual mistress is housed.

As a love-nest, the place had its points. (Chandler, 1943)

See LOVEMAKING and MAKE LOVE TO.

love that durst not speak its name (the) male homosexuality.

A 19th-century use, but still seen:

... stiff collar and tie, always formal, even when declaring the love that durst not speak its name. (Burgess, 1980)

loved one the corpse.

American funeral jargon, which makes an unwarranted assumption in many cases:

As for the Loved One, poor fellow, he wanders like a sad ghost through the funeral men's pronouncements. (J. Mitford, 1963)

Evelyn Waugh entitled his 1948 novel about the Californian funeral industry *The Loved One* but it is otherwise almost entirely free from euphemism, like most of his writing; the dedication turned out to be to the wrong Mitford sister, Nancy instead of Jessica.

lovemaking copulation.

Originally it implied only courtship:

Christopher, in lovemaking, as in most things, would pursue methods unknown to her. (Somerville and Ross, 1894—Christopher was someone who would not have heard of, let alone put into practice lessons from, the Kamasutra)

Nowadays it normally refers to copulation:

Rachman's love-making was clinical and joyless. (Green, 1979)

lover a woman's regular extramarital partner in copulation.

In a marriage, if the lover begins to be bored by the complaisant husband, he can always provoke a scandal. (G. Greene, 1978)

It is remarkable that the language has evolved no specific word for so common a need, and this use invites us almost to assume that husbands do not love their wives. *Lovers* means 'both partners in unmarried heterosexual copulation on a regular basis':

Soon, however, everybody knew that they were lovers. (Harris, 1925, of Parnell and Mrs O' Shea)

Today *lovers* can also be homosexual partners in a physical relationship:

'Are you and she lovers?' asked Treece. 'No; she's never done *anything* to me,' said Viola. (Bradbury, 1959)

low see HIGH[2].

low-budget cheap.

The intention is to banish the association between cheapness and nastiness. *Low-budget* is used mainly of films or television programmes, *low-cost* of purchases or services, *low-key* of publicity, *low-income* of living accommodation:

... based in a low-income unit made of cinder blocks and unpainted pine planks. (McInerney, 1992)

Low girls were prostitutes of the meaner kind:

The most of the low girls in this locality do not go out till late in the evening, and chiefly devote their attention to drunken men. (Mayhew, 1862)

low flying far exceeding a speed limit in a motor vehicle.

As distinct from FLYING LOW, which is having a trouser zip inadvertently undone. The two are not used interchangeably.

low profile the avoidance of publicity.

The imagery is from tank warfare, where you try to keep your hull down to reduce the target. A usage of politicians, corporations, etc. doing nothing when they should be acting, or hiding what they are doing.

lower abdomen the genitalia.

Of males, it is a useful evasion for sports commentators when a player has suffered a disabling blow. *Lower stomach* is used of both sexes:

... caressed the hair of her lower stomach affectionately. (Bradbury, 1976)

lower ground floor a cellar or basement.

Part of the armoury of deception in the world of real estate. On 25 January 1995, after talking about euphemisms to a lively audience in Daventry, I was directed by my hotel to take breakfast in the 'Lower Ground Floor', the former cellar.

lower the boom on to arrest.

The assumption has to be that the victim is already in harbour, which suggests inverted imagery:

We lowered the boom on Ross Minchen. He's behind bars right now, with his lawyer fighting to get him out. (Sanders, 1986)

Also commercially of putting a customer on the stop list for non-payment.

lubricate to bribe.

A rare variant of OIL. But when James Boswell *lubricated* a female, he referred to his seminal discharge.

lubricated drunk.

A variation of OILED. It can also mean 'being maliciously plied with intoxicants by others'.

Lucy in the sky with diamonds lysergic acid diethylamide.

A phrase formed on the initials *LSD*. Many adults who affected admiration for pop music, especially that of the Beatles, tended to miss the references to illicit narcotics and the implicit sanctioning of drug abuse. This was a title of a Lennon/McCartney song of 1967.

lulu a sum fraudulently claimed as expenses by an employee.

Probably punning on the American slang *lulu*, something exceptional, and a payment *in lieu*.

lumber *?obsolete* to copulate.

It has the air of British rhyming slang—say *lumber and lump*, hump or thump—but that is speculation:

> Zoë lumbers for a fiver. (Kersh, 1936)

lump a corpse.

American criminal jargon which is probably short for *lump of meat*:

> The lump is on the way down now. The big problem... is whether to do a cut 'em-up before lunch or after. (Sanders, 1973, of a medical postmortem examination)

lumpy *obsolete* pregnant.

You would think of obvious imagery, but there is an old dialect meaning, AWKWARD, which invites wider speculation. *Lumpy* also meant 'drunk'.

lunchbox the male genitalia.

Mainly American homosexual jargon of doubtful provenance.

lunchtime engineering bribery.

Excessive hospitality, where a vendor plies the customer's buyer with intoxicants or rich food.

lungs a woman's breasts.

Viewed sexually by a male:

> '... it's not a bad piece.' 'Good lungs,' Eddie admitted. (Sanders, 1982, and not of a singer)

lush intoxicants.

Probably from the meaning 'succulent':

> We gets in some lush, and 'as some frens, and goes in for a regular blow-hout. (Mayhew, 1862)

A *lush* is still a drunkard:

> He was a lush. He got the sack. (Theroux, 1973—he was dismissed for drunkenness and not given some dry white wine)

Lushy or *lushed* means 'drunk':

> ... on a bench by a railing of the boat, lushed to the gills. (L. Armstrong, 1955)

All these images were once recalled by referring to Alderman Lushington, who was a London brewer.

Lydford law arbitrary punishment.

This is a sample entry of the many local geographical euphemisms which have now largely been lost except to students of local history or, in the British Isles, to those who trawl through the EDD. In this example, special courts in the tin-mining districts of Devon and Cornwall known as the Stannary, held jurisdiction under which a judge at the small Devon border town of *Lydford* caused a tin-miner to be hanged in the morning and then sat in judgement on him during the same afternoon.

lying-in see LIE IN.

M anything taboo beginning with the letter *M*.

Especially the initial letter of *marijuana* in addict use.

machine the erect penis.

An old but rare variant of ENGINE OR TOOL:

> ...that machine whose touch has something so exquisitely singular in it, to make its way into me. (Cleland, 1749)

mackerel a pimp.

The French *maquereau* is more widely used. Shortened in America to *mac, mack*, or *macko*.

madam the female keeper of a brothel.

> 'What can I do for you Madam?' 'Miss,' she said. 'In my country a lady doesn't like being mistaken for a madam.' (Deighton, 1978)

Rather a come-down for 'my lady' but the honorary title of any bawd from Shakespeare's *Madam Mitigation* (*Measure for Measure*) to *Madam Mitchell*, who kept a noted Second World War brothel in Madras. For the French *madame* was also the guillotine.

made at one heat stolen.

This is a sample entry from the days when the smithy was more essential to society than the garage today. In forge-work, you only make an article by reheating it and reworking it many times. If there were only one operation, it had to be thievery. Somerset.

Magdalene a prostitute.

Christ's disciple, Mary, was supposed to have been a prostitute who repented and she gave her name, mainly in the 19th century, to others who had yet to renounce their way of life:

> After that our Magdalenes were left alone. (Fraser, 1982, writing in 19th-century style of prostitutes)

magic word (the) please.

Not *abracadabra* and magic at least for those children whose parents, or grandparents, try to teach them manners.

mail a letter to urinate.

A common and transparent American male excuse for leaving others. But see also POST A LETTER.

mail cover the interception and unauthorized reading of letters.

It should mean no more than an envelope. The term was used by the US Post Department for its clandestine post-Second World War perusal of letters from American citizens to Communist countries. (Jennings, 1965)

main thing (the) an act of extramarital copulation.

It appears to have been that for Pepys in his encounters with females:

> ...there finding Mrs Lane, took her over to Lambeth where we were lately, and there did what I would with her but only the main thing, which she would not consent to. (Pepys, 1663, with *but* meaning 'except')

mainline illegally to inject a narcotic intravenously.

> He made himself a fix... and he mainlined it. (Pereira, 1972)

The *main line* is the blood vessel in the arm, from railroad imagery, and the effect is almost immediate:

> A high-wire performer who hit the main line in his own office. (Chandler, 1953)

make[1] to copulate with extramaritally.

Normally the male *makes* the female:

> The team made eight hits
> And a girl in the bleachers called Alice.
> (*Playboy's Book of Limericks*)

To *make a play for* is to seek to copulate with extramaritally, usually of a male:

> 'Don't make a play for me, Peter.' 'I wasn't planning to.' (Sanders, 1983)

However either sex can *make it with* the other:

> Georges Simenon, who says he made it with ten thousand different women. (Hailey, 1979)
>
> This old meat made it with Bernard Shaw. (Bradbury, 1976)

Make out is American:

> I know you were making out with that German maid. (Mailer, 1965)

Also as *make whoopee*:

> I heard two people in the next room making whoopee—the old man's archaic term for fornication. (Styron, 1976)

A *make* can be a single act of copulation and the woman described as an *easy make*. A place where individuals habitually go looking for this kind of relationship is a *make-out joint*:

> ...the bar was the best make-out joint in Fort Lauderdale. (Sanders, 1982)

(Etymologists will observe that *make* and DO[1] carry their subtle distinctions into euphemism also.)

make[2] to steal.

Army usage whence, in criminal jargon, any fraudulent act. A *make* is a robbery:

> 'It's not a make,' I said. 'You're in trouble.' (Chandler, 1939—the intruder was not just a thief)

make a call to urinate.

Perhaps punning on a visit to the lavatory and the CALL OF NATURE:

'I just want to make a call,' said Willoughby, and he disappeared into the toilet. (Bradbury, 1959)

Common use by both sexes, but rarely of defecation.

make a decent woman of to marry (a woman whom you have impregnated).

You ought to hear Hope when she gets scared he'll never come back and make a decent woman of her. (Stegner, 1940—the putative father was at sea in the navy)

A less common version of MAKE AN HONEST WOMAN OF.

make a hole in the water *?obsolete* to kill yourself by drowning.

I suppose you have to jump from a height:

Why I don't go and make a hole in the water I don't know. (C. Dickens, 1853)

make a mess to urinate or defecate involuntarily.

Nursery, sickroom, and geriatric use of humans and generally of domestic animals, especially indoors:

If he makes another mess... I'll have him destroyed. (N. Mitford, 1945, of a dog)

make a suggestion to propose casual copulation.

It is what men *suggest* to women generally, or prostitutes to men:

... if anybody had made a suggestion to her then, she would have slapped his face... But look at her: she'd sleep with any Tom, Dick or Harry for two or three pounds. (Kersh, 1936)

To *make an improper suggestion* is more explicit and also used of homosexual approaches.

make an honest woman of to marry (a woman you have impregnated).

HONEST used to mean 'not ready to copulate outside marriage' and this phrase was once used seriously:

It was your son made her sae, and he can make her an honest woman again. (W. Scott, 1822, of a pregnant woman)

Now only used humorously:

But if you're really so old-fashioned... it's called 'making an honest woman of me'. (Price, 1970, of a man proposing to marry his sexual mistress)

make away with[1] to kill.

The victims are usually domestic animals, if unwanted or ill. Of humans usually reflexive and referring to suicide:

Ready to make away with themselves. (Burton, 1621)

To *make an end of* is explicit.

make away with[2] to steal.

From the act of physical removal.

make babies together to copulate.

Usually within marriage and not really anticipating a multiple birth. It is perhaps my least favourite euphemism. To *make a child*, slightly less cloying, means 'to become a parent':

Aren't you ever sad... that we haven't made a child? (G. Greene, 1932)

make little of to copulate with outside marriage.

Usually with the woman as the object, after she has been made large by impregnation:

You let *David Power*, the doctor's son, make little of you, and get you into trouble? (Binchy, 1985)

Also perhaps of either party:

So don't talk about people *making little* of other people, or of him *disgracing* me. I was just as eager, all the time, as he was. (ibid.)

make love to to copulate with.

Formerly, it meant no more than 'to court':

... generally they had made love to her, and, if they did not, she presumed they did not care about her, and gave them no further attention. (Somerville and Ross, 1894, of a flirt)

Today to *make love to* is to engage in heterosexual copulation:

He should make love to her, or, in the parlance, screw her (Masters, 1976)

or homosexual activity:

The allegation that he had ever made love to Maclean. (Boyle, 1979, of his fellow-traitor Burgess)

To *make love to yourself* is to masturbate:

She sometimes made love to herself on the bath mat. (McCarthy, 1963)

make off with to steal.

A commoner variant of MAKE AWAY WITH[2]. It is never your own property that you take with you. In obsolete English use it also meant 'to kill'.

make old bones to live long.

Euphemistic in the negative in which it is normally used:

I feel I shall never make old bones. (N. Mitford, 1945)

make out see MAKE[1].

make room for tea to urinate.

A jocular and almost genteel usage, although based on somewhat flawed physiology. You may also claim to be making room for another beer etc.:

'Knock that back and have another.' 'I'll make room for it first if you don't mind.' (Amis, 1986)

make sheep's eyes at to show sexual interest in (another).

From the unintelligent staring of the wide-eyed beast. In former times you might *cast sheep's eyes at*:

> I swear I have often seen him cast a sheep's eye out of a calf's head at you. (Swift, 1738 —*calf* too implies youthful longing, as in *calf love*)

make the beast with two backs see BEAST WITH TWO BACKS.

make the (bed) springs creak to copulate.

The usual *bed* imagery, within or outside marriage:

> We'd been married a long time and made the springs creak times without number. (Fraser, 1971)

The springs may also *squeak* under the same provocation:

> 'It would improve everyone present if the bedsprings squeaked a bit more often.' 'Let's leave sex until after tea,' said Treece. (Bradbury, 1976)

make the chick scene to copulate with a woman.

Of an American male homosexual, for whom heterosexuality is taboo:

> ... that roaring faggot ... He makes the chick scene from time to time. (Mailer, 1965)

make the supreme sacrifice to be killed.

On war service but not necessarily in action:

> Fellow members who had made the supreme sacrifice ... (Boyle, 1979, of war dead)

make time with *American* to copulate with extramaritally.

The male is usually the maker:

> It doesn't help when they go into the bar and find a couple of guys trying to make time with them. (Sanders, 1983, of a club for women)

make up to to attempt to court.

Of either sex to the other:

> ... me mother would have a fit if she thought I was making up to you. (Cookson, 1967)

make water to urinate.

> Heave up my leg, and make water against a lady's farthingale? (Shakespeare, *Two Gentlemen of Verona*)

Discharge it would be slightly more accurate and see also WATER.

make yourself available to to be ready to copulate with outside marriage.

The woman indicates the *availability*, usually without payment:

> He ... would have toyed with her and cast her aside ... if she had been callow enough to make herself immediately available to him. (W. Smith, 1979)

However the politician who *makes himself available* for nomination or reselection thinks to gain votes from a show of reluctance.

maladjusted naughty or stupid.

In standard usage this describes a child who is emotionally unbalanced or suffers from bad home influences. But mainly educational or 'social worker' jargon in a world where there are no bad or simple-minded children, only a sick society. Whence *maladjustment*, mental illness, in children or adults:

> I was good at diverting myself, and others, from the deeper causes of my 'maladjustment'. (Irvine, 1986—she was in an institution for the insane)

malady of France *obsolete* syphilis.

The FRENCH ACHE:

> My Moll is dead i' th' spital
> Of malady of France.
> (Shakespeare, *Henry V*)

male[1] a lavatory for the exclusive use of males.

In factories or office blocks, with its *female* counterpart.

male[2] homosexual.

As in *male* movies for those who are *male* identified or oriented. *Female* can be, but isn't usually, a similar signal for women homosexuals.

male beast *American* a bull.

Also as *male brute* or MALE COW, in 19th-century prudery. For a list of similar euphemisms see BIG ANIMAL. *Male hog*, a domestic pig which had not been gelded, would be tautological were it not that most hogs raised for a slaughter are less fortunate.

male cow *American* a bull.

Another example of 19th-century prudishness in this regard:

> Even bulls became male cows. (Bryson, 1994)

male parts the male genitalia.

But not the manly breast or other physical indications of masculinity:

> His hair and beard hung in untidy yellowish ropes over his bronzed body, almost as far as his male parts. (Farrell, 1973)

See also PRIVATE PARTS.

malt relating to beer-drinking, whence to drunkenness.

Not from drinking malt whisky but because malt is a constituent of beer. Thus to be *malty* was to be drunk and a *malt-worm*, a habitual drunkard, might be described as being *troubled with a malt sucker inside*. The obsolete Scots *malt above the water* or *malt above the meal* meant 'drunkenness':

> When he was riding dovering hame (wi' the malt rather abune the meal). (W. Scott, 1814 —*dovering* meant 'drowsy')

Malt today is used mainly as a shortened form of *malt whisky*.

Malthusianism *obsolete* the prevention of conception during copulation.

Malthus was a curate in Surrey who published anonymously in 1798 his *Essay on the Principal of Population as it affects the future Improvement of Society*. This enunciated the theory that population grows in a geometrical, and resources in an arithmetical, ratio—population must therefore be controlled by wars, disease, and famine:

> He has been described as a timid bird in the sociological aviary and was an inappropriate person to have a euphemism for birth control —Malthusianism or neo-Malthusianism —fathered on him. (Pearsall, 1969)

mama bear *American* a policewoman.

A variant of the GOLDILOCKS joke. Also as *mama smokey*.

man[1] a male with whom a female regularly copulates extramaritally.

Occasionally in this sense *tout court*:

> He is not my man, he is my husband. (*Evesham Journal*, 1899, quoted in EDD)

Man friend is more explicit and see also FRIEND.

man[2] a policeman or warder.

The American Citizens' Band *man with a gun* is a patrolman with radar. *The Man* is an American prison governor:

> If he went to The Man to complain about it, you got him alone someplace, more places to ambush a man in prison... (McBain, 1981, of buggery in jail)

man[3] a personal valet.

Rather more than merely a shortened form of *man-servant*. The black *man* as a method of male address comes from the need to assert the dignity of the individual which had been insulted by the servile *boy* used by generations of whites.

man cow *American* a bull.

For a list of this particular American form of prudery, see BIG ANIMAL.

man friend see MAN[1].

man-root see ROOT[1].

managerial privileges promiscuous copulation with a female entertainer.

Producers and directors are supposed to be the particular objects of favour:

> Tammy gave what we call 'managerial privileges' to agents, impresarios and the rest of the gang. (Allbeury, 1980)

Also as *management privileges*:

> On the bed upstairs, Julie had let him enjoy what are known in show business as 'management privileges'. (Allbeury, 1981)

manhood the male genitalia.

> ... tying a handkerchief round the remains of his once-proud manhood. (Sharpe, 1979—he had snagged his penis on a rosebush)

Literally, the state of being an adult male. Much figurative use for male copulation:

> I was oblig'd to endure one more trial of his manhood. (Cleland, 1749, of a prostitute)

To *eliminate manhood* is to castrate:

> I know what you mean about eliminating manhood—even in animals. (Hailey, 1979)

manual exercise masturbation.

Usually that of a male by himself and punning on the worthy *manual work*. Also formerly as *manual pollution*:

> TOSS OFF. Manual pollution. (Grose)

manure see HORSE APPLES.

maracas *American* the breasts of an adult female.

From the Mexican rattle, which is shaped like a gourd.

marble orchard *American* a cemetery.

From the tombstones.

marbles the testicles.

From the glass spheres or alleys, which used to be made of marble. Mainly used figuratively of a mental state. To *have (all) your marbles* is to be sensible or alert:

> Burgoyne has got all his marbles. (C. Forbes, 1992—he was not about to embark upon a game)

and to *lose your marbles* is to be mentally unstable:

> ... now openly saying that Sir Ian has lost his marbles. (*Private Eye*, August 1980)

marge a female homosexual taking the female role.

In 19th-century England it meant 'an effeminate youth', either from an abbreviation of the name *Margery*, or from the meaning 'edge', or punning on both. Still in American use.

marginalized being different from the majority.

Living on the edges of society and resentful of the mores accepted by or the conditions enjoyed by the majority:

> Most are sincere in their concern for society's 'marginalised', as they call them. (*Daily Telegraph*, 20 April 1992, of clerics)
>
> ... the political drive for 'empowerment' of 'marginalised' groups such as blacks, women and gays. (Mary Kenny, *Sunday Telegraph*, 30 January 1994)

Maria Monk the male semen.

Rhyming slang for SPUNK. From *The Awful Disclosures of Maria Monk*, a scurrilous anti-

Roman Catholic pornographic book published in the 19th century and still in print, rather than from the poetess who died in 1715. (My thanks to Mr Sean Barker for this correction.)

marine residence a dwelling near the sea.
Real estate jargon, but not a houseboat.

marital aid an instrument to use in seeking sexual pleasure.
In fact, the object is unlikely to feature in any exchange within wedlock, often merely adding zest to masturbation or other solitary activity:

> ... in their bedroom drawers I would find what the dirty shops called 'marital aids'. (Theroux, 1983)

marital rights copulation by a man with his wife.
They were demanded from a reluctant wife in an age when both lay and ecclesiastical law held that it was a woman's duty to copulate with her husband on request, even at the cost of debilitating, dangerous, and unwanted pregnancies. Today used only by husbands with willing wives and a dated sense of humour. To *exercise your marital rights* with other than your wife is to copulate extramaritally.

mark[1] a swindler's victim.
Marked or 'watched', for his suitability, whence the obsolete punning *walk penniless in Mark Lane*, to have been swindled but not necessarily in that London street.
The obsolete Scottish *mark* was an invulnerable spot on the body of a wizard or witch, which played an important role in detecting them as such:

> ... through which mark, when a large brass pin was thrust till it was bowed, both men and women, neither felt a pain, nor did it bleed. (Ritchie, 1883, describing how you might unmask a witch or wizard)

mark[2] to injure in custody with bruising.
American police jargon, and euphemistic only in the negative:

> You told me not to mark him. (Macdonald, 1952—a jailer was speaking of a prisoner who had been assaulted)

mark someone's card to single someone out for reprimand or punishment.

> She marked his card on the spot, and he's going to take early retirement, having 'had it conveyed to him' that he will never make Permanent Secretary. (A. Clark, 1993—a senior British bureaucrat had failed to show respect to the Prime Minister and realized that his career prospects were thereby diminished)

From the practice in horse-racing, 'to tip someone off', whence 'to put someone right'; or possibly from scoring at golf, from the record which deters the player from relying on his powers of recall; or from *marking* a playing card. Early usage pre-dates the disciplinary cards shown by referees, but contemporary usage may also allude to these.

marriage joys copulation.
But what of shared children, companionship, warmed slippers, or a cooked meal? Shakespeare meant none of these in:

> The sweet silent hours of marriage joys. (*Richard III*)

Now too denoting extramarital copulation.

marshmallows *American* the testicles.
From the confection of a similar size and shape. Also referring to the breasts of an adult female, from the softness, you might think, rather than the size.

martyr to (a) suffering from.
The death or persecution is only figurative. The *Daily Telegraph* of 7 September 1978 hesitated to call the British Prime Minister a liar, a 'martyr to selective amnesia' being a more telling and memorable indictment.

Mary[1] a homosexual playing the female role.
Either male or female, from the name.

Mary[2] marijuana.
An abbreviation much used in pop songs for oblique reference to narcotics. Because some English speakers pronounce the *J* in *marijuana*, sometimes as *Mary Jane* or *MJ*. *Mary Anne* is rare.

Mary Jane the vagina.
A rare 19th-century female example of the common male practice of calling their genitals by given names. Also as *Lady Jane*.

Mary Palm male masturbation.
Perhaps she was the original FIVE-FINGERED WIDOW:

> KGB men never go out with girls, they just live with Mary Palm. (de Mille, 1988)

massage[1] to bribe.
Literally, to apply friction to muscles. It refers to the use of excessive hospitality etc. to soften up industrial buyers. Also used as a noun denoting such conduct.

massage[2] to assault violently.
American police jargon for the use of force to obtain information:

> 'Shellacking', 'massaging' ... and numerous other phrases are employed by the police ... as euphemisms to express how they compel reluctant prisoners to refresh their memories. (Lavine, 1930)

massage[3] masturbation.
Often as the provision of EXTRAS:

> 'You want a massage?' she says. I says forget it. They don't mean massage. (Theroux, 1975)

Sometimes also denoting copulation; see also MASSAGE PARLOUR.

massage[4] to overstate or wrongly increase (profits or value).

Brokers' jargon for a rise in prices that is brought about by stimulating interest in a security through unwarranted puffs short of overt fraud. Also of manipulating or presenting accounts so as to give an unduly favourable result:

> The massaging of profits came at a 'vital time' for the company, which was floated publicly by Walker in 1985. (*Daily Telegraph*, 3 June 1994, of a quoted company where some officers were later accused of falsely inflating profits)

massage parlour a brothel.

The friction applied is not to tone up the muscles:

> Whether we worked in a Massage Parlour or were rich ... we were still the same to you. Easy women. (Bogarde, 1978)

masses (the) those ruled by Communist autocrats.

Marxist jargon for the great body of humanity whose communal will is revealed to those who have won power over them:

> 'Look at Lenin,' she said. 'Did he think about himself?' 'He thought about the masses.' (McCarthy, 1963)

masseuse a prostitute.

Usually working in a MASSAGE PARLOUR:

> Accompanied by my personal assistant-cum-masseuse Miss Rita Chevrolet. (*Private Eye*, February 1980)

masterpiece a competent piece of artistic work.

Originally, the article submitted by the apprentice to show that he has acquired the skill of the master. An advertising cliché, used especially for quite banal work, the author of which may previously have turned out something of merit.

masters (our) the politicians in government.

A sarcastic and resentful use by senior civil servants who consider themselves better fitted to govern by virtue of their intellect and experience than the elected amateurs who head their ministries:

> ... the very archetype of everything our masters have told us to avoid. (le Carré, 1980, of a civil service gaffe)

mate to copulate.

Of animals this is a standard, non-euphemistic, use:

> Mating pythons are a very rare and a very strange sight. (Richards, 1936)

In humans *mating* means 'marriage' but is used of copulation within or outside marriage. The man *mates with* the woman:

> He'll never be able to mate with a woman again (West, 1979—but with what, if not with a woman?)

and a *mating* is an act of copulation:

> ... half a dozen mamas enjoyed unexpectedly vigorous matings later that evening. (Erdman, 1974)

material part the penis.

Perhaps as much circumlocution as euphemism and see also PRIVATE PARTS:

> Mrs Booth ... danced 'so deliciously' on stage ... that she threatened to throw young Thompson's 'material part' into an erection. (R. Holmes, 1993, quoting Thompson, 1725)

matron an old woman.

In standard usage, any married woman, with a presumption of sobriety and godliness.

mattress (in compounds and phrases) relating to copulation.

The common association of beds and copulating in a variety of compounds such as *mattress drill* or *beating the mattress*, copulation, and rare uses such as *mattress extortion*, sexual blackmail by a female:

> So you con him into moving to sunny Florida. Maybe a little mattress extortion there. (Sanders, 1982)

mature[1] old.

> ... the high payers at the front wind up with some of the more mature girls. (Moynahan, 1983—older stewardesses work in the 1st class section in aircraft)

Literally, fully developed. In educational jargon, a *mature student* is someone who returns to class after years outside the educational system. See also OF MATURE YEARS.

mature[2] fat.

The language of those who seek to sell clothes to obese women. The *maturer* figure is no more mature than the *mature*.

Maud a male homosexual.

Usually paid for his services. From the girl's name and not the Scottish striped shepherd's plaid.

maul to caress (a reluctant female).

Literally, to handle roughly, but to an unwilling partner, any fondling is excessive:

> Because you give me the occasional meal ... doesn't mean you have the right to maul me. (Archer, 1979)

Female use.

mausoleum crypt a drawer for a corpse, facing on to a corridor.

Funeral jargon for the slots facing into the building, which are harder to sell, due to the absence of a view:

> The crypts facing the corridor are called 'Mausoleum crypts'. (J. Mitford, 1963)

A far cry from the mausoleum of Mausolus's widow who, with the help of a few thousand slaves, ran up the famous sepulchre at Halicarnassus around 353 BC to keep his memory alive.

mayday an international distress call.

A corruption of the French *m'aider*. Euphemistic in the negative, where *no mayday* in American hospital jargon means that the patient should be allowed to die.

me-too (of goods and services) slavishly copying.

Commercial use where a product is launched similar to that of a competitor to attempt to exploit a market he has developed:

> Everybody knows there are 'me-too' drugs... But they sometimes lead to new discoveries. (Hailey, 1984)

meaningful (of a relationship) not sexually promiscuous.

In the sense 'being significant', because everything that occurs has a meaning. Of adult heterosexual behaviour which is considered to be other than transitory or profligate, and a *meaningful relationship* between two adults is expected to have a degree of permanence.

measure for the drop to dismiss from employment.

From the duties of the hangman, who knew that weight was more important than height if the job were to be performed properly, and punning on DROP[6]:

> Time to move you on... Time to measure you for the drop. (le Carré, 1989)

meat a person or the genitalia of a person viewed sexually.

If a female by a male, usually all of her:

> Away, you mouldy rogue, away. I am meat for your master. (Shakespeare, *2 Henry IV*)

If a male by a female, his erect penis:

> A lot of them look like they need... a hot meat injection. (Styron, 1976, of young women)

If denoting a male homosexual, the partner playing the female role in buggery:

> Together, he and Jimmy had shared some of the choicest meat inside the prison. (McBain, 1981)

Whence many compounds: *fresh meat*, a young prostitute; *stale meat*, an experienced or older prostitute:

> ... since to the accustomed rake the most prized flesh is the newest, some now counted her stale meat (Fowles, 1985, of an experienced prostitute)

meat and two veg, the penis and testicles:

> ... carrying a carving knife with which she planned, she shrieked, 'to cut off his meat and two veg' (Monkhouse, 1993)

the obsolete British *meat-house*, a brothel, and the modern American *meat-rack*, a meeting place for male homosexuals:

> The meat racks, the quick sex, the beatings... (Collins, 1981, of a male homosexual's life)

a bit of meat, an extramarital sexual partner:

> I don't want you coming round here after my little bit of meat (Richards, 1933, of a sexual mistress)

tube of meat, the erect penis:

> All because of that lousy tube of meat. I want to hump every woman I see. (Sanders, 1982)

Meat is also American underworld slang for the corpse of someone killed violently, with the *meat wagon* an ambulance or a hearse:

> They have the meat wagon following him around to follow up on the business he finds. (Chandler, 1943, of Marlowe, his corpse-prone private eye)

meathead a fool.

> Rev, in this town, with this Administration? Don't be a meathead. (M. Thomas, 1987)

All of us have meat of sorts in our heads, if you think about it.

medal showing (a) a visible undone trouser button.

Pre-zip warning from one male to another and see also ABYSSINIAN MEDAL.

medical correctness the avoidance in speech of direct reference to a taboo condition or illness.

Not diagnosing patients accurately and treating them wisely:

> Medical Correctness is motivated by compassion, but seized by a dangerous illusion: that if you change words, you change reality. (M. Holman in *Financial Times*, October 1994)

See also POLITICALLY CORRECT.

medicine spirituous intoxicants.

This substance is seldom ingested to treat disease:

> ... fond of taking their medicine. (Mayhew, 1851, of drunkards)

The pretence that you drink spirits for your health is not new, nor does it take anybody in.

medium small.

Literally, between big and little, but not in the grocery business.

meet with an accident to be murdered.

From the American underworld, or those who write about it:

> 'He met with an accident. He's dead.' ... 'Yeah, he's dead. You shot him.' (Chandler, 1939)

meet your Maker to die.

This and similar expressions are used by those with no confidence that the rendezvous will be kept. Similarly, a Muslim might, if so favoured, *meet the Prophet*:

> He intended to meet the Prophet shod, smiling, and at peace. (M. Thomas, 1980)

meeting (at/in a) where you claim to be when you do not wish to talk to someone.

The standard rebuff to an intruder by telephone or in person.

mellow slightly drunk.

Literally, ripe, but a euphemism since the 17th century:

> Two being 'half-drunk', and the third 'just comfortably mellow'. (Bartram, 1897)

melons *American* the breasts of an adult female.

Either *tout court*, or as *honeydew melons*, or *watermelons*, perhaps for those with a larger bust.

melt *?obsolete* to ejaculate semen.

The imagery is from liquefying under heat:

> ... made me soon sensible of his melting period. (Cleland, 1749)

member[1] the penis.

> Affection and the erect male member tend to go hand in hand, if you'll pardon the expression. (Amis, 1978)

Literally, any limb of the body. The obsolete British *member for Horncastle*, a cuckold, was a complex vulgar pun on the Lincolnshire parliamentary constituency.

member[2] see CLUB[3].

membrum virile the penis.

Literally, in Latin, the male member:

> And not a bad label for his membrum virile either. (Sanders, 1980)

memorial *mainly American* relating to death.

Literally, maintaining the memory of anything. An American *memorial association* or *society* is a rather grander version of the burial club:

> 'Funeral societies' or 'Memorial Associations'. (J. Mitford, 1963)

> Memorial societies ... constitute one of the greatest threats to the American ideas of memorialization. (ibid.—*memorialization* is funeral jargon for spending as much money as can be extracted from the bereaved on smartening up the corpse, a fancy casket, etc.)

An American *memorial counsellor* is a salesman of plots in a cemetery:

> A cemetery salesman (identified on his card as a 'memorial counsellor') (ibid.)

and a *memorial park* is a cemetery:

> ... not in a graveyard or cemetery, but rather in a 'memorial park'. (ibid.)

A *memorial home* is a structure on the walls of which you can pay to have a tablet fixed recording the death, and a *memory garden* is an open space within the confines of a cemetery or crematorium.

men a lavatory for male use only.

Usually in a place where nobody is trying to sell you anything, such as a government office. Often expanded to *men's toilet* and in America to *men's room*, which latter could just as well be a dormitory, but isn't:

> His first destination, the men's toilet. (Bradbury, 1976)

> I went into the men's room, just to look in the mirror. (Theroux, 1973)

The sexual counterpart is WOMEN.

men in blue see BLUE[1].

ménage à trois three people living together in a sexual relationship.

Literally, domestic arrangements for three and see also À TROIS. To *maintain a clandestine ménage* is to keep a sexual mistress:

> Although he was indeed married, he also maintained a clandestine ménage. (R. V. Jones, 1978)

men's magazine a pornographic publication for male readers.

> He even had a little stash of mens magazines in an old hatbox at the back of his clothes closet. (Bryson, 1989)

Formerly filled with repetitive heterosexual titillation, but perhaps now also for homosexuals.

men's room see MEN.

mense(s) menstruation.

Literally, in Latin, month(s), but long standard English:

> He would say ... 'I'se glad to see ye after yer mense', before beginning the churching. (Linton, 1866—the *churching* was a rite of supposedly cleansing a woman after childbirth)

Nearly always in the plural:

> A woman does not get gout unless her menses are stopped. (Condon, 1966)

mental mad.

> Non-U mental/U mad. (Ross, 1956, who here and elsewhere reminds us that greater security or better education tend to reduce or eliminate the use of euphemism in speech)

Literally, pertaining to the mind. Because *mental* was once used to describe any illness from mild eccentricity to lunacy, those with a mild condition could find themselves locked away with severely ill people, while the latter might not be receiving the treatment and attention which they needed. In lay use, *mentally handicapped* still lumps together those with widely differing degrees of affliction. *Mentally retarded*, of low attainment or intelligence, may misleadingly imply a wilful holding back (see RETARD). To be *mentally challenged* is not to try to solve a difficult puzzle:

> ... the general-for-specific euphemism *sick* is frequently used to describe someone who is *mentally challenged*. (Allan and Burridge, 1991)

mental disease syphilis.

From the days when patients with third degree syphilis and dipsomaniacs formed the majority of those in lunatic asylums:

> ... even in 1966 Winston's son Randolph referred to his grandfather as suffering from a 'severe mental disease'. (Massie, 1992—Lord Randolph Churchill had contracted syphilis either from a prostitute at Oxford or from a maid at Blenheim after his wife's confinement with Winston)

merchandise any illicit possession.

Often narcotics, but otherwise synonymous with the more frequent *goods* (see GOODS (THE)).

mercy a mild oath.

American Citizens' Band radio use for illegal profanities. The obsolete Scottish *mercy*, whisky, brought warmth and comfort:

> The Bailie requires neither precept nor example wi' his tumbler when the mercy's afore him. (Galt, 1826)

mercy killing the murder of a terminally ill patient.

Often by the deliberate administration of excessive medication and for the victim it may be a *mercy death*. The Nazis also adopted the expression:

> ... the Gestapo is now systematically bumping off the mentally deficient people ... the Nazis call them 'mercy deaths'. (Shirer, 1984)

In modern use:

> Mercy death suspected in hospital. (*Daily Telegraph*, headline 30 December 1994)

merger accounting the false statement of subsequent profitability.

Literally, the creation of provisions against the cost of assimilating an acquisition:

> By the alchemy of merger accounting, some of that 'cost' could be recycled into profits. (*Daily Telegraph*, 16 November 1990—Burton, having acquired Debenhams, was alleged to have arranged for £80 million to trickle into profits during a period of two years following the deal)

merry drunk.

Cheerful, but not offensive. The obsolete *merry-begot* or *merry-begotten* was a bastard, conceived I suppose in pleasure rather than in drink:

> That Joe Garth is a merry-begot. (Caine, 1885)
> A love or merry-begotten child, a bastard. (Grose)

meshugga mad.

From the Yiddish *shagag*, to go astray or wander (OED):

> 'They say he's meshugga.' 'No sign of that today.' (Deighton, 1988)

mess[1] to copulate casually.

> I got a decent wife. I don't go messing any longer. I just don't have the energy. (Sharpe, 1977)

Probably a shortened form of *mess about*, to act in a sloppy, unconventional, or disorganized way. A *messer* is a part-time prostitute with a reputation for spoiling the market for full-time workers.

mess[2] turds or urine in an unwanted place.

Made by household pets, but also occasionally by humans and other animals:

> ... the goat which was for ever trotting in and making a mess on the fireplace. (W. S. Moss, 1950)

message an advertisement.

American television jargon:

> We'll return after these messages. (ABC announcer, September 1983)

Mexican brown *American* marijuana.

> That's what speed and Mexican brown does to ya. A hardballer. (Wambaugh, 1983)

From the country of origin and the colour. It can also be *Mexican green* or *red*. A *Mexican mushroom* is the hallucinogenic *Psilocybe Mexicana*.

Mexican raise a promotion with no increase in pay.

Many Mexicans working in the USA without permits are subject to exploitation. A *Mexican promotion* means the same thing.

Mexican toothache see MONTEZUMA'S REVENGE.

mickey the penis.

An Irish example of an international practice —see WILLY:

> Mister Quigley couldn't get his mickey to go hard. (R. Doyle, 1993)

In Australian slang the *mickey* is the vagina, but whether this comes from the association with *mouse* (see MOUSEHOLE) or is the result of seeing things upside down I cannot say.

Mickey Mouse fraudulent.

From the cartoon character via the meaning 'bogus or ineffective':

> It was the Revenue who made the first breach of Fleet Street's 'Spanish practices' by exposing

the 'Mickey Mouse' payments to printers. (*Daily Telegraph*, 11 August 1994)

micky to taunt or mimic.
From *take the piss* and the rhyming slang *Mike* or *Micky Bliss*:
Look at Bill wobbling his belly... mickying her, he is. (Cookson, 1967)
See also TAKE THE MICKEY.

Micky Finn a drugged intoxicant given to an unsuspecting victim.
From a late 19th-century Chicago innkeeper of evil reputation. The commonest additive was chlorine, which reacts with alcohol to form chloral hydrate, with dire effects. Occasionally abbreviated to *MF*:
Had I been slipped an MF? (Burgess, 1980, of a drink which proved highly intoxicating)

microwave club a place where married women may meet men for extramarital copulation.
Usually an American dancehall or bar. The microwave oven allows a wife to serve her husband with hot food despite an afternoon's dalliance.

mid-job see ON THE JOB.

middle-aged old.
Halfway to three score years and ten is thirty-five but no man under forty-five or woman under fifty would admit to having reached middle age:
...in that advanced stage of life that we optimistically call middle age. (Deighton, 1982)
Middle-aged spread is obesity around the waist:
Middle-aged spread is a genuine fact of life... The flesh can resist the pull of gravity for so long. (Matthew, 1983)

middle leg see THIRD LEG.

middle passage the west-bound transatlantic voyage of a ship carrying slaves.
The *passage* was the sea voyage from Africa to the Americas in the *triangular trade* (see TRIANGULAR TRADE (THE)).

midnight baby *American* a bastard.
From the mysterious time of conception rather than of birth:
I never knew who my daddy was. I was what they called a 'midnight baby'. (Sanders, 1984)

migraine a case of sub-acute alcoholic poisoning.
It is the same word as *megrim*, which meant 'depression' or 'a fad' as well as 'a bad headache'. This medical condition is much called in aid by women who do not wish to acknowledge having been drunk, the taboo against female drunkenness being greater than against male:
She had stayed at home with a hangover that she called a migraine. (Manning, 1978)
The affliction may also be contracted at short notice by a female who does not want to copulate with her regular sexual partner.

Mike Bliss see MICKY, TAKE THE MICKEY.

military intelligence spying.
It could mean no more than knowing how to fire a gun:
Foreigners have spies; Britain has Military Intelligence. (Follett, 1978)
But nearly everybody has *military attachés*, whose main function is to uncover the military secrets of the country to which they are accredited.

militia an armed body operating outside normal military regulations.
Literally, a body supplementing and under the control of regular forces. It is usually raised to support an autocracy, or, like the notorious *milice* in the Second World War, to assist the suppression by a conqueror:
He more than anyone else knew that the Militia existed in order to betray. (Genet, in 1969 translation)
...the *Service de l'Ordre Légionnaire*—which is now the Milice—the scum of the scum. (Price, 1978)

milk regularly to defraud.
By taking small amounts from a till, persistent pilfering, etc. Now especially of illegally syphoning fuel from motor vehicles.

milk run a comparatively safe wartime flight.
Second World War fliers' jargon, usually of a mission over enemy territory taken regularly but sometimes of a single occasion:
We'll be over the sea most of the way... Another lousy milk run. (Deighton, 1982)
From the daily doorstep delivery common in the British Isles.

milk yourself to masturbate.
Of a male, from *milk*, the semen. Occasionally, too, of ejaculation during copulation.

minder a bodyguard.
Often for the protection of a criminal but no chaperon, as in the British television series so named. Also of a soldier accompanying a journalist in a war zone etc., to ensure his safety and to see that nothing is seen which the army wishes to conceal.

mingle bodies to copulate.
A purist might say that only a limited portion of each does the mingling. Usually of casual copulation:
...in the eight times their 'bodies had mingled' since that first evening. (Boyd, 1982)

Ministry for Internal Affairs see INTERNAL AFFAIRS.

Ministry of Defence see DEFENCE.

minor function (the) urination.

As distinct from defecation, which I have yet to see called the *major function*:

> ... going to the W.C. (Generally for the minor function.) (Franklin, DRS)

minority group non-whites in a country where the majority is white.

The usage avoids a direct reference to skin pigmentation:

> I used to be coloured, right? Then I was a Negro. And then I turned into an Afro-American. After that I was just a member of a Minority Group. Now, I'm black. (Theroux, 1982—he's probably back to being *coloured* again, the way these things go)

The members of the *group* can be *minority types*:

> What's the deal here—you don't let minority-type people sit at your front booths? (McInerney, 1992, of seating in a restaurant)

Minorities may also include those with less prevalent religious or sexual views:

> ... the minorities ran the risk of losing others' sympathy and support. (Jennings, 1965)

minstrel a black person in a predominantly white-occupied country.

From *nigger minstrel*, a performer who blacked his skin to emulate or caricature a negroid appearance:

> ... he was responsible for admitting the Minstrels in such numbers in the first place. (*Private Eye*, 1981, of Enoch Powell)

Especially in derogatory use by British whites of immigrants from the West Indies.

minus buttons of low intelligence.

Also as *minus marbles*, *screws*, etc., implying incompleteness:

> ... he'd throw down his pen and admit the fellow was minus some buttons, crazier than a bed bug. (Burgess, 1980)

misadventure the consequence of negligence.

In normal usage, bad luck causes misadventures, but not in medical jargon where a *therapeutic misadventure* means the patient died from receiving the wrong treatment and a *surgical misadventure* tells us that the scalpel slipped.

misconduct extramarital copulation.

Usually of a woman within marriage, from the literal meaning 'bad behaviour'. To *commit misconduct* is synonymous in legal jargon with committing adultery. Standard English

misfortune a bastard.

Literally, ill-luck, which it certainly was for mother and child:

> ... had 'had a misfortune'—in the shape of a bouncing boy. (Bartram, 1897)
>
> To light of a misfortune is the ordinary euphemism. (EDD)

A *misbegot* or *mishap* was a bastard. A *mishap* was also premature delivery of a foetus, in which case an animal or woman was said to *misgo*:

> 'Tis a thousand pities her should'a miswent. (EDD)

miss[1] see MISTRESS.

miss[2] to fail to menstruate at due time.

Shortened form of *miss a period* and often with overtones of unwanted pregnancy:

> 'Has'er missed then?' 'No, but us've 'ad some worryin' times.' (Conversation in south Devon between two males in 1948)

Mis(s) is a common abbreviation too for miscarriage.

Miss Emma *American, ?obsolete* morphine ingested illegally.

Probably from the old phonetic alphabet, in which *Emma* signified the letter *M* (as in the now rare *pip emma*, p.m. or *post meridiem*).

miss Nancy see NANCY.

missionary position a position for copulation in which the male lies atop the female.

European missionaries brought this fashion to the Polynesians, who had preferred the quadripedal approach:

> 'The guy's on top and the girl's on the bottom, and they're—well, you know, screwing?' ... 'Not the missionary position.' (Theroux, 1973—but it was)

Sanders (1982) probably invented the *Ms-sionary position*—the female astride the male.

misspeak to lie.

Originally, to speak evil or to speak incorrectly:

> ... do they bar him for his 'misspeakings', or do they just take over and appoint someone else as candidate? (*Private Eye*, October 1986)

This is one of Richard Nixon's contributions to the language from the days of Watergate.

mistake[1] an unwanted pregnancy.

Usually within marriage, and also denoting the resultant child:

> Told him he was rubbish, a mistake. (Francis, 1987, of a bastard)

mistake[2] urination or defecation other than in a prepared place.

By young children in their clothing and by animals indoors other than in prepared trays etc. Also denoting the resulting urine and turds:

That was enough to make her father overlook the chewed shoes and occasional mistakes with which the dog was littering the house. (Clancy, 1987)

mistress a man's regular extramarital sexual partner.

Originally, the female head of the household, but now always used improperly except when shortened to *Mrs* or in girls' schools:

My mistress is my mistress. (Shakespeare, *Titus Andronicus*)

Kept mistress is explicit:

It's not fair to the girl, this life as a kept mistress. (Harris, 1925)

Miss was formerly used in the same sense:

Priests, lawyers, keen physicians, kept misses. (Galloway, 1810)

misuse *?obsolete* to copulate with outside marriage.

But not the opposite of USE[1]:

Did you ever misuse my Ephie... Did you ever have her? (Keneally, 1979)

mitotic disease cancer.

From *mitosis*, the process whereby a cell splits into two identical parts:

The label used by many Australian doctors in place of 'cancer' is 'mitotic disease'. (Allan and Burridge, 1991)

Now common medical jargon.

mixer *American* a prostitute who finds customers in bars.

In standard usage, a fluid added to a spirituous intoxicant to make it more palatable.

mob an association of criminals.

The 19th-century English use came from the Latin *mobile vulgus*, the rabble, and a *mob* might also be a prostitute. Now mainly of an American criminal gang:

Wasn't it enough he had to pay protection on his place to the mob? (Collins, 1981)

A member of such a gang is a *mobster*:

A mobster newly acquitted from a charge of swindling the city. (Ustinov, 1971)

mobility impaired crippled.

The *impairment* is the result of disease etc., and not through being locked away somewhere. A common American usage. In Canada, you can be said to be *impaired* if you drive a vehicle while drunk or under the influence of narcotics.

model a prostitute.

Shortened form of *model girl*, a mannequin. *Model* is also widely used non-euphemistically and most girls describing themselves as *models* do so without any implication of sexual activity. However prostitutes who advertise their availability by defacing telephone booths etc. profess to being so employed, as do high-class prostitutes who have no need for promotional activities:

Miss Keeler, 20, a freelance model, was visiting Miss Marilyn Rice-Davies, an actress. (*Daily Telegraph*, December 1962, quoted by Green, 1979)

modern (of weapons) nuclear.

A military usage to differentiate from any old way of killing people:

... the power range and prospective development of 'modern' weapons—a frequent euphemism—would favour a surprise attack against the United States. (H. Thomas, 1986, quoting US Chiefs of Staff paper of September 1945)

modern convenience a lavatory indoors.

This does not imply that a *convenience* (see CONVENIENCE[1]) is necessarily ancient. British real estate jargon, shortened in small advertisements to *mod cons*. In similar code *all mod cons* means you can expect hot and cold running water and a bath as well as an indoor lavatory.

mole a conspirator or spy within an organization.

Espionage and labour union jargon, from the burrowing habits of the mammal and perhaps its blackness, but not its blindness:

There were no 'moles' at large in Washington: 'Indifference, not treachery, was at the root of America's attitude'. (Boyle, 1979)

molest to assault sexually.

Originally, to inconvenience, but so pervasive is the euphemism that a female may be reported as having been brutally assaulted but not *molested*, unless her assailant's motives were sexual as well as predatory:

I revived her by threatening to carry her into the bushes and molest her. (Fraser, 1975)

Also of any sexual assault by an adult against a child, as in the phrase *child molester*.

moll a prostitute.

Probably from the common girl's name; in obsolete use a *moll-shop* was a brothel. Previously, a *moll* was merely a sweetheart, as she remains when the female companion of the American gangster. The obsolete British *Moll Thompson's mark* was nothing more than emptiness of a bottle of intoxicant, punning on the initials *MT*.

Molotov cocktail a simple petrol bomb.

Molotov was the Russian foreign minister in the Second World War whose name has thus been not inappropriately perpetuated. He it was who concluded Stalin's peace pact with Hitler in 1939, among other perfidious acts in a long and often sinister career. But even this claim to fame was wrongly acquired, the device having been invented by the Finns for use against their Russian aggressors.

mom-and-pop *American* staid and old-fashioned.

Like your aged parents:

> Are you gonna be a mom-and-pop camcorder with Kuralt-ian notions of 'on the road', or are you up to heavyweight digital effects and dazzling graphics? (*Fly Rod & Reel*, March 1991)

Usually used of a small retail business:

> ... a small-time mom-and-pop dope store would be allowed to flourish unmolested. (McBain, 1981)

momentary trick (the) *obsolete* copulation.

Referring to the duration of a casual encounter:

> ... for the momentary trick
> Be perdurably fined.
> (Shakespeare, *Measure for Measure*)

Monday man *American* a stealer of clothes from washing lines.

Clothes were traditionally washed and hung out to dry on Mondays. As modern appliances have released women from the drudgery of the old copper boiler, so the washing may be done, and stolen, any day. A *Monday man* did not ride in a *Monday car*, which, like the *Friday car*, was likely to have been sloppily assembled by inattentive workers, nor would he be welcome in the British *Monday Club*, an association of hard-line conservatives.

Monday-morning quarter-back *American* a fantasist who judges by hindsight.

The spectator who watches a weekend game may take his criticism to work with him on Monday:

> ... the Monday morning quarterback who could have won the ball game if he had been on the team. But he never is. He's high up in the stands with a flask on his hip. (Chandler, 1958)

monkey[1] see SUCK THE MONKEY.

monkey[2] (the) *American* addiction to illegal narcotics.

Probably from *having a monkey on your back* which you cannot shake off:

> You think it's the monkey that's killing you. (Macdonald, 1971, of a heroin addict)

monkey's (a) an obscenity.

Shortened form of *monkey's fuck*, a matter of trifling importance, and usually in the phrase *not give a monkey's*:

> ... doesn't it worry you that ninety-nine point nine per cent of the population couldn't give a monkey's? (Lodge, 1988)

For Grose, a *monkey's allowance* was more kicks than halfpence. In the 19th century to *have a monkey on your house* was to owe money on a mortgage, probably from the animal's ability to adhere or from its long tail. That kind of monkey might also alight elsewhere:

> Oh yes, there's a monkey setting on his chimney. (EDD—he has a mortgage)

monkey business casual copulation.

Literally, any mischief which a monkey might get up to. Used sometimes as a warning from a girl to her suitor but more often from her mother to them both:

> 'No monkey business,' he agreed. 'Shit, I won't touch her.' (Sanders, 1977—an artist is speaking to a young model's mother)

monosyllable the vagina.

The taboo *cunt*, viewed sexually by a male. Grose says 'A woman's commodity' and see also DSUE for a learned disquisition.

Montezuma's revenge diarrhoea.

Usually but not necessarily contracted in Mexico by US visitors. Montezuma II was the Aztec emperor when Cortes invaded and was killed by his own people in 1520 when he told them to submit to the invader. Also as the AZTEC TWO-STEP, *Mexican toothache*, *Mexican two-step*, *Mexican foxtrot*, etc.

monthly period menstruation.

From its regularity and incidence:

> ... her monthly period. We call it menstruation. (Sharpe, 1978)

Shortened to *month's*:

> ... my wife ... gone to bed not very well, she having her month's upon her (Pepys, 1662)

or more often to *monthlies*:

> Molly was easily excited, especially about the eighth day after her monthlies had ceased. (Harris, 1925)

Also as *monthly courses*, which is obsolete, or *monthly blues*, from the associated depression:

> 'You all right?' 'Yes.' 'Monthly blues?' (de Mille, 1988)

mooch to pilfer.

In standard usage, to hang about, whence to beg and so to steal:

> I don't mean to say that if I see anything laying about handy that I don't mooch it. (Mayhew, 1851)

I include this obsolete British entry as a rare example of a word which has reverted from its euphemistic to its proper use in modern speech.

mood freshener an illicit narcotic.

From the stimulus:

> It was enough to send you racing to the bathroom for a discreet puke or a quick blast of mood freshener. (McInerney, 1992)

moon to expose the buttocks to (another person) by lowering clothing in public.

Presumably from the roundish expanse of flesh revealed:

...the Chinese soldiers provoking incidents by dropping their pants and presenting the bare bums northward, mooning the Soviet border guards. (Theroux, 1988, of Manchuria)

moonlight[1] smuggled spirits.

Obsolete smugglers' usage, from the time when the business was best done; a habitual smuggler was said to have been *bred in the moonlight*:

Thirty 'crack' hands, who had been bred in the 'moonlight' from boyhood. (Vedder, 1832)

In many other uses *moonlight* was synonymous with smuggling and a *moonraker* was a smuggler, from the practice of throwing contraband into a pool if detected, recovering it later:

Getting ready for the moonrakers at the great pool. (Verney, 1870)

moonlight[2] to wound.

The agrarian disturbances of 19th-century Ireland took place at night, some warnings about threatened arson etc. being signed *Captain Moonlight*. For all the publicity, relatively few threats were carried out:

He had deposed to his experience of being moonlighted in the thigh. (*Daily Telegraph*, November 1888)

Whence the *moonlighter* who made such threats etc.:

...ranged on the side of the moonlighters and the hillside men. (Flanagan, 1988, writing of the 19th century)

moonlight[3] to work at a second job without paying taxes etc.

Moonlight because the work is often done in the evening, but it can be done at any time:

A joiner who 'moonlights' at weekends for his mates... (Shankland, 1980)

A worker who continues to draw unemployment monies from the state is also said to be *moonlighting*.

moonlight flit the clandestine departure of a tenant in arrears with his rent.

As distress could be levied on household chattels in the premises for arrears of rent, but not on the same chattels if moved elsewhere, you had to take all with you when you vacated the property, and this was best done at night:

He has e'en made a moonlight flitting. (W. Scott, 1822)

You might also have made a *moonlight flight, march, touch*, or *walk*; or you might *bolt* or *shoot the moon*:

Nobody was allowed to shoot the moon. (Besant and Rice, 1872)

The phrase was also, and occasionally still is, used of a debtor fleeing his creditors:

He was fain to make a moonlight flitting, leaving his wife for a time to manage his affairs. (Galt, 1821)

See also FLIT[1] (DO A).

moonraker see MOONLIGHT[1].

moonshine whisky.

From an illicit still, which is operated at night to avoid detection:

...made their living by odd ends of trade, from moonshine, from cutting lumber... (Keneally, 1979)

moose a prostitute.

Neither a corruption of *mouse* nor punning on the deer, but an American Korean war usage from the Japanese *musume*, a girl (DAS).

mop up to kill or capture (surviving opponents).

Military jargon. The imagery is from cleaning up spillage and the activity is *mopping up*:

Franco ruled. It was all over bar the mopping up. (Boyle, 1979)

moral immoral.

Another of the opposites, like DEFENCE. To be in *moral danger* is not to be at risk from an excess of goodness, and the police *Morals Squad* is after vice, whatever the rectitude of its members.

more than a good friend a person with whom you regularly copulate outside marriage.

Another kind of FRIEND:

It would have taken no special investigation to establish that they were more than good friends. (Price, 1971)

morning after (the) sub-acute alcoholic poisoning.

Shortened form of *the morning after the night before*, when excessive or adulterated alcohol had been consumed.

most precious part the male genitalia.

Valued for copulation rather than urination:

Corporal Brownlee was hit in the most precious part of his body. (Farran, 1948)

moth in your wallet (a) stinginess.

The *Tineola bisselliella* doesn't normally go for leather, although it favours an undisturbed site for its eggs:

Symington would pick up the tab... there were no moths in his wallet. (Sanders, 1985)

mother[1] an elderly male homosexual.

A modern American use which replaced the obsolete British meaning 'a bawd'.

mother[2] *American* a term of vulgar abuse.

Shortened form of *mother fucker* but those who use it are unlikely to know that Oedipus was said to have sired four children by Jocasta in a complex saga which includes blinding and suicide. Used as an insult but an inanimate object may also be so castigated:

I remember back in Quang Tri we had an A. P. C. was a real mother. Always throwing tracks, breaking down. (Boyd, 1983, of an armoured personnel carrier)

mother five fingers male masturbation.

Of the same imagery as FIVE-FINGERED WIDOW:

Always looking for something better. Know what I mean? Then I end up with Mother Five-fingers. (Sanders, 1981)

mother's blessing a narcotic administered to a baby.

The *blessing* was the peace which came from silencing a crying child:

Give the babies a dose of 'Mother's Blessing' (that's laudanum, sir, or some sich stuff) to sleep 'em when they's squally. (Mayhew, 1862)

The usage and practice survived to the Second World War but are now obsolete.

mother's ruin gin.

Its 19th-century cheapness led to wide female addiction and consequent demoralization. Now only humorous use:

... struggling to get his arms around a Europack of litre-sized Mother's Ruin. (*Private Eye*, April 1980)

Franklin in DRS says it is rhyming slang, adding 'the phonetics are poor'—too poor, I suspect, when another etymological explanation is simpler. Occasionally as the punning *mother's milk*.

motion (a) defecation.

Medical jargon, from the movement of the bowels and generally in the plural, where *motions* means 'faeces':

She had dreams of cooking by perpetual motion, or rather by perpetual motions. (Sharpe, 1976, of Mrs Wilt's 'biological' lavatory that was supposed to generate heat for domestic purposes)

motion discomfort airsickness.

Airline jargon, in support of the pretence that any passenger actually enjoys air travel:

'I am still suffering from motion discomfort' ... 'It means air sickness.' (N. Mitford, 1960)

The *motion discomfort bag* you may find on an American airline is for you to be sick in.

mount to copulate with.

Standard English of animals; the occasional use of humans puns on the action of mounting a horse:

Like a full-acorn'd boar, a German one,
Cried 'O!' and mounted.
(Shakespeare, *Cymbeline*)

A male may describe his complaisant sexual partner as a good *mount*—it remains a mystery where the *bad mounts* get to.

mount a corporal and four to masturbate.

Of a male, punning on the constitution of a British army guard and the thumb and four fingers.

mountain dew whisky.

In modern use synonymous with MOONSHINE, from the process of distillation and the place where it is done:

A 'greybeard' jar of the real Glengillodram mountain dew. (Alexander, 1882)

mousehole the vagina.

Not necessarily viewed sexually:

Scissored her legs open—and pulled a length of magician's scarves, knotted end to end, out of her mousehole. (Theroux, 1978)

Perhaps also punning on *mouse*, a sexually attractive female:

... tempt you again to bed;
Pinch wanton on your cheek; call you his mouse.
(Shakespeare, *Hamlet*)

mouth *obsolete* to kiss lecherously.

Literally, to utter. Also as *mouth with*:

He would mouth with a beggar, though she smelt brown bread and garlick. (Shakespeare, *Measure for Measure*)

move to steal.

Mainly American, with the common imagery of shifting the stolen article.

move your bowels to defecate.

Medical jargon and perhaps circumlocution rather than euphemism, as in many cases we are merely recording cause and effect:

He lay in bed, reading nothing; he moved his bowels. (Bradbury, 1959, writing of a stay in hospital)

Whence *good bowels* as an absence of constipation:

Good bowels were beyond price (Keneally, 1979, of military service)

and *movement* as an act of defecation:

Observe the time of day when he has his movement. (McCarthy, 1963)

movement[1] see MOVE YOUR BOWELS.

movement[2] an ossified institution or association of institutions.

Those who affect this title are often remarkable for their rigid and unchanging attitudes and behaviour, although to be fair the British Building Society *movement* appears to have moved from its former deep conservatism in recent years.

Mozart drunk.

From the rhyming slang *Mozart and Liszt*, pissed. BRAHMS is more common.

Mr Priapus *obsolete* an erection of the penis.

Priapus, the Pan of Mysia, was (and still is) depicted in that condition:

> ... as I write and describe them cause Mr Priapus to swell in my breeches. (Pearsall, 1969, quoting from 19th-century pornography, *them* being the 'lips of her plump parting cunny')

See also PRIAPISM.

Mrs Chant a lavatory.

British rhyming slang for 'aunt' (see AUNT[2]). Female use.

Mrs Duckett a mild oath.

Rhyming slang for 'fuck it'. Franklin says: 'A comment rather than an expletive: the workman who hits his thumb with a hammer uses no euphemism' (DRS).

muck a mild oath.

Used for *fuck* figuratively in all inflections. DAS says that *muck up*, 'to make a mess of' ' = fuck up, a euphem.' but I'm not so sure.

mud-kicker a poor prostitute.

She may lie down al fresco, and she is liable, in American use, to bilk or rob her customer.

mudlark *obsolete* a scavenger or thief.

Used in London to denote both those who frequented the exposed banks of the River Thames at low tide to recover anything of value lost or discarded from ships, and those who picked up stolen goods which an accomplice had tossed over the side:

> The mudlarks are generally known as thieves. (Mayhew, 1862)

muff the female pubic hair.

But not used for keeping the hands warm:

> I had a photograph of that sanctimonious prick Merriman with his nose in some call-girl's muff. (M. Thomas, 1980)

A *muff diver* indulges in cunnilingus:

> ... the tufts of facial hair known as *bugger's grips* can also be described as muff diver's depth marks. (Jolly, 1988)

mug to rob by violence in a public place.

In obsolete British use it was to bribe with an intoxicant, from the container:

> Having... mugged as we say in England, our pilot. (Ingelo, 1830)

Specifically in 19th-century London of robbery by garrotting but today any violence suffices for a *mugger*, who so robs; from *mug*, a dupe.

mug-shot a frontal photograph taken by the police for identification.

From slang *mug*, the face. The subject seldom sits voluntarily.

muggy drunk.

Literally, moist, and usually of the weather:

> They're raything muggy oft. (Clark, 1839, of drunkards)

Muggy also means 'stupid'.

mule[1] whisky.

Usually from an illicit American still and with a strong kick.

mule[2] a carrier of illegal narcotics in bulk.

From the smuggling on rough mountain tracks in Central America:

> Some smuggle for their own use, but most are 'mules', paid $1,500 or so a trip. (Moynahan, 1983, of airline narcotics smuggling)

multicultural in which people of different skin pigmentations are expected to mix harmoniously.

Multicoloured would be deemed offensive as well as inaccurate. The usage makes the patronizing assumption that first- or second-generation immigrants wish to be like the indigenous inhabitants:

> All-black schools in multi-cultural Brent would be a form of apartheid. (*Daily Telegraph*, October 1983)

An attempt to impose or achieve such a harmonious outcome (or the outcome itself) is called *multiculturalism* or *multiculturism*:

> ... 'multiculturism' ... provides certain minorities with a way through the university, and little fiefdoms within the curriculum for those on the vocal left lucky enough to identify themselves with them. (*Daily Telegraph*, 23 February 1991—the standards of entry were not uniform for all applicants)

municipal farm a prison.

As distinct from the *funny farm* (see at FUNNY[4]):

> A striker caught with a slingshot was sentenced to the municipal farm. (Lacey, 1986)

Murphy (of a prostitute) to cheat (a customer) by unfulfilled promises of copulation.

An American and not an Irish trick, but perhaps from the simplest of *Murphy's Laws*, if something can go wrong it will. Whence the *Murphy game*:

> ... there were rooms for hire above the bar and that Star's specialty was the Murphy game —rolling drunk customers. (Maas, 1986)

muscle *American* to assault criminally.

From the force used:

> You couldn't muscle anyone, Peter. You're a softy. (Sanders, 1983)

Whence a *muscleman*, who is no keep-fit enthusiast but employed to carry out criminal acts of violence:

... kind of muscleman for a big protection gang in Tokyo. (West, 1979)

Muscleman is also shortened to *muscle*:

Not so much between the ears, but he was a good muscle. (Sanders, 1980)

mush to rob (householders) while professing an itinerant trade.

Shortened form of slang *mushroom*, an umbrella. Umbrella-men went from house to house, itinerant crooks found it good cover, and the trade got a bad name. *Mush* is still a British mode of male address, importing no ill-will or accusation of dishonesty.

musical (of a male) homosexual.

British 1930s usage:

In Harry's estimation they were both homosexual—or 'musical', as the Noel Coward set would say. (Follett, 1991)

muslin *?obsolete* a female viewed sexually by a male.

She used to wear it in her dress or skirt; the whole was described usually as a *bit* or *piece of muslin*.

muster your bag to be ill.

British naval usage, from having to take your kit to the sick bay if you reported sick. However in the Second World War British army you were too weak to carry your kit bag if you dared to report sick.

mutate to dye.

Of women's hair and certainly not, as in the literal sense, to change genetically and permanently through natural processes:

She 'mutates' or 'colour-corrects' her hair. (Jennings, 1965)

mutilate *mainly American* to castrate.

Originally, in this sense, to cut off a limb. Formerly of humans and animals but now mainly of tom-cats. Standard English.

mutt *?obsolete* deaf.

From the rhyming slang *Mutt and Jeff*, better known as the British First World War service and victory medals than for the comic cartoon characters, but no longer well-known as either.

mutton a person viewed sexually by another.

As you may expect, it is normally the man who so views the woman:

The duke ... would eat mutton on Fridays. He's now past it. (Shakespeare, *Measure for Measure*)

Mutton, a prostitute, is obsolete, but *in her mutton* still means 'copulating', and not necessarily for payment by the male. *Mutton-monger*, a male profligate, is obsolete. To COME YOUR MUTTON is to masturbate, of a male. *Mutton dressed as lamb* is a derogatory description of a woman affecting the dress or style of someone much younger:

'Youthful excess is one thing,' said the Dean, 'but mutton dressed as lamb is another.' (Sharpe, 1974)

mutual abuse/pleasuring see ABUSE[1] and PLEASURE.

muzzy drunk.

Literally, dull and overcast (of the weather). Quite common female use of themselves when slightly drunk.

my word a turd.

British rhyming slang, mainly of canine deposits on pavements etc.

N

nab to steal.

Literally, to catch or arrest:

They ha' nabb'd my gold. (Clark, 1839)

Nab is found in many 18th-century compounds such as *nab the stoop*, to stand in the pillory, *nab the snow*, to steal linen from a line, etc. Often corrupted to *nap*, in which form it was used of venereal disease, from the 'catching' of it. The First World War *napoo*, death, may have been punning on *nap* and a corruption of *n'y a plus*, but I suspect it was merely from the French.

naff off go away.

A vulgarism brought into greater prominence after being used by the outspoken British Princess Anne in 1982. Partridge suggested *naf*, the vagina, was back slang for *fan*, a shortened form of FANNY, and also tells us that *naff* was prostitutes' slang for 'nothing'. The OED gives 'a euphemistic substitution for fuck'. Dr Wright has 'to be frivolous or to argue snappishly... esp. of children and persons of diminutive appearance'. Dr Johnson gives 'A kind of tufted sea-bird'. The modern slang meaning is 'dated' or 'unfashionable' and might derive from 'gone away' but probably does not. None of this gives us any positive etymological answers, I fear.

nail[1] *American* to copulate.

Of a male, and perhaps from the slang *nail*, the penis. But we also have the rhyming slang *hammer and nail*, TAIL[1], which would be an expected etymological development; and, with even less charm, the concept of the male entry into the female; or, equally charmless, an analogy with *screw* (see SCREW[1]).

nail[2] a cigarette

Abbreviation of *coffin nail* from the adverse effect of health, but referring usually to inferior tobacco:

Smoke if you want to... I thought you were desperate for one of those East German nails. (Deighton, 1994)

nameless crime (the) *British* buggery.

A common use when homosexuality was illegal although openly practised among some intellectuals, especially at the ancient universities, which were largely male societies. The charge sheet for those detected and prosecuted would refer to 'the abominable crime of buggery', although murder and rape earned no such descriptive embellishment.

nancy a male homosexual.

From the female name. Originally as *Miss Nancy* and then as *nancy boy*:

He looked a bit of a nancy boy to me. (Matthew, 1978)

Also shortened to *nance*.

nanny *obsolete* a prostitute.

The female partner of a GOAT rather than from the nursery form of *nurse*. A *nanny-house* was a brothel:

... speech smacking of grogshop or nanny-house. (Graves, 1940, writing in archaic style)

Napoleon's revenge diarrhoea.

As suffered by British tourists in France:

A lady friend, travelling through France with her family, was stricken with a rather severe attack of 'Napoleon's revenge'. (*At Your Convenience*, 1988)

napoo see NAB.

nappy an infant's towel to contain excreta.

Probably merely a shortened form of *napkin*, a small *nap* or piece of linen, but it might also pun on the catching function—see NAB.

narrow passageway to the unknown (the) death.

You are given no chance to turn aside:

The narrow passageway to the unknown which everyone must cross. (J. Mitford, 1963)

The obsolete British *narrow bed* was a grave.

nasty[1] **(the)** a spirituous intoxicant.

Unpleasant to the teetotaller but now humorous use only:

'What you need is a wee bit of the old nasty.' I uncorked the Armagnac. (Sanders, 1983)

Also in the fuller form *the nasty stuff*:

How about a bit of the old nasty stuff before we turn in? (Sanders, 1977)

nasty[2] drunk.

From taking too much of *the nasty* (see NASTY[1] (THE)) or the way drunks behave:

I shared a car back to London with Peter and we sat in the back getting thoroughly nasty on a clutch of freebie bottles of Hine or Martell. (Fry, 1994)

nasty complaint (a) a venereal disease.

Not just telling the waiter about a hair in your soup:

After a business trip to the Middle East, Brown found that he was suffering from a nasty complaint. (*Private Eye*, February 1989—it was being implied that he had venereal disease)

national assistance see ASSISTANCE.

national indoor game see GAME[2].

national security guard an instrument of civil repression.

A phenomenon of any totalitarian state because autocracy can brook no overt opposition:

> The shark pool ... was established by Nassir's feared henchmen from the National Security Guard. (*Daily Telegraph*, August 1980)

See also SECURITY SERVICE.

national service compulsory conscription into the armed forces.

Civil servants, politicians even, may consider their service to the nation no less meritorious than spending a limited period on and off the barrack square. The usage conceals the military nature of the engagement, although some pacifists in the British Isles were given the alternatives of going down the mines, felling trees, or attending the sick.

National Socialist a member of a chauvinistic and totalitarian political party.

The *National Socialist German Workers' Party* was better known as the Nazi party:

> ... we shall settle accounts with them in the manner to which we National Socialists are accustomed. (Hitler, speech of 20 July 1944 in translation, of the Jews)

nationalize to appropriate.

Denoting the appropriation of foreign resources, businesses, and the like, such as the seizure by the Iranians in May 1951 of the British refinery at Abadan. It is also in standard use to denote the taking of undertakings from the private sector into state control with payment of compensation. The subsequent inefficient operation of many such enterprises brought the term into disrepute. See also PUBLIC OWNERSHIP.

native a black.

Literally, as in Dr Johnson's definition, 'an original inhabitant', but extended in the colonial era to all non-whites:

> 'He admits to having abandoned twenty men to their deaths.' Vera said: 'They were only natives.' (Christie, 1939)

A white expatriate's black mistress was known as his *native comfort*.

Native American a person with North-American Indian ancestry.

For the transatlantic observer, a rather harsh usage which appears to disparage the greater part of those who were born in the US and look upon it as their native land:

> 'An Indian,' I said ... 'I mean, a Native American.' (Theroux, 1993)

natural[1] *?obsolete* an idiot.

Probably a shortened form of *natural (born) fool*, an expression which antedated this use by a century (OED):

> We had oor naiteral. He was known as Daft Jamie. (Inglis, 1895)

natural[2] bastard.

Originally describing a child who was sired by the father of the family as distinct from an adopted child, it came from the late 16th century to mean 'sired outside of wedlock'. A *natural son* was a bastard (and therefore unable to inherit), sired by a *natural father*:

> Edward VII, a most wide-ranging man in his attraction to ladies, was his natural father. (Condon, 1966)

Today it is used non-euphemistically of the biological parents of an adopted child, a usage which recalls the earlier sense.

natural break the intrusion of advertisements in a television broadcast.

The British licensing authority stipulates that the interruptions should not spoil the continuity of a programme. Many viewers welcome the opportunity to make tea or urinate without missing anything. Whence the humorous *natural break*, an invitation in a meeting etc. for those involved to go and urinate.

natural functions (the) urination and defecation.

Eating, sweating, and breathing are equally natural, to name a few:

> ... reaching peaks of embarrassment whenever he wished to fulfil one of his natural functions. (R. V. Jones, 1978)

Occasionally as *natural purposes*. The obsolete *naturals*, a shortened form of *natural parts*, meant the genitalia of humans and animals and to *be in your naturals* was to be naked.

natural vigours copulation.

By a male in the days when it was thought to come less naturally to females:

> I have my natural vigours, like any man. (Fowles, 1985, of a man excusing a sexual approach to a woman)

nature's garb nudity.

No clothes at all, as worn by a *naturist*, in pursuit of *naturism*, a penchant for individual or communal nakedness.

nature's needs urination or defecation.

A variant of *natural functions* (see NATURAL FUNCTIONS (THE)):

> For another of nature's needs I also inserted a large rubber bag. (Theroux, 1975)

The American *nature stop* is taken by the highway to urinate rather than to admire the scenery.

naturist/naturism see NATURE'S GARB.

naughty copulating extramaritally.

Usually by a female, as it was not considered wickedness in a male:

> She had been naughty as a girl, she said, especially with one boy. (Harris, 1925)

A *naughty girl* or *lady* is a prostitute:

> ... to which, incidentally, came many of the naughty ladies of Paris to improve their complexions. (Fingall, 1937)

The *naughty-house* was a brothel:

> This house, if it be not a bawd's house, it is pity of her life, for it is a naughty house. (Shakespeare, *Measure for Measure*)

naughty bits (the) the male or female genitalia.

A usage of adults with children which makes a number of unmerited assumptions and might even implant unfortunate inhibitions.

nautch girl a prostitute.

Literally, a professional Indian dancing girl:

> She kept a troupe of nautch-girls who were also prostitutes. (Richards, 1936, of India)

A *nautch*, vagina, might just have come from the Hindi but I favour derivation from the 18th-century American *notch*, a 'narrow opening or defile through mountains, a deep narrow pass' (SOED).

Neapolitan bone-ache *obsolete* syphilis.

The disease you caught from the Italians, if not the French or the Spanish:

> Vengeance on the whole camp! or, rather, the Neapolitan bone-ache! for that, methinks is the curse ... (Shakespeare, *Troilus and Cressida*)

Syphilis was also known sarcastically as the *Neapolitan favour*.

near[1] stingy.

From the literal meaning 'close', itself a shortened form of *close-fisted*:

> Some were beginning to consider Oak a near man. (Hardy, 1874)

near[2] imitation.

Mencken gives: 'near-silk, near antique, near-leather, near mahogany, near-silver and near-porcelain' (1941). Consumer protection legislation has thinned the list, but *near-beer*, served in an unlicensed den called a *near-beer club*, survived the Second World War as a way of circumventing British liquor licensing laws:

> Near-beer costs two shillings a glass: call it just beer—forget the 'near'. (Kersh, 1936)

necessary a lavatory.

The Italian *necessario* or the French *nécessaire*, which latter survives in literary use:

> ... this unlucky medicine chest having played the same part that Marie Antoinette's nécessaire did in the escape to Varennes. (N. Mitford, 1945)

Also in obsolete use as *necessary house*:

> A contrivance for emptying every Necessary House in the City of London ... (Monsarrat, 1978, writing in archaic style)

Whence the *necessary woman*, who was not the mandatory female member of any committee but the lavatory emptier:

> Trott the Necessary Woman, who stalked the house at all hours ... to empty and then clean the several privies. (ibid.)

necessities urination or defecation.

Perhaps a survival from NECESSARY:

> Only let him out in the garden for necessities. (Herriot, 1981, of a dog)

neck to kiss and caress amorously.

Short of copulation, from the placing of an arm around the other's neck at some stage:

> To copulate or at least neck, in the relative comfort of a parked sedan. (Ustinov, 1971)

neck-oil an intoxicant.

Army use of beer:

> We were fond of a drop of 'neck-oil', which like 'purge' was a nickname for beer. (Richards, 1936)

necklace to murder by igniting a rubber tyre placed on the shoulders of the victim.

A method used by some blacks, mainly of the African National Congress party, against other blacks, for crime or being of a different political persuasion:

> ... some stone throwing, petrol bombing and necklacing of innocent people. (BBC News, 30 August 1989, reporting black rioting)

necktie party *American* a lynching.

The *necktie* was the noose:

> Its solitary bent branch enough to tell any Western fan that it would eventually be used for a necktie party. (Deighton, 1972)

A *necktie sociable* is also a lynching. The victim was *measured for a necktie*:

> ... then he knew he was being measured for a necktie. (Price, 1985, of someone facing death by lynching)

need a desire to urinate or defecate.

Short for *need to urinate* etc.:

> Mostyn ... had flitted from the room to cope with a nervous need. (Price, 1970)

needle to strengthen (an intoxicant) by adulteration.

Originally by introducing an electric current through the rod shaped like a needle, whence any form of lacing:

> The smell of needled beer ... (Longstreet, 1956)

negative is used, in the same way as the adverbial prefix *less*, to avoid precision or as an evasion in many phrasal euphemisms, of which a sample only appear below.

negative containment a dangerous leak of radiation from a nuclear reactor.

Quoted in *Daily Telegraph*, 9 March 1994.

negative contribution a sale at a loss.

Manufacturers' and commercial jargon. The *contribution* is that part of the price left over after deducting the cost of labour and materials. A *positive contribution* indicates that some or all overhead and selling costs have been recovered. A *negative profit contribution* means that you have lost money after deducting all your costs.

negative employee situation *American* the dismissal of staff.

Not just someone failing to answer in the affirmative.

negative growth a decline.

Politicians so speak of the national product, businessmen of turnover or profits:

> With International Leisure somewhat becalmed at 112p having shown no negative growth in two years... (*Private Eye*, September 1986)

It is taboo to use words which expose recession or failure.

negative incident *American* an event which may cause adverse publicity.

The language of PUBLIC RELATIONS:

> 'Will they have a representative on the train?' 'To minimize negative incidents... I'm using their jargon, dammit.' (Francis, 1988)

negative (income-)tax state payment to the poor.

The proposition seems to have been first expounded by Milton Friedman under the title *negative tax*. The object is to give the rich less to spend (as at present through income tax) and the poor more, cutting out the present cumbersome channels of individual assessment and distribution.

negative patient care outcome death.

The phrase could be understood to mean no more than that a test has proved the absence of infection.

negative stock holding orders which cannot be delivered.

This is how your computer describes empty shelves in the warehouse when you have overdue orders and clamant customers. Generally computers are programmed to deduct orders from unallocated stocks and to throw up re-order or remanufacture schedules.

negatively privileged poor.

Sociological jargon and a correct statement only of those who have elected to lead a life of monastic asceticism. See also PRIVILEGED and UNDERPRIVILEGED.

negotiable we do not expect to receive the asking price.

Estate agents' jargon, often shortened to *neg.* in small advertisements. Of an overseas commercial sale to a corrupt buyer, a *negotiable* price means that it can be inflated by the amount of bribery needed to win the order.

negotiate to yield or appease.

The language of diplomacy where bullies or appeasers are involved:

> Halifax... had urged the Polish Foreign Minister, Beck, to negotiate, (i.e. yield) upon Hitler's demand to annex Dantzig. (Crossman, 1981—it was not the last time the arch-appeaser was to urge *negotiation* with Hitler)

negotiation an illegal act in espionage.

> He was an extraordinary asset to the American intelligence community, a veteran of twenty-two years of the most complicated 'negotiations'. (Ludlum, 1979)

Whence *negotiator*:

> The State Department's small band of 'negotiators'... (ibid.)

In standard usage a *negotiation* is a business transaction involving haggling.

negro *obsolete* a slave.

Grose gives 'A black-a-moor: figuratively used for a slave. I'll be no man's negro; I will be no man's slave'. Rawson (1981) has an erudite note on how *negro*, with or without an initial capital, moved into and out of fashion. It is now definitely out, which is why I chose *black* and *white* to describe people of different ancestry, remaining conscious of the inaccuracy of those labels.

neighbourhood connection *American* a receiver of stolen property.

Thieves' jargon.

Nelly a homosexual.

Either male or female, although a *nelly fag* is male. From the meaning 'a weak-spirited or silly person' (SOED), which could just have evolved from *not on your Nelly Duff* via *duffer*, but this is only guesswork. A duffer is a useless person but see FLUFF YOUR DUFF for a possible link to this usage.

Nelson's blood rum.

The corpse of the Admiral was returned from Trafalgar via Gibraltar in 1805 for burial in London. The preservative was probably brandy and not rum. Tradition has it that it was depleted on the voyage because sailors syphoned it off and drank it.

neo-Malthusianism see MALTHUSIANISM.

neoplasm a cancer.

Literally, a fresh growth. Mainly medical jargon.

nerve agent a noxious gas.

British military jargon. It could mean anything which excites one of the senses and so stimulates a nerve, not excluding a woman's perfume.

nervous breakdown a severe mental illness.

Not paralysis, where some of the nerves really do break down:

> The man before him had similarly had a nervous breakdown and had had to be brought South by an Indian sub-assistant surgeon. (Allen, 1975)

The expression may in lay use describe any condition from depression to madness, temporary or permanent.

nest the vagina.

With visual imagery:

> ... in your daughter's womb I'll bury them:
> Where, in that nest of spicery, they shall breed.
> (Shakespeare, *Richard III*)

19th-century British slang but now perhaps obsolete.

Netherlands (the) the male and female genitalia.

Probably punning on the Low Countries, itself a convoluted punning 16th-century joke:

> The Netherlands?—O, sir, I did not look so low. (Shakespeare, *Comedy of Errors*)

Netherlands is obsolete but *nether parts* or *regions* in the same sense persist.

networking using social contacts for political or financial purposes.

Probably less from the British *old-boy network*, the acquaintance formed at private schools and fostered in later life, than from the jargon of the world of telecommunications and management information systems:

> I hate the word 'networking', but I love parties and clubs. (*Sunday Telegraph*, 24 July 1994, quoting an international businessman; his statement is reminiscent of 'As I told the queen, I hate namedropping')

neutralize to kill.

Much more than rendering neutral or inert:

> It means they don't know he's been... neutralized. (Follett, 1978, of a killing)

never-never (the) a contract for hire-purchase.

A common usage in the pre-inflation days of the 1930s, from the ethic that you should *never* enter into a contract for something for which you could not pay cash or you would *never* finish paying. Goods so acquired were said to be *on the never-never*. Whence figurative adjectival use:

> Critics rebuked it for rousing expectations it could not satisfy, and for fostering a never never mentality among a public who now irritatedly refused to pay the price of profligacy. (Lee, 1989, of the 1979 Irish budget, which had been so reckless as to suggest a 2 per cent levy on farmers, who at that time paid only 1 per cent of their gross income in tax against 16 per cent for the average industrial worker)

new age travellers vagrants.

British itinerants who reject conventional attitudes to employment and trespass:

> In addition [to gypsies] there are estimated to be 2,500 to 5,000 'New Age Travellers' (*Daily Telegraph*, 19 August 1992)

and see also TRAVELLER.

New Australian an immigrant.

After the Australians decided to welcome immigrants who were neither British nor white, a phrase had to be coined to avoid any reference to their skin pigmentation or country of origin, even if it meant linguistically denying antipodean babies their birthright.

New Commonwealth a group of countries in which the majority of the inhabitants are not white.

After the Second World War *Empire* had too many overtones of conquest and white supremacy and a new name was needed for the agglomeration of former British colonies etc. which continued to consult with each other:

> At the Commonwealth Prime Ministers' Conference in September [1962] it was clear that neither the 'old Dominions' nor the 'new Commonwealth' were happy about recent developments in the negotiations. (Crossman, 1981—the negotiations were for British entry into the EEC)

Whatever the relative length of your association with Britain, you belong to the *New Commonwealth* if the majority of your citizens are black, although the readmitted South Africa may be an exception.

new cookie see COOKIE.

new economic zones the barren places to which you exile your opponents.

They are too busy trying to stay alive to cause trouble, or they die. Thus the victorious Communist Vietnamese eliminated those who were unable to get hold of a boat:

> Vietnam's 'New Economic zones' (in fact areas of internal exile where many starve and perish). (*Daily Telegraph*, February 1980)

New Order a chauvinist tyrannical autocracy.

The rule of the Nazis:

> ... told them there was a new order in Europe. (Keneally, 1982, of Nazis in Second World War Poland)

But *order* for some was disorder for most. The Philippine variant was called the *New Society* and a great many Filipinos looked back with regret to the old.

Newgate a prison.

Denoting other British prisons than the notorious one in London. There were many compounds to do with jail, hanging, etc. such as *Newgate bird*, a thief, *Newgate solicitor*, a corrupt lawyer, etc. Now obsolete except in literature.

news management the official suppression of information.

For American military or political purposes. The *management* embraces delay and manipulation rather than attempts to get lies published.

nibble an act of casual copulation.

Literally, a small bite:

> She makes a damn pretty widow ... 'Wouldn't mind a nibble myself.' (Lyall, 1972)

Occasionally also used within marriage by a husband.

nice time a single act of copulation with a prostitute.

Prostitutes' jargon often used when soliciting:

> You've given me the ticket, and I've given you a nice time. (G. Greene, 1932)

Nice-nice is copulation:

> I should have made nice-nice with Martha. (Sanders, 1983)

Nick[1] the devil.

After one of the Nordic evil spirits or monsters:

> O thou! Whatever title suit thee,
> Auld Hornie, Satan, Nick, or Clootie.
> (Burns, 1785)

Today usually as *old Nick* but also as *Nickie* and *Nicker*.

nick[2] to steal.

In standard usage, cutting or catching unawares, whence originally only of pilfering:

> We dinna steal. We only nick things whiles. (Crockett, 1896)

Today of any stealing:

> He was caught nicking shavers. (N. Mitford, 1960)

The obsolete *nicks* meant 'stolen property'.

nick[3] a police station or prison.

A common usage, probably from being caught, or *nicked*.

nick[4] to castrate.

From the cutting:

> Through mist or fog to nick a sturdy hog. (Dickinson, 1866)

Now used also of vasectomy.

nickel and dime *American* to short-change or cheat.

From the pre-Second World War stores such as Woolworth, which offered goods to the value of 5c and 10c:

> The kind of guy who'll nickel-and-dime his own mother. (M. Thomas, 1987)

night bucket a pot for urine.

Usually in communal male sleeping quarters where it can avoid the ingress of cold air through repeated opening of a door in winter.

night club a place in which to meet prostitutes.

In proper usage, a restaurant open to the public for entertainment until late at night, and some of them are indeed properly conducted, but:

> A night-club or dance-hall hostess ... are the modern equivalents of the old-time disorderly house and of the street-walker. (Lavine, 1930)

The obsolete British *night house* also had no facilities for copulation on the premises so that prostitutes took their customers elsewhere, but it provided no entertainment other than eyeing the females:

> These generally resort to night-houses, where they have a greater chance of meeting customers. (Mayhew, 1862, of prostitutes)

night job a contract in which a prostitute devotes the entire night to a single customer.

Interchangeable with ALL-NIGHTER:

> They ran to wake up mama, who was sleeping after a night job. (L. Armstrong, 1955)

night loss the involuntary ejaculation of semen during sleep.

Mainly female usage, usually referring to the soiled bedlinen. Also as *night emission* or NOCTURNAL EMISSION.

night physic copulation.

The 'medical' treatment taken, especially by the male, at night. Also as *night exercise* or *nocturnal exercise*:

> ... if I'm not down to twelve stone by the time we reach Calcutta, it won't be for want of nocturnal exercise. (Fraser, 1975)

Other variants are the American punning *night baseball*, which is usually of casual copulation away from the home, and *night games*:

> 'He was too old for games.' 'What kind of games?' 'Night games,' she said softly. (Theroux, 1992)

night stool a portable lavatory.

Sickroom use. It looks like a square seat.

nightcap a drink of an intoxicant.

You don't place it on your head but drink it before retiring to an unheated bedroom:

> A 'nightcap', which consisted of a stoup of mulled claret, well spiced and fortified with a glass of brandy. (Lowson, 1890)

Now of any intoxicant drunk in the evening:

> May I please offer you a nightcap? (M. Thomas, 1980—the offerer was trying to pick up a stranger)

nightingale a police informer.

From the 'singing' properties of man and bird. In obsolete British use a *nightingale* was also a prostitute, from her hours of work; or a soldier who cried out while being flogged, to avoid which a victim would *chew*, or *bite*, a *bullet*, thus further enriching the language.

nightsoil human faeces.

Soil has meant 'excrement' since the 16th century and primitive lavatories were cleaned at night, sometimes by a *nightman* in an operation called, in London at least, a *wedding*:

> ...thrust our ragged clothes, with a stick deep into the night soil at the necessary house. (Graves, 1940, writing in archaic style)

Now mainly jocular figurative use as *in the nightsoil*, a variant of *in the shit*. *Night water*, of urine, is obsolete:

> You try to tell us that the might of this great army rests upon goddam night water? (Keneally, 1979, of a Confederate Army forbidden to make noise at night)

nightwork *?obsolete* copulation.

Punning on nocturnal labour:

> Ha, 'twas a merry night. And is Jane Nightwork alive?... She was then a bona-roba. (Shakespeare, *2 Henry IV*)

nineteenth (the) the bar at a golf club.

Shortened form of the humorous *nineteenth hole*; the first eighteen involve striking a ball and walking after it.

nip[1] to steal.

A variant of PINCH[1] and much rarer. It used to appear in many compounds to do with pilfering, cutting purses, etc. and also meant 'to give short measure':

> Ye was set aff frae oon for nipping the pyes. (Ramsay, 1737)

A *nipper* was a thief.

nip[2] a measure of spirituous intoxicant.

Originally, a *nipperkin*, the eighth part of a pint:

> Down to the bar to snatch a furtive 'nip'. (Doherty, 1884)

nip[3] to castrate.

From the action of the tool employed:

> It was to 'nip' some calves... or more correctly to emasculate them by means of the Burdizzo bloodless castrator. (Herriot, 1981)

no better than she ought to be sexually promiscuous.

Usually of a younger woman by an older. Also as *no better than she should be*:

> ...dissolute young Guards officers dining and spending the night with women no better than they should be. (S. Hastings, 1994)

no chicken old.

A *chicken* is the young of a domestic fowl, whence a child. Always of a female:

> And Caroline is twenty-seven. No chicken. (Bogarde, 1981)

Also as *no spring chicken*:

> She's old enough in the picture. 'I'm no spring chicken myself.' (Macdonald, 1976)

no comment I admit nothing.

Much political and business use in reply to journalists. It is the only defence of those who know that, when scandal is in the air, to be quoted is to be misquoted, and selectively.

no longer with us dead.

Especially used of a former associate:

> None of us could believe that the charming Deborah... was no longer with us. (Mailer, 1965)

No more is more direct:

> Mrs de Moleyns, a loving wife, a tender mother, and a good and true friend to the poor in her village, is now no more. (Dunning, 1993, quoting from a 19th-century source)

No longer in service comes from the jargon of espionage:

> Fensing is no longer in service... officially we're calling it suicide. (Hall, 1988)

no mayday see MAYDAY, ROUTINE (NURSING) CARE ONLY.

no more see NO LONGER WITH US.

no right to correspondence (have) to be dead.

Russian Communist usage in various similar forms—the dead cannot read letters:

> 'No right to correspondence'—and that almost for certain means: 'He's been shot'. (Solzhenitsyn, 1974, in translation)

no scholar dull.

Parental description of a child who may then be said to have other talents such as being 'good with his hands'; if you are a manufacturer he may be recommended to you as a potential industrial manager, being thought unsuited for more intellectual callings. (This is a sample entry, of many expressions starting with *no*, which seek to soften or understate a weakness or infirmity.)

no show[1] the fraudulent entry of a name on a pay sheet.

Of a person who fails to report for work but, usually with the connivance of another, continues illegally to collect his pay; or of the entry on a pay sheet of a worker who does not exist or is not employed, the pay being drawn by a third party. See also GHOST[1].

no show[2] a person who reserves an aircraft seat and fails to travel.

A curse of the airlines, which in turn encourages them to overbook and then to *bump* passengers (see BUMP[5]).

no-tell *American* used for extramarital copulation.

It is the manager who keeps quiet:

> He was found in one of those no-tell motels. (Wambaugh, 1983)

nobble to kill.

Literally, to tamper with a horse illegally, whence to do a lot of other evil deeds such as bribing and stealing in the 19th century:

> Ah thowt ah'd tak a wauk an nobble a few specimens foe me-sen. (Treddlehoyle, 1892)

Whence the modern meaning:

> 'I saw a bloke nobbled here,' she said. 'I mean killed.' (Theroux, 1976)

nocturnal emission an involuntary ejaculation of semen.

Spitting, vomiting, or ejaculation during copulation are not so described. A rather grander version of NIGHT LOSS, with a suggestion of erotic dreaming:

> He got a great deal of pleasure from nocturnal emissions. (Sharpe, 1978)

nocturnal exercise see NIGHT PHYSIC.

nocturne a prostitute.

Literally, a night scene in a painting or a dreamy musical composition. I give this obsolete British use to allow me to quote George Sand's apocryphal pun to Chopin: 'One nocturne deserves another.'

noddy *British* a policeman.

By transference from PC Plod (see PLOD) whose exemplary behaviour graced the Noddy books?

> ... hardly worth the shoe leather of the luckless noddy taking statements. (Blacker, 1992)

noggin an intoxicating drink.

Originally, an eighth of a pint of any liquid:

> Only share of two noggins wid my brother. (Carleton, 1836)

Now used of any type of beer or spirits, but not usually of wine.

nominal extortionate.

Literally, token or minimal in relationship to what might be reasonable. Used by professional advisers, especially merchant bankers, of fees which are breathtaking and ruinous to their industrial clients, the implication being that they are doing you a favour by their moderation.

non-aligned vacillating in allegiance.

The representatives of small countries which so described themselves met in Belgrade in 1961, claiming with more or less sincerity that they favoured neither Washington nor Moscow, whence Jennings' 1965 description of them as 'no more than potential parasites'.

non-Aryan Jewish.

The Nazis adopted de Gobineau's theories of *racial purity*, which class Slavs and gypsies with Jews under this heading. As Himmler had limited numbers of gypsies and no Slavs to persecute before 1941, Poles apart, *non-Aryan* became generally associated with Jewishness, as dysphemism rather than euphemism.

non-heart beating donor a corpse.

It must sound better than *cadaver* to the recipient of an organ transplant:

> Their proposed alternative for dead... 'non-heart beating donor'. (*Daily Telegraph*, 11 March 1994, quoting *The British Medical Journal—they* were transplant surgeons)

non-industrial poor and relatively uncivilized.

The latest in a long line of euphemisms adopted to avoid offending post-colonial rulers:

> 'Civilized' and 'primitive' were to be replaced by 'industrial' and 'non-industrial'. (*Daily Telegraph*, 12 May 1993, quoting a document issued by Leeds Metropolitan University)

non-penetrative sex mutual sexual activity short of copulation or buggery.

Not merely kissing with closed lips or the fertilization of eggs after spawning:

> Robyn and Charles were into non-penetrative sex these days. (Lodge, 1988)

non-performing asset a loan on which interest is not being paid.

Bankers' jargon, disregarding the fact that it is the borrower who is not performing.

non-person a person without civil rights.

Communist jargon used to denote those whose former fame or achievements embarrass the present rulers:

> Krupsky was banished twenty years ago. He became a non-person. (Ludlum, 1979)

non-profit avoiding taxation.

Not any old loss-making enterprise. An American business device whereby the eventual beneficiary passes the profit through a tax-exempt charity:

> The profits that are now routinely extracted by the promoters on 'non-profit' cemeteries are spectacular. (J. Mitford, 1963)

non-traditional casting putting a black in a role written for a white.

Stage jargon of a practice which may give greater satisfaction to the director than to the audience:

> But the logic doesn't apply, and the term 'non-traditional' is inadequate. What we really have here is theatrical PC. (*Daily Telegraph*, 6 February 1993, of casting a black without make-up in the role of a white New Englander)

non-white a person whose ancestry is not entirely white.

Particularly of those with any black ancestry:

> Non-whites are even more overwhelming in their desire for work. (Pei, 1969)

Howard writes of *non-white*: 'the latest silly extremity into which we have been forced by euphemism' (1977), but this taboo about skin pigmentation will go on spawning euphemisms until a person with only one white great-grandparent is described as *non-black*, or found unworthy by society of separate notice.

nonsense casual sexual activity outside normal courtship.

Literally, an absurdity:

> He was a calm, down-to-earth creature who brooked no kind of 'nonsense'. (Bogarde, 1981, of the proprietor of an erotic photographic studio)

Genteel and perhaps obsolete usage.

nookie copulation.

A *nook* is a small orifice or corner, whence the vagina viewed sexually:

> You might even have enough free time to get you a little nooky. (Styron, 1976)

Now very common, within or outside marriage, in popular speech.

North Britain Scotland.

More common when Scotland was seen as relatively backward, not least by many Scots people themselves:

> Near to this Marble are deposited the Remains of Hugh Campbell Esq[re] of Mayfield in the County of Ayr North Britain 5 Jan 1824. (Memorial in Bath Abbey)

Often abbreviated to *NB*, especially on mail. Whence *North British*, Scots or Scottish:

> The poet Burns wrote in the North British dialect. (Wodehouse, 1930—Jeeves was correcting Wooster's enunciation)

Also as *North Briton*, a Scot, although John Wilkes, who used the nom-de-plume, was a Londoner. See also WEST BRITON.

nose habit see HABIT.

nose job cosmetic surgery on a nose.

Denoting surgery mainly for females, who tend not to be satisfied with the feature with which nature endowed them:

> Turns out that she always wanted a nose job. (Clancy, 1989)

nose open eager for copulation.

Bulls and stallions flare their nostrils when sexually excited. Of humans the phrase is used of the male or female although the physical symptoms are different from those of animals:

> 'I seen her mooching around upstairs.' Murf licked his lips. 'She's got your nose open?' (Theroux, 1976)

not a great reader illiterate.

You still hear this among country folk in south-western England, and no doubt elsewhere. This is a sample entry of many phrases prefixed *not* which seek to minimize or conceal a weakness or infirmity.

not all there of low intelligence.

Of a mental state and not the amputation of a limb:

> That poor creature who's not quite all there. (Christie, 1940)

Atypically, *all there* means 'keenly intelligent'.

not as young as I was old.

None of us is as young as we were, even as the eye crosses the page. We use this expression as an excuse for our own failing powers, but of another might say that he was *not in his first youth*.

not at home at home but unwilling to speak to a caller.

The converse of *At home*, a specific invitation to visit at a set time:

> 'Want to see Mrs Morny.' 'She's not at home.' 'Didn't you know that when I gave you the card?' ... 'I only knew when she told me.' (Chandler, 1943)

Also as *not in*:

> Weren't you told she was not in? (ibid.)

not available to comment unwilling to risk being compromised.

This is how journalists describe a victim who has denied them the chance of confronting him with an accusation or scandal.

not dead but gone before dead.

Christian usage, implying both that the dead person will be reunited with a spouse still alive, and that meanwhile he is enjoying eternity in a state not dissimilar to life on earth. Less often as *not lost but gone before*, with *before* meaning 'ahead'.

not in see NOT AT HOME.

not inconsolable ready to copulate with other than your usual but absent sexual partner.

For *consolation* see CONSOLE:

> It is feared she waited for him in vain. Not that the Lady Frances, a creature of some resilience, proved inconsolable. (Boyle, 1979 —the *him* was the spy Kim Philby)

not interested in the opposite sex homosexual.

A genteel usage of adult males.

not invented here we reject and denigrate all ideas other than our own.

The main defensive mechanism of the employees in any establishment whose duty it is to think or innovate. Your existence is no longer justifiable if outsiders discover or invent something which you have been paid to discover or invent:

> They didn't think of it, so they'll piss all over it. *Not invented here*! (M. Thomas, 1980)

Often abbreviated to *NIH*.

not sixteen annas to the rupee of low intelligence.

British Indian army use, for the old currency in which 4 pice = 1 anna, 16 annas = 1 rupee. And the British *not sixteen ounces in the pound* meant the same thing. But see also TWELVE ANNAS IN THE RUPEE.

not very well very ill.

Hospital and valetudinarian jargon which ignores the presumption that *very well* implies perfect health. A hospital may use *not at all well* or *not doing well* to signal that a geriatric patient is near death.

notice dismissal from employment.

Shortened form of *notice of dismissal*, which is given or received. *Notice* as a verb is obsolete:

> Notice me as much as ever ye like, I'll not clean them pigs out. (M. Francis, 1901)

nouvelle cuisine small portions of food sold at high prices.

The presentation on the plate is often tasteful, but there is plenty of room for an elaborate layout:

> She says that both her cooking and its presentation are more voluptuous than nouvelle cuisine. (*Country Homes*, June 1990)

See also HAUTE CUISINE.

nuggets *American* the testicles.

Literally, anything small and valuable. Also used of the breasts of an adult female and of capsules of illegal narcotics.

nuisance see COMMIT A NUISANCE.

nullification killing.

A form of cancelling out, I suppose:

> They have also been reported to have been used to kill enemy divers, in the case of the US, as part of a 'swimmer nullification' programme. (*Sunday Telegraph*, 29 March 1992, of captive whales being trained by the CIA)

number is up (your) you are about to be killed.

First World War usage, from the game of *house*, where each player has a numbered card, and perhaps referring to each soldier's individual army number. It indicates the fatalism of the trenches, with a deity selecting his faithful or victims on a chance basis:

> It's all right, you laughing, but I know my number is up. (Richards, 1933, of the First World War)

number nine a laxative.

The standard British army purgative. Some figurative use as when a sluggard might be told he needed a *dose of number nines*.

number one(s)[1] urination.

Mainly nursery usage, in the phrase *do number one(s)*.

number one[2] self-interest.

As in the Second World War British army adage *Number One comes first*, act selfishly; now usually in the phrase *look after number one*:

> ...he believes trade policy should be founded on protection. Look after Number One. (A. Clark, 1993)

number one, London *British, ?obsolete* menstruation.

Any etymological link with the town house of the Duke of Wellington, which had that address and is now a museum overlooking Hyde Park Corner, is obscure.

number two(s) defecation.

Mainly nursery usage, in the phase *do number two(s)*; but also of adult activity:

> Stand over him and, as he put it 'do number two—oh lots of it—all over me'. (Theroux, 1973)

numeracy see COMPREHENSIVE.

nun *obsolete* a prostitute.

The religious orders provided many allusive words for sexual subjects before and immediately after the dissolution of the English monasteries, partly because of the supposed dissolution of their members and partly because the economic power of the Church had engendered that mixture of envy and outrage in 16th-century England without which Henry VIII's Reformation would have failed. Thus when Hamlet says to Ophelia 'Get thee to a nunnery' (Shakespeare, *Hamlet*) he is telling her he thinks she is a prostitute, a *nunnery* being a brothel.

nurse to suckle (a baby).

Probably from the name given to a woman who was paid to suckle another's child:

> Priss...was nursing her baby...'I never expected a breast-fed grandson,' said Priss's mother. (McCarthy, 1963)

The obsolete English *nurse-child* was a bastard, because it was raised away from its mother.

nursing home an institution for geriatrics.

Literally, a hospital for any sick person, in which sense it is still widely used in the British Isles.

nut a lunatic.

Nut is slang for the human head and this must be a shortened form of *gone in the nut* or some such expression:

It was the laugh of a nut. (Chandler, 1940)

A *nut college, farm, hutch,* or *house* is an institution for the insane:

... round up of nut-houses, likely nutters on parole. (Davidson, 1978)

The FBI list of mad or unstable people likely to attack a public figure is called the *nut-box*. A *nutter* is a lunatic, who may be *nutty, nuts,* or *off his nut,* mad.

nuts testicles.

Also as, and perhaps even a shortened form of, *nutmegs*, but see *cobs* under COBBLERS for the same imagery:

... the new government ... will cut our nuts off. (M. Thomas, 1980—the threat was figurative only)

nymph a prostitute.

In standard usage, a mythical semi-divine and beautiful maiden. More explicitly as *nymph of darkness, of delight, of the pavement,* etc.

NYR an airman lost in action.

Second World War usage, as an abbreviation of *not yet returned* from a mission over enemy territory:

'We've got a lot of NYRs, Lester.' 'Not Yet Returned doesn't mean dead.' (Deighton, 1982—but it meant shot down or crashed, with death a probability)

O

O opium.

Used by addicts.

oats copulation.

Usually by a male, within or outside marriage, with an implication of regular need, as in the daily nourishment of horses:

> I have to go out later, so you'll have to wait even longer before you get your oats. (B. Forbes, 1986—a woman was refusing not to make porridge for but to copulate immediately with a man)

The *oats* in this case are cultivated but see SOW YOUR WILD OATS.

obligatory appointed other than on merit.

Of membership of committees, boards, etc. where it may be thought expedient or politically necessary to have other than those chosen from a male dominant group:

> ... she's my recommendation ... for our obligatory female. (Price, 1985, of such an appointee)

See also STATUTORY and TOKENISM.

oblige[1] to copulate with.

Literally, to gratify, which a female may do in this regard with or without payment:

> Does Mrs Hagerty oblige or are we reduced to the habits of our youth? (le Carré, 1989—Mrs Hagerty was a housekeeper to whom an alternative, if she proved chaste, was masturbation)

oblige[2] to work as a domestic servant.

The British usage, by implying that the employee (always a female and often elderly) is conferring a favour on the employer, seeks to remove the supposed stigma of doing someone else's housework on a regular basis for money. See also HAND and HELP[1].

oblique homosexual.

The common imagery of divergence from what is STRAIGHT:

> ... whether she has unmasked his disguise, or because her tastes were oblique, or because she is a man who thinks she is a woman ... (Bradbury, 1983)

obtain to acquire illegally.

Usually of stealing but also of forbidden or other embargoed goods:

> '... many small pleasures ... not the least of which is obtaining Cuban cigars.' 'Obtaining' was the Director's favourite euphemism. (van Lustbader, 1983, of an American espionage agency)

occupational dose an exposure to radiation in excess of normal.

The amount of additional radiation suffered by those employed in the nuclear industry or as radiotherapists and falsely represented as not constituting a health hazard.

occupied defeated and annexed.

Not all conquerors depart:

> Let us hope the Administration will not be foxholed by Beijing, and will stand with Congress, which unanimously passed a resolution declaring Tibet an occupied country. (*New York Times*, 13 April 1993)

Much pejorative use, but the British *armies of occupation*, scattered around the globe after the Second World War, were only too happy to go home.

occupy to copulate with (a female).

From the physical entry:

> These villains will make the word as odious as the word 'occupy'; which was an excellent good word before it was ill sorted. (Shakespeare, *2 Henry IV*)

An *occupying house* was a brothel. The rare modern use seems to come rather from *occupied*, or engaged, with:

> Karl was not ready, having been occupied with a Negro girl in his tent. (Harris, 1925)

odd homosexual.

From the meaning 'out of the ordinary' but as more people find nothing out of the ordinary in homosexuality, so this euphemism is falling into disuse:

> The successful challenges that have been made to the popular media images of lesbians as 'odd girls' and 'twilight lovers' ... (Faderman, 1991)

odorously challenged smelly.

Noteworthy as an ingenious extension of those before whom that traditional symbol, the gauntlet, is thrown down—see CHALLENGED for other examples:

> The list of minority victim-groups with special rights is getting longer every year, and now includes Hispanics, Asian-Americans, women (all 51 per cent of the population), the obese, and finally the smelly (odorously challenged). (*Sunday Telegraph*, 20 November 1994—but the list is far from exclusive)

of mature years old.

We do not mean a female of nineteen or a male of twenty-one; it is not full development we see but incipient decline:

> A good many of my students were civil servants, some of them of mature years. (Forsey, 1990)

See also MATURE[1].

off[1] *American* to kill.

Perhaps a shortened form of *bump off* and hardly from the meaning 'in a state of decay':

> Maybe he stiffed the waiter and the guy followed him down here and offed him. (Sanders, 1973)

DSUE also gives 'to die' as a First World War meaning and today to *off yourself* is to commit suicide:

> I just don't wanna off myself like so many cops do. (Wambaugh, 1975)

off[2] to copulate with.

American male use, perhaps from the slang *have it off* (see HAVE IT).

off[3] with its implications of departure and decay precedes many phrases describing types of mental illness:

Off at the side, of a mild condition, is obsolete:

> Not 'all there'—'off at the side'. (Linton, 1866)

Off your head may imply that you are mad, with many slang variants for *head*, including *chump*, *gourd*, *napper*, *nut*, *onion*, *turnip*, etc.:

> I must be going off my chump. (Wodehouse, 1930)
>
> He feared she had gone off her gourd, and he was scared. (Sanders, 1982)
>
> The fixture was scratched owing to events occurring which convinced the old boy I was off my napper. (Wodehouse, 1930)
>
> When ... she informed him one day that she was engaged ..., he went right off his onion. (Wodehouse, 1922)
>
> Unless he had gone off his turnip, I suppose. (le Carré, 1980—he was not refusing vegetables)

Off your rocker and *off your trolley*, indicating perhaps a temporary derangement, come from the apparatus, or *trolley*, which picks up the overhead electric supply for a tram or trolley bus:

> I think he was really off his rocker for a bit. (Amis, 1988)
>
> There are moments when I wonder if I'm tipping off my trolley. (Deighton, 1985)

Off the rails means 'mad', again with transport imagery:

> You could see that a very unstable personality placed in this environment would go off the rails. (*Maclean's Magazine*, 9 November 1993)

Off the wall means anything from 'eccentric' to 'mad', from the unpredictable bounce off the fence in baseball:

> ... it was a crazy cackle, and maybe she really was off the wall. (Sanders, 1982)

Off your tree probably comes from the slang meaning 'axle', the imagery being of a wheel coming away, although there is an American meaning of *tree*, the mind:

> Who the hell is she? She's off her tree. (le Carré, 1989)

off-colour[1] vulgar or offensive.

The colour is not necessarily BLUE[2]:

> I don't want any of your off-colour stuff from the Drones' smoking-room. (Wodehouse, 1934)

off-colour[2] ill.

Of a temporary affliction where the victim may be paler than usual. Also of menstruation.

off duty menstruating.

A female use, to explain why she cannot copulate.

off games menstruating.

A variant of OFF DUTY, punning on *fun and games* and a schoolgirl's minor indisposition:

> ... errant husbands who have looked to her for corrective therapy during periods when their wives have been in the country/abroad/off games. (*Private Eye*, December 1983)

off the chandelier bogus.

Of bids at an auction, where the auctioneer is trying to run up the price by pretending there are active bidders in the room:

> ... the bidding, which moved slowly from $4 to $6 million, proved to be all 'off the chandelier'. (*Daily Telegraph*, January 1990)

Depending on the decor, bids may come off the walls, pictures, or whatever catches the auctioneer's eye.

off the payroll dismissed from employment.

Joining the rest of humanity which was never on that particular payroll in the first place:

> So the old boy hadn't known I was 'off the payroll'! (Shirer, 1984—an American publisher had contacted a journalist who had been dismissed from the newspaper)

off the peg inferior or ill-cut.

In standard usage this describes garments which are bought ready-made rather than tailored:

> In an off-the-peg dress ... she did not look her best. (Ellmann, 1988)

Most people of normal dimensions now buy such clothes, their quality has improved greatly, and the condescension which the usage implied is now seldom met.

off the rails[1] see OFF[3].

off the rails[2] being detected in reprehensible conduct.

Criminal or sexual, of a person hitherto considered above reproach, and implying thereafter a continued pattern of such behaviour:

> Johnny Depp is a dream as the bad boy tempting a nice girl off the rails. (*TV Quick*, 19 December 1992)

off the voting list dead.

Voting lists containing the name of every adult are regularly updated but the dead do not enjoy the franchise, except in Northern Ireland, Chicago, and any other constituency where living voters are enjoined to 'vote early and vote often'.

off the wagon see ON THE WAGON.

off-white wedding the marriage of a pregnant bride.

She may or may not eschew the pleasure of a virginal white dress:

> I married Pauline hastily—a quiet off-white wedding at her parish church. (Lodge, 1962 —Pauline was pregnant)

offer yourself to ask a man to copulate with you.

Usually extramaritally and free of charge:

> She tracked me down to my rooms in Oxford and offered herself to me. (Amis, 1978)

In obsolete use either sex might, it seems, *offer kindness* to the other in the same sense:

> Offred her such Kindnes, as sticks by her ribs a good while after. (Wilson, 1603, quoted in ODEP)

oil to bribe.

A variant of GREASE[1]. In the same sense, palms and wheels may be *oiled*.

oiled drunk.

Things may for a time seem to run more smoothly:

> Phipps, described by Yakimov as 'a trifle oiled', had attacked the Major. (Manning, 1965)

The commonest cliché, whatever the state of inebriation, is *well oiled*:

> He was well oiled by the time the coffee waiter returned. (Deighton, 1988)

The obsolete British *oil the wig* meant 'to become drunk' and the obsolete Scottish *oil of malt* was whisky.

old a' ill thing the devil.

Our ancestors really believed that if you talked of the devil, he might appear, but with a certain naivety assumed that he could be fooled by allusive references. Many of these 'nicknames' and evasions were preceded by *auld* or *old*, and include *bendy, blazes, bog(e)y, boots, boy, chap, child, cloutie, dad, davy, driver, gentleman, gooseberry, Harry, hornie, lad, mahoon, man, Nick, one, poger, poker, Roger, ruffin, Sandy, scratch, serpent, smoker, sooty, thief, toast*, etc. Only a few are euphemistic; others were dysphemisms and some downright insulting. If nothing else did, the variety must have confused him. Elsewhere I have dealt with the various names, ignoring the prefix *old*. And I have referred to the rather sad practice of farmers who left a patch of ground untilled for the devil's use, in the hope that it would placate him and he would leave the rest of the farm alone (see GOOD MAN):

> The old man's fold, where the druid sacrificed to the demon. (EDD—I wonder if the farmer had to pay tithe on it too)

old Adam (the) a man's lust.

From the unregenerate character of our common ancestor before life became complicated for him and he passed on to us, with St Paul's help, our sexual complexes:

> I felt the old Adam stir at the sight of her. (Fraser, 1973)

old bill see BILL.

old boots a sexual mistress who has been discarded by another male.

Less often as *old shoes*. The later incumbent may be said to *keep* or *ride in* such footwear, his predecessor having had the first use.

old faithful menstruation.

By coming back regularly, anxiety about pregnancy is lifted.

old-fashioned derelict.

Real estate jargon:

> When applied to houses old-fashioned means a draughty ruin. When applied to clubs it means bad food and no women. (Theroux, 1982)

old Joe venereal disease.

Stalin was so known in a brief period of Second World War popular enthusiasm, which seems to have no relevance at all to this American usage.

old maid an unmarried woman who is unlikely to marry.

A *maid* was an unmarried girl and, after the 17th century in the normal linguistic progression, an unmarried female of any age:

> There will the devil meet me, like an old cuckold, with horns on his head, and say, 'Get you to heaven, Beatrice, get you to heaven; here's no place for you maids.' So deliver I up my apes, and away to St Peter. (Shakespeare, *Much Ado*—see LEAD APES IN HELL for the simian allusion; in those days there was thought to be no place for virgins in hell)

Now standard English:

> I'm able to keep myself, and to wait as long as I choose till I get married; *I'm* not afraid of being an old maid! (Somerville and Ross, 1894)

For a dissertation on the plight of unmarried females a century or less ago, see WITHOUT A HEAD.

old man[1] see OLD A' ILL THING.

old man[2] the penis.

Male usage, usually denoting your own, possibly from OLD MAN[1], the devil with his licentious conduct, although the phrase is used of the penis in its flaccid state:

> His old man needed to set it trying to haul itself up into his abdomen. (Amis, 1978—he had difficulty in achieving an erection)

To *give the old man his supper* is a vulgarism for 'to copulate'.

old man[3] a man with whom a woman cohabits and copulates extramaritally.

The phrase is widely used of a husband in speech:

> She had an 'Old Man'—the name we used to have for a common-law husband. (L. Armstrong, 1955)

old man's fold see OLD A' ILL THING.

old man's friend pneumonia.

It was the illness which allowed the elderly to die quite quickly without much pain. Penicillin may now preserve them for more lingering, painful, and degrading deaths.

old soldier (the) a pretence of hardship or physical incapacity.

Acted, or *come*, by beggars seeking alms and sympathy in the days when wounded soldiers were returned to society without pensions or other relief. We still use the phrase of a child who seeks consolation without special apparent cause.

older woman (the) an elderly female.

Advertising jargon which omits to state what her age is compared with. Similarly the advertisers' *larger woman* is not merely bigger than a midget, but raw-boned and very tall, or very fat.

oldest profession see PROFESSION (THE).

on[1] drunk.

In a mild state:

> I shouldn't like to zay how he was drunk... he was a little bit on like. (EDD)

That use is obsolete but we still hear *half on*, which as usual is no less drunk than the whole.

on[2] pregnant.

The obsolete usage differs from the modern:

> I doubt she's on again, poor lass (EDD)

meant 'I think she's pregnant again' whereas we might take it as no worse a circumstance than perhaps missing a train to work. But we have no doubt about the status of a lady who is *four months on*. EDD also gives 'Of a female "Maris appetans"' but, exceptionally, without an example and it may just have been a a shortened form of *on heat*.

on[3] *American* habitually using illegal narcotics.

A shortened form of *on drugs*:

> But a woman like that living a life like that, has *got* to be on. (Sanders, 1977)

on[4] available for casual copulation.

On in the sense 'happening' or 'going ahead':

> Those legs at the corner table might be on, but then they could just be here for conversation. (Blacker, 1992—he was alluding to the person and not just the limbs)

on a cloud under the influence of narcotics.

From the floating feeling. The cloud is sometimes numbered *nine* after the cumulonimbus, which may reach 30,000 to 40,000 feet according to the measurement of the US Weather Bureau.

on heat[1] able to conceive.

Of female animals other than humans, from the increased temperature associated with the condition. *In heat* is less common.

on heat[2] anxious for copulation.

From ON HEAT[1], no doubt, and usually of women:

> Those bloody women! Like a lot of randy she-cats. And there's that bitch back again, on heat as usual. (Manning, 1962, of a princess and not a female dog)

Of a male it may mean no more than lusting after:

> Are you on heat for her, Reverend? (Cornwell, 1993)

In heat is less common:

> 'I'm no bitch in heat,' she said between tight teeth, 'take your paws off me.' (Chandler, 1958)

In the heat means 'copulating':

> ... make love to her afterwards. Would you like to hear tapes [of] Mike Santos in the heat? (West, 1979)

on her way pregnant.

The destination is unstated:

> She is two months on her way. (Shakespeare, *Love's Labour's Lost*)

Now rare.

on jawbone see JAWBONE.

on the bash see BASH.

on the beach see BEACH.

on the bend see BEND.

on the bottle see BOTTLE[1] (THE).

on the chisel see CHISEL.

on the club ill and absent from work.

The British *club* was an association of workers paying money weekly into a communal fund from which they received help when ill. *On the panel* meant the same thing, from the *panel*, or list, of doctors willing to treat patients unable to pay their fees (although the Scottish *on the panel* meant 'being accused in court'):

Mr James Mitchel was upon the panell at the criminal court for shutting at the Archbishop of St Andrews. (Kirkton, 1817)

The Scottish *on the board*, indigent, came from the parochial *board* which dealt locally with paupers (EDD). *On the box* formerly meant 'destitute', being helped from the donations made in church:

Fifteen got assistance from the Poor's Fund; or as it was generally expressed... fifteen... were on the box (Pennecuik, 1715)

but latterly *on the box* was synonymous with *on the club*, the weekly contributions being collected in a box.

on the coal see GO ON THE COAL.

on the couch engaged in casual copulation.

The same imagery as the *casting couch* (see CASTING COUCH (THE)), although offices are seldom so furnished:

My wife thinks I have endless lines of big-titted girls trying to get me on the couch. (Deighton, 1972)

A person described as being *on the couch* may be the patient of a psychiatrist and so, by implication, may be mentally ill.

on the cross engaged in robbery as a profession.

Not punning on the ordeal suffered by Jesus and others, but derived from the *cross* of *double-cross*:

The hostile glare of the decent did not prevent men and women 'on the cross' from constructing pecking orders. (Hughes, 1987, of convicts in New South Wales)

See also *cross girl* at CROSS.

on the grind engaged in prostitution.

Prostitutes' jargon punning on GRIND, to copulate, and the daily *grind* of honest toil.

on the Hill engaged in bribery.

The *Hill* is the Capitol Hill in Washington DC, where lobbyists may offer more than verbal persuasion to legislators and their staffs.

on the hoist see HOIST[1].

on the job copulating.

A common pun on being engaged in work:

'We told him you'd been on the job continuously'... He paused fractionally as the implications of that statement flashed through his mind. (Price, 1970)

Whence *in mid-job*, so engaged:

If he could snap his fingers and boof, there he was in mid-job, very pleasant. (Amis, 1978)

on the labour see LABOUR[2].

on the loose see LOOSE[1].

on the left illegally.

Describing operating without a permit etc. and using the common sinister connotation:

... a small shop whose manager made more money selling drink 'on the left' than he did by dry-cleaning. (Clancy, 1988)

on the make *American* seeking an extramarital sexual relationship.

Literally, overly ambitious or greedy in an impatient way. This euphemism is used of both sexes:

Once in a while... a man and a woman talk without dragging bedrooms into it. This could be it, or she could just think I was on the make. (Chandler, 1953)

on the needle addicted to illegal narcotics taken by self-injection.

From the hypodermic syringe. *Needle Park* in New York City is a haunt of drug addicts who may discard needles after use.

on the nest *American* pregnant.

From the sedentary behaviour of a broody hen.

on the pad in receipt of regular bribes.

American police jargon, from the *pad* on which the particulars of the arrangements are noted:

Everybody's on the pad then... The pimps, the barkeeps, they just put up the dough. (Turow, 1987—but not the bakers)

on the panel see ON THE CLUB.

on the parish destitute.

Historically every British parish had to provide for the needy poor within its boundaries, either on its own or in conjunction with neighbouring parishes. The poor might be housed in the CHURCH HOUSE, the parish house, or the *union* (see UNION[2]). The system was financed by a 'parochial rate', which meant that to be *on the parochial* was synonymous with being *on the parish*:

They did their very best to get him gang on the 'parochial'. (*Aberdeen Weekly Free Press*, March 1901, quoted in EDD)

In urban areas, the destitute might be described as *on the town*.

on the peg see ON THE SHELF.

on the pill see PILL[2] (THE).

on the piss engaged on a drunken carouse.

Usually drinking beer, where the bulk induces frequent urination. It does not mean that, like the former Indian premier Desai, you drink your urine for medicinal purposes.

on the pull seeking an extramarital sexual partner.

Of males or females, but not necessarily prostitutes, or even desiring copulation:

She wasn't on the pull that night and, even if she had been, my public profile was too low to grace her boudoir. (Blacker, 1992)

(I am indebted to my teenage Crossley granddaughters for clarification of this entry.)

on the ribs *?obsolete* indigent.

I suppose because of the protrusion of the ribs of an undernourished person:

'How's life, Duke?' 'On the ribs.' 'You skint?' 'Dead skint.' (Kersh, 1936)

on the road to reform subject to harsh treatment for a political prisoner.

In Communist Russia, *reform* was the crushing of a prisoner's individualism by a combination of starvation, banishment to a remote region, brutality, and excessive manual work. For those subjected to it, the *road* was a hard one.

on the rockpile *?obsolete* in prison.

From the breaking of stones as a convict's chore rather than a reference to the British prison at Portland, where the famous stone is quarried:

Anyone who's fool enough to invite ten years on the rock-pile for his superstitions deserves all he gets. (Fraser, 1975, writing in 19th-century style)

on the roof see ON THE TILES.

on the seat defecating.

Not normally used of female urination:

Tell them I'm on the seat, my compliments. (Seymour, 1977, indicating a slight delay before being available)

on the shelf unmarried and unlikely to marry.

Usually but not always of females and derogatory in modern use. The imagery is from retailing. Occasionally, of women only, as *on the peg*, where an unused garment may stay.

on the side (of a benefit) enjoyed illegally or immorally.

Such as a bribe; undeclared and untaxed income; and extramarital copulation, especially where a man keeps a sexual mistress, or *bit on the side*. The imagery is from the additional food on a separate plate but served with the main dish.

on the skids failing.

Of a commercial enterprise. A *skid* is a piece of wood on which an object is placed to facilitate unstoppable movement, such as the launching of a ship:

Its current affairs flagship World in Action is on the skids. (*Private Eye*, May 1981)

on the square living honestly.

Criminal jargon in a society where it is reprehensible to be law-abiding:

Going on the square is so dreadfully confining. (Mayhew, 1862)

The members of the Freemasons' secret society so describe their participation among themselves, not because they lead honest lives but from the set square used in building.

on the street(s) see STREET (THE).

on the stroll engaged in prostitution.

From the leisurely walk while seeking custom:

Hello, Mayann. What in the world are you doing out on the stroll tonight? (L. Armstrong, 1955—I don't think Mayann bothered to elaborate)

on the take see TAKE[1].

on the tiles engaged in a carouse.

In the nocturnal company of the tom-cats; usually describing a male taking part in any leisure activity outside the home and involving a late return:

I saw you sneaking up the stairs. Been having a night on the tiles, have you? (Sharpe, 1975)

The American *on the roof* is rarer:

I was on the roof last night and I've got a hangover. (Chandler, 1944)

on the town[1] engaged in a carouse.

Literally, on a rare visit to a city's theatres etc. without much thought of expense and used of both sexes without any implication of the debauchery imported by ON THE TILES. The obsolete British *on the town* meant 'engaged in prostitution as a regular way of life':

She had been on the town for fifteen years. (Mayhew, 1862, of a prostitute)

on the town[2] see ON THE PARISH.

on the trot[1] escaped from prison.

From the running:

I'm looking for someone, and if he's here, he's probably told you he's on the trot. (Follett, 1978)

on the trot[2] see TROTS (THE).

on the wagon refraining from drinking intoxicants.

Wagon is a shortened form of *water wagon*, in which potable supplies may be distributed. Of a single case of abstinence, as someone about to drive a car; or of a former alcoholic who is trying to cure himself:

On the wagon now, of course, and what he drunk was with a wink and shake of the head. (Longstreet, 1956)

Off the wagon indicates backsliding by a former addict:

When a man like that goes off the wagon, he bites dust. (Kersh, 1936)

on top of copulating with.

Of a male, from the common posture and implying a casual single act:

Isn't there anything else to interest you, except twenty minutes on top of a girl? (Kersh, 1936)

on your back copulating.

Of a female, from the position commonly adopted:

One way to travel. On my back. (L. Thomas, 1977—the speaker is a prostitute who thus paid for her transport)

The Second World War British slogan to encourage the growth of vegetables, *Back to the Land*, was lewdly attributed to the *Women's Land Army*, the young females directed to work on farms, who had an undeserved reputation for profligacy.

on your bones *?obsolete* indigent.

Starvation has consumed the flesh:

Give us a chance, constable; I'm right on my bones. (Galsworthy, 1924)

on your ear bankrupt.

Your ear is also what you are traditionally *thrown out on*, presumably from the uncomfortable landing. In America it may also mean 'drunk'.

on your shield dead.

The shield doubled for a stretcher if you were killed in battle:

...the only way out was on your shield. (Keneally, 1982, of trying to resist the Nazi police)

on your way out dying.

The common imagery of departure:

...a pretty little nurse to special him on his way out. (Price, 1979—he was a dying patient)

onanism masturbation.

Onan spilled his seed on the ground, for which he was slain by the Lord (Genesis 38:9,10):

One night I got thinking of E... and for the first time in months practised onanism. (Harris, 1925)

The biblical evidence points to the withdrawal method of contraception rather than to masturbation, of which Harris was fully aware:

Very soon I played Onan and like the Biblical hero 'spilt my seed upon the ground'. (ibid. —he had been copulating)

Occasionally of female masturbation, which was not free from taboo and medical scaremongering:

Those poor girls, he went on, were dying by the thousands from consumption, but really from self-abuse or onanism, as it was often called. Masturbation would also arrest growth, distort the pelvis, and prevent the development of the breasts. (Pearsall, 1969, quoting MacFadden's *The Power and Beauty of Superb Womanhood*, 1901)

one-armed bandit see FRUIT MACHINE.

one foot in the grave near death.

Through old age or terminal illness, the other foot supposedly functioning normally. Some figurative use of an ineffective person and in both cases dysphemism rather than euphemism.

one for the road an extra drink of an intoxicant before leaving company.

From the warming, or stirrup, cup taken before cold winter journeys on horseback or in an unheated coach.

one-night stand a single night of copulation with a chance partner.

Punning on the travelling show which plays a single performance before moving on but not also punning, I think, on *stand*, an erection of the penis or the copulation of a stallion:

An opportunity for extracurricular sex occurred... Afterwards there had been still more opportunities—some the usual one-night stands. (Hailey, 1979)

A *one-nighter* is such a chance partner:

This little lady is a born one-nighter. (Francis, 1982)

one o'clock at the waterworks (it's) your trouser zip is undone.

An American warning from one male to another. From the hour when an employee might leave his office and go for lunch?

one of those a male homosexual.

Used by those who are not homosexuals, but for male homosexuals *one of us*:

When you asked him whether he knew any girls—the shadow of homosexuality, is he one of those? (le Carré, 1986)

Less often, sober and godly matrons may refer to a prostitute as *one of those*; for these, *one of us* would mean a social equal, usually in the negative and derogatively:

...he's not what Aunt Fenny calls one of us. (P. Scott, 1968, of a policeman commissioned into the army)

one off the wrist an act of masturbation.

Not a lost wristlet watch:

I'm afraid Mother was enjoying a quick one off the wrist. I'm a noisy lover when it comes to myself. (Fry, 1994—'Mother' was a homosexual male)

one over the eight an excessive intake of intoxicants on a single occasion.

There are eight pints in the gallon, which was considered a sufficient amount of beer for a regular drinker:

'Had one over the eight,' diagnosed Mr Blore accurately. (Christie, 1939, of a person who was drunk)

one-parent family a parent living alone with dependent offspring.

There are normally two parents, of whom one in this usage is permanently absent from the home, or, in the case of many young females, was never there at all:

> The one-parent family is going to be the big social problem of the 1980s, with the present rate of divorce. (Price, 1979—and an even bigger one of the 1990s)

Also as LONE PARENT and SINGLE PARENT.

one thing copulation.

There can be few wives who have not, at least once in their marriage, expressed the view that a male's interest in a female is solely sexual:

> I'd really—only—wanted—one—thing. She told me so this morning. (Amis, 1978)

one too many an intoxicant taken to excess.

Whence *had one too many*, became drunk:

> ... had one too many in a bar somewhere. (McCarthy, 1963)

one way ride a murder carried out by habitual criminals.

Where you go if you are *taken for a ride*:

> Charlie Luciano—now nicknamed Lucky Luciano on account of a one way ride that he came back from ... (Collins, 1981)

one way street *American* a heterosexual person.

Homosexual jargon for a person who is not homosexual. See also SWING BOTH WAYS.

one wing low see FLY ONE WING LOW.

open access needing no qualification.

Denoting entry to a university course without passing previous examinations:

> But both courses are 'open access'. (*Daily Telegraph*, October 1983, of degree courses in two London Polytechnic Colleges with Marxist bias, in 'social sciences'; the intention was to make it easy for blacks to enrol, and students could have REMEDIAL lessons in the English language before embarking on their further studies)

open housing a housing policy allowing no restriction on new residents in a district.

American white Christians are compelled to allow non-whites and non-Christians (especially blacks and Jews) to set up homes in the locality which they have kept to themselves. The objection, apart from snobbery and prejudice, is usually economic because *open housing* may drive down values of real estate.

open-legged see OPEN YOUR LEGS.

open marriage a marriage in which neither spouse hides extramarital copulation.

The *openness* consists in not lying to each other about lying with others:

> A groovy couple with an open marriage... (Bradbury, 1976)

An *open relationship* refers more often to a couple cohabiting and regularly engaging in mutual sexual activities but not to the exclusion of third parties:

> You and I have had an open relationship, with no strings. (Lodge, 1988, of such an arrangement)

open your bowel(s) to defecate.

A *bowel* is literally an intestine, whence any internal organ, and was so used by Cromwell:

> The enemy in all probability will be in our bowels in ten days. (Letter, 1643)

The euphemism is medical jargon:

> 'Have you had your bowel open?' he had asked Carfax. (Bradbury, 1959)

open your legs to copulate with a male.

Of obvious derivation:

> I'll teach her not to open her legs for bloody Germans. (Allbeury, 1978, of a Frenchwoman in the Second World War)

Open-legged describes a promiscuous woman as inviting copulation. The woman need not be a prostitute:

> ... the risks to my health, in being so open legg'd and free. (Cleland, 1749—but she did in fact make a charge)

opening medicine a laxative.

Not the first dose in a series, but *opening bowels*:

> Any pukka old soldier would have much preferred a dose of opening medicine. (Richards, 1933—to compulsory Church Parade)

operant conditioning payment by results.

American management jargon, with *operant* meaning 'having an effect' or 'capable of being measured'. In some circles *piece-work* has similar undesired overtones.

operational difficulties the ostensible reason why your journey will again be late.

The excuse given by transport operators to cover up breakdowns or incompetence. Also as *operational reasons*:

> The Aeroflot flight was eight minutes late. For 'operational reasons' the girl at information explained. (Seymour, 1982—for most airlines eight minutes late is early)

Also in the same sense as *operating difficulties*:

> 'Operating difficulties', I assume, which is BR-speak for some ASLEF slob, having drunk fourteen pints of beer the previous evening, now gone 'sick' and failed to turn up. (A. Clark, 1993—BR was the state-owned British

railway network and ASLEF the main union for engine drivers)

operator a swindler.

Literally, anyone who carries out an operation but, whatever his fee-structure, do not so describe a surgeon in his hearing:

> 'What does that mean—operator?' 'Well, I've done a bit of villainy.' (L. Thomas, 1978)

A politician, businessman, etc. described as an *operator* may use unconventional or questionable tactics to achieve his ends; of illegal narcotics, he is a dealer.

optically challenged blind.

The blind, or those with very bad eyesight, may also be *optically inconvenienced, marginalized*, or otherwise insulted by a succession of similar faddish usages.

oral sex cunnilingus or fellatio.

Passionate kissing is not so described:

> He preferred oral sex, something that obviated the need for a bed. (Green, 1979, of Rachman)

In the same sense, *oral service* is not what your dentist provides.

orbs a female's breasts viewed sexually.

There is no suggestion that they are twice the size of HEMISPHERES:

> ...her open night dress displayed all the splendours of her mature bosom's magnificent orbs. (Pearsall, 1969, quoting a 19th-century usage)

orchestras the testicles.

From the rhyming slang *orchestra stalls*, balls:

> ...catching one a direct bullseye in the orchestras, thus putting one completely *hors de combat* for at least a week. (Matthew, 1983)

order of the boot (the) summary dismissal from employment.

After the ancient *orders* of chivalry. There is no actual kicking, nor shoving if you get the *order of the push*.

orderly marketing price fixing between competitors.

The customers are all quoted much the same price, or the markets are shared out on a geographical basis between the fixers. *Orderly progress* means either 'price fixing' or 'the retention of a monopolistic position', both being dear to the hearts of those managing British state industries:

> Last week he sang of 'orderly progress' as 'preferable to the dangers of unbridled competition'. (*Sunday Express*, May 1981)

ordure excreta.

Literally, filth. Of faeces:

> Barbarians! The place is covered in...human ordure (Boyd, 1982—troops had defecated in every room)

and of vomit:

> But it's hard enough...without havin' that ordure there atop ye. (Keneally, 1979—soldiers were being sick)

organ the penis or vagina.

Shortened form of *sexual organ* or *organ of sex*. Usually of the penis:

> He displayed the organ, the secondary function of which is the relief of the bladder (Manning, 1965)

with *organs* meaning 'the penis and testicles':

> You've got to have a healthy view of your organs. (Bradbury, 1976—you do it with mirrors?)

Occasionally of the vagina:

> ...that organ of bliss in me, dedicated to its reception. (Cleland, 1749)

organization (the) a band of criminals.

Literally, a body of people working in concert, as in some pompous business titles such as *The Rank Organization*. Underworld jargon:

> It's the business of the Organization, and I don't know anything about that. (Seymour, 1992)

organize to induce to join a trade union.

Trade union jargon, from the *organization* of a branch etc. You organize both those who wish to participate and those who have to be coerced. In this language a company without a union, however well run, is said not to be *organized*.

orientation homosexuality.

We have moved a long way from the Christian desire to site a building so that it faces towards the east. In this use, a shortened form of *sexual orientation* meaning SEXUAL PREFERENCE:

> Trent had made no secret of his orientation, had gone public six years before. (Clancy, 1988)

oscar a male homosexual.

Not a successful actor but from the late Mr Wilde. More common in America and Australia than in the British Isles or his native Ireland.

ossified *American* drunk.

Instead of being STONED, you are turned into bone.

other arrangements repeated adultery.

Particularly where the male has been long absent from home:

> Many PoWs came home to discover that their wives had made 'other arrangements'. (Hastings, 1987, of the Korean war)

other place a British house of parliament.

For reasons of pedantry, it is not done for the member of one of the legislative bodies to refer to the other directly in the course of debate. Also as *another place*, which in a saner world would include everywhere beyond the confines of the chamber.

other side (the) death.

In spiritualist jargon, across the barrier between this world and the next. For others, the far bank of the Styx or Jordan on the way to the Elysian Fields or to life eternal. For a spy it is the enemy:

> He thought the Other Side was maybe savouring the tourist attractions... (Price, 1985)

other side of the tracks (the) the poor section of town.

When the American railroad arrived, it was often located on the edge of town where property was cheaper and it could be placed downwind of houses to avoid smoke, noise, and fire hazards. Eventually the town would develop around the station, with the richer inhabitants staying in the cleaner area and the poorer on *the other side of the tracks*. Now some figurative use too.

other way (the) homosexual.

Of either sex and a departure from the normal:

> He wouldn't look at his servants. His inclinations, if she knew it, are all the other way. (G. Greene, 1932—the servants were female)

other woman (the) a sexual mistress.

The usage ignores the fact that all womankind is *other* than the wife or other permanent partner:

> If Polly were not the 'other woman', she would advise Gus to go back to her. (McCarthy, 1963)

The other for a male is not a homosexual encounter but chance extramarital copulation with a female—see BIT.

ounce man *American* a dealer in illicit narcotics.

He buys in bulk and sells, often after adulterating the product, in small lots.

our masters see MASTERS (OUR).

out[1] available for marriage.

In the days when debutantes had their 'season' and were held out as being of marriageable age. See also COME OUT.

out[2] overtly homosexual.

From being *out of the closet* (see CLOSET[2] and COME OUT[3]). Whence to *out*, to acknowledge publicly that you are homosexual, and to *be outed*, to be thus exposed:

> Militant activists claim that they are now 'negotiating' with five other bishops (who, it is said, are being urged to admit to homosexuality or be 'outed'). (*Sunday Telegraph*, 12 March 1995)

Such acknowledgement, whether under duress or voluntary, is called *outing*:

> It is here that Outrage's tactics, particularly its threatened 'outings' of individual clergymen, are likely to cut sharpest. (ibid.—Outrage is the title of a body urging openness in homosexual matters, and equality of treatment between homosexuals and heterosexuals)

out of circulation menstruating.

Female usage, often to a male, with imagery from the lending library.

out of context said inadvisedly.

A use by politicians when they have forgotten what exactly they said, wish they had never said it, or were unaware that anyone was recording it. As journalists are known to select unfairly if it improves the story, this defensive manoeuvre is often effective.

out of town *American* in prison.

Suggesting perhaps that the convict might be away on business. Some humorous use.

out of your mind mad.

As distinct from momentarily overlooked. In various phrases, such as *out of your head, senses, skull*, etc.:

> He's out of his skull... ready for certifying. (Bogarde, 1981)

The usages are also of eccentric behaviour, as is *out of the envelope*, which comes from the design parameters within which an aircraft is expected to perform, as to rates of climb, stall, turn, etc.:

> He's somewhat out of the envelope, to use an old test pilot's phrase. (BBC Radio 5, 26 June 1994)

out to grass dismissed from employment.

Or in retirement, from the horse which is too old to work but has been spared the knacker.

out to lunch mentally unstable.

The American imagery is of a short absence from home, whence a mild and perhaps temporary mental affliction:

> His wife died about two years ago and he's been somewhat out to lunch ever since. (Diehl, 1978)

outhouse a lavatory.

From its separation from the dwelling house, across a court or down the garden. Whence to go *out the back*, to the lavatory. *Outdoor plumbing* is an American humorous use for a lavatory which is no more than a shed, a seat, and a hole in the ground.

outrage to copulate with a woman forcibly.

Literally, to offend her; an *outraged* female is someone who has been, or considers that she has been, raped:

> She complained to him that... some British soldiers had assaulted and outraged her... She

could have identified at least forty men who had outraged her. (Richards, 1933, of a First World War French village prostitute)

outstanding expensive.

Estate agents' puffing, from the meaning 'exceptional'. Literally, all construction is outstanding, with the exception of in-ground swimming pools, crypts, subways, and fallout shelters.

oval office the vagina viewed sexually.

Punning on the office of the President of the US:

> Ace, he was looking for a girl... 'Gone visit the oval office?' asked a man. (McInerney, 1992)

over-active naughty.

But beware of raising an eyebrow when a parent so describes her offspring:

> ...'we do have a special course for the Over-active Underachiever,' continued the headmaster. (Sharpe, 1982)

A child described as *hyperactive* is likely to show the same ill-discipline.

over-civilized decadent.

Nazi dysphemism in a culture where to appreciate beauty was to be effete:

> They are nearer to France, Europe's most over-civilised country. (Goebbels, 1945, in translation, of the Rhinelanders, whose supine reception of the Anglo-American invaders contrasted with the fanatical defence on the Eastern Front. There was a simpler explanation for the contrast, of which those facing capture by the Russians were keenly aware.)

over-familiar see FAMILIAR WITH.

over-gallant see GALLANT.

over-geared insolvent.

Gearing is the ratio of borrowings to assets. An *over-geared* company is too deeply in debt but, to maintain confidence and avoid a writ for defamation, it is taboo to talk of insolvency.

over-invoicing the payment of a bribe in a place selected by the recipient.

In some markets, where corruption is endemic but to be caught is to have your hands or your head cut off, the selling price of goods may be inflated by an amount which will in due course be placed to the credit of the person bribed somewhere beyond the jurisdiction of his masters. The invoice sent with the goods is the sum of the true cost plus the bribe. See also UNDER-INVOICING.

over Jordan dead.

Those Christians who have reached the *other side* (see OTHER SIDE (THE)). Whence the punning film title *Johnson over Jordan*. Occasionally as *over the creek*.

over-refreshed drunk.

There are several phrases prefixed by *over* which, sometimes humorously, seek to attribute drunkenness to a socially acceptable cause. Some may have taken a REFRESHER too many:

> ...post-prandial euphoria that Harry Woods euphemistically termed 'over-refreshed'. (Deighton, 1978)

Others may have *over-indulged*:

> ...thought for a moment I might have been over-indulging. (*Private Eye*, July 1981)

Politicians especially seem to be prone to *overtiredness* (see at TIRED[2]):

> He turned up to the first production meeting —in the morning—in an advanced state of over-tiredness. (*Private Eye*, 1980)

Others become *over-excited*. *Overcome* may be no more than a shortened form of *overcome by drink* etc. For *over the bat* see BAT[2].

over the broomstick cohabiting and copulating outside marriage.

This common 19th-century imagery is examined in JUMP[2]:

> ...this woman in Garradstreet here, had been married very young, over the broomstick (as we say), to a tramping man. (C. Dickens, 1861)

over the top achieving sexual orgasm.

Usually of a female:

> She made love to herself on the bath mat... She always felt awful afterwards... especially when she took herself...'Over the Top'. (McCarthy, 1963)

I suspect no link with attacking from a trench in the First World War:

> Darling, you can't really imagine ONE going over the top? (N. Mitford, 1960—the homosexual Cedric is explaining why he declined to participate in the Second World War and I suspect the delightful Miss Mitford knew both meanings)

over the wall escaping from prison.

Of obvious imagery but the British navy has to be different with its jargon meaning 'in prison':

> The Court Martial sentenced him to six months over the wall and he got dismissed from Service as well. (Jolly, 1988)

over there engaged in trench warfare in France in the First World War.

When the extent of British casualties encouraged evasion:

> [Peter] was seventeen and a half; next year would see him fighting. He had learned much of what it was like over there from his brother. (Hastings, 1994, quoting a fragment of E. Waugh—his brother Alec was serving on the Western Front)

For the American military, it meant service in Europe and the First and Second World Wars.

over your time late in menstruating.

Female use, with an implication of unwanted pregnancy. It may also refer to a delayed birth.

overcome see OVER-REFRESHED.

overdo the Dionysian rites to become drunk.

Dionysus discovered the art of wine-making and travelled widely to pass on his important knowledge. Being of catholic tastes, he engaged in sexual orgies, plays, human sacrifice, and flagellation, in addition to drinking wine. The obsolete *overdone*, drunk, probably owes nothing to this classical libertine.

overdose an attempt at suicide by self-poisoning.

Medical jargon, whether the protagonist fails or succeeds, and abbreviated to *OD*. It may refer to the attempt or to the person who makes it:

> *She's* a person, *not a goddamned OD*! (Clancy, 1989, of an attempted suicide patient in hospital)

overdue[1] pregnant.

From failing to menstruate at the expected time but not necessarily of an unwanted pregnancy.

overdue[2] in difficulty or crashed.

Aviation jargon, of an aircraft which has failed to report by radio as required during flight, or not landed as expected:

> *Overdue* connoted something quite different from *late* in airline parlance. (Block, 1979)

overfly to spy from an aircraft over enemy territory.

Literally, to cross a country in the course of a commercial flight by a recognized and agreed path. The American government in May 1960 described Gary Powers' shooting down in a U2 aircraft over Russia as an *overflight*. In 1962 they exchanged Powers for the Russian spy Rudolf Abel.

oversee *obsolete* to bewitch.

To *oversee* is to inspect or supervise, but one glance was enough for a true witch:

> It have brought all kind of disaster along with it. I must have been overseen when I took it. (Gissing, 1890)

To *overlook* was just as dangerous:

> Wha kens what ill it may bring to the bairn, if ye overlook it in that gate? (W. Scott, 1819)

And as *overshadow*:

> The last witness said deceased had been 'overshadowed' by someone. (*North Devon Herald*, 1896, quoted in EDD)

In Scotland friends and relatives at the bedside giving comfort were said to *oversee* a dying person.

own goal an accusation or campaign which damages the originator.

From the soccer player who inadvertently scores against his own side. Much loved of political commentators.

oyster an animal's testicle used for food.

From the shape and size. Also in America as *mountain oyster* and see also PRAIRIE OYSTER[1].

P

P anything taboo beginning with the letter *P*.

Usually *piss* and spelt PEE. Used figuratively in the common *P off*, go away.

PC see POLITICALLY CORRECT.

PG see PAYING GUEST.

pacify to conquer.

Literally, to bring peace to:

> ... the unsettled areas where we are still engaged in pacifying the Tajicks, Uzbecks and Khokandians. (Fraser, 1973—the areas had long been 'settled' by the nations named, but not by the Russian invaders)

Pacification is such conquest. For the British in Africa, their colonial rule was the *era of pacification* (Allen, 1970) and apologists in Washington were to use the same word:

> Pacification, for example, was hardly anything more than a swollen, computerized tit being forced upon an already violated population. (Herr, 1977, writing about Vietnam)

For invaders everywhere, a *pacification camp*, *centre*, etc. is a political prison:

> ... concentration camps are 'pacification centers'. (Commager, 1972)

pack it in to die.

Literally, to desist:

> That's where Jack's mate from Hong Kong packed it in. (Theroux, 1973, of a death)

In the First World War to *pack up* was to be killed in battle, from the meaning 'to cease to function'.

package on (a) drunkenness.

From the *load* (see LOAD[1]) which you have taken aboard and not from the obsolete *pack*, a rum-like liquor named for the unfortunate English General Pakenham, who was killed in the Battle of New Orleans (fought two months after the signing in Europe of a peace treaty between the combatants).

package store *American* a place which sells intoxicants.

The shopkeeper will hand them to you in a sealed container which you may not break open on the premises.

packet[1] a serious wound or death.

Literally, a small pack, whence an article sent by post, as in the *packet-boat* which carried the mails. In the First World War this particular missive was despatched at you by the enemy and the recipient *copped* or *caught* it.

packet[2] venereal disease.

The article *copped* or *caught* by careless or unwary servicemen in the Second World War. But today to *catch a packet* may mean no more than having a number of bills descend on you at once.

pad[1] dishonestly to inflate (a claim).

From *padding* clothing etc., to cause an apparent increase in size:

> The surcharges, padding and fictitious costs that were an inevitable part of every account. (Deighton, 1972)

There is no etymological link with the obsolete *pad*, to rob, as in *footpad*, a person who robbed on foot, which was from a *pad*, a path (see also HIGHWAYMAN).

pad[2] a cloth worn by a menstruating woman.

Euphemistic only as *wearing a pad*, which does not indicate preparedness to play hockey or baseball. In obsolete English use, such a towel was a *padlock*, punning on *pad* and on *padlocking the gentleman's pleasure garden*—the vagina.

Paddington relating to hanging.

And not just hanging about for a train at the London terminus:

> Tyburn being in the parish of Paddington, execution day was known as Paddington Fair, the hood drawn over one's head was the Paddington spectacles, and in dying one danced the Paddington frisk. (Hughes, 1987, of the 19th century)

paddy wagon a police vehicle.

From the preponderance of Irishmen in New England police departments and not from the national origins of those incarcerated:

> McCord and the other burglars being led out of the building and into a paddy wagon. (Colodny and Gettlin, 1991—the building was called Watergate)

pagan *obsolete* a prostitute.

Literally, one who does not worship God, and prostitution was no occupation for the upright:

> *Prince Henry* What pagan may that be?
> *Page* A proper gentlewoman.
> (Shakespeare, *2 Henry IV*)

paint the tape *American* fraudulently to record financial deals at fictitious prices.

The *tape* was the ticker tape on which market information would be diffused:

> Some of the amazing prices you read of in auctions are created by the owner selling to himself—what is called 'painting the tape' on Wall Street. (Train, 1983)

paint the town red to carouse.

Usually of a single session of celebratory drunken debauchery. It has been suggested, without a great deal of conviction, that the phrase originated in the American west, where a drunken spree might start in the brothel area, and then move up-town.

painted woman a prostitute.

Not an artist's model but one who painted her face before the practice became in succession permitted, normal, and then obligatory. Now perhaps obsolete.

painters are in (the) I am menstruating.

From the disruption and discoloration. Dated female use.

pair a woman's breasts.

Viewed sexually by a male. A female said to have a good or magnificent *pair* is neither an identical twin nor is she being complimented on her eyes or ears.

palm[1] an indication of bribery.

From the upturned hand, palms and venality have long gone together:

> You yourself
> Are much condemn'd to have an itching palm.
> (Shakespeare, *Julius Caesar*)

Whence many punning terms for bribery —*palmistry, palm soap, palm oil, palm grease*, etc.:

> It would be hard to dispute that a little such palm-grease must, upon occasion, have found a compliant hand. (Monsarrat, 1978)

To *anoint a palm* is to bribe.

palm[2] to cheat by prestidigitation.

You conceal cards in the palm of the hand. Also used figuratively of other forms of cheating or sharp practice, as in the phrase to *palm off with*, to give (someone) something of lesser value than agreed.

pan a pedestal type lavatory.

Literally, any bowl, but a *bedpan* is not used for cooking or washing. Whence *down the pan*, irretrievably lost. A *pancake*, a round, flat pile of cow shit, comes from its appearance.

panel (the) see ON THE CLUB.

panel-house a brothel.

I would like to think this American use came from the obsolete eastern English dialect *panel*, a prostitute:

> Panels march by two and three,
> Saying, Sweetheart, come with me.
> (Old ballad quoted in EDD)

However the more likely derivation is merely a description of a place where there are cubicles divided by panels. Occasionally too as *panel-joint*.

pansy a male homosexual.

From the delicate flower, *viola tricolor*:

> You're just a filthy pansy! No wonder your marriage has failed. (Masters, 1976)

panther sweat *American* whisky.

DAS suggests: 'May have originally been a euphem. for "panther piss"':

> 'Ran alky through here,' he said, 'in a beatup truck, white lightning, panther piss —whatever you want to call it.' (Sanders, 1980)

But where did *panther piss* come from, in the absence of panthers?

paper[1] a consideration other than in cash.

The jargon of the stock markets, when securities in the prospective buyer's firm are offered to holders of stock in the company they seek to acquire, rather than cash.

paper[2] a quantity of illegal narcotic.

From the folded paper in which it is passed to the buyer, and especially of methylamphetamine. *Paper acid* is LSD.

paper[3] a commercial agreement to murder.

Another American form of *contract* (see CONTRACT[2]), usually in the phrase *have paper out on* a person:

> 'It wasn't no amateur hit.' 'Are you tellin' me there was paper out on her?' (Diehl, 1978)

paper aeroplane a project to construct a new type of aircraft.

Usually drawn in outline, with a draft specification, in the hope of securing backing for development costs from a potential customer. The pun is less obvious with *paper helicopter*:

> There is a heavy health warning about assuming that paper helicopters always fly. (*Daily Telegraph*, 2 March 1994, of a document prepared by a hopeful manufacturer)

paper-hanger[1] *American* a policeman punishing a motorist for speeding.

From the *ticket* which is handed out on such occasions. With similar punning humour, the officer may be described as *doing his paperwork*.

paper-hanger[2] a person who passes false negotiable instruments.

Usually the instruments are cheques which have been stolen or are knowingly uncovered by assets:

> I've been stung too many times by the summer people. Paperhangers, I call them. (Theroux, 1974—an innkeeper was bemoaning his losses from cashing cheques for holidaymakers)

Also as *hang paper*, to issue cheques or other securities fraudulently:

> Jimmy gave me some good skinny on how to hang paper with minimum risk. (Sanders, 1990—*skinny* was originally 'a course or class in chemistry' (DAS) whence slang for any instruction)

paper the house to fill a theatre by giving tickets away.

Punning theatrical jargon, the *house* being the audience.

Paphian *?obsolete* a prostitute.

Paphos, or Cyprus, was sacred to Venus, the goddess of love:

> Cyprians of the better sort ... well acquainted with its Paphian intricacies. (Mayhew, 1862, of London prostitutes)

parallel parking *American* extramarital copulation.

The car illegally left beside the permitted one at the kerb. See also PARK[1].

parallel pricing the operation of a cartel.

The quotations or sale prices of competing manufacturers are held in an agreed relationship. Common commercial use.

paralysed very drunk.

And immobile:

> Dead drunk, paralysed, spifflicated. (Chandler, 1953)

paralytic very drunk.

Immobile, but not from paralysis or palsy:

> We had a marvellous wedding, Jerry and me. I was paralytic. (Theroux, 1983—but what did Jerry think about it?)

paramour a person with whom you regularly copulate extramaritally.

Originally, a suitor, acting 'through love', and of both sexes, although women seem to have more *paramours* than men:

> Married women go there with their paramours, for they are sure of secrecy. (Mayhew, 1862)

parboiled drunk.

The common culinary imagery. This time, thoroughly boiled, whence overheated.

parity the achievement of the best in any aspect of conditions of employment.

Trade union jargon. The equality you seek is only with the best, whether in terms of wages, hours of work, holidays, pensions, sick pay, etc.

park[1] to kiss and embrace in a parked car.

The American use, supposedly from parking your car in a secluded spot. There is no etymological link with the 19th-century London prostitutes or *park women*:

> Park women, properly so called, are those degraded creatures, who wander about the paths most frequented after nightfall in the Parks, and consent to any species of humiliation for the sake of acquiring a few shillings. (Mayhew, 1862)

park[2] to transfer (stocks) for an accomplice so as to conceal ownership.

Using the same imagery as WAREHOUSE:

> Last year he also, on five occasions, arranged with Keith Place of Natwest, to 'park' stock with each other to reduce their book positions —with an understanding to repurchase. (*Daily Telegraph*, 25 June 1994)

parliament *British* a lavatory.

19th-century slang, punning on 'sitting'. The obsolete British *parliament*, whisky on which excise duty had not been paid, reflected the widespread smuggling of spirits before the Railway Age:

> It's as good parliament as ever a gentleman tasted. (Croker, 1862)

parlor house *?obsolete* a brothel.

The American parlor, with the kitchen and bedroom, was where you expected to meet a woman:

> The parlor houses, cribs, brothels and bagnios had disappeared ... and a thousand prostitutes had been thrown out of work. (Gores, 1975)

parsley bed the place whence newborn babies came.

EDD defines it as 'A euphemism for the uterus' but the ensuing quotation and dissertation do not support the definition (vol. IV p. 427):

> How do babies come? What is the parsley bed the nurses and doctors say they come out of? (Pearsall, 1969, quoting from 1879)

There were taboos in Somerset against transplanting parsley, which would involve a fatality in the household. Girls tended to favour *parsley* while boys preferred *gooseberry bushes* for their ante-natal accommodation. Both crops are noted for their ability to thrive without being carefully weeded or cultivated.

part furnished without security of occupation.

For almost seven successive decades from 1914, British tenants were given the right to occupy unfurnished premises without fear of eviction while, in the majority of cases, paying uneconomic rents. Landlords resorted to the device of leaving some sticks of furniture in the premises in an attempt to evade what they considered unfair legislation, advertising such accommodation as *part furnished*.

part with child to abort involuntarily.

Or, in the Scottish dialect, to *part with Patrick*:

> Or he wan back she parted wi' patrick. (Graham, 1883)

partially sighted nearly blind.

If baldness were a subject of pity instead of humour, we would talk of the *partially haired*.

participate in to work for.

The usage seeks to play down the tedium of menial employment:

> Secretaries who are seen to participate in the fast-moving oil business... (*London Times*, May 1976, quoted in DDE)

partner a person cohabiting and copulating with another.

Usually heterosexually but also of homosexuals:

> Maternity nurses at the Royal United Hospital in Bath have been told to call fathers of newborn babies 'partners' rather than 'husbands', so as not to upset single mothers. (*Sunday Telegraph*, 20 March 1994)

It is apparently of no consequence if such abuse of language upsets husbands and wives.

parts see PRIVATE PARTS.

party[1] a battle.

Usually denoting short unsuccessful or violent fights:

> Dutch civilians weeping...for the few returning guests departing from what someone on the staff had chosen to call 'a party'. (Bogarde, 1978, on the Battle of Arnhem)

party[2] *American* an act of casual copulation.

Not, I think, from the social meeting for pleasure but because *party* is slang for a person, whence occasionally for a person who is sexually promiscuous:

> A smart little party of sixteen or so. (Overheard in Connecticut, 1982)

A *party girl* is a prostitute, presumably from her attending social events where she can meet customers:

> There were some snide references to what had befallen her, including a mention that she was known as a 'party girl'. (Sanders, 1986)

party member a Communist.

The usage dates from the pre-Second World War time when you kept quiet about being a Communist because many would consider you to be a traitor with revolutionary tendencies:

> That's why people convert to Catholicism, or become party members. (Bradbury, 1959)

pash a homosexual desire.

A shortened version of *passion*. Formerly much used to denote such feelings among schoolgirls for each other, or for a female teacher:

> Are you getting a pash for that little thing? (G. Greene, 1932, the question being asked by one woman of another about a third)

Less often of one-sided heterosexual feeling:

> Janet seems to be getting a pash for this Savory man. (ibid.—but people normally have pashes *on* rather than *for*)

See also CRUSH.

pass[1] to die.

The passage is from this world to the next:

> Harriet laughed. 'You have only to let them pass and they lose their importance.' 'You may pass with them, of course,' David said with a wry, sombre smile. (Manning, 1962 —Harriet was speaking of events)

More often in a phrase, such as *pass away*:

> Flora must have felt she was going to die, for just before she passed away... (L. Armstrong, 1955)

pass into the next world:

> He was the first to pass into the next world (Richards, 1933)

pass on:

> She murmured something sensitive just before she passed on (Bradbury, 1976)

and *pass off* the earth etc.:

> I haven't got the wind up, but some strong healthy men have been unlucky enough to pass off this Ball of Clay in double-quick time since we have been at this station. (Richards, 1936, of India)

Christians and spiritualists may *pass over*, arriving on the far bank of the Styx, the Jordan, the Great Divide, etc. To *pass in your checks* is a rarer variant of CASH IN YOUR CHECKS. To *pass out* only rarely means 'to die', usually meaning 'to faint'. A *passing* commonly describes a death.

pass[2] an unsolicited sexual approach.

Usually by a male to a female whom he does not know well, perhaps from the reconnaissance made by an aircraft overflying the target before attacking:

> Too many passes had been made at it and it had grown a little too smart in dodging them. (Chandler, 1943, of a woman's face)

Occasionally of homosexuals:

> Burgess sought Rees out later earning a mild rebuff for 'making a tentative pass' at him. (Boyle, 1979)

Although normally *made*, *passes* can apparently also be *thrown*, as in football:

> Threw a pass. Yes, as a matter of fact he did. (Amis, 1988, of such an approach)

pass air *American* to fart.

You may also be said, if so moved, to *pass gas* or *wind*.

pass water to urinate.

The phrase is so common that we give it no thought but see WATER:

> The nurse took him into a little cubicle and asked him to pass water into a bottle. (Bradbury, 1959)

passing see PASS[1].

past its sell-by date outmoded or useless.

A common cliché of the mid-1990s, based on the marking of perishable foodstuffs etc.:

Socialism: the package that's passed its sell-by date. (*Daily Telegraph*, 13 March 1987)

past (your) anything shameful or secret about your past life.

Of criminal activities or failure in business, and especially of extramarital copulation, in which case your *past* can be your general promiscuous conduct or an attachment to a single person:

'Part of your past, I presume?' 'No. At least, not as you mean it.' (Manning, 1965)

pasted *American* drunk or under the influence of illegal narcotics.

Probably from the common 'beating' imagery.

pasture to copulate.

Of a male, grazing as it were:

Fielding thought of Hecht pasturing in that thick body. (le Carré, 1962)

patient see AGENT.

patriotic front an organization of terrorists.

They cease to be described as terrorists if they win power.

patron[1] a man who keeps a sexual mistress.

Originally, he who stands in the relationship of a father, whence the concept of protecting:

An impotent and unkind man will produce a woman predisposed to fall in love instantly with her succeeding patron. (Chandler, 1944)

patron[2] a customer.

The *pater* has here moved from being a protector to being a supporter, as of the arts. The shopkeeper who so describes his customers seeks to put a finer gloss on a commercial relationship.

pause[1] the natural permanent cessation of menstruation.

Shortened form of *menopause*. Literally, a *pause* is a cessation for a short time of something that will be resumed.

pause[2] a statutory restriction on increases in pay.

One of a series of terms aimed at disguising British political attempts to hold down wage increases on a national basis:

In 1961 ... Selwyn Lloyd introduced what he euphemistically described as the Pause, to combat growing inflationary problems. (Green, 1979)

The implication was that increases would be resumed after a brief delay.

pavement artist an expert at clandestinely following another.

Espionage jargon, punning on the beggar who chalks on paving stones:

Rostov had commandeered all Petrov's best pavement artists and most of his cars. (Follett, 1979)

pavement princess *American* a prostitute.

Found at the roadside where the trucks pull up.

paw to fondle (a woman) sexually.

Punning on *paw*, the hand, and perhaps on the vigour with which an impatient horse strikes the ground with its hoof:

When you ask any of the men here, they just want to paw you. (Chandler, 1953)

pax Britannica the British colonial rule.

Unlike the golden years of the Pax Romana, its injustices clouded by time, the two centuries of British empire, which petered out in apathy in the 1960s, saw continuous fighting somewhere or other. The Romans enjoyed the advantage that their subjects saw no acceptable alternative to the Imperial system. We should not however be too dismissive:

The Pax Britannica gave both British and Indians a quite remarkable degree of safety in their daily lives. (Allen, 1975)

pay a visit to urinate.

Shortened form of *pay a visit to the lavatory* and punning on a social call. Sometimes expanded, as *pay a visit to the old soldiers' home*, but seldom used of defecation.

pay nature's debt to die.

From the necessity of death in the natural order. This, and to *pay nature's last debt*, imply natural death. To *pay the extreme, final*, or *supreme penalty* is to die violently, whether by being murdered, executed, or killed in war. To *pay your debt to society* is to be killed judicially, usually for murder. A British soldier who *paid with the roll of a drum* did no more than fail to pay his debts. It was forbidden to arrest a soldier for debt when he was on the march.

pay off *American* to kill.

Probably from the standard English *paying off* a ship's crew at the end of a voyage, their service being concluded.

payoff a bribe or illicit reward.

The action is clothed with commercial propriety:

Ezra is still in the saddle, even after that payoff business in Malawi. (M. Thomas, 1980)

paying guest a stranger lodging for payment in a private house.

A standard English usage which is thought to add gentility to a commercial transaction and abbreviated to *PG*. A *guest, lodging*, or *rooming house* is in business only to provide for such lodgers.

peace a preparation for violence.

First noted in Hitler's notorious *Peace Speech* of 17 May 1933, which heralded his assaults upon his neighbours. As so often, the concept and language were adopted by Communists etc. with *peace councils, campaigns, offensives, initiatives*, and so on aimed at destabilizing their opponents and inciting violence. The British *peace women* at the American nuclear base at Greenham Common were noted, among other things, for their readiness to resort to trespass. A *peace-keeping action* in another country is an invasion, using a *peace-keeping force*:

> First they call the army a peace-keeping force. (Sharpe, 1979)

peace at last death.

A tombstone and obituary favourite, referring to the dead person and not the survivors.

pecker the penis.

Literally, an instrument for making a hole by pecking. Common American use:

> ... caution a feller about despairing of his poor engine and perhaps hitting his pecker with a hammer. (Theroux, 1973)

The British *pecker* was the nose, whence the expression *keep your pecker up*, keep cheerful, an exhortation which an American might find impracticable as well as impertinent.

peculiar homosexual.

A variant of QUEER[3]:

> The idea came to her that Dick was, well, *peculiar*. (McCarthy, 1963)

For Webster in 1833 the *peculiar members* were the testicles and in obsolete British use, a *peculiar* was a sexual mistress, being something for your exclusive use.

peculiar institution (the) *American* slavery.

19th-century usage when slavery was normal in the American South but also continued in some Unionist states for much of the Civil War:

> ... it was unthinkable that the American flag should impose the South's 'peculiar institution' on new lands won by Americans from every part of the country. (G. C. Ward, 1990—dispossessing American Indians was all right, it seems)

peddle your arse to be a prostitute.

From *peddle*, to offer for sale, and see also ARSE:

> I'm too old to peddle my ass. (Sanders, 1981, of a female)

Some homosexual use also.

pee to urinate.

The usual spelling of P:

> During the next few days I peed endlessly into containers which were duly transported to the laboratory and analysed. (Oakley, 1984)

A *pee* is an act of urination:

> The Brigadier, on his way back from a quick pee in the bushes... (Bogarde, 1978)

Pee-pee is either urination, although still only on a single occasion, or the penis:

> I happen to be magnificent as a fucker; does that surprise you—don't lie, you've secretly wondered if my pee-pee is dead too. (Goldman, 1984—a paraplegic was speaking)

peel a banana *American* to copulate.

Of a male, perhaps from the movement of the prepuce or the removal of clothing.

peeler a policeman.

After the original BOBBY, Sir Robert Peel:

> If they'd been tipped every peeler in London would have been there in plain clothes waiting for us. (Clancy, 1987—the police would have received clandestine notice, not gratuities)

Whence perhaps the American slang *peel*, to arrest.

peeper a private detective.

From their supposed frequent involvement in the observation of adultery:

> 'Merely an ex-cop trying to hustle a living'... 'That's tall talk for a peeper.' (Macdonald, 1952)

peeping Tom a voyeur.

Leofric, the Anglo-Saxon Lord of Coventry, agreed to postpone an increase in taxes if his wife, Godiva, rode naked through the streets. The townspeople were forbidden to watch and how anybody would have known about it if Tom hadn't peeped is a matter for conjecture. Standard English.

peg an intoxicating drink, usually of spirits.

British Indian usage; a shortened form of CHOTA PEG. Northern English communal drinking bowls had *pegs* which marked each person's share, but the Indian use is probably from *a peg in your coffin*, which each brandy or whisky in that climate probably was:

> We had our pegs on the verandah. (Fraser, 1977)

To *peg* is to drink intoxicants to excess:

> What with rum and pepper—and pepper and rum—I should think his pegging must be nearly over. (C. Dickens, 1861—punning on to *peg*, to knock on the floor for attention)

peg out to die.

From the scoring at cribbage, where the first to finish moves his peg to the end of a row of holes on a board.

pellets rabbit shit.

Standard English. The rabbit, having no second stomach, eats the *pellets* a second time, to recover full nourishment from them.

pencil[1] the penis.

From the shape and construction rather than the shared Latin ancestry. Partridge gives 'pencil and tassel, a (little) boy's penis and scrotum: lower-class euphemism' (DSUE—*tassel* being British rhyming slang for 'arsehole'). See also LEAD IN YOUR PENCIL.

pencil[2] not legally binding.

Attributive use, from the ability to erase what is written in pencil:

> Book studio space and make it firm, no pencil deals. I want it in dry ink. (B. Forbes, 1972)

A busy or self-important person who *pencils* an appointment in his diary is likely to cancel it or fail to keep it.

penetrate to copulate with.

Again sharing its etymological stem with *penis*. It has overtones of lack of (female) consent, just as the espionage jargon *penetrate* does, meaning 'to enter a building without permission'.

penman a forger.

Literally, a skilled writer with a pen. Underworld use:

> Then there are the 'blanks', the unfilled identity cards on which the penman can work at will using the originals to produce forgeries of superb quality. (Forsyth, 1994)

penny *American* ?obsolete* a policeman.

A pun on COPPER from the US cent which is called a *penny*. Although pre-decimal British coins of low value were called *coppers*, the pun failed to bridge the Atlantic.

people of/with those patronized as being inferior.

The language of 'social science' and the POLITICALLY CORRECT. Thus *people of colour* are black:

> Black people *may* be black, but many now prefer 'African American' or 'people of colour' —though *never* 'coloured people'. (*Daily Telegraph*, 23 February 1991)

Less often in the singular as *person of colour* etc. A *person with AIDS*, or *PWA*, is someone infected with that condition, the usage being less harsh than 'AIDS victim' or 'sufferer'.

People with impaired hearing are deaf, *people with learning difficulties* are of low intelligence:

> ... the deaf shall be described as 'people with impaired hearing' and the mentally handicapped as 'people with learning difficulties'. (*Daily Telegraph*, 1 October 1990)

People with differing abilities are subject to a physical disability (the phrase was coined by one B. F. Freeman, winning him $50,000 in a 'Create a New Word' competition (Beard and Cerf, 1992)).

People of size, which might be thought to include all of us, does not refer to stature but to girth:

> ... mainstream society should shed its prejudices against those known in the current politically correct jargon as 'people of size'. (*Sunday Telegraph*, 13 November 1994)

people's imposed by autocracy.

The language of totalitarianism in various compounds, as follows:

people's army an army pledged to the support of a regime when the former non-political and professional army has been disbanded.

people's car a device for financing Nazi rearmament.

In 1938 any German who had paid 750 marks at a rate of not less than 5 marks a week received an order number, but none received a car. Today *Volkswagen* has long shaken off its dubious beginnings.

people's court a tribunal supporting the regime without trained judges and juries, and without justice or mercy.

There is however a certain irony in the fact that the three Communists acquitted before the courts in 1934 of involvement in the Reichstag Fire should have been the first victims of the Nazi people's court, or *Volksgerichtshof*.

people's democracy a Communist autocracy.

Newspeak at its best, since its citizens enjoy no effective voting rights.

people's justice summary killing without trial.

Not even the legalistic routine of a PEOPLE'S COURT to delay the process:

> Spare them after all? When they should be punished according to the people's justice! (Kyle, 1983, of the Czar and his family)

people's militia an armed force supporting those who have recently seized power.

It may be institutionalized to keep a watch over and counterbalance the remains of a professional army.

people's republic a Communist autocracy.

Slightly less offensive than a *people's democracy* although the *people* are unlikely to notice the difference:

> ... fatuous violation of language that in our day terms the grotesque dictatorship a 'People's Republic'. (Theroux, 1979)

Percy see PERSON (YOUR).

perform[1] to defecate or urinate when required.

Nursery usage when a child is being taught to avoid doing so involuntarily:

> On the rare occasions when—by pure chance —he 'performed', she moderated her pantomime of approval. (McCarthy, 1963)

To *perform a natural function* is more specific:

> Temple felt an urge to perform a natural function. (Boyd, 1982)

Used too of domestic pets.

perform[2] to copulate.

Chiefly of a male:

> You see ... he can perform, or he wants to, anyway he does. (Amis, 1978, of copulation)

Whence *performer* as someone who so *performs*:

> ... the writer or artist ... is a better performer in love's lists than the navvy (Harris, 1925)

or as someone who engages in sexual activity in public for a payment:

> He's the performer. In return, he'll get it for nothing. (Manning, 1978, of a couple copulating before an audience)

Perform is also used of homosexual activity or sexual deviation.

period[1] the time of menstruation.

Shortened form of *monthly period* of menstrual flow:

> 'Next Monday?' asks Howard. 'No good,' says Flora. 'That's my period.' (Bradbury, 1975—they were planning copulation)

period[2] old and dilapidated.

Literally, a passage of time, but this attributive use is perhaps no more than a shortened form of *colonial period* etc. Estate agents so describe a residence when decay and disrepair are so apparent that they cannot be glossed over:

> Impressive stone-built period house (available for the first time for 50 years). Ideal for renovation. (*Western Daily Press*, May 1981)

periodic rest a term in prison.

Usually of a habitual criminal. I particularly recall it being used of the incarceration of Hoffa, the former boss of the Teamsters' Union, whose stay in jail was not accompanied by great physical hardship. He had the distinction of going inside through the efforts of Robert Kennedy and coming out in 1971 through the clemency of Richard Nixon.

permissive less constrained by custom in personal conduct.

As in the British *permissive society*, which was said to have emerged in the 1960s with the reforms initiated by Roy Jenkins as Home Secretary, including that which stopped acts of male homosexuality constituting a criminal offence.

perpetual care fund a supplemental burial charge.

American funeral jargon. The customer is persuaded to pay a capital sum which will supposedly keep the grave in good order for ever:

> Another idea used by virtually all cemeteries is the 'perpetual care fund' ... a surcharge of ten to twenty per cent for future care. (J. Mitford, 1963)

person (your) the male genitalia.

Probably shortened from *personal parts*, which also describes the vagina, as though your nose or ears are not equally personal to you. Shortened to *Percy*, punning on the male name.

person of/with see PEOPLE OF/WITH.

persona non grata someone caught spying.

Literally, any unwelcome person, but used specifically of a diplomat caught spying on the host nation. The phrase is used equally of those subjected to a retaliatory expulsion. Occasionally as an initial abbreviation and used as a verb:

> They're already PNG'd, and they're going on the next Pan Am. (Clancy, 1988, of two expelled spies with diplomatic status flying out on a now extinct airline)

personal assistant a secretary.

The usage enhances the vanity of the employer and the salary of the employee. Sometimes reduced to *assistant*:

> ... two remarkably pretty girls, dark-haired, upright of carriage, secretaries perhaps, assistants rather. (Amis, 1988)

personal hygiene the paraphernalia of menstruation.

Such as towels, tampons, etc., and whether clean or used; also as *hygiene* or *feminine hygiene*. All of these terms overlook the fact that menstruation is a natural process rather than an illness; *hygiene* originally meant 'knowledge or practice that relates to the maintenance of health'. Containers in lavatories in hotes and workplaces for the disposal of towels and tampons after use are labelled *personal hygiene* and their final disposal is known as *personal hygiene services*. In astronauts' jargon, a *personal hygiene station* is a lavatory in a spacecraft.

personal parts see PERSON (YOUR).

personal relations copulation.

Often in the phrase *have personal relations with*, copulate with, although in literal terms you do this with anyone you meet, of either sex, by shaking hands with or talking to them:

> Personal relations, as they used to say. But what's personal about relations? ... Two victims sharing groins. (Bradbury, 1965)

Personal relations is similarly used for homosexual encounters:

> Burgess had ample opportunity to indulge his fetish for 'personal relations' under cover of the rigidly enforced nightly blackout. (Boyle, 1979)

personal representatives those who administer the estate of a person who dies intestate.

Or who appoints no living executor. But the *person* they *represent* is dead.

personal services extramarital sexual activity.

Usually copulation by a female for payment:

> Recruiting 'a lady of my acquaintance' for personal and espionage services... (Boyle, 1979)

Personalized messages, delivered over the telephone or by cassette, are usually pornographic aids to masturbation.

personality a nonentity.

Literally, the fact of being a person with individual characteristics. Entertainment jargon, in a milieu where the article on sale is individuality:

> He wouldn't allow the *TV Times* to describe him as a TV personality. That's just for jokeless comics who wished they could sing and dance. (Deighton, 1972)

persuade to compel through violence or threats.

Literally, to convince by argument:

> No less than 260 of our illustrious legislators were vulnerable to KGB 'persuasion'. (*Private Eye*, April 1981, suggesting that membership of the House of Commons does not immunize from adultery, homosexuality, pederasty, chicanery, extravagance, etc.)

Police, criminal, and espionage use; for some villains a *persuader* is a handgun or other violent weapon:

> ... pistols, whips, blackjacks, lengths of rubber hose called 'persuaders'. (Lacey, 1986)

pet[1] to caress physically during courtship.

Probably from the stroking of the domestic animal:

> ... held in his gentle brutal mitts for a petting session. (Ustinov, 1971)

There is unhappily no etymological link with the obsolete northern English *petting-stone*, a rock which a bride had to negotiate at her wedding, to leave all her *pets*, or 'ill-humours', behind her.

pet[2] a sexual mistress.

The American imagery is of the domestic animal kept for its owner's pleasure, or pleasuring:

> Cynical as a Park Avenue pet after her butter and egg man goes home. (Chandler, 1958)

peter *mainly American* the penis.

Apart from the common use of male names for the penis, the etymology is obscure and I don't think much of the suggestion that it comes from a *petard*, a mine:

> 'Twas the peter of Paul the Apostle. (*Playboy's Book of Limericks*)

Peter Funk *American* an auctioneer's stooge.

Probably from the automatic repeating device, or *peter*, used in telegraphy. It has the sound too of rhyming slang but the available rhymes are not appropriate.

petite (of a female) very small.

Not merely a child. Much used in the garment trade.

petite amie a sexual mistress.

The little female friend, but not normally French. Less often as *petite femme*:

> Time the *petite femme* got herself into a *negligée*. (N. Mitford, 1945)

Petit ami as the partner for a male homosexual is even rarer:

> Your *petit ami* was calling me a horrid baggy little man. (Sharpe, 1977)

petrified drunk.

Showing no sign of movement, as if turned into stone. Mainly American; in the British Isles it still normally means no more than 'frightened'.

petticoat dominated by a female.

Thus *petticoat government* is the situation in which a wife makes the decisions for the household, including how the husband shall spend his leisure time and his money:

> Adair's idea of 'petticoat government' included the power of the Women's Council of the Cherokee. (Allen, 1992)

petty house see LITTLE HOUSE.

phantom *American* a person paid while not working or a non-existent employee whose wage is drawn by another.

The victim is usually a large public employer. Either the person on the payroll exists but as a friend of the foreman; or a politician gets paid without working; or the payroll numbers are inflated by a person who does not exist, with the foreman stealing the wages.

pharmaceuticals illicit narcotics carried personally.

Along with your aspirins and toothpaste:

> ... whom Caryn still saw, but now only as a matter of form and pharmaceuticals. (M. Thomas, 1982—she got her supply from him)

pharmacy a private store of illegal narcotics.

Literally, a place where drugs are dispensed:

> Barney convoying personal pharmacies through airports. (M. Thomas, 1980)

physic a laxative.

Literally, any medical treatment:

> The physic will clean him out real good. (L. Armstrong, 1955)

I have not met the further meaning of 'copulation', given by DSUE, except as NIGHT PHYSIC.

physical involvement copulation outside marriage.

Not just shaking hands, which is all the words imply, except of a team game where there would be overtones of unsporting play:

> Her solicitors have been instructed to sue any hack who dares to suggest a physical involvement. (*Private Eye*, March 1981)

pick to steal.

OED notes a use in 1300, which makes it one of the oldest euphemisms in the language, and in regular use since then:

> A charge of picking and unlawfully intromitting with his neighbour's goods. (Hector, 1876)

The obsolete forms of *pickle*, and Scottish *pike* meant the same:

> Ye pykit your mother's pouch o' twal-pennies. (W. Scott, 1818)

An animal which *picked* gave premature birth, from the dialect meaning 'to throw':

> ... produces a calf prematurely—in local phrase, 'picks her cau'f'. (Atkinson, 1891)

pick a daisy to urinate.

A punning female use, perhaps from the bending down and the *daisy*, or chamber pot, with the common floral motif around its rim. To *pick a pea* punned with less subtlety whereas to *pick a rose* brought to mind a nozzle producing a fine spray. These, and other flowers might be also *gathered*, *plucked*, or *pulled*.

pick-me-up a drink of an intoxicant.

Literally, a medicine taken as a tonic, whence jokingly of spirits:

> If I had any more of these pick-me-ups I'd be under the table. (Theroux, 1979)

pick off to kill.

Choosing whom you aim at:

> Go ahead. You can pick him off. (Genet, in 1969 translation, of a killing)

pick up to acquire (a sexual partner) casually.

Of either sex, often at a first meeting:

> Rachman continued to pick up other girls. (Green, 1979)

A woman who regularly *picks up* men may be a prostitute and a man who does the same with women may be seeking the services of a prostitute. A person you so meet is a *pick-up*:

> You don't think they make me look like a tart? ... I'll go up the Broadway looking for pick-ups. (Theroux, 1976)

pick up a nail to contract gonorrhoea.

The discomfort felt by the male when urinating or undergoing a pre-penicillin cure was akin to lameness in a horse. The obsolete *pick up a knife* was to fall off a horse in the days when to be a bad rider was like being a bad driver today, it being jokingly suggested that the faller left the saddle on purpose.

pickled drunk.

The common culinary imagery but this time with a reference to the preservation of anatomical specimens in alcohol:

> ... you were a bit pickled at the time and so not to be blamed for what you did. (Wodehouse, 1930—the action was to have knocked down a pedestrian and driven on, considered less reprehensible then than now)

pie-eyed drunk.

Unable to focus rather than with eyes like pies:

> Brother Yank doesn't believe in getting his nose in the trough before 10 p.m., by which time one and all are absolutely pie-eyed. (*Private Eye*, April 1981)

piece[1] a female viewed sexually by a male.

Literally, a part of something and a synonym of BIT:

> The greatest little piece in the business, and for half a page in your rag—she'd do it. (Deighton, 1972)

More often as a *piece of arse* or *ass, buttered bun, crackling, crumpet, gash goods, muslin, rump, skirt*, etc., most of which are dealt with under those headings:

> I was day after day closeted with this choice piece of rump, and not so much as touching her, let alone squeezing or grappling. (Fraser, 1975)

A *piece of trade*, a prostitute, and a *piece on a fork*, copulation, are both obsolete.

piece[2] a handgun.

Perhaps a shortened form of *fowling-piece* but so used of both cannons and pistols since the 16th century, and of cross-bows before that:

> 'You carry a piece?' her asked suddenly ... 'Oh no,' I said. 'I don't believe in violence.' (Sanders, 1980)

piece[3] *American* a small quantity of illicit narcotics.

Insofar as there are standards in this business, a *piece* is an ounce.

piece of paper a speeding ticket.

Apart from the nuisance of having to pay a fine, this American Citizens' Band usage minimizes the indignity of having been caught.

piece of the action see ACTION[1].

piece off *American* to bribe.

In general use, to buy silence or favours, from the actual or figurative peeling bills off a bankroll. Specifically it refers to bribing a foreman etc. to give you a job, in return for handing over a part of your wages.

piffled *American* drunk.

Piffle is nonsense, but that may not lead us very far. Also as *piffed*.

pig a police officer.

An ancient term of abuse, being noted by Grose. *Pig-feet* may prove to have been ephemeral:

> ... they'd tell the pig-feet if they came asking around. (Lyall, 1982)

In a pig's ear, meaning 'under no circumstances', is not from information passed to a policeman, or to the animal, but from the receptacle so named on the bridge of a ship which a watch-keeper might use for urination without leaving his post. I'm not sure whether that has an etymological link with the Scottish *pig*, a pot for urine:

> Into my putrid channel
> At night each wifie tooms her pig.
> (Ogg, 1873—to *toom* is to empty)

pigeon[1] the dupe of a criminal.

Old enough for Grose but still modern criminal use. In a *pigeon-drop* the victim pays money to thieves for a share in a bogus bank-roll which they profess to have found. But see also FLY THE BLUE PIGEON.

pigeon[2] to steal.

Probably from the slang American sense of passing off a losing betting slip etc. as a winner rather than from what the wood pigeons do to the brassica in my vegetable garden every winter.

pigeon[3] a vehicle caught speeding.

From the speed of the bird:

> The Smokey at milepost 116 has a pigeon. (CBSLD)

American Citizens' Band slang.

piggyback to use another's reputation for your financial or social ends.

Literally, a *piggyback* is a ride given to a child on the back of an adult:

> You're doing me the very same way. You're piggybacking. (Theroux, 1978)

pile into to copulate with.

Of a male, with the common violent and penetrating imagery:

> I'm 'bout worn out pilin' inter that li'l darlin'. (Fraser, 1971—a husband was speaking of his wife's sexual desires)

piled with French velvet *obsolete* infected with syphilis.

A complex pun on the *pile* of shorn cloth, *velvet*, the vagina, and the MALADY OF FRANCE:

> Thou art piled, for a French velvet.
> (Shakespeare, *Measure for Measure*)

pill[1] the penis.

This obsolete Scottish/northern English use, from Norwegian dialect, survives in the diminutive *pillock*, which is commonly used figuratively as a mild insult.

pill[2] **(the)** a contraceptive taken orally by women.

Not just any medicament prepared for swallowing:

> In the pre-Pill world of our youth...
> (Bradbury, 1976)

Whence *on the pill*, taking such contraceptives regularly and by implication able to copulate without impregnation.

pillow partner a person with whom you often copulate outside marriage.

Of either sex, but not a spouse:

> I can usually make use of a native pillow partner. (Fraser, 1971)

pills the testicles.

From *pill*, a ball, rather than the medical tablets but, unlike BALLS, not used figuratively or as an oath. In obsolete Irish use, to be *pilled* was to be blackballed from membership of a club etc.:

> After someone he had put up for the Kildare Street Club had been pilled, he never entered the doors of the Club again. (Fingall, 1937, of her husband. The Kildare Street Club was the rendezvous in Dublin of the Protestant gentry, especially prior to the First World War.)

pin the penis.

Of the same tendency as PRICK but much less common.

pin-money the receipts from casual prostitution.

From the standard English meaning 'funds which a woman can spend on personal needs', this income being earned without her husband's exertions, and perhaps also punning on PIN.

pin-up an erotic picture of a female.

In the Second World War titillating and often crude photographs etc. of women were displayed in barrack-rooms etc., secured by drawing-pins. A woman so depicted, or considered worthy of having that distinction, was also known as a *pin-up*. Now also used to denote erotic pictures of males.

pinch[1] to steal.

In standard usage, to nip between the fingers:

> He had spent most of his life in clink for pinching anything from a roll of linoleum to a hurricane lamp. (Bogarde, 1972)

pinch[2] to arrest.

From the grasping of the suspect:

> He got acquitted for that there note after he had me 'pinched'. (Mayhew, 1851)

In modern American use too as an arrest:

> Maybe he knows something that could hang a pinch on her. (Chandler, 1958)

pine overcoat a coffin.

Worn if you are killed but not if you die in bed. For other examples of this type of macabre humour, see WOODEN BOX.

ping-ponging *American* passing a rich patient from one specialist to another.

A medical version of the long rallies in table tennis.

pink panther an unmarked police car.

Perhaps from the plain-clothes detective in the television cartoon. American Citizens' Band slang.

pink pound the purchasing power of homosexuals.

A version of the 'coloured' currencies which trade at a rate outside that dictated by the open market, such as the *green pound*, which was once used for agricultural computation. In this case the *pink* is perhaps from the colour of the boudoir and the allusion is to the higher spending power of those without families to support:

> The pink pound is going from strength to strength. (*Daily Telegraph*, 13 March 1995)

pink slip a notice of dismissal.

If that is the message, the paper on which it is written is *pink*, whatever the colour. Also of normal retirement:

> I'm forty-seven hours and fifty-five minutes from owning my own pink slip. (Wambaugh, 1983, of a policeman about to stop working)

To be *pink-slipped* is to be summarily dismissed:

> The first month, eleven of the twenty-three staffers were pink-slipped. (Sohmer, 1988)

pioneer[1] a soldier sent to intervene in a foreign war.

Originally, a soldier who clears the way for his own following troops:

> China had sent several fresh brigades of 'volunteers' and 'pioneers' into the fray. (Ustinov, 1966)

pioneer[2] a person who has undertaken not to drink intoxicants.

An Irish phenomenon, for an ebullient race to whom moderation is often inimicable.

piped drunk.

Probably not from the obsolete *pipe*, to put liquor into a cask or to drink it, and perhaps from the notion that any vessel overfilled needs plumbing. *Pipe*, to cry, is from the rhyming slang *pipe your eye*.

piran *?obsolete* drunk.

This entry is for my English West Country readers:

> St Piran is the patron saint of tinners, popularly supposed to have died drunk. (EDD)

Found in Cornwall.

piss pins and needles to be infected with gonorrhoea.

Of a male, from the sensation while urinating. In rarer usage, to *piss pure cream*. The obsolete *piss your tallow* was to ejaculate before vaginal entry:

> Send me a cool rut-time, Jove, or who can blame me to piss my tallow? (Shakespeare, *Merry Wives of Windsor*)

pissed drunk.

From the need to dispose of beer drunk to excess:

> I am not introspectively drunk. I am merely pissed. (Sharpe, 1977)

But you can equally become *pissed* on wine or spirits. The figurative *pissed off* means no more than 'disaffected', drunk or sober. Occasionally abbreviated to to *P.O.'d*:

> I think the president was very angry... In fact royally P.O.'d might be a very good word for it. (*Washington Post*, March 1987, quoting Maureen Reagan)

pistol *obsolete* the penis.

Of obvious imagery. Whence Shakespeare's punning character.

pit-stop an occasion for urination.

With somewhat inverted logic, from the pause for the replenishment and repair of a car during a race:

> The hiatus allowed the control-room crew to ... make necessary pit-stops. (Clancy, 1989—they left to urinate)

place a lavatory.

Probably the location given over to a special use rather than a shortened form of a *place to urinate*. You may often hear a male enquiry: *Where's the place?*

place-man a spy.

Originally, someone who holds a responsible place in government service, but in this use punning on having been *placed* as a spy by his masters:

> Soviet officials had access to a variety of French political and military secrets through experienced 'place men' such as Burgess's associate. (Boyle, 1979)

place of safety an inhumane prison.

Himmler's favoured term for his concentration camps. The obsolete British *place of correction*, a common jail, was at least named in honest, albeit usually unfulfilled, expectation:

> Your places of correction would be as quiet as Chelsea Hospital. (Ustinov, 1971)

plank to copulate with (a female).

Almost certainly the usual violent imagery from the echoic of the strip cartoons, despite the etymological attractions of lying on hard ground or hardening hats by felting:

When I see a pretty girl waiting for a bus I always get horny... I can plank her, because she wants a ride. (Theroux, 1989)

planned unexpected and unwelcome.

A common usage when we prefer not to face up to the fact that we have been wrong or lacking in foresight. Any corporate statement claiming that an event occurred *as planned* should put the analyst on notice that all is far from well. The *planned reduction* of a workforce indicates summary dismissals to avert a crisis and the *planned withdrawal* of an army signifies defeat:

Surprise and mobility, coupled with overwhelming air support, turned 'planned withdrawals' into creeping rout. (Boyle, 1979)

A *planned termination* is either a suicide or an abortion. A bureaucrat styled as a *planner* is primarily concerned in preventing any change in the status quo, *planning*, or in America *zoning*, being a mechanism used.

planned parenthood *American* induced abortion.

The antithesis of what is normally categorized as *planned*:

A rash of violence and killings at abortion centres throughout the United States (or Planned Parenthood Clinics as they are delicately called). (A. Waugh in *Daily Telegraph*, 14 January 1995)

Normally *planned parenthood* refers to contraception, where the *planning* is aimed at not becoming a parent.

plant[1] to bury (a corpse).

The imagery of horticulture, without the crop:

Y'wouldn't want to be planted without ceremony. Why not put Baptist? (Manning, 1962)

Planted means 'dead and buried'. The obsolete *plant a man* was to copulate, again using horticultural imagery but this time anticipating that the seed might bear fruit.

plant[2] falsely to place incriminating material on.

Again with horticultural imagery:

With the evidence you'd arranged for him to find... Or to put it bluntly, planted. (Crisp, 1982)

A *plant* can also be a story introduced into a periodical for the purpose of advertising or promotion, and not necessarily false.

plant the books see BOOK.

plasma an intoxicant.

Literally, the substance in the blood in which other elements are suspended:

And speaking of the old nasty—it's past noon, and you could use some plasma. (Sanders, 1985)

plastered very drunk.

Literally, covered with a substance that sticks to you, as does the smell of intoxicants, but perhaps only from the immobility of a limb in plaster:

You could tell by his eyes that he was plastered to the hairline. (Chandler, 1953)

plater a person who engages in fellatio on another.

Often of a prostitute who performs the act. From the concept of eating meat, presumably ham, as fellatio is also known as a *plate of ham*.

play to copulate or masturbate.

The imagery of *sport* (see SPORT (THE)):

As well a woman with an eunuch play'd
As with a woman.
(Shakespeare, *Antony and Cleopatra*)

There are many compounds and phrases on this theme, exemplified in the entries that follow. See also *make a play for* at MAKE[1].

play around to copulate casually.

Usually with more than one person contemporaneously:

Not with the chauffeur... I don't have to dig down that far if I want to play around. (Chandler, 1939)

play at hot cockles to masturbate.

Of a female, the *cockles* being the vulva.

play away to copulate outside marriage or a stable sexual relationship.

From the team game played on the opponent's ground.

play games to copulate casually.

Usually of a woman and perhaps with a single partner:

She was playing games with Vannier. (Chandler, 1943)

play hookie to copulate extramaritally with one partner.

Of either sex, with the imagery of staying away from school:

The safest racket in the world is to rob a married man or woman who is playing hookie. (Lavine, 1930)

play house a brothel.

Punning on but different from a *playhouse*, or theatre.

play in the hay to copulate casually.

But not necessarily al fresco, and see also IN THE HAY:

If every girl who's ready to play in the hay was to get married, we'd have damned few spinsters. (Fraser, 1969)

play mothers and fathers to copulate.

Usually outside marriage, with a British version *play mums and dads*:

> And at a moment like this my wife has to play mothers and fathers with that bastard. (C. Forbes, 1985)

play on your back to copulate.

Of women:

> Lulls him whilst she playeth on her back. (Shakespeare, *Titus Andronicus*)

play the ace against the jack to copulate.

Of a woman, punning on a game of cards and JACK[1].

play the beast with two backs see BEAST WITH TWO BACKS.

play the field to be sexually promiscuous.

Of either sex, from betting on several or all the runners.

play the goat to be sexually promiscuous.

Of a male. But *play the giddy goat* means merely 'to act stupidly'.

play the organ to copulate or masturbate.

Of either sex, punning on the musical instrument and see also ORGAN.

play the skin flute to engage in buggery.

The imagery is as in FLUTER:

> He looks like a guy who plays the skin flute. (Sanders, 1984)

play tricks to copulate outside marriage.

Usually of a female and often in the past tense where a single woman becomes pregnant, but not, I think, punning on the prostitutes' jargon TRICK.

play with to masturbate or sexually excite.

Heterosexually or homosexually, often as *play with yourself*:

> All the time we were playing with ourselves, I kept thinking of Mary's hot slit. (Harris, 1925)

play your harp to be dead.

A jocular usage, but not by those who believe in a posthumous harp-playing paradise, who tend not to be sacrilegious. But see also STOKE LUCIFER'S FIRES.

playmate someone of either sex with whom you copulate outside marriage.

Or *playfellow*:

> To seek her as a bed-fellow,
> In marriage-pleasures play-fellow.
> (Shakespeare, *Pericles*)

A *playboy*, apart from being a wealthy hedonist, can be any promiscuous male, although not to be confused with the obsolete English/Irish *playboy*, the devil:

> The divil sitting cheek be jowl with him in his own chimbley corner... an' himself an' the playboy sloughed out o' the same pipe. (MacManus, 1898)

please yourself on *obsolete* to copulate with.

Of a male in the days when females were not meant to take much pleasure in it:

> Perhaps they will but please themselves upon her. (Shakespeare, *Pericles*)

pleasure to copulate with.

Normally of the male, the activity being *pleasuring*:

> Not the most joyous pleasuring I have taken part in... (Fraser, 1969, after copulation)

but sometimes of the female:

> Three doe-eyed, heavy hipped women pleasuring one man. (Masters, 1976)

Also as *take pleasure with*:

> Later, stirred by the curry, he took pleasure with his second wife. (Sanders, 1977)

Pleasures may be copulation:

> ...afternoon pleasures are exchanged for a few days' work. (B. Forbes, 1972, of a producer casting female roles in a film)

A *pleasure house* is a brothel:

> It was a pleasure house, where those rich ofay (white) business men and planters would come. (L. Armstrong, 1955)

Mutual pleasuring can be either copulation or masturbation of each other by any combination of males and females. The obsolete *pleasure-garden* was the vagina viewed sexually by a male, with its *pleasure-garden padlock*, the towel worn during menstruation.

pledge (the) an undertaking never to drink intoxicants.

A feature of the campaign of freechurchmen against the 'demon drink':

> He felt the Band of Hope has been worthwhile when some of the old boys came to see him during a holiday in the village. It warmed his heart to be assured 'I've kept the pledge'. (Tyrrell, 1973)

The pledge might also be *signed*, *taken*, or *broken*:

> If they ever start off the Little White Ribboners in Russia, all the members will have to be boys, for they'll never get the women to take the pledge. (Fraser, 1973)

Now sometimes used humorously by those expressing remorse after drunkenness.

plod a policeman.

At second remove from his measured gait via Enid Blyton's creation and the friend of NODDY:

> Why hasn't it been given straight to us? Why are the...'plods' involved? (Seymour, 1989—an investigator from a narcotics squad was denigrating the local police)

plot to rob.

Originally a Scottish and northern dialect word meaning 'to plunge in boiling water', whence 'to scald (a fowl) before plucking' and then 'to pluck':

> When old John Knox and others some
> Began to plott the bags of Rome.
> (Henderson, 1856)

The devil's name *plotcock* came from the symbolic *cock* and his penchant for flaying people alive:

> Seven times does her prayers backwards pray,
> Till Plotcock comes with lumps of Lapland clay. (Ramsay, 1800—all genuine witches pray backwards and Lapland was their fabled homeland)

plough[1] *obsolete* to copulate with.

Of a male, punning on the entry of the share into the furrow and the chance of issue:

> He plough'd her and she crop'd. (Shakespeare, *Antony and Cleopatra*)

plough[2] to fail (a candidate) in an examination.

Almost always in the passive and of uncertain origin. It may also mean drunk, usually as the American *plowed*, the derivation not being from a failure to pass a sobriety test.

plough under needlessly to cause the death of.

From the way a farmer disposed of an unsaleable crop. In 1940 Wendell Wilkie, opposing Roosevelt's third term as President, appealed to isolationists and pacifists by saying that his opponent was determined to 'plow under every fourth American boy'. Because Wilkie lost, we forget how close he came to winning.

ploughman's lunch bread and cheese.

Originally, an advertising gimmick thought up by the trade association for British cheesemakers, but enthusiastically adopted by British innkeepers who were then able to charge excessively for what had once been a cheap snack, and even more if 'garnished' by a lettuce leaf and a slice of tomato:

> ... the cricket pitch being watered,
> ploughman's lunches being served in the Barley Mow... (*Daily Telegraph*, 30 July 1994)

pluck to copulate with.

Of a male. DAS says: 'Rhyming euphem. for the taboo "fuck".' But to *pluck a rose* is to copulate with a female virgin and the imagery may merely be from the gathering of a flower.

pluck a daisy see PICK A DAISY.

plucked from us unexpectedly or prematurely dead.

In floral imagery, the deity being credited with choosing the choicest blooms:

> The most heavenly girl in the whole glorious world has been plucked from us. (Mailer, 1965)

In British universities, *plucked* formerly meant 'failing to graduate'; an unpaid tradesman had the right to pluck the gown of the Chancellor when the name of his debtor was read out at a degree congregation, and the degree was not conferred until the debt had been settled.

plug[1] to kill or wound by shooting.

Literally, to stop a hole, which a bullet may do, after first making it:

> I'd plug you as soon as I'd strike a match. (Chandler, 1943)

plug[2] to copulate with.

Of a male, again from the hole-filling. Mailer uses *plug* also of buggery:

> There was a high private pleasure in plugging a Nazi... she was loose... as if this was finally her natural act. (1965)

plum(p) pudding club see IN THE CLUB.

plumb to copulate with.

Literally, to sound a depth:

> There once was a plumber of Leigh
> Who was plumbing a girl by the sea.
> (Vulgar limerick)

plumber a presidential staff member acting improperly.

His function, after the Ellsberg disclosures in 1971, was to trace or stop any *leak* (see LEAK[2]):

> Young and Krogh were later dubbed the Plumbers, because they were assigned to stop news leaks. (Colodny and Gettlin, 1991, reporting the cross-examination of Admiral Welander on 22 December 1975)

plumbing[1] a lavatory.

From the ancillary piping:

> Unless you've shifted the plumbing around here, I can find it. (M. Thomas, 1980—he was looking for a lavatory)

plumbing[2] the parts of the body concerned with urination and defecation.

A genteel and rather coy use, likening the body mainly internally with that aspect of domestic construction:

> Helena had known about sex from a very early age but treated it as a joke like what she called her plumbing. (McCarthy, 1963)

plunge away to copulate.

Of the male, from the movement:

> Flashy is plunging away on top of the landlord's daughter in the long grass. (Fraser, 1977)

pocket to steal.

Usually of money or trifles, small enough to go into it and without premeditation.

pocket pool male masturbation.

From the cue and balls used in the game:

> You're playing with yourself. Lay off the pocket pool. (Theroux, 1978)

The more common British form is *pocket billiards*. A *pocket job* implies ejaculation:

> ...reduced to performing pocket jobs. (Styron, 1976)

The obsolete British *pocket the red* meant copulation by a male, with punning imagery from snooker or billiards.

poetic truth lies.

A translation of Goebbels' expression:

> Convenient lies ('poetic truth') as he once called them. (Trevor-Roper, 1977)

Before we start preening ourselves, we should remember that our public figures can also be ECONOMICAL WITH THE TRUTH.

point Percy at the porcelain to urinate.

Of a male. The *porcelain* is the stall or bowl in a urinal and see also *Percy* at PERSON (YOUR).

poison a preferred intoxicant.

A jocular reference to the possible harmful effects:

> 'What's your poison?' Dundridge said he'd have a gin and tonic. (Sharpe, 1975)

poison pill the deliberate assumption of corporate liabilities to deter or repel an unwanted predator.

A tactic of the defended bid, with success leaving a sour taste in the mouth, or worse:

> 'Poison pill' meant that AbCom would issue a dilutive new stock...and that would double or triple the cost of AbCom to an unfriendly enquirer. (M. Thomas, 1985)

poison the blood of a nation as a prostitute to transmit venereal disease.

An insight into 19th-century British morality with its implication that the woman was the only guilty party:

> A woman...only yesterday had two buboes lanced...and yet she was...poisoning the blood of the nation with the most audacious recklessness. (Mayhew, 1862)

poke[1] (the) summary dismissal from employment.

Punning on the meaning 'to push', but a *poke* is also a sack, as the phrase *a pig in a poke*:

> He's gi'en him t'poake (*Leeds Mercury Supplement*, April 1896, quoted in EDD)

would mean the same thing today, but the passive might convey a different meaning to modern ears:

> I wor poked afore eleven o'clock t'next morning. (Burnley, 1880, of a dismissal)

poke[2] to copulate with.

Of a male, with common imagery:

> Don't get to poke too many women too often. (Bradbury, 1976)

A *poke* is either a single act of copulation:

> Nice trouble-free way of victualling your girl-friend between pokes (Amis, 1978)

or the female participant as seen by the male as in the phrase *a rattling good poke*; in male gallantry or conceit there are no bad *pokes*. Also some homosexual use of buggery, whence the American *pogey bait*, candy, from the 'inducement held out by old sailors for the favours of fair-cheeked, smooth-bottomed young cabin boys' (Styron, 1976). The obsolete British *pole* merely meant 'to copulate' without these other refinements.

poke[3] a prison.

I suggest from the sack—see POKE[1]—rather than being thrust or *poked* inside:

> He just got out of poke a few months ago. (Sanders, 1970)

polar bear a policeman.

A rare but not necessarily wintry elaboration of *bear* (see BEAR[2]). American Citizens' Band slang.

pole[1] see POKE[2].

pole[2] the erect penis.

An obvious vulgarism.

police action a war.

Normally attributed to Harry Truman at the time of the Korean War, and now common when a participant seeks to play down a conflict:

> Truman agreed with a reporter who asked 'Would it be correct to call it a police action under the United Nations?' This was a phrase which would later haunt Truman. (Hastings, 1987)

Dean Rusk was more precise when he called the Korean conflict a *limited war*, which fortunately it remained.

polish the mahogany to urinate.

From the wooden lavatory seat. Heard on BBC television on 6 February 1994.

political and social order internal repression.

The Brazilian version of a familiar autocratic device:

> The Department of Political and Social Order, a bland title for the administration of terror and thumbscrews. (Simon, 1979, of Brazil)

political change a humiliating defeat.

Kissinger's contemporary description of the final conquest of South Vietnam by the Communists and the humiliating American withdrawal.

political education the arbitrary imprisonment of critics.

A Communist phrase to describe and justify the suppression of political opponents. Also as *political re-education*.

political engineering *American* using government patronage to engender electoral support.

It specifically involves organizing defence procurement projects so that contracts can be placed in as many congressional districts as possible, regardless of expense or efficiency.

politically correct conforming in behaviour or language with dogmatic opinions.

The chosen subjects of dogmatism are wittily and provocatively enunciated in *The Official Politically Correct Dictionary and Handbook* (Beard and Cerf, 1992). What set out to be a campaign to deal sensitively and respectfully in both word and deed with issues like women's rights, skin pigmentation, age, obesity, chronic illness, homosexuality, incarceration, and many other often taboo subjects, was before long to assume McCarthy-like overtones in its stridency and intolerance. Thus a professor in California, learning that household pets should be called *animal companions*, had a sexual-harassment charge filed against him for wondering if the nude models in *Penthouse* should be called *Penthouse animal companions* instead of the customary *Penthouse Pets* (ibid.). Whence *political correctness*, adopting or supporting such action:

> Many men now consider themselves to be the victims of political correctness and pluralism that leaves them at a disadvantage in competition for jobs. (*Independent*, 14 January 1995)

Also shortened to *PC*:

> PC holds that Western civilization is the product of racial and sexual hierarchies which should be unseated. (*Sunday Telegraph*, 21 July 1991)

pollute to affect in a taboo manner.

Literally, to corrupt or make dirty. To *pollute* yourself is to masturbate, to *pollute* another is to copulate with her extramaritally, *polluted* is drunk or under the influence of illegal narcotics, etc.

polygraph a lie detector.

Literally, a machine giving a number of simultaneous print-outs:

> What we used to call a lie detector, sir. A Polygraph. (le Carré, 1989)

pony an act of defecation.

From the rhyming slang *pony and trap*, crap, and usually as *have a pony*, to defecate. Some figurative use:

> The voice must have realised that I was giving him a load of old pony. (McNab, 1993—he was lying during his interrogation by the Iraqis)

pooftah see POUFF.

poontang *American* casual copulation.

A corruption of the French *putain*, a prostitute, and formerly used in the south of copulation between a white male and a black female:

> A growin' Southern boy's got to have his poontang. (Styron, 1976)

poop[1] to defecate.

Nursery usage, and also of domestic pets, whence the *pooper scooper* for removing your dog's turds from the sidewalk. As happens when this topic arises, the word has been used for vulgarities associated with buggery, so that a *pooper scooper* is also a male homosexual.

poop[2] to fart.

Mainly used by and of children, and probably onomatopoeic.

pooped drunk.

Originally, flooded by the sea coming over the stern, but not only of sailors:

> ... seldom sober by seven and almost always pooped by eight. (Sharpe, 1978)

poor-mouth to ignore or refer to in unfavourable terms.

Of a more consistent practice than the occasional denigration when you BAD-MOUTH:

> Naturally the Chinese have always poor-mouthed the foreign-built railways' contribution to their economic well-being. (Faith, 1990)

poorly[1] very seriously ill.

Hospital jargon, replacing the normal meaning 'mildly unwell'. The patient described as *very poorly* is unlikely to survive.

poorly[2] menstruating.

From the meaning 'unwell' and often in the female phrase, *my poorly time*.

pop[1] to ingest (a narcotic) illegally.

Either from *popping*, injecting into, a vein or from *popping* the pills into the mouth. Whence *popper*, such a narcotic:

> The ammoniac aftersmell of poppers hung in the air. (M. Thomas, 1982)

pop[2] an act of copulation.

Perhaps from the sensation of orgasm:

> Azalo figured she'd be lucky to get twenty bucks a pop. (Sanders, 1985, of a prostitute)

pop[3] to pawn.

In what was also known as the *pop shop*:

I had to pop the silver; you know what I mean. (Guinness, 1985)

And in the old song:

Up and down the City Road,
In and out the Eagle.
That's the way the money goes —
Pop goes the weasel!

The 'Eagle' was a London public house of which a former landlord was the father of one of my aunts by marriage; the *weasel* is from the rhyming slang *weasel and stoat*, overcoat.

pop[4] to kill.

Causing another to POP OFF?

We don't pop people any more. We've learned from the Argentines. People just disappear. (Sanders, 1984)

pop off to die.

From the slang meaning 'to depart', probably with the imagery of a cork leaving a bottle, whence to die of natural causes:

Look here, Hugh, I'm afraid Percy has popped off. (Matthew, 1978)

You may also be said to *pop your clogs*:

It's either join us or pop your clogs. (Fraser, 1983—he was to be killed if he refused to join the pirate crew)

pop the question to propose marriage.

The question used to be asked by the male, but when *popped*, related to nothing other than wedlock:

Just heard yesterday that my divorce comes on today so was elated and popped question to Dutch girl and got raspberry. So that is that, eh. Stiff upper lip and dropped cock. (E. Waugh, July 1936, quoted in S. Hastings, 1994—he had been divorced for some years but wished to have also a papal annulment)

Pope's telephone number (the) *?obsolete* Vat 69 whisky.

The London telephone system formerly used the first three letters of a place name as part of the telephone code for subscribers in that district. Thus 'Vatican' would have been reached by dialling VAT followed by an individual number.

popping up the daisies dead.

The corpse is supposed to provide sustenance for the common churchyard wild flower. Some jocular use, even of those cremated.

popsy a prostitute.

Originally, and still used as, a term of endearment to a girl, whence 'an attractive young female'. The euphemistic use is usually generic:

...enough popsy to satisfy an army. (Fraser, 1977)

The variant *poppet* implies a less sexually-charged regard.

population transfer forcible resettlement.

Not the natural movements which take place on a surprising scale in a civilized country but the forcible uprooting of a group of the same race for political reasons, as practised by the Germans under Hitler, the Russians under Stalin, the South Africans under apartheid, etc.

porch climber a thief from houses.

A convenient mode of access to an upstairs window:

He was a two-bit porch climber with a few small terms on him. (Chandler, 1939—a *term* is a prison sentence)

pork[1] *American* a Federal benefit diverted to local political purposes.

From the richness of the meat, it can come in the form of funds, jobs, or patronage. A *pork barrel* is a scheme funded federally with pickings for local politicians and *to pork-barrel* is to secure votes by getting expenditure in your constituency:

America's production of space centres... symbolise an ancient discipline which lies at the heart of politics here: pork-barrelling. (*Private Eye*, July 1983)

A *pork chopper* gets a job as a sinecure in return for past favours.

pork[2] the penis viewed sexually.

With the usual MEAT imagery:

I've known greater beauties, and a few that were just as partial to pork. (Fraser, 1982—they liked copulating not eating)

A *pork puller* is a male masturbator:

...he's one a these pissy pork pullers. Takes a leak and beats off. (Wambaugh, 1975)

The *pork sword* is, in vulgar speech, the erect penis:

'She isn't getting any'...'Any what?' 'Cock. The old pork sword.' (B. Forbes, 1989)

pork[3] to copulate with.

Perhaps an unusual verb form of PORK[2]:

Larren's porkin her and takin the money to keep her in style. (Turow, 1987—he was copulating with a woman and taking bribes to finance his activities)

pork pies lies.

British rhyming slang, often as *porky pies* or *porkies*. Very common among both children and adults.

porridge prison.

Partridge suggests a pun on STIR but it is also a staple item of breakfast fare in British prisons.

positive militaristic and aggressive.

How tyrants like to see and describe themselves:

...was in tune with Japan's increasingly aggressive or, to use the euphemistic Japanese term, 'positive' foreign policy. (Behr, 1989)

possess to copulate with.

Historically the male *possessed* the female, despite the physical contradiction:

> I have bought the mansion of a love,
> But not possess'd it.
> (Shakespeare, *Romeo and Juliet*)

If used at all today, *possess* would imply extramarital chance copulation as our culture now rejects the concept of the wife as a chattel.

post a letter to defecate.

Punning on an excuse for absenting yourself from company and the process of defecation, a letter being commonly pushed through a small aperture to fall in a receptacle below.

post-viral syndrome a debilitating disease of uncertain origin.

A lay term denoting myalgic encephalomyelitis or ME. Some journalists prefer put-downs such as *yuppie flu, executive flu,* or *Royal Free disease.*

posterior(s) the buttocks.

Originally, later in time, from which 'behind', and after that the etymological progression is obvious:

> Her posteriors, plump, smooth, and prominent. (Cleland, 1749)

pot[1] to kill by shooting.

From hunting for the cooking pot but also used of attempts to kill or wound:

> ... wasn't anything much else to shoot at so I took to potting them. (Sharpe, 1978)

A *pot* is such a shot, usually taken without premeditation.

pot[2] a habitual drunkard.

From the drinking vessel rather than the slang term for belly and to *pot* is to drink intoxicants to excess. *Pot-walloper* had two meanings, 'a drunkard', and someone entitled to vote under the British Reform Act of 1832, because he had boiled, or *walloped,* his own pot in the parish for the previous six months and was therefore a bona fide householder. *Potted* (see POTTED[2]) is drunk and *pot-valour* is drunken courage.

pot[3] a receptacle for urine.

Literally, any container for liquids:

> I had taught him to use a pot. (N. Mitford, 1960, of a child's urination)

To *pot* is to sit a young child on such a container in the hope that it will urinate or defecate. Also as the diminutive *po*:

> Eeny-meeny, miney-mo,
> Sit a ... on a po. (Old rhyme)

or *potty*:

> She's on the potty. (Goldman, 1984—a child was explaining why her mother could not come to the telephone)

In American Citizens' Band slang a *potty mouth* is someone who uses foul language on the air. *Potty training* involves teaching children to urinate or defecate in the required place. Military aircraft without lavatories carry, for male use, *Port-a-Pots*:

> ... he even managed to use a Port-a-Pot without missing. (Clancy, 1989, of a man urinating in a helicopter)

pot[4] marijuana.

Either from the American Indian *potaguaya* or from the container in which the leaves and stalks are brewed—the shortening of *pot liquor* to *pot* favours the latter:

> ... to graduate to student parties to smoke pot. (Bradbury, 1976)

pot hunter an egoist seeking public recognition.

Not an archaeologist or drunkard but a sycophant conducting himself so as to achieve RECOGNITION.

potation an alcoholic drink.

Literally, the act of drinking, from which any drink:

> ... returned next day only partially recovered from the potation that had celebrated the event. (Somerville and Ross, 1894)

potboiler a repetitive or facile work by an established artist or author.

Originally, of the person who perpetrated it but now of the product:

> Then, when I got in the swing of things and began turning out four potboilers a year ... (Sanders, 1980)

Potomac fever *American* a desire to be elected to Federal office.

And specifically to the Presidency, from the river flowing through Washington DC:

> Baxter contracted a terminal case of Potomac Fever. He started to dream of the White House. (M. Thomas, 1980)

Potsdam *obsolete* a prison for captured British soldiers.

In the First World War you might talk of *dining with the Kaiser in Potsdam* if captured:

> ... so this was 'Potsdam', this moist foul-smelling cell. (Grinnell-Milne, 1933)

potted[1] dead.

Punning perhaps on being shot by a *pot* (see POT[1]) and on the imagery of *plant* (see PLANT[1]). British rather than American.

potted[2] drunk or under the influence of illegal narcotics.

In the one case from the drinking vessel, in the other from ingesting *pot* (see POT[4]).

potty[1] mad.

Usually in a harmless or eccentric way:

> It was only a question of time before the Goat-Major would go stone potty. (Richards, 1936)

Perhaps from slang *gone to pot*, destroyed.

potty[2] see POT[3].

pouch to steal.

A Scottish version of POCKET:

> I had given Master Boy Scout a fair amount of money ... doubtless he had merely pouched it. (Fergusson, 1945, of paying for assistance behind the Japanese lines in Burma in the Second World War)

pouff a male homosexual.

Not from the English dialect meaning 'a big stupid person' (EDD) but probably in both cases from the exclamation, implying a lack of substance or value:

> Don't tie the tapes under your chin ... or they'll think you're a pouff. (Francis, 1978)

The meaning 'a round footstool', is almost obsolete:

> ... sitting animatedly forward on what used to be called a pouf or pouffe but obviously couldn't be these days. (Amis, 1978)

Also as *pooftah*:

> If Prince Charles shows no interest, he *must* be a pooftah. (A. Waugh, *Private Eye*, July 1980)

pound to copulate with.

Male usage, with the common violent imagery, although of chance extramarital occasions rather than those involving coercion or roughness. Whence the punning *pound the keys*:

> ... hoped the little bubblegummer had been well pounded by the piano-tuner so she could get ... to the home for unwed mothers. (Wambaugh, 1975)

pound off *American* to masturbate.

Of a male. Also as *pound the meat*.

pound salt go away and leave me alone.

A shortened form of the vulgar American invitation to *go pound salt up your arse* and owing nothing, I think, to POUND which is not normally used figuratively or abusively. DAS says: 'This euphem. much more common than the full term'. Also, but less often, as *pound sand*.

pourboire a bribe.

Significantly more has to change hands than would pay for a drink:

> And he'll need to make cash transfers to someplace that doesn't have an extradition treaty with the US, and where government officials are not insulted by the offer of a small pourboire. (Sanders, 1977)

powder a narcotic taken illegally.

In the form in which it is often marketed:

> Why would any fool use powder for pleasure when he can have a woman? (Clancy, 1989)

powder (take a) see TAKE A POWDER.

powder room a lavatory for the exclusive use of females.

It used to be that part of a warship adjacent to the guns where the gunpowder was stored. Children were used to pass it thence to the gundeck so that the size of the passage could be restricted and danger from enemy fire or flashbacks minimized. Today the *powder* is the scented talc which women put on their faces. To *powder your nose*, to urinate, is usually of women:

> Back in the Long Gallery some of the women went upstairs to 'powder their noses'. (N. Mitford, 1949)

and occasionally, jocularly, of men:

> Peter laughed, 'I've had enough of the bloody sun. I'm going to powder m'nose.' (Manning, 1978)

To say you are going to *powder your nose* is genteel, but the rarer *powder your puff* is not, punning on the powder puff and the pubic hair. For a male, to *powder your hair* in the 18th century (when wigs were commonly powdered) was to become drunk.

powder your nose see POWDER ROOM.

pox (the) syphilis.

Literally, any disease which brings up pustules on the skin but, as Dr Johnson reminds us, 'This is the sense when it has no epithet':

> I couldn't be sure she hadn't got the pox. (Archer, 1979)

Poxed is so infected and *poxy* a term of abuse. See also BLACK POX.

practice development advertising and marketing.

British lawyers and accountants, conditioned by decades of touting for clients but forbidden to advertise, still feel a need to conceal their hypocrisy:

> Professions cringe at the word. A more acceptable term ... is 'practice development'. (*Financial Times*, November 1987—the 'word' was marketing)

prairie oyster[1] the testicle of a calf.

Eaten in America as a delicacy. Calves' testicles are also eaten in the British Isles but described as *sweetbreads*, the term for the pancreas in standard usage.

prairie oyster[2] a pungent alcoholic drink with raw egg in it.

Not to be confused with the American *prairie dew*, an illegally distilled drink.

pre-arrangement the payment for a funeral before death.

American funeral jargon for selling burials and their trappings in advance to the morbid and lonely:

> The cemetery industry has found an answer to high cost through pre-arrangement. (J. Mitford, 1963)

Also as *pre-need*:

> A 'pre-need memorial estate'; in other words, a grave for future occupancy. (ibid.)

pre-dawn vertical insertion an invasion by armed parachutists.

Neither clocking on for the first shift nor starting the day with copulation, but how the American invaders of Grenada on 27 October 1983 described their mission.

pre-driven *American* (of a car) not new.

The jargon of the used-car dealer, implying that new models had not even been road tested. See also PRE-OWNED.

pre-emptive strike an unprovoked military attack without warning.

Pre-emption is buying first, whence denying the purchase to others. If you are the one so attacked, you may call the action a *pre-emptive offensive*:

> It would be important ... for the forces of the Pact to be fully prepared ... for the more likely contingency of a pre-emptive offensive. (Hackett, 1978)

Pre-emptive self-defence is killing:

> He had written a legal opinion asserting that pre-emptive action would be no more an assassination than would a case in which a policeman gets off the first shot at the man who is pointing a gun at him. 'Pre-emptive self-defense' he called it. (Woodward, 1987)

pre-owned (of a car) not new.

Advertising jargon of the motor trade, conveniently forgetting the initial ownership by the manufacturer and dealer and 'the modern euphemism for "second hand"' (Pei, 1969). Two decades later *previously owned* was preferred:

> Buyers looking for a 'previously owned' motor car (to use the current trade euphemism) tend to be very selective. (*Daily Telegraph*, October 1987)

By 1990 prospective customers were being invited in advertisements to buy *owned* Rolls-Royces, as though there were also a store of abandoned vehicles to draw from, if they so chose.

precautions contraception.

Shortened form of *precautions against pregnancy*, usually *taken* or *neglected*:

> She hoped she might be pregnant, since she had taken no precautions. (McCarthy, 1963)

precocious spoilt and ill-mannered.

Originally, developing early. Used of children other than your own, out of earshot of their parents.

predilection homosexuality.

Literally, a tendency or preference for anything, even for heterosexual encounters:

> 'Predilection?' he said, giggling. 'What a sensitive way of putting it!' (Sanders, 1986)

preference being homosexual.

Shortened form of *sexual preference*, I suppose, but not used of heterosexuals:

> Names and addresses; sweethearts and wives; habits and preferences. Complete with photos and medical sheets. (Deighton, 1994)

pregnancy interruption an induced abortion.

An *interruption* is literally a disturbance with an assumption of resumption. This medical jargon is sometimes enlarged to *voluntary pregnancy interruption* or *VPI*.

preliterate uncivilized.

Anthropological and social science jargon of societies which remain illiterate. The usage denotes a touchiness on behalf of savages who can't read what they would be touchy about.

premature conceived before marriage.

The taboo about pre-marital sex in western society no longer appears to exist, which makes this fiction to explain a birth of a normal baby less than nine months after a wedding no more than a reminder of the recent past.

premium costing more.

An attributive use of a noun which originally meant 'an award or prize', whence something worth more than its face value. Advertising jargon and much loved by estate agents.

prepare *American* to embalm.

For viewing by the survivors rather than St Peter:

> So the worst racket of all was built up: the embalming or 'preparing' of the 'loved one'. (E. S. Turner, 1952)

Whence the *preparation room*, where the corpse may lie:

> He suggests a rather thorough overhauling of the language ... 'preparation room not morgue'. (J. Mitford, 1963)

See also AT-NEED.

prepared biography *American* an updated obituary.

A delicate expression masking the inevitability of death:

> In America, incidentally, an obituary held in reserve for future use is ... described as a 'prepared biography'. (John Gross in Enright, 1985)

preparedness *American, obsolete* the military help given to the British in the Second World War prior to Pearl Harbor.

American isolationism was so widely supported even by those who were not pro-Nazi, such as Lindbergh, that support for the embattled British Empire had to be disguised by euphemism:

> ...he had financed an expensive advertising campaign in the country's largest newspapers, savagely attacking 'preparedness'... (Lacey, 1986, of Henry Ford, whose anti-Jewish paranoia cannot have failed to see attractive features in Nazism)

present arms to have an erect penis.

Punning on the military drill in which the rifle is held vertically in front of the body:

> ...by the time she was done I would be ecstatically ruined, and certain sure I'd never be able to present arms again. (Fraser, 1971)

The obsolete English *presenter* was a prostitute, who *presented* herself to you, but you still had to pay.

preserved drunk.

An American variant of the more common PICKLED with alcohol the agent of preservation.

press conjugal rights on to copulate with (a reluctant wife).

The *pressure* in this phrase is usually within marriage:

> Some fear that he might have been pressing his 'conjugal rights' could have accounted for it. (Kee, 1993—Parnell was afraid that his mistress, Katie O'Shea, might be having to copulate also with her husband and see also CONJUGAL RIGHTS)

Also as *press your attentions on*.

pressure torture.

Exerted on someone in custody:

> '...he's trained to withstand pressure.' 'Interesting usage, pressure.' (Seymour, 1989, of a prisoner being subjected to torture)

pressure of work an excuse for any neglect, inefficiency, or discourtesy.

Not used by business-like people:

> I feel an awful worm, not having written to you for so long, but a genuine pressure of work stopped me. (P. G. Wodehouse in a letter of 1930—in Donaldson, 1990)

prestigious expensive.

Originally, concerned with juggling, or prestidigitation, but latterly commonly used as conferring prestige. Estate agents love it:

> City of London's most prestigious fully-serviced apartment block. (*Times*, May 1981, but not Buckingham Palace)

Also, unhappily, in other commercial puffs.

preventative a contraceptive sheath.

Army usage; in America syphilis and gonorrhoea were the *preventable diseases*. The obsolete British *preventative man* neither sold nor wore such articles, but was a coastguard.

preventive detention arbitrary imprisonment.

In a totalitarian state, without process of law. In the British Isles it means a long sentence for a dangerous or hardened criminal.

preventive war an unprovoked attack without warning.

The theory is that you are *preventing* your opponent getting the drop on you. Today it is thought slightly effete for a nation to declare war formally and in this respect, as in so many others, the Nazis set a pattern of conduct which has been widely emulated.

previously owned see PRE-OWNED.

prey to (a) suffering from.

The victimization is only figurative as with those who describe themselves as a *prey to dyspepsia*. Not so the obsolete British *prey to the bicorn*, a cuckold. The *bicorn* was a mythical two-horned beast which devoured men whose wives dominated or deceived them. Its counterpart, the *chichevache*, which ate obedient wives, was reputed to feed but rarely.

priapism an erection of the penis.

From Priapus, the Pan of Mysia. For medical practitioners:

> **Priapism**, a condition caused by sudden obstruction of the blood vessels so that the blood cannot flow away from the erect penis. (T. Smith, 1986)

In lay use a *priapism* or *priapus* is a more temporary and less dangerous condition:

> He threatened her with a priapus that had already once inflicted upon her an almost mortal wound. (Nabokov, 1968)

See also MR PRIAPUS.

price adjustment an increase in price.

In retail trading, all *adjustments* and *revisions* are upwards—a reduction needs no euphemism.

price crowding a price increase not authorized by the proprietor.

Mainly supermarket jargon. The device is normally used by a manager to create a reserve which can be used to make good losses or shortages for which he might be held responsible.

prick the penis.

Once standard English but now common slang, of obvious origin:

What did in for him
Was a prick in the skin,
When the prick should have been in Ophelia.
(*Playboy's Book of Limericks*, of *Hamlet*)

Also used figuratively as a term of mild abuse or rebuke among males.

pride the erect penis.

Shortened form of *pride of the morning*, an erection of the penis upon waking which comes from the proper meaning 'mist or a shower heralding a fine day':

Said a just-wed professor named Ted,
To a redhead coed in his bed...
Won't you swallow my pride dear instead?
(*Playboy's Book of Limericks*)

prima donna see PRINCESS.

prime saleable.

Literally, first, whence implying of first quality. Commonly used of perishable foodstuffs, especially meat.

prime the pump deliberately to cause inflation.

Of government spending when the political consequences of higher unemployment etc. outweigh the damage caused by short-term inflation—few politicians concern themselves with the consequences longer term caused by inflation because that will be somebody else's problem:

The new administration coming into power in just two weeks would have no choice but to 'prime the pump' through massively increased government expenditure. (Erdman, 1986)

primed *Scottish/English* drunk.

Like a pump, and perhaps also alluding to an explosive charge:

When he was 'primed', was Nathan's wont to pass,
No licensed house without another glass.
(Doherty, 1884)

Prince of Darkness the devil.

Not the eldest son of King Edward III but yet another way of avoiding talking of the devil—see OLD A' ILL THING.

princess an expensive prostitute.

From the meaning 'a classy type of female' or 'one who affects airs':

Willy goggled at a couple of painted princesses swaying by in all their finery... 'Whores' says I. (Fraser, 1973, writing in 19th-century style)

The use seemed to be a synonym of *prima donna*:

By lorettes I mean those I have touched on before as prima donnas (Mayhew, 1862)

but survives in modern speech, as the American PAVEMENT PRINCESS.

private enterprise illegal trading by an employee.

Literally, trade or industry not financed by or under the control of the state, but some pejorative use by socialists etc. The euphemism is of government servants who exploit their position for personal gain, of smuggling by transport workers, etc:

But there was a great deal of what you might call private enterprise on that run. (Price, 1970, of smuggling by airline staff)

private office see PRIVY.

private parts the human genitalia.

Those not ordinarily exposed to the public gaze:

'No more private selves, no more private corners in society, no more private properties, no more private acts.' 'No more private parts,' says Barbara. (Bradbury, 1975)

Less often of domestic animals, where there is no concealment:

Buller was licking his private parts with the gusto of an alderman drinking soup. (G. Greene, 1978)

Shortened either to *privates*:

He had not let Oliver in until his privates were covered over with water (Bradbury, 1976)

or to *parts*:

'You find the model ugly?' 'No not at all. I mean her... parts.' (Amis, 1978)

In obsolete use as *privy parts*:

He moved their privy parts to the front (Plato, in translation—Zeus was the mover)

or as *privities*:

...felt great pain in her privities, as if her swooning had not spared her and some rude forcing had taken place. (Fowles, 1985, writing in archaic style)

private patient a person paying personally for medical care.

All British citizens pay indirectly for medical services provided by the state but some choose to make further personal arrangements, usually through insurance. The usage disregards the fact that each *patient* is *private*, by whatever means the bills are paid.

privateer *British, ?obsolete* a prostitute.

She works on her own, usually part-time like the pirate who combined *private man of war* and *volunteer*.

privates see PRIVATE PARTS.

privatize to pass from the control of politicians and bureaucrats to that of directors.

The fate or destiny of utilities or businesses formerly operated without commercial constraints and therefore increasingly inefficient. The expressed intention is to enable any individual to own shares but in Britain the greed of most directors has diverted attention from other achievements.

privileged rich.

Sociological jargon and not really implying that they have honourable distinctions. Indeed in the eyes of those who use this dysphemism the opposite is true. The converse is UNDERPRIVILEGED.

privy a lavatory.

From the privacy:

> Hadjimoscos, sick in a privy, had spewed out his false teeth. (Manning, 1960)

Private office is now obsolete in this sense, having been adopted as a rather grand social distinction for those whose status demands individual secretarial arrangements. The obsolete *privy stool* was a portable combination of lavatory seat and bucket:

> ... chairs and privy-stools necessary for a royal visit. (Monsarrat, 1978, writing in archiac style—I hope the two puns were unintentional)

For *privy parts* see PRIVATE PARTS.

pro a prostitute.

A shortened form of *professional* or of *prostitute* —or both:

> You the bloke that floated them pros out to the *Everett*? (Theroux, 1973—some prostitutes had been sent out to a ship)

I suppose the Second World War military *pro-pack*, a soldier's contraceptive kit, came from this shortened form or from PROPHYLACTIC.

pro-choice in favour of abortion on demand.

Not the selection of a prostitute, or even suggesting that there is the option of remaining celibate if impregnation is not desired. The opposition describes itself as *pro-life*, suggesting that those not subscribing to its views are at best callous.

probe to copulate with.

Of a male, but not with a blunt-ended exploratory surgical instrument:

> Says Barbara frankly 'I was probed.' 'That's true,' says Howard. 'At the purely external level you got screwed.' (Bradbury, 1975)

problem an unwanted and often irreversible condition.

The word is used in many phrases to conceal truth or inadequacy. Thus a *cash problem* in an individual is a shortage of money, and not a lack of pockets in which to keep it. In a company, a *cash flow problem* means it is overtrading or insolvent. A *communication problem* means that nobody understands us or we don't understand them. A *crossword problem* means we cannot do the crossword (a *problem problem*?) although a problem crossword is one we should be able to solve. A *drink problem* is alcoholic addiction for a *problem drinker*:

> ... the fact that she was a 'problem drinker' (Styron, 1976),

although a *drinks problem* at a party would indicate that you are running out of supplies. A *heart problem* is a malfunction or failure of that organ, with other organs or bodily zones being similarly identified according to your disability. A *husband* or *wife problem* indicates that the problem husband or wife is a drunkard, takes illicit drugs, copulates extramaritally, spends too much, or lacks some common marital virtue. In addition, a *woman's problem* may be the onset of menstruation, which she may describe as her *problem days*, or it may be a disorder of the womb. Hitler referred to the *so-called Austrian problem* when looking for justification of his 1938 invasion. To have a *pigmentation problem* is to be black in a mainly white country:

> ... wants to send anyone with a pigmentation problem back to Islamabad. (Sharpe, 1979)

A *weight problem* is obesity, and not starvation:

> If you are destined to be fat, food makes you fat. But I have never had a weight problem. (Murdoch, 1978)

procedure any taboo or unpleasant act.

Literally, a method of acting. In the language of doctors and dentists, it is anything which will cause pain. For police and lawyers, it is a civil or criminal legal action. For pregnant women, it may be an induced abortion. For the Nazis, it was mass murder:

> Schindler heard rumours that 'procedures in the ghetto' were growing more intense. (Keneally, 1982, of Second World War Poland)

process the penis.

Literally, anything which sticks out:

> ... washing my process and asking me if I've got the clap. (Theroux, 1979)

procure to arrange (prostitution) on behalf of another.

Literally, no more than to obtain, of anything, but legal jargon in this sense:

> ... she had never heard of my sister, but she would undertake to procure her for me for seventy-five dollars. (Fraser, 1971)

Formerly a *procurer* did nothing more harmful than to arrange affairs or collect taxes for another but today he and a *procuress* are pimps:

> A middle-aged man doing the same thing was a dull dirty procurer. (Theroux, 1973)

prod to copulate with.

Of a male with the same imagery as *poke* (see POKE[2]) but much less common.

product a service.

Jargon of bankers and other moneylenders which seeks to suggest that their activities actually *produce* something, and now standard English:

> Beginning with the M & S Chargecard, followed by personal loans and a number of investment products. (*Daily Telegraph*, 3

March 1994—M(arks) & S(pencer) is a retailer with multiple outlets)

profession (the) prostitution by females.

Prostitutes' jargon:

... containing some bitter observations by an old member of the Profession (Londres, 1928, in translation)

and the *oldest profession* with its Biblical references is a cliché:

It was maybe the oldest profession ... but New Orleans was proud *and* ashamed of its cathouses. (Longstreet, 1956)

A *professional* or *professional woman* is a prostitute, the latter perhaps punning on the propriety of the learned professions:

He cannot afford to pay professional women to gratify his passions. (Mayhew, 1862)

See also PRO.

professional car a hearse.

American funeral jargon. *Processional* would be more appropriate.

professional foul *British* a deliberate infringement.

Soccer jargon for unsporting and often dangerous violence against an opponent to deny him an advantage—the *foul* of one who 'plays' as a *professional*.

progressive opposed to restraints on behaviour.

Literally, moving towards improvement but socialist etc. jargon for those whose conduct is not restrained, on occasions, by the law, morality, or good manners:

Day Release Apprentices had to have their weekly hour of progressive opinions. (Sharpe, 1979)

For a Communist, *progressive* means 'Communist'.

proletarian Communist.

The *proletariat*, from the Latin *proletarius*, 'the lowest class in the Servian arrangement' (Smith), first indicated those in feudal service and then anyone who worked for a wage, among whom middle-class revolutionaries traditionally seek support. Whence a *proletarian democracy*, a Communist autocracy, *proletarian internationalism*, Russian imperialism, etc.

promised engaged to be married.

From the days when a man (and very occasionally a woman) might be sued for *breach of promise* if he changed his mind before the wedding:

Loud, of course, and facetious were the lamentations that Francie had not returned 'promised' to one or other of these heroes of romance. (Somerville and Ross, 1894)

promoted to Glory dead.

A usage of the Salvation Army, whose members live as closely as any may get to the Christian ethic, and deserve any glory that may be going.

promotion bribery.

As with the *promotional* competitions among dealers arranged by oil companies, with prizes of holidays for two in far off places which can be awarded ahead of the event; or of the excessive expenditure devoted by drug and medical supply firms to the British National Health Service to ensure the wasteful selection of branded products in preference to cheaper equivalents.

prong to copulate with.

Of a male, with the common 'forking' imagery:

I hear she's some kind of guru to the old man ... Think he's pronging her? (M. Thomas, 1985)

prophylactic a contraceptive sheath.

Literally, the prevention of any disease but widely used in the Second World War of any process to reduce venereal disease.

proposition to suggest engaging in a sexual act to.

Of both heterosexuals and homosexuals:

He might feel like hitting the first one who propositioned him. (Davidson, 1978, of homosexuals)

Whence a *proposition*, such a suggestion:

I didn't take her up on a proposition she made to me ... a bodily proposition. (Masters, 1976)

proposition selling the use of misleading hypotheses to confuse a buyer.

A device of the con-man which a statutory cooling-off period may blunt but will not negate:

His technique is old-style American 'proposition selling'. The salesman puts forward a series of numbskull propositions with which you have NO CHOICE BUT TO AGREE. (*Daily Telegraph*, August 1989, of a TIME-SHARING scam)

protection extortion.

The practice of selling immunity from your own depredations is well documented from Anglo-Saxon Danegeld in England to 20th-century Mafia activity in America:

He was supplying Rachman's clubs with protection. (Green, 1979)

Just as being *in the rag-trade* is making money in the clothing business, so being *in protection* is living by such extortion:

I'm going into protection ... scare the shopkeepers silly. (Murdoch, 1977)

protective custody arbitrary imprisonment.

The Nazi *Schutzhaft* of February 1933, the subject of much imitation by Communists and others:

> *Schutzhaft* (Protective Custody) a catch-all word whereby men, women and children disappeared and were never seen again. (Deighton, 1978)

The pretence is that the victims are incarcerated to prevent ill befalling them.

protective reaction *American* bombing.

A Vietnam usage (Commager, 1972).

protector[1] *?obsolete* a man keeping a sexual mistress.

From the 19th-century convention that an unmarried woman living alone should have a male to look after her:

> They are dismissed... and set once more adrift. They do not remain long... without finding another protector. (Mayhew, 1862, of such women)

protector[2] a contraceptive sheath.

From the avoidance of impregnation or the transfer of a virus.

protectorate a conquered territory.

Widely used by the European colonizers of Africa, who were anxious to *protect* themselves against seeing a rival grab the territory ahead of them, although they did not use the word in that sense. More recently of Bohemia and Moravia under Nazi Germany:

> The Anglo-Americans intend to reach the Protectorate before the Soviets. (Goebbels, 1945, in translation. Unhappily for the Czechs, and for Europe, his political judgement was superior to his forecasting; the Western powers held back their armies and so condemned Czechoslovakia to prolonged Russian domination.)

Thus for the Nazis and others, to *protect* was to incorporate and rule by force:

> He had warned that Germany would know how to 'protect' the ten million Germans living on its borders... Everyone knew what Hitler meant by 'protect'. (Shirer, 1984)

provision an arbitrary adjustment in figures to be publicly reported.

Provisions are literally reserves made against contingencies, to avoid a misleading statement of assets or profits. Reserves and deductions made on a subjective basis to satisfy a client by reducing taxable income or otherwise understating the financial position are also called *provisions*. Public accountants have to be flexible because they lose audits if they upset clients.

provocation the statement of a contrary view.

Literally, a hostile act calculated to excite a reaction. Hitler and other tyrants profess to construe any opposition as such, even when a neighbour dares to point out it might defend itself if attacked.

prune-juice a spirituous intoxicant.

From the colour and the laxative effect? American *pruned*, drunk, may come from *prune-juice* but a more likely derivation is from feeling like a tree which has lost its appendages and extremities.

psycho a lunatic.

From the Greek, it means 'relating to the breath', whence 'of the soul' or 'mind'. This common use is probably no more than a shortened form of *psychopath*:

> 'Keep that psycho away from me,' Wade yelled, showing fear for the first time. (Chandler, 1953)

psychological warfare the dissemination of lies or half-truths.

Usually waged as an adjunct to bombing, shooting, etc.:

> ... the Foreign Office's Information Research Unit, responsible for what had once been termed psychological warfare. (N. West, 1982)

psychologically disadvantaged under the influence of narcotics.

Ignoring the physical ill-effects:

> Wilson, who won a lawsuit in 1992 claiming that his father, Murray Wilson, had bullied him into giving away the publishing rights to his songs while he was 'psychologically disadvantaged' (spaced out on drugs)... (*Daily Telegraph*, 7 October 1994, of one of the Beach Boys)

public assistance see ASSISTANCE.

public convenience see CONVENIENCE[1].

public house an establishment where intoxicants may be sold and drunk.

Not a British version of the *long house* of the inhabitants of Borneo. Now commonly and internationally shortened to *pub*, the 19th-century *public* being obsolete:

> Being also a public, it was two stories high. (W. Scott, 1814)

public ownership managed under the control of politicians and bureaucrats.

Of commercial businesses, utilities, etc. which have no individual owner or shareholders:

> Various failings, real or imaginary, in state-run undertakings created the need for fresh euphemism, and 'public ownership' was promptly produced. (Simon Hoggart in Enright, 1985)

public relations the presentation of yourself or your client in a favourable light.

Literally, making relevant facts or opinions known to the public. The commercial or political euphemism refers to conscious distortion or selection of truth. A *public relations officer* is paid to do this for others. Abbreviated respectively to *PR* and *PRO*:

> We're in the same racket... and you're the kind of PRO I like. (Deighton, 1982)

public sector borrowing requirement government overspending.

The *public sector*, that part of a mixed economy which is directly controlled by government, has no sanction of bankruptcy to inhibit profligacy:

> A series of heavy expensive settlements has piled up that debt, euphemistically called the Public Sector Borrowing Requirement. (*Daily Telegraph*, December 1980)

public tranquillity internal repression.

In China internal political control is the function of the Department of Public Tranquillity, one of the softer terms for that necessary totalitarian function.

pudding club see IN THE CLUB.

puddle the result of involuntary urination.

From the shallow and temporary pool of rainwater, and usually made in this sense by small children or domestic pets:

> ... my foot landed in the middle of Telek's puddle. (Butcher, 1946—Telek was Eisenhower's dog)

puff (powder your) see POWDER ROOM.

pull[1] to cause a horse to lose a race fraudulently.

Racing jargon, from the jockey's handling of the reins. To *pull up* is also used in the same sense, although it literally means 'to bring to a halt'.

pull[2] to copulate extramaritally.

From drawing the object of your lust towards you, perhaps, although it sounds a bit thin:

> If someone does recognize me, word will go back that the brigadier's pulling outside duty. (Ludlum, 1984—he was meeting a woman in a truck stop)

pull a daisy see PICK A DAISY.

pull a train *American* to copulate in immediate succession with a number of males.

The imagery is from coaches behind an engine:

> ... trying to persuade her to pull the train for a few of the choirboys. (Wambaugh, 1975—the *choirboys* were off-duty policemen)

The male participants successively *board a train*:

> I just can't board the train like horny old Spencer. (ibid.)

pull his trigger to cause to ejaculate.

Of obvious punning imagery:

> I know how to pull his trigger. His wife doesn't. (Sanders, 1981)

pull in to arrest.

Police jargon, often expanded to *pull in for a chat* etc.:

> What do you say to a man from SAVAK when he says ... 'We'd like you to replace Barnheni as office manager, because we'll be pulling him in for a chat very soon'. (West, 1979)

pull off[1] to masturbate.

Of a male. Also as *pull the pudding, pull the wire, pull yourself*, etc.

pull off[2] *American* to refrain improperly from prosecuting a criminal.

From the meaning 'to draw away from':

> The detectives who were offered all kinds of inducements to pull off. (Lavine, 1930)

pull out of the air fraudulently to invent.

Usually of figures in prospectuses and other published accounts. To *pull figures out of a hat*, meaning the same thing, takes its imagery from conjuring:

> The *Veterinary Record* said he 'pulled figures out of a hat to fit his arguments'. (*Private Eye*, May 1981)

pull rank to use seniority to secure an unfair advantage.

Of those in hierarchical employment such as sailors or civil servants and euphemistic only when not used of normal commands or orders. Standard English.

pull the long bow see DRAW THE LONG BOW.

pull the pin[1] to desert your wife.

The male imagery is from the uncoupling of rolling stock on a railroad with the engine running free, and not from activating the primer on a hand grenade, although for many husbands the latter might seem more appropriate.

pull the pin[2] to retire.

With the same imagery:

> ... he wondered if he could afford to pull the pin when he got twenty-five years in. (Wambaugh, 1983)

pull the plug on to kill by withdrawing mechanical life support.

Punning on the electrical connection to life support machinery and the flushing of a lavatory. Also of killing:

> Hubby Luther pulled the plug on her. (Sanders, 1986, of a wife murderer)

pull the pudding *British* to masturbate.

Of a male, often shortened to *pull the pud*.

pull the rug to render bankrupt.

From causing a person standing on a rug to fall when you jerk it:

He thinks the United Nations peacemongers could pull the rug. (Forsyth, 1994, referring to the period preceding the Gulf War)

Normally used of a banker or creditor who refuses to give further time to pay, thereby crystallizing other indebtedness.

pump ship to urinate.

Of a male, from the jet of water expelled over the side. Also as *pump bilges*:

See if you can put a Martini together while I pump bilges. (Clancy, 1989)

pump up to copulate with.

By a male, from the motion involved:

If you work for a big corporation, the head of the firm is always pumping up the secretary. (*Sunday Telegraph*, 20 March 1994—the secretary of a big corporation is usually a male, as is its head, which makes the generalization as inaccurate as it is incorrect)

pump your pickle to masturbate.

Of a male. The *pickle* in this case is a gherkin, whence the penis. Also as *pump your shaft*:

So there he stood, pumping his turgid shaft. (Sanders, 1973)

punch *American* to copulate with.

Of a male, with the common violent imagery:

Danny introduces Angel to this broad which Danny has been punchin' since high school. (Diehl, 1978)

punk has so many modern connotations with adolescent excesses that it is easy to forget its venerable history. A 16th-century *punk* was a prostitute:

She may be a punk, for many of them are neither maid, widow, nor wife. (Shakespeare, *Measure for Measure*)

A 20th-century American male *punk* was a homosexual playing the female role, although the word is now used both as a term of abuse and to acknowledge comradeship. In the sense 'worthless', *punk* means 'low-quality marijuana', although *punk pills* are any narcotic illegally swallowed. *Punk rock* was a fashion of the late 1970s and early 1980s; the music was a loud tuneless noise with a strong beat, performed under flashing coloured lights. Followers of the fashion, who were known as *punks*, wore ritual decorated leather clothing and affected impractical coiffures.

punter *English* an inexperienced visitor who can be overcharged or robbed.

Literally, someone who bets on horses, whence a habitual loser:

Many airport taxi-drivers strongly object to driving their fellow-countrymen, motivated... by the prospect of picking up a 'punter', someone who can safely be overcharged. (Moynahan, 1983)

pup to impregnate (a woman).

Canine imagery, although there need be no suggestion of bitchiness:

I want all these wenches pupped. (Fraser, 1971)

In coarse speech, to *pup* may also mean 'to be delivered of a child', and see *in pup* at IN CALF.

puppy fat obesity in a child.

Usually of a young female, with the implication that the plumpness will vanish as the child grows up without any change in its eating habits.

pure not having copulated outside marriage.

The proper meaning 'free from adulteration', becomes 'free from adultery'. An 18th-century *pure* was a sexual mistress, although not noted by Grose or Dr Johnson, from her freedom from disease rather than her chastity or modesty.

purge[1] *?obsolete* beer.

Probably from its laxative effect:

We were fond of a drop of 'neck-oil', which like 'purge' was a nickname for beer. (Richards, 1936)

purge[2] to attack violently.

Literally, to cleanse from some kind of defilement:

The next day what they euphemistically call a 'purging operation' was effected. In this instance they purged Fatah. (Price, 1971)

purification of the race the systematic killing of Gypsies, Jews, and Slavs by the Nazis.

Those living in Germany had to conform to the Teutonic ideal. We sometimes forget mad, crippled, or deformed Germans were also killed by the Nazis with the same justification.

purple heart a narcotic in pill form.

Usually morphine or a barbiturate of that colour and punning on the American medal for the wounded:

...the notorious so-called Purple Heart tablets which were a source of drug addiction in teenagers in Britain from the early 'sixties. (Foster, 1968)

pursue other interests see LOOK AFTER YOUR OTHER INTERESTS.

push[1] to copulate with.

The usual thrusting imagery but also from the rhyming slang *a push in the truck*, a fuck:

'You pushing her?'... 'Every chance I get.' (Sanders, 1970—the lady was not confined to a wheelchair)

In West Africa, *push* is copulation:

Sing, dance, cook, plenty push. (Sanders, 1977—a female servant was being extolled to a bachelor)

A *pushing academy, school*, or *shop* was a brothel, punning on the other meaning 'a fencing school':

...for the income of the whores of the so-called *pushing academies*. (Keneally, 1987)

I mean he spent an hour a day at the pushing shop down near the railway, rooting himself stupid. (Keneally, 1985)

push[2] (the) peremptory dismissal from employment.

Not always *given* by the employer; until at least 1951 the employee might have taken the initiative:

It is conceivable that not all employers relished the idea of encouraging ambitious young men to give their firms 'the Push'. (E. S. Turner, 1952)

push[3] a sustained attack.

Second World War military jargon:

The gen is that the jerries are preparing a push on Alam Halfa. (Manning, 1977)

push[4] to distribute (narcotics) illegally.

Literally, to sell energetically. A *pusher* is any dealer in illegal narcotics, especially by retail:

He was on the weed. I pretended to be a pusher. (Chandler, 1958)

A *share pusher* fraudulently sells bogus or over-valued securities.

push the button on *American* to cause to be killed.

The act of setting in motion or of turning out a light:

You never gonna get the guys who pushed the button on him. They too big for you. (Sohmer, 1988—an FBI agent's partner had been murdered)

push up the daisies to be dead.

If buried, you may nourish the common churchyard flower:

If I'd been born fifty years sooner I'd have been pushing up the daisies by now. (N. Mitford, 1960)

pussy the vagina.

A commoner version of *cat* (see CAT[2]):

She could not even get her forefinger into her pussy. (Harris, 1925)

Pussy-whipped means besotted with a female who takes advantage of the infatuation, punning on *horse-whipped*:

An old man like that. Our father. Pussy-whipped. (Sanders, 1980)

An American *pussy lift* is an operation to tighten the vagina and so enhance sexual enjoyment:

...Piper with the happy illusion that pussy lifts were things cats went up and down in. (Sharpe, 1977)

Pussy is also an act of copulation viewed by a male:

...Brancusi,
Unafraid of black pussy,
Walked under the ladder and had her.
(*Playboy's Book of Limericks*—the sculptor was using a black model)

put an act of copulation.

In common speech, *had* or *done* by the male, from the meaning 'a push or thrust'. In American use also as a verb meaning 'to copulate':

...you been put-putting with blondie here, my wife. (Mailer, 1965)

Put and take is the joint act by male and female. To *put a man in your belly*, to copulate, of a female, puns on the male ingress and the conception:

So you may put a man in your belly. (Shakespeare, *As You Like It*)

To *put it about* is to copulate promiscuously, usually of a female:

Certainly not some blonde tart who undoubtedly put it about if the mood took her (C. Forbes, 1987)

and a *putter about* is a male profligate:

There are *rumours* to the effect that Michael is a consistent and conscientious putter-about. (Fry, 1994)

Put it in and *put it up* are explicit in male use:

They thought it would save their kids or their daddies, letting me put it up them. (Allbeury, 1980, of a German prison guard)

To *put out* is to copulate casually, of a female, from the meaning 'to act in an extroverted way':

Any girl...is caught in a sexual trap. If she won't put out the men will accuse her of being bourgeois. (Lodge, 1975)

To *put to* is to copulate, of women or men, from the proper meaning 'to start work':

As rank as any flax-wench that puts-to,
Before her troth-plight.
(Shakespeare, *Winter's Tale*)

put a move on *American* to make a sexual approach to (a stranger).

Usually by the male and possibly also of a successful outcome, such as extramarital copulation:

...too sore and shaken to put a move on her. (Wambaugh, 1983)

Also as *put the moves on*:

He doesn't seem to comprehend the etiquette of putting the moves on a woman. (de Mille, 1988)

put a person's lights out to kill.

Lights means 'eyes' but the phrase also puns on extinguishing a lamp:

All men who were lucky at gambling very soon had their lights put out. (Richards, 1933, of First World War trench life)

put away[1] to kill.

Of old, diseased, or unwanted domestic animals:

> I have left instructions for Buller to be put away—as painlessly as possible. (G. Greene, 1978, of a dog)

In obsolete use, *put away* might mean 'dead', from the burial of the corpse:

> Some poor comrades undertook to see her put away. (Hartley, 1870)

To *put yourself away* is to commit suicide or, in obsolete use, to *put hands in (or on) yourself*:

> Belus... put hand in himself and became his own executioner (Brand, 1789)

and:

> Who being to be tryed, put hands on himself at the devil's instigation. (Maidment, 1844)

put away[2] to confine to a lunatic asylum or prison.

From the involuntary removal from society:

> He was a bit 'tropo'... They put him away in the end. (Simon, 1979)

put away[3] to consume (intoxicants).

Not putting the bottle back in its rack:

> ... it was astounding to see [her] put away the booze. (Styron, 1976, of an alcoholic)

If you *put it away*, you regularly drink intoxicants to excess:

> ... the walking wounded of the day watch *really* put it away. (Wambaugh, 1983)

put daylight through to kill by shooting.

Mainly First World War usage from figuratively making a hole through the body:

> He wouldn't have given him that chance, but soon put daylight through him. (Richards, 1933)

put down[1] to kill.

Formerly of execution:

> The most... accomplished lady... was suffered to be put down as a common criminal (Hogg, 1822)

and of murder:

> I am going to be forced to put down the first hostage. (W. Smith, 1979)

Normally of killing old, diseased, or unwanted domestic pets:

> ... an old smelly Border Terrier which Uncle Matthew had had put down. (N. Mitford, 1945)

But if you *put down* pheasants, you encourage them to breed so that you can kill them at your pleasure.

put down[2] to denigrate or oppress.

Often by a dominant group but also of an individual snub:

> The majority keeps putting down the minority. (*Daily Telegraph*, 1 March 1995, of obese men who complained they were the 'butt of lewd jokes by women')

put in a bag killed.

Of soldiers, from the days when soldiers were buried in sacking and see also BODY BAG.

put in your ticket to die.

A ship's officer gives up his current licence, or *ticket*, on retirement. Mainly marine use.

put lead in your pencil see LEAD IN YOUR PENCIL.

put off to kill.

Of animals and perhaps obsolete:

> Ir ye gaun to pit aff da auld koo? (*Shetland News*, 1900, quoted in EDD)

put on to deceive.

I suppose from the imposition:

> ... if he's putting us on I'm going to pull his arms off. (Forsyth, 1994—the speaker is referring to an informer)

put on the spot to kill.

American underworld jargon, the *spot* being a place of danger:

> Youthful killers on the East Side can be hired to 'knock off' or 'put a guy on the spot'. (Lavine, 1930)

put out[1] to kill.

Perhaps a shortened form of PUT A PERSON'S LIGHTS OUT.

put out[2] see PUT.

put out a contract see CONTRACT[2].

put out to grass to cause to retire prematurely.

Usually for early senility, inefficiency, etc. with imagery from a horse which escapes the knacker:

> If you think you are going to be put out to grass, you are mistaken. (Price, 1970, of a man being moved from his job prior to normal retiring age)

put the arm on to extort money etc. from (a person) by threats or violence.

The imagery is from wrestling:

> Other guys roll over and lie still the moment you put the arm on them. (le Carré, 1980)

Also as *put the scissors on*, again from wrestling:

> ... if I don't get it then in one-pound notes, I'll put the scissors on you. (Kersh, 1936)

Put the muscle on refers to the *muscle*, or powerful person, used for such errands:

> I was looking for a job, no question about it. But I wasn't trying to put the muscle on them. (Colodny and Gettlin, 1991—he was being accused of attempted blackmail)

And as *put the black on* where *black* is a shortened form of BLACKMAIL, *put the burn on* (see BURN[3]), and *put the bite on* (see BITE).

put the boot in to cause injury or damage.

From the conduct of ruffians when they have floored their victim. Used figuratively of any harmful activity or hurtful words against a party in difficulty:

> Mrs Lupey says living successfully in a family is largely a matter of timing, and, I must say, I picked exactly the right moment to put the boot in. (Fine, 1989)

To *put the clog in,* in the jargon of professional sportsmen, means 'deliberately to injure an opposing player':

> There were many who thought the Dutch had put the clog in on the Saudi striker. (*Daily Telegraph*, 22 June 1994—the Dutch were playing the Saudis at soccer, not presenting them with their national footwear)

put the clock back fraudulently to alter the reading of a mileometer.

Punning car trade jargon of a practice which tricks buyers of used cars, the *clock* being the mileometer. Today done circumspectly although no less frequently as a British criminal offence. Also as *turn the clock back*, or, of the operation, to *clock* a vehicle.

put the file in order to conceal a mistake or omission.

Bureaucratic jargon with the file-mentality cocooning the civil servant. So long as the documentation is in order, all are above criticism.

put the finger on see FINGER[1].

put the skids under wilfully to cause to fail.

From the method of launching a ship or getting treetrunks to a mill. Once on the *skids*, the motion cannot be voluntarily arrested.

put to to cause to mate with.

Of mares etc.:

> We put her to Sandcastle yesterday morning. (Francis, 1982, of a mare)

See also PUT.

put to bed with a shovel dead.

Or with a *mattock* or a *spade*, of natural death or a killing. To *put underground* is to kill, usually leaving the burial to others:

> If you don't keep quiet for ten minutes, I'll put you underground too. (G. Greene, 1932)

Put under the sod means no more than 'dead and buried':

> Charlie, who was put under the sod, poor chap, a year come Michaelmas. (Pease, 1894)

put to rest dead.

The corpse does not necessarily have also to be buried:

> ... didn't expect things to change much until she was put to rest. (Sanders, 1986)

See also AT REST.

put to sleep to kill (a domestic animal).

Commonly of old, ill, or unwanted pets:

> 'I'll have it put to sleep!' he shouted... 'Oh, darling,' she pleaded, 'he's only a puppy.' (Ustinov, 1966)

put to the sword to kill.

Usually of a large number of helpless victims, by any form of violence:

> ... took Siakat by storm and put not only the Egyptian garrison, but every man, woman and child in the place to the sword. (Harris, 1925)

put yourself about to be sexually promiscuous.

Mainly of males, from circulating freely:

> By all accounts our friend put himself about a bit. (Blacker, 1992)

put yourself away see PUT AWAY[1].

python see SYPHON THE PYTHON.

quail *obsolete* a prostitute.

Not from the Celtic *caile*, a young girl, but the common avian imagery, this time from the reputedly amorous game bird:

> Agamemnon, an honest fellow enough, and one that loves quails. (Shakespeare, *Troilus and Cressida*)

Quaker gun a decoy.

A usage from the American Civil War because it wouldn't fire, and the Quakers are still pacifists:

> After a whole battery of Quaker guns was discovered at Centerville. (G. C. Ward, 1990)

quaker's burial ground see BURY A QUAKER.

qualify accounts to throw doubt on published figures.

Literally, *qualify* means 'to modify in some respect', and there are indeed some technical *qualifications* which do not indicate that the directors are a bunch of crooks and the company is headed for receivership. The *qualification* is a modification of the standard form of words in an auditor's report.

quantitatively challenged fat.

But not Sumo wrestlers:

> Without some such ordinance the fate of the quantitatively challenged teenager in the United States—and there are many of them—is sad to contemplate. (A. Waugh, *Daily Telegraph*, 4 October 1993)

See also CHALLENGED.

quarantine a military blockade.

Originally, the period of forty (*quarante*) days in which a widow might stay in her deceased husband's house, whence any period of isolation against disease etc. Kennedy used the phrase of his 1962 blockade of Cuba.

queen a male homosexual.

Usually an older man playing the female role or affecting effeminate manners or dress:

> He won't hold your hand and ask for your autograph like that old Harley Street queen you normally see. (Deighton, 1972)

An obsolete form was *cotqueen*, which really meant 'the wife of a manual labourer':

> You cotqueen, you. (Harris, 1925)

The obsolete *queen* or *quean*, a prostitute, came from its meaning 'any female animal, especially a cat':

> To call an honest woman slut or queen. (W. Scott, 1820)

A *queen-house* was a brothel.

Queen's evidence *British* betraying a fellow malefactor.

Or *King's evidence*, depending on the occupant of the throne. The derivation is from the convention that the Crown prosecutes in British criminal cases:

> But a suspect may, if he refuses to co-operate, perhaps by 'turning Queen's evidence' or becoming a 'supergrass' (David Pannick in Enright, 1985)

queer[1] drunk.

Originally, not in your normal state of health, and still occasionally used of a drunkard, with a suggestion that his condition may have been caused by something else. The meaning 'to make drunk' is obsolete:

> Queered in the drinking of a penny pot of malmsey. (W. Scott, 1822)

queer[2] of unsound mind.

Perhaps a shortened form of *queer in the head*. In this usage, people tend to be *a bit queer*, implying a harmless and mild condition.

queer[3] homosexual.

Almost always of males. It is used adjectivally:

> I'm not, um, queer. Well, you know, I don't like boys (Theroux, 1975)

and as a noun:

> Three or four queers talking together in queertalk. (From a poem of 1947, in Ginsburg, 1984)

Queertalk is not gobbledegook. Whence too *queercore*:

> This east London gay/lesbian foursome are noisy advocates of queercore, the movement of aggressively out young gay people. (*Guardian*, 25 February 1994—for *out* see OUT[2])

queer[4] **(the)** forged banknotes.

American criminal usage:

> He was all for printing the queer. (Sanders, 1990)

question[1] to arrest.

British police jargon, much used when publicizing particulars of a suspect because of the legal implications of a direct accusation of guilt. If the police announce that they would like to *question* someone corresponding with your description, you should take an overnight case to the interview.

question[2] a persistent problem to which there appears to be no answer.

Common political usage:

> I have always expressed my belief that the present Parliament and Government would fail to settle the Irish land question. (Kee, 1993, quoting Parnell from 1881)

Such a *question*, in French and German as well as English, may also concern matters to which allusive reference may be thought preferable:

> One of his friends... held a leading position in the Paris office of the *Commissariat-General aux Questions Juives*, the Vichy agency charged with hunting down Jews, listing them for deportation and, in due course, looting their property. (*Sunday Telegraph*, 2 October 1994, of Mitterand)

questionable immoral or illegal.

Literally, which should be inquired into, but now almost always in a derogatory or euphemistic sense. A *questionable motive* is concealed or dishonest, a *questionable act* offends the law or propriety, a *questionable payment* is a bribe, etc.

quick *?obsolete* pregnant.

From its first standard English meaning 'animate', and used of pregnancy after the foetus has started kicking:

> She's quick; the child brags in her belly. (Shakespeare, *Love's Labour's Lost*)

quick one a drink of an intoxicant.

Not necessarily drunk by an addict:

> His short sharp nose looked as if it had hung over a lot of quick ones in its time. (Chandler, 1943)

Less often as a *quickie*:

> And maybe we'd better break open the bottle for a quickie. (Sanders, 1980)

quick time a single act of copulation with a prostitute.

The jargon of prostitutes who have a time-based tariff:

> Want a quick time, long time, companionship, black leather bondage? (graffito quoted in Rees, 1980)

The common *quickie* can be with any female:

> Stone had never been fastidious about where he'd take his girls for a quickie. (Deighton, 1972)

quietus death.

Literally, a legal discharge from an obligation, whence removal from an office:

> When he himself might his quietus make
> With a bare bodkin
> (Shakespeare, *Hamlet*)

and in modern use:

> It looks as if Armstrong has got his quietus. (Christie, 1939)

quit to die.

From the departure. Also as *quit the scene* etc. To *quit cold* is to be killed and not to perish through hypothermia:

> Quit cold—with a slug in his head. (Chandler, 1939)

Those who *quit breathing* might not have died of natural causes:

> Tafoya asked if there was anybody 'that should quit breathing permanently'. (Maas, 1986)

quod prison.

It was formerly spelt *quad*, a shortened form of *quadrangle*, the area in which students were confined as a punishment:

> He has got two years now. I went to see him once in quod. (Mayhew, 1862)

To *quod*, to send to prison, is obsolete:

> ... been quodded no end of times! She knew every beak as sat on the cheer. (ibid.—the *beak*, or magistrate, sat on the chair)

quota appointed to fulfil a quota rather than on suitability or qualification.

American employers of more than 15 people are required to reflect in their workforce the local mix of race to a minimum ratio of 80 per cent regardless of qualification or aptitude:

> Quota employees have become a standard office joke. (*Sunday Telegraph*, 20 November 1994)

R and R see REST AND RECREATION.

RD see REFER TO DRAWER.

racial displaying prejudice against or hostility towards an ethnic group.

Originally, referring to humanity in its entirety, as when Marie Stopes was the President of the *Society for Constructive Birth Control and Racial Progress*, or distinguishing between people of different races. For the Nazis *racial science* concerned itself with showing that the Germans were better than anyone else—the *master race*—and *racial purity* required that there be no Gypsy, Jew, or Slav among your ancestors since 1750. This would then qualify you to apply to join the SS, who, by 1940:

> had already done sterling work in matters of racial purification. (Keneally, 1982—they had killed lots of unfit people, Gypsies, Jews, and Slavs)

Today *racial discrimination* is no longer the anthropological ability to tell one group from another but an assumption, or the actuality, of prejudice against anyone who is not white. To ensure conformity with the ethic, individual behaviour is policed, publicized, and punished by *race relations officers*, the whole supervised by a *Race Relations Board*, and fostered by the *race relations industry*, an expression used by those who think the problem has become an obsession with others. For many people *racist* or *racialist* is a term of abuse and those suspected of *racism*, the real or supposed preference for their own ethnic type, suffer calumny and even obligatory 're-education':

> ... the Catholic bishops, too, have excitedly discovered 'racism awareness courses'. (*Daily Telegraph*, 20 April 1992)

Race-norming is fixing lower pass standards in examinations for blacks than for whites:

> Race-norming is an unfair practice. (*Chicago Sun-Times*, 14 May 1991—unfair both to those discriminated against and those patronized)

racked *American* drunk or under the influence of illegal narcotics.

Not tortured on a rack, but laid out, you might suppose.

racy prepared to copulate extramaritally.

A variant of FAST; a *racy girl* is not a sprinter or jockey but a prostitute:

> The Eden Hotel ... where the racy girls hung out, was entirely rubble. (Shirer, 1984)

radical supporting extreme political or social change.

Literally, going back to the roots, from which comes the meaning 'ignoring accepted mores'. In obsolete English dialect a *radical* was 'an impudent, idle, dissipated fellow' (EDD)—readers of Bradbury's *The History Man* may think little has changed. For the Nazis, to be *radical* was ruthlessly and fanatically to disregard the conventions of civilized conduct:

> Had we proceeded in a more radical fashion in our treatment of prisoners of war the numbers of German soldiers ... surrendering ... would have been smaller. (Goebbels, 1945, in translation)

Today *radical* is often used pejoratively:

> ... avid, punitive, radical ladies ... enlisting my support for experimental sex-play in the nursery schools. (Bradbury, 1976)

And before the Second World War the New York police had a *Radical Squad* whose duties were mainly breaking up Communist rallies.

rag (the) a brothel.

British Indian Army use, perhaps from the slang name for the London Army and Navy Club:

> In this brothel, or Rag as it was called by the troops ... (Richards, 1936)

The obsolete *rag water* was gin, because those who became addicted ended up dressed in rags.

rag(s) on menstruating.

Usually in the phrase *have the rag(s) on*:

> That stupid little cunt ... is refusing to work because she's got the rags on. (B. Forbes, 1989—the female was an actress, not a prostitute)

The American *ragtime*, menstruation, puns on the type of music.

ragged drunk.

The way you feel. A venerable slang usage. However, *ragging* is not getting drunk, but menstruating, from RAG(S) ON.

railroad *American* to treat in a ruthless and unfair way.

The imagery is from the track which you cannot leave. Denoting wrongful imprisonment:

> ... railroaded to jail in an incredibly short time (Lavine, 1930)

or summary dismissal from employment:

> His father, in real life, had been framed and railroaded out of his position (McCarthy, 1963)

or merely pressing for an immediate decision.

railroad bible a pack of playing cards.

From the gambling which was prevalent on long railroad journeys:

> In the United States a pack of cards became known as a 'railroad bible'. Some 300 card sharks operated the Union Pacific. (Faith, 1990—for the sake of the passenger safety, I think he meant 'operated on the Union Pacific')

raincoat[1] *American* a male contraceptive sheath.

Perhaps punning on the RUBBER.

raincoat[2] a private investigator.

From the clothing they are supposed to wear:

> It will be interesting to see if Lloyds is prepared to use the raincoats (private investigators). (*Daily Telegraph*, 6 August 1994—the insurance body was pressing defaulting members to cover their losses)

rainmaker a person paid primarily for his contacts.

He brings in to an American financial house what his African counterpart brings down from the skies, both activities intended to generate growth without subsequent effort:

> Thanks to his mother and the Wallenberg connection, young Paul Mayer could bring in big business. In the United States they would have called him a 'rainmaker'. (Erdman, 1986—Mayer was operating in Europe)

raise a beat to have an erect penis.

Also as *have a beat on*, from the observable pulse. Commonly used figuratively in the negative, when an exhausted male will claim that he *cannot raise a beat*. To *raise a gallop*, with its equine imagery punning on GALLOP, is rarer.

raise a belly to impregnate a woman.

From the subsequent swelling:

> He raised so many bellies in the gay capital that the registrar of births had to increase his staff owing to the way he had exercised his. (Pearsall, 1969, quoting 19th-century pornography—the *gay capital* was London, not San Francisco)

raise the wind to secure funds to meet an obligation or need.

The imagery is from setting in motion a becalmed sailing boat. Still widely used, especially when you have no resources of your own and the funds are hard to obtain.

raisin a male homosexual.

I suspect from the French meaning 'lipstick'; *fruit* (see FRUIT[1]) is probably a later variant:

> He had more wrinkles than Auden, that other amazing raisin. (Theroux, 1978, of Maugham)

rake-off a payment made under bribery or extortion.

Usually on a regular basis from an illegal operation, such as gambling. The imagery is from removing with a rake, with perhaps the roulette table in mind:

> I'll give you a third, as I gave Curtis. The 'rake-off' don't hurt anyone. (Harris, 1925—the inverted commas show the then novelty of the usage)

ram to copulate.

Of a male, from the fecund animal and perhaps too the common violent imagery:

> Flirting and ramming with white women... (Fraser, 1975)

A *ram* is a promiscuous male:

> Must 'ave been quite a ram in 'is day. (Ustinov, 1971)

ram-riding public humiliation.

A hen-pecked husband might be compelled to mount a sheep in this ancient ceremony:

> I found the stairs full of people, there being a great Riding there today for a man, the constable of the town, whose wife beat him. (Pepys, 1667)

In Cornwall an adulterous wife was so punished:

> They had seized the woman... and were hauling her along in a Ram Riding. (Quiller-Couch, 1891)

ramp to cheat.

Originally, to snatch. Commonly of robbing, tricking, overcharging, giving wrong change, etc. A *ramp* is anything which causes you to be cheated, deceived, or overcharged. In obsolete British army use, the *ramps* was a brothel, perhaps because you often paid dearly for your pleasures, then or later.

randy eager for copulation.

A *ran-dan* was a carouse:

> Is the laird on the ran-dan the night? (Tweeddale, 1896)

and *randy* is a corruption of it. In the late 19th century 'A randy sort o' a 'ooman' (EDD) was one who enjoyed a good party, but the association with intoxicants has now gone:

> I want you just as you are. Final. Got it? I'm randy now. (Bogarde, 1981)

A British use which makes both sexes look with disfavour on the American shortened form of the name Randolph.

Rangoon itch a fungal infection of the penis.

Burmese prostitutes were notoriously disease-ridden:

> The houses you come away from with the fungus on your pecker known as 'Rangoon itch'... (Theroux, 1973)

Rangoon runs diarrhoea.

Not journeys to and from the city. One of the geographical alliterations inspired by cities where disease is commoner than cleanliness; see also *runs* at RUN[4].

rank capable of impregnation.

Literally, lush or strong-smelling. The use of women is obsolete:

> ...the ewes, being rank,
> In the end of autumn turned to the rams.
> (Shakespeare, *Merchant of Venice*)

rap the accusation of a criminal offence.

Literally, a rebuke or slap:

> I'd rather be under a murder rap—which I can beat. (Chandler, 1951)

rap club *mainly American* a brothel.

To *rap* is to talk or chatter (whence also to perform *rap music*). *Rap clubs, parlors,* and *studios* offer more than conversation alone:

> In the face of a crackdown on street prostitution many of the girls... are taking shelter in 'rap clubs'—which have replaced massage parlors in the sex-for-sale world. (*New York Post*, 22 June 1973)

raspberry[1] a fart.

From the rhyming slang *raspberry tart* but almost always used figuratively of a mild admonition or reproach. Oddly, people understand the association and vulgarity of *blowing a raspberry*, making a sound like a fart through pursed lips.

raspberry[2] a lame person.

From the rhyming slang *raspberry ripple*, cripple.

rat a mild oath.

Shortened form of *drat*:

> Rat the town, I say (Fielding, 1729)

but today used, if at all, in the plural. *Rabbit* was once used in the same way but not, I think, euphemistically.

rather exceedingly.

Many expressions introduced by *rather* are on the borderline of understatement and euphemism, so that a *rather naughty* child is a badly behaved brat. Similarly *rather a handful* denotes worse trouble with a spouse, child, or horse than a *handful*, as much as you can handle, and *rather poorly* in hospital jargon means that the patient is very ill.

rationalize arbitrarily to reduce.

Literally, to think in a rational manner, whence to deal sensibly with a problem. Often of employment, where the manager who *rationalizes* the workforce dismisses summarily many of the employees. Other resources too can be *rationalized*:

> Ever since the Government... encouraged local authorities to 'rationalise' their recreational areas, school pitches have been lucratively sacrificed for houses and supermarkets as a way of keeping down the rates. (*Daily Telegraph*, 3 March 1994)

rattle[1] to copulate with.

Of a male normally, from the shaking about which may be involved:

> All I'd done was rattle Mandeville's wife. (Fraser, 1971)

A *rattle* can also be the woman involved, always with a laudatory adjective; or a single act of copulation:

> It was her thinking she was the thinking man's rattle. (Amis, 1978)

rattle[2] to urinate.

From the American rhyming slang *rattle and hiss*, piss, with probably something of the common serpentine imagery.

rattled *mainly American* drunk.

I suspect, from its antiquity, that the derivation is from the Scottish meaning 'to beat', with the common violent imagery.

raunchy lustful or pornographic.

It originally meant 'sloppy' whence, with unusual rapidity of progression, 'poor', then 'cheap', then 'drunk' and now the sexual uses:

> ... importuning me with words delectably raunchy and lewd. (Styron, 1976)

And you still meet it of any wild behaviour:

> But then things got a little raunchy. They wanted to go down to Greenwich Village and see the freaks. (Sanders, 1981)

ravish to copulate with (a woman) against her will.

Originally, to seize or carry off anything:

> The ravish'd Helen, Menelaus' wife,
> With wanton Paris sleeps
> (Shakespeare, *Troilus and Cressida*)

and, in modern use:

> I don't know why, but that ravishing of Lily made her dear to me. (Harris, 1925)

The dated female exclamation of delight *How ravishing!* came from the meaning 'ecstatic', and not from any Freudian fantasies.

raw naked.

The undressed state:

> But screw the pyjamas; I sleep raw. (Sanders, 1983)

Usually as *in the raw* (see IN THE SKIN).

razor to wound or kill by cutting.

The cut-throat razor is not here used for shaving:

> ... razored in barrelhouses and end up being shot in a saloon. (Longstreet, 1956)

re-educate to extinguish a former political allegiance in (a person).

The Communists, who use harsh imprisonment, are better at it than the Americans who relied on logic and bribes:

> ... turn every deserter into a defector by 're-educating' him in a camp. (McCarthy, 1967, writing about Vietnam)

re-emigration the return of black immigrants to their birthplace.

Enoch Powell's idea was to pay them to volunteer to return whence they emigrated to Britain before their numbers increased through children being born with rights of citizenship:

Repeating that repatriation (which he called 're-emigration') was also a vital part of Conservative policy. (Cosgrave, 1989)

reading Geneva print drunk.

This is a sample entry of several obsolete literary puns on the city renowned for its piety and its printing, and on gin which is also called *Geneva* from the French *genièvre*, the juniper berry:

You have been reading Geneva print this morning already. (W. Scott, 1816)

realistic not inflated.

Of prices generally, but especially used by estate agents when they wish to signal that a house is being offered for sale at something near its true value.

ream to bugger.

Originally, to enlarge a hole by inserting a metal tool:

... maybe a night in the slammer where the boogies will ream you. (Sanders, 1985—a policeman was threatening a male homosexual)

reaper (the) death.

Father Time carries a scythe as well as an hourglass. A *grim* was a death's head or skeleton in northern English dialect, and the more common *grim reaper* may come from this as well as the figure's remorseless mien:

The goal was to outmanœuvre the Grim Reaper. (J. Mitford, 1963)

rear¹ the buttocks.

Not the heels or shoulder blades. Females, especially in America, may be said to have a *rear end*:

... her sumptuous rear end. (Styron, 1976)

rear² to defecate.

The etymology suggested by DHS of soldiers falling out to the rear seems unnecessarily complex. British *rears*, lavatories, were so named from their location behind and away from houses, and the usages are probably linked.

receding nearly bald.

A shortened form of *receding hairline* although, when the expression is used, that is usually only one symptom of incipient baldness. Men use *receding* of themselves, or of those whom they wish to flatter. In others, male baldness is a subject for frankness and humour.

receiver a dealer in stolen property.

From his willingness to 'receive anything brought' (Mayhew, 1862) and standard English. The obsolete British *receiver-general* was a prostitute, punning on the office of a senior official and her *reception* of men generally.

recent unpleasantness a war.

A version of LATE DISTURBANCES with its variants and showing the same wish to play down or forget the horror.

recognition the receipt of a medal from government.

Not just knowing a likeness. The assumption is that the recipient has done something more noteworthy than support the ruling party financially or otherwise:

... someone who hopes that it may result at some future date in their 'recognition'. (A. Clark, 1993—he was as caustic about others who sought such rewards as he was anxious to secure the appointment as a Privy Councillor for himself)

recreational drug an illegal narcotic.

As opposed to one taken for medical purposes:

Sloth, gluttony, recreational drugs were out. (McInerney, 1992)

rectification of frontiers the annexation of territory by force.

The party which seeks the *rectification*, from Hitler onwards, is never minded in turn to divest itself of territory.

red devil an illegal narcotic.

In many compounds, from the colour of the pill. A *red devil* is a barbiturate or seconal. The two exceptions are *red dirt marijuana*, the wild plant, from the Southern American deserts where it grows; and *red cross*, morphine, because it may be stolen from a first aid kit.

red eye bad whisky.

From its effects on the drinker, and also of other inferior American concoctions. However the *red-eye* or *red-eye special* is any overnight flight from the west coast to the east of the USA, or less often a night flight to Europe:

I'm on the redeye back to the Big Apple. (M. Thomas, 1980, of a return to New York City)

red lamp a brothel.

From the traditional sign:

There was a Red Lamp in Bethune situated about five yards off the main street. (Richards, 1933)

Occasionally as *red-lighted number*:

... also featured at the red-lighted number of the brothel area of a town. (Longstreet, 1956)

A *red-light area, precinct, district*, etc. is a brothel quarter:

They paid for promotion or detail to the red-light precinct. (Lavine, 1930—a New York policeman would expect regular bribes from bawds)

red rag (the) menstruation.

A variant of RAG(S) ON and punning on the cliché, a *red rag to a bull*. In speech either *waved* or *worn. The red flag is up* puns on the danger signal rather than the Communist emblem.

Red Sea is in (the) I am menstruating.

Alluding perhaps to the adventures of Moses and others recorded in Exodus and punning on the sea which covered the channel of escape.

red-squad (the) police concerned with subversion.

An American use, from fear of Communist influence:

> The New York Police Department has a Red Squad. They change the name every two years or so—Radical Bureau, Public Relations, Public Security. Right now they call it the Security Investigation. (M. C. Smith, 1981)

redhaired visitor (a) menstruation.

A rare version of the common *visitor* theme (see VISITOR (A)).

redistribution of property looting.

Mainly Second World War use of soldiers in Europe:

> He didn't call it stealing though, 'redistribution of property' he called it. (Price, 1978, writing of the Second World War)

redistribution of wealth punitive taxation.

Despite much evidence that an optimum rate of tax (to achieve the highest return) is lower than the maximum (to punish the rich), the myth persists that the latter will somehow put money in the pockets of the poor:

> ... wilful and cruel disruption of the economic fabric that was called the redistribution of wealth. (Allbeury, 1976)

redlining *American* refusing credit solely because of the place of residence of the applicant.

You highlight the address in a list, figuratively or in fact:

> ... entire areas of the city, poor areas, humble areas, were beyond the credit ... the inhabitants of those districts were exiled from creditworthiness. That foul practice was called redlining. (M. Thomas, 1987)

reds (the) menstruation.

A common female use of obvious derivation.

reduce your commitments involuntarily to leave employment.

Not just paying off your debts or moving to a cheaper house:

> ... a former finance director of Mirror Group Newspapers facing charges of false accounting and conspiring with Robert and Kevin Maxwell, has reduced his commitments ... (*Daily Telegraph*, 2 March 1995)

reduction in force see RIF(F).

redundant dismissed from employment.

Originally, in superabundance, which an individual *made redundant* can hardly be:

> 'And now they've turned you out?' he asked. 'Who said they had?' ... 'I thought you said something about being made redundant.' (Sharpe, 1974)

See also RIF(F).

reefer a marijuana cigarette.

The etymology is uncertain, although some suggest it comes from the method of hand-rolling the cigarettes:

> A two-time loser sneaking home from a reefer party. (Chandler, 1943)

refer to drawer this cheque is unpaid through lack of funds.

Banks use this phrase because it is dangerous to dishonour a cheque by mistake and thereby imply that the drawer has written it fraudulently. Commonly abbreviated to *RD*.

referred failed.

Originally, put back. English university jargon.

refresh your memory to extract information through violence.

Police usage and quite different from consulting an aide memoire:

> They compel reluctant prisoners to refresh their memories. (Lavine, 1930, of violence by the New York police)

Also used by criminals in their transactions with their victims. If you *refresh* your memory or recollection, you are recalled to the witness stand after having previously perjured yourself:

> ... after the indictment they'll give her a chance to 'refresh her recollection'. He made the quotation marks in the air. (Turow, 1990)

refresher a drink of an intoxicant.

From the supposed bracing effect:

> He marches out, with his hat on one side of his head, to take another 'refresher'. (Jefferies, 1880)

refreshment a drink of an intoxicant.

If offered on a Sunday school outing, a *refreshment* is likely to be tea or lemonade but on most other male social occasions, you would expect beer or spirits.

regular[1] in the habit of daily defecation or menstruating at a predictable time.

Laxative advertisements enshrined the defecatory use:

> I've always been regular as clockwork, all my life, and then, bingo. (Ustinov, 1971, of defecation)

Irregular is used as the converse in both senses:

> Irregularity was one of my problems these days, so I was unusually prepared. (P. Scott, 1975, of menstruation)

regular[2] small.

In the jargon of packet sizes—*jumbo, family, economy*, etc.—*regular* is the little one and often the worst value.

regularize[1] to make good (a bureaucratic omission or mistake).

Civil service jargon. So long as the impersonal file has been amended to comply with regulations, it matters little what delays and blemishes it hides.

regularize[2] to invade.

The intended implication is that the political situation is being restored to normal. It took one Polish, one East German, and twelve Russian divisions to *regularize* the position in Czechoslovakia in 1968.

relate to copulate.

Literally, to be connected in any way:

> 'Can't you just say "fuck" once in a while?' But Piper wouldn't. 'Relating' was an approved term. (Sharpe, 1977)

In obsolete use you might be *in relation with* another:

> This is of course Weguelin, so she must have been in relation with both. (Kee, 1993—Weguelin had stayed at the O'Shea residence before Parnell and, after the divorce, Parnell's Irish enemies alleged that Katie had copulated with him as well)

For a further disquisition see HAVE RELATIONS and RELATIONSHIP.

relations see HAVE RELATIONS, HUMAN RELATIONS, and *sexual relations* at SEXUAL INTERCOURSE.

relations have come (my) I am menstruating.

From the limited duration and inconvenience of the visitation, or in some cases, the relief at seeing them. The visitors are sometimes *country cousins*, from their ruddy complexions.

relationship a connection which involves copulation outside marriage.

In fact we have a *relationship* with anyone we deal with, as buyer or seller, friend or enemy:

> For just over three months Jeanie has had a relationship with a Russian. (Allbeury, 1982—he was copulating with her)

Used alone or with adjectival embellishment, where the *relationship* may be close, long-term, special, or as the case may be. Now common homosexual use also.

relative deprivation the fact that some people are richer than others.

Sociological jargon which avoids any reference to poverty. See also DEPRIVED.

release[1] to dismiss from employment.

The employee has not in fact been held against his will:

> ...since released (not surprisingly) to pursue 'other business interests', the banking euphemism for goodbye. (*Private Eye*, April 1988)
>
> The pilot's release from the team is a result of an administrative action. (*Daily Telegraph*, January 1987—it could hardly have been implemented without an administrative action, you might suppose, whatever the causation. It appears the pilot was dismissed from an aerobatic team because he had crashed an aeroplane.)

Of single and multiple dismissals.

release[2] a death.

The soul has left the body for more congenial climes. Much used after a painful terminal illness in the cliché *happy release*. *Released* means 'dead', usually of those killed by violence:

> Let these serve as a sacrifice to those dead innocent spirits so cruelly released at Jhanoi. (Fraser, 1975, writing in 19th-century style)

release[3] the ejaculation of semen.

By whatever means, under the theory that unrelieved sexual tension in a male is bad for him:

> ...indulged in this pastime night after night as much to give him some 'release' (she actually uses the odious word). (Styron, 1976, of masturbation)

released see RELEASE[2].

relief[1] public aid given to the indigent.

Originally, a feudal payment to an overlord on coming into an estate:

> The parish granted no relief and even if it had done so it is very doubtful whether the strikers or their wives would have accepted it. (Richards, 1936)

This is an early, but still current, example of the string of euphemisms which we use to conceal the charitable nature of state aid to the poor.

relief[2] urination.

You usually *need* or *obtain* it:

> Archie had needed immediate relief in the bathroom. (Davidson, 1978)

Occasionally denoting defecation or vomiting, although an American *relief station* is a lavatory, which provides for all three functions. See also RELIEVE YOURSELF.

relief[3] copulation.

With the male's tensions subsequently quietened, I suppose:

> ...the Euphoric Spring has heated your blood to the extent that you're prepared to fly me

six thousand miles to obtain relief. (Lodge, 1975)

relief[4] masturbation.

Again from the supposed relaxation of pressure:

He played blue movies in his head featuring himself and Robyn Penrose, and crept guiltily to the *en suite* bathroom to seek a schoolboy's relief. (Lodge, 1988)

relieve to dismiss from employment.

This American use is of the same tendency as LET GO, as though the dismissal were a kindness to the employee by letting him get on with something else. The standard English *relieve of duties* is usually of an official for misbehaviour or dereliction of duty.

relieve of virginity to copulate with (a female virgin).

Usually extramaritally and perhaps no more than circumlocution:

Dottie had wanted to be 'relieved' of her virginity. (McCarthy, 1963)

relieve yourself to urinate.

You obtain *relief* (see RELIEF[2]):

He felt a sudden urge to relieve himself. (Diehl, 1978, of urination)

Rarely of defecation, except in a fuller form such as *relieve your bowels*:

They went in the dawn, brass lotah in hand, to relieve their bowels in the spaces between the houses. (Masters, 1976)

relinquish to be dismissed from (employment).

The implication is that the giving up was voluntary:

Mr Barker 'relinquished' these roles in May last year on the same day Hartstone issued its second profits warning. (*Daily Telegraph*, 16 July 1994)

relocation killing.

A Nazi use for the despatch of Jews etc. to the death camps:

In Berlin, they wrote 'relocation' and believed themselves excused. (Keneally, 1982, of orders for the rounding up of Jews for murder)

relocation camp a place for the imprisonment of enemy aliens.

American Second World War usage when the presence of Japanese-Americans on the possibly threatened West coast posed a considerable dilemma:

... most of them were interned at the time in 'relocation camps'. (Jennings, 1965, of Japanese-Americans)

remain above ground not to die.

The presumption is that the dead are buried. The following quotation shows what can happen when a euphemism is interpreted literally:

Mrs Van Butchell's marriage settlement stipulated that her husband should have control of her fortune 'as long as she remained above ground'. The embalming was a great success. (J. Mitford, 1963—Mr Van Butchell showed more enterprise than taste)

remainder[1] to kill.

From the resultant corpse:

He did not feel pity often, but he almost felt it for whoever was to be remaindered there. (Goldman, 1986, of an assassin)

remainder[2] to dispose of (surplus stock of a book) by selling cheaply.

The jargon of the publishing trade and the humiliation of the author.

remains a corpse.

Funeral jargon; *mortal remains* is specific:

Today, though, 'body' is Out and 'remains' or 'Mr Jones' is In. (J. Mitford, 1963)

remedial applicable to the dull and the lazy.

Literally, helping to cure something, as *remedial education* should be, that is to say special instruction to overcome a specific weakness in an otherwise normal child. In educational jargon, *remedial* is used to avoid referring to any fundamental inadequacy of which mention would be taboo:

... the staff even have to lay on a remedial English course for students with a 'less than adequate mastery of the English language'. (*Daily Telegraph*, October 1983, of a former polytechnic)

As with lunacy, the blurring of levels of disability is no kindness for those who require long-term help.

remittance man an unsuccessful, embarrassing, or improvident member of a wealthy family sent to reside in a distant country.

Money is remitted to him so long as he stays away:

Remittance man—a form of Kenya settler said to depend on remittance from UK to stop him returning. (Allen, 1975)

remove to kill.

But not necessarily making off with the body. DSUE says of *removal*: 'Ex a witness's euphemism in the Phoenix Park assassination case', but Dr Johnson gives both 'dismissal from a post' and 'going away', both of which are common 'death' images. The Scottish *removed*, dead, probably came in those pious times from the separation of the soul from the body:

When a person has just expired, the Scotch people commonly say, he is removed. (*Monthly Magazine*, 1800, quoted in EDD)

rent a payment extorted illegally or for an illicit transaction.

Not necessarily on more than one occasion, as from a tenant to a landlord. In obsolete use, it included cash paid to a highway robber. Also of regular extortion and, in homosexual jargon, money paid to a male by another, the recipient being a *rent boy*:

Moreover, when the *Sunday People* published its front-page 'caring' shocker about Harvey and the ex-public school rent boy. (*Private Eye*, October 1986)

A *renter* is a prostitute, male or female, who works on a part-time basis.

repose to be dead and buried.

An American extension of the common 'resting' imagery:

The companions will repose one above the other in a single grave space. (J. Mitford, 1963)

A *reposing room* is a morgue:

Reposing room or slumber room, not laying-out room. (ibid.)

reproductive freedom *American* the freedom to abort a healthy foetus.

Also used to denote the effect on a woman's life of the availability of contraceptives. See also PRO-CHOICE for other examples of how the participants in this debate manipulate words.

resettlement mass murder.

Literally, voluntary or involuntary movement of residence. The Nazi *Unsiedlung* took Jews from ghettos to their death:

...the huge 'resettlements' from the Warsaw ghetto...were coincident with the establishment of...Treblinka and its gas chambers. (Styron, 1976)

residential provision a place in a boarding institution.

This British usage is more than inelegance or circumlocution, because in sociological jargon you must use words which avoid the taboo *board school*, a prison for young criminals, or *boarding school*, to which the rich send their children in the hope of a more thorough and intensive education than that usually provided free by the state. A *resident* may be a homeless geriatric, a lunatic, a chronic invalid, or a prisoner.

resign to be dismissed from employment.

Usually said of and by senior employees, to avoid losing face:

I worked as a personal secretary in London until I was fi...until I resigned. (Bradbury, 1976)

resign your spirit to die.

An obsolete use which seems to discount the prospect of reincarnation:

Resigned her Spirit to Him who gave it on the 13th day of March 1818. (Memorial in Bath Abbey)

resisting arrest in custody.

Police usage to explain the wrongful wounding or killing of a prisoner:

I like it better you get a slug in the guts resisting arrest. (Chandler, 1939)

Of the same tendency as TRYING TO ESCAPE.

resolved without trial involving the acceptance of a guilty plea.

Part of the American process of plea-bargaining, but not implying that the accused was acquitted for want of prosecution:

...it should be 'resolved without trial', an oblique reference to a guilty plea. (Turow, 1990)

resources control *American* the destruction of crops.

Vietnam usage. It should mean no more than rationing:

...bombing, defoliation, crop-spraying, destruction of rice supplies, and what is known as 'Resources Control'. (McCarthy, 1967)

rest and recreation extramarital copulation.

Originally, a short period of leave during wartime. Often abbreviated to *R and R*:

The Russians had probably been a patrol team, and had chosen the farm for a little informal R & R. (Clancy, 1986—they had raped a female occupant)

rest home an institution for geriatrics.

Not punning on the fact that most of them will stay there for the rest of their lives:

A ninety-two-year-old who died in a rest home. (J. Mitford, 1963)

The *Convalescent Home* in East Haddam, Connecticut, masked its purpose with a less accurate title.

rest room *American* a lavatory.

Wide use by both sexes, usually to denote those in public use and not excluding the positively unrestful cubicles on long-distance buses:

...asked where the bathroom was. The restroom was filthy. (Diehl, 1978)

The attempt by the American funeral industry to use *restroom* for morgue not surprisingly found few takers.

resting unemployed.

Theatrical jargon which seeks to imply that the idleness is voluntary:

The demoralization of so many of my out-of-work companions. 'Resting' is one of the least restful periods of an actor's life. (Murdoch, 1978)

restorative a drink of an intoxicant.

Now more common than the REFRESHER which uses the same imagery:

> ... a gin and tonic or a shot of whisky ... at the end of a busy day or tiresome week. That is not thirst. To enjoy it, we do not need to be dry; just in need of a restorative or relaxant. (*Independent*, October 1989)

restorative art the cosmetic facial treatment of corpses.

American funeral jargon:

> ... transferred from a common corpse into a Beautiful Memory Picture. This process is known in the trade as embalming and restorative art. (J. Mitford, 1963)

A *restorative artist* is a beautician to the dead:

> ... features remoulded by the hand of a Restorative Artist into unfamiliar expressions of benign sweetness. (ibid.)

restore order to invade and conquer a country.

The excuse used by the Communists for the invasions of Hungary and Czechoslovakia, but still current:

> This has involved moving in masses of arms and men of the ANC's 'armed wing', the MK, to step up the violence—thus creating an excuse for the South African Defence Force (SADF) to be sent in to 'restore order' and to topple Buthelezi. (*Sunday Telegraph*, 27 March 1994—similar tactics had already achieved the desired result in Ciskei and Bophuthatswana)

restorer *obsolete* an agent for a thief who negotiated ransoms.

In the 19th century, especially in London, dogs were commonly taken from their owners, this not being considered theft—dogs, like blackberries and field mushrooms today, could not be stolen under common law. Their return would be the subject of negotiation:

> A restorer ... undertook 'to restore the dog if terms could be come to'. (Mayhew, 1851)

See TRANSPORTED for the story of the Duke of Beaufort's dog's collar.

restraint[1] an attempt to limit wage increases.

One of a series of euphemisms used by governments which seek to curb the inflation generated through their own profligacy by limiting wages and salaries:

> Workers agreed to a policy of wage restraint and profit maximization in order to wipe out a large budget deficit. (*Dissent*, Winter 1991)

restraint[2] a recession.

A usage of politicians who wish to show that they are in command of events and seek to avoid using words with bad connotations:

> The country (under Harold Wilson) was going through a period of severe economic restraint. (Mantle, 1988)

restricted growth dwarfishness.

Restricted derives from a Latin verb meaning 'to hold back deliberately', 'to bind fast'. A BBC programme broadcast on 15 January 1987 was devoted to *people of restricted growth*.

restructured presented in a dishonest or misleading way.

Of financial reports, where the method of presentation introduces falsehoods or hides truth:

> When the Saudis take a look at some of these 'restructured' balance sheets, they're going to need about ten seconds to figure out what pushing oil back down to ten bucks a barrel would do to a twenty-to-one debt to equity ratio at Texaco. (M. Thomas, 1987—and did, as it turned out, although other factors came into play too)

resurrection man *obsolete* a stealer of corpses.

When it was widely supposed that those who died in Christian belief would in due course undergo a resurrection of the body, few wished to risk having their corpses dissected in pursuit of medical knowledge for fear of a dismembered or partial return to earth. The market had therefore to be supplied, especially in Scotland, by raiding churchyards. This punning usage may first have been applied in 1829 to Burke and Hare, who carried the business a stage further by murdering future supplies when a paucity of natural deaths caused fresh corpses to be in short availability. Also as *resurrection cove* or *resurrectionist*.

retainer a series of payments made regularly to an extortioner.

Literally, a sum paid to book the services of a lawyer etc.:

> I can afford a substantial retainer. That's what you call it, I've heard. A much nicer word than blackmail. (Chandler, 1958)

retard a simpleton.

Literally, anything delayed or held back:

> How long is the old girl going to take? No one said she was a fucken ree-tard. (Theroux, 1978)

Retarded is educational jargon used to describe a person with a congenital etc. inability to learn.

retire[1] to kill.

The victim certainly stops working:

> I just retired a junkman. (Diehl, 1978)

(The French *en retraite*, retired, led a Belgian friend, kindly talking to me in English, to refer to an old soldier as a 'general in retreat'.)

retire[2] to urinate.

From temporarily leaving somewhere. A *retiring room* is any old lavatory in America but something rather grander in the British Isles, being provided or reserved for royalty and honoured guests at functions. When the monarch *retires* on such occasions, she does not abdicate.

retire[3] to dismiss from employment.

The victim rarely ceases to work:

> George Owen was 'retired' from Mercury by Lord Young, C & W's well-rewarded chairman. (*Daily Telegraph*, 6 December 1994—announcing that Owen then chaired the board of Energis, a rival of Cable and Wireless)

retiring room see RETIRE[2].

retreat a lavatory.

Punning on the solitary religious activity. In various forms such as the doubly punning *beat a retreat*.

retrenched *mainly American* dismissed from employment.

Literally, reduced in the interest of economy, but illogically used of those who have had to leave rather than those who form the continuing workforce:

> Factories closed. Retrenched workers committed suicide. (Naipaul, 1990)

returned empty *obsolete* (of a woman returning to Britain after a visit to India) neither married nor engaged to be married.

Thus were described the unsuccessful members of those who had been on the FISHING EXPEDITION[1]. The imagery would be familiar in the days when affluent families were accustomed regularly to return used soda syphons etc. to a supplier, and may just have punned on the unimpregnated condition of the young marriageable women involved.

returned to unit failed.

British army jargon, often abbreviated to *RTU*, to describe those who fail to complete a special course to qualify for an elite corps, to become an officer, etc.:

> They would be conditionally accepted or RTU'd. Returned to their original units. (Allbeury, 1982)

revenue emolument increased taxes.

The language has moved some way since the *emolument* was the fee paid to the miller, and even today it should be no more than what you earn. The equally evasive *tax enhancement* means the same thing.

reverse a defeat.

Literally, the opposite of anything, but used in this sense since the 16th century, especially if it is your side which loses. Similarly too of a pecuniary or business misfortune.

reverse discrimination the failure to appoint a qualified applicant because he is white.

Discrimination, tout court, might seem to have covered the concept:

> White men have scored two major victories in reverse discrimination rulings by the US Supreme Court, confirming that the mood in America is turning sharply against race-based 'affirmative action'. (*Daily Telegraph*, 19 April 1995—or confirming a change in the political make-up of the court; see also DIVERSITY)

reverse engineering unauthorized copying.

Not the gear which propels backwards. You obtain your competitor's product, take it apart, and then use the technological improvements, suitably disguised, in your own.

revision of prices see PRICE ADJUSTMENT.

revisionist anybody who questions the current policy of an autocracy.

A Communist dysphemism, with much pejorative use:

> ... illegal cheap hotels that seem to have a collectively revisionist attitude regarding official papers. (Ludlum, 1979)

reviver a drink of an intoxicant.

From its supposed ability to liven up the drinker, but not used only, as you might suppose, of the first potation.

revolving-door unduly lenient and ineffective.

Of the treatment of criminals who, soon after capture, are released to continue their former activities, figuratively entering the court or jail through such a door:

> The people of California are sick of revolving-door justice. (*Daily Telegraph*, 4 March 1995)

rib joint *American* a brothel.

Probably from the obsolete *rib*, a woman, after the manner of Eve's creation. DAS says 'from "tenderloin" reinforced by "crib joint"', which might be right although I favour less complex etymologies.

rich friend see FRIEND.

Richard a turd.

From the rhyming slang *Richard the Third*. He had a bad press from the Tudors and Shakespeare, which is why he is commonly considered more of a shit than Edward, William, Henry, or George, of whom there were also three or more. See also DICK[1, 2].

ride to copulate with.

Usually of a man, with the common equine imagery:

> You rode like a kern of Ireland, your French hose off. (Shakespeare, *Henry V*)

But also of a woman, especially if above the man:

> Gabby groaned as she rode him at a little under a canter. He lay easing himself up to her. (L. Thomas, 1979)

A *ride* is either a female viewed for copulation, or the act:

> Reckon you'll count it a pretty dear ride you had, friend. Was she good? (Fraser, 1971 —Flashman had copulated with his master's wife)

To *ride St George* was to copulate with 'The woman uppermost in the amorous congress, that is, the dragon upon St George' (Grose—it was said to be the way to beget a bishop). However to *ride abroad with St George but at home with St Michael* imported neither sexual variety nor custom at Marks and Spencers but that the man was a braggart away from home but henpecked indoors. A *riding master* was a woman's extramarital sexual partner:

> I was the Queen's current favourite and riding-master (Fraser, 1977, writing in 19th-century style)

and *riding time* was the season for impregnation, used of sheep and vulgarly of women:

> Warn him ay at ridin time
> To stay content wi' yowes at hame.
> (Burns, 1786—*yowes* means 'ewes')

The imagery may also be of cycling; see TOWN BIKE.

ride the red horse to menstruate.

Also as *ride the rag*. In America the *horse* may be white, from the colour of the absorbent cloth worn.

ride the wooden horse *obsolete* to be flogged.

From the *horse*, or stool, over which the victim was strapped.

ride up Holborn Hill *obsolete* to be killed by hanging.

Holborn Hill was on the road between Newgate Prison and the Tyburn gallows in London:

> I shall live to see you ride up Holborn Hill. (Congreve, 1695)

This is a sample entry, many cities making similar use of a geographical feature which the condemned man passed between prison and the scaffold. Of general use was *ride backwards*, because they sat you that way in the cart for the journey. To *ride the mare* was less common, and I have not traced the etymology. To *ride out* was to be a thief, particularly on the borders of England and Scotland, where riding and robbery were almost synonymous:

> Ride, Rowlie, hough's i' the pot. (Nicholson and Burn, 1777—*hough* was the last remaining piece of beef, and it was time to rustle some more)

rif(f) *American* to dismiss from employment.

The initial letters of *reduction in force*, usually in the passive voice:

> Ask any Federal Government employee what it means when he receives his Reduction in Force letter, and he will say 'I've been riffed'. (Letter in *New York Times Magazine* quoted in Wentworth and Flexner, 1975)

rifle *British, ?obsolete* to copulate with.

Of a male. You have a choice of three derivations—from the technical term for the copulation of hawks, from the plundering of booty, or from the concept of screwing.

right-sizing *American* dismissing many employees.

Right for the management or owners, perhaps:

> 'We enter 1995 with the bulk of our right-sizing behind us,' Lou Gerstner, chairman of IBM, on last year's 35,000 redundancies. (*Daily Telegraph*, 28 January 1995)

See also DOWNSIZE.

right thing see DO THE RIGHT THING.

right-wing chauvinist and totalitarian.

A euphemism for some of those extremists who so describe themselves, but much derogatory use by Communists etc. in reference to those holding more moderate socialist views than themselves.

ring[1] the vagina.

Viewed sexually, from its supposed shape:

> ...I'll fear no other thing
> So sore as keeping safe Nerissa's ring.
> (Shakespeare, *Merchant of Venice*)

A woman who had copulated before marriage was said, with punning vulgarity, to be *cracked in the ring*. Now also used of the anus of either sex, especially in male homosexual use.

ring[2] a cartel.

From the concept of meeting in, and making complete, a circle. Wide commercial use, referring in particular to dealers who combine to buy cheaply at auction or firms which agree to submit high tenders:

> Wellington City Council, which recently protested strongly against the submission of equal tenders by a number of British firms, has now decided to accept a tender for electric

cable which is £3,000 below the 'ring price'. (*Times*, 13 May 1955)

ring eight bells to die

The American *end of watch* (see at END) adopts the same imagery. Jolly (1988) draws our attention to the punning Alastair Maclean novel title, *When Eight Bells Toll*.

ring the bell to impregnate a woman.

From the fairground trial of strength which involves a sledge-hammer and a moving object in a vertical column. The use implies intent.

ringer a racehorse etc. fraudulently substituted for another.

In early 20th-century slang, a *ringer* was a person who closely resembled someone else, often in the cliché *a dead ringer*. There may be some connection with *ringing the changes* in campanology.

rinse a dye applied to the hair.

Literally, a cleaning by water. Mainly female hairdressing jargon. Some old women with white hair have *blue rinses*:

> ...married the Buick dealer on the adjacent lot, and got a blue rinse. (Bradbury, 1976)

rip off to cheat.

Either by stealing or overcharging, with imagery perhaps from tearing paper off a pad or notes off a roll:

> We get ripped off for half a million, and we respond with free psychiatric treatment and maintenance for the villain's family. (West, 1979)

Also of plagiarism; a *rip-off* is any example of such cheating:

> Such rip-offs of their material are strictly banned by the GTV hierarchy. (*Private Eye*, May 1981)

A *rip-off artist* or *merchant* is a habitual swindler:

> ...there are plenty of ripoff merchants around in this game. (*Private Eye*, September 1981)

rip off a piece of arse to copulate with a female extramaritally.

Neither by flenching the lady nor by refusing payment. As usual, *piece* and *bit* are interchangeable, as are *arse* and *ass*:

> ...picks up a hooker and rips off a bit of ass. (Theroux, 1973)

ripe drunk.

And ready to fall? This American use has, I fear, no connection with the picturesque obsolete English *ripples on*:

> ''E 'ad the ripples on,'—drunk he was not, though he had exceeded his rightful allowance. (EDD—*ripples* are the attachments to the sides of a cart to enable it to carry more than its normal load)

ripped *American* drunk or under the influence of illegal narcotics.

Feeling torn perhaps:

> Last night, you got ripped on tequila. (*Midnight Zoo*, 1991)
>
> Dave Gilbert... told Min once he'd been ripped on LSD and put the top of a hamburger bun on in place of a distributor cap. (Lawrence, 1990)

ripple to experience a succession of orgasms.

In a female, presumably from the succession of waves on a shore:

> A bird sang low; the moonlight sifted in;
> The water rippled and she rippled on.
> (Roethke, 1941)

Whence *rippling*, such an experience:

> ...sometimes a whole series of orgasms followed (wasn't 'rippling' the word used?). (Hailey, 1984)

rise an erection of the penis.

In America you call an increase in pay a *raise*, to avoid misunderstanding.

river (the) intoxicants.

From the rhyming slang *River Ouse*, booze.

rivet *?obsolete* to copulate with.

Of a male, usually extramaritally on a single occasion, from the metal fastener passed through a hole in engineering:

> When I was an undergraduate you got sent down if you were caught riveting a dolly. (Sharpe, 1974)

roach[1] the butt of a marijuana cigarette.

Perhaps it looks like a beetle. Used either as a synonym of *butt*:

> The marijuana cigarette which he smoked down to the roach (Longstreet, 1956)

or as the entire object:

> The waitress took the roach, sniffed it, and said, 'Thank you, dear. Just what I need.' (Saunders, 1986)

roach[2] *American* a cockroach.

A shortened form of *cockroach* and a rare survival from the days when any mention of *cock* was considered impolite (see ROOSTER). *Rooster-roach* proved an unsatisfactory substitute. In modern use a *roach* is also an offensive name for a policeman.

road apples *American* horse shit in the street.

From the way it may be piled naturally and its value as free manure.

road is up for repair (the) I am menstruating.

A multiple pun on the red warning light, the closing of the passage, and the temporary nature of the affliction.

rock see ROLL[1].

rock and roll a regular payment by the state to the involuntarily unemployed.

British rhyming slang for DOLE.

rock crusher *American* a convict.

From the activity in which prisoners were traditionally engaged.

rocks the testicles.

A variant of the standard English STONES; see also GET YOUR ROCKS OFF.

rocky[1] mad.

Unstable, like an unbalanced chair:

> I guess you're a bit rocky. You haven't escaped from anywhere, have you? (G. Greene, 1932)

rocky[2] *American* drunk.

Again from the lack of balance.

rod[1] to copulate with.

Of a male, from the insertion of a rigid probe as when cleaning a chimney.

rod[2] a handgun.

Literally, a straight bit of wood:

> I don't never let Frisky carry a loaded rod. (Chandler, 1939—a craftsman who at least knew when he was writing incorrect English)

Rodded is so armed:

> The derby hat saw if I was rodded. He took the Lüger. (ibid.)

rod[3] the penis.

From its propensity to rigidity:

> The liveliest parts of his body became spiritualized, and his rod itself. (Genet, in 1969 translation)

roger[1] to copulate with.

Of a male, usually naming the female. Most authorities trace the derivation to a name commonly given to a bull, but it may have come from the obsolete name used by shepherds for their rams. It seems most likely, however, that it comes from the name for the penis (see ROGER[2]).

> ...find oneself being rogered by one of his libidinous heroes. (Bradbury, 1976)

Also spelt *rodger*.

roger[2] the penis.

Rare. I suspect not from its role in *rogering* (see ROGER[1]), but from the *Jolly Roger*, or pirate flag.

roll[1] to copulate with.

Of either sex, from the movement:

> A beautiful blonde virgin from Boulder
> Swore no man on earth had yet rolled her.
> (*Playboy's Book of Limericks*)

A *roll* is an act of copulation:

> ...our last meeting had been the monumental roll in her pavilion. (Fraser, 1975)

The cliché *a roll in the hay* is an act of casual copulation, but not necessarily in an agrarian setting:

> A hotel room rented...for a roll in the hay. (Chandler, 1953)

The obsolete *roll the linen*, to copulate, punned on the ironing of sheets:

> ...in my absence she'd been rolling the linen with any chap who'd come handy. (Fraser, 1977, writing in 19th-century style)

In rare use to *rock*, to copulate, probably comes from the association of *rock* and *roll*. See also GET YOUR ROCKS OFF.

roll[2] *American* to rob violently.

Often applied to a drunkard who is knocked, or *rolled*, over before being robbed, and in general use of street theft:

> ...rolled by a tough hackie and dumped out on a vacant lot. (Chandler, 1953)

roll[3] to kill.

After violent assault, and perhaps only an extension of ROLL[2]:

> ...both now dead. James 'rolled' by Rough Trade in Blackheath. (A. Clark, 1993)

Roman sexually orgiastic.

From the fabled orgies of the ancient Romans, which involved intoxicants and general abandon, rather than the depravities of the modern city or its church. It is a shortened form of *Roman culture* or *Roman way* in American advertisements offering access to sexual depravity. A *Roman spring* is lust in the elderly.

Roman candle a failure of a parachute.

Failing to open fully, it resembles the firework:

> ...we were all well acquainted with details of a Roman candle. (Farran, 1948, of parachuting)

romance copulation outside marriage with one person.

In standard usage, a courtship, from the *romance*, or tale of chivalry, which was set down in vernacular French rather than in Latin:

> I am distressed to see the old French word 'romance' used as a code name for East African activities. (A. Waugh, *Private Eye*, December 1980)

Also as a verb:

> Stratford Court, where he'd romanced another highly recognisable blonde star, Frances Day. (Monkhouse, 1993—it had been the venue for a single case of extramarital copulation)

romantic entanglement an extramarital sexual relationship.

Generally more sordid than romantic; see also ROMANCE:

Half of fashionable London has its ... romantic entanglements. (Flanagan, 1988, writing of the 19th century)

Also as *romantic affair, relationship*, etc.

romp to copulate.

Literally, to frolic:

What these Indians don't know about the refinements of romping isn't worth knowing. (Fraser, 1975)

A *romp* is also an act of copulation, or a partner in such an act:

I'd rather think of her as the finest romp that ever pressed a pillow. (Fraser, 1970)

room and board with Uncle Sam imprisonment.

An American refinement of GUEST:

Using narcotics without a licence can get you room and board with Uncle Sam. (Chandler, 1953)

rooster a cock.

A survival from 19th-century prudery:

... engine noises clinging to the trees, the rooster crowing. (Theroux, 1993)

Now the standard term for a male fowl in American English. However, we no longer have infestations of *rooster-roaches*; nor do we drink *rooster-tails* or call our helmsman a *rooster-swain*. See also DARK MEAT[1].

root[1] the erect penis.

From the source of procreation or from the shape of *root* vegetables?

... a thicket of curling hair that spread from the root all round thighs and navel. (Cleland, 1749)

Man-root is explicit:

... moving her pussy the while up and down harshly against my man-root. (Harris, 1925)

root[2] to copulate.

With porcine imagery probably, rather than from ROOT[1], and with a suggestion of unwonted enthusiasm:

... he spent an hour a day at the pushing shop ... rooting himself stupid. (Keneally, 1985)

Also as *rootle* or *root about*:

Where did you ever learn to root about like that? Didn't know such things went on outside a Mexican whorehouse. (Mailer, 1965)

A *root rat* is a male profligate:

They're supposed to be so holy but some of them are unbelievable root rats. (Theroux, 1992, of male Mormon missionaries in Polynesia)

rope[1] **(the)** death by hanging.

Noose and all:

We're dealing with big violent organized gangs. Comes of scrapping the rope. (Kyle, 1975)

rope[2] *American* marijuana.

From the association with *hemp* (see HEMP[2]).

roses (your) menstruation.

From the colour of blood:

Such a bad headache. Had her roses probably. (Joyce, 1922)

The obsolete *rose-coloured* was a euphemism for *bloody* as a mild form of swearing.

rosy drunk.

From the facial glow. The meaning 'wine' may have been merely the anglicizing of *rosé*:

... finished the rosy, and applied himself to ... another glassful. (C. Dickens, 1840)

rotten *American* drunk.

You certainly tend to feel it later.

rough trade an uncouth male in a sexual role.

Aggressive and often badly-dressed, he may be the consort, with whom she regularly copulates, of a wealthy or cultured woman:

... being admonished ... for her public Ugandan activities with her 'rough trade' boyfriend. (*Private Eye*, April 1981)

Much homosexual use, both of the uncouth person and of the practice of consorting with him:

I don't do chickenhawks and I don't do rough trade and I don't work men's rooms. (M. Thomas, 1980)

round objects you are wrong.

A derisive riposte, punning on BALLS, but not used of the testicles medically.

round the bend mad.

Mentally going out of sight:

'Keitel also is going round the bend,' Jodl observed. (C. Forbes, 1983)

See also HARPIC.

roundheels *American* a sexually promiscuous woman.

From the unsuccessful boxer who spends much of his time in the ring on his back, the shape of his heels facilitating his frequent falls:

Little roundheels over there ... she's a blonde. (Chandler, 1951)

routine (nursing) care only allow to die.

Hospital jargon for the procedure in a situation where extra medication would only prolong suffering.

roving eye a tendency towards extra-marital sexual activity.

Usually of males but also of women, and not referring to the ceaseless vigilance of a naval officer on watch:

> This was a predator, a huntress. Artemis for pants. Old Cap'n Hawley called it a 'roving eye'. (Steinbeck, 1961)

A *rover* is a promiscuous person, probably from the hunt for partners rather than the optical movement:

> He is single, but he is no rover. (Turow, 1987)

rub off to masturbate (another person).

Usually with a male as object. Also as *rub up*:

> Lucy was standing between his legs and rubbing him up. (Sanders, 1982—she was not applying embrocation)

To *rub yourself* is to masturbate, of either sex:

> ... he rubbed himself and the orgasm came. (Harris, 1925)

The corresponding noun is *rub-off* and in obsolete use a *rub off* was a single act of copulation. See also GROIN for *rub groins together*.

rub out to kill.

The act of erasing:

> Somebody rubbed him out this afternoon with a twenty-two. (Chandler, 1939)

rub the bacon to copulate.

One of the common 'meat' images:

> If [they] really did have the hots for each other, maybe Scoggins walked in on them while they were rubbing the bacon. (Sanders, 1979)

Also as *rub the pork*:

> As long as you and I keep rubbing the pork. (Sanders, 1982, of a man and his sexual mistress)

rubber a male contraceptive sheath.

The American word for what in the British Isles is an inoffensive article of stationery. The word comes from the material and not from the *rubbing* of copulation:

> Inside my valise
> Are some rubbers and grease.
> (*Playboy's Book of Limericks*)

Whence the advertisers' *rubber goods*:

> A druggist with a *Rubber Goods* sign taped to his window. (Theroux, 1973)

The American *rubber cookies* is rare.

rubber heel a police detective.

Also too of American private detectives, from the walking around quietly.

rubber tire see SPARE TYRE.

ruddy a mild oath.

For 'bloody'. Literally, glowing with a pink hue:

> You ask for the impossible. You ask for the ruddy impossible. (Hemingway, 1941)

rude noise a belch or fart.

Which a child may say it has made, or be reprimanded for making.

rug a wig worn by a male.

The covering over a bare area:

> Your hair. It's beautiful. Is it a rug? (Sanders, 1973)

Whence the American figurative *pull your rug in handfuls* etc., to be exasperated, but see also PULL THE RUG.

ruin to copulate with (a female) outside marriage.

The implication was that her marriageable worth had been lowered by the experience:

> I've often heard the boys boasting of having ruined girls. (Mayhew, 1851)

Whence *ruined (in character)*, a woman known to have so copulated:

> ... seduced by shopmen, or gentlemen of the town, and after being ruined in character ... (Mayhew, 1862)

rumble to steal.

Probably from the improvised seat at the back of a carriage from which servants might toss purloined goods to an associate on foot, and certainly a *running rumbler* was 'A carriage-thief's confederate' (DSUE). Modern airline jargon too for cabin staff stealing consumable stores:

> Methodically, the stewards first 'rumble' the dry stores. (Moynahan, 1983)

run[1] to smuggle.

As in the current *gunrunning*. The OED devotes over 14 pages to *run*, which gives the etymologist a wide choice of derivations for any euphemism; this seems to have developed from the single voyage or excursion:

> You can lay aground by accident and run your goods. (Slick, 1836)

A *runner* was a smuggler and a *run* is still an attempt to smuggle:

> A fine clear run ... all the goods snugly stowed away. (Ainslie, 1892)

run[2] to flee in defeat from a battlefield.

The motion is away from the enemy, not towards him, and the usage is by the winners:

> What? Do they run already? Then I die happy. (Wolfe, 1759, referring to his victory at Quebec)

run[3] an unexpected and sustained series of demands on a bank for repayment.

The phenomenon occurs when depositors lose confidence in a bank which has borrowed 'short' and lent 'long':

> ... if the run persisted, cash reserves would be exhausted and FMA obliged to close its doors. (Hailey, 1975)

run[4] to urinate or defecate.

Usually with urgency, needing to move fast. The *runs*, diarrhoea, is of the repeated urgency and perhaps the composition of the faeces:

> ...don't eat any of those goddamn grapes —they'll give you the runs. (Price, 1978)

In genteel use, a *runny tummy* is diarrhoea.

run[5] **(the)** peremptory dismissal from employment.

From the supposed speed of your final departure. A mordant wit may give you your *running shoes*.

run around with to copulate with extramaritally and regularly.

Literally, no more than to consort with socially:

> Gus had walked out on her because she had been 'running around' with a Party organiser. (McCarthy, 1963)

run away permanently to leave the matrimonial home.

Usually of a wife but not always with or for another male:

> The fact that she did not even take her handbag with her is proof... that she was not 'running away'. (Murdoch, 1978)

To *run off* means the same thing, but is used of either sex, usually when going to a new sexual partner:

> Rita's third husband had run off with a male dancer. (ibid.)

run into a bullet to be killed.

Often used where there is a pretence that the killing was accidental:

> If it develops that a rival ran into a spare bullet while someone was practising target-shooting, that's just too bad. (Lavine, 1930)

run off the bathwater to urinate.

As emptying a tub. A *run-off* is urination.

run on (a) menstruation.

Common female use of obvious imagery.

run out of steam to be sexually impotent.

Of a male, with a suggestion of previous virility. The imagery is of an engine improperly fired:

> ...normal except they've run out of steam and can't make it with a woman any more. (Hailey, 1979)

run round the Horn *American* repeatedly to mislead or deceive.

Usually of prolix evasion, from the fluctuating winds of the Cape which might frustrate the progress of a sailing ship:

> 'I won't run you around the Horn,' Sandecker spoke quickly, 'but I can't tell you any more than I already have.' (Cussler, 1984)

Also of the movement of a newly-arrested person from one police cell to another to prevent access to a lawyer:

> By the time the lawyer finds out, we've moved him again. We waltz him 'around the horn'. It's an old routine. (Sanders, 1973)

runner a policeman.

Not merely in Bow Street or London, where many Victorian *runners* were remarkable for their old age and immobility. Both they and the criminal classes took unkindly to Peel's transfer of his successful innovations in Dublin to the streets of London. (As with RUN there are many euphemistic meanings for *runner*—'smuggler', 'fugitive', 'conveyer of illegal bets', 'journalist who makes irregular contributions', etc.)

running bear *American* a mobile highway policeman.

A Citizens' Band amplification of BEAR[2].

runny tummy see RUN[4].

runs see RUN[4].

rural construction *American* the policing of conquered territory.

A Vietnam usage, and not referring to cottages in the countryside:

> Rural Construction is the old name for Revolutionary Development—the 'workers' are paramilitary elements. (McCarthy, 1967)

Similarly *rural development* was the establishment of a community which you hoped did not support the Communists, often abbreviated to *RD*:

> We sterilize the area prior to the insertion of the R.D. teams. (ibid.)

rush job the marriage of a pregnant woman.

The hastily arranged wedding is often to the putative father.

rush the growler see GROWLER-RUSHING.

rusticate to banish.

Standard English of dismissing British students from university for a while, even if they continue to reside in a town. The Chinese Communists take things more literally:

> His parents had been rusticated—sent shovelling. (Theroux, 1988, of city dwellers banished to the countryside)

rusty rifle (a) *?obsolete* syphilis.

British army usage, likening an impediment in the urethra to rust in the barrel. A soldier found to have either was likely to be severely disciplined.

S and M a sexual deviation involving violence or humiliation.

The initials of *sadism and masochism*:

> The chap was into S and M. Well, 'S' really. Very keen on spanking. (Theroux, 1982)

ST a cloth worn during menstruation.

A common female abbreviation of SANITARY TOWEL.

sack (the) dismissal from employment.

A workman who had to provide his own tools kept them in a bag or sack at his employer's workshop. To be given it by the master was a token of dismissal:

> ... sacked by a British bank for interfering with a woman in Fixed Deposits. (Theroux, 1973)

An unsatisfactory member of the Sultan of Turkey's harem who *got the sack* received more peremptory and drastic treatment: she was stitched up in one and thrown into the Bosphorus.

sad homosexual.

> ... a giveaway, this time specially directed at the 'Sad' community. (*Private Eye*, February 1984)

Of males who are presumably no longer *gay*, but see GAY[2].

saddle soap flattery.

From its qualities of making the seat more comfortable by softening it:

> ... he pointed out he would save the saddle soap in future and come up with easier missions. (Coyle, 1987—a soldier had been flattering his commander but still had been allotted the hard tasks)

saddle up with to copulate with.

Of a male, with the common equine imagery:

> He had been saddling up with all the wenches on his estate and breeding bastards like a buck rabbit. (Fraser, 1970)

In modern use perhaps more as *get in the saddle*:

> Just before they get in the saddle they say, 'Okay, put your clothes on—you're under arrest'. (Theroux, 1973)

A *saddle-broken* woman is used to copulation:

> ... too bad she had a husband, of course, but at least she'd been saddle-broken. (Fraser, 1973)

Specialist dictionaries tell us that a *saddle* is the vagina but I have seen no literary use.

safe see SAFETY.

safe house a refuge for spies.

Not merely a dwelling which is unlikely to collapse:

> The Russian spy master had a 'safe house' for a time at 3 Rosary Gardens. (Boyle, 1979)

In espionage jargon, *secure house* is a synonym.

safe sex copulation or buggery in which a male uses a protective sheath.

No longer, as once upon a time, merely for the avoidance of pregnancy or a curable disease:

> Whether this is a covert invitation to 'safe' sexual relations is unknown. (Pei, 1969)

Now recommended to prevent the spread of AIDS:

> She brushes back Gina's badly braided hair and tells her to get hip to safe sex. (*Tribune* (Oakland), 11 March 1991)

safety *American* a male contraceptive sheath.

Protecting from venereal disease and a paternity suit. Less often as a *safe*.

St Colman's girdle has lost its virtue there has been extramarital copulation.

The mythical but magical garment encircled fully only those who were chaste. The euphemism was used in 1890 when Parnell's adultery with Katie O'Shea, which had been widely known about but not publicized, was exposed in a court hearing over several days before a jury. Whether history might have been changed if the Catholic church and other Irish nationalists had not turned against him is a matter for speculation.

salami tactics the gradual elimination of non-Communists from a coalition.

First used of the post-Second World War method through which non-Communists were sliced away from the Hungarian coalition government formed under Russian occupation:

> Why should the Russians try to annex the whole of Europe... If they try anything it will be salami tactics. (Lynn and Jay, 1989)

saloon a place where intoxicants are drunk after purchase.

Literally, it is no more than a hall. An American *saloon* is almost any bar but the British *saloon bar* was a better furnished, dearer, and often drearier room than the public bar.

salt to cheat by improper addition.

From adding to food, to improve or disguise taste. The object may be a mine, where valuable minerals are introduced to misrepresent the worth; or an account, where wrong items are charged for.

salt and pepper a black consorting heterosexually with a white.

From the contrasting pigmentation. In this American use, the male is usually black.

salute upon the lips a sexual kiss.

19th-century reticence when those who were 'free of their lips' were said also to be 'free of their hips':

> ...he repeatedly subjected me to the *assault* of his *salutes* upon my lips. (Fraser, 1977, writing in 19th-century style)

salvage to steal.

Literally, to save from fire, shipwreck, or other disaster. In the First and Second World Wars *salvage* became synonymous with looting, from bombed or deserted buildings. Today it covers any stolen articles, particularly if they are later offered for sale in a market *ouvert*, where in the British Isles a buyer always acquires a good title even against the rightful owner.

Sam[1] a policeman.

Especially if he is on counter-narcotics duties for *Uncle Sam*.

Sam[2] a black male who is submissive to whites.

American black slang, from *Little Black Sambo* and the common slave servant name.

same gender oriented homosexual.

It appears in America as *SGO* for short, but let us hope not for long.

samizdat see SELF-PUBLICATION.

sample a quantity of urine.

Medical jargon and a shortened form of *sample of urine* required for tests:

> We think of taking 'samples' of urine in little bottles to the doctor's surgery. (Peter Mullen in Enright, 1985)

San Quentin jail bait see JAIL BAIT.

sanction an assassination.

Literally, no more than a penalty:

> ...he had performed a half-dozen counter-assassinations ('sanctions' in the crepuscular bureaucratese). (Trevanian, 1973)

Espionage jargon.

sanctum sanctorum see HOLY OF HOLIES[2].

sand-box *American* an area within a room in which a cat is expected to urinate and defecate.

So called even when the medium introduced in the hope of reducing the smell or spillage is other than sand.

sand rat a cheap prostitute.

British Indian army use, from the squirrel-like mammal which infested *bashas*, or sleeping huts:

> The few cases that were contracted were with the Burmese and Chinese sand-rats. (Richards, 1936, of venereal disease)

sanguinary a mild oath.

A rather laboured and uncommon form of BLOODY[2].

sanitary man a cleaner of lavatories.

Sanitary means 'pertaining to health', but to avoid confusion and loss of face the formerly styled British *sanitary inspector* now calls himself a *public health inspector*:

> ...latrine buckets introduced which the sanitary men emptied every night. (Richards, 1933)

The American *sanitation man* remains a *dustman* in the British Isles.

sanitary towel a cloth worn during menstruation.

> She sold sanitary towels to the younger women in the pension, passing them over, wrapped in plain paper, with a secrecy that suggested a conspiracy. (Manning, 1977)

Once again health and cleanliness are confused. *Sanitary napkin* is a more dated form:

> Don't block the toilet with sanitary napkins. (Bradbury, 1959)

sanitary treatment embalming a corpse.

American funeral jargon which ignores the fact that fresh corpses are aseptic:

> The use of the word 'embalming' is best avoided... other terms as 'Temporary Preservation', 'Sanitary Treatment', or 'Hygienic Treatment.' (J. Mitford, 1963)

sanitized[1] cleaned.

You read it on the irritating paper strips across lavatory bowls and tooth mugs in certain types of hotel which need to convince you that they tidy rooms between customers.

sanitized[2] rendered harmless.

Of files etc. from which damaging evidence has been weeded out:

> Erlichman says he never received that material, and doesn't know whether he got all of what Welander had turned over to Haig, or if the batch was sanitized by either man. (Colodny and Gettlin, 1991)

sapphic a female homosexual.

Sappho was the poetess who lived in Lesbos:

> I never picked you for a sapphic... were you always that way? (McCarthy, 1963)

A *Sappho* is a female homosexual, *sapphism* is female homosexuality, and a *sapphic attachment* is such a relationship:

> A lady gym teacher with whom she had formed a Sapphic attachment... (Theroux, 1979)

> One of the fillies started an affair with a lady passenger... I had to make up to an emigrant to tempt my Sappho back to me. (Londres, 1928, in translation)

sartorially challenged badly dressed.

An extension of the CHALLENGED theme which has added a new dimension to the world of euphemism:

> The sartorially-challenged Sir John Harvey-Jones... (*Daily Telegraph*, 30 March 1994)

satin gin.

From the supposed smoothness. Now rare except in brand names.

saturated *American* drunk.

Thoroughly soaked in intoxicants.

sauce[1] (the) *mainly American* intoxicants.

Usually whisky and implying excessive drinking. Someone *on the sauce* is an alcoholic:

> I had been on the sauce and behaving badly. (Theroux, 1978)

A version of GRAVY[1].

sauce[2] to copulate with.

Of a man, although I can think of no rational derivation:

> Said as if the name was a reason for my never having sauced her. (Fry, 1994)

sauna parlour a brothel.

Public wash-houses have catered for other masculine needs than cleanliness since antiquity:

> ...more magazines restrict advertisments for 'saunas' or 'escorts' to a few pages. (*Sunday Telegraph*, 28 August 1994, on attempts to control advertising by prostitutes)

You are however more likely to be offered a sauna in a *sauna parlour* than a massage in a MASSAGE PARLOUR.

sausage the penis.

Mainly nursery use, without sexual connotations. In the same society it also may mean a turd.

save spend.

A commercial inducement to buy something you do not need because of a price reduction.

save it to refuse to copulate.

Of a woman who permits passionate embraces before marriage, but no more:

> A wet tongue kiss, a few minutes in their arms ...but—she was saving it for her husband. (Longstreet, 1956)

say a few words to make a speech.

Of indefinite duration. The pretence of an impromptu performance is often belied by the furtive production of notes.

say Kaddish for to mourn the death of.

Kaddish is a Jewish prayer 'specially recited also by orphan mourners' (OED):

> He had said Kaddish for so many of his own generation. (Forsyth, 1994, referring to an octogenarian Jew)

scald to infect with venereal disease.

From the burning sensation, especially in the male, and you were likely to be so infected in the obsolete British *scalding-house*, or brothel.

scalp to kill.

Originally the scalp was the skull, and then the skin and hair covering it. The verb form arose from the practice of the American Indians in which the skin and hair was removed from the head of their victims, both to prove their success and to retain a trophy. Successful braves were popularly supposed to have collections of dozens of scalps. In figurative use, a *scalp* may be a man enamoured of a popular and attractive woman who may or may not return his feelings; someone with whom a profligate person claims to have copulated; or a famous person persuaded to assist a third party in some appeal or enterprise.

scalp dolly a wig.

The *dolly* is probably from the child's toy, rather than the spreading of the head of a rivet or a corruption of *doily*. You use this American term only if someone else is wearing the wig.

scandal sheet an expense claim.

From the newspaper which is also likely to contain exaggerated or fictional episodes.

scare (the) criminal extortion.

From the intimidation of the victim and usually as *put the scare into*.

scarlet fever *obsolete* lust for soldiers.

I cannot exclude this rare treble pun, on the disease, the colour of the uniform, and the activities of the *scarlet woman*:

> Nursemaids are always ready to succumb to the 'scarlet fever'. A red coat is all powerful with this class, who prefer a soldier to a servant. (Mayhew, 1862)

scarlet woman a prostitute.

From the woman 'arrayed in purple and scarlet colour...THE MOTHER OF HARLOTS' (Revelations 8)—but sometimes any adulteress will do:

> The Colonel evidently objected to its presence in the house at the same time as his Scarlet Woman. (Sharpe, 1978)

Our Protestant ancestors found it an abusive and useful epithet for the Roman Catholic Church.

scatters (the) diarrhoea.

In humans and animals. It probably comes from *scate*, 'a dysenterical disease in sheep' (EDD) although to *scatter* was to urinate in obsolete British dialect.

scheduled classes *Indian* those condemned by birth to menial employment.

Indian society retains gradations which would provide scope for those anxious to eliminate social barriers:

> ...the Dulits (or scheduled classes or harijans or untouchables, to take the wounding nomenclature back through its earlier stages). (Naipaul, 1990)

schmear see SMEAR[1].

school *American* a prison.

The *big school* is for men, the *little school* for women and children.

scorched *American* drunk or under the influence of illegal narcotics.

From BURN WITH A LOW (BLUE) FLAME? Even if not, it uses the same imagery.

score[1] to copulate with a woman outside marriage.

Usually of a single episode and without payment:

> Brunton was all set to score with a Moral Philosophy student in his rooms—a female student. (Price, 1979)

The punning *know the score* is to be sexually experienced, of men and women.

score[2] *American* to commit a successful crime.

Mainly of crime committed by those addicted to illegal narcotics:

> At first... we thought it was a junkie looking to score. (Sanders, 1985, of a murder)

Whence to buy such narcotics:

> There were drive-up windows in garages to which white kids could come to score. (Turow, 1987)

score adjustment giving higher marks to blacks.

A device for trying to conceal lower scholastic achievement or to compensate for inadequate schooling etc.:

> The little-known practice also is referred to in certain government and employment circles as 'within-group norming' or 'score adjustment strategy'. (*Chicago Sun-Times*, 14 May 1991, of *race-norming*; see at RACIAL)

Scotch mist drunk.

British rhyming slang for PISSED, from the drizzling cloud which blots out the landscape.

scour to administer a laxative to; to purge.

Literally, to clean thoroughly the inside of anything. A beast with *scour* has diarrhoea, which humans also caught from bad beer, or *scour-the-gate*:

> There's first guid ale,
> And second ale and some,
> Hink-skink and ploughman's drink,
> And scour-the-gate and trim.
> (Chambers, 1870)

scratch[1] the devil.

Because of his propensity to 'seize rapaciously' (OED). Usually as *old scratch*:

> Give over action to like old Scratch. (Slick, 1836)

See also OLD A' ILL THING.

scratch[2] a wound.

A brave soldier sought to minimize the extent of his injury:

> She gave a little scream. 'You are wounded! Your arm!' 'It's a scratch, nothing more.' (Fraser, 1970)

scratch[3] *American* to kill.

Literally, to retire from a contest, and in America also to eliminate the name of a candidate from a list:

> I scratch the colonel in Hong Kong, Corrigon shows up. I scratch Corrigon, there's the dame. (Diehl, 1978, of killings)

screw[1] to copulate with.

Of a male, from the entry into a fitting aperture:

> 'Well you, Howard,' says Flora, 'who did you screw last night?' (Bradbury, 1975)

But either sex may be said to *screw around*, to copulate indiscriminately:

> Blokes who screw girls who screw around a lot are usually blokes who screw around a lot. (Amis, 1978)

A *screw* is a male's sexual partner, always with a laudatory adjective—as I note elsewhere, in male vanity or fantasy, there are no *bad screws*. *Screwing* means 'copulation':

> Everyone gets laid, too, but that doesn't eliminate screwing as a subject. (Theroux, 1975)

Also much figurative use as a synonym for *fuck*:

> She was drowned out by a chorus of 'Screw the profiteers'. (Hailey, 1979)

screw[2] a prison warder.

He turns the key:

> ...known as a hard-boiled 'screw'. (Lavine, 1930)

The British derivation is probably from the tightening of the screw on an apparatus on which a prisoner underwent forced exercise, or *hard labour*, in his cell.

screw[3] to cheat.

A venerable standard English usage, from the accentuated application of force implicit in the process. It is the victim who usually so refers to his plight in the passive tense:

> Your chances of being screwed by a Canadian factory owner then were just as good as your chances of being screwed by an American factory owner now. (*Saturday Night* (Toronto), 12 February 1974)

screwed drunk.

Probably a pun on TIGHT[1]:

> ...a glance sufficed to show even Philippa... that he was undeniably screwed. (Somerville and Ross, 1897, of a drunkard)

To be *half-screwed* is to be no more sober.

screwed down dead.

As the coffin is sealed after the last peep at the corpse:

> Then don't talk as if I'd been screwed down. (Cookson, 1967)

screwy having an abnormal mental condition or behaving in an eccentric manner.

The abnormality may be temporary and may exist only in the eyes of the person who uses the expression:

> 'The girl is screwy,' I said. 'Leave her out of it.' (Chandler, 1958)

The American *screw factory* is an institution for the mentally ill:

> ...had to be taken to the screw factory. (Wambaugh, 1975, of a lunatic)

To be *screwed up* is to be confused or upset, while to *screw up* is to handle a situation badly. See also HAVE A SCREW LOOSE.

scrub *American* to remove hesitation or error from.

Cleaning the tape of a pre-recorded speech so that the broadcast version does not say too much, or tell all the truth.

scrubber a prostitute.

Of the meaner sort and perhaps from the posture of the floor cleaner:

> Not all of them were scrubbers. Jane Wentworth wasn't... Marilyn would have fitted into that list of likely pick-ups. (Price, 1979)

A London *Times* 1972 headline 'Heath's Whitehall Scrubbers' party' was changed in the second edition, before the office cleaners had time to instruct their lawyers, to 'Celebrating a whiter Whitehall'.

scrump to steal.

Literally, to collect windfalls or other small apples to make *scrumpy*, rough cider. It is still used in Somerset today in this sense:

> To goa a scrumpin, that is fetchin apples off somebody's trees. (Hallam, 1866)

sculpted manufactured.

Advertisers' pretentiousness, seeking to imply that a forged, cast, turned, moulded, or other machine-manufactured article has been produced through a sculptor's individual skill:

> Each is sculpted in a classic traditional pattern. (*Aspect*, October 1983, making an 'offer' of 'FREE! Solid Brass Candlesticks')

scupper a prostitute.

American naval use, and it sometimes embraces amateurs. Literally, the *scuppers* are that part of the ship through which waste water is washed.

scuppered killed in battle.

The derivation from *scuppers* seems inappropriate because those so killed die involuntarily, while the essence of a *scupper* is its planned egress.
In standard usage plans, arrangements, or hopes may be figuratively *scuppered*:

> We're here to raise money for a very important charity, and we're not going to let that be scuppered. (*Daily Express*, 8 June 1992)

sea food whisky.

An American Prohibition use 'to mislead the police or strangers' (DAS). Most of it came by sea or via the Great Lakes. *Sea food* today is snob catering jargon to avoid the downmarket connotations of *fish and chips*. You may expect to order fish, molluscs, and marine crustaceans, but not plankton or seaweed.

sea gull a prostitute.

She follows the American fleet from port to port. Also used of similarly peripatetic wives and sweethearts. See also GULL.

sea lawyer see BARRACK-ROOM LAWYER.

season (the) the fixed period in which marriageable girls were publicly displayed.

When a British girl with rich parents had to COME OUT:

> 'The Season' being a sort of ritual marriage market to which every parent then subscribed anxiously. (Blanch, 1954)

seat the buttocks.

A transference from the thing you sit upon to the part on which you sit. With BOTTOM, a very familiar coy evasion.

seat cover a nubile female in a car.

Punning American Citizens' Band slang:

> Lay an eyeball on that seat cover comin' up in that show-off lane. (CBSLD)

See also CHECK THE SEAT COVERS.

secluded inconveniently isolated.

Estate agents' jargon to describe a house with no access to public transport, utilities, shops, etc.

seclusion involuntary solitary confinement.

The subject does not seek his privacy. Jargon for the treatment of violent criminals and lunatics.

second eye *American* the anus.

Homosexual use. See also BRONZE EYE.

second strike retaliation.

Nuclear jargon, and not a further blow from the party making the FIRST STRIKE. A *second-strike capability* is your ability to reply in kind after such an attack, inflicting *second-strike destruction*:

> Both superpowers have to bear in mind the high probability of second-strike destruction. (Hackett, 1978)

secret parts the human genitalia.

Those not normally revealed in company:

> *Hamlet* Then you live about her waist, or in the middle of her favours?
> *Guildenstern* Faith, her privates we.
> *Hamlet* In the secret parts of Fortune? O, most true; she is a strumpet.
> (Shakespeare, *Hamlet*)

secret (state) police an instrument of civil repression.

The full phrase is a literal translation of *Geheime Staatspolizei*, usually shortened to *Gestapo*. Today a *secret police* is a necessary adjunct to any tyranny.

secret vice masturbation.

By either sex, although normally by a male:

> ... the various lubricants I had used while practicing the Secret Vice. (Styron, 1976)

Also as the *secret sin* or *indulgence*.

secretary the sexual mistress of an employer.

Often a proximity relationship but it is also a favoured title, especially where the parties are travelling together:

> Wives, daughters and mistresses too —documented as secretaries. (Deighton, 1978)

In a large firm, the formal post of company secretary is usually held by a man.

sectioned detained involuntarily in a mental hospital.

From the part of the Act which empowered such confinement:

> Should she be sectioned under the Mental Health Act and forced back into hospital? (*Times*, 19 October 1991)

The appropriate sections of the current British legislation are numbers two and three. Under American service regulations of the Second World War, the section was numbered eight:

> You hold on ... Or you get shipped home on a Section Eight. (Deighton, 1982, of Second World War airmen)

secure house see SAFE HOUSE.

security risk anyone you disagree with.

Espionage jargon of those who cannot be completely trusted. The expression was used by Senator Joseph McCarthy to attack those in public life, and especially actors, who did not share his pathological hatred of communism.

security service a department for internal espionage and repression.

Literally, a body charged with the prevention of crime:

> There was no sign of a smoking pistol pointing to ministerial knowledge of past illegal acts by the RCMP Secruity Service. (*Maclean's* (Toronto), 9 April 1979)

For the Communists, it was concerned with the security of the rulers and not the ruled. The system was exported to client states through *security advisers*:

> Shehu made the way easy for the rapid growth at the end of 1945 of a Soviet military mission to which 'security advisers'—dull euphemisms for torturers ... —were already attached. (H. Thomas, 1986, of Albania)

seduce to copulate with (a woman) extramaritally.

Originally, to persuade a vassal to break vows of loyalty. In the days of chaperoning, seductions were mainly limited to wives:

> By long and vehement suit I was seduced
> To make room for him in my husband's bed.
> (Shakespeare, *King John*)

The word is now used even if the woman consents without 'long and vehement suit'.

see[1] to copulate with extramaritally.

Of either sex, from the sense 'to visit':

> What would you say if I told you I'd been seeing someone? (Theroux, 1989—the questioner was a wife who had been copulating with a third party and not just visiting a friend)

A prostitute who *sees* a client copulates with him, in the same sense that a lawyer *sees* a client, or a dentist *sees* a patient. To *see company* is explicit.

see[2] to satisfy by bribery.

Usually as the American *see the doorman, the cops*, etc.:

> ... doing business without 'seeing the cops'. (Lavine, 1930)

Lavine also uses *see* for passing part of a bribe to a superior:

> Woe to the cop who collects anything ... and doesn't 'see the sergeant'. (ibid.)

see a man about a dog to go to any place that is the subject of taboo.

Probably from dog fancying. Your destination depends on the company you are with—a lavatory, in mixed company; an inn, in the presence of your family at home; home, if parting from friends who are staying together drinking.

see the rosebed to urinate out of doors.

Of a male at night, usually in mixed company when the lavatory is reserved for females. Others may *see the compost heap, the view, the vegetable garden*, or whatever. But to *see your aunt* may also be to defecate, in female use, punning on *aunt*, the lavatory (see AUNT[2]).

seed the male semen.

That which is sown:

> She that sets seeds and roots of shame and iniquity (Shakespeare, *Pericles*, with his penchant for vulgar puns)

and, in modern use:

> I felt my seed coming. (Harris, 1925)

The American *seed-ox* was a bull in the days when words such as *bull*, *cock*, *ram*, and *stallion* were taboo in polite speech.

seen better days (that has) poor.

Usually describing people who command sympathy because their poverty has succeeded affluence. Of machinery it means 'worn out'.

seepage the amount stolen from a retail store.

Literally, the liquid which has slowly escaped from a container. Now the jargon of supermarkets.

segregation the availability of inferior facilities for a minority ethnic group.

Literally, no more than separating one thing from another. A dysphemism for giving whites better conditions than blacks.

segregation unit a cell for solitary confinement.

American prison jargon.

select capable of being offered for sale.

Shopkeepers' jargon for perishable commodities such as tomatoes, which are unsaleable if rotten. Things so described have rarely been subjected to any process of selection. For an estate agent, *select* also means 'better than average'—you can reject any implication that there has been any discrimination in their choice of what they will sell.

selected out *American* dismissed from employment.

Sam Goldwyn, famous for such contradictory catchphrases as 'include me out', would have been proud of it.

selective indiscriminate.

Of various military activities, where you wish to play down the horror. *Selective ordnance* is usually napalm, less widely destructive than a nuclear blast but hardly discriminating. A *selective strike* or *response* is one where you don't wipe out all your enemy. *Selective facts* are lies.

self-abuse masturbation.

Usually by a solitary male, from the supposition that it may damage his body or his soul:

> ... two of them being pretty hopeless cases through self-abuse. (Richards, 1936, of white soldiers serving in India)

Self-gratification is less common:

> Nor would loutish self-gratification quell this imperious, feverish desire. (Styron, 1976)

Self-manipulation is rarer:

> I have started to become obsessed by sex. I have fallen to self-manipulation quite a lot lately. (Townsend, 1984)

Self-indulgence literally means giving yourself an undeserved treat:

> Pandora says she is not going to risk being a single parent ... So I shall have to fall back on self-indulgence. (Townsend, 1982)

Self-pollution and *self-pleasuring* are obsolete. *Self-love* usually refers to female masturbation and does not imply that the person who practises it is also narcissistic.

self-catering you buy and prepare your own meals.

Jargon of the tourist industry for a contract for accommodation only. In addition to catering, you also wash up which is often not much of a 'holiday' for those who have to do it.

self-defence an unannounced military attack.

Specifically the explanation by Iraq for its September 1980 attack on Iran.

self-deliverance suicide.

Deliverance is the preferred usage of those who advocate euthanasia, but you may also hear *self-destruction*, *-execution*, *-immolation*, *-termination*, or *-violence*, the choice of words revealing the opinion of the speaker.

self-publication the illegal circulation of uncensored writings.

A translation of the Russian:

> They called it *samizdat*, 'self-publishing', a spoof on the name of the state publishing house, Gosizdat. (Moss, 1985)

sell out to betray.

But not necessarily for cash:

> You'll sell me out fast. And you won't have any five thousand dollars. (Chandler, 1958)

A *sell-out* is such a betrayal or, to an extremist, any settlement of a labour dispute.

sell yourself to copulate for payment.

Correctly viewed, the transaction is at best one of hire, lease, or licence:

> This woman went out on the streets ... to keep them both alive ... so she sells herself. (Bradbury, 1959)

Others may *sell* their *back*, *body*, or *desires*:

> A housewife that, by selling her desires,
> Buys herself bread and clothes.
> (Shakespeare, *Othello*)

Today too some homosexual use.

For a candidate or politician, to *sell yourself* means to persuade others of your worth:

He emphasizes his business experience—to sell himself as a manager and penny-pincher. (*Esquire*, February 1994, referring to Ross Perot; a *penny-pincher* is not a thief but a person who is careful with money. In fact Perot spent millions of dollars on his abortive presidential candidacy.)

semi-detached sharing a party wall.

The usage avoids direct mention of the fact that the house is not separate from its neighbour:

And the novel's title was the first recorded use (in 1859) of the word 'semi-detached'. 'Double cottages' built with a shared party wall had been common in the eighteenth century. (Muir, 1990)

Standard English.

send ashore to dismiss from the navy.

For misconduct on land or sea.

send down[1] to dismiss from university.

The opposite of *up*, in residence. Usually done because of misconduct or failure to achieve minimal results:

When I was an undergraduate you got sent down if you were caught riveting a dolly. (Sharpe, 1974)

send down[2] see DOWN THE LINE[1].

send down the road to dismiss from employment.

Usually at short notice, with one or many employees being affected. Used by employees rather than employers but only heard in the British Isles today among older people.

send home to kill.

Home meant, for many people, heaven:

If it would please the Lord to take it home. (EDD)

A Christian might also be *sent* to *heaven, his long account*, or *the skies*; an American Indian to his *happy hunting ground*; a Chinese to *the happy land*:

The only successful way to get rid of a competitor... is to send him to the happy land of his forefathers by having him 'put on the spot' (Lavine, 1930)

or you might:

send him to the land of the lotus blossom. (ibid.)

So too for other religions and cultures. The phrases are also used with reference to third parties regardless of their origins or allegiances:

My faithful Jasper has gone to the happy hunting grounds. (du Maurier, 1938)

send in your papers (of an officer) to retire prematurely.

From the figurative return to the sovereign of the commission once addressed individually to each British officer:

... I've put up a fearful black? I'm not sure I shan't have to send in my papers. (P. Scott, 1975)

send over the edge to drive mad.

The *edge* of sanity, I suppose:

A mental hospital would send him over the edge. (Bradbury, 1959)

send to the showers see TAKE AN EARLY BATH.

send up to pass a prison sentence upon.

A shortened form for *send up the line* or *river*. That is where the prisons were and where convicts in New York and New Orleans ended up. To *send up* is also to ridicule and, as you may also *deflate* those whom you *send up*, the derivation is elusive.

senior citizen an old person.

In most of America those over 55; in Britain a woman over 60 and a man over 65:

I told them to send half a dozen senior citizens who look a bit sad and just a little threadbare. (L. Thomas, 1979)

But do not so describe a High Officer of State, a General, or even a geriatric President. Also shortened to *seniors*:

Discover Tunisia in the Luxury of our air-conditioned Coach. Seniors a Speciality. (le Carré, 1986—and suckers too, it would seem, unless I have missed a new meaning of *discover*)

sensible ugly or unfashionable.

Of women's shoes and clothes, perhaps with supposed transference from the wearer or from the lists of clothing laid down for girls attending boarding schools:

Her breasts, neatly harnessed under a dark sweater, did not swing as she walked. She wore the ultimate in 'sensible' shoes. (Irvine, 1986)

sensitive payment a bribe.

So described because of its impropriety and probable illegality in the hands of the recipient. As the meaning has become explicit among exporters THIRD PARTY PAYMENT has become a preferred usage.

sent *American* drunk or under the influence of narcotics.

The subject enjoys another state of consciousness, if not unconscious.

separate[1] to dismiss from employment.

Literally, to cause to part. Now rare.

separate[2] (of a couple) to cease living together.

As distinct from what happens when they go about their respective daily business. To *be separated* is to be living apart from your spouse, perhaps with no intention of cohabiting again, but without being divorced. *Separation* represents the status of either spouse or both collectively, and may often be without mutual consent:

> Since her separation from a drunken husband some years ago, Sheila's friend Maureen Bowler had become a noted feminist. (Aldiss, 1988)

separate development the suppression of blacks by whites.

The Afrikaans *apartheid*, which sought to disregard the economic interdependence of the three main categories in South Africa, apart from any unfairness and so on. In 19th-century America, blacks after the abolition of slavery were said to be *separate but equal*, which also had the effect of ensuring that they were separate and unequal.

seraglio a brothel.

Originally, the palace of the Turkish sultan in the Golden Horn, of which a part only was the 'harem', or secret spot.

sergeant see TOP SERGEANT.

serious credibility gap see CREDIBILITY GAP.

serpent see STUNG BY A SERPENT.

serum an intoxicant.

Literally, the fluid part of the blood after clotting. Second World War army usage.

servant a slave.

An example of 19th-century American euphemism which reminds us that these conditions obtained not so very long ago.

serve to copulate with.

In standard usage, of male animals, but a fruitful ground for innuendo as in the British television comedy series set in a store and entitled 'Are you being served?':

> It was a pity there wasn't time and leisure, or I'd have served her as I had once before. (Fraser, 1969, of copulation)

Specifically as *serve your lust*:

> I would we had a thousand Roman dames
> At such a bay; by turn to serve our lust.
> (Shakespeare, *Titus Andronicus*)

service[1] to copulate with.

A rarer form of SERVE:

> Aldo had walked in while he was servicing the cigarette girl over his desk. (Collins, 1981)

Whence the punning American *service station*, a brothel. In standard English *service* is arranged copulation by a male mammal, usually a stallion or bull.

service[2] a charge additional to the cost of the goods supplied.

Levied on you in restaurants etc. regardless of the quality of the attendance. The roadside *service station* is a misnomer as the motorist is usually expected to attend to his own needs, except where he finds the apparently tautological announcement *Attended Service*.

service[3] to kill or destroy.

American military jargon:

> But in 'servicing' these 'things', they had killed sixteen men and had lost one of their own. (Coyle, 1987—the *things* were tanks)

service of military investigation an instrument of repression and torture.

Specifically the 1937 Spanish Communist *Servicio de Investigacion Militar* which provided in its prison cells:

> ... lights to dazzle, noises to deafen, baths to freeze, clubs to beat. (H. Thomas, 1961)

services no longer required dismissal from employment.

The blow is supposed to be softened by an implication that the function no longer exists:

> I was given a discharge, ostensibly on the grounds that my services 'were no longer required', this being a curious euphemism. (R. V. Jones, 1978)

set back to cause (a person) to pay a cost that cannot easily be afforded.

From the proper meaning, 'to cause a reverse or relapse':

> Then luncheon, that set me back considerably. (N. Mitford, 1960)

set up[1] to provide living accommodation for (a sexual mistress).

From the meaning 'to establish or initiate':

> When Christine refused to leave Ward and be set up in a flat, he ceased to meet her. (Green, 1979, of Profumo)

set up[2] to incriminate falsely.

As with skittles, for the purpose of knocking them down again:

> They 'set up' MacLennan in an attempt to discredit him. (*Private Eye*, July 1980)

set up shop on Goodwin Sands *?obsolete* to be shipwrecked off the Kent coast.

A low-lying island of some 4,000 acres lying in the English Channel was taken from the Anglo-Saxon Earl Godwin by the Norman conquerors and handed to clerics who neglected the sea walls, a great storm overwhelming it in the year 1100. Since then it has remained a hazard to shipping, emerging above the waves to a varying extent at each low tide.

settle[1] to kill.

Literally, to reach a conclusion:

> Jack Plenty had settled the Balagnini with a lovely back-hand cut. (Fraser, 1977, of a killing)

settle[2] to conquer and appropriate.

The language of aggression and imperialism. Whence the *settler* who goes to live in conquered territory:

> Rubin resists call to evict settlers. (*Daily Telegraph*, 7 March 1994, of Jews who had taken over part of the Arab city of Hebron)

Such communities among an indigenous population may be termed *settlements*:

> The settlements are usually built on hilltops outside Arab towns and villages. (ibid.)

settled unlikely ever to marry.

An Irish use in a society where religion, remoteness, and tribalism often combined to limit the catchment area:

> Being generally regarded as 'settled' in the expressive Irish phrase, into single blessedness, he sprang it on all of us that he was going to be married to a schoolteacher. (Fingall, 1937)

seven (chuck a) *Australian* to die, or collapse in a faint.

There is no seven on a dice cube. You might also be said to *throw a seven*:

> If she sees the thing she won't scream and throw a seven. She'll shoot. (Upfield, 1932)

Found also in the catchphrase *Threw a seven, went to heaven*.

seven year itch a wish for extramarital sexual variety.

> There's something called the seven-year itch ... middle-aged men quite suddenly cutting loose. (Moyes, 1980)

Seven years is the classic period of change:

> Time's pace is so hard that it seems the length of seven year. (Shakespeare, *As You Like It*)

See also ITCH.

sewage see EFFLUENT.

sewn up[1] *British* pregnant.

Perhaps from the meaning 'finally arranged', or from the appearance of a bale packed in sacking.

sewn up[2] *American* drunk.

Like a corpse prepared for burial in sacking?

sex[1] copulation.

Heterosexually, of the male or female:

> I could have asked to wash after sex (Green, 1979)

and also now used of homosexual activity. The obsolete *sex love* was explicit:

> Katie told him in 1891 ... that 'sex love' between herself and Willie was 'long-since dead'. (Kee, 1993—Katie was Mrs O'Shea, Willie was Captain O'Shea and he was Parnell)

sex[2] the penis or vagina.

From the reproductive functions:

> I rubbed my hot sex against her little button. (Harris, 1925)
>
> 'Oh, how lovely your sex is!' I exclaimed ... my left hand drew down her head for a long kiss while my middle finger still continued its caress. (ibid.)

sex appeal the attraction felt by males for an adult female.

But not used of the equally potent feelings in women induced by certain males or the mutual attraction of other animals:

> Tommy had lashed out in defence of his mother's sex appeal. (E. S. Turner, 1952—according to the advertiser his father had sought other female company because she had neglected to use the correct toothpaste)

Perhaps more euphemistic when abbreviated to *SA*.

sex worker a prostitute.

American usage which seeks to remove any stigma from the calling. I have no literary reference for the delightful alternative form, *sex care provider*, whose therapy is strictly non-medical.

sexual act (the) see ACT (THE).

sexual ambiguity having bisexual tastes.

Ambiguity in this phrase does not usually imply doubt or uncertainty—rather it indicates an excess of catholicism:

> ... over-stressing his sexual ambiguity, even his deviance with regard to drugs. (Davidson, 1978)

sexual assault an unsuccessful attempt at rape.

Nowadays no longer a euphemism but:

> 'Sexual assault' is the euphemism for the rape that fails ... Sexual assault depended on the time and place. (Pearsall, 1969, of 19th-century usage)

sexual intercourse copulation.

Not dealings or conversation between individuals. Now standard English:

> If he gets pinched with a girl in a hotel room, stop sexual intercourse. (Chandler, 1953)

Sexual commerce is archaic and there was no suggestion that anybody got paid for their services. *Sexual congress* does not refer to activities on or around Capitol Hill:

> Eight days later in the little summer-house sexual congress takes place. (Boyd, 1987)

Sexual knowledge, which is usually *had* by the male with an under-age girl, does not mean that she has been told about the birds and the bees. *Sexual relations* may also imply familiarities short of copulation and *sexual relief* refers to what the male obtains, implying that his health might suffer from an excess of celibacy. *Sexual liaison* is sometimes used in this sense:

> [Mao] believed, as some Chinese emperors had believed, that sexual liaison with young virgins enhanced the chance of longevity in an old man. (Cheng, 1984)

These concepts are further explored at COMMERCE, CONGRESS, CRIMINAL CONVERSATION, INTERCOURSE, KNOW, etc.

sexual preference homosexuality.

Not of men who prefer blondes to brunettes or of women who fancy moustache-laden kisses although it may sometimes refer to heterosexuality or bisexuality:

> ...impossible to ask questions about (as they said on the current affairs programmes) Ron's 'sexual preference'. (Keneally, 1985)

Also as *sexual irregularity*, which does not imply random couplings:

> She spoke of your sexual irregularities (Burgess, 1980)

sexual orientation, which excludes heterosexuality:

> But my sexual orientation was the true instigator of apostasy (ibid.)

sexual proclivity:

> She discovered her boyfriend's, uh, sexual proclivities (Sanders, 1986—he was also homosexual)

and *sexual tropism*:

> ...it is replacing your former militancy on behalf of the sexual tropism you and I both represent. (Burgess, 1980—*tropism* is normally a vegetable rather than an animal response to a stimulus)

sexual variety promiscuity.

It does not mean having both male and female forms rather than being hermaphrodite.

shack up to cohabit and copulate extramaritally.

A *shack* is a roughly built rural residence but the arrangement usually has a degree of permanence:

> Why not shack up together?...I don't got to drive so far, you got fun. (Bradbury, 1965—his New York cabbie, in suggesting that they shared a hotel room, misused both the phrase and the verb to *get*)

Also as *shack up with*:

> Since she had shacked up with Joe, the youth had kicked over many traces. (Allen, 1971)

shade1 to reduce in price.

Shopkeepers' jargon, making the asking price a *shade less than it was*. You meet the use in the kind of store where haggling is taboo.

shade2 a dealer in stolen goods.

In American underworld speech, he provides some cover for the thief.

shaft1 to copulate with.

Of a male, from the insertion of a spindle in a bore:

> ...he was out drinking or shafting someone older and uglier than she was. (Sanders, 1977)

Less often as a noun:

> Well, it was clear enough that the old thing had no trouble, even across the dividing decades, in spotting him as a king of shaft. (Amis, 1988)

shaft2 the penis.

In its erect state, from the handle of a tool:

> As you thrust your shaft in and out of me, I felt a strange sort of pleasure. (Harris, 1925)

A rare meaning, 'the vagina', comes from a space into which an object may be inserted and moved, such as an elevator shaft.

shag1 to copulate with.

Perhaps from the old meaning 'to shake' or 'wrestle with'—the cormorant is certainly not a renowned sexual performer. Men usually do the *shagging*:

> Out shagging some quiff.... (Sanders, 1982)

although they may also be heard to say that a woman *shags like a rattlesnake* in a phrase which employs daunting imagery.

shag2 to masturbate.

Usually of boys, again from the shaking. The 19th-century English dialect *shag-boy* was a ghost:

> Fairies and shag-boys! lasses are often skeart at them. (EDD)

shake1 to rob.

By violence or trickery:

> How much you shake him for? (Chandler, 1953)

To *shake down* is to cheat or rob through deceit, usually without violence:

> Find out what they're all trying to shake us down for (Bradbury, 1976)

and a *shakedown* is a scheme which defrauds a victim:

> It was a shakedown. For a two-hundred-dollar camera Sony made a hundred and the girl made a hundred. (Theroux, 1973)

shake2 an arrest.

In American police jargon, usually on trivial grounds to show activity or to fill a quota:

> We ain't got no shakes yet today...Maybe we better write a couple of F.I.'s? (Wambaugh, 1981)

shake hands with the bishop to urinate.

Of a male, whose uncircumcised penis may resemble the chessman:

> Help me to the toilet...I have to go and shake the bishop's hand. (Theroux, 1979, quoting Borges)

Others may *shake hands* with their *best friend*, their *wife's best friend*, or *the unemployed*.

shake the pagoda tree *obsolete* to make a rapid fortune in India.

As a foreign colonist and perhaps by dishonest means, punning on the *pagoda*, an Indian gold coin:

> ... won handsome fortunes by 'Shaking the Pagoda Tree', by the private trade that then was permitted to John Company's servants. (*Spectator*, 1912, quoted in ODEP—*John Company* was the East India Company which conquered most of India before handing it over to the British crown, demonstrating the truism that the flag follows trade)

shakedown see SHAKE[1].

shakes (the) a symptom of chronic alcoholism.

Euphemistic, if at all, in so far as it seeks to minimize the shameful delirium tremens.

shame extramarital copulation by a woman.

What disgraced the female party did not seem to disgrace the male:

> Is't not a kind of incest, to take life
> From thine own sister's shame?
> (Shakespeare, *Measure for Measure*)

The shame was once the devil:

> The shame be on's. (Beattie, 1801)

shanghai to kidnap.

Originally, to render senseless and carry on board ship as a crew member from the crime-ridden Chinese city, because some of the crew you arrived with might be absent when you came to sail:

> He'd rue the day he shanghaied me aboard his lousy slave-ship (Fraser, 1971)

but also used of any forceful removal:

> ... shanghai'd might be a more accurate description for all that had happened to her during the last 24 hours. (Price, 1982)

share favours with to copulate with extramaritally.

In this case the *sharing* is between the donor and another:

> And who does she pick to share her favours with? (Bogarde, 1981)

share pusher see PUSH[4].

share someone's affections knowingly to copulate with someone who is contemporaneously regularly copulating with another.

Of either sex:

> The mistress even suggested that his wife should contemporaneously share his affections. (*Daily Telegraph*, 1979)

share someone's bed to copulate with someone.

Usually extramaritally, when it is assumed that such a proximity will always overcome chastity or disinclination:

> I say you share his bed—*puta*. (Deighton, 1981—*puta* is vulgar Spanish for 'prostitute')

sharp and blunt the vagina.

Rhyming slang but used only, I think, in the phrase *a bit of sharp and blunt*, a single act of copulation.

sharp with the pen inclined to overcharge.

Of lawyers or retailers who charge 'what the traffic will bear' and, in the latter case, especially when old price tags are changed to reflect inflation.

sharpen your pencil[1] to distort figures in published accounts.

This falls short of criminal falsification but involves taking that view on stock valuation, bad debt reserves, etc. which tends towards the desired outcome, showing higher profits or a stronger balance sheet.

sharpen your pencil[2] to alter your stance in bargaining.

A commercial usage and injunction to a seller who is asking too much, or a buyer who is offering too little.

sharpener a drink of an intoxicant.

Usually whisky and gin, which are supposed to liven you up:

> I managed to escape from Colditz for a sharpener or twain with the Major at the RAC Club. (*Private Eye*, May 1981)

sheath a contraceptive worn by a male.

Literally, the covering in which a blade is kept:

> It was typical of Murray to call it a sheath, he thought. (Boyd, 1981—Murray was a rather pompous Scot)

sheath the sword to copulate.

Of a male, with obvious imagery. Literally, to cease to fight.

shed a tear to urinate.

Of either sex, I suppose from the bodily secretion of fluid.

sheep buck a ram.

A buck is literally the male of several quadrupeds, but not of the genus *Ovis aries*. I confess to a weakness for these examples of 19th-century American prudery about farmyard animals.

sheet a record of criminal convictions.

Police jargon, from the paper on which they are written although today we probably are more likely to have a disk:

A sheet that might include gambling arrests, maybe some boosting. (Sanders, 1973)

sheet in the wind mildly drunk.

A *sheet* is a rope tying a sail to a spar, and not the sail itself. If one or more are loose, the vessel is in some disarray:

A thought tipsy—a sheet in the wind. (Trollope, 1885)

A drunkard may also be *three sheets in* (or *to*) *the wind*:

An American lady who was three sheets in the wind and said I looked like a movie actor (Theroux, 1973)

or *four*, or *several*:

There were French seamen at the next table... all several sheets to the wind. (Moss, 1987)

sheets copulation.

On the marriage, or any other, bed:

Happiness to their sheets! (Shakespeare, *Othello*)

sheila an unattached nubile female.

A common term in Australia for any young woman:

Oh for Christ's sake, an Aussie spots a sheila at a party, or what he takes to be a sheila. (Amis, 1988)

Also used in the British Isles.

shell shock the inability to continue fighting.

For many in the First World War a severe neurosis induced by exposure to horror and danger but for others a justification for not again risking their lives:

John Bowen Colthurst had been posted to Dublin to recover from shell-shock; the events of the past two days had proved too much for his disordered mind. (Bence-Jones, 1987 —downtown Dublin did not turn out to be the ideal spot for recuperation over Easter in 1916)

shellacked *mainly American* very drunk.

Literally, covered with shellac, a varnish which is stoved to give a glazed appearance.

sheltered for those unable to look after themselves.

Of accommodation where staff are available to supervise and help geriatrics or invalids who are no longer self-sufficient and whose families cannot or will not help them.

shelved dismissed from employment.

Normally describing those asked to retire early because of their declining powers:

...so that men who lack drive and imagination can, without undue cruelty, be shelved. (Colville, 1976)

sheriff's hotel *American* a prison.

Now obsolete in the British Isles where the post of sheriff is mainly honorary. If you *danced* at a *sheriff's ball*, you were killed by hanging.

sherpa a senior civil servant.

He carries the figurative burden for his British political boss.

shield *American* a police badge.

And, by transference, a policeman or his authority.

shift[1] an act of defecation.

From the sense 'a movement' and usually as *do a shift* in male use.

shift[2] to copulate.

Again I suppose from the movement involved:

Let we shift... You give baby me. (Theroux, 1971)

ship to dismiss from employment or to expel from college.

American commercial imagery, from the dispatch of goods.

shipped home in a box (be) to be killed.

Usually of a soldier etc. killed abroad:

Shelley had to get him out, or he'd be shipped home in a box. (C. Thomas, 1993—a prisoner was about to be murdered)

Used also, though less often, of anyone who dies abroad.

ship's lawyer see BARRACK-ROOM LAWYER.

shirtlifter a bugger.

Ignoring the daily occasions on which normal and upright men usually lift their shirts or shirt-tails:

...when you sup with a shirtlifter you should use a very long spoon. (*Private Eye*, January 1987)

Whence *shirt-lifting*, buggery:

...an opposition to what the good old-fashioned 'bloke' sniggeringly refers to as shirt-lifting. (*Sunday Telegraph*, 4 September 1994)

Also shortened to *lifter*:

Earlier this year Tasmanian 'lifters'... handed themselves over to the police. (ibid.)

shit stabber a bugger.

British army usage:

Arab men are very affectionate with each other, holding hands and so on. It's just their culture, of course, it doesn't mean they're shit stabbers. (McNab, 1993)

shoot[1] to kill by a firearm.

Literally, to discharge a projectile. This standard usage, which may also mean 'to wound', implies an accurate aim by the person who does the shooting:

He was condemned to death and shot within two hours. (Goebbels, 1945, in translation)

Thus a target may be *shot at*, and perhaps missed, or *shot dead*, about which there is no ambiguity:

Saddam invited the Minister into a side-room, pulled his side-arm, shot him dead and returned to resume the cabinet meeting. (Forsyth, 1994)

In the British Isles, unlike America, hunting with a gun is also called *shooting, hunting* being reserved for chasing on horseback after foxes or deer:

Charles Edouard is shooting, which he calls hunting. (N. Mitford, 1960)

shoot[2] peremptory dismissal from employment.

Perhaps a pun on discharge—see FIRE—or from the velocity with which you finally leave your place of work.

shoot[3] *?obsolete* diarrhoea in cattle.

From its expulsion:

It piss'd the bed, and shute the bed. (Graham, 1883)

shoot[4] to inject an illegal narcotic intravenously.

From the direct passage into a vein. To *shoot gravy* is to inject after mixing the narcotic with your own blood and heating both. A *shooting gallery* is where you may obtain illegal narcotics.

shoot a line to boast.

With imagery from whaling? Nowadays you are more likely, if of that mind, to *shoot the bull*:

No one lingers, no one sits down and shoots the bull. (Theroux, 1988, of the aftermath of a Chinese banquet, not of a Spanish *corrida*)

Shooting the breeze is usually of male flirtation:

Inside, oblivious of all this, are the two highway patrolmen, sitting at the counter eating apple pie with ice-cream and shooting the breeze with the waitress. (Bryson, 1989)

The cliché to *shoot yourself in the foot* means 'to act in a way which harms you':

Following the latest fiasco, Eurotunnel was accused of 'shooting itself in the foot' by the AA. (*Daily Telegraph*, 13 March 1995—the firm was obliged to abandon a promotion after two weeks owing to lack of trains)

shoot a lion to urinate.

Of a male, usually out of doors. In America you may also say you are going to *shoot a dog*.

shoot among the doves *obsolete, Scottish* to boast or exaggerate.

From the ease with which tame birds might be brought down:

A lady... had heard her husband mention... that such a gentleman... was thought to shoot among the dows. She immediately took the alarm and said to him with great eagerness... 'My husband says ye shoot among the dows. Now as I am very fond of my pigeons, I beg you winna meddle wi' them.' (EDD)

shoot blanks to be sexually impotent.

Of a male and see also SHOOT OFF:

That's pretty big talk for a man shooting blanks. (Garner, 1994, and not of one using a starting pistol)

Common use of a man who has had a vasectomy.

shoot off to ejaculate semen.

Usually prematurely under intense sexual excitement:

I had to change my underwear when I got back here. That's right. I shot off in my drawers. (Diehl, 1978)

The punning *shoot over the stubble* is to ejaculate in a woman's pubic hair. To *shoot your roe* refers to any ejaculation. A *shot* is an ejaculation:

It's the only one where you get three shots for your money: The shot upstairs (fellatio). The shot downstairs (vaginal copulation). And the shot in the room (whisky). (Longstreet, 1956, of a brothel)

The obsolete *shoot between wind and water* was to infect with venereal disease, punning on the crippling shot to a sailing ship.

shoot the agate *American* to seek out a woman for copulation.

From an affected form of strutting used in black American parades which was so named.

shoot the cat *British* to become drunk.

Originally, to vomit from any cause. There were many phrases linking cats, vomiting, and drunkenness.

shoot the moon see MOONLIGHT FLIT.

shoot with a silver gun *British* to be unable to provide meat by hunting.

In season, a gentleman was supposed to be able to keep his family well, and his retainers occasionally, supplied with game birds:

Shooting with a *silver gun* is a saying among game eaters. That is to say, *purchasing* the game. (Cobbett, 1823)

See also CATCH FISH WITH A SILVER HOOK.

shooting gallery see SHOOT[4].

shop[1] *American, ?obsolete* to dismiss from employment.

I would have shopped the fellow in an instant ...He was most impertinent. (Wilson, 1915)

shop[2] to give information which leads to the arrest of.

You would think from selling it, but a lot of *shopping* is done out of malice or self-protection:

> [He] volunteered for a fiver to 'shop' his pals. (*Tit-Bits*, 20 May 1899)

A *cop shop* is a police station.

shop door is open (the) your trousers are unfastened.

An oblique warning to another male, of buttons or zip.

shoplift see LIFT[1].

short[1] a measure of spirituous intoxicant.

Shortened form for *short drink*, as distinct from beer:

> A pint of beer, or a glass of 'short' (neat gin). (Mayhew, 1851—but today *shorts* are any kind of spirits)

short[2] a pistol.

As distinct from a *long*, a rifle. Army jargon:

> We had no shorts (*pistols*), they were all longs, and it was going to be almost impossible to bear them if we were compromised. (McNab, 1993)

short-arm inspection an examination for venereal disease among men.

Punning on the *small arms inspection* of rifles etc. and the *short-arm*, or penis:

> Periodical medical checks, known as short-arm inspections. (Allen, 1975)

short hairs the pubic hair.

Even though they may be more luxuriant than those on the legs etc. The use is almost always in a figurative cliché:

> I think I've got them by the short hairs. (Sharpe, 1974)

The *short and curlies* is specific.

short illness (a) see LONG ILLNESS (A).

short time a single act of copulation.

Prostitutes' jargon for a contract with few preliminaries and no sequel:

> The price for a short time with massage stayed the same. (Theroux, 1973)

Less often as a *short session*:

> She's short sessions. Never lets a man stay for more than half an hour. (Archer, 1979)

If the hotel clerk asks you whether you need a room for a *short time* or a *short-term* occupation, he suggests you will use the room for such activity and he will charge you accordingly:

> An overnight stay, sir? Or a short-term residency? (Keneally, 1985)

shortening the front line[1] retreating under pressure.

Soldiers thus excuse a defeat by implying that a salient is being voluntarily abandoned.

shortening the front line[2] slimming.

Punning on the military euphemism (above) and usually of fat men.

shortism a supposed prejudice against small adults.

Another category of those discriminated against:

> Small step in battle to end shortism. (*Daily Telegraph*, 12 April 1994, and see also VERTICALLY CHALLENGED for further elucidation)

shorts (the) indigence.

Usually of a temporary nature, short of cash until the next pay day:

> ... if you get the shorts, don't be bashful about asking me for help. (Sanders, 1986)

shot[1] see SHOOT OFF.

shot[2] a measure of spirituous intoxicant.

I think from the method of discharging, or *shooting*, it into the glass. Frequently in America, but less so as the years pass, an imprecise measurement depending on your relationship with the bartender.

shot[3] a narcotic taken illegally.

By injection:

> The keepers could sell the balance ... to other prisoners in need of a shot. (Lavine, 1930)

shot[4] drunk.

Probably from the slang meaning, 'finished' or 'exhausted' although the variant *shot away* suggests derivation from wounding. However drunk, you are unlikely to be ever more than *half shot*:

> ... unlimited wine being dispensed in all the public buildings. The whole population seemed to be half-shot. (Fraser, 1970)

shot in the tail pregnant.

A rather tasteless multiple pun.

shotgun marriage the marriage of a pregnant woman to the baby's father.

The man is supposed to come to the altar under duress:

> Princess Caroline of Monaco is finding it impossible to secure an annulment of her 1978 marriage ... made even more difficult following a shotgun marriage last December to Italian Stefano Casiraghi. (*Private Eye*, August 1984)

Less often as *shotgun wedding*.

shout[1] **(the)** peremptory dismissal from employment.

Employees still say they have *had the shout*, even if dismissed *sotto voce* or by letter.

shout[2] an obligation to pay for a round of intoxicants.

Not euphemistic if you take your turn but perhaps so of someone said *not to pay his shout*—he cadges from others. To *shout yourself hoarse* is to be drunk, from having ordered too many drinks.

shove[1] to copulate.
Of a male, from the common pushing imagery.

shove[2] **(the)** peremptory dismissal from employment or courtship.
No physical ejection or rejection can be assumed.

shove over *American* to kill.
Originally, to overthrow:

> Did you—did anybody—have any idea that she was gonna get shoved over? (Diehl, 1978)

shovelled under dead.
Whether buried or not:

> My last day in the Fourteenth Army will be the day they shovel me under. (M. Fraser, 1992, of the Second World War in Burma)

show[1] to menstruate.
Usually of animals and especially used of mares when breeding is planned. In humans the noun a *show* indicates vaginal bleeding which may be caused by an imminent miscarriage, the onset of labour, or other reasons.

show[2] a battle.
Mainly First World War use, minimizing the danger by reference to a theatrical production:

> 'I am watching the show over on our right.' Some of our new divisions... had advanced through a gap. (Richards, 1933)

show your charms (of a prostitute) to seek a customer.
She may in public reveal more than chaster women but less than the term might suggest, until terms have been agreed:

> A woman was showing a man her private charms, and inviting him to enjoy them. (Masters, 1976)

showers[1] deviant sexual activity.
A code word in prostitutes' advertisements, from the penchant of some males for sexual antics involving the urine of another, a *golden shower* (see also WATER SPORTS). A *brown shower* is offered for participants who prefer faeces. The slang 'showercap' is either a contraceptive sheath or diaphragm.

showers[2] see TAKE AN EARLY BATH.

showers[3] gas chambers for prisoners.
Part of the Nazi pretence that prisoners were not about to be gassed:

> His first job was to work in one of the ante-rooms where prisoners had to remove their clothes before going through a door to the 'showers'. (Christopher Booker in *Sunday Telegraph*, 29 January 1995, of Auschwitz)

shrink see HEAD-SHRINKER.

shrinkage the amount stolen from retail stores.
Literally, a reduction in weight or volume of packed goods due to settlement or dehydration. Retailers' jargon, especially of thefts or embezzlement in supermarkets.

shroud waving a device for safeguarding or augmenting expenditure on medical projects.
The sponsor is threatened, usually with more publicity than veracity, that deaths will result if funds are withheld:

> She noted that shroud waving had 'quite a high success rate'. (*Sunday Telegraph*, 29 March 1992)

A *shroud waver* is a doctor or politician, or frequently a combination of the two, who so acts.

shuffle off this mortal coil to die.
The Bard said it first, through the voice of Hamlet:

> ...left a hundred grand when he shuffled off his mortal coil. (Sanders, 1986)

sick menstruating.
A slightly rarer variant of ILL[1].

sick-out a strike by public service employees.
Those prevented by American law or contract from going on strike may pretend to be absent due to illness.

side orders sexual practices of an unusual or depraved nature.
From the dishes served additionally to a main course:

> Alvin C. had been having no side orders of sex; no arguments either, or drink or drugs. (Davidson, 1978—although the phrase here could refer to his not copulating outside marriage)

sides pads worn to accentuate a woman's figure.
From the days when men seemed to be attracted to wider hips:

> She pulled off a pair of 'sides', artificial hips she wore to give herself a good figure. (L. Armstrong, 1955)

sight deprived blind.
Perhaps no more than circumlocution despite:

> The blind are now 'sight-deprived' as if to refute any suspicion that they got that way voluntarily. (Jennings, 1965)

sigma phi syphilis.
Medical jargon from the Greek letters used in shorthand, which also conceals the diagnosis from some of the patients.

sign the pledge see PLEDGE (THE).

silk (the) a parachute.

Euphemistic in the phrase *on the silk*, referring to military aircrew obliged to abandon an aircraft in flight:

> ... you've got to stick to your own air space or ride down on the silk. (Hackett, 1978—if you collide, you will crash)

Whence the figurative *hit the silk*, to seek to escape from or to avoid a calamity:

> In markets like this, if that happens, everyone'll try to hit the silk at once and no one'll get out the door. (M. Thomas, 1987)

silver an unskilled black worker.

Obsolete American local usage in the Panama Canal Zone:

> Race was expressed by the Panama Canal Company not in terms of black and white but by the designations gold and silver. The euphemism was derived from the way workers were paid: the unskilled workers, most of them black, were paid in silver, the skilled workers, nearly all white Americans, were paid in gold. (Theroux, 1979)

simple of small intelligence.

Not just lacking knowledge and experience, as in Simple Simon's commercial exchange with the Pieman. *Simple* is now widely used of idiots considered fit to remain in society.

sin to copulate extramaritally.

Literally, to commit a forbidden act but, since St Paul's obsession with that kind of wrongdoing, now Christian use for any activity which is taboo sexually:

> Most dangerous
> Is that temptation that doth goad us on
> To sin in loving virtue.
> (Shakespeare, *Measure for Measure*)

Sinful then means relating to such copulation, as in *sinful commerce* and see also *live in sin* at LIVE AS MAN AND WIFE

sing (of a criminal) to give information to the police.

From the imagery of the song-bird in the cage, of your own misdeeds, or of other criminals:

> ... had him under the lights all fuckin' night ... and about nine this morning he starts singin' like Frank Sinatra. (Diehl, 1978)

single parent a parent living alone with dependent offspring.

A variant of LONE PARENT and no longer for the most part referring to someone who has been widowed.

singles describing a place etc. where people divorced, unmarried, or apart from a spouse can meet another person heterosexually.

From *single*, unmarried, although women tend on these occasions to hunt in pairs. Thus in America you find *singles bars*, *nights*, and *joints* where the participants look for anything from a drinking or dancing partner to copulation or marriage:

> Used to be a singles joint but lately it's turned really rough. (Deighton, 1981)

sink *?obsolete* a lavatory.

Originally, a drain or cesspit:

> Usuph pretended to wander off to the regimental sinks. (Keneally, 1979, writing in 19th-century style)

sip a drink of spirituous intoxicant.

Literally, anything drunk in small quantities:

> By the time they had had a few sips there was damned little left for us. (Richards, 1933, of a rum ration)

The Scottish and northern English *siper*, a drunkard, came from *sipe*, to soak:

> The Hivverby lads at fair drinking are seypers. (Anderson, 1808)

siphon off see SYPHON OFF.

siphon the python see SYPHON THE PYTHON.

sissy a male homosexual.

An American spelling of CISSY:

> Little teeny sissy with gold hair. Looks enough like a girl to be a queen. (Wambaugh, 1983)

sister a prostitute.

In the Far East pimps claim this kinship:

> ... pimps accosting you ... with promises of their sister. (Fraser, 1977)

The dusky lad who invites you to copulate with his 'sister, very white, very clean', makes three false assertions. Occasionally elsewhere as *sister of charity* or *sister of mercy*, both of the same tendency as NUN.

sit-down job an act of defecation.

Usually of a male, who does not need the western pedestal seat for urination:

> Oh, a sit-down job, is it? (Higgins, 1976, of defecation)

sit-in a trespass to draw attention to a supposed injustice.

Usually by a body of people, without violence. A *sleep-in* continues overnight and during a *love-in* the participants while away the hours in casual copulation.

sit-upon the buttocks.

More common in the British Isles than America where *sit-upons* in 19th-century prudery were trousers, the equivalent of the British *sit-in-'ems*. Also as *sit-down-upon*, which should be a chair. An obsolete western English form was a *sitting*:

> She had a tumour going from her sitting. (EDD, from 1887)

sitting by the window under-employed.

A phenomenon of Japanese industrial society where paternalistic attitudes deter the dismissal of employees whose jobs have been superseded by computerization etc.:

> Either more and more underworked employees are left, as the Japanese say, 'sitting by the window', or else these jobs get vaporised in the white heat of the technological revolution. (*Daily Telegraph*, 4 April 1995)

six feet of earth death.

The length of a grave rather than its depth:

> Six feet of earth make all men equal. (Proverb)

six feet underground dead and buried.

The regulation coffin depth:

> I'm glad his father's six feet underground. (G. Greene, 1978)

sixty-nine see SOIXANTE-NEUF.

sizzle to be killed by electrocution.

One of several American culinary images.

skewer to copulate with.

Of a male, from the action of transfixing meat:

> The crooked shadow of Harvey skewering Hornette... (Theroux, 1978—they were copulating during a public performance)

skidmarks the stains of shit in underpants.

Mainly British boys' use.

skim to embezzle or extort.

As cream from milk, on a regular basis:

> ...the two brokers set up the 'skimming' operation mainly dealing in overseas shares through overseas brokers and charging the Kuwaiti organisation inflated prices. (*Daily Telegraph*, 16 April 1994—they were alleged to have stolen some £2 million)

It is also American gambling jargon:

> Skimming is the term used to describe the removal of gambling revenues before they are counted for state or Federal taxes. (*Daily Telegraph*, September 1979)

A *skim* is a bribe or other sum so regularly received:

> A skim of a hundred and eighty was damned thin for a bull lieutenant. (Weverka, 1973)

skin *American* a male contraceptive sheath.

From the shape and texture. Whence the punning *skin-diver*, a male who copulates using such a contraceptive.

skin- pornographic.

From the implication of nudity. A *skin-flick* is a pornographic film:

> ...bought the rights on this new Swedish skin-flick (Deighton, 1972)

which may be shown in a *skin-house*, or cinema specializing in pornography. A *skin-magazine*, often shortened to *skin-mag*, contains erotic pictures, mainly for male edification or whatever.

skin off all dead horses to marry your sexual mistress.

A *dead horse* is something now of small value but at one time useful. In obsolete Irish use, to *work on a dead horse* was to complete a job for which you had already been paid and when your task was over you were said to have *skinned* it.

skinful an excessive quantity of intoxicants.

Usually of beer, which suggests a distended human belly or bladder rather than derivation from a wineskin:

> Take it easy, Larry. You've got a skinful. (Chandler, 1958)

skinny-dip to bathe in the nude.

The subject of far greater taboo in America than Europe:

> I'm going skinny-dipping... Who's game? (Sanders, 1982)

skippy a male homosexual.

Taking the female role and using an affected walk. American black slang.

skirt a woman viewed sexually by a male.

The garment is worn only by females—males call them kilts:

> He's got a nice skirt all right... I wouldn't say pretty, but a good figure. (G. Greene, 1932, and not of a transvestite)

A *bit* or *piece of skirt* is either a male's partner in copulation or the act itself:

> He enjoyed nothing better in the world than a nice bit of skirt. (Richards, 1936)

skivvy a prostitute.

In standard usage, a female domestic servant. The American *skivvie-house* is a brothel:

> Little chickie workin' the skivvie houses... (Herr, 1977)

sky piece a wig.

Worn by a man. It used to mean a hat. Also in America as a *sky rug*.

slack to urinate.

Of a male. The variant *slack off* suggests derivation from loosening or relieving pressure.

slack fill delivering less than the customer thinks he has bought.

Commercial jargon for the manufacture of glass etc. containers which look as if they hold more than they do. Also of not filling them to full capacity.

slag a sexually promiscuous woman.

Often young but not content to form a pair bond with a single male. Partridge (DSUE) suggests 'perhaps ex slagger', which was an obsolete term for a bawd but I just wonder, bearing in mind the social background of many of those so described, whether it isn't back slang for *gals*. See also YOB.

slake your lust to copulate.

Of a male, usually extramaritally, from *slake*, to quench or satisfy:

> ... let him slake his lust on one of his own serf-women. (Fraser, 1973)

In obsolete Westmorland dialect, a *sleck-trough* was a prostitute—the cooling place into which a smith plunged his red-hot iron.

slammer a prison.

Either from the *slamming* of the door as you go in, or the rough treatment you receive inside:

> 'You'll turn her into an addict. And she's —what? Sixteen? Jesus.' 'She's already been in the slammer.' (Theroux, 1976)

Occasionally shortened to *slam*:

> Now kin we jist wrap this up and take me to the slam. (Wambaugh, 1983)

slap and tickle casual copulation.

> And what sells this year's new royal books but the same slap and tickle? (*Esquire*, December 1993)

Literally, no more than what might occur in any courtship, which is all the phrase normally implies.

slash an act of urination.

Originally, a splashing or bespattering:

> All I was doing was quickly relieving myself or, in plain language having a slash. (Sharpe, 1979)

Common use by both sexes.

sledge unsportingly to harass (an opponent).

Jargon of professional cricket where the rewards may become more important than the game. OED suggests the origin may come from a sledgehammer rather than the conveyance used to pull a man to his execution. The practice and the phrase originated in Australia and it is unfortunate perhaps that both have escaped.

sleep to be dead.

When Christians await the resurrection of the body. Often in compounds according to the circumstance. Thus to *sleep in your leaden hammock* or *in Davy Jones' locker* was to have died and been buried at sea:

> Though Drake their famous Captain now slept in his leaden hammock (Monsarrat, 1978)

and to *sleep in your shoes* was to be killed in battle:

> The dreary eighteenth day of June
> Made mony a ane sleep in their shoon;
> The British blood was spilt like dew
> Upon the field of Waterloo.
> (Muir, 1816)

By the same token, *sleep* is death:

> Anyone who went to sleep in a dug-out where there was not much air with one of those fires going... would soon drop into a sleep from which there would be no awakening. (Richards, 1933)

sleep around to copulate promiscuously.

Of either sex, supposedly in various beds:

> ... sleeping around with a lot of West Indians. 'I never approved of Christine's lust for black men.' (Green, 1979)

In this and certain following entries, *sleep* is synonymous with copulation, as though the latter involved something akin to somnambulation.

sleep over to copulate overnight with an extramarital partner at the partner's residence.

Not involving, as you might suppose, the occupation of bunk beds:

> He wanted her to sleep over that night. (Sanders, 1982)

sleep together (of a couple) to copulate.

Usually of extramarital copulation except in the negative where it refers to the cessation of copulation between spouses. Now too of homosexual activity.

sleep with to copulate with.

One of the commonest uses, normally of extramarital copulation of either sex. Of a woman:

> One couldn't accept a fur coat without sleeping with a man (G. Greene, 1932)

and, of a man:

> East African (European) officers as a whole maintained a very much stricter code in the matter of sleeping with African women —sometimes referred to as 'sleeping dictionaries', from their obvious advantages as language instructors. (Allen, 1975)

sleeper a spy infiltrating an organization but staying inactive.

Politically usually of espionage, industrially of disruption where a union activist is introduced into a company to foment unrest. The latter *sleepers* tend to wake up as soon as their probationary period, during which they can be dismissed without complaint, has been passed:

> Philby, Burgess and Maclean had little option but to lie low. As 'sleepers' they could secretly console one another. (Boyle, 1979)

sleeping dictionary see SLEEP WITH.

sleeping partner someone with whom you regularly copulate outside marriage.

Punning on the partner who takes no active part in the running of a business:

> ...the service of a Somali girl-friend or sleeping partner. (Allen, 1975)

sleepy time girl a prostitute.

Again the 'sleeping' imagery for a wakeful occupation. She may also be a sexual mistress:

> Seems like the bim was one of his sleepy time girls. (Chandler, 1953—a *bim* is a female friend)

sleighride the condition of being under the influence of narcotics.

Punning on, and usually from, *snow*, cocaine.

slewed drunk.

Not going straight:

> Mr Hornby was just a bit slewed by the liquor he'd taken. (Russell, c. 1900)

Also as *half-slewed* where, as in other cases, the half equals the whole.

slice to cheat (a customer).

Retailers' jargon for overcharging or under-delivery, from removing a sliver from cheese etc. which has been weighed for delivery. The obsolete *shave*, using the same imagery, was more colourful, especially in the punning phrase to *shave the gentry*.

slice of the action see ACTION[1].

slight chill a pretext for not keeping an engagement.

An indisposition which the draughts of royal palaces seem to induce:

> 'What shall I tell them? A slight chill?' 'That sounds a deal too much like Buckingham Palace. Just say I'm out.' (Ustinov, 1971)

Royal personages are also martyrs to *slight colds* and *indispositions*. However in one case at least the phrase was used to conceal the gravity of an illness:

> Every other paper reported that Attlee is now getting better from a slight indisposition. (Crossman, 1981—he had had an attack of cerebral thrombosis)

slip[1] to give premature birth to.

Usually of domestic animals:

> Cows slipped their calves, horses fell lame (Hunt, 1865)

but not for the great diarist:

> Fraizer is so great with my Lady Castlemain and Steward and all the ladies at Court, in helping them slip their calfes when there is occasion. (Pepys, 1664—Fraizer was a court physician and royal abortionist)

Of humans to *slip a foot* or *slip a girth* was to give birth to a bastard, both with imagery from a fall at riding:

> Slipping a foot, casting a leglin-girth or the like. (W. Scott, 1822)

slip[2] to die.

The concept of gliding easily away:

> The kid's 'slipping'. Dying. (Londres, 1928, in translation)

More often in compounds. To *slip away* is an easy natural death, usually of old age:

> To 'slip awa' within sight of ninety. (Maclaren, 1895)

Old people may also *slip off*. With nautical imagery you may *slip* your *breath*, *cable*, *grip*, or *wind*:

> He was going to slip his cable with all the good scandal untold. (Fraser, 1971)

To *slip to Nod* is obsolete:

> He the bizzy roun' hath trod,
> An' quietly wants to slip to Nod.
> (Taylor, 1787—later in the verse his fate is to 'trudge on Pluto's gloomy shore')

slippage mental illness or decline.

Not an ability to skate nor even physical deterioration:

> I learned all this much later from my mother who, after my father's death, had begun to show signs of slippage. (Desai, 1988)

slit the vagina.

Literally, a narrow straight incision:

> Her first movement brought her sitting down on the step above me and at once my finger was busy in her slit. (Harris, 1925)

slopped drunk.

Spilling over the edge figuratively, and perhaps also handling your glass maladroitly.

slops the police.

Back slang indicating disrespect:

> ...sent out a girl for the slops. (Sims, 1902—she was asked to fetch a policeman)

sloshed drunk.

To *slosh* is to be a glutton but there is also the imagery of an over-full container:

> ...her career of piss artistry, when she could still pretend she got sloshed out of not knowing about alcohol. (Amis, 1986)

Often as *half-sloshed* which means no less drunk.

slot to kill.

Probably from the imagery of piercing:

> If the ragheads had me tied down naked and were sharpening their knives, I'd do whatever I could to provoke them into slotting me. (McNab, 1993—the *ragheads* were the Iraqis)

Whence *slotted*, killed:

> It would be a pity for him to get slotted after you guys have gone to so much trouble. (M. Smith, 1993)

slow stupid.

Mainly educational jargon of children, but also of adults of low mental capacity. Also as *slow upstairs*:

> He's the Irish version of a street hood, very good with weapons but a little slow upstairs. (Clancy, 1987, repeating a common but fallacious myth concerning the intelligence of the Irish)

slowdown *American* a deliberate failure to do work for which you are being paid.

A version of the British GO SLOW, a bargaining tool in a labour dispute, especially when, as for Federal employees, striking may be illegal:

> ... air controllers or postal workers staged 'slowdowns'. (*Daily Telegraph*, August 1981 —for the controllers the *slowdown* shortly became a full stop)

slug[1] a bullet.

In the 17th century, when the usage arose, bullets had much the same shape and colour as the gastropod:

> ... felt that a .38 slug could save wasting a lot of time and the taxpayer's money. (Allbeury, 1976)

To *get a slug* is to be killed or wounded by a bullet but *slugged* means hit by any agency including a fist or an excess of alcohol.

slug[2] a quantity of spirituous intoxicant.

Probably punning on SHOT[2] rather than from the rare meaning 'to swallow'. Usually in the cliché *a slug of whisky*.

slugged drunk.

From the hitting, the swallowing, or the measure?

sluice[1] to copulate with.

Literally, to flush:

> ... she has been sluic'd in's absence,
> And his pond fish'd by his next neighbour.
> (Shakespeare, *Winter's Tale*)

This may well be a spurious entry based on a single metaphorical use which neither Dr Johnson nor Sir James Murray recognized but still more worthy of notice than the American *sluice*, to shoot eagles from a helicopter.

sluice[2] a lavatory.

From the controlled flow of water:

> He's in the sluice. (Bradbury, 1959—he was in a lavatory and not a mill-stream)

slumber room a morgue.

American funeral jargon, using the common 'sleeping' imagery:

> Lavish slumber rooms where the deceased receives visitors for some days before the funeral. (J. Mitford, 1963)

A *slumber cot* or *box* is a coffin and a *slumber robe*, a shroud.

slush bribery.

Originally, 'a soft mixture of grease or oil' (WNCD). A *slush fund* is a corporate bribery budget:

> A non-existent British Leyland 'slush' fund... (*Private Eye*, May 1981)

smack illegal heroin.

A corruption of the Yiddish *schmeck*, to sniff, rather than a derivation from the nickname (and physical impact) of the bandleader James Fletcher Henderson (1898–1952):

> 'Hey, Johnny, you want smack?' (Simon, 1979)

small folk the fairies.

From their stature in the days when they were real to countryfolk and, with their vicious natures, not to be trifled with or talked about directly. Also as *small men*:

> The small men. I mean the pixies (Mortimer, 1895)

and as *small people*:

> The small people are believed by some to be the spirits of the people who inhabited Cornwall many thousands of years ago. (Hunt, 1865)

Mainly western English.

smallest room the lavatory.

Even if, by geometric computation, it isn't:

> **smallest room, the** The bathroom; restroom. *A facetious euphem.* (DAS, in an interesting case of defining one euphemism by two others)

smashed drunk or under the influence of illegal narcotics.

Your consciousness, if nothing else, is destroyed by the substance consumed:

> I was smashed last night. Some of the guys at this party were on methedrine with their acid. (Deighton, 1972)

The obsolete *smash the teapot* was to resume regular drinking of alcohol after a period of abstinence.

smear[1] to bribe.

Literally, to spread, whence to spread largesse:

> A little smearing of the right palm... (Longstreet, 1956—he didn't mean to imply that the left palm wouldn't have served an equal purpose)

The American *schmear* comes from the German *schmieren* via Yiddish in the same sense, and also occurs as a noun:

> I got the feeling that a schmear changed hands somewhere along the way. (Sanders, 1977)

smear[2] a test for cervical cancer.

A cervical sample is taken for laboratory analysis; an abnormal *smear* indicates that there might be pre-cancerous cells present:

Course I did ask once when I went to the family planning for a smear. Well, you wonder if all is well. (Lochhead, 1985)

smear[3] to attempt to bring into disrepute.

You hope the dirty stain will remain:

You'll have to do better than try and smear me. (Crisp, 1982, of such an attempt)

smear out to kill.

A variant of WIPE OFF:

The opposition had twice tried to smear me out. (Hall, 1969)

smeared *American* drunk or under the influence of illegal narcotics.

Using the same imagery as the slang *blotto*?

smell of to be tainted with.

When the supposed disability is taboo. Thus to *smell of the counting-house* was, for the landed gentry, to be contaminated by trade:

If she thought that any of her newcomers smelt of the counting-house, she would tell her friends 'Have nothing to do with them.' (Bence-Jones, 1987—if the Protestants in southern Ireland had been as commercially-minded as their counterparts in the north, they might better have survived the loss of their farms which the tenants had the right to buy with government aid in the late 19th century)

smell the stuff *American* to sniff cocaine as an addict.

Of a general propensity and not a single experiment.

smoke to inhale (narcotics) illegally.

I am assuming that inhaling and expelling the smoke from tobacco is no longer taboo. Drug users usually *smoke* marijuana:

Clinton confessed that he had smoked marijuana... at Oxford in the late '60s. (*Time*, 20 April 1992)

A whole lot of girls, real sweet chicks. And we'll be smoking too. (Bradbury, 1965)

The smoke may be opium:

There isn't much record he went for tea-sticks or the smoke. (Longstreet, 1956)

Prior to the 17th century you *smoked* only if you were cremated, burnt to death, or went to hell. The American *smoke*, a black person, was derogatory and may be obsolete:

Smokes and a white gal... Lousy. Crummy. (Chandler, 1939)

smoke it to kill yourself.

From putting a handgun in your mouth:

I hear some detective from West L.A. smoked it. (Wambaugh, 1983, of a suicide)

smoke screen a police radar trap for motorists.

An American Citizens' Band warning to approaching drivers obliquely punning on SMOKEY.

smoker *south-western English, ?obsolete* the devil.

The fire, brimstone, etc.

smokey a policeman.

The DAS suggests that it comes from *Smokey the Bear*, the US Forestry Service symbol, and see also BEAR[2]:

The only enemies are the weather and the occasional lawman, known as 'Smokey Bear'. (*Daily Telegraph*, 4 March 1995, writing about American truckers)

In Citizens' Band slang there are many compounds—*smokey beaver*, a policewoman; *smokey on four legs*, a policeman on horseback; *smokey with camera*, police with radar; *smokey on rubber*, police moving in a car; and *smokey with ears*, police with Citizens' Band radio; see also CBSLD.

smoking gun conclusive evidence of guilt.

From the emission from the barrel immediately after it has been fired:

... the tape is a 'smoking gun', that is, in police and prosecutorial slang, direct evidence of criminal guilt. (Colodny and Gettlin, 1991, of a White House tape of June 1973)

smooth to distort (published accounts).

You conceal or try to even out fluctuations by carrying forward exceptional movements up or down. This soothes stockholders and financial commentators.

smother to copulate.

Of a male, from his supposed attitude on the female:

I've smothered in too many hall bedrooms. (Chandler, 1939)

smut house *American* a place where pornographic programmes are screened.

Smut as in *smutty jokes* rather than a boiler-room:

He had never watched queer movies before, and after this night he had no plans to watch another one. This was his third such smut house in the last ninety minutes. (Grisham, 1992)

snaffle to steal.

Originally, to saunter, as many chance thieves do:

He cud snaffle the raisins an' currins away. (Bagnall, 1852)

snake pit a lunatic asylum.

From one of the common delusions of the mentally ill:

The old man was always threatening to stash her away in a snake-pit. (Macdonald, 1971)

snake ranch a brothel.

From the dangers met in such a place and punning on American slang *snake*, the penis.

snapper an ampoule of amyl nitrate.

The drug is used in the treatment of heart disorders but, as it is popularly supposed in America to be an aphrodisiac, it commands a strong illegal market. You use it by snapping the top off an ampoule, and sniffing:

> ... a box of snappers in plain view on a dresser top. (Sanders, 1977)

snatch[1] a single act of copulation.

Usually extramaritally. The derivation might be from any one of several standard English meanings of *snatch*—a snare, an entanglement, a hasty meal, a sudden jerk—or the slang SNATCH[2], the vagina. Shakespeare could have been using it in either sexual sense:

> ... it seems some certain snatch or so
> Would serve your turns
> (*Titus Andronicus*)

but there is no equivocation in:

> I could not abide marriage, but as a rambler I took a snatch when I could get it. (Burton, 1621)

The modern use survives in America more than in the British Isles.

snatch[2] the vagina.

Perhaps from the *snatch*, or portion of hair:

> ... if the number of the vaginas ... were lined up orifice to orifice, there would be a snatch long enough ... (Styron, 1976)

snatch[3] to kidnap or steal.

From the seizing in either case:

> Snatching Steven was going to be one big piece of chocolate cake. (Collins, 1981, of kidnapping)

The snatch is the particular execution of the crime.

snatch[4] to arrest (a criminal).

Either singly, or by taking a ringleader from a mob. Whence a *snatch squad* of police etc., whose duty it is to make such mob arrests.

sneak to steal.

In standard usage, to move furtively (whence too the children's use, to inform against). It was applied particularly in many phrases to 19th-century robbery from houses:

> He saw Seth Thimaltwig snake hawf a pahnd o' fresh butter. (Treddlehoyle, 1893)

Today we use only the tautological *sneak thief*.

sneezer *American* a prison.

Possibly a corruption of FREEZER—which may have come into the language because the typist couldn't read Chandler's writing:

> ... tossed in the sneezer by some prowl car boys. (Chandler, 1953)

sniff to inhale narcotics or stimulants illegally.

Usually cocaine:

> Department wives who drink, analysts who are screwing their secretary, translators who sniff. (Deighton, 1994, of possible blackmail targets—it is to be hoped that the analysts had the services of more than one secretary)

Also of glue by juveniles:

> ... an increasing number of children ... have adopted glue-sniffing. (*The Practitioner*, 1977, quoted in Hudson, 1978)

sniff out to kill.

Perhaps a corruption of *snuff out* (see SNUFF (OUT)) because to *sniff out* is normally no more than to detect:

> ... before some busybody at the top sniffs out Sniffers. (Manning, 1977—of killing and not detection but Miss Manning liked to play with words)

To *take a long sniff*, your last deep breath is more logical:

> Half a dozen horsemen galloped past, firing six-guns into the air. The young cowboy said, 'Seems like you might be taking yourself a long deep sniff.' (Deighton, 1972)

snifter a drink of spirituous intoxicants.

Literally, a sniff, whence a small portion of brandy etc. offered so that the aroma can be tested. Now of any intoxicant:

> He turned, snifter in hand. (Wodehouse, 1934)

snip a vasectomy.

Medical jargon, punning on the surgical cut and the easy way to earn a fee. *Snib* and *snick* are English dialect words for the castration of lambs and both are likely to come into use of vasectomy.

snort[1] a drink of spirituous intoxicant.

Perhaps it makes you exhale noisily:

> There's a pint in the glove compartment. Want a snort? (Chandler, 1958)

snort[2] to ingest an illegal narcotic.

Usually by sniffing and a *snort* is a portion:

> 'I'm not worried about it,' she said with a half-smile as she casually spooned two snorts. (Robbins, 1981)

snout a police informer.

The nose of the PIG:

> I know all about snouts. And I didn't have to pay for this. (James, 1986, of police information)

snow[1] cocaine.

In its crystalline or powdered form, from the colour and coldness:

> Not all jazz-players smoke marijuana or opium, or sniff snow. (Longstreet, 1956)

Whence many derivatives, such as *snowball*, a quantity of cocaine; *snowbird*, an addict who takes illegal narcotics; *snowed in, under* or *up*, under the influence of narcotics; *snow-storm*, a gathering where cocaine is ingested illegally.

snow[2] deliberately to obfuscate (an issue) or deceive (a person).

As the landscape may be obscured by a snowfall. To *snow* an issue is to produce masses of print-outs and other documentation which make it difficult for the recipient to pick out and understand the relevant points. The person subject to such treatment may be *snowed*:

> Little job? Don't let them snow you, old friend. (Price, 1970)

The operation is known as a *snow-job*:

> A lie, a cover-up, a snow-job was fatal. (Allbeury, 1980)

snowdrop a military policeman.

From the white spats worn by American Second World War police:

> '... we've even put the 787th Military Police Company into the Junior Constitutional Club.' ... 'Your snowdrops, you mean.' (Deighton, 1982)

snowing down south (it's) *American, obsolete* the hem of your petticoat is showing.

An oblique warning to the wearer. Petticoats are usually white.

snuff (out) to kill.

From extinguishing a candle:

> You mean you make sure he doesn't go off like a mad dog, snuffing people left and right. (van Lustbaden, 1983)

Snuff out is the older version:

> I'd have snuffed out every life in India. (Fraser, 1975)

To *snuff it* is to die:

> An' Ray Tuck's been running Lippy's errands —or was, until Lippy snuffed it. (Price, 1982)

snug drunk.

Literally, comfortable. Many English inns have *snugs*, shortened form for *snug bars* but any pun is, I think, unconscious. British estate agents also use *snug* to gloss over inconveniently restricted space:

> Now he knew 'snug' meant tiny. (Theroux, 1974)

so *?obsolete* pregnant.

From the meaning 'in such a situation':

> A euphemism for pregnant ... 'Mrs Brown is so.' (EDD)

Less often as *so-and-so*, which may derive from the obsolete meaning 'in poor health'.

so-and-so a mild insult.

Each *so* being a substitute for an abusive epithet.

so-so[1] unwell.

Literally, mediocre, and usually how you describe yourself rather than another.

so-so[2] drunk.

The common imagery of being unwell. Of another when he is drunk, or of yourself if suffering from a hangover.

so-so[3] menstruating.

Again, the *unwell* imagery. DSUE also gives *so*, but I have not heard or read it elsewhere.

soak a drunkard.

In modern use, and *soaked* is drunk. Formerly, and logically to *soak* was to drink excessively without showing symptoms of drunkenness:

> A 'slug for the drink' is a man who soaks and never succumbs. (Douglas, 1901)

social disease a venereal disease.

Mainly 19th-century usage, which you might contract from the *social evil*, prostitution, which, especially in 19th-century England, prospered as a result of delayed marriage by men and the fear of repeated pregnancy by women during an era of falling infant mortality:

> 'He has contracted a social disease, which makes it impossible that he marry.' 'You mean he's got a dose of clap?' (Fraser, 1970, writing in 19th-century style)

A sufferer might also have a *social infection*:

> ... contracting certain indelicate social infections from—hem hem—female camp-followers. (Fraser, 1975, again in 19th-century style)

social evil see SOCIAL DISEASE.

social justice an imprecise dogma based on envy rather than the rule of law.

Its appeal lies in its suggestion that wealth should be shared rather than earned:

> The robbery of the rich is called social justice. (Roberts, 1951)

We must always beware any political programme which incorporates *justice* in its title, and even more any institution which seeks to impose such dogma, such as Haverford College, in Pennsylvania, where:

> ... students must complete a 'Social Justice Requirement' in order to graduate. This means taking courses in such things as Feminist Political Theory. (*Daily Telegraph*, 23 February 1991)

social ownership expropriation.

The 1986 version of *nationalization* (see NATIONALIZE) when it had become unpopular with the electorate:

> ... the substitution of phrases like 'social ownership' for nasty brutal words like 'nationalization'. (*Daily Telegraph*, August 1986)

social security the payment of money by the state to the indigent.

It was the morning most people went to collect their social security. (L. Thomas, 1979)

The latest and most durable of a line of evasions by which we seek to mask the plight of, and charity to, the poor.

social worker a public employee helping the poor, ineffective, sick, or old.

Not an ant, or similarly organized creature. The use does not mean, though it may imply, that work such as manufacturing or compiling a dictionary is anti-social.

sodomite a bugger.

Sodom, the Biblical Dead Sea city from which this derives, had a reputation for evil:

She once made a fearful gaffe about Sodomites, mixing them up with Dolomites. (N. Mitford, 1949)

Sodomy is buggery:

It often led to downright sodomy. (Harris, 1925—*it* is masturbation)

The shortened form *sod* is often a mild insult.

soft[1] of low intelligence.

Shortened form for *soft in the head*:

She's saft at best, and something lazy. (Burns, 1785)

soft[2] inflicting less harm than an alternative.

The alternative to HARD in pornography, illegal narcotics, etc. A *soft drink* is non-alcoholic and will harm your teeth more than your liver. For the soldiers, a *soft target* is one which is easily attacked. A *soft option* is a simple solution, with overtones of laziness or cowardice.

soft commission a bribe.

Paid in addition to any overt commission for the introduction of business:

It is the first time in Imro's history for a breach of rules on 'soft commissions'. (*Daily Telegraph*, 25 June 1994—the British Investment Management Regulatory Authority fined Unit Trust managers who had accepted £50,000 of travel expenses from stockbrokers in return for placing business with them)

soft soap flattery.

Originally, what is now generally known by a word borrowed from Hindi, shampoo. See also SADDLE SOAP.

soil[1] *obsolete* extramarital copulation.

Of either sex, from the sullying of reputation:

Who is as free from touch or soil with her
As she from one ungot.
(Shakespeare, *Measure for Measure*)

soil[2] human excrement.

From the days of the earth closet or deep-trench latrine, when it had regularly to be removed and, if by night, became NIGHTSOIL. To *soil yourself, your pants*, etc. is to defecate involuntarily:

He sometimes soiled his pants here in the park. (McCarthy, 1963)

soixante-neuf simultaneous fellatio and cunnilingus.

From the reversible numbers *6* and *9*, and the position adopted by the participants. This French form is normal in the British Isles—I speak etymologically—with *six-à-neuf* being rare:

... six-à-neuf meaning a slightly contortive sexual diversion. (Jennings, 1965)

The American usage, more direct or less Francophone, is *sixty-nine*:

... every act from masturbation to 'sixty-nine' was indulged in. (ibid.)

The participants may also be described as *sixty-nining*.

solace extramarital copulation.

Supposedly consolation for the absence of a spouse:

'... seeking that well-known solace.' 'From two men at the one time?' (Keneally, 1979)

Usually of women.

soldier *American* a mobster.

At the boss's orders, he threatens, assaults, or kills:

I lend you a couple of soldiers—you frighten the crap outta number one on the list. (Collins, 1981)

solicit to offer sexual services for money.

Literally, to request or entreat of anything, for example the British *solicitor*, who pleads for you in court, or the American *solicitor*, who is barred from trying to sell you anything in your place of business except between 8.30 and 9.00 a.m. on Monday mornings. Legal jargon of prostitution where the prostitute does the *soliciting*:

She was soliciting to cover her air fare. (Gardner, 1983)

The person who does the asking is a *solicitor* rather than a *solicitress* or *solicitrix*, and this leads to much ribald male humour, except among British lawyers. Of homosexuals, it can refer to either party:

The defendant was accused of having improperly solicited another man in a public lavatory. (Boyle, 1979, of Burgess)

solicitor general the penis.

Punning on male promiscuity and the British high office of state.

solid waste human excrement.

Civil engineering jargon. The term does not include empty tins or potato peelings.

solidarity participation in a strike on behalf of others.

In the 19th century it meant, as in modern Poland, the coming together of workers in a single bargaining unit. Now trade union jargon for support given to other unionists engaged in a dispute with a third party.

solitaire *American* suicide.

The game that can be played by one person.

solitary sex masturbation.

Not hermaphroditism. Also as the *solitary sin* or *vice*:

> Carter had 'seen young unmarried women, of the middle-class of society reduced, by the constant use of the speculum, to the mental and moral condition of prostitutes; seeking to give themselves the same indulgence by the practice of solitary vice'. (Pearsall, 1969, quoting from 1853)

something an intoxicant.

Usually in the cliché *Would you like a little something? Something damp* is more explicit and usually indicates whisky. *Something short* is any spirit:

> She pulled out a bottle of gin, asking me if I would have a drop of something short. (Mayhew, 1862)

Also as *something moist*:

> I doubt if he were quite as fully sensible of that gentleman's merits under arid conditions, as when something moist was going. (C. Dickens, 1861)

something for the weekend a contraceptive sheath.

Or packet, from the days when the main purveyors were barbers and men had their hair cut more often:

> Condoms weren't called condoms, the euphemism was 'something for the weekend'. (Monkhouse, 1993)

something on you a damaging piece of knowledge about you.

Not the clothes you are wearing. In this sense, *on* means 'against':

> He's got something on her and she's afraid of him. (Chandler, 1958)

something-something a mild expletive.

You invite the hearer to choose his own profanities.

somewhere in... the location is secret.

Wartime usage to prevent identification of the location of specific troops:

> As it was, most already had their soldier 'somewhere in France'—that delightful euphemism of the censors. (Horne, 1969)

Soldiers used great ingenuity to inform their families where they were, although the one who told his wife he 'bumped into Dicky Grenville the other day' during the illegal British occupation of the Azores was punished by court-martial, the censor also recalling that 'At Flores, in the Azores, Sir Richard Grenville lay'.

somewhere where he can be looked after off our hands.

Used of aged, ill, or burdensome dependants, implying that they, not you, will benefit from the care of paid strangers:

> Get him out of here as soon as possible, to somewhere where he can be looked after. (Bradbury, 1959)

son et lumière the clandestine recording of an indiscreet act.

Espionage jargon, from the public display which involves lights and music:

> ...a British MP been leaping into bed with a Czech agent, in Prague itself, with full sound and camera coverage, *son et lumière* as the professionals say. (Lyall, 1980)

son of a bachelor a bastard.

Born outside wedlock, as was the more common *son of a bitch*, often abbreviated to *SOB*, where the mother was also insulted. Those terms are now fairly mild insults without any implication as to parentage. For a dissertation on *son of a gun* see GUNNER'S DAUGHTER.

song and dance a male homosexual.

Rhyming slang for 'nance' and perhaps punning on the supposed tastes of men who dance for a living.

sop a drunkard.

Literally, something dipped in liquid or the liquid in which it is dipped. I suspect it is confused with the common SOT.

sore a carcinoma.

The symptom, in this case an ulcer, is used for the dread affliction:

> Her own mother had died of a 'sore'. (Mann, 1902)

sot a drunkard.

Acting like a fool, perhaps. To *sot* was to become drunk:

> Drover blades, who drink and sot. (Nicholson, 1814)

sought after expensive.

Estate agents' puff, when they want to imply a buyer will have plenty of competition. But any property, however humble, is likely to be sought after, at a price, if advertised.

soup illegal explosive materials.

Possibly from *jelly*, a shortened form of gelignite.

souper *Irish* a Roman Catholic seeking to convert or converted to Protestantism.

In the periods of 19th-century food shortage, even before the famines, Irish Protestant clergy fed their own flocks, including converts from Roman Catholicism:

> Proselytizers, or soupers, from their offering soup to starving people... (Carleton, 1836)

Those who swallowed the bait and the soup were also *soupers*:

> I'll turn souper this day for the male. (Barlow, 1892—a *male* was a meal)

I heard a Protestant so describe his ancestor in West Cork in January 1994.

souse a drunkard.

The common culinary imagery, this time from soaking in vinegar or the like:

> That much would just get a real souse started. (Chandler, 1953)

Soused means 'drunk':

> I could see that mother was getting soused. (L. Armstrong, 1955)

south (the) the poorer or less industrialized countries.

From the geographical location of many of them relative to Europe and North America. This 1982 version is a happy break from the *developing* theme but cannot expect much currency in South Africa or the Antipodes.

South Chelsea Battersea.

An example of what happens in most big cities where a fashionable area is bounded by an unfashionable:

> Battersea... South Chelsea, the snobs call it. (Theroux, 1982—and perhaps, too, the estate agents)

Southern Comfort masturbation.

Punning on the brand of whisky:

> I usually wind up giving myself another kind of Southern Comfort, you know what I mean? (Lodge, 1980)

souvenir[1] a bastard.

Certainly a lasting memory for the mother:

> I expect in some cases [the troops] had left other souvenirs which would either be a blessing or a curse to the ladies concerned. (Richards, 1933)

souvenir[2] a turd.

A reminder of a canine presence:

> ... not many people looking up, preferring instead to study the sidewalks for souvenirs of its vast dog population. (McBain, 1981)

souvenir hunting looting.

Second World War army usage, now used by the thieves to refer to robbery after a natural disaster or riot.

sow your wild oats to behave wildly or irresponsibly.

With extravagance or promiscuous seminal distribution like the persistent weed *Avena fatua*, or wild oat:

> We all have to sow our wild oats at some time or another. (Sharpe, 1974)

Standard English.

sozzled drunk.

Originally, splashed:

> 'We were all rather sozzled that night.' 'I wonder if he was drunk when he killed himself.' (Murdoch, 1977)

Also as *half-sozzled*.

space a grave.

American funeral jargon:

> As for other euphemisms... 'space' for 'grave'. (J. Mitford, 1963)
>
> The 'space and bronze deal' as it is called by the door-to-door sales specialists. (ibid.—you buy your plot and casket in advance)

spaced out *American* under the influence of illegal narcotics.

From the floating sensation, especially after taking a hallucinogen:

> The doctor arrived, but to our dismay he was totally incompetent. I mean, he was spaced out on drugs or something. (Peck, 1987)

Spaced out may also be used to describe an abnormal physical or mental state not induced by narcotics:

> She looked sick, exhausted, depressed, and spaced out. Someone had slipped her a mickey. (Greeley, 1986)

Less often as *spaced*.

spade a black.

Derogatory white use, from the colour of the playing card, and now regarded as extremely offensive:

> And if that little spade said something different, she's lying. (Macdonald, 1976)

Spanish fly a supposed aphrodisiac.

No dipterous insect but the cantharis beetle which, by inducing vaginal itch, is said to be an aphrodisiac in women:

> ... they'd put women in a barrel ready mixed with goddam Spanish fly to make 'em saucy. (Keneally, 1979)

Spanish gout syphilis.

British sailors thought that Spanish girls must have infected them, if not French, Italians, or other inferior breeds.

Spanish practices the regular use of dishonest devices by employees fraudulently to reduce hours of attendance and increase pay.

Such as those exposed in London's Fleet Street in the 1980s; here the scandal was not so much that the employees customarily and systematically cheated their employers but that the employers tolerated it for so long and the tax authorities turned a blind eye to it:

> A year ago, as well as the overmanning, the exploitative 'Spanish practices' and the interrupted production... (*Times*, January 1987)

(Because of long acclimatization to the absolute protection enjoyed in Great Britain by trade unions against damages in a civil action, British-born managers tended to acquiesce in the continuation of some notorious abuses of the power thus given. Eddie Shah, Robert Maxwell, 'Tiny' Rowland, Rupert Murdoch, and Conrad Black did not share the same inhibitions.)

Spanish tummy diarrhoea.

The British equivalent of the American *touristas* (see TOURISTAS (THE)).

spare tyre obesity at the waistline.

Usually of a male, from the roll of fat overhanging his belt:

> I longed to melt away that spare tyre before it was too late. (Matthew, 1983)

In America sometimes as *rubber tire*.

speak to to propose marriage with.

This is a rare reminder of the 19th-century reticence about marriage:

> When Jamie 'spoke to' Janet Carson, who told her people at once, having no opposition to expect... (Strain, 1900)

The Scottish *speak for* and *speak till* are obsolete.

speak with forked tongue to dissimulate.

> Owners and players act as if there were no fans, as if the fans were a myth invented by sportswriters for days when there is no... multi-millionaire owner to scold for speaking with forked tongue. (*Guardian*, 11 August 1994)

Serpents again get unfair metaphoric treatment in this phrase which, if not used by American Indians, was at least attributed to them in numerous Westerns.

speakeasy an establishment selling intoxicants without a licence.

Mainly American Prohibition use, although noted in 1889 (OED). You had to lower your voice to avoid advertising the transaction.

special[1] ruthless and extra-legal.

Usually in phrases concerned with espionage or repression; see the various entries which follow SPECIAL[4].

special[2] mentally or physically inferior.

A *special pupil* going to a *special school* may be suffering from a mental or physical abnormality; *special games* are for those with restricted mobility. *Special* is seldom used in this sense of those of superior attainments, although they may be equally distinctive.

special[3] *British* a part-time policeman.

Shortened form of *special constable* but without political overtones in England, Wales, or Scotland. The Northern Irish *B Specials* were a para-military force which supported Protestant ascendancy.

special[4] nuclear.

Thus nuclear weapons are known to servicemen as *special stores* or *special weapons*:

> ...a considerable number had been in Special Weapons Stores overrun by the offensive. (Hackett, 1978, of nuclear warheads)

special action the rounding up and murdering of Jews by Nazis.

> ...the incredible *numbers* involved in these Special Actions... These Jews, they come on and on. (Styron, 1976)

Special Branch the branch of the British police force concerned with subversion.

> Here you call your political police the Special Branch, because you English are not so direct in these matters. (Deighton, 1978)

special court a tribunal.

Established by the executive of an autocracy to overrule and supersede an independent judiciary; the Nazi *Sondergericht* of March 1933 was a good example.

special detachment an army or police unit used to terrorize dissidents etc.

As might be expected, the Nazis perfected the concept by, in Poland, so naming a police force of Jews who worked under control of the SS to harass other Jews:

> Even the Jews of the Special Detachment were reluctant to pick the children up. (Styron, 1976, writing of Poland in the Second World War)

special duty an extra-legal or illegal act performed under authority.

The phrase is now widely used, and not only by the Communists, but again the concept came from the Nazis whose anti-Jewish street bands were called *Einsatzgruppen*:

> 'Special duty groups' is a close translation. But the amorphous word 'Einsatz' had another shade of meaning... knightliness. (Keneally, 1982—the squads were usually recruited from the *Sicherheitsdienst*, or *SD*)

special education a prison regime calculated to kill, cow, or indoctrinate dissidents.

Mainly Communist practice and usage.

special fuzz *British* police concerned with controlling subversion.

This is used only of those who come directly in touch with the public:

A hairy hitchhiking student had only recently complained to him that the special fuzz were becoming hard to pinpoint. (Price, 1971)

special investigation unit malefactors for political purposes.

... the work of the Special Investigations Unit (Plumbers). (Colodny and Gettlin, 1991, writing about Watergate)

special police the police seconded to the control of subversion and civil disorder.

The word *police* seeks to imply a degree of fairness and solicitude for the feelings of the public which seldom characterizes such bodies. A London variant is the *special patrol group*, a riot squad with an occasional penchant for individual weaponry.

special regime a treatment intended to kill or destroy the health of a prisoner.

A prisoner on *special regime* in the former Soviet Union would be required to do heavy manual work on 800 calories of food daily, so long as he survived. The other *regimes* are *general* (the mildest), *intensified*, and *strict*.

special services and investigations the covert monitoring of lawful citizens.

A New York police section was so described:

Caulfield had been a member of the NYPD and its undercover unit, the Bureau of Special Services and Investigations (BUSSI) ... known for its ability to penetrate and keep track of left-wing and black groups. (Colodny and Gettlin, 1991)

special squad a unit established by a government to murder or harass opponents.

A phenomenon of American autocracies where the courts have retained some independence of the executive. For the Nazis the *Sonderkommandos* were used for rounding up and oppressing Jews:

... the Sonderkommandos were unleashed to advance with an appropriate sense of racial history and professional detachment into the old Judaic ghettos. (Keneally, 1982, of Poland in the Second World War)

special treatment the torture and killing of your opponents.

This was the treatment accorded by the Nazis to Jews, Gypsies, Slavs, and other assorted enemies of the state:

... what Sonderbehandlung means, that though it says *Special Treatment*, it means pyramids of cyanosed corpses. (Keneally, 1982)

A 1983 British Airways advertisement in Germany was a literal translation of 'You fly frequently. Don't you deserve a little special treatment?' Many people felt the use of *Sonderbehandlung* was a Freudian slip.

specimen a sample of urine.

Medical jargon, sometimes confusing to patients:

He should show his *specimen* privately to his family doctor. (T. Harris, 1988, of urine and not some physical attribute)

speed an illegal drug that excites.

Usually amphetamine despite the fact that a *speedball* is a mixture of narcotics. To *speed*, punning on driving a car above a legal limit, is to take such a drug:

They were speeding and tripping at the same time. (Deighton, 1972)

spend to ejaculate (semen).

Usually in copulation:

Spending his manly marrow in her arms (Shakespeare, *All's Well*)

or, in modern use:

I could after the first orgasm go on indefinitely without spending again. (Harris, 1925)

spend a penny to urinate.

Normally referring to urination by either sex, although only females were required to produce the coin formerly needed to operate the lock of a British public lavatory turnstile or cubicle. Occasionally denoting defecation, although in that case men had to *spend a penny* too.

spend more time with your family to be dismissed from employment.

Often of a senior employee who has been peremptorily dismissed:

... he has not resigned ... He will be preparing for the trial and 'would like to spend some more time with [his] family'. (*Daily Telegraph*, 2 March 1995, referring to someone accused of false accounting while employed by one company; he had been working for another company of which the head was himself facing criminal proceedings for rape)

spend the night with to copulate with casually.

Of either sex usually in a transient relationship:

She wanted me to go and spend the night with her. (L. Armstrong, 1955)

There is also a legal presumption that a male and a female, if not married to each other, cannot spend a night in each other's company without copulating.

sphere of influence a foreign territory which can be controlled politically, militarily, and economically without interference from a third party.

The end product of the colonial cartel, such as the 1907 Anglo-Russian Agreement concerning Persia or of the deals done at Yalta:

> In a stroke, we clear the British from India, and extend your majesty's imperial influence from the North Cape to the isle of Ceylon. (Fraser, 1973, of 19th-century Tsarist expansionism)

spicy salacious.

Figuratively highly-flavoured, whence HOT[1]:

> ...she would be talking about a sexual episode —the man in the Norman Mailer story sodomizing his girlfriend, for example—and she would call it 'spicy'. (Theroux, 1989)

spiflicated drunk.

Originally, beaten up, which is how you may feel in either event. Occasionally in the slang shortened form *spiffed*.

spike[1] to adulterate (an intoxicant).

Perhaps from *spiking*, or destroying, a gun by driving a metal object through the touch-hole, or possibly merely from the practice of inserting a hot piece of metal in the glass of fluid. It may denote the addition of alcohol to a non-alcoholic drink:

> When I complained that it was my first day and I was afraid to drink, Mary reluctantly bought me an orange juice and then spiked it with vodka when my back was turned. (Bolger, 1991)

Also the addition of illegal narcotics to a drink:

> A couple of hours later Beano spiked their tequilas with angel dust, which was his idea of a good New Year's joke. (O'Connor, 1991)

It is normally done without the knowledge or consent of the victim.

spike[2] to reject for publication.

Editorial jargon, from the metal spike on which rejects were filed:

> The chances are that no sub-editor is going to spike the story. (Deighton, 1982)

spike[3] *American* a hypodermic needle.

A specific sharp-pointed piece of metal:

> It was for the spike he held out toward her in his open hand... Her eyes never left the needle, or the loving smile her face. (Crews, 1990)

spill to give gratuitous information.

Normally to the police and a shortened form of *spill the beans*, to pass on damaging information:

> If Hench shot somebody, she would have some idea... She would spill if she had. (Chandler, 1943)

spill yourself to ejaculate semen.

In masturbation or copulation:

> Ulf who is nothing and has no career had spilled himself on their precious sheets. (Seymour, 1980—they had used the girl's parents' bed)

spin doctor *American* a person employed to monitor, correct, suppress, and interpret pronouncements by a candidate for election.

Spinning from the bias and *doctoring* from mending. First noted in print in the *New York Times* in October 1984, after a debate between Ronald Reagan and Walter Mondale (ODNW), and brought into wider prominence during the Clinton presidential campaign. Also as *spinner*.

spirits[1] a man's semen.

In obsolete use, the essence of maleness, whence the symbol of courage:

> Much use of Venus doth dim the sight... The cause of dimness of sight is the expense of spirits. (Bacon, 1627)

The modern SPUNK has the same duality of meaning.

spirits[2] a spirituous intoxicant.

Literally, no more than any liquid in the form of a distillation or essence:

> He gave me a piece of an honey-comb, and a little bottle of Spirits. (Bunyan, 1684)

Now generically of whisky, gin, rum, vodka, etc.:

> 'Spirits don't seem to agree with you.' 'They differed from me sharply this time.' (Amis, 1978)

spit feathers to have a hangover.

From the dryness of the mouth. I am sorry that this useful and graphic phrase seems to have passed out of use.

splash to crash into the sea.

Of an aircraft and also used transitively:

> So, if Bronco... does have to splash the inbound druggie, nobody'll know about it. (Clancy, 1989—Bronco was a fighter pilot)

splash your boots to urinate.

Usually of a male, but not necessarily out of doors or even wetting your footwear:

> I was up splashing my boots. (Theroux, 1971, of urination)

splice the mainbrace to drink intoxicants.

The mainbrace was the rope which held the mainsail in position and a vessel was in peril if it broke. In rough weather *splicing*, or mending by rejoining it, was a hazardous operation and the seamen deserved a reward when it was done. For them, it was a double tot of rum, as it still is to celebrate some national event in the navy, but for the rest of us usually mundane whisky or gin and tonic.

split to copulate with.

Of a male, with obvious imagery:

> If you want to split the black oak... then you'll find it great—down Macpherson Road or among the taxi dancers at the Great World. (Barber, 1981, of copulating extramaritally with a black woman)

In obsolete use to *split a woman's shape* was to impregnate her. Whence too *split-mutton*, the penis, and other vulgarities.

split on to inform against.

Some children's use, probably from the sense of dividing:

> It's the meanest thing out—that splitting on a pal. (Trollope, 1885)

spoken for retained as a sexual mistress.

Literally, engaged to be married:

> You can often spot these spoken-for girls in the public trucks, sitting and smiling a lovely white smile. (Theroux, 1992—of the South Pacific where teeth of the indigenous population decay early. French expatriate soldiers may provide their sexual mistresses with dentures, which they repossess to indicate ownership when they go on leave to their wives and families in France.)

sponge a habitual drunkard.

Punning perhaps on the soaking up of intoxicants and his willingness to accept free refills from others. The British *sponging-house* was not an inn but a temporary prison for debtors, where they were relieved, or *sponged*, of cash and valuables before passing into a long-stay debtors' prison.

sponsor an advertiser.

Originally, a godparent, whence one who supports a candidature or activity, but now standard usage of paying for broadcast advertisements by financing other productions:

> Sponsors didn't write the programmes any longer, but they did impose a firm control on the contents. (Bryson, 1994)

spoon with to caress heterosexually.

Perhaps from the phrase to *lie spoons*, to nestle closely with the convex side of one against the concave side of the other; or from the old Welsh custom of giving your sweetheart a suitably carved wooden spoon as a token of your interest. A 19th-century British use of *spooning* was of homosexual relationships between males:

> 'Spooning' between master and boy was a subject for cruel jest. (Pearsall, 1969, of 19th-century English boarding schools for boys)

sport (the) copulation.

Usually viewed as such by the male:

> He had some feeling of the sport; he knew the service. (Shakespeare, *Measure for Measure* and not of a player of real tennis)

In literary use it becomes the *amorous sport*, *sport for Jove*, etc. To *sport* is to copulate:

> Now let us sport us while we may. (Marvell, c. 1670)

In modern use, any *sporting* copulation is usually on a commercial basis. Thus a *sport-trap* is the brothel area of a town:

> Storyville became and stayed the biggest tourist and sport-trap in the nation. (Longstreet, 1956—and it so remained until 1917 when it was shut down to protect American servicemen from temptation and disease)

A *sporting-house* is a brothel:

> She was like a lot of sporting-house landladies I've known through life. (L. Armstrong, 1955, of a bawd)

There you might find *sporting* girls or women —not hockey-players—who inhabited the *sporting section* of the town, punning on the part of the newspaper given over to ball games etc.:

> You came to the sporting-section, the cathouses around 22nd street. (Longstreet, 1956)

sports medicine illegal drugs.

As administered to athletes in Communist East Germany:

> ...in order to win, everything possible must be done, and... sports medicine had its part to play. (*Sunday Telegraph*, 27 February 1994 —the use of such drugs was called *laufende Versüche*, or continuing experiments, and if the athlete cheated, that was acting *mit Abstrichen*, or with exceptions)

sportsman a gambler.

The modern equivalent of GAMESTER[2]. Usually of regular or spectacular punters only.

spot[1] a drink of a spirituous intoxicant.

I suppose a shortened form for a *spot* of whisky etc.:

> I think I could do with a spot. (E. Waugh, 1955)

spot[2] to kill.

From the entry mark of the bullet, perhaps, or a shortened form of to PUT ON THE SPOT:

> That's enough to spot a guy for. (Chandler, 1939)

spot[3] a tubercular infection.

Usually of pulmonary tuberculosis where there is a hole in the lung which appears as a spot on the x-ray plate. Whence the description of such disease as 'a spot on the lung'.

sprain your ankle to copulate with a man before marriage.

Usually in the past tense, especially where the woman is pregnant. British women also suffered similar injuries to their knees, elbows, and thighs —see BREAK YOUR ELBOW.

spread for to copulate with.

Of a woman, usually willing, once, and outside marriage. More explicitly to *spread your legs*:

> They must both be paid, cash on the barrel-head, before she would spread her legs (Monsarrat, 1978)

or vulgarly to *spread your twat*:

> Spreading that twat of yours for a cheap, chiseling quack doctor... (Styron, 1976)

spring to secure the release of (someone) from prison before the end of a sentence.

Of legal pardon or escape, and occasionally of bail before conviction, from the unexpected and positive action of a released coil:

> The proprietor knew how to 'spring' them, that is, get them out of jail. (L. Armstrong, 1955)

sprung slightly drunk.

Like a ship which leaks but hasn't sunk:

> How's a chap to get sprung, much less drunk? (Westall, 1885)

Half-sprung is no less drunk.

spunk a man's semen.

Originally, courage, and still so used in some innocent circles:

> ... a term Lady Maud found almost as offensive as Colonel Chapman's comment that she was full of spunk (Sharpe, 1975)

but for the less innocent:

> ... right off there, with my fresh spunk in her. (Keneally, 1979)

See also SPIRITS[1]. *Spunk* is occasionally used too of the vaginal sexual discharge. The obsolete Scottish *spunkie* was whisky:

> Spunkie ance to make us mellow. (Burns, in an undated letter)

spur of the moment passion unpremeditated extramarital copulation.

Not momentary anger or other forms of suffering:

> ... spur of the moment passion with a married woman. (*Daily Telegraph*, April 1980)

squash to kill.

Of humans, treating them as we do insects:

> 'At best? Two busted kneecaps.' 'And at worst?' 'They'll squash me.' (Sanders, 1980)

squashed *American* drunk.

From the way you feel, and not from drinking fruit squash.

squat[1] to defecate.

From the posture and perhaps too from the dialect meaning 'to squirt':

> The authorities were trying to teach the people not to squat behind their huts. (McCarthy, 1967)

For females, a *squat* may mean urination only. Some figurative use, as:

> ... the S2 hasn't told me squat about the enemy now facing me. (Coyle, 1987—an S2 is an American staff officer responsible for obtaining and disseminating information about the enemy)

squat[2] to occupy (a building or land) by trespass.

Squatters' rights is an English legal concept dating from the social and economic need in the Middle Ages to see land and buildings, vacated and ownerless through plague, brought back into productive use. The modern use originated in America. The verb is used both transitively:

> Hobo punks hop trains, squat abandoned buildings, collect welfare, and dumpster food (Esquire, January 1994)

and intransitively:

> She was working... to identify and locate people who are homeless or squatting in abandoned buildings. (*Philadelphia Inquirer*, 17 December 1989)

A *squat* is the action or the property in which it happens:

> ... they eventually discovered his body in some squat. (B. Forbes, 1989)

A *squatter* is someone who so acts:

> Squatters of empty, unused houses may be evicted after a summary hearing at which they cannot defend themselves and may be imprisoned if they refuse to move within 24 hours. (*Kindred Spirit*, Autumn 1994)

squeal as a criminal to give information to the police.

There is an implication of duress, with the *squeal* indicating pain. Of informing on others or confessing your own guilt:

> ... loath to 'squeal' or turn him in. (Lavine, 1930)

squeeze to extort money etc. from illegally.

From the pressure applied:

> The Red Eleven would stick by him and fight anyone who tried to squeeze him. (Theroux, 1973)

The squeeze is such extortion and especially the developed and endemic version in the Far East:

> Perhaps this Englishman, like the French, wanted his squeeze. (Moss, 1987, of bribery)

A *squeeze* may be no more than a tip:

> Brooke nodded to the little chauffeur, then handed him some money as he had seen Jeremy do. *Squeeze*, they called it. (Reeman, 1994)

squib off to murder.

Usually by shooting, from the noise made by a firework:

> The night Joe got squibbed off. (Chandler, 1939)

squiffy intoxicated.

Literally, uneven or lopsided:

> 'The man was squiffy,' said Aunt Agnes. 'It was written all over him.' (E. Waugh, 1933)

squirrel *American* a psychiatrist.

Because he feeds on a NUT, or a diet of them. Whence a *squirrel tank*, or institution for the insane:

> ... the perpetrator went nuts after the accident and is now in the squirrel tank. (Wambaugh, 1975)

squirt to defecate.

Normally of diarrhoea, and much used as a mild insult, neither the giver nor the receiver being aware how rude it is:

> ... a very coarse name, which we can change euphemistically into—squirts. (Vachell, 1934)

Diarrhoea is also *skeet, squit, squitters*, or *skitters*:

> 'Skitters,' I said. 'That'll wait for no man. Run for it. I'll wait.' I dashed for the toilet. (Steinbeck, 1961)

> ... the senile Labrador that drools and squitters all over the stairs. (Theroux, 1982)

stab to copulate with.

Of a male, using the common imagery of violence and of pushing:

> He'd stabb'd me in mine own house ... he will foin like any devil. (Shakespeare, *2 Henry IV —foin* means 'thrust')

Being *stabbed with a Bridport dagger* was not copulating with a native of the Dorset town but death by hanging, Bridport being famous for rope-making because of a climate in which flax flourished.

stable the prostitutes who work for a single pimp.

The common equine and riding imagery. The pimp may be called a *stable-boss* but a *stable horse* was a stallion in 19th-century American prudery.

staff the erect penis.

A rarer version of ROD[3]:

> ... the registrar of births had to increase his staff owing to the way he had exercised his. (Pearsall, 1969, quoting 19th-century pornography)

stag pornographic.

> But you can get late-night stag movies piped into our place. (C. Forbes, 1983)

From the meaning 'male', as in *stag party*:

> All weddings are the same, and Vye wanted everything done the way everyone else did it, including the stag party—where they showed four skin-flicks. (D. A. Richards, 1988—but I think most such occasions, although boisterous and rowdy, do not involve pornography)

stag month see STEG MONTH.

stain *obsolete* to copulate with outside marriage.

Of a male, who pollutes the female morally rather than seminally:

> Give up your body to such sweet uncleanness,
> As she that he hath stain'd.
> (Shakespeare, *Measure for Measure*)

staining minor bleeding.

Medical jargon, from the seepage of blood into bandages.

stake (the) killing by burning.

The victim was tied to a pole. The significance of this form of death for heretics was that nothing remained to reappear and cause further trouble at the Resurrection.

stake-out a police trap.

From the prior exploration of the site of a suspected future criminal attempt:

> ... he was running a stake-out ... over in the meat-packing district. (van Lustbaden, 1983)

stale *obsolete* a prostitute.

Someone already used sexually by others:

> ... poor I am but his stale. (Shakespeare, *Comedy of Errors*)

Stale was also urine, possibly from its common retention for laundry and other purposes:

> The dung and stale of cattle. (Marshall, 1817)

stale meat see MEAT.

stand[1] the erect penis.

Of obvious derivation:

> When it stands well with him, it stands well with her. (Shakespeare, *Two Gentlemen of Verona*)

The penis may also be said to *stand to attention*:

> She finished ... posing as a nude Britannia with helmet and union jack. I wondered how many men in her audience would be standing to attention. (Monkhouse, 1993)

stand[2] to be available for breeding.

Of a male quadruped, although *mount* might seem more appropriate:

> ... the stallion had stood for three seasons and therefore covered a hundred and twenty mares. (Francis, 1982)

Standard English.

stand before your Maker to die.

It would be presumptuous, I suppose, to sit:

> ... none should expect Gratitude until it is his turn to stand before the Father of us all. (le Carré, 1986)

stand down to be dismissed or prematurely retired from employment.

Literally, to end a term of duty or to revert to a lower state of preparedness after an alert. The term is used to protect the self-esteem of a departing employee.

standard small or poor quality.

No longer the level of size, quality, etc. against which judgement of other products can be made. The supermarket *standard pack* is small and comparatively expensive; a *standard* durable product is the one without accessories or elegant styling.

standstill an attempt by government to restrict pay increases.

See also FREEZE[1], GUIDELINES, PAUSE[2], and RESTRAINT[1] which called in aid the same imagery with the same lack of success:

> Thus, in the House of Commons on 6 November 1972, when Heath announced a standstill on wages and prices, thereby introducing the kind of incomes policy which he had always sworn to eschew... (Cosgrave, 1989)

star in the east an undone fly-button.

An oblique warning to a male which has not survived the zip age.

stardust cocaine.

With the same imagery as ANGEL DUST. Addict slang.

stark naked.

Stark means 'quite' and this is probably merely a shortened form of the idiom *stark naked. Stark as the day they were born* (Buchan, 1898) might be read 'quite as the day' etc. The obsolete meaning 'dead' came from a dialect meaning 'stiff', often in the tautological *stiff and stark*.

start bleeding to menstruate for the first time.

The female concerned will certainly have bled from her nose or a wound on previous occasions:

> Yes, I matured early... I started bleeding at eleven. (Sanders, 1970)

starter home a small house.

Not you remember, you remember, the house where you were born, but the first you may be induced to buy (with apologies to the English poet Thomas Hood (1799–1845)).

stash a supply of illegal narcotics.

Or the place where it is hidden. Both as verb and noun, a venerable usage until taken over in the jargon of drug addicts:

> Never carry when you can stash. (Addict adage)

state farm an institution for involuntary detention.

Where you keep prisoners, children, geriatrics, or lunatics. Also in America as *state homes, hospitals, (training) schools*, etc.

state of nature nudity.

The imagery of NATURE'S GARB, but the opposite is not an unnatural state:

> Charles Boon, who scorned pyjamas and was often to be encountered walking about the apartment... in a state of nature. (Lodge, 1975)

state of sexual excitement having an erection of the penis.

Also of female manifestations such as hardening of nipples:

> ...scenes in which she caresses the naked figure of Christ in a state of sexual excitement. (*Daily Telegraph*, August 1989)

state protection the preservation of tyranny.

As in the Communist *Department of State Protection*, which controlled political prisons, publication, spied on citizens, etc. in the USSR, or the Ugandan tyrant Amin's *State Research Bureau*, which had the same function.

states' rights the continuation of discrimination against blacks.

Literally, the powers reserved to each individual American state as distinct from Federal jurisdiction. The South professed to see Federal moves against white supremacy as an attack on the individual rights of each state.

status deprivation being thought badly of.

Educational jargon of a child who does badly at school or is objectionable, and therefore not rated highly by teachers or other pupils.

statutory appointed other than on merit.

Of membership of committees, boards, etc. where those perceived as oppressed or belonging to a category not properly recognized secure appointment other than on merit:

> I realised that the government would wish to include certain 'statutory members' such as representatives of the trade unions and the Co-operative movements—though not, to my regret, a Statutory Lady. (Cork, 1988)

See also TOKENISM and OBLIGATORY.

statutory rape copulation by an adult male with a consenting girl below a specified age.

A criminal offence so described in various states of the USA, with differing ages. The first legal proscription of such activity came in medieval times, when British males were punished if they were caught copulating with girls under the age of 10: they must have been in the habit of doing so to render legislation necessary. The minimum age has risen slowly over the years, remaining at 12 in most places until the present century. In 19th-century American legal jargon a *statutory offense* was rape against a female of any age.

steady company a person with whom you consort and copulate regularly.

Usually in the phrase *keep steady company*:

> We've been keeping steady company for the past five years now. (McBain, 1981, of a man and his sexual mistress)

See also COMPANY[1] and KEEP COMPANY WITH.

steaming see TAX.

steer dishonestly to influence the placing of business.

By pretending to give disinterested advice in the selection of an adviser or service when you are receiving a commission or a reciprocal benefit:

> ... bribery of hospital personnel to 'steer' cases. (J. Mitford, 1963, describing how funeral firms secured business)

steg month *obsolete* the period around childbirth when a husband might copulate extramaritally with relative impunity.

From being a gander in northern English dialect, a *steg* became an aimless male, wandering about while the goose is hatching the goslings. See also GANDER-MOONER for a further dissertation and explanation. During this time the unavailable wife was known as a *steg-widow*. I suspect the *stag month* and *stag widow* of other lexicographers, including Partridge, are mistaken corruptions.

stem the erect penis.

Not in this case the opposite of the stern but of the same tendency as ROOT[1]:

> Gently she tugged, guiding my stem between her sleepy breasts. (L. Thomas, 1989)

step down to be dismissed from employment.

Normally, to retire of your own volition:

> Saunders must step down. (*London Standard* headline, January 1987, of a Chairman about to be dismissed)

step-ins *American* women's underpants.

Not a bath tub or a pair of shoes. This usage survived most of the evasions used for trousers—see UNMENTIONABLES[1].

step off to die.

Clearly a shortened form but I'm not sure of what. The obsolete Scottish *step away* also meant 'to die':

> Garskadden's been wi' his Maker these two hours; I saw him step awa. (Ramsay, 1861)

step on to kill.

Again drawn from our habitual treatment of insects:

> Jack and Hyme talk so casually about killing and death. 'Should I step on him?' 'We should have killed the cock-sucker.' Like that. (Sanders, 1980)

step out on *American* to deceive (a regular sexual partner) by copulating with another.

Of either sex:

> Do you think Haveabud and your mother had a sexual relationship? Do you think I ever stepped out on her? (A. Beattie, 1989)

To *step out with* someone and *step out together* is to be courting:

> Before long they were stepping out together and although Thea was strictly chaperoned, they had soon become very close. (M. Clark, 1991)

stepney *British, obsolete* a pimp's favourite prostitute.

The one he treated in other respects as a wife out of a number under his control. A *stepney* was the fifth wheel carried on the step, or running-board, of a car and only brought into use when one of the other four was out of service.

sterilize to destroy.

Literally, to render barren, whence to purify or make clean. Of obliterating tapes, removing documents from files, etc. In Vietnam jargon, it meant dropping bombs and killing suspects:

> We sterilize the area prior to the insertion of the R.D. teams. (McCarthy, 1967—*RD* was *rural development*; see RURAL CONSTRUCTION)

stern the buttocks.

Naval imagery in general use. The punning *stern-chaser* may have homosexual or heterosexual preferences.

stewed drunk or under the influence of narcotics.

The common culinary imagery:

> ... most of the time in camp ... poor old Abel was stewed. (Keneally, 1979—Abel was not a prisoner taken by cannibals)

You are no less drunk if *half-stewed*. The narcotics use is rarer:

> They kept piling the old hashish into the shisheh ... He's totally stewed. (Deighton, 1991—a *shisheh* may be a bowl made from shisham wood, or *Dalbergia sissoo*)

stews (the) *obsolete* a brothel.

Originally, a bath house which often doubled as a brothel:

> An I could get me but a wife in the stews. (Shakespeare, *2 Henry IV*)

stick[1] to kill.

Supposedly with a pointed weapon, of cattle in an abattoir and of wild pigs in hunting. It used also to mean 'to wound':

> The black thief has sticket the woman. (Carrick, 1835)

stick[2] a spirituous liquor added to another drink.

Perhaps you have simply placed, or *stuck*, one liquid inside another:

> Coffee, if you like, with a 'stick' in it. (Praed, 1890)

Still used in northern England and Australia.

stick[3] to copulate with.

Of a male, perhaps from the slang *stick*, the penis:

> Said he, with a snicker,
> As he watched the guy stick her...
> (*Playboy's Book of Limericks*)

Also as to *give stick*, punning on the offering of violence.

stick[4] a marijuana cigarette already rolled by another.

Perhaps a shortened form of *stick of tea*, a thin form of cigarette. Often in compounds such as *dream-stick* and the punning *joy-stick*.

stick[5] a handgun.

From its shape, but ROD[2] is more common:

> He hit some East Side apartment for a bundle. Ice, mostly. Never carried a stick. (Sanders, 1970, and not of a someone with a limp sliding into a building)

stick[6] a cluster of bombs dropped from an aircraft.

Perhaps because they fall in a straight line.

stick it into[1] to extort money etc. from through threats.

From figuratively wounding with a weapon:

> They had pictures, who the hell knows what else? But they stuck it into him. (Diehl, 1978 —the *pictures* are photographs)

stick it into[2] to copulate with.

Of the male and barely euphemistic:

> Brother was sticking it into sister every night. (Mailer, 1965—they were committing incest)

The American *stick it on* is rarer:

> Men liked to think they were sticking it on some kind of technical virgin. (McBain, 1981)

stick up to rob with a threat of, or with actual, violence.

From the command to *stick up your hands*, and a *stick-up* is such a robbery:

> 'You'll hold me up, I suppose!'...'I'm a stick-up artist now, am I?' (Chandler, 1939)

sticky a spirituous intoxicant.

Usually a liqueur, from its tacky properties:

> I spend the next two hours...with a litre bottle of some colourless but potent sticky at my elbow. (*Private Eye*, August 1983)

sticky-fingered thieving.

Other people's property adheres to the fingers. Usually of embezzlement or chance pilfering.

sticky stranger a clandestine electronic listening device.

Espionage jargon—you have to fix the apparatus by glue:

> You'll want to look around for a sticky stranger. If they think you've got something to hide, they'll plant another ear. (Francis, 1978—an *ear* is also such a device)

stiff[1] a corpse.

From the rigor mortis and a shortened form of the 19th-century *stiff one*:

> Would she stick it till she was a stiff'un. (Mayhew, 1862)
>
> When anyone was killed they piled the stiffs outside the door. (*Scribner's Monthly*, July 1880, quoted in EDD)

The abusive *stiff*, a moron, may come from this use.

stiff[2] drunk.

You often feel and look like a corpse:

> I was quite stiff by the time we got to the burial ground. (Styron, 1976—he was drunk)

stiff[3] having an erection of the penis.

Of obvious derivation:

> ...she approached me where I lay, stiff as a dagger. (Styron, 1976)

stiff[4] to fail to meet your financial obligations.

It is one form of death, I suppose. Also as the American *stiff out*.

stiff-arm to compel through threats or violence.

From a disabling hold in wrestling:

> One more attempt to stiff-arm him occurred at 8.30 p.m. (Colodny and Gettlin, 1991—the White House was seeking to make the Attorney-General suppress the Nixon tapes)

stiffener a drink of a spirituous intoxicant.

A variant of the common BRACER and not because it may make you STIFF[2]:

> ...careless riders would fall away, in search of a few stiffeners. (Flanagan, 1988)

stimulant a spirituous intoxicant or an illegal narcotic.

Not a bribe, a kiss, a gift, an encouraging word, or any of the other things you might find stimulating:

> ...if ever there was a man who needed a snappy stimulant, it was he. (Wodehouse, 1934, referring to an intoxicant)

Their main source of revenue is from trafficking in stimulants, especially crystal methamphetamine (known as ice). (*Economist*, 29 February 1992)

sting to deprive by trickery.

Of robbery, overcharging, cheating, or any other kind of knavery:

He has completely dead eyes, and looks at you with all the warmth of one deciding how much he can sting you for your bridgework. (L. Barber, 1991)

The sting is the ultimate coup in an elaborate confidence trick or complex police operation set up to catch criminals:

The sting resulted in the serving of 198 arrest warrants for fewer than 100 individuals. (*Law and Order*, May 1990)

stinking very drunk.

Probably not from your *stinking of drink* but from the meaning 'exceedingly', as in the cliché *stinking rich*. At one time corrupted to *stinko*:

Are you stinko? (Chandler, 1953)

stir a prison.

Probably from the Romany and not from what you do to your breakfast PORRIDGE:

A friend of mine who's in stir. (Chandler, 1939)

To be *stir-wise* is to be experienced in prison life:

He's too stir-wise for me. (ibid., of a convict)

stitch up to fabricate evidence against.

From the securing of a canvas bag:

'Someone *else* did it, I tell you!' 'Who? *Why?*' 'To stitch me *up*.' (C. Thomas, 1993, of a prisoner falsely accused of murder)

stitched drunk.

Mainly military use, using the common death imagery as a soldier's corpse was sewn up in canvas before burial. Unhappily not from the obsolete Irish *stitch in your wig*, mild drunkenness, where to *stitch* was to rumple and the imagery is of a wig slightly askew.

stoat a libertine.

It is unclear why the European ermine should have acquired such a reputation:

He fancied everyone really. By way of being a stoat. (le Carré, 1989)

Forster, in 1971, used the same animal to represent homosexual lust.

stock beast *American* a bull.

Also as *stock animal*, *brute*, or *cow* in 19th-century prudery.

stockade a military prison.

Literally, a strong fence forming an enclosure:

... you fly or you go to the stockade. (Deighton, 1982, of American Second World War fliers)

stoke Lucifer's fires to be dead.

Usually of one who has led a sinful life for which he is presumed to be doing penance into eternity on the end of a shovel:

There was a rumour of his death, or he's probably been stoking Lucifer's fires these thirty years. (Fraser, 1970, writing in 19th-century style)

Lucifer, for the Christians, was inseparable from Hell (*New Larousse Encyclopedia of Mythology*, 1968).

stomach (a) obesity around the waist.

Usually of a male and incorrectly specifying the internal chamber through which food passes for digestion. *A bit of a stomach* also implies obesity, and not post-surgical deprivation.

stomach cramps see CRAMPS (THE).

stoned drunk or under the influence of illegal narcotics.

It is hard to see what the discomfort of St Stephen and others had to do with this common use. Of drunkenness:

The day Butler's Military Cross was gazetted they both got stoned out of their minds (Price, 1979)

and of narcotics:

He did his best work half-stoned. When you stare at motels for a living, you need to be stoned. (Grisham, 1992, referring to an investigator who habitually smoked cannabis)

stones the testicles.

Of man and animals:

A philosopher, with two stones more than's artificial one (Shakespeare, *Timon of Athens*)

and an East Anglian farmer will say that a castrated sheep is *two stone lighter*. The obsolete *stoned-horse-man* was not a heroin addict but the groom who took a stallion—*stony*—around farms to impregnate mares.

stool pigeon a police informer.

Pigeons were tied to stools to lure other pigeons for capture. To *stool* is to inform against:

... stooled on a bank job in Michigan and git me four years. (Chandler, 1939)

stoop your body to pollution *obsolete* to copulate extramaritally.

Of a female, although she is more likely to have been recumbent:

Before his sister should her body stoop
To such abhorred pollution.
Then, Isabel, live chaste.
(Shakespeare, *Measure for Measure*)

stop one to be killed or wounded.

A common First World War use:

We old ones aren't lucky enough to stop one that way. (Richards, 1933—he was referring to a BLIGHTY)

To *stop a slug* is more specific, except for a keen gardener:

> I wasn't hired to kill people. Until Frisky stopped that slug I didn't have no such ideas. (Chandler, 1939)

To *stop the big one* is to be killed:

> The guy stopped the big one. Cold. (ibid.)

stoppage[1] an inability to defecate.

Medical jargon and also used of nasal and other bodily malfunctions.

stoppage[2] a strike by employees.

Trade union jargon which is still used even if the factory concerned remains in production.

story a lie.

Nursery usage although the punning *story-teller* may also be used of an adult. A *tall story* implies exaggeration and a *cock-and-bull story* is a fabrication.

straddle to copulate with.

Of a male, with the common riding imagery:

> I felt a moment's pang at the thought that I'd straddled her for the last time. (Fraser, 1985)

straight conventional, heterosexual, or honest.

In the sexual sense, used both by heterosexuals of themselves and homosexuals of heterosexuals:

> And he wasn't going to ask his queer friends ... and there were none of his straight friends he reckoned he could ask except us. (Amis, 1988)

Few of us would say of ourselves that we are *straight*, meaning 'honest', although we might hope that others might say it about us, even if it would be less of a compliment than is implied by the French *homme droit*. Criminals so describe honest people, until they themselves may decide to *go straight*.

straighten out to bribe.

You induce another to follow the line which you indicate. The phrase is also used of our forceful, but inevitably unavailing, correction of someone with different opinions from our own.

strangle to cause (a horse) to run badly in a race.

From pulling the bridle surreptitiously:

> Sandie had 'strangled' a couple at one stage. (Francis, 1962, of a crooked jockey)

strategic capability the possession of nuclear weapons.

The jargon of nuclear armour, it should mean, at most, your ability to work out a plan. From our experience of the Second World War, anything described as *strategic* in the way of bombing etc. shows small discrimination between civilian and other targets and a *strategic nuclear war* delivering *strategic warheads* from *strategic submarines* is unlikely to prove any different.

strategic withdrawal a retreat under pressure.

Any military strategy which involves moving away from the enemy is hardly likely to lead to victory:

> We've admitted a strategic withdrawal ... the Jerries are coming hell for leather down the coast road. (Manning, 1965)

A *strategic movement to the rear* seeks to imply, falsely, that some of your side are still in front of you. A *strategic retreat* is even less of a face-saver:

> The Germans announced an Allied retreat. Merely a strategic retreat, said the British News Service. (Manning, 1960)

stray off the reservation to reveal the truth.

Another contribution to the language from the Watergate conspirators:

> ... if Jeb 'strayed off the reservation'—the phrase had come to be used in the Nixon inner elite to mean refusing to adhere to the approved story of the burglary and the cover-up—Dean would not have remained at liberty himself. (Colodny and Gettlin, 1991)

stray your affection *obsolete* to copulate extramaritally.

To *stray* is to cause to wander:

> Stray'd his affection in unlawful love. (Shakespeare, *Comedy of Errors*)

streak to run naked in a public place.

In this phenomenon, which started in the mid-1970s, the speed was supposed to restrict the visibility. A *streaker* so behaves.

streamlining the simultaneous dismissal of a number of employees.

I suppose in the expectation that the rest will go faster.

street (the) prostitution.

The place where customers are picked up:

> 'You're the only person who can save us.' 'How?' 'Why, the street, of course.' (Londres, 1928, in translation, of prostitution)

A *street-walker*, a prostitute, is dated:

> The modern equivalents of the old-time disorderly house and of the street-walker. (Lavine, 1930)

A *street girl* is also a prostitute:

> ... her wretched career from housewife to street girl. (Green, 1979—were both occupations wretched?)

The American *street tricking* is finding customers as a prostitute on the streets:

> This old campaigner we call Mabel the Monster, been street trickin' must be ten years now. (Diehl, 1978)

On the street(s) is engaged in prostitution:

> She fell in love with Mary Jack's pimp, who put her on the street. (L. Armstrong, 1955)

The obsolete Scottish/northern English *Street and Walker's place* meant 'unemployment', from walking the streets looking for work.

street drugs narcotics, hallucinogens, etc. sold illegally.

As distinct from those bought in a pharmacy. Prior to the legalizing of British off-course gambling, *street bets* were those placed illegally with local bookmakers.

street money *American* electoral bribes.

Wherever paid:

He claimed Mrs Whitman's campaign paid what is known as 'street money' to black clergy and elected officials to dissuade them from getting out the black vote. (*Daily Telegraph*, 23 November 1993)

stretch a period of imprisonment.

A shortened form of *stretch of years*:

The bosses always get the longest stretch in the penitentiary. (L. Thomas, 1979)

stretch the hemp *?obsolete* to kill by hanging.

From the material of the noose:

Molly Maguire stretching the hemp in the last act. (*Pearson's Magazine*, October 1900, quoted in EDD)

Also as *stretch the neck*.

stretch your legs to urinate.

Why we say we have intervals in meetings or stops on long journeys:

Another five or ten minutes, and you'll be able to stretch your legs. And then after that I fancy you'll be able to travel more comfortably. (Price, 1978)

stretcher a lie or exaggeration.

From *stretching* your credulity and the truth:

Is ole Wheat still telling Gus back there them stretchers regarding his gran'daddy? (Keneally, 1979)

Whence the punning *stretcher case*, a habitual liar and to *stretch*, to lie:

I ain't about to start stretching at this time of day. (Fraser, 1985—Flashman isn't often so reticent)

strike out to die.

From baseball and therefore confined to America.

string up to kill by hanging.

Usually in a makeshift manner on the limb of a convenient tree which always seems to be to hand in cowboy films.

stripper a thief.

Especially of radios etc. from American cars and houses:

...our motherfucking car stripper is halfway to Watts. (Wambaugh, 1975)

In standard English a *stripper* removes paint or takes off her clothes in public in a manner calculated to entertain male spectators.

stroke *mainly American* to attempt to persuade by flattery.

As you might comfort a pet:

He asked himself over a glass of vodka whether Pokryshkin had handled—he didn't know the Western expression 'stroked'—him enough to create a false impression (Clancy, 1988)

and the Watergate team reported:

We are giving him a lot of stroking. (Colodny and Gettlin, 1991—the team were trying to gain the silence of a witness)

A *stroke job* is such flattery:

'I want to be as open and candid as I can.' The stroke job's starting, Barcella thought. (Maas, 1986)

stroke off to masturbate.

Usually of the male. And sundry vulgar compounds, as *stroke the bishop, dummy, lizard*, etc.

stroller *Irish* a habitual itinerant.

But not ON THE STROLL:

You'll not trick me, stroller. I saw you pull up and there's no-one with you. (O'Donoghue, 1988—addressing a lone gypsy)

strong-arm to steal.

With the use or threat of force:

If he had not strong-armed that money out of me I would have given him lots more. (L. Armstrong, 1955—his own surname originated in the Scottish/English borders where for centuries such activity was endemic)

strong waters spirituous intoxicants.

Not a fast flowing stream:

...it does not one-tenth of the harm that strong waters cause among the poorer classes. (Fraser, 1985, writing in archaic style of opium)

In Ireland, a *strong weakness* was dipsomania:

Bob would be marked as a man with what our countryside calls 'a strong weakness'. (Flanagan, 1988)

strop your beak to copulate.

Of a male, from the movement in sharpening an open razor and the slang *beak*, the penis. Occasionally of masturbation by a male.

structured arranged as a cartel.

The concept is of something being put together in an orderly manner but the American *structured competition* covers up illegal agreements on price and other competition.

struggle a political campaign.

For most people most of the time a figurative usage to make what is trivial seem important but see ARMED STRUGGLE. The Nazi *struggle for national existence* was not the fight against the Anglo-Americans or the Russians but the extermination of Slavs, Jews, and Gypsies:

> ... a struggle for national existence meant racial warfare. (Keneally, 1982)

strung out addicted to illegal narcotics.

From the haggard appearance? Also of anyone under the influence of narcotics.

stubble see TAKE A TURN IN THE STUBBLE.

stuck cheated.

Probably a shortened form of *stuck with a poor bargain*:

> I experienced that peculiar inward sinking that accompanies the birth of the conviction that one has been stuck. (Somerville and Ross, 1897, of a horse deal)

stuck on infatuated with.

No doubt from the desire to enjoy propinquity:

> Archer, are you stuck on the girl or something? (Macdonald, 1976)

stud a male viewed sexually.

Usually of a man who is thought to be ready to copulate promiscuously:

> Sex? ... No stud in the world is worth two million dollars (West, 1979)

but also of homosexuals:

> I don't go to no leather joints lookin' for some stud to fistfuck. (M. Thomas, 1980)

The derivation is probably from the place where stallions are kept for breeding, their availability, their fee, etc.—*stud farm, at stud, stud fee*, etc. —rather than the imagery of a projecting lug.

studio a small apartment with no windows at eye level.

House agents' jargon, trading on the traditional indirect illumination from a skylight in which some painters choose to work.

stuff[1] any taboo or forbidden substance.

Literally, any substance or material. Of semen:

> ... put stuff
> To some she-beggar
> (Shakespeare, *Timon of Athens*)

of contraband spirits:

> A considerable amount of 'stuff' finds its way to the consumers without the formality of the Custom House (Stoker, 1895)

and of illegal narcotics:

> ... he smokes too much, and 'stuff'. (Bogarde, 1981)

stuff[2] to copulate with.

From the physical entry rather than impregnation despite:

> A maid, and stuf'd! there's goodly catching of cold. (Shakespeare, *Much Ado*)

Also some use of buggery. Much vulgar figurative use:

> As for the flute, he knew where he could stuff that. (Davidson, 1978)

Much use, again figurative, as a synonym for *fuck*, as in *get stuffed* and *stuff that*.

stump liquor *American* an illegal spirituous drink.

Probably made by a *stump-jumper* or hillbilly:

> People in these hills still made moonshine, or stump liquor as they call it. (Bryson, 1989, of eastern Tennessee)

stung drunk.

From the debilitating effect. To be *stung by a serpent* was to become pregnant, punning on the *serpent*, or penis, and implying something both unexpected and unwanted.

stunned drunk.

Common American slang with obvious imagery.

stunt a limited battle.

Much more than a trick, but these horrors were played down in the First World War:

> If he don't get the Victoria Cross for this stunt I'm a bloody Dutchman. (Richards, 1933)

stunted hare a rabbit.

For seamen, the mention of a *rabbit* is taboo although it is a long time since British chandlers substituted salted rabbit meat, which decays quickly, for the conventional salted pork.

stupid drunk.

From the drunkard's behaviour rather than the folly of getting like it. Common in Scotland as *stupid-fou*:

> He was na stupid-fou, as was his wont on market days. (Strain, 1900)

subdue to your will to copulate with extramaritally.

Males do it, overcoming female fears or scruples. The woman has to be royal or rich to *subdue* a man:

> ... the queen has only two uses for foreign men—first to subdue them to her will, if you follow me ... (Fraser, 1977)

submit to to copulate with.

Of a woman, usually extramaritally and with an implication of reluctance:

> They refuse to submit to his pleasure, and will not return him the money. (Mayhew, 1862, of cheating prostitutes)

subsidy publishing the publication of a book at the author's expense.

Not a House Journal or the products of Her Majesty's Stationery Office.

substance an illegal narcotic.

Literally, any matter and normally in compounds such as *illegal substance* or *substance abuse*, which means regularly ingesting illegal narcotics or sniffing glue or solvents:

> ... she'd been a nurse too long, had too often seen the results of substance abuse. (Clancy, 1989)

succubus a prostitute.

Originally, a female demon who copulates with men in their sleep, thus for the fastidious providing an excuse for involuntary nocturnal seminal ejaculation:

> 'Yes, thou barbarian,' said she, turning to Wagtail, 'thou tiger, thou succubus!' (Smollett, 1748)

Succuba would seem the correct gender, but is wrong:

> 'She's a witch. She'll destroy everything!' 'A succuba, is she? I'd like to meet her.' (Cornwell, 1993)

succumb[1] to die.

Literally, to give way to anything, and usually of natural death:

> Hibbert... succumbed to a heart attack at his desk. (Condon, 1966)

succumb[2] to copulate outside marriage.

Another form of giving way, by either sex:

> I'm willing to bet you five dollars she doesn't succumb even to the charms of William. (Archer, 1979)

suck daisy-roots to be dead.

The common imagery of the churchyard flower seen from below.

suck off to practise fellatio or cunnilingus on.

Of obvious derivation:

> One American GI is forcing a Vietnamese woman to suck him off. (*Guardian*, 27 September 1971, referring to fellatio)
>
> Equilibrists suck each other off deftly. (Burroughs, 1959—an *equilibrist* is someone who 'performs feats of balancing')

suck the monkey *obsolete* to steal rum.

British naval usage, from the insertion of a straw in the cask. Also of illicit drinking of rum inserted in a coconut for subsequent consumption on board ship.

suffer the supreme penalty to be killed.

Usually of a convict, flogging, imprisonment, and fines being lesser sanctions:

> As for... the murder of her Indian subordinate ... eventually one or two men suffered the supreme penalty. (P. Scott, 1973)

sugar[1] a bribe.

The common imagery of *sweetening* and a *sugared* deal is one which involves corruption.

sugar[2] a mild oath.

Common genteel usage, for the taboo *shit*.

sugar[3] an illegal narcotic.

In two specific uses: of any white narcotic in crystalline form; and of LSD deposited on a lump of sugar to make it palatable.

sugar daddy a man with a much younger sexual mistress.

Daddy from the age gap and *sugar* from the sweet things of life which she may expect from him:

> Kathy's Sugar Daddy Evicted. (Headline in *Western Daily Press*, May 1981)

Sometimes shortened to *daddy*.

suggestion the improper revealing of privileged or confidential information.

With *insider dealing* (see INSIDER) a crime nowadays, new linguistic evasions have to be evolved:

> He'll get a commission of five percent of all profits generated by his 'suggestions', which I guess is standard procedure. (Erdman, 1987, of the improper passing of share information)

sun has been hot today there are signs of drunkenness.

Probably from the old British habit of putting weak cider or ale in harvest fields. On a hot day a reaper would get progressively more drunk as he slaked his thirst. A drunkard might also be said to *have the sun in his eyes* or *have been in the sun*:

> We guessed by his rackle as he'd bin i' the sunshine. (Pinnock, 1895—*rackle* was riotous conduct)

sun has gone over the yardarm let us drink intoxicants.

The sun, in temperate latitudes, sinks in the late afternoon below the *yardarm*, a spar running horizontally from the mast to support a sail:

> Ah well, sun is over the yardarm, so down to work. (*Private Eye*, May 1981, of drinking intoxicants)

Sunday traveller *Irish* an illegal drinker of intoxicants at an inn.

Only a bona fide traveller could be served with intoxicants on Sundays in Ireland:

> ... a door consecrated to the unobtrusive visits of so called 'Sunday Travellers'. (Somerville and Ross, 1897)

sundowner an intoxicating drink taken in the evening.

Of spirits, which, in the tropics, only an alcoholic would regularly drink during the day because of the effects of additional dehydration. The hard drinking starts at dusk:

> We're all right now, the sun has gone down and we can have a whisky. (Allen, 1975)

As he sits there on a hot evening swilling his sundowners... (G. Greene, 1978—*there* was in Zaire)

sunset years old age.

Trading off the beauty of sunsets rather than the darkness to follow. But still less sickly than the *golden years* (see GOLDEN YEARS (THE)).

supercharged drunk or under the influence of illegal narcotics.

From the increase in pressure and see also CHARGE[2].

supporters' club the employees of a potential customer improperly favouring a vendor.

Punning on an association of sports fans. Commercial use of those who have been influenced by bribes in the form of cash, gifts, or lavish entertainment.

supportive anxious to intermeddle.

Literally, ready to support, but the use may imply a deep commitment to or obsession with a cause and a criticism of those who may not share the same emotions:

...if the caring and supportive wanted a political focus, it was necessary to drive...to meet others with similar ambitions for the use of the planet. (*Daily Telegraph*, May 1990—the writer was deploring the absence of like-minded environmentalists close to her home; perhaps her argument would have been more convincing if she had left the car in the garage. Note the absence of the noun after the two adjectives.)

sure thing a woman reputed by men to copulate promiscuously.

Probably from the racehorse so described by the tipster, although there are no certainties in either sport:

...hardly at all like someone who in her time had been one of the surest things between Bridgend and Carmarthen. (Amis, 1986)

surgical appliance see APPLIANCE.

surgical strike a bombing raid.

Supposedly as accurate as the first incision of the scalpel:

...precision bombing is 'surgical strikes'. (Commager, 1972, writing about Vietnam)

surplus *American* to dismiss from employment.

Perhaps from the noun phrase *war surplus*, supplies offered for sale when no longer needed, whence the verb to *surplus*, to dispose of unwanted equipment. Sometimes of a single dismissal with an implication that there was one employee too many, but:

IBM has reportedly 'surplused' 25,000 jobs corporate-wide. (*Computer Shopper*, July 1993)

surrender to to copulate with extramaritally.

Of a female with the common imagery of male dominance:

Girls seemed to prefer the story of her surrendering to Koolman in exchange for a leading role. (Deighton, 1972)

surrendered personnel Japanese prisoners of war.

An evasion used by the British 14th Army after the Second World War to encourage its defeated opponents not to obey their martial code and continue fighting until death:

By October, thousands of Japanese Surrendered Personnel (as a salve to their dignity they were never referred to as prisoners)... (M. Clark, 1991)

surveillance spying.

Literally, no more than keeping a watch over, as a ward or a prisoner. Now police and espionage jargon for clandestine observation, and *electronic* or *technical surveillance* is the use of hidden microphones or other gadgetry of spying.

survivability the extent to which it is thought that a nation can survive a nuclear attack.

Military jargon. Mario Pei also gives *termination capability*, the ability to bring to an end a nuclear war, presumably by destroying the other side too.

swallow the anchor to retire from a naval career.

A British naval usage also adopted by yachtsmen etc.:

At sixty-three, their painful knees and hands were making it increasingly difficult to work the foredeck, but at the same time neither of them relished the prospect of swallowing the anchor. (M. Clark, 1991)

swallow the Bible *American* to perjure yourself.

From the taking of an oath in court:

They will stick together, stretch conscience and at times 'swallow the Bible'. (Lavine, 1930)

See also EAT THE BIBLE and SWITCH THE PRIMER.

swamped drunk.

Sunk by too much liquid. In the British Isles in the 19th century it also meant 'bankrupted'.

sweat it out of to obtain information from by coercion.

American police jargon; the *sweat-box* was the room where you did it. The obsolete English *sweat-box* was a cell in any police station.

Sweeney see FLYING SQUAD.

sweep *American* to kill.

Vietnam usage where a *sweeping* operation was not the cleaning of a barracks and to *search and sweep* was to go over hostile territory and kill anyone you found there.

sweet momma a man's sexual mistress.

Originally, for an American black, any woman of a kindly disposition regardless of age or motherhood. A *sweet man* is a black woman's regular sexual partner outside marriage.

sweet tooth an addiction to illegal narcotics.

An American pun on CANDY.

sweetbreads animal glands used as food.

Literally, the thymus or pancreas, but also the testicles. See also the American VARIETY MEATS and PRAIRIE OYSTER[1].

sweeten[1] to bribe.

The common bribery imagery:

> Now-a-days ane canna' phraise,
> An' sooth, an' lie, an' sweeten,
> An' palm, an' sconse.
> (Lauderdale, 1796—all the activities were *sui generis* except *sconce*, to trick)

A *sweetener* is such a bribe, often in gift form rather than in cash:

> Giving big commissions, sweeteners, call it bribery if you like... (Lyall, 1980)

sweeten[2] improperly to force up bidding at an auction.

Auctioneers' jargon for purporting to accept spurious or non-existent bids.

sweetheart an arrangement which improperly benefits two parties at the expense of a third.

In various phrases such as the American *sweetheart contract* between an employer and a labour union which benefits the firm and the union officials at the expense of the workers; or a *sweetheart price* which cheats stockholders:

> And at a real good sweetheart price, too. Less than $6 billion over four years. (M. Thomas, 1980)

swell to be pregnant.

Of obvious imagery:

> Unless it swell past hiding, and then it's past watching (Shakespeare, *Troilus and Cressida*)

and in tasteless modern American use.

swill to be a habitual drunkard.

Literally, to rinse out but long standard English of drunkenness. The usual stream of derivatives, such as *swilled*, *swiller*, *swill-pot*, etc. seem to have passed into disuse.

swim for a wizard *obsolete* to test for magical powers of evil.

Witchcraft was a fruitful subject of taboo and euphemism, and I include this sample entry to remind us of the social behaviour and beliefs of our recent ancestors:

> So late as 1863, an old man was flung into a mill-stream... being what is called 'swimming for a wizard'. (Harland and Wilkinson, 1867—presumably, he drowned if he was human and you killed him if he proved himself a wizard by not drowning)

The custom obtained in Lancashire.

swing[1] to be killed by hanging.

From the rotation of a suspended corpse:

> On high as ever on a tow
> Swing'd in the widdie.
> (Anderson, 1826—*tow* was hemp, from which a noose; and *in a widdie* was swinging around)

Still used figuratively of receiving punishment.

swing[2] to engage in any taboo act.

From a slang meaning 'to act in a modern and unrestrained manner'. Of taking illegal drugs, copulation, homosexual activity, and anything else which may offend decency or propriety:

> *Thomas* Did you ever swing with her?
> *Cynthia* Twice. No more.
> *Thomas* Bent—isn't she?
> (Sanders, 1970—of homosexuality)

swing around the buoy to have an easy job.

British naval imagery from a ship at anchor, moving with the tides, and the consequent period of relaxation for the crew.

swing both ways to have both homosexual and heterosexual tastes.

Probably from *swing*, to be in the fashion as the *swinging sixties*, when open promiscuity and vulgarity were popularized:

> You swing both ways, huh? (Sanders, 1982—the male was copulating with a female homosexual)

swing off to die.

Not by hanging or even by other violence. The imagery is avian or possibly from jazz music:

> She placed flowers on his grave on the day he swung off. (Longstreet, 1956—he meant the anniversary)

swing the lamp to boast.

British naval usage and imagery, probably from the action of a signaller passing a message between ships at night rather than from the movement of a suspended lamp below decks:

> There were several groans and Andy Laird, the chief stoker, shouted, 'Swing the bloody lamp, somebody!' (Reeman, 1994—another crew member had been reminiscing)

swing the lead to pretend unfitness to avoid work or duty.

The association with the calling of the leadsman is unclear, unless he is claiming to be grounded while still in clear water:

> The majority were swinging the lead and would do anything to prevent themselves being marked A1. (Richards, 1933—*A1* was an active service category)

swipe to steal.

In the original American usage, perhaps a corruption of *sweep* and not from *swipe*, a blow. Usually of chance pilfering. The obsolete Norfolk meaning 'to raise lost anchors for an admiralty reward', sounds like a corruption of *sweep*.

swish to flaunt your homosexuality.

Of an American male who walks in a manner recognized by fellow homosexuals.

switch-hitter a person with both heterosexual and homosexual tastes.

From the American ambidextrous baseball player. In obsolete British use to *switch* was to copulate, along with to *swinge* or to *swive*.

switch on to excite.

A less common version of TURN ON using the same imagery.

switch-selling dishonest advertising of cheap goods designed to induce a customer to buy something dearer.

Not offering riding whips or false hair but a practice condemned to death in 1962 by the British Code of Advertising Practices, but too useful to be allowed to die:

> ... there must be no 'switch selling', namely advertising one article at a cheap price in the hope of persuading the customer to switch to a more expensive one. (E. S. Turner, 1952)

switch the primer *Irish* to perjure yourself.

The *primer* was a prayer-book and a Roman Catholic would have small regard for the mana of the Protestant bible produced for him to swear upon:

> He switched the primer himself that he was innocent. (Carleton, 1836)

swordsman a male profligate.

With common imagery:

> 'Bit of a swordsman, was he?' ... 'The post-mortem suggests there was sexual activity on the night of the murder.' (Blacker, 1992)

sympathetic ear a self-righteous person who forces his attention on those suffering a misfortune.

Literally, someone prepared to listen with sympathy:

> No tragedy is too immense and no personal anxiety too insignificant to be absorbed by Britain's vast emotional sponge of psychotherapists, social workers, trauma experts, do-gooders, and assorted sympathetic ears. (*Daily Telegraph*, 31 March 1994)

syndicate *American* an association of powerful criminals.

> 'When we talk about the rackets, are we talking about the same guys?' 'We're talking about the syndicate.' (Ustinov, 1971)

Literally, any group of business etc. associates, but in this use a shortened form of *crime syndicate*.

syndrome any taboo medical condition.

Originally, a set of symptoms of which the cause was conjectural or unknown, but now denoting established afflictions such as DOWN'S SYNDROME, *Acquired Immune Deficiency Syndrome* (AIDS), *Korsakow's Syndrome* (delirium tremens), and the recently noted *School Phobia Syndrome* (see EDUCATION WELFARE MANAGER).

syphon off to steal.

Usually by embezzlement:

> No way he could have spent more than half of what was coming in ... The best guess was that Birdsong ... was siphoning it off. (Hailey, 1973)

Also describing the stealing of petrol from motor vehicles.

syphon the python to urinate.

Of a male, with the common serpentine imagery.

syrup a wig.

From the rhyming slang *syrup of figs*. Usually one worn by a male, for whom the taboo remains strong.

T

TB see CONSUMPTION.

tablet a contraceptive taken orally by women.

As distinct from an aspirin, say. *The pill* is more common (see also PILL[2] (THE)).

tackle the penis and testicles.

Seen sexually, from the meaning 'equipment', and humorously as *wedding tackle*.

tactical done involuntarily under pressure.

Originally, relating to the deployment of troops, but a *tactical re-grouping* is a retreat. A *tactical nuclear weapon* is smaller than a *strategic* one, for use primarily against soldiers rather than cities.

tagged[1] hit by a bullet.

Literally, labelled, and perhaps from the superstition among soldiers that the bullet which hits you *had your name on it*:

> 'Tagged!' he realized. There was no mistaking it, he had been hit before. (W. Smith, 1979)

tagged[2] *American* detected in the commission of a crime.

> Ralph got tagged for stealing stamps. (Steinbeck, 1961)

Again I suppose from the identification when you want to remain anonymous.

tail[1] a woman viewed sexually by a male.

From *tail*, the buttocks:

> It's tail, Lew. Women. (Bradbury, 1976)

More common as a *bit* or *piece of tail*:

> She was a piece of Scandinavian tail that he'd picked up. (Matthew, 1978)

See also *flash-tail* at FLASH[1].

tail[2] to follow surreptitiously.

From staying close behind, whence a *tail* who so follows and a *tail job*, the act of following:

> You can do a tail job on him. (Allbeury, 1976)

tail-pulling the publication of a book at the author's expense.

Publishers' punning usage, from the meaning 'teasing'.

take[1] to steal.

Almost too venerable to be accepted as a euphemism—the OED gives an illustration in this sense from 1200. In modern use, *on the take* is receiving bribes:

> You're on the take from one of the mobs. (Deighton, 1978)

Also as *take your end*:

> Chicago was a right town then. The fix was in. The dicks took their end without a beef. (Weverka, 1973)

take[2] to copulate with.

Usually of the male:

> To take her in her heart's extremest hate (Shakespeare, *Richard III*)

and in modern use:

> It didn't stop the waves of lust as he took her. (Allbeury, 1976)

Rarely, although with rather more logic, the female *takes* the male:

> Chandra ... had been the cause of his love affair ... for she had taken him just to forget Chandra. (Masters, 1976)

take[3] to kill.

The victims are animals, by culling or hunting:

> And many of the creatures she allowed to escape. 'You take him,' she would say. (Mailer, 1965)

take[4] to conceive.

Of domestic animals, as of cuttings or grafts:

> Some mares won't 'take'. (Francis, 1982, of horse breeding)

take[5] to overcome or master.

An omnibus use which describes anything from passing another vehicle on the highway to any kind of villainy:

> He had no doubts he could 'take' the apartment at Fontenoy House; he was, after all, one of the best cracksmen in London. (Forsyth, 1984)

take a bath to suffer a heavy financial loss.

Your boat is capsized:

> His old man took a bath in real estate about ten years ago, got in the shower, and emptied his brains out with a .45. (Diehl, 1978)

In 18th-century England to *go to Bath* meant much the same, so adroit were the frauds and beggars in that spa.

take a bit from to copulate with extramaritally.

Usually of a female on a regular basis:

> Margot Dunlop-Huynegin is taking a little bit now and then from her husband's valet. (Condon, 1966)

take a break to allow the intrusion of advertisements.

British television jargon; statutorily, the advertisements are meant to appear when the continuity of the programme will not be interrupted. This poses a problem for a producer who wants to hold the attention of his audience without unduly contravening the regulations of his licence to transmit.

take a hike to be dismissed from employment.

A variant of the more common WALK[2]. Also used of any peremptory ejection:

> They told him to take a hike, because it was so gross. (Theroux, 1993)

See also HIKE[2].

take a leak see LEAK[1].

take a liberty see TAKE LIBERTIES.

take a long walk off a short pier to drown.

The assumption is that the pedestrian cannot swim. Used figuratively as a mild riposte or insult.

take a point to listen unsympathetically to a contrary view.

A device of politicians who eschew contradicting constituents:

> She received us most cordially and took all the points we made. (*Daily Telegraph*, February 1980, of Mrs Thatcher)

take a powder to leave hastily to avoid an obligation or publicity.

From the rapid departure necessitated after taking a laxative. Of checking out of a hotel without paying; of deserting a spouse:

> ... she's the one who took the powder. I didn't ask her to leave (Turow, 1987)

of running away in battle:

> ... your guys took a powder and the Krauts just came rolling over our support areas (Deighton, 1981)

or of going into hiding:

> Dean commented that it would be a good idea ... for Hunt to take a powder. (Colodny and Gettlin, 1991, of Watergate)

take a turn in the stubble to copulate.

The *stubble* is the female pubic hair as in the punning *shoot over the stubble*, to ejaculate before vaginal entry. According to Grose, a man might take other punning *turns, in Cupid's Corner, Love Lane, Mount Pleasant*, and other suggestive locations in and around London; and a woman might *take a turn on her back* in any part of the kingdom.

take a view to publish a misleading statement.

Accountants' jargon and self-justification when they modify published figures by taking a subjective view of reserves, inventory, etc. to please a client.

take a walk[1] *American* to resign from employment.

Or less often, to be dismissed:

> I think he should take a walk. Who needs this shit? (M. Thomas, 1985—he referred to a troublesome affair and not to the employee)

take a walk[2] to defect.

And not return:

> Years ago—before Fiona took a walk... (Deighton, 1988—Fiona had defected to Russia)

take a walk[3] to be stolen.

The usage may avoid a direct accusation of theft:

> If half a million pounds took a walk... (Deighton, 1988)

See also GONE WALKABOUT.

take a wheel off the cart to bankrupt someone.

Bankers' jargon for demanding repayment of a loan on terms with which the borrower cannot comply. In 1977 a banker who told me he didn't intend to do it, did.

take advantage of to copulate with casually.

Of a male, alluding to the female's weakness and his ungentlemanly conduct:

> My later behaviour in taking advantage of her did no more than damage her self-respect. (Amis, 1978—but how did he know?)

Also in the obsolete form to *take vantage*:

> 'I fear her not, unless she chance to fall'... 'God forbid that! for he'll take vantages.' (Shakespeare, *3 Henry VI*)

take an early bath to be sent off the field for foul play.

In the British Isles, too, occasionally of being replaced by a substitute after playing badly or being injured. In similar vein, an American sportsman may be *sent to the showers*.

take-away where no facilities for eating on the premises are provided.

Of catering establishments, whence an implication of inferiority as, for example, in the figurative:

> We'll be able to advertise Take-Away Degrees. (Sharpe, 1979)

take care to take steps to avoid impregnation during copulation.

As in CAREFUL[2]:

> I won't hurt you and I'll take care. (Harris, 1925)

The converse is the non-euphemistic *take a risk*, to copulate casually without contraception, risking unwanted pregnancy or disease.

take care of[1] to kill or render impotent.

Literally, to look after, whence to account for:

> He might have been afraid you would have him taken care of. (Chandler, 1943)

Clearly, the commissionaire or the night-watch could easily be 'taken care of'. (Forsyth, 1994)

take care of[2] to bribe.

Again from the meaning 'to look after':

Osborne had always known which officials needed to be taken care of. (Archer, 1979)

take for a ride[1] *American* to murder.

You bundled your victim into a car and killed him in a secluded place:

...takes him for a ride. His death is attributed ... (Lavine, 1930)

You fared no better if you were *taken for an outing* or *out into the country*.

take for a ride[2] to cheat.

Damaging, but not terminal.

take home to die of natural causes.

The devout, for whom heaven is *home*, are taken there by Jesus. Occasionally of an animal:

If it would please the Lord to take it home. (EDD)

take in your coals to contract venereal disease.

American male naval usage, punning on the burning sensation and the heat of a coal-fired steamship.

take leave of life to die.

Evasive circumlocution as much as euphemism, although it does suggest a voluntary decision where dying is concerned:

He could eat nothing, not rally his strength; and within ten days he took leave of life. (Monsarrat, 1978)

take liberties to make an unwanted sexual approach to a woman.

It covers anything from attempted fondling to copulation:

...[the licentious monk] proceeded to take still further liberties. (Lewis, 1795—the entry of her mother saved the girl from rape)

Also as *take a liberty*:

Nobody ever tried to take a liberty with her. (McCarthy, 1963)

take little interest in the opposite sex to be a homosexual.

Or *no interest*. The phrase should be an object lesson in the danger of resorting to euphemism, especially when making important statements. One of the British traitor Vassall's referees said that:

...he took very little interest in the opposite sex. (N. West, 1982)

If she had said he was a homosexual, and therefore a potential blackmail victim, he would not have been given access to top secret information, and various other unfortunate events would have been avoided, not least those which happened to Vassall himself.

take needle to inject narcotics illegally.

Not the action of a sempstress:

About to take the needle. (Mailer, 1965, referring to a drug user)

See also ON THE NEEDLE.

take off[1] to ingest illegal narcotics.

From the intended effect. In America to *be taken off* is not to be imitated but to be robbed of drugs or the cash to buy them by a user of illicit drugs. A *take-off artist* is one who thus secures supplies.

take off[2] *Scottish, obsolete* to die.

Before any visible manifestation of wings, however:

You were in the house at the time of his taking off. (Beatty, 1897, referring to a death)

take out[1] (of a male) to court (a female).

The action may take place in the front room, if secluded enough.

take out[2] to kill.

Perhaps short for *take out of circulation*:

If a KGB agent named Talaniekov appeared on the scene he was to be taken out as ruthlessly as Schofield. (Ludlum, 1979)

To *take out* opposition is to overpower it by violence:

Japanese counter-terrorist people had decided to take out the headquarters of the fanatical ultra-left Red Army Faction. (Forsyth, 1984)

take precautions see PRECAUTIONS.

take someone's name away *obsolete* to copulate with someone casually.

In such a case, the woman's name was unlikely to be changed by marriage to her partner but her *good name*, or reputation, could be lost:

The captain of the football team spent a whole year trying to take my dear name away from me. (Mailer, 1965)

take someone's shirt off to reduce someone to penury.

He will figuratively be reduced to parting with his apparel. Also as *take someone's pants off*:

What about a game of poker...I'm going to take the pants off you. (C. Forbes, 1992—the threat was financial and not sexual)

take something to drink an intoxicant or use an illegal narcotic.

In various forms, including the polite enquiry *What will you take?*, which offers a choice of intoxicants. It also refers to narcotics:

'Have you taken anything?' (This meant drugs.) (Murdoch, 1977)

Drink taken implies a degree of drunkenness:

I had only three bottles o' porther taken. (Somerville and Ross, 1908)

take steps to initiate proceedings for divorce.

Legal jargon, perhaps short for *take steps to end the marriage*—or 'to secure as much alimony as you can'. Also used of other kinds of litigation.

take stock of the situation *British* to decide whether or not to abandon a case.

Legal jargon, using the imagery of shopkeeping:

[The judge] indicated to prosecuting counsel that it would be a good idea if he 'took stock of the situation'—which means in judicial language, that he should throw in the towel. (*Private Eye*, May 1981)

take the air to urinate.

From the days when you had to leave the house:

Danny rose and said he needed to take the air, a gentlemanly statement of his desire to use the outhouse. (Keneally, 1979)

take the can (back) see CARRY THE CAN.

take the drop to be killed by hanging.

From the scaffold:

He's as good as taken the drop already. (G. Greene, 1934)

To *take a drop* is merely to drink intoxicants, although implying habitual excess, in which case the victim dies more slowly.

take the mickey to taunt or make fun of someone.

How can you be too old to be a Kylie fan? All our friends take the mickey but we all think she's gorgeous. (*Fast Forward*, 17 January 1990)

From the rhyming slang *Mike Bliss*, piss, adapted for figurative use. Also as *take the micky*, *Michael*, *Mick*, or *Mike*. See also MICKY.

take the pledge see PLEDGE (THE).

take to bed to copulate with.

What does it matter to me if she lets a man take her to bed? (G. Greene, 1932)

Used of either sex. See also BED[2].

take to the cleaners to rob or cheat.

The process thoroughly removes all surplus matter:

Dantzler's sporting a new Ferrari, braggin' on the street how he took some cowboy to the cleaners. (Diehl, 1978)

take to the hills to escape or go away.

Fleeing captivity, real or figurative:

I really thought seriously of taking to the hills with our little Laura. (B. Forbes, 1983—a married man was thinking of leaving his wife for another woman)

take too much to be drunk.

The precise intoxicant is unspecified:

I very much fear he has taken too much. (E. Waugh, 1933—she thought he was intoxicated)

Referring to a single drinking bout or regular intemperance.

take under protection (of a pimp) to control (a prostitute).

The jargon of prostitution:

When a Pole has chosen a Jewish girl he calls it 'taking her under his protection'. (Londres, 1928, in translation)

take up with to cohabit and copulate extramaritally with.

Literally, no more than to consort with or support:

After a quarrel too, a lad goes and takes up with another young gal. (Mayhew, 1851)

take with you to kill.

Used of people who are themselves about to be killed:

... a few desperate wretches taking as many Sioux with them as they could. (Fraser, 1982)

take your departal see DEPART THIS LIFE.

take your end see TAKE[1].

take your leave of to bereave.

The final parting:

... so absolutely unlike the way Frank would have wished to take his leave of us. (M. Thomas, 1982, referring to a death)

take your life to kill yourself.

As distinct from *take your life in your hands*, to risk your life rashly, or just to *take life as it comes*, to live life in a casual way:

Beautiful Young Society Matron Takes Life in Plunge. (Mailer, 1965, referring to a suicide)

To *take life*, to kill, is explicit.

take your snake for a gallop to urinate.

Of a male, with the common serpentine imagery, the slang *snake* being the penis.

taken dead.

Not being killed, as in TAKE[3], but being taken from this world to another or as the case may be:

He was taken with leukaemia. (Ustinov, 1971, giving a cause of death)

The dialect *took* is more emphatic:

Took he was—in the pride o' his prime. (Ollivant, 1898)

A *taking* or, less often, a *taking hence* is a natural death:

I was present at her taking, and though I be partial to death-beds... (Zack, 1901)

The early days before the taking hence of her brother John. (Jane, 1897)

taken-away *obsolete* a changeling.

The Scottish fairies took your bonny breast-fed child and replaced it by a sad puny creature, weaned on to an inadequate diet, but no corpse was left in the cot for scientific enquiry because 'Whoever lives to see him dee will find in the bed a henweed or a windlestrae, instead o' a Christian corpse'. (Galt, 1823—referring to a *ta'en awa*. A *windlestrae* is a stalk of withered grass.)

taken short needing to urinate at an inconvenient time or place.

I suppose from the days when coaches and trains without corridors used to stop at staging posts or stations, but not between them:

> We used empty bully-beef tins for urinating in. If a man was taken short during the day, he had to use the trench. (Richards, 1933, writing of the First World War)

taking death. See TAKEN.

talent women viewed sexually by a man.

> He had no plans to get trapped by just any piece of gash. The talent in the place had to be seen to be believed. (Collins, 1981)

Presumably he hopes one of them has a talent for copulation, although he wouldn't put it that way.

The punning *talent-spotting* is male searching for such females.

talk to the old gentleman to die.

> I reckon thar be a few Pawnees talkin' to th' old gennelman 'fore long. (Fraser, 1982)

The *old gentleman* was the devil; see also OLD A' ILL THING.

tall under the influence of illegal narcotics.

In drug users' slang usually with reference to marijuana, and a pun on HIGH².

Tampax time the period of menstruation.

Of obvious derivation:

> When it's Tampax time, the lady is a tramp. (B. Forbes, 1989)

tank a prison cell.

From its size and shape. In America it is usually held in reserve for the accommodation of newly arrested prisoners.

tank fight *American* a fraudulent boxing match.

One of the fighters *dives* into a figurative *water tank*—collapses voluntarily to the floor—whence the pun on the contest between armoured vehicles.

tanked (up) *American* drunk.

Motoring imagery, which may owe something to the German *tanken*, to fill with fuel:

> He got tanked up one night and stood on his chair and sang. (Theroux, 1973)

tanquem sororem *British* without copulation.

The literal meaning of the Latin is 'as a sister'. 19th-century legal jargon in a plea for nullity on the ground that the marriage has not been consummated by copulation but was a relationship as between brother and sister.

tap¹ to drink intoxicants.

From piercing a cask with a bung, to draw off liquid:

> I got the square bottle out and tapped it with discretion. (Chandler, 1939)

To *tap the admiral* is reputed to come from the naval habit of sucking liquor furtively through a straw, not excluding brandy from the cask in which Admiral Nelson's corpse was being preserved on its return from Trafalgar for a state funeral in London.

tap² a clandestine listening device.

Espionage jargon for something drawing off sound rather than liquid.

tap³ to obtain an advantageous loan or other finance from.

Again the imagery of the faucet:

> He's invested in movies, I believe, though being a chum I've never tapped him. (C. Forbes, 1983)

tap⁴ the constant availability of stock from willing sellers.

The implication is that they know more than the prospective buyer. Whence the Stock Exchange saw: 'Where there's a tip, there's a tap.'

tap⁵ see DO-LALLY-TAP.

tap a kidney to urinate.

Of either sex, from the renal function:

> I tapped a kidney in the ladies' room. (Theroux, 1978)

tapped *American* indigent.

Like a cask which has been broached and emptied. Especially of someone who has lost a lot of money gambling.

taps (the) *American* death.

Military use, from the roll of a drum at a funeral.

tar an illegal narcotic.

Formerly opium, from the colour and tackiness perhaps. Now denoting heroin produced in Mexico; a shortened form of *black-tar heroin*, with the same derivation.

tarbrush (the) partial descent from a non-white ancestor.

Usually as a *touch* or a *lick of the tarbrush*, from the difficulty in eradicating every trace of tar from a brush which you may later wish to use for applying white paint, genes controlling black pigmentation being equally persistent:

> ... her body was slightly darker than could be expected even by a rich girl's sunburn, her breasts were brown. ('Touch of the tarbrush there,' murmured Pinn.) (Murdoch, 1974)

The use is offensive.

target of opportunity random bombing.

The common instruction to Second World War bomber crews who might fail to reach the assigned targets but had to jettison their load if they wanted to improve their prospects of a safe return:

> They bombed 'targets of opportunity'... shutting your eyes, toggling the bombload, gaining height, and getting the hell out. (Deighton, 1982)

tarry-fingered *?obsolete* thieving.

Sailors in Scottish ports had a reputation for pilfering—the *tar* came from the tarpaulins they wore:

> To prevent 'tarry-fingered' customers, all the hobs were hooked in unison. (Gordon, 1880)

tart a prostitute.

From *jam tart*, rhyming slang for 'sweetheart':

> Young lady indeed. She's a tart. (G. Greene, 1932)

Also used to denote a young woman who is willing to copulate casually.

taste to copulate with.

Of a male, extramaritally and once only:

> If you can make't apparent
> That you have tasted her in bed.
> (Shakespeare, *Cymbeline*)

You might also more specifically *taste her body*. The use survives in the modern American *taste*, a single act of extramarital copulation by a male.

taste for the bottle an addiction to alcohol.

From BOTTLE[1] (THE):

> A letter from her daughter Norah to Henry Harrison delicately hinted at a taste for the bottle. (Foster, 1993, of Katherine (or Katie but never Kitty) Parnell in later life)

taters the testicles.

A shortened form of *potatoes*, and figuratively still 'lost' by English Midlanders who act stupidly.

tax *American* to steal with a threat of violence.

Our contributions to central and municipal funds, involuntary and onerous though they may be, are not made under threat to our person:

> The principle of 'taxing'—mugging to steal shoes—is well established in the tough cauldrons of America's inner cities. (*Daily Telegraph*, June 1990)

Let us hope the use proves as ephemeral as the equivalent British *steaming*, where a crowd of youths runs through a crowd assaulting people and thieving from them.

taxi drinker a prostitute.

She introduces herself to potential customers by accepting a drink from them at a bar. Punning perhaps on the *taxi dancer* whom a solitary male may pay to dance with.

tea marijuana.

From its likeness, when chopped, to tea leaves:

> ... marijuana; he called it tea. (Styron, 1976)

Texas tea is specific and *tea sticks* or *sticks of tea* are marijuana cigarettes:

> There isn't much record he went for tea-sticks or the smoke. (Longstreet, 1956)

> Three highballs and three sticks of tea. (Chandler, 1940)

American *tea heads* may smoke marijuana illegally at a *tea party*.

tea leaf a thief.

> Or go and be a straightforward tea-leaf —thieve, rob. (Kersh, 1936)

Rhyming slang.

tear off a piece of arse to copulate extramaritally.

Of a male, with no actual physical damage to the partner but with overtones of haste and a transient relationship. As is frequently the case with such vulgarisms, *bit* may be substituted for *piece*, and *ass* for *arse*.

tearoom *American* a public lavatory frequented by homosexuals.

Denoting a known meeting place, as one frequented for the purposes of refreshment and gossip. Whence *tearoom trade*, those who seek out other homosexuals in such a place:

> The Tea Room Trade they call it in America; in England, Cottaging. (Fry, 1991)

technical adjustment a sudden fall in stock-market prices.

It covers anything from an absence of buyers, bad economic news, and stock dumping, to panic selling. The phrase seeks to imply that jobbers are merely covering their positions by marking prices down without undue selling orders. Be wary equally of a *technical correction* or *technical reaction*.

technical surveillance see SURVEILLANCE.

technicolor yawn vomiting due to drunkenness.

Of obvious imagery:

> No sooner was Lord Matey allowed back in than he failed to stifle a technicolour yawn and swamped the entire bar. (*Private Eye*, February 1988—but I don't think the colour

film process was ever Anglicized as this spelling suggests)

tee in to avoid payment for telephone calls.

Used in electronic jargon to mean 'to connect one circuit to another':

> It was suspected someone was 'teeing in', which is a method of getting direct access to the phone line without paying for the calls. (*Daily Telegraph*, 31 March 1994—the device cost £10, the thief was fined £327, and the amount the telephone company lost was £44,586.90)

temperance see INTEMPERANCE.

tempered in the forge of labour imprisoned.

A Communist phrase, usually of those shut up for expressing political or nationalistic dissent.

temple (of health) *?obsolete* a lavatory.

A genteel usage, based on the supposed medical advantages of regular defecation and the need so to *worship* daily.

temporarily abled see ABLEISM.

temporary permanent and embarrassing.

The American army in Vietnam was especially prone to *temporary setbacks* etc.:

> It caused heavy casualties to be announced as light, routs and ambushes to be described as temporary tactical ploys. (Herr, 1977, of the American news services)

For Harold Macmillan, a *temporary local difficulty* was a major political crisis. For a firm, a *temporary liquidity problem* is insolvency:

> Your old man's got a temporary problem of liquidity. (le Carré, 1986)

ten commandments (the) scratches by a woman's fingernails.

When she says to a man 'Thou shalt not':

> Could I come near your beauty with my nails,
> I'd set my ten commandments in your face.
> (Shakespeare, *2 Henry VI*)

In occasional modern use too, of punches by either sex.

ten o' clock girl *British, obsolete* a prostitute.

After being arrested in London as a prostitute and bailed overnight, she had to attend court at ten o' clock the following morning.

ten one hundred *American* urination.

Slang based on the CB code for stopping to urinate beside a road. A *ten two thousand* is a seller of illegal narcotics.

ten years' turmoil the period of philistine anarchy under Mao.

The Cultural Revolution (see CULTURAL) necessitated all manner of evasions to mask its true objective, which was the perpetuation of Mao's rule:

> 'Yes, because the Ten Years' Turmoil'—he used the current euphemism—'went too far.' (Theroux, 1988—the speaker was Chinese)

tender a fool *obsolete* to give birth to a bastard.

To *tender* is to attend or wait upon, whence to offer or present. So the punning Polonius:

> Tender yourself more dearly;
> Or ... you'll tender me a fool.
> (Shakespeare, *Hamlet*)

tenure a job for life.

University jargon for a teacher confirmed in a post. The security of employment until retirement age is said to be there to encourage fearless thinking but it also protects the idle, the ageing, and the incompetent at the expense of students, other teachers, and research:

> He set up his tents in various different universities, from all of which he was tactfully evicted. He never achieved 'tenure'. (Murdoch, 1983)

term *obsolete* the period of menstruation.

In standard usage, any specific period:

> My wife, after absence of her terms for seven weeks ... (Pepys, 1660)

terminate[1] to kill.

Literally, to end:

> The people he terminated died for specific reasons. (M. Thomas, 1980)

When killing illegally, the CIA *terminates with extreme prejudice*:

> I'm afraid the project's been terminated. There was prejudice, extreme prejudice. (Lyall, 1980, of such a killing)

terminate[2] *American* to dismiss from employment.

Another ending:

> ... they had been sent home and demoted or else fired—'terminated' was our word. (Theroux, 1982)

terminate[3] to induce an abortion.

Yet another cessation, usually referring to the ending of unwanted pregnancies but also to those ended on medical advice:

> A nice girl from a nice home ... the thought of termination was unthinkable. (Seymour, 1980)

terminological inexactitude a lie.

British parliamentary usage, to circumvent the convention against calling another member a liar:

...half-lies, or as Erskine May finds more acceptable, terminological inexactitudes. (Howard, 1977)

test the mattress to copulate.
Usually extramaritally, with common imagery:
I returned unexpectedly...and found you testing the mattress with our dear old friend Johnny. (Deighton, 1981)

Texas tea see TEA.

thank to bribe.
In many places, verbal expressions alone are not enough:
'Have you thanked the captain?' 'I always thank everybody,' I replied naively. (Simon, 1979, describing passing a North African frontier with a British motorcycle)
To *thank* can also be to tip a servant, and a trap for a British visitor to America. I once handed a scantily clad female in a Philadelphia bar a $5 bill for $1.60 drinks—it must have been long ago—saying in the English fashion 'Thank you'. 'Thank yew, sir,' replied the harpie, keeping the change.

that and this urination.
British rhyming slang for 'piss'.

that way[1] homosexual.
Of either sex:
I never picked you for a sapphic...were you always that way? (McCarthy, 1963)
Like that usually refers only to male homosexuals.

that way[2] pregnant.
Female use, often referring to unwanted impregnation.

the worse drunk.
A shortened form of *the worse for drink* etc.:
She had never known him the worse for liquor. (Mayhew, 1862)

them a woman's breasts viewed sexually by a male.
A similar evasion to IT[3]:
...clothing disarranged to reveal a, to him, rare glimpse of 'them'. (Muir, 1990, quoting K. Amis's *Jake's Thing*)

thick stupid.
A synonym for *dense* and a shortened form of *thick in the head* but with no imputation of lunacy.

thick of hearing *obsolete* deaf.
A western English variant of the common HARD OF HEARING:
Doubtless I may be thick o' hearing. (Quiller-Couch, 1890)

thief *mainly Irish* the devil.
From his evil ways. He is usually *old*, or *black*, or *the thief of the world*:
May the thief o' the world turn it all into whishky an' be choked wid it. (Bartram, 1898)

thing any taboo object to which you refer allusively.
Such as a ghost, for which:
'Summut' or 'Things' is preferred. (*Spectator*, February 1902, quoted in EDD)
Or the penis:
She that's a maid now...shall not be a maid long, unless things be cut shorter. (Shakespeare, *King Lear*)
In modern use:
Measured my 'thing'. It was eleven centimetres. (Townsend, 1982—the speaker is a youth)
The *Watergate thing* was an unfortunate series of events for Richard Nixon.

thing going a sexual relationship between two people.
We did have a thing going in London, but nothing serious. (Reeman, 1994—the male speaker was married to a third party)

thingy the penis.
Nursery use. Also as *thingamajig*:
You stand there with your thingamajig in my toothmug. (Sharpe, 1979)

third age (the) senescence.
As in the *University of the Third Age*, a British lecture and discussion group for elderly people.

third degree (the) police violence to extract information.
Probably from the scale of seriousness of burns, of which the third degree is the worst:
A veritable catalogue of police third-degree methods is contained in a recent [February 1930] issue of *Harvard Law Review*. (Lavine, 1930)
Occasionally shortened to *third*:
He's giving me a third about a gun. (Chandler, 1934)
Etymologists who are also Freemasons may reject the association with BURN[3] in favour of their own experience at the third stage of their induction into the 'craft', when the ordeal and questioning are reported to be daunting.

third leg the penis.
Sometimes also as the *middle leg*:
He had to learn to live with the fact that his third leg had proved faulty. (Goldman, 1984—referring to someone who was impotent)

third party payment a bribe.
The favoured commercial euphemism of the 1990s. A *third party* is literally someone with a casual connection to the matter in hand.

third sex (the) homosexual males.

Formerly, eunuchs were so described. Now rare.

third world poor and uncivilized.

The other two were America and Russia and their respective allies:

> ... a wealthy Bostonian, from a family of some distinction, adventuring in Third World philanthropy. (Theroux, 1980)

A welcome change from the DEVELOPING routine. See also FIRST WORLD.

thirst (a) an addiction to alcohol.

Usually in the phrase *have a thirst*:

> There's a man that had a thirst, as the Irish would say. (Follett, 1991, and not just the Irish)

thread the needle *?obsolete* to copulate.

Of a male, with obvious imagery.

three-letter man a male homosexual.

Punning, I suppose, on *four-letter man* (see FOUR-LETTER WORD). The three letters are 'f-a-g'. (DAS). In earlier days, when Latin was more widely acclaimed and homosexuality less, he was a thief, the letters being 'f-u-r'.

three letters the Russian internal apparatus of repression.

A usage in Russia prior to the Second World War during Stalin's indiscriminate terror:

> ... the Three Letters (a pseudonym used in conversation to avoid mentioning the dreaded GPU). (Muggeridge, 1972)

three-point play *American* the recruitment of a black woman.

The imagery is from basketball. The employer gets a point for taking on another worker, so reducing unemployment; a second point if the worker is female, to show that he is not prejudiced about employing women; and a third point when he contributes to his quota of non-whites. You get a bonus if the recruit has American Indian ancestry.

three sheets in the wind see SHEET IN THE WIND.

threepennies (the) *British, obsolete* diarrhoea.

From the rhyming slang *threepenny bits*, shits, but now perhaps as rare as the coins themselves.

thrill a single act of copulation.

Formerly a euphemism for the sexual orgasm (DSUE), but now used in the prostitute's invitation *Can I give you a thrill?*

throne a pedestal lavatory.

From the shape, elevation, and solitary situation. Whence *enthroned*, of a person sitting on it:

> ... she looked along the vista and saw, at the far end Lord Doneraile enthroned playing the violin. (Bence-Jones, 1987, writing about an Irish mansion where the lavatory had been sited in the conservatory leading from the hall)

throw[1] to copulate with.

The common violent imagery:

> And better would it fit Achilles much
> To throw down Hector than Polyxena.
> (Shakespeare, *Troilus and Cressida*)

Today a male may *throw a leg over* or *throw a bop into* his sexual partner.

throw[2] to give premature birth to.

Usually of cattle, and still used at least in western England:

> Sight o' yoes've a-drow'd their lambs. (EDD —a *sight o' yoes* is many ewes)

throw[3] to lose deliberately or fraudulently.

A shortened form of *throw away*:

> I heard you were supposed to throw it. (Chandler, 1939, referring to a boxing match)

See also THROW IN THE TOWEL.

throw in the towel to concede defeat.

Boxing imagery, where the second does it for his principal. The implication can be that a braver or more determined person would have held on longer.

throw the book at to charge with every feasible offence.

Mainly police jargon, the *book* being the manual setting out criminal offences:

> You'll just have to throw the book at me... I don't sell out—even to good police officers. (Chandler, 1958)

throw up to vomit.

The oral expulsion is in fact usually downwards:

> I got so mad I actually threw up. Puked! (Theroux, 1982)

An Australian may be said to *throw a map*. To *throw up your toenails* is to be violently sick.

thud *American* an aircraft crash.

Vietnam use, and later some civilian use, from the dull distant noise.

thump to copulate with.

Of the male, using the common violent imagery:

> Jump her and thump her. (Shakespeare, *Winter's Tale*)

And in modern use:

> Well, if I'd had my way, he'd still have been thumping her every night. (Fraser, 1973)

thunderbox a portable lavatory.

The thunder is presumably from the noise of defecation:

When it rained the clients had to row themselves to the thunder-box at the bottom of the yard. (Simon, 1979)

The less common *thunder-mug* is a portable container for urine:

Have a water pitcher, wash-basin, fancy soap dish, and a thunder-mug. (Butcher, 1946, referring to Second World War camping equipment)

tick a person clandestinely following another.

From the parasitic arachnid, which sticks to your skin:

He saw his tick come in through the revolving doors, look around and, spotting Kim, make for the elevator. (van Lustbaden, 1983)

ticker the heart.

But you only speak of it in this way if you have reason to suspect it is likely to cease ticking:

'In any case I have a bad heart.' '*My* ticker was none too good,' said Mr Flack. (Theroux, 1974—the conversation was between valetudinarians)

ticket punched see GET YOUR TICKET PUNCHED.

tickle to copulate with.

Perhaps from the preliminary caresses, or the association with TICKLER[1]:

When the swollen little girl told her father the name of the man who had been tickling them—and I defy you to find a more revolting terminology.... (Condon, 1966—the girl was pregnant)

tickle your fancy a male homosexual.

British rhyming slang for NANCY. It is also a pun on *tickle your fancy*, a dialect term for the flower *Viola tricolor*, which most of us know better as the pansy.

tickler[1] the clitoris.

Perhaps from its significance in sexual arousal:

I went back to caressing her tickler. (Harris, 1925)

tickler[2] a contraceptive sheath.

...a wall-mounted vending machine that dispensed rainbow-coloured French ticklers. (Sanders, 1983)

tiddly slightly drunk.

From the rhyming slang *tiddly wink*, a drink:

I poured her wine carefully. 'Ma, you'll get tiddly.' (Bogarde, 1983)

A *tiddly wink* or *kiddlywink* was an unlicensed inn or a pawnshop before it came to mean the game played with counters, no doubt by drinkers.

tie a can on *American* to dismiss from employment.

Probably from the attachment of a can to the tail of an unwelcome stray cat, to drive it away. See also CAN[2].

tie one on *American* to go on a carouse.

The etymology of this phrase is uncertain:

We could tie one good one on, two days, three days, five empty bottles at the foot of the bed. (Mailer, 1965)

tied up unwilling to see or speak to a caller.

If a secretary says her boss is *tied up*, have no fear of bondage or a criminal incursion, nor even that he is moderately overworked—most of the time he just can't be bothered to speak to you. In obsolete English use it would have meant that he was constipated:

I be terrible a-tied up in my inside. (EDD)

tiger sweat an inferior intoxicant.

Referring to bad beer and spirits (but with no aspersions cast on 'Tiger' beer from Singapore):

King Kong is not a movie, it's cheap alcohol, also known as Tigersweat. (Longstreet, 1956)

Also as *tiger juice, milk*, or *piss*.

tight[1] drunk.

The OED suggests a connection with SCREWED, but which is a pun on which?

Well, he got in at last, and he lit a candle then. That took him five minutes. He was pretty tight. (Somerville and Ross, 1897—to give an example of so common a usage may seem redundant but it allows me to record a more picturesque euphemism in the same story: 'Let it not for one instant be imagined that I had looked upon the wine of the Royal Hotel when it was red.')

tight[2] *American* not capable of interception.

Espionage jargon, the opposite of *loose*, as in *loose talk*:

And, for the good Lord's sake, use a tight phone. (M. Thomas, 1980)

tightwad a miser.

His roll of bills is seldom loosened for the extraction of a banknote:

Cost him a hundred bucks to cancel which must have killed the old tightwad. (M. Thomas, 1987)

Tight-fisted is standard English for a mean person.

Tijuana bible *American* a pornographic book or picture.

Visitors to the Mexican border town find erotica openly displayed.

Tijuana taxi *American* a police car.

The external lights and signs on some American police vehicles call to mind the decoration of Mexican cabs.

time the happening of something subject to a taboo.

Of childbirth:

Elizabeth's full time came that she should be delivered. (Luke)

My wife—she be near her time wi' the eleventh. (M. Francis, 1901)

Of death:

Mr Ralph wur to die, his toime had coom. (Antrobus, 1901)

Of imprisonment:

'Listen,' he said, still softly. 'I did my time'. (Chandler, 1939)

time of the month menstruation.

Common female usage:

Could be that time of the day, that time of the month. (Bradbury, 1965)

time-sharing a compounded annual rental paid in advance.

A number of people are induced to advance capital in respect of the same property, with access limited to each for a stated period in each year. Whence the satirical:

Don't say two timing, sweetie, think of it as time sharing. (Young woman to old man in Smirnoff Vodka cartoon, *Private Eye*, June 1981)

tin handshake a derisory payment on dismissal from employment.

A less desirable outcome than the *golden handshake* (see at GOLDEN):

He's sacked, given a tin handshake and left to rot. (Allbeury, 1981)

tincture[1] a partial descent from other than white ancestry.

Literally, a pigment, and used of those whose darker skin pigmentation is noticeable:

She had a tincture herself or she would not have mentioned their race. (Theroux, 1977)

tincture[2] an intoxicant.

Literally, in pharmacy, a medical solution in alcohol:

So while I was shunted off for tinctures with a lot of silly women in leotards.... (*Private Eye*, February 1981)

tinker a gypsy.

The trade at which they were once adept and still a common use in Ireland. Occasionally shortened to *tink*:

I've had more than one tink woman to *chavver*... I'll take a bet a big girl like you's been *chavvered* by half the gyppos in Ireland. (O'Donoghue, 1988—to *chavver*, to copulate with, which has not yet made the OED, is a variant of the slang *chauver*, probably from the Romany *charver*, to touch)

An Irish synonym for *tinker* is *itinerant*, which is not euphemistic, and see also TRAVELLER. Many of these people are not of Romany descent but live with their families, their horses, and their vehicles on the *long acre*, the roadside verge.

tinkle to urinate.

Onomatopoeic nursery usage, from the noise of urine against a mild-steel receptacle (a 'tin-pot' would be enormously expensive):

If you sprinkle when you tinkle
Be a sweetie, wipe the seatie.
(Graffito in lavatory)

tinpot pretentiously assuming the trappings and manner of authority.

The etymology comes not from any lack of value in pots made of tin, but from the days when tinkers were wont to pass off inferior articles as such. Now only pejorative and figurative use:

...give away every scrap of Empire that remains to any tinpot potentate that asks for it. (*Private Eye*, July 1981)

tint to dye.

Literally, to colour slightly, but now barbers' jargon for dyeing hair:

...we drove sixty miles to Banbury to get her hair dyed—'tinted' they said in the shop. (Kyle, 1988)

tip[1] *American* to copulate with someone other than your regular partner.

A *tip* was a ram and in obsolete Scottish use to *tip* was to serve ewes with the ram (EDD):

Tip where you will, you shall lamb with the leave. (Old proverb)

However, I suspect this slang usage has a different origin.

tip[2] to die.

Often in the phrasal verb to *tip off*, with the common avian imagery:

They all tipped off an' deed. (Binns, 1889)

tip[3] to drink intoxicants to excess.

From the motion of tipping the container:

You're tipped, darling. You're hurting. (Steinbeck, 1961)

To *tip the bottle* is specific:

If she 'tips the bottle' he knocks her about a little more to teach her to keep sober. (Burmester, 1902)

Tipped and *tipsy* mean drunk:

'Was he tipsy?' 'I dare say... now you mention it, I think he was.' (E. Waugh, 1933)

A *tiper*, *tipper*, or *tippler* is a drunkard. *Tip* was ale once sold in Scotland at 2d. a pint, but a *tipple* is any intoxicant:

Helpers had brought in the drinks and bits. 'Do dig into the tipple,' said Serena. (Bradbury, 1976)

A *tippler* used to be an innkeeper, who kept a *tippling-house*:

No vyattler nor tipler to sell any ale or beer brewed out of town. (Lincoln Corporation Records, 1575)

tip[4] to warn or inform.

A shortened form of *tip off*, with a suggestion of betrayal or breach of confidence:

'Who tipped you?' he said, smiling... 'If I find him... I'll have his balls.' (Sanders, 1983)

tip over to rob.

From upsetting a stall and stealing some of the goods in the confusion, rather than from knocking down your victim. In America it applies to any theft; also to a sudden police raid.

tipple an intoxicating drink. See TIP[3].

tired[1] unwilling to copulate with a regular partner.

A female excuse which may or may not have to do with weariness:

...a kind of marital signal, looking to her for sexual encouragement, the unspoken suggestion that they would make love, 'I'm tired' or 'I'm not tired'. (Theroux, 1976)

tired[2] drunk.

The symptoms can be the same. The sarcasm is stronger with *overtired* (see OVER-REFRESHED). The British minister George Brown was perhaps the first to be described as *tired and emotional*:

Mr Brown had been tired and overwrought on many occasions. (*Private Eye*, 29 September 1967—he was reputed to be a drunkard)

to one side of the truth untrue.

Of political evasion:

'Nothing asked and nothing taken', was how Gladstone put it which, if not strictly falsehood, was certainly to one side of the truth. (Kee, 1993)

to the knuckle devoid of resources.

All the meat is gone, as with ON YOUR BONES:

It's to the knuckle. It's not MGM or anything. There's no money. (Bogarde, 1983)

toilet a lavatory.

Originally, a towel, whence washing and the place where the washing was done. And we use *toilet paper* for wiping rather than washing.

tokenism an appointment made other than on merit.

Often in relation to a black:

There was evidence of 'tokenism', employing black staff purely for their colour. (*Daily Telegraph*, June 1984)

Whence a *token appointment*, to placate a pressure group or our social conscience:

The token black, Dr Clifton R. Wharton, Jr. had gone in 1975. (Lacey, 1986, of the Ford Board of directors)

Tokyo trots diarrhoea.

Not a Japanese race but one of the many alliterative geographical attributions.

tolbooth *Scottish, obsolete* a prison.

Originally, the town hall, where tolls were paid. The jail was often in the same building:

How many gypsies were sent to the tolbooth? (W. Scott, 1815)

tom[1] an act of defecation.

From the rhyming slang *tom tit*, shit, but never used as an insult. The full expression is also used of turds:

All that Tom Tit blown up in the air. (B. Forbes, 1986—a sewage works had been bombed)

tom[2] to act as a prostitute.

I suppose from the sexual reputation of the cat, although *queen* might seem the correct gender.

tomboy a prostitute.

From the feline reputation and perhaps also a corruption of *tumble* (see TUMBLE[1]):

A lady
So fair... to be partner'd
With tomboys.
(Shakespeare, *Cymbeline*)

Today it means no more than a young girl with the athletic and other tastes of a boy, although some tell us that boys and girls should not have different interests.

tomcatting regular extramarital copulation by a male.

Again from the cat's reputation:

The tomcatting made history in the form of songs. (Longstreet, 1956, of New Orleans)

tommy the penis.

Rarer than DICK[1], commoner than *Harry*:

She... had to use her hand to get my Tommy in again. (Harris, 1925)

too many sheets in the wind drunk.

An uncommon variant of SHEET IN THE WIND and its variants in numerical form.

took see TAKEN.

tool the penis.

Literally, any instrument, whence the punning:

'Draw thy tool'... 'My naked weapon is out.' (Shakespeare, *Romeo and Juliet*)

and in modern use:

No accountability could be apportioned anywhere for how his tool behaved, or failed to behave, while he slept. (Amis, 1978)

Grose has:

TOOLS. the private parts of a man.

tooled up carrying a gun.

A *tool* as a weapon is obsolete standard English and this use by modern thieves when about their business puns on the processes of mass production.

toot a carouse.

Perhaps from the noise, although *toot* is one of those words with many slang meanings for taboos down the centuries, from the devil, lunacy, defecation, and farting to illegal ingestion of narcotics:

Her husband was off on a toot. (Chandler, 1953—he was on a drunken spree)

top[1] *obsolete* to copulate with.

Either a corruption of the standard English *tup*, or from the position adopted by the male, or from his supposed dominance:

Behold her top'd? (Shakespeare, *Othello*)

top[2] to kill.

The imagery is from silviculture or horticulture:

Just who did top Ambassador Mobuto? It came as a great relief to all concerned to find he had topped himself. (*Private Eye*, March 1980)

To *top* was much used of executions and hangings:

These fellows you are topping in batches. (Flanagan, 1979, of public hangings)

The obsolete *topping fellow* was a public hangman, and a gruesome pun.

top and bottom see HALF AND HALF.

top and tail to clean up a baby.

Nursery usage with imagery from preparing gooseberries or root crops, which need trimming at either end. The baby may have vomited as well as defecated.

top-heavy drunk.

And unable to stand upright without swaying:

We kept on drinking and yarning until stop-tap. At that time we were getting a little top heavy. (Richards, 1933)

top sergeant a female homosexual taking the male role.

American soldiers in that rank have a reputation for roughness and toughness. Sometimes shortened to *sergeant*.

top up to conceal inferior goods below those of higher quality.

Usually of fruit sold by weight where only part of the purchase is visible:

... a few tempting strawberries being displayed on top of the pottle. 'Topping up,' said a fruit dealer. (Mayhew, 1851)

topless exposing your breasts in public.

Beach, bar, and entertainment usage:

As one of the show-girls who had to strut around the stage topless... (Green, 1979)

Thus a *topless bar* is not one which is open to the heavens. However a *topless* person used to be, and occasionally still is, bareheaded.

torch of Hymen (the) copulation only within marriage.

Hymen, the god of marriage, was depicted as carrying a torch:

The torch of Hymen burns less brightly than of yore. (Mayhew, 1862)

torpedo *American* a hired assassin.

As a submarine may fire a single missile from its hidden place, so is he recruited to undertake a specific murder away from his normal base.

toss[1] an act of copulation or ejaculation.

Perhaps alluding to hay—see IN THE HAY—or from TOSS OFF:

He had a toss in the hay with his tootsie tonight. (Sanders, 1981)

Whence the common vulgarity *I don't give a toss*.

toss[2] to search (another's property).

Usually without consent and looking for illegal narcotics, but also of any incursion:

'How did you find out the apartment had been searched?'... 'She... knew where everything was kept. She swears the place was tossed.' (Sanders, 1986)

The imagery is from throwing things into the air.

toss down to drink (an intoxicant).

From the movement of the container, but it could be hay off a stack:

'We need to talk,' he said, 'and toss down a few before you go.' (Shirer, 1984)

toss off to masturbate.

Of a male, from the ejaculation:

I could have another whisky, toss myself off in the loo. (Theroux, 1973)

The figurative *tosser* is a term of abuse:

What would they know? Bunch of tossers. (C. Thomas, 1993)

toss out to feign withdrawal symptoms from narcotics.

A trick by an addict to induce a doctor to prescribe a fresh supply of narcotics.

tot a drink of intoxicant.

Literally, anything small, whence a small drinking vessel or measure, which used to be from quarter to half a pint. Formerly to *tot* was to drink intoxicants to excess:

An' th' women folk... can tot
That Dunville's Irish whiskey.
(Doherty, 1884)

totty a prostitute.

Normally it meant 'of bad character':

> I tyell yü bestways 'ave nort tü dü wi' she; er's nort but a totty twoad. (Hewett, 1892)

I think the suggestion that the usage comes from a corruption of the name Dorothy is wrong (DSUE). Still used in English dialect.

touch[1] to copulate with.

Of the male, despite the mutuality:

> ...you have touch'd his queen
> Forbiddenly
> (Shakespeare, *Winter's Tale*)

and still heard in Devon:

> I asked her, Did I touch you in the night? (Reported conversation, 1948, after an unplanned pregnancy)

For Grose 'To touch up a woman is to have carnal knowledge of her' but see TOUCH UP[1]. To *touch yourself* is to masturbate, usually of a female:

> You want to know whether I have touched myself. Sure; all girls have. (Harris, 1925)

touch[2] a theft.

Usually by stealing from a pocket, from the physical contact. Today normally of borrowing money which you do not intend to repay:

> A quick ten or twenty dollar *touch*, which of course was never intended to be returned. (Lavine, 1930)

The obsolete British *touch-crib* was the kind of brothel in which you were liable to be robbed as well.

touch of the tarbrush see TARBRUSH (THE).

touch signature a fingerprint.

Bankers' jargon when they want positively to identify their customers without accusing them of being crooks:

> The practice is known by the euphemism 'touch signature', an approach which one banker described as 'part of our back-up security system'. (*Daily Telegraph*, September 1980, of fingerprinting)

touch up[1] digitally to excite the genitals of (another).

Usually the male does it to the female, unless they are homosexuals:

> ...it would be ridiculous to keep you from your work just because you touched up some Jewess. (Keneally, 1982)

The punning *touch-hole*, the vagina, is obsolete.

touch up[2] to dye (hair).

Barbers' jargon, implying a partial application where in fact the whole is treated.

touched[1] of unsound mind.

Shortened form of *touched in the head*:

> ...an uncle who had a passion for concrete dwarves...who his mother said was a bit touched in the head. (Sharpe, 1974)

touched[2] *?obsolete* drunk.

And usually only mildly affected by the contact:

> In respect of her liquor-traffic, she was seen 'touched' about once a week. (Tweeddale, 1896)

tourist inferior.

Of hotel accommodation and especially public transport. Airlines use verbal ingenuity in calling the dearer seats for richer travellers something else, such as *club*, *sovereign*, *executive*, or *clipper*.

touristas (the) *American* diarrhoea.

Suffered by many a tourist, or *turista*, whose digestion is unprepared for the diet, climate, want of cleanliness, and excessive alcohol to which it may be subjected on a Mexican vacation.

tout a police informer.

From the tipster who covertly observes racehorses in training; now the jargon of Irish terrorism:

> ...if there's a tout on the mountain and he's dead you won't find tears on me. (Seymour, 1992)

town bike a prostitute.

So called because she is available for all the men to RIDE.

town house a modern dwelling built on a small plot.

Mainly British estate agents' jargon for a modern terrace house, which need not be located within the urban centre. I don't think there is any implication that the occupant is rich enough to have a country house as well.

town pump a prostitute.

Every man went there for satisfaction, in the days when there was no piped water to individual houses.

toy boy a man consorting sexually with a much older woman.

She may not pay him money for his services but is more likely to ply him with gifts:

> At 48 she is like a teenage girl again—raving it up with four different lovers including a toyboy of 27. (*News of the World*, 15 November 1987)

tracks the scars left by repeated injection of illegal narcotics.

Like American railroad lines:

> Russell inconclusively scanned her arm for tracks. (McInerney, 1992)

trade (the) prostitution.

Prostitutes' jargon everywhere. In America the *trade* is also a prostitute's customer:

> She doesn't like the trade, she packs it in and goes home. (Diehl, 1978)

A *trader* is a prostitute and *living by trade* is so working:

Oh, there's no doubt they live by trading. (EDD)

See also ROUGH TRADE.

traffic with yourself *obsolete* masturbation.

Of either sex:

For having traffic with thyself alone,
Thou of thyself thy sweet self dost deceive. (Shakespeare, *Sonnets*)

trainable of low intelligence.

Educational and sociological jargon, to distinguish from an idiot, who cannot be trained at all. Jennings (1965) also quotes from *Today's Health* 'educable, corresponding to moron'.

tramp a prostitute.

Originally, from her walking the streets, but now of any promiscuous woman:

When it's Tampax time, the lady is a tramp. (B. Forbes, 1989)

transfer the forcible deportation of a population.

Not voluntarily going from one location to another:

Ze'evi, 62, is an advocate of transfer, the euphemism employed by its supporters for the removal from Israel and the Occupied Territories of the Arab population. (*Daily Telegraph*, October 1988)

transfer pricing false understatement of prices in order to avoid tax.

A practice where goods are sold from one subsidiary to another which is not within the same customs union or tax regime:

This could be achieved by the delicately contrived device of transfer pricing, by which [companies with branches in Ireland] understated the cost incurred by their Irish enterprises, and exaggerated their earnings. (Lee, 1989—Irish manufacturing firms paid little or no tax)

translate the truth to lie.

A British parliamentary evasion to comply with the convention about not calling another member a liar in the House of Commons.

translated *obsolete* drunk.

Literally, transferred from one state or place to another, as from life to death or just from one clerical living to another:

Bless thee, Bottom, bless thee! thou art translated. (Shakespeare, *Midsummer Night's Dream*)

transport death.

In the Second World War the *transport* of the Jews at the hands of the Nazis was to their death.

... of labour and transport lists, of the lists of living and dead. (Keneally, 1982, of Second World War Jewish forced labour in Poland)

transported sentenced to exile for a criminal offence.

Not merely carried from one place to another. In this use, you went by sea:

One old offender, who stole the Duke of Beaufort's dog, was transported, not for stealing the dog, but his collar. (Mayhew, 1851—under English common law dogs were not capable of being stolen)

trash unsportingly to harass (an opponent).

From the meaning 'rubbish'. An American version of the Australian SLEDGE:

They are fast and noisy and they 'trash' their opponents while playing. (*Sunday Telegraph*, 20 March 1994, of regular chess players in Washington Square Park, New York)

travel to cohabit and copulate extramaritally.

Of the homeless poor in 19th-century London:

He could not remember a single instance of his having seen a young Jewess 'travelling' with a boy. (Mayhew, 1851)

The female *travelled* with the male.

travel expenses bribes.

Paid whether you made the trip or not, or at first-class rates if you rode second:

Owen, a former miner, had been recruited during a 1957 visit to Czechoslovakia and had been supplied with his 'travel expenses'. Thereafter he received regular cash payments from the Czechs. (N. West, 1982—the Czechs had been ordered by the Russians to cultivate and suborn British trade unionists. Owen was one of three members of parliament named by the defector Forlik as being in their pay. Nobody was more surprised than the accused when he was later acquitted of charges of spying.)

traveller a habitual itinerant.

Often gypsies, although it is a way of life for many Irish families without Romany blood:

... there must have been fifty or sixty travellers crammed in the back of the close, malodorous cave. (O'Donogue, 1988, of gypsies)

Also as the *travelling people*:

News was passed on with the speed of Morse among the travelling people. (ibid.)

In obsolete British use to *travel the road* was to be engaged in highway robbery; see also *commercial gentleman* at COMMERCE, and NEW AGE TRAVELLERS.

tread to copulate.

Of birds, from their foot movements:

The cock that treads them shall not know. (Shakespeare, *Sonnets to Sundry Notes of Music*)

treasure the vagina.

Viewed sexually by the male, especially when free access is denied to him:

I fall crazy in love ... and she keeps her sweet treasure all locked up. (Styron, 1976)

treat to bribe.

Especially at 19th-century elections, where the activity was known as *treating*:

... the emollience with which the established Radical election agent offers treating at the polls. (Foster, 1993—the limited franchise made possible the buying of votes by bribing individuals, which was economically less harmful than the modern practice of bribing the entire electorate with promises incapable of fulfilment)

treatment the use of violence to extract information.

Far removed from the medication which cures sickness:

I guess if this was a KGB operation we should get Leggat out and then give him the treatment. (Allbeury, 1977—Leggat's real name was Pyatokov, which was why they were prepared to be beastly to him)

tree-rat a prostitute.

The small mammal infests the bashas used, among others, by troops in India as billets:

... any man who availed himself of the 'tree rats' or 'grass *bidis*' was properly dealt with. (Allen, 1975—a *grass bidi* was also a prostitute; see BIBI)

tree suit a coffin.

A modern American variant of the obsolete *timber breeches*:

He'll get a good settin'-down some day, afore he gets into his timber breeches. (*Cornhill Magazine*, August 1902, quoted in EDD)

triangular situation a situation in which two people wish to cohabit and copulate with a third, but not by sharing.

As distinct from a MÉNAGE À TROIS:

... not only was much left intentionally unresolved on the political scene, but also much in the triangular situation at Eltham. (Kee, 1993, of a conversation in 1885 between Parnell and O'Shea. Mrs O'Shea lived at Eltham.)

triangular trade (the) trading in slaves.

On the first leg, manufactured goods went from England to Africa; on the second leg, slaves went from Africa to America; on the third leg, commodities went from America to Europe. It was also known as the *African trade*.

trick *mainly American* a prostitute's customer.

From the limited turn of duty rather than any deception, I suspect:

Lots of women walking the streets for tricks to take to their 'pads'. (L. Armstrong, 1955)

To *call the tricks* was to solicit as a prostitute:

They weren't allowed to call the tricks like the girls in Storyville (ibid.)

and to *trick* the customer was to copulate with him:

And I never tricked him. He never asked for it. (Wambaugh, 1981, of a prostitute)

A *trick-babe* is a prostitute but to *do the trick* is merely to copulate with someone:

I dare say would have done the trick if this clown Yei hadn't come. (Fraser, 1985 —Flashman had been alone with a woman)

trim to copulate with.

In modern American use of a male, probably from the concept of cutting into shape. To *trim your wick* is explicit of a male, punning on WICK and what you used to do to candles:

'You're just getting old. Lucky to be able to —' 'Ah, shut up. I got my wick trimmed all right.' (Lyall, 1975)

The obsolete British *trim the buff* came from *trim*, to beat, and *buff*, nudity. *Trim* is also used of a woman's sexuality, although I have not worked out the derivation—perhaps the imagery is naval.

trip a condition induced by the ingestion of illegal hallucinogens.

The departure from your normal situation:

The kind of thing that hippies switch into when the trips turn sour. (Bradbury, 1975)

To *trip* is to hallucinate as a result of taking the drugs:

They were speeding and tripping at the same time. (Deighton, 1972)

triple entry fraudulent.

Of bookkeeping, where *double-entry bookkeeping* (see DOUBLE ENTRY) is a procedure which is meant to make deception more difficult:

... carried with him, like bad breath, the reek of the back-streets—of furtive deals and triple-entry accountancy. (R. Harris, 1992)

trolley a single act of copulation.

From the rhyming slang *trolley and truck*.

trot *obsolete* a prostitute.

The common equine imagery, whence the punning:

Marry him to ... an old trot ... though she have as many diseases as two and fifty horses. (Shakespeare, *Taming of the Shrew*)

The British *trot on your pussy*, which is also obsolete, meant 'to copulate', of a female, and embraced a number of vulgar puns.

trots (the) diarrhoea.

The need is too immediate for walking:

I'd already got the trots. They're supposed to cement you up. (P. Scott, 1975, of pills)

A sufferer is said to be *on the trot*.

trouble any unpleasant or unwanted experience.

Euphemistic when the subject is taboo. Of an unwanted pregnancy to a spinster:

She got into trouble. Through an old white fellow who used to have those coloured girls up to an old ramshackle house of his. I do not have to tell you what he was up to. (L. Armstrong, 1955)

Of childbirth:

When I'm over my trouble I'll come to see you. (M. Francis, 1901)

Of menstruation and of any illness of a persistent and unpleasant nature, especially varicose veins and piles:

I was confident it was nae rheumatics, though what his trouble was I couldna juist say. (Service, 1890)

The *troubles* are fighting or violence in Ireland against the British or between groups whose differences are more tribal than religious:

The 'troubles'—that quaint... word for murder and mayhem. (Theroux, 1983)

In obsolete south-western English dialect *troublesome* meant 'haunted':

Th' old 'ouse up to Park's troublesome 'pon times. (EDD)

truant with your bed *obsolete* to copulate extramaritally.

Not necessarily while you are absent from school without consent:

Tis double wrong to truant with your bed,
And let her read it in thy looks at board.
(Shakespeare, *Comedy of Errors*)

true not copulating with anyone other than your regular sexual partner.

The opposite of FALSE and UNTRUE:

She was true to me all the time we lived together. (L. Armstrong, 1955)

The American funeral jargon *true companion crypt*, a common grave, suggests posthumous sexual fidelity:

True Companion Crypt—permits husband and wife to be entombed in a single chamber without any dividing wall to separate them. (J. Mitford, 1963)

trull a prostitute.

Am sure I scared the Dauphin and his trull,
When arm in arm they both came swiftly running.
(Shakespeare, *1 Henry VI*)

trunk *American* falsely to conceal.

Referring to the hiding of evidence etc., as if in a voluminous item of luggage:

And so you gave her that file to trunk. (Turow, 1987)

truth raid dropping leaflets on an enemy.

Almost certainly no more the truth than it was a raid:

So the R.A.F. was employed in showering tons of non-lethal leaflets on Germany: ...'truth raids', Sir Kingsley called them: ignominious 'confetti warfare' was the view of another M.P., General Spears. (Horne, 1969)

truth-shader *American* a liar.

Not someone who tells only part of the truth:

The second Republican choice, businessman John Lakian, has shown himself to be a truth-shader impressive even by the generous standards of Massachusetts. (*Sunday Telegraph*, 14 August 1994)

trying to escape in custody.

An excuse given for the murder of prisoners:

Codreanu had not been shot while trying to escape. He had been assassinated by order of the king. (Manning, 1962—the report had been that he was trying to escape from custody)

tube buggery.

Usually *had* or *laid*:

...about eight of them's gonna lay more tube than the motherfucking Alaska pipeline... (Wambaugh, 1983—of the prospect facing a male prisoner and the discomfort which would ensue)

tuck away in earth to kill.

From the burial, but not usually of natural death:

He was going to be quietly tucked away in earth at the frontier station after dark. (G. Greene, 1932)

To *tuck under the daisies* implies burial after natural death, the *daisy* being a symbol of the churchyard:

After me poor old man was tucked under the daisies... (MacDonagh, 1898)

tuft-hunter a vainglorious person.

From the huntsman who wishes to display evidence of his prowess:

An unashamed tuft-hunter, he faithfully followed the Jesuit tradition established in England of concentrating on the upper echelons of society. (S. Hastings, 1994, of a Jesuit priest who was selective in his choice of souls to be saved)

tumble[1] to copulate with.

Of either sex, from the alacrity of the move into a prone position:

Quoth she, before you tumbled me,
You promised me to wed.
(Shakespeare, *Hamlet*)

Modern use is also intransitive:

I'm not a regular girl and you expect me to tumble. (Weverka, 1973)

A *tumble* is a single act of copulation:

A discreet visit in a tri-shaw for a tumble at Dunroamin. (Theroux, 1973)

tumble[2] a drink of an intoxicant.

From the rhyming slang *tumble down the sink*, and sometimes used in full:

Afterwards, Dickie Leeman ... surmised that I'd had 'a tumble down the sink' at lunchtime. I never drink before six p.m. (Monkhouse, 1993)

tumescence an erection of the penis.

Not just any old swelling and usually in a medical or pompous rather than a sexual sense; also *tumescent* as an adjective:

I don't in the least mind letting girls see my penis. I suppose it is because I fear ... becoming lightly, or indeed heavily, tumescent and attracting the attention of other men. (A. Clark, 1993, explaining why he was reluctant that men also should be so favoured)

tummy ache menstruation.

One of the symptoms, but not the most obvious of them.

tummy bug diarrhoea and sickness.

Usually of a mild condition, not typhoid, cholera, etc.

tummy-tuck a cosmetic operation to reduce frontal flab on the trunk.

Felicia Dodat was going to have a tummy-tuck. (Sanders, 1986)

Not what Billy Bunter sought in the tuck-shop.

tumour (a) cancer.

Originally, any swelling; for Dryden the *tender tumour* was the erect penis.

tuned drunk.

Electronic imagery, from the heightened sensitivity perhaps.

tup to copulate with.

Dr Johnson coyly says 'To but like a ram', but the use in reference to ovine copulation remains standard English, being euphemistic only when borrowed for humans:

... but then he cruelly upped and tupped a PR girl leaving Patricia simply squelching in misery. (Fry, 1994)

turf accountant *British* a person who accepts bets professionally.

The *turf* is racing and *accountants* are supposed to be honest professional men.

Turk *American* a bugger.

In standard English he was a fierce man, who proved more than a match for the Crusaders, or a ragged boy.

turkey shoot *American* a business easily concluded.

From the size and relative immobility of the bird, which originated in the Americas and not the Levant. Of a deal bringing easy profits:

... a chance for a real turkey shoot just turned up (M. Thomas, 1982—a wealthy customer had appeared)

or of killing an easy victim:

Already there was mounting criticism in the Press that the battle had turned into a turkey shoot. (de la Billière, 1992, of the Gulf War)

But a *turkey farmer* is a businessman who finds profits hard to come by:

... at least I'm not a turkey farmer. My last three films made money. (B. Forbes, 1983)

Turkish ally an unreliable supporter.

From their supposed cowardice and treachery although, etymologically, the Greeks fare little better:

... the rock was a Turkish ally, ready to change sides if the going got rough. (Trevanian, 1972)

Turkish medal *British, obsolete* an inadvertently exposed trouser fly-button.

A warning in the pre-zip days from one male to another, from the casual way in which some Turks wear Western-style dress:

Their flybuttons were undone, and now I could understand why these buttons were called 'Turkish medals' by British soldiers in the First World War. (Theroux, 1975)

turn[1] an act of copulation.

Male usage, with imagery from the music hall perhaps; to *turn up* a woman is to copulate with her extramaritally.

turn[2] to subvert from allegiance.

Espionage jargon:

The case might be a textbook Soviet attempt to 'turn' an American military officer. (*Daily Telegraph*, February 1981)

In America too to *turn round* or to *turn around*:

'Why does a feller earning a handsome salary in the American State Department decide to chuck it all in and join a bomb factory?' 'I got turned around.' (Theroux, 1976)

turn[3] a sudden illness.

Anything from dizziness to a cerebral haemorrhage. This is a 19th-century extension of meaning from a word which merits fifteen pages in OED.

turn[4] *American* to have predominantly black residents.

Of a neighbourhood formerly occupied only by whites.

turn in to betray to authority.

Literally, to hand over to another, as a piece of work to a tutor:

> ...fearing the other might reveal something or even connive to turn in the other. (Sanders, 1980, of criminals)

In everyday slang, to *turn in* means 'to go to bed'; it used also to mean 'to be killed by hanging'.

turn it in to die.

Not voluntarily, of suicides, as might be supposed.

turn of life the menopause.

An obsolete variant of the *change* (see CHANGE[1] (THE)).

turn off[1] to kill.

Usually by hanging and probably with imagery from a lamp rather than from association with the *turning tree*, the gallows on which the corpse rotated:

> ...it gives a man a wonderful appetite for his breakfast to assist at turning off a dozen or more rebels. (Richards, 1936)

turn off[2] not to excite sexually.

The opposite of TURN ON. In the 19th century it meant 'to dismiss a sexual mistress':

> He can turn a poor gal off, as soon as he tires of her. (Mayhew, 1851)

turn on to excite.

Through any taboo or illegal agent. Of sex:

> He left bruises! I suppose he thought he was —what's the expression—turning me on. (Theroux, 1977)

Of illegal narcotics:

> 'Hey, want to turn on with me? Here, I'll make you one.' He fumbled with his cigarette papers and took out his stash. (Theroux, 1976)

And of anything else which has the same effect, from whips to carrot juice.

turn out upon the streets to become a prostitute.

The common *street* imagery (see STREET (THE)):

> Another young girl...advised me to turn out upon the streets. (Mayhew, 1862)

turn south to collapse.

Of stock markets, from the way the graphs move:

> But dealers believe that when bond prices turned south, hedge funds may have had to dump as much as $60 billion of bonds. (*Sunday Telegraph*, 6 March 1994)

turn the clock back see PUT THE CLOCK BACK.

turn up[1] see TURN[1].

turn up[2] to give the police information about.

From exposing what formerly was hidden, but in American English used of a criminal betrayal:

> He would be set free if he 'turned up the gang'. (Lavine, 1930)

turn up your little finger to be a habitual drunkard.

From a way of holding the glass, although many hold a teacup the same way:

> Ye maun keep unco sober, and no be turnin' up yere wee finger sae aften. (Ballantine, 1869)

In Scotland you might in the same sense *turn up your pinkie*:

> So very fond was Tam of 'turnin' up his pinkie' that he latterly lost both his credit and his character. (Murdoch, 1895)

turn up your tail to defecate.

Al fresco:

> ...it being very pleasant to see how everyone turns up his tail, here one and there another, in a bush, and the women in their Quarters the like. (Pepys, 1663—the facilities at Epsom were insufficient for those moved by the spectacle and the famous salts)

turn up your toes to die.

Most people are buried on their backs:

> I'll turn merrier toes to th' sky nor thee, lad, when it comes to deeing. (Sutcliffe, 1899)

Occasionally as *turn up your heels*, implying face-down interment.

turn your bike around to urinate.

Of a male out of doors. The imagery of this perhaps obsolete phrase is uncertain. Was he going the other way to seek privacy, or is the expression in the SHOOT A LION category?

turn your coat dishonourably to desert a cause.

From the days when distinctive clothing facilitated recognition and indicated allegiance:

> Perhaps wisely they turned coat and told us where he was. (Allen, 1975—Ali Dinar's spies betrayed him)

Some modern figurative use, especially of politicians, and *turncoat* is standard English.

turn your face to the wall to die.

Not from the reversal of a picture of a disgraced person but from the privacy sought by the dying:

> Sahib turns his face to the wall and all is up with him and us. (P. Scott, 1977)

turning tree see TURN OFF[1].

twelve annas in the rupee of mixed Indian and white ancestry.

British Indian derogatory use of those of mixed race, especially if they pretended to be European —there were sixteen annas in the rupee:

> I took the conventional attitude ... of making jokes about 'blackie-whitie' and 'twelve annas in the rupee'. (Allen, 1975)

But see also NOT SIXTEEN ANNAS TO THE RUPEE.

twenty-four hour service we have a telephone recording device.

A misleading advertisement—annoying when used by a plumber and you have a burst pipe in the early hours.

twilight home an institution for the unwanted old.

Not a summer house facing the west but from the cliché *twilight of your life*:

> ... arranged for her mother to be packed off to a comfortable and expensive 'twilight home'. (Murdoch, 1978)

At least it holds out no promise of rejuvenation (see REST HOME).

twin tracking *British* sinecures reciprocally given to each other by sympathetic politicians in neighbouring administrations.

Thus the political masters of one district are paid (albeit absentee) employees of another, to the masters or mistresses of which they provide similar situations. The politicians can then devote all their energies to retaining office without the chore of having to earn a living:

> ... the Bill will seek to limit the politicisation of local authorities ... ending so-called 'twin tracking', where councillors are offered well-paid posts by sympathetic neighbouring councils. This has been used by left-wingers to build up a power base. (*Daily Telegraph*, June 1989)

twisted[1] killed by hanging.

From the subsequent gyration of the corpse:

> You'll be the first Christian twisted in this awful place. (Keneally, 1987, writing of Australia)

twisted[2] drunk or under the influence of illegal narcotics.

An American version of TIGHT[1].

two-backed beast see BEAST WITH TWO BACKS.

two-by-four a prostitute.

Rhyming slang for 'whore', from the rag used as a pull-through to clean the barrel of the .303 army rifle, although we called it a *four-by-two*.

two fingers a vulgar riposte.

The British equivalent of the Latin gesture using a single digit:

> I must find something else first before I give the Captain the two-fingered farewell. (B. Forbes, 1989—he was seeking other employment)

two-on-one two people sexually using a third.

Usually of two prostitutes with a single male, although it is to be hoped that in practice they take turns:

> If you'd be interested in a two-on-one ... (McBain, 1981—two prostitutes were propositioning a man)

Also of male homosexuals:

> Enjoyed more damn two-on-ones with Jimmy up there in Castleviews ... (ibid., of convicts)

two-time to copulate regularly with two people contemporaneously.

Originally in slang it meant 'to cheat', but see also DOUBLE TIME:

> Lonsdale ... who is the latest escort of the gracious Princess Margaret, is reputed to be still two-timing with his old flame. (*Private Eye*, December 1981)

Tyburn *British, obsolete* appertaining to death by hanging.

The gallows were located near the modern Marble Arch in London; the district, named after the two burns, or streams, has shed its unhappy connotations in its modern name, St Marylebone. *Tyburn* was to be found in many phrases. The *Tyburn dance, hornpipe*, or *jig* was a hanging, of which there were about 2,000 a year at this one location in the 16th century, and twelve a year in the 19th century. A *Tyburn tippet* was a noose and the *Tyburn tree* or *triple tree*, a gallows, with its *Tyburn blossom*, a young thief 'who in time will ripen into fruit borne by the deadly never-green' (Grose). The *King of Tyburn* was the hangman and to *preach at Tyburn Cross* was to be killed by hanging:

> That souldiers sterne, or prech at Tiborne crosse. (Gascoigne, 1576, quoted in ODEP)

A *Tyburn ticket* was a certificate of exemption from the payment of all taxes in the parish in which the felony was committed, given to the person who secured a conviction and hanging or, as an alternative, a reward of about £20, which was forty times the minimum value of the stolen property in respect of which a thief could be hanged. A *Tyburn top* was no more than a wig worn 'in a knowing style' (Grose).

U

Uganda extramarital copulation.

Private Eye use in many forms based on an alleged incident in which an African princess, detected in compromising circumstances, explained that she and the man had been 'discussing Ugandan affairs'. Now also, in *Private Eye*, of male homosexual activity:

> One second-year student called 'Elsie' offers to discuss Uganda with anyone as an act of Christian love. (*Private Eye*, May 1981—he was a candidate for ordination as a priest)

ultimate intentions the extermination or banishment of Jews.

A Nazi evasion:

> How did you know this? About ultimate intentions? (Keneally, 1982—the question was asked of a Polish Jew in the Second World War)

un-American differing from an accepted or assumed standard.

The phrase was introduced in 1844 to deride the 'Know Nothing' movement and has been found useful by bigots ever since. In politics, conservatives find anything other than conservatism *un-American*:

> It would be regarded as un-American and therefore rejected. (Goebbels, 1945, in translation, of Bolshevism)

Anyone who stood up to Senator McCarthy was un-American, along with numbers who failed to stand up to him. Even losing your accent by living too long in Europe puts you at risk:

> They'd be branded for ever as un-American. (N. Mitford, 1960, of those who did so)

unacceptable damage that degree of nuclear destruction which involves surrender.

Nuclear jargon, assuming that there will be something left to surrender or, again in the jargon, that you retain SURVIVABILITY.

unavailable[1] unwilling to accept a caller.

Social and business jargon whether the call is on the telephone or in person. The expression falsely implies a frustrated willingness to be helpful.

unavailable[2] menstruating.

Female usage, especially to a normal sexual partner.

unavailable[3] evading arrest.

Police and underworld jargon:

> Ray Tuck is 'unavailable' at the moment. And we've got a three-line whip out on him. (Price, 1982)

unbalanced mad.

Not just dizziness:

> We have to accept the position that Ed was unbalanced. (Condon, 1966—Ed was mentally ill)

unbundling asset stripping.

The word chosen to explain the designs on BAT, a conglomerate, of certain financiers who sought to acquire it:

> This would be a highly-geared company. Our purpose is unbundling, and the proceeds would be used immediately to repay debt. (*Daily Telegraph*, July 1989, quoting James Goldsmith)

Whence *unbundler*, an asset stripper.

uncaring see CARING.

uncertain economically depressed.

The jargon of economists who fear that to talk openly of disaster may bring it about:

> ... the economic situation in the UK remains uncertain. (M. Thomas, 1980)

uncle a pawnbroker.

Perhaps punning on the Latin *uncus*, the hook on his scale, and the supposed benevolence of a brother of one of your parents. In the same sense, although without a classical allusion, the French call a pawnbroker an aunt and further to confuse things, in obsolete London usage, an *uncle* was a lavatory. See also AUNT[2].

Uncle Tom a black who defers unduly to whites.

> ... kissed the right asses, moved on up there. Fuckin' Uncle Tom shit. (Diehl, 1978)

From Stowe's *Uncle Tom's Cabin, or, Life Among the Lowly* of 1851. Some of those who now use the term derogatively do not know that the novel was a tract against slavery.

uncomfortable urgently needing to urinate.

By implication there is no lavatory immediately available.

uncover nakedness to copulate.

Tautological, it would seem, except to those who prepared the Authorized Version of the Bible:

> Frequently the words used to cover the sex act are 'uncover nakedness' (another example of the literal transcription of a Hebrew metaphor). (Peter Mullen in Enright, 1985—I commend his delightful essay *The Religious Speak-Easy* to all students of language or clerical vandalism)

under-arm an armpit.

Literally, any of the under part of the arm; also a method of lobbing a ball, service at tennis, or anything carried in that manner, such as the obsolete Yorkshire *under-arm bairn*, a dead child carried by its mother to its grave to save the expense of a coffin. Advertisers especially like to avoid the taboo body-hair and pocket of sweat, or *under-arm wetness*.

under hatches dead.

British nautical usage, from the closing of the hold, and of the coffin. The obsolete naval *under sailing orders* meant that you were dying.

under-invoicing a fraudulent device to avoid import duties.

Where the importing country imposes high import tariffs and the importer has access to external funds, it is common for the documentation to show a price below that agreed between the parties, the balance of the agreed price being paid by informal transfer and thus not subject to duty. See also OVER-INVOICING.

under the counter illegal.

The physical reality with many goods in war-torn countries, and for many years thereafter:

> This gave him access to what extras were being kept under the counter. (Teisser du Cros, 1962, writing of Second World War Paris)

Now used figuratively of criminal transactions involving stolen goods, wages paid without deduction of tax, etc.:

> ...called for an end of 'shamateurism', the nudge-nudge, wink-wink under-the-counter payments and perks to leading players. (*Daily Telegraph*, 5 February 1994)

under the daisies dead.

You may also be *under the sod, undersod, underground*, or *under the grass* even though you have been cremated:

> If he dhraws thim mountainy men down on me, I may as well go under the sod. (Somerville and Ross, 1908)
>
> Small wonder that th' ghosties stir up an' dahn, time an' time, when them as lig undersod... (Sutcliffe, 1900—*lig* means 'lie')
>
> You can live there when I'm underground, which will be any day now. (Murdoch, 1983)

under the influence drunk.

Shortened form of the legal jargon *under the influence of alcohol or drugs. Half under* is no less drunk.

under the screw in prison.

And doing hard labour, from the screw which tightened the treadmill, or the slang term for a warder, derived from it.

under the table[1] very drunk.

You are supposed to end up there after dropping senseless from your chair. Normally today used figuratively of someone who is able to consume less intoxicant than his companion:

> I'll drink you under the table, Max. Be warned. (Deighton, 1981)

under the table[2] involving bribery.

From the surreptitious passing of the money. Occasionally too of any transaction arranged in such a way that tax is avoided.

under the weather[1] drunk.

In standard usage, unwell and therefore also used of someone with a bad hangover.

under the weather[2] menstruating.

Again from the meaning 'unwell'.

underachiever an idle or stupid child.

Literally, a child mentally capable of doing better, especially in examinations, but failing through nervousness or ill-health. This educational jargon seeks to excuse wilfulness under a cloak of misfortune:

> ...'we do have a special course for the Over-active Underachiever,' continued the Headmaster. (Sharpe, 1982)

underdeveloped poor.

Both of states:

> The use of underdeveloped is a clue to a state of mind, that of the international do-gooders (Pei, 1969)

and of communities:

> All big cities have these little under-developed areas in them. (Theroux, 1982)

underground see ABOVE GROUND.

underground production see BLACK ECONOMY.

underground railroad the protection of runaway American slaves organized by philanthropists in the North.

Perhaps punning on its hidden nature and its then illegality:

> The escape route for runaway slaves was known as the 'underground railroad' because it was so reliable. (Faith, 1990—someone was not familiar with London Transport)

underprivileged poor.

Literally, lacking honourable distinctions, so that it embraces us all, unless we are royalty, Nobel prize-winners, or have been decorated for gallantry:

> One righted the balance by being more than fair to the underprivileged. (Bradbury, 1959)

underweight a young prostitute.

The jargon of the WHITE SLAVE trade:

> ... women from seventeen to twenty years old. These are *underweight* and must be provided with false papers. (Londres, 1928, in translation)

undiscovered country (the) death.

The

> Undiscover'd country, from whose bourn
> No traveller returns
> (Shakespeare, *Hamlet*)

and in modern use:

> I shall have entered the great 'Perhaps', as Danton I think called 'the undiscovered country'. (Harris, 1925)

undo to copulate with outside marriage.

Of a male, from the loss of reputation rather than the removal of clothing:

> Thou hast undone our mother. (Shakespeare, *Titus Andronicus*—today the children might have said 'thou hast done our mother')

undocumented illegal.

Of Hispanic Americans entering the USA illegally, especially when working without a permit.

uneven bad.

Company chairmen continue to transmit their messages in code even though the cypher was broken long ago:

> Shares in Coats Viyella ... yesterday slipped 4 to 163p as chairman Sir James Spooner told the annual meeting that trading conditions remain uneven. (*Daily Telegraph*, June 1989)

unfailing in marriage vows not guilty of adultery.

A euphemism in the negative only:

> I counted to ten, debating to ask her if she's been unfailing in her marriage vows. (Deighton, 1994)

unfaithful having copulated with someone other than your regular sexual partner.

Of either sex, usually within marriage:

> 'She's been unfaithful to me' ... 'He thinks it's a violation of our marriage because it was someone he didn't like.' (Bradbury, 1965)

And of homosexual relationships:

> ... the person he loved was being unfaithful to him in Paris. (N. Mitford, 1949—the person was male)

unfallen not yet guilty of any sin.

> ... he felt unfallen and did not yet understand how wickedness began. (Murdoch, 1983)

The *falling* is from grace, and not a bicycle. Of either sex, normally, but not necessarily, referring to the possibility of sexual adventures.

unforced error a mistake.

The jargon of the tennis commentator, who fails to understand that most of us only play as well as our opponent allows us to.

unfortunate engaged in prostitution.

A common 18th- and 19th-century use, especially by women who earned their living in other ways, or not at all:

> ... those unfortunate young women, who ... were the juster objects of compassion. (Cleland, 1749, of prostitutes)

unglued mad.

Perhaps too of a temporary affliction:

> She was completely unglued. You know, I tried to reassure her. (Turow, 1990)

unhealthy homosexual.

Those who use the phrase do not necessarily regard heterosexual activity as healthy:

> Hattie heard one of the mistresses, talking about her and Pearl, say, 'It's an unhealthy relationship'. (Murdoch, 1983)

unheard presence someone dismissed from employment.

Of a character WRITTEN OUT OF THE SCRIPT:

> The failure of his engagement to Lizzie Archer sealed the fate of Nigel Pargeter, who will become an 'unheard presence'—radio terminology for sacked. (*Daily Telegraph*, February 1990—but both the relationship and Nigel were restored in due course)

unhinged mad.

The common gate imagery:

> Gordon Masters is quite unhinged—has taken to coming into the Department wearing his old Territorial Army uniform. (Lodge, 1975)

unilaterally controlled Latino assets see ASSET.

union[1] copulation.

Of humans and animals, the making into one:

> The union of your bed. (Shakespeare, *Tempest*)

Commonly used too of marriage.

union[2] an institution for the homeless poor.

Shortened form of *union house*, set up by a Poor Law Union and still around until after the Second World War in the British Isles:

> We used to ... tramp it from one union to another. (Mayhew, 1862)

Union Jack for the death of.

Army usage, from the custom of draping the national flag over a soldier's coffin:

> I could see it was the Union Jack for this one, no error. His frame was wasted and yellow. (Fraser, 1975)

I have not met an American equivalent featuring Old Glory, but I imagine it exists.

unique unusual.

Advertising jargon:

> Unique tranquil location adjacent to the Law Courts. (London *Times*, May 1981)

Every location is unique nor are there degrees of uniqueness, despite the description of a putting course in Woolacombe, Devon as the *most unique in Britain*.

uniquely *American* (in compound adjectives) suffering from a defect.

The language of the POLITICALLY CORRECT. Thus the *uniquely abled* are crippled, *uniquely co-ordinated* clumsy, *uniquely proficient* incompetent, etc.

united dead.

With your *Maker* etc. or with a spouse who has predeceased you. Monumental usage.

University of ... a prison for political prisoners.

First used of Napoleon III, who developed economic theory alongside his romantic attachments while a prisoner in the Castle of Ham:

> At the height of his career as Emperor, he was fond of saying... 'I took my honours at the University of Ham'. (Corley, 1961)

Now used of the former places of confinement of President Mandela and others.

unknown to men not having copulated.

And a man might be *unknown to women* in the same sense:

> I am yet
> Unknown to woman.
> (Shakespeare, *Macbeth*)

unlace your sandal *?obsolete* to copulate.

Of a woman, from the first step in the act of undressing:

> When a *Casita* woman unlaces her sandal from thirty to thirty-five times a day, you can compliment her on being a good worker. (Londres, 1928, in translation)

unlawful involving extramarital copulation or bastardy.

Our forefathers seemed to be more concerned about the bastardy than the adultery:

> ... in his unlawful bed, he got
> This Edward.
> (Shakespeare, *Richard III*)

A bastard was *unlawful issue*:

> ... the unlawful issue that their lust
> Since then hath made between them
> (Shakespeare, *Antony and Cleopatra*)

being *unlawfully born*:

> I had rather my brother die by the law than my son should be unlawfully born.
> (Shakespeare, *Measure for Measure*)

But an *unlawful purpose* covers many vices apart from its criminal connotation, extramarital copulation being only one:

> May be the amorous count solicits her
> In the unlawful purpose.
> (Shakespeare, *All's Well ...*)

unlimber your joint see JOINT[2].

unmarried homosexual.

Most bachelors are not homosexual and as ever the euphemism depends on the context:

> Neighbours of unmarried Mr Hamilton contacted police six months ago... a male model and a tenant at Mr Hamilton's house... is acting as Mr Hamilton's agent. (*Sunday Telegraph*, December 1986)

In obituaries, *he was unmarried* can now be a code for 'he died from AIDS'.

unmentionable crime buggery or sodomy.

Once one of the great taboos:

> The practice of bedding the men by threes and not in pairs was supposed, optimistically, to reduce unmentionable crime. (Hughes, 1987, of convicts)

unmentionable disease venereal disease.

Still not spoken of in polite circles:

> ... adding an unmentionable disease to the old lady's dossier of Wilt's faults. (Sharpe, 1979)

unmentionables[1] trousers or undergarments.

19th-century prudery which in its extremest form extended to the sexual implications of table legs:

> She had vowed never to change her unmentionables until her husband, Archduke Albert, took the city of Ostend by siege. (Jennings, 1965—as it held out for three years, she must have kept her vow at the expense of her friends and her marriage)

Also as *unexpressables*, *unspeakables*, *untalkaboutables*, *unutterables*, *unwhisperables*, *ineffables*, *indescribables*, and *inexpressibles*:

> They wear all manner of pantaloons and inexpressibles. (James, 1816)

See also *benders* at BEND.

unmentionables[2] haemorrhoids.

A female way of talking about piles, or, in modern slang *Emmas*, which might just be a shortened form.

unnatural homosexual.

Legal jargon, as in *unnatural crime*, *practice*, *vice*, or, less often, *filth*:

> ... the severe penalties imposed on unnatural practices in our own country by an Act of 1886 have merely had the effect of advertising them. (Richards, 1936)
>
> ... seeing a Turk severely whipped and his beard singed for attempting unnatural vice.

(Ollard, 1974—what a Turk was doing in St Helens in 1683, apart from the attempt, is not known)

Also of bestiality and buggery of a female:

> ... trying to sort out which portion of anatomy fitted the next ... in what ... appeared to be a series of extremely unnatural acts. (Sharpe, 1975)

unofficial action a strike in breach of agreement.

The *action* is inaction. Of stoppages, sometimes encouraged by a trade union which might be sued if its support were overt:

> Was it another day of 'unofficial action'? Had an epidemic of sunstroke decimated the staff of London Transport? (Blacker, 1992—the trains were not running)

unprotected sex copulation without using a condom.

Not batting against a hard ball without a box. The *protection* is against venereal infection and AIDS, but not pregnancy.

unripe fruit young virgins.

A 19th-century evasion:

> Reference in the scurrilous periodicals to men with tastes for 'unripe fruit'. (Pearsall, 1969)

unscheduled caused by accident or necessity.

Airline jargon which seeks to avoid any implication of loss of reliability or safety:

> Engineers have a nice phrase for engine breakdowns. An 'unscheduled engine removal'. (Moynahan, 1983)

unscrewed mad.

The eventual state of one having a *screw loose* (see HAVE A SCREW LOOSE):

> ... this is pure banana oil! You've come unscrewed. (Wodehouse, 1934)

unsighted blind.

Literally, prevented from seeing by an intervening obstruction.

unslated mad.

Perhaps obsolete northern English imagery from the loss of slates from the roof of a house:

> He's gone clean off his head, unslated. (Brierley, 1886)

unsound not to be trusted.

Bureaucratic jargon, of judgement rather than honesty:

> '... Tyler was unsound.' 'And you can't say worse than that in Whitehall.' (Lyall, 1980)

Perhaps from a ship which is not in good condition.

unstanched *obsolete* not having copulated.

A *stanch* is something which stops blood and I think the imagery is from the cessation of menstruation during pregnancy, but

> As leaky as an unstanch'd wench (Shakespeare, *Tempest*)

could as well refer to the absence of a protective towel.

unstoned castrated.

Of animals rather than St Stephen. See STONES.

untouchables the lowest caste in Indian society.

They are given the job of cleaning the latrines and so on. See also SCHEDULED CLASSES.

untrimmed *obsolete* not having copulated.

The imagery is from a wick rather than from the meaning 'to put in order':

> In likeness of a new untrimmed bride. (Shakespeare, *King John*)

untrue having copulated outside marriage.

The reverse of TRUE, of either sex:

> The thought that you might have been untrue ... would have broken my heart. (Fraser, 1975)

unwaged involuntarily unemployed.

It sounds more like a war which didn't take place:

> Claire is trying to get her father to give cheap food to the unwaged. (Townsend, 1982—he was a greengrocer)

unwell[1] menstruating.

From the meaning 'ill':

> ... all's well that ends unwell. (Harris, 1925, quoting a woman who had thought she was pregnant)

unwell[2] drunk.

Covering up the taboo condition with one of its symptoms:

> 'Our Mr Fellowes' had been 'very unwell at the time of the move.' 'He wasn't unwell,' said my sister. 'He was drunk.' (Bogarde, 1983)

up[1] to copulate with a woman.

Rare, except as *upped* which is used mainly of rape. The phrase *up a woman* is explicit:

> 'When you're up who, Barbara's down on whom?' asks Flora. 'Flora, you're coarse,' says Howard. (Bradbury, 1975)

There are many vulgar puns, such as *up her passage*, *up her way*, etc.

up[2] under the influence of illegal narcotics.

Especially amphetamines. Whence *ups*, *uppers*, or *uppies*, such narcotics:

> I knew one 4th Division Lurp who took his pills by the fistful, downs from the left pocket

of his tiger suit and ups from the right. (Herr, 1977)

up[3] *American* having forgotten your lines.

Theatrical slang.

up along old.

An English dialect usage which is still heard in the south-west. I think the Scottish *up in life* is obsolete:

Though up in life, I'll get a wife. (Boswell, 1871)

up in arms having an erection of the penis.

The common military imagery:

I'd never have thought to be still up in arms when Susie ... was hollering uncle. (Fraser, 1982)

up the creek in severe difficulties.

The *creek* is *shit creek*, or the anus; if you were *up the creek* you were a bugger and liable to severe penalties if found out. The shortened form and the full phrases, of which for most people the provenance is mercifully lost, are now only used figuratively:

... telling them that if they'd followed her this far up shit creek it's long way to walk back. (*Private Eye*, July 1981)

up the loop mad.

Army usage, with imagery perhaps from railway shunting practice, but perhaps not:

A lot of us believed he was really up the loop for having played at it so long. (Richards, 1936, of a soldier feigning madness to secure discharge)

Whence the more common *loopy*, mad or eccentric:

The reason is typically Muriel Spark, both down-to-earth practical and mildly loopy at the same time. (Barber, 1991)

'Ah,' said the Bishop, 'and suppose one of your children were sick in some way?' 'Loopy?' 'If you like.' (Fry, 1994)

up the pole pregnant.

Punning on the meaning 'in trouble', and on *pole*, the penis:

'We've planned this for a long time.' 'When you discovered she was up the pole.' (Binchy, 1985—*this* was marriage)

up the river in prison.

From the location of Sing Sing and other American jails relative to their neighbouring cities. See also SEND UP.

up the spout pregnant.

The imagery is from loading a shell into a rifled barrel from which, the copper band being engaged, it can be extracted from the breech only with danger and difficulty:

The chorus, four times repeated, was: 'She was up the bleeding spout'. (Richards, 1936)

Also, in obsolete usage, of bankruptcy.

up the stick pregnant.

Perhaps punning on *stick*, a gun barrel and the penis.

uppish drunk.

From the sometime feeling of elevation, and perhaps too the cheekiness.

upstairs[1] an allusion to a taboo act or place.

She's gone upstairs implied that a pregnant woman was about to give birth. Of an invalid *he's been upstairs two months* indicates the extent of his infirmity. Socially *would you like to go upstairs?* invites urination. *Upstairs* too is where the bedrooms are, for copulation:

Was he going to haul her off upstairs, leaving first-years honours [students] to riot away among the cakes below while he satisfied his passion? (Bradbury, 1959)

For the devout morbid, *upstairs* is death, where heaven is, while to go *upstairs out of this world* was to be hanged, punning on the climb up the scaffold.

upstairs[2] in authority.

The senior staff occupy the higher floors:

And now the pressure put on from upstairs to put the clamp on the case ... (van Lustbaden, 1983)

Uranian a homosexual.

I hope from the Latin *Urania*, another name for Aphrodite, rather than a coarse pun on an inhabitant of the planet Uranus, but sad experience says we can never be quite sure:

Many of the Uranians or Urnings (favourite term among the literati), were disgusted by the physical manifestations of their tendencies. (Pearsall, 1969)

O child of Uranus, wanderer down all times ...
Thy Woman-soul within a Man's form dwelling.
(ibid., quoting Carpenter, c. 1895)

urban renewal slum clearance.

Not a tidied up business district:

The abandoned warehouse was in a depressed area long overdue for urban renewal. (Bagley, 1982)

use[1] to copulate with.

Of a male, normally outside marriage:

Be a whore still: they love thee not that use thee (Shakespeare, *Timon of Athens*)

and in modern use:

The fact that her father had used her killed my liking for Kätchen. (Harris, 1925)

There are many explicit phrases such as *use for his vile purposes*, *use to sate his lusts*, and *use as a woman*:

Do you suppose that slaver captain has been ... using her ... as a *woman*? (Fraser, 1971, and not referring to a mixed tennis partnership)

use[2] to be addicted to illegal narcotics.

A shortened form of *use drugs* etc. and of the jargon *use some help*:

'I think we can use some help,' ... he said, passing the vial and the gold spoon to her. (Robbins, 1981)

An American *user* is an addict.

use[3] **(in)** capable of conception.

Of animals:

... none of the mares he covered three weeks or more ago has come back into use. (Francis, 1982)

use a wheelchair to be physically incapable of walking.

An example of the healthspeak which forbids any direct speech about certain physical conditions:

You should not say that someone 'cannot walk', instead say 'uses a wheelchair'. (M. Holman in *Financial Times*, October 1994, also reporting that the Spastics Society had decided, at a cost of £750,000, to restyle itself 'Scope', which 'carries a degree of weight and a feeling of progress and some other positive associations')

use of Venus *obsolete* copulation.

By a male, Venus being the goddess of love:

Much use of Venus doth dim the sight. (Bacon, 1627—Shakespeare would never have written that)

use paper to defecate.

Hospital jargon, and not of writing a letter home.

use the facilities to urinate.

Facilities such as the bathroom or the outhouse:

... a gentlemanly statement of his desire to use the outhouse. (Keneally, 1979)

use your tin *American* to identify yourself as a policeman.

From the badge:

I'd be in civilian clothes ... Could I use my tin? (Sanders, 1973—the speaker is a policeman)

used second-hand.

To relieve the stigma of prior ownership, especially of cars. But for the British 18th-century warrior General Guise *used up* meant 'dead or wounded'—he sent for reinforcements after he had *used up* his grenadiers in an attack on Cartagena.

useful fool a dupe of the Communists.

Lenin's phrase for the sincere pacifists and shallow thinkers in the west whom the Communists manipulated:

... the Judas goats leading what they call 'the useful fools' up the garden path to the knacker's yard—the brave sons of Ireland in the IRA and the honest pacifists in CND. (Price, 1982)

V-girl a female who might copulate extramaritally without payment.

To enhance the status of work in Second World War American munitions factories, female workers were called *Victory Girls*. In the changed environment and absence from home, many became sexually promiscuous, as did their British sisters. Shortened to *V-girl*, servicemen used the expression of any woman who *volunteered* to copulate without payment. Later, when many of the women became infected with disease, the *V* came also to stand for *venereal*.

vacation *American* a prison sentence.

> ... won a twenty years' vacation in the Big House. (Lavine, 1930)

Literally, a holiday which involves any absence from home.

valentine a notice of dismissal from employment.

Punning on the *cards* received by some on 14 February (see CARDS (YOUR)). It was also used in America of a warning to an employee of conduct which was unsatisfactory and might lead to dismissal:

> The captain ... may distribute a few complaints or 'valentines' for dereliction of duties. (Lavine, 1930)

vanity publishing the publication of a book at the author's expense.

The assumption is that the venture is not commercially attractive to a professional publisher. Less often denoting the encouragement of favourable coverage in the media:

> And persuading his friend, Sir Roland Smith, to interview him on his career must count as an exercise in vanity publishing. (*Daily Telegraph*, 16 April 1994, on Swarj Paul)

vapours (the) *obsolete* menstruation.

Originally, in standard English, the unhealthy exhalations occurring in the body, whence any illness, especially if it affected the stomach.

variety meats *American* offal.

Lungs, liver, testicles, and all the bits you would rather not spell out with precision. See also SWEETBREADS.

Vatican roulette the use of the 'safe period' method of contraception.

Punning on the Roman Catholic dogma against contraception and Russian roulette—in either case you cannot be quite sure that there isn't one *up the spout*:

> But it seems that Vatican roulette has failed them again and a fourth little faithful is on the way. (Penguin blurb for Lodge's *The British Museum is Falling Down*)

vault[1] to copulate with.

Of a male, predating the modern *jump* (see JUMP[2]):

> Whiles he is vaulting variable ramps. (Shakespeare, *Cymbeline*)

Whence the obsolete punning *vaulting-school*, a brothel.

vault[2] a cupboard for the storage of a corpse.

American funeral jargon. Literally, any structure with an arched roof, which is how many early tombs were built. However:

> That vault we are describing here is designed as an outer receptacle to protect the casket and its contents from the elements during their eternal sojourn in the grave. (J. Mitford, 1963)

velvet[1] the vagina viewed sexually by a male.

The fabric with a smooth, thick, luxurious pile:

> ... pitiless calculation of a female with velvet to sell. (Mailer, 1965)

velvet[2] a payment for which there is no consideration.

Either a bribe:

> Money is deposited in the 'velvet-lined' drawer of my desk ... (Lavine, 1930)

or an exceptional profit:

> ... to get back his original investment in order to be able to work in 'velvet'. (ibid.)

Again from the properties of the cloth.

venereal see VENUS.

Venus appertaining to copulation.

The Roman goddess of love appears in many compounds and variations:

> ... his heart
> Inflam'd with Venus.
> (Shakespeare, *Troilus and Cressida*)

Venereal, now used only of sexually-transmitted diseases, once meant 'beautiful or lustful' and *venery* once meant the pursuit of women as well as of deer. The *venerous act*, copulation, is obsolete:

> ... it did afford him some pleasure to see the venerous act performed. (Fowles, 1985, using archaic language of a voyeur)

verbally deficient illiterate.

Not merely having a restricted vocabulary. Jennings (1965) points out how odd it is that those who cannot read should need a written euphemism to conceal their ignorance.

vertically challenged of short stature.

My god-daughter Rebecca Stephens was not *challenged* in this way when she climbed Mount Everest:

> A better deal for the vertically challenged was urged yesterday by Dr David Weeks, a consultant psychiatrist, who said that 'shortism' was as pernicious as sexism and racism. (*Daily Telegraph*, 12 April 1994—Dr Weeks should know, being himself 5ft 2in. tall)

See also CHALLENGED.

very poorly see POORLY[1].

vibrator an electrically operated device for producing sexual stimulation.

> The biggest-selling sex aid by far is the vibrator. (Cauthery & Stanway, 1983)

See also CORDLESS MASSAGER.

vicar of Bray a cowardly or opportunistic trimmer.

A cleric held this living in the 16th century during the reign of four English monarchs, two of whom were Roman Catholic and three Protestant. (Henry VIII was both.) When the vicar was accused of being of a changeable turn he replied:

> No, I am steadfast, however other folk change I remain Vicar of Bray. (Reported by Alleyn, Bishop of Exeter)

Elsewhere in the country, livings were lost and regained with each turn of the tide, as reference to the names of incumbents in many parish churches will illustrate.

victualler the keeper of a brothel.

He provides the MEAT:

> *Falstaff* ... suffering flesh to be eaten in thy house contrary to the law; for the which I think thou wilt howl.
> *Hostess* All victuallers do so.
> (Shakespeare, *2 Henry IV*—note the two sexual puns)

Whence the obsolete punning *victualling-house*, a brothel.

vigilance informing to the authorities on fellow-citizens.

The usage and practice of totalitarian states, to deter or detect any incipient dissident:

> ... everyone informs right from the nursery... They call it 'vigilance'. (M. C. Smith, 1981, of Russia)

violate to copulate with extramaritally.

The usual violent imagery and used in this sense since the 15th century where a male has used force or blandishments:

> With unchaste purpose, and with oath to violate
> My lady's honour.
> (Shakespeare, *Cymbeline*)

viper a person who sells narcotics illegally.

> Vipers selling a smoke-pot to school kids. (Longstreet, 1956)

American dysphemism although we have to point out that the serpent's evil reputation is undeserved.

A *viper* is also an addict to marijuana, which is sometimes known as *viper's weed*.

Virgin Mary see BLOODY SHAME.

virtue the property of not having copulated extramaritally.

Literally, conformity with all moral standards, but in this use of women since the 16th century:

> Their triumphs over the virtue of girls. (Mayhew, 1851)

Whence *virtuous*, not having so copulated:

> Betimes in the morning I will beseech the virtuous Desdemona. (Shakespeare, *Othello*)

visible minority a black.

In a society where the majority is white. It can also include indigenous North Americans:

> An Ad from Concordia University in Montreal, Canada, says it especially encourages applications from women, 'visible minorities', and the disabled. (*Daily Telegraph*, 17 February 1992)

visit see PAY A VISIT.

visiting card an act of urination or defecation in a public place.

Of domestic pets:

> He's left his visiting card. (Ross, 1956, of a dog)

visiting fireman[1] a boisterous reveller.

Especially at a convention etc. some distance from home:

> ... a visiting fireman in search of a cheap thrill would get mugged and robbed. (McBain, 1981)

Oddly, one of the most disturbed stays I can recall in an American motel was occasioned by the night-long antics of firefighters attending such a meeting.

visiting fireman[2] an investigator or manager sent from head office to a subsidiary.

Usually after poor performance by the local management. Fighting generals have their critics too:

> He should not get into any arguments or debates with visiting firemen who take his time. (Butcher, 1946, of Eisenhower in Algiers)

Also of a team sent to correct a mistake:

> When visiting 'firemen' move in, the bodel has to move out... (Forsyth, 1994—a *bodel* is a young Israeli living in a foreign country and assisting the Israeli secret service MOSSAD)

visitor (a) menstruation.

Common female usage, often with punning extensions. In America the *visitor* may come from a place called *Redbank*.

visually challenged ugly.

You would think, only when they look in a mirror:

> Margaret Beckett, visually challenged Deputy Leader of the Labour party... (A. Waugh, *Daily Telegraph*, 4 October 1993)

See also CHALLENGED.

vital statistics the measurements of a woman's chest, waist, and buttocks.

Here, as so often, *vital* means no more than important, which the information is in the world of entertainment.

vital statistics form a death certificate.

American funeral jargon which seems singularly inappropriate to death:

> A death certificate would be referred to as a 'vital statistics form'. (J. Mitford, 1963)

vitals the testicles.

Literally, the parts of the body essential to the continuation of life, whence usually the organs located in the trunk:

> ...him so bad with the mumps and all, so that his poor vitals were swelled to pumpkin size. (Graves, 1941)

void water to urinate.

An obsolete form of PASS WATER:

> When, at the end, they went too far, she voided her water on the deck. (Monsarrat, 1978, writing in archiac style)

voluntary done under duress or compulsion.

Such as an admission of guilt obtained under duress:

> ...denied that any coercive measures had been used in obtaining the 'voluntary confession' (Lavine, 1930)

or unpaid work undertaken through compulsion, such as chores in the army.

voluntary patient a patient in a psychiatric hospital supposedly free to leave.

Those who are confined through legal process, or SECTIONED, cannot discharge themselves. The expression *voluntary patient* is not used of those in hospital for the treatment of physical illness.

voluntary pregnancy interruption see PREGNANCY INTERRUPTION.

volunteer a person instructed to fight or work for a third party.

Used originally of those who intervened in military formations for the Nazis and Communists during the Spanish Civil War, and now of any organized military interference where you wish to help an ally without declaring war against his enemy:

> ...intervention on the enemy's side of overwhelming reinforcements of Chinese 'volunteers'. (Boyle, 1979, of the Korean War)

voyeur a person who enjoys watching the sexual activity of others.

Literally, a watcher of anything:

> Hamilton had been an enthusiastic voyeur... In one home, microphones had been installed throughout the bedrooms. (Green, 1979—an écouteur too, it seems)

To *voyeurize* is so to act:

> That's a hell of a way to get experience... voyeurising ancient broads. (Sharpe, 1977)

Vulcan's badge *literary* an indication of cuckoldry.

Venus, while married to Vulcan, committed adultery with Mars. Thus, in literary use, to *wear Vulcan's badge* is to be a cuckold.

vulnerable poor or inadequate.

None of us is incapable of being wounded, but there are some more at risk of harm from the slings and arrows of modern life than others.

W/WC see WATER CLOSET.

wad-shifter a person who never drinks intoxicants.

A *wad* is a doughy bun, often taken with *char*, tea. British army usage in a society where temperance was taboo:

> If a teetotaller he was known as a 'char-wallah', 'bun-puncher' or 'wad-shifter'. (Richards, 1933)

wages of sin (the) premature death.

The original exposition suggested perhaps that the good people would survive, even though they died, while the bad had no such prospect:

> For the wages of sin is death; but the gift of God is eternal life. (Romans 6:23)

Today it is often taken literally:

> I could have mentioned that the wages of sin are death—that the Union captain's carnal desire for the powdered, rouged, weeping old woman we'd left that evening had brought him a well-deserved end. (Baldwin, 1993)

wagon see ON THE WAGON.

waiting for employment unemployed.

A Chinese communist usage:

> He told me he had plenty of time since he was 'waiting for employment'—an expression used by the People's Government for 'unemployment' which was not supposed to exist in a socialist state. (Cheng, 1984)

wake to watch over a corpse.

Literally, to keep awake, whence to stay awake by night to ensure that a body is not molested before burial:

> For nobody cared to wake Sir Robert Redgauntlet like another corpse. (W. Scott, 1824)

Your vigilance had sometimes to extend after the interment:

> Wauk the kirkyard... to prevent the inroads of resurrection-men. (EDD)

Whence our modern *wake*, the feast which may follow a death:

> 'There's a wake in the family,' an euphemistic expression for death. (ibid.)

wake a witch *Scottish, obsolete* to force a woman to confess to witchcraft.

As with SWIM FOR A WIZARD this entry is included to illustrate the behaviour of our recent ancestors. Here an iron hoop was placed over the victim's face, with four prongs in her mouth. Chained to a wall so that she could not lie down, she was kept awake by relays of men until she admitted she was a witch, after which she might be ducked or burnt to death.

walk[1] to be a prostitute.

Seldom *tout court*, but if so used, the confusion may be considerable, as in the case in 1891 of Daisy Hopkins, who was sentenced to 14 days in prison by the University Court of Cambridge after being accused of *walking with a member of the university*. This was held, on appeal to a wiser body, to imply no offence. Usually expanded to *walk with* or *walk the streets* (but see also WALK OUT[1]):

> Women walking the streets for tricks to take to their 'pads'. (L. Armstrong, 1955)

walk[2] to be dismissed from employment.

This usage wrongly implies a voluntary departure:

> Thing is, I give you maybe three, four years, you'll walk. (Diehl, 1978, of such dismissal)

The figurative *walking papers* are what may be handed to you when you are dismissed:

> I should give you your walking papers (Theroux, 1989, of such a threat)

or when your spouse or sweetheart unilaterally brings cohabitation or courtship to an end.

walk[3] to be stolen.

Of small tools, items of army kit, etc. This ironic usage avoids accusing your mates of theft by attributing the power of locomotion to inanimate objects.

walk[4] to acknowledge dismissal before an umpire's adjudication.

Cricket jargon:

> Gooch's initial movement suggested that he was going to walk, which might have deceived the umpire. (*Daily Telegraph*, 27 January 1995—he was given out incorrectly)

Euphemistic only in the negative, where *not to walk* is an imputation of bad sportsmanship in a batsman who knows he has been dismissed but hopes for an umpiring error in his favour.

walk[5] to escape deserved punishment.

A shortened form of *walk free from court* etc. and only used of the guilty:

> 'Havistock is going to walk, isn't he?' 'Sure he is,' Al said. 'What could we charge him with?' (Sanders, 1986—Al was a policeman)

Also used in the sense 'to be released from prison', especially when the time served has seemed too short to an observer:

> ... the most they'll get is twenty years, walk in seven or eight. (Clancy, 1989—policemen were deploring the prospective treatment of rapist/murderers)

walk out[1] to court.

The usage has survived the days when preliminary courtship was a pedestrian affair:

Caleb was 'walkin' a maid out'. (Agnus, 1900)

The maid would more often *walk out with* her swain:

...the colonel's daughter, who was walking out with Mike Seymour (M. Clark, 1991)

or *walk with* him:

You'll dance at the hops with me, ride with me, but you won't walk with me. (Cookson, 1967—a girl was complaining at the limits set by a man to their friendship)

Courting couples were said to *walk with* or *walk along of* each other. See also WALK[1].

walk out[2] to go on strike.

This does not refer to the departure on foot of workers at the end of a shift; in a *walk-out*, a strike, the majority today leave by car. Usually of cessation of work at short notice, contrary to agreed conditions of employment, and without recourse to negotiating procedures. Common until the 1980s, the walk-out is now almost unheard of.

walk penniless in Mark Lane see MARK[1].

walk the plank to be murdered by drowning.

Favoured by pirates for the disposal of their captives.

wall-eyed drunk.

Literally, strabismic, with difficulty in focusing, and drunkenness can cause that too.

wallflower week menstruation.

The period during which a woman will not copulate, from the female who sat on the periphery of a hall waiting in vain for a man to ask her to dance:

Suddenly came the sweet green age of chlorophyll, offering new hope for wallflowers and old maids. (E. S. Turner, 1952, referring to an advertising fad)

wallop the mattress to copulate.

To *wallop* is to beat. Of either party with familiar imagery:

She'd never have walloped the mattress with me like that if she'd been false. (Fraser, 1975)

wander to philander within marriage.

Usually of the male:

...her pain, particularly with her husband's wandering, was sometimes intense. (Turow, 1990)

wander off to defecate.

From the seeking after seclusion and sometimes mentioning the destination:

The following morning, after we had all wandered off into the appropriate field, and washed at the pump, and breakfasted... (Simon, 1979)

wandered *Scottish, ?obsolete* mentally confused.

From the inability to concentrate:

...sick in mind as in body. He seemed, as my wife's relatives would have said, to be 'wandered'. (Fraser, 1969, writing in archaic style)

wank to masturbate.

Of a male, from the meaning 'to beat or thrash', as in the dialect *wanked*, exhausted. Sometimes as *wank off*:

He himself felt only guilt and depression, like as a lad he used to feel when he wanked off. (Lodge, 1988)

Also as a noun:

He seems to be recording, in his own graceful way, a wank in the woods. (Fry, 1994)

A *wanker* is both a term of abuse and of SELF-ABUSE:

Harrison's are a load of wankers. (Sharpe, 1982, of schoolboys)

Wankery is pornographic literature etc. catering for males:

...locking himself in with a load of new-bought wankery. (Amis, 1978)

Wank meaning 'to copulate' appears only in the corruption *wang*, as in *wang-house*, a brothel:

I had expected the opium parlour to be something like a wang-house filled with sleepy hookers. (Theroux, 1973)

want[1] idiocy.

Shortened form for *want of understanding* etc.:

I had a want and been daft likewise. (Galt, 1826)

Whence the modern *wanting*, of low intelligence. In Scottish English an idiot might *want some pence in the shilling*:

...of rather a wild frantic nature, and seem to want 'some pence in the shilling'. (Mactaggart, 1824)

want[2] to wish to copulate with.

When you *want* a man or woman, it is not for social intercourse:

Yet he wanted my mother, his half-sister, and in trying to get his way with her caused her untold agony of mind. (Cookson, 1969)

Specifically you may *want sex* (which does not mean you are a hermaphrodite) or *want relations, intercourse, love, a body*, etc.:

Since she was fifteen, men had wanted her body. (Allbeury, 1976)

want out to wish to kill yourself.

Literally, to wish to extract yourself from a deal or arrangement:

'Does the letter signify anything to you?' 'Only that he wanted out.' (B. Forbes, 1983, of a suicide note)

war criminal a leader of the enemy.

This dysphemism is usually only applied to the losers, although Goebbels described Churchill and Roosevelt as *the two war criminals* (1945, in translation), but in fact his companions were so arraigned later that year. A *war crime*, when used euphemistically, is the act of having led a country which has lost a war.

warehouse to hold (stocks) for a principal seeking to conceal his interest.

Stock Exchange jargon for an illegal or clandestine operation, with imagery from holding goods in a store belonging to another:

> It is even suggested that the diminutive legal person could have 'warehoused' some of the Howard shares. (*Private Eye*, March 1981)

warm sexually aroused.

And sometimes no cooler than HOT[1]:

> The warm effects which she in him finds missing. (Shakespeare, *Venus and Adonis*)

The obsolete *warm one* was a prostitute whom you might find in a *warm shop*, a brothel. I am sorry that we seem to have lost the descriptive phrase, to *warm up old porridge*, to renew a discontinued sexual relationship, with its gentle reminder that the taste is never quite the same as before.

warm someone's backside to beat (someone).

Not to stand before an open fire on a cold day:

> Please don't think I don't know how to warm your backside. (Theroux, 1993—of a threat to a child)

warm someone's bed to copulate with (someone).

Usually extramaritally, with the normal assumption that a male and female in such proximity will copulate:

> It was equally possible she was warming another man's bed. (Moss, 1987, and not by using a hot-water bottle)

warning termination of employment.

An older form of NOTICE, but given rather than received:

> If respectable young girls are set picking grass out of your gravel, in place of their proper work, certainly they will give warning. (Somerville and Ross, 1897)

warpaint facial cosmetics.

Punning jocular female usage, from American Indian adornment before battle:

> Baby was down with a fresh dressing of warpaint. (Sharpe, 1977)

wash[1] *obsolete* urine.

As commonly used in the laundry:

> Dochter, here is a bottle o' my father's wash. (Graham, 1883—it was for medical examination)

A *wash-mug* was a piss-pot.

wash[2] to deal unnecessarily in securities to obtain commission.

British Stock Exchange jargon, from handling something which remains the same after treatment. The American CHURN is more elegant.

wash[3] to bring into open circulation (money obtained illegally).

A variant of LAUNDER:

> We *must* wash the money... If that money isn't broken down, Kalenin... just won't cross. (Freemantle, 1977)

wash the baby's head to drink intoxicants in celebration of a birth.

There is probably some connection with the rite of christening:

> To wesh ther heeads e bumper toasts. (Treddlehoyle, 1846)

In modern use more as *wet the baby's head* (see WET[2]).

wash your hands to urinate.

This is what arriving guests are commonly invited to do, or suggest doing, from the proximity of the lavatory and the hand washbasin. A *wash and brush up* means much the same thing and *washroom* is widely used in America for lavatory.

wash your hands of to dissociate yourself from (anything unpleasant).

Like Pilate, who 'took water, and washed his hands before the multitude, saying, I am innocent of the blood of this just person: see ye to it' (Matthew 27:24).

washroom see WASH YOUR HANDS.

waste[1] *American* to kill.

From one of the many meanings 'to destroy', 'to use up', 'to expend needlessly', 'to spill' (as in south Devon you may *waste* your milk if you overturn the cup), etc.:

> You wanted a photo of Roger Kope, the cop who got wasted. (Sanders, 1973)

waste[2] urine or faeces.

In America particularly of canine excreta, where *dry waste* is dog turds. *House waste* is what is emptied from the EARTH CLOSET. A spacecraft has a *waste management compartment* rather than a lavatory. The *waste reception centre*, on the other hand, in Oxford and elsewhere, is a refuse dump.

waste time to masturbate.

A common American usage which happily never reached the British Isles.

wasted drunk.

Not from spilling the liquid or the *wasting away* of the body of the drinker:

> To an American, the word *bar* suggests a place to get either happily squiffed or unhappily wasted. (*Travel and Leisure*, June 1990)

watcher a person who observes surreptitiously.

Espionage jargon, and not just someone switching on the television (who enjoys the rather grander appellation of *viewer*):

> He should distribute as many watchers as he could muster. (le Carré, 1989, of espionage)

water urine.

Used in this sense since the 14th century even though the liquid described differs significantly from our concept of the clear and potable compound of hydrogen and oxygen:

> Sirrah, you giant, what says the doctor to my water? (Shakespeare, *2 Henry IV*)

A male who goes to *water* his *garden, roses, nag,* etc. will urinate, usually out of doors. *Watershaken* is an obsolete term for involuntary urination. *Waterworks* is a punning reference to the human urinary system, especially when it is malfunctioning (but see also WATERWORKS[2]):

> ... busily at work cauterizing her waterworks. (Sharpe, 1979)

water closet a lavatory with a flush mechanism.

Abbreviated internationally to *WC* and occasionally in the British Isles to *W*:

> The W is a frequent non-U expression for 'lavatory' (W.C. is also non-U). (Ross, 1956)

water cure a form of torture.

A far cry from attending a spa to relieve your rheumatism. The water is applied in persistent drips externally or in excessive quantities orally. The torture still persists in modern tyrannies because it is simple, cheap, and effective.

water of life whisky.

The Scottish/Irish *usquebaugh* rather than the French *eau de vie*:

> 'Uisgebeatha?' Murdoch said in Gaelic. 'The water of life.' (Higgins, 1976)

I suspect in 19th-century Scotland it may have been used of any spirituous intoxicant:

> A glass of brandy or usquabae. (W. Scott, 1824)

water sports sexual activity involving urination.

Not swimming, diving, etc., but a form of sexual activity in which one partner urinates on the other:

> ... they're interested in leather and water sports. (Theroux, 1990, of sexual deviants)

See also SHOWERS[1].

water stock to render securities less valuable by constant dilution.

The practice was perfected by Daniel Drew with the securities of the Erie railroad. In his work as a drover, he fed salt to his cattle as they were being driven to market, causing them to drink a lot and put on weight but not flesh. He took his system, and the same name for it, with him when he started financing railroads (Faith, 1990).

watering hole a place licensed to sell intoxicants.

Punning jocular usage:

> A blinking sign I took to be a watering hole... (Theroux, 1979)

He would have been affronted if it offered only water.

waterlogged very drunk.

Literally, saturated with water, whence heavy and sluggish, unable to absorb more liquid.

watermelon an indication of pregnancy.

It is the shape of the swelling that is referred to in such American phrases as *have a watermelon on the vine* or *swallow a watermelon seed*.

watermelons see MELONS.

waterworks[1] see WATER.

waterworks[2] tears.

Especially those of a woman or child thought to be producing them to obtain sympathy:

> It's impossible to talk reason with Ma; she just turns on the waterworks. (Seth, 1993)

wax to remove unwanted hair from (a part of the body).

Mainly female usage and practice:

> Mumsy and I are motoring up to London to have our legs waxed at Fortnums. (*Private Eye*, April 1981)

See also BIKINI WAX.

way of all flesh (the) death.

Quoting the Biblical

> I am going the way of all flesh (Joshua 23:14; Douay Bible)

and made a cliché by Samuel Butler's novel of the same title, published posthumously (1903).

way out under the influence of illegal narcotics.

In standard usage, showing any wide deviance from a norm, whence too the narcotics-induced elation in which some performers consider they work best.

weakness a tendency towards self-indulgence.

Often *tout court* of drunkenness:

> ... their Mr Fellowes *did* have a weakness (Bogarde, 1983—he was a drunkard)

and for the paradoxical *strong weakness* see STRONG WATERS. Often qualified, as, in a woman, a *weakness for men*, profligacy; in a man, a *weakness for boys*, homosexuality.

weapon the erect penis.

Punning on the shape and the use:

> My naked weapon is out (Shakespeare, *Romeo and Juliet*)

or, less obscurely:

> ... my weapon sheathed itself in her naturally. (Harris, 1925)

wear a bullet to be killed or wounded by shooting.

Underworld American slang:

> 'Who's wearing the bullet?' I asked her. (Chandler, 1958)

wear a smile to be naked.

A smile and nothing else. American rather than British use.

wear away to die a lingering death.

Usually being *consumed* by pulmonary tuberculosis, the common scourge until the Second World War, which was also known as *wearing*:

> Sickened, Took the bed, an' wear awa'. (Grant, 1884)

wear Dick's hatband see DICK[1].

wear down *Scottish, ?obsolete* to grow old.

Physically accurate and a nice allusion to the burdens of long life:

> I and my Jenny are baith wearin' down. (Rodger, 1838)

wear green garters *Scottish, obsolete* to remain unmarried after a younger sister's wedding.

By Scottish tradition, the unmarried elder sister wore green or yellow garters at the wedding. The taboos which surrounded spinsterhood arose from the plight of women who failed to obtain the support of a husband and were forbidden by convention to support themselves.

wear Hector's coat to be a traitor.

The *Hector* referred to was Hector Armstrong in whose house Thomas Percy, Earl of Northumberland, took refuge after the failure of his rebellion against Queen Elizabeth I in 1569. Armstrong betrayed him for money to the Regent of Scotland, but died in penury.

wear iron knickers to refrain from copulation.

Women only are figuratively so clothed:

> Her Italian father ... wanted her to wear iron knickers until she was twenty-one. (Follett, 1979)

wear the breeches to be the dominant partner in a relationship.

Usually used of women, alluding to the fact that at one time only men wore *breeches*, *trousers*, or *pants*:

> Helpmate, a thick, stubborn-looking lady of 40, childless, and most likely wearing the breeches. (*Century Magazine*, July 1882)
>
> Eddie Murphy's boastful ladies' man [falling] madly in love with Robin Givens's gorgeous, ruthless executive, who is even more predatory than he is ... This film's brassy flouting of money, power, and sex appeal would appear naive no matter who wore the pants, as they used to say. (*New York Times*, 12 July 1992)

And they 'used to say it' long ago, too:

> That you might still have worn the petticoat,
> And ne'er have stol'n the breech from Lancaster. (Shakespeare, 3 Henry VI)

It may also now be used of dominant men.

wear your heart upon your sleeve to fail to conceal heterosexual longing.

At one time men might advertise their intentions or desires by displaying some keepsake from the woman:

> But I will wear my heart upon my sleeve
> For daws to peck at.
> (Shakespeare, *Othello*)

wee to urinate.

We have a choice of derivation. *Wee*, meaning 'small', may refer to LITTLE JOBS, or it may be a corruption of the French *eau*; urine or WATER are also known as *wee*. A *wee-wee* indicates neither greater nor repeated urination:

> 'Just a minute,' said Viola, 'I want to wee-wee.' (Bradbury, 1959)

Nursery and some genteel use.

wee drop *mainly Scottish/Irish* a drink of whisky.

Despite the *wee*, the portion is usually substantial:

> Manis was always fond of the wee dhrap. (MacManus, 1899)

See also DROP[2]. A *wee half* is less common:

> ... a 'wee hauf' held my heart in cheer. (Murdoch, 1873)

Wee dram remains common. These phrases are the Scottish and Irish equivalents of a LITTLE SOMETHING.

wee folk the fairies.

The British Christmas pantomime tradition makes us forget that especially in Ireland fairies were pretty unpleasant creatures whose malevolence made them the subject of fear and taboo:

> The belief in the 'wee folk', or 'gentry', is very much more widely spread. (*Cornhill Magazine*, February 1877, quoted in EDD)

Also as the *wee people*:

> ... they attribute it to the wee people. (Mason, c. 1815)

You had to appease those you feared by speaking kindly of them.

weed (the) marijuana.

Formerly tobacco, whence a cigarette or cigar when smoking was taboo:

> ... a man whose private worth is only to be equalled by the purity of his milk-punch and the excellence of his weeds. (Bradley, 1853)

In modern use from the leaf of the pistillate hemp plant:

> ... opened the door and sniffed the weed. (Chandler, 1958—he could smell cannabis)

A *weedhead* is a habitual smoker of marijuana.

weed killer *American* a chemical defoliant used in warfare.

Vietnam jargon for the substance which also destroyed crops, and perhaps the health of those who administered it:

> ... defoliants are referred to as weed-killers. (McCarthy, 1967, writing about Vietnam)

weekend dishonestly to use a customer's money after the close of business on a Friday.

Banking jargon and practice. By delaying putting the credit to a customer's account on a Friday —or even a Thursday—the bank gains for itself interest on the amount until at least the following Monday. All banks do it, but some are more blatant or greedy than others, especially with remittances from abroad and documentary credits. Paying an inflated charge for telegraphic transfer and advice affords some protection to the customer but I have known a Middle Eastern bank to *weekend* for upwards of three weeks.

weekend warrior a part-time prostitute.

Punning on the derisive term for National Guardsmen and members of the Territorial Army who hold civilian jobs and engage in military activities only at weekends.

weenie the penis.

Possibly from the German *wienerwurst*, Vienna sausage, whence *wienie* or *weenie*, a frankfurter. To *step on* or *shoot your weenie* is a variant of the cliché, to shoot yourself in the foot:

> So long as I don't step upon my weenie. (Clancy, 1989)

weigh the thumb deliberately to overcharge.

From the practice of depressing the scales to give a heavier reading but now used figuratively of any overcharging.

weight problem see PROBLEM.

weirdie a male homosexual.

Also used of anyone whose tastes, dress, politics, etc. differ from our own, as in the dated cliché *beardies and weirdies*.

welfare state aid to the poor.

It originally meant prosperity, which is not how its recipients today see it:

> ... his girl friend threatened to call the cops when he took half of her welfare money. (Wambaugh, 1983)

The British *welfare state* purported to provide every citizen with free medical attention and schooling; and a home and subsistence for the poor.

well not menstruating.

The opposite of UNWELL[1]:

> ... soon after I am well each month. (Harris, 1925)

well away very drunk.

> The Colonel ... overcomes his resistance to vodka to such an extent he is soon well away and sings songs of Old Kentucky. (Carter, 1984)

You can also be *well bottled, well in the way, well oiled, sprung, corned*, etc.:

> I'll nut say drunk, but gay weel cworn'd. (Whitehead, 1896)

Some forms are obsolete.

well built obese.

Of men and women, and of children also because manufacturers know better than to describe somebody's little darling as fat:

> But, importantly in this case, there is a well-built girl attendant who is chased about the stage. (*Daily Telegraph*, 31 October 1972)

Formerly as *well fleshed*:

> Well-fleshed men could niver stand up long agen an ale-pot. (Sutcliffe, 1901)

well corned see WELL AWAY.

well endowed having large genitalia or breasts.

The person so described may have a tiny dowry, or none at all:

> ... she was probably as pretty, if considerably less well endowed physically. (Price, 1970 —she had smaller breasts)

Amply endowed means the same thing, although a *large endowment* sounds like boasting:

> Exceptionally good-looking, personable, muscular athlete is available. Hot bottom plus large endowment equals a good time. (Advertisement which Representative Frank answered, later appointing the personable prostitute his personal aide: quoted in *Sunday Telegraph*, September 1989)

well hung having large genitalia.

Used critically of bulls, stallions, and rams, and lewdly of men:

> The blowen was nutts upon the kiddey because he is well-hung. (Grose)
>
> He had a deep voice and looked from his tight pants to be fairly well hung. (Phillips, 1991)

well in the way see WELL AWAY.

well-informed sources a friend or confidant of the reporter.

Literally, an authorized or knowledgeable spokesman. The vague attribution gives weight to a thin or speculative story, or seeks to cover a *leak* (see LEAK[2]).

well oiled see WELL AWAY.

well rewarded overpaid.

With a hint of greed in a corporate world where backs are regularly and mutually scratched:

> ... Lord Young, C & W's well-rewarded chairman. (*Daily Telegraph*, 6 December 1994)

well sprung see WELL AWAY.

welly a contraceptive sheath.

A shortened form of *Wellington boot*, which is also made of rubber for protective reasons:

> **wellies from the Queen** are condoms held by the QM at the brow during foreign port visits. (Jolly, 1988—the *brow* is the gangway)

wench *archaic* a prostitute.

Originally, a girl, whence a promiscuous woman:

> Let my lord take wenches by the score. (Blackhall, 1849)

To *wench* is to use a prostitute (of a male), and *wenching* a tendency in that direction.

West Briton an Anglicized Irishman.

Often Protestant and educated in England, affecting the manner and speech of the British professional classes. Long used derogatively:

> Those on the other side, he said, were mere 'West Britons'. (Kee, 1993—the speaker was C. S. Parnell, himself a Protestant educated at Cambridge, a cricket enthusiast who spoke with a standard upper-class English accent)

Whence *West Britonism*, the continuance of rule by the Protestant ascendancy in Ireland:

> The O'Conor Don is a sample of West Britonism in Ireland—he is a sample of the rights of England and Englishmen to rule Ireland. (ibid.—Charles O'Conor affected the more glamorous appellation)

Thus too *West British*:

> After a short time the paper's policy could no longer with any justice be called 'West British'. (Fleming, 1965, of the *Irish Times*, formerly the mouthpiece of the southern Irish Protestants and maintaining a Unionist stance long after the creation of the Free State)

Still in common use, often shortened to *West Brit*; see also *North Briton* at NORTH BRITAIN.

wet[1] to urinate.

Literally, to damp through any agency, but now standard English when we *wet the bed*:

> Boys and girls who steal, vandalize, or wet the bed. (Bradbury, 1976)

To *wet yourself* is to urinate in your clothing, perhaps through fear:

> Grooters felt her legs almost doubling underneath her and she wet herself (Davidson, 1978)

as is to *wet your pants* etc.:

> Merriman thought he was going to wet his pants. (M. Thomas, 1980)

The obsolete *wetting* was stale urine used in laundry or cloth manufacture:

> I slat a pot of wettin in his feace. (Wheeler, 1790)

wet[2] a drink of an intoxicant.

Occasionally on its own:

> Bring me a wet, I'm near parched. (Cookson, 1967)

Also used as an adjective, in the sense 'serving or consisting of intoxicants'. The British *wet canteen* was the one where alcohol was served:

> We spent a very pleasant evening, the First Battalion having a wet canteen, and when we started back we were three sheets in the wind. (Richards, 1933)

Wet goods or *stuff* were intoxicants, especially in American Prohibition use:

> The wet goods flowed. You couldn't move all of it. (Longstreet, 1956)

A *wet hand* was a drunkard who might be said to *wet* his *mouth, beard, quill, whistle*, etc.:

> Simply must wet m'whistle. (Manning, 1960, who can never be forgiven for Yakimov's untimely death)

You *wet a baby's head* when you celebrate its birth—another rite of christening—and you *wet a bargain* when you drink to seal it:

> ... and be dam we'll wet our bargain. (Somerville and Ross, 1908)

To *wet the other eye* is to drink one glass of an intoxicant after another (OED). A *wetting* used to be an intoxicating drink:

> The young chaps bring their bottles oot,
> And ilk ane gets a wettin'.
> (Lumsden, 1892)

wet-back an illegal Mexican immigrant into the USA.

From supposedly swimming across the border:

> A lot of them were wet-backs. (Macdonald, 1971, of Californian orange pickers)

wet deck a prostitute who copulates with one man immediately after another.

Nautical usage, from her condition and the sequential action of waves on a ship:

And who would have the first bout, in any case? I'll not take your wet decks. (Monsarrat, 1978, writing in archaic style of sailors sharing the services of a prostitute)

The obsolete *wet hen* was a prostitute.

wet dream an involuntary seminal ejaculation while asleep.

The experience may be accompanied by an erotic dream:

Any dreams, wet or non-wet... (Amis, 1978)

General and genteel use.

wet for wishing to copulate with.

Of a woman, from the vaginal secretion and usually of extramarital lust:

I am rotten-ripe, soft and wet for you. (Harris, 1925)

wet-job a murder.

But not necessarily by drowning:

If anyone fancied the idea of doing a 'wet-job' on me then the bomb would go off in hours. (Allbeury, 1983)

Also as *wet operations*:

Max was an expert at what the chekists tactfully described as mokrie dela, 'wet operations'. (Moss, 1987)

wet your wick to copulate.

Of a male, usually extramaritally:

And Carlo had tried to wet his wick, because in Oregon that was no big deal, and before the sun was up her father had opened his throat for the ants to have a drink. (Seymour, 1984)

wet weekend (a) *Australian* a period of menstruation.

An occasion perhaps when the opportunity for sport is curtailed.

wetness sweat.

Genteel female usage:

The competent, knowledgeable people with public lives which transcend choices about bathroom bowl cleaners and products to prevent underarm 'wetness' have been males. (Mackie, 1983)

wetting see WET[1, 2].

whack to kill.

The common hitting imagery:

Joe, you know when Geoff got whacked, don't you? (Sanders, 1977, not on his bottom but pushed under a train at Union Square station)

whack off to masturbate.

Of a male—to *whack* is to pull, among other meanings:

Zoona—who was eventually thrown out of school for whacking off in full sight of three mothers during parents day. (Collins, 1981)

whacked drunk.

In standard usage, in slang 'very tired' after being beaten, I think, rather than after sexual activity:

...a very wet party. Everyone got whacked out of their skulls. (Sanders, 1982)

whanger a penis.

The common hitting imagery:

It couldn't have been more killingly awful if he had taken out his whanger and stuffed it in Lady Draycott's ear. (Fry, 1994)

what the traffic will bear an excessive but obtainable fee.

A principle used by most lawyers etc. when billing corporate, careless, or worried clients. The imagery is from transport pricing policy.

what you may call it any taboo object.

The lavatory to many females, or the penis or vagina, although not necessarily in a sexual sense.

whatsit any taboo object.

A shortened form of WHAT YOU MAY CALL IT but usually of a lavatory only:

The whatsit is through there if you want it. (B. Forbes, 1983—a woman was telling a man where the lavatory was)

Whatzis is rare:

...you'll probably use it to shoot off your whatzis. (Sanders, 1982—the *it* was a handgun which might hit the firer's penis)

whelp to give birth to a child.

A *whelp* is literally the cub of a bitch, a lioness, or a tigress:

...she was so close to what she called 'whelpin' that she couldn't be moved. (Keneally, 1979)

whiff *?obsolete* to kill.

Originally, in slang, 'to hit out at':

He wasn't alone when you whiffed him. (Chandler, 1939)

whiff of associated with something taboo.

From the odour:

...we got a definite whiff of march hare. (Monkhouse, 1993—somebody was acting like a lunatic)

Carlyle's *whiff of grapeshot* was the firing on the Paris mob by Napoleon whereby he established order and his own reputation.

whiffled drunk.

To *whiffle* was to be unsteady and a *whiffler* a person who talked wildly:

'I did thirty days without the option for punching a policeman in the stomach on Boat-Race night.' 'But you were whiffled at the time.' (Wodehouse, 1930)

whip to steal.

Usually of small objects, probably from the concept of moving a distant article quickly with the use of a whip.

whip off *American* to masturbate.

Of a male, but not involving sadism. Also as the vulgarism *whip your wire*. In obsolete slang use, the *whip* was the penis.

whip the cat to be drunk.

Cats are associated with vomiting, and vomiting with drunkenness.

whistle the penis.

Nursery usage, from the shape in a young boy.

whistleblower a person who reveals confidential information.

From the action of stopping a game, but some pejorative use because confidential information may be made public in breach of the terms of employment:

> But the marginalising of local government, and giving powers and public funds to unelected, unaccountable quangos (with rules that punish 'whistleblowers'...). (*Daily Telegraph*, 5 February 1994)

See also BLOW THE WHISTLE ON.

whistled drunk.

The derivation was probably from the obsolete *whistle-shop*, a British inn in which you *wetted your whistle*, your *whistle* being your mouth or throat. *Whistlers* were the unlicensed sellers of spirits:

> The whistler, otherwise the spirit-merchant. (Moncrieff, 1821)

whistling *American* poor.

Of a narcotics addict unable to buy supplies, from the wind which whistles through an empty house.

white elephant an unwanted or onerous possession.

The King of Siam, also titled the King of the White Elephant, was said to present such a beast to any courtier he wished to ruin. Unable to sell or work the animal, the courtier had to provide for it with no return:

> The £2,000 million white elephant. (*Private Eye*, March 1981, of Concorde)

white eye inferior whisky.

Made out of colourless barely potable alcohol and no doubt punning on the rolling of the eyeballs of those who drink it. The *white stuff* is gin or vodka:

> He was drunk... He'd been on the white stuff all day long and he was drinking it like water. (le Carré, 1989)

white feather cowardice.

Such a feather in the plumage of a fighting cock was said to indicate poor breeding whence less aggressive behaviour:

> There's a white feather somewhere in the chield's wing, for all he's so big and buirdly. (Hamilton, 1898—*buirdly* means 'fine-looking')

In the First World War chauvinistic British women took the imagery literally, handing out white feathers to young men in mufti, secure too in the knowledge that they were themselves ineligible for service in the trenches.

white-knuckler *American* a small aircraft used on a scheduled service.

From the anxious grip of its passengers in poor weather:

> You take a white-knuckler... from Hyannis Airport through sea-fog to Logan. (Theroux, 1978—Logan is Boston's international airport)

Various small carriers enjoy the title of *White-knuckle line* conferred by their regular passengers.

white lady heroin.

Also known as *white stuff*, which can include cocaine and morphine; and as *white powder*:

> He was still getting $100,000 a year... and that bought a goodly amount of the sweet white powder. (M. Thomas, 1982)

white lightning a spirituous intoxicant or LSD.

> Ellen... unfolded some tinfoil which she said contained three tabs of Owsley's original 'white lightning', the Mouton-Rothschild of LSD. (*Village Voice*, 1 June 1972)

The intoxicant is illegally distilled and has no added colouring:

> ...'white lightning', 'white mule', or just plain 'corn', as the local moonshine whiskey is called. (*Double Dealer*, July 1921)

white man's burden the privileged status of a white in a colonial territory inhabited by blacks.

Kipling invented the phrase but not the concept of the chore of 'civilizing' the natives:

> ...impress the natives with what wonderful things the British were doing for them; the whole idea of the White Man's Burden. (Allen, 1975, writing of the colonial period)

You may recall that in the pictures of intrepid marches in the interior of Africa by white men, the black men always seem to be doing the porterage.

white marriage a marriage in which the parties do not copulate.

Perhaps from the virginal colour, although that concept seems dated too:

I don't think there's much sex in poor Tom. What's known as a white marriage. (Burgess, 1980)

white meat[1] the breast of poultry.

Especially chicken; see DARK MEAT[1] for a dissertation. Now standard English.

white meat[2] a white woman viewed sexually by a black man.

The converse of *dark meat* (see DARK[2]).

white money *American* funds improperly acquired made capable of open spending.

BLACK MONEY recycled.

white nigger a white person who favours blacks.

Mainly derogatory use. It formerly meant 'a poor Southern white' in the days when the blacks in the South were all poor and poor whites were despised:

Ingrid You were poor also ... *nein?*
Anderson Yes. My family was white niggers. (Sanders, 1970)

white plague (the) pulmonary tuberculosis.

A 19th-century scourge and the source of much euphemism:

When scarlet fever, cholera, typhoid fever, and the 'white plague' (tuberculosis) took such a toll of young ladies in their prime, there was a considerable body of fiancés who mourned for the rest of their lives. (Pearsall, 1969, writing of the 19th century)

white rabbit-scut *?obsolete* cowardice.

The *scut* is the short white erect tail, the sign of a fleeing rabbit:

What, leave Marsh and show the white rabbit-scut to Nicholas Ratcliffe? (Sutcliffe, 1900)

white sale an occasion when concessions are freely given.

Presumably from the recurrent discounting of *white goods*—bed linen or domestic appliances:

I got him everything. It was a white sale at the U. S. Attorney's Office. (Turow, 1990)

white satin gin.

From the colour and the smoothness:

White satin, if I must know, was gin. (Mayhew, 1862)

white slave a white prostitute working outside Europe.

Usually under a pimp's strict control in the Middle East or South America. Whence *white slavery*:

White slavery—the seduction and selling, and of course buying, of women for immoral purposes (Londres, 1928, in translation)

and *white slaver*, the finder or the pimp:

I'm not a white slaver in case they exist. (James, 1972, referring to an invitation to a young white woman to go on a journey with a stranger)

white tail a completed but unsold civil aircraft.

The manufacturer leaves it in its white undercoat until he can find a buyer who will stipulate the colour scheme. For him, such a stock is a double disaster; the finance charges continue and the existence of unsold aircraft spoils the market.

white top a geriatric.

In reality many of the men will be bald and the women blue:

The problem with 'white tops', old folks with failing reflexes, impaired facilities or the effects of prescription drugs, let loose on the highways, is causing concern in Florida. (*Daily Telegraph*, December 1988—and not just in Florida)

whites the vaginal secretion of a sexually aroused female.

Also used of other vaginal secretions due to ill health. Occasionally as *white flowers*:

Fleurs-blanches ('white flowers', flowers being the colloquial expression for the menses) were caused by erotic reading. (Pearsall, 1969—French novels were found very potent in the 19th century)

whitewash an attempt to hush up an embarrassing or shameful event.

The compound of lime and water, or similar non-permanent materials, quickly and liberally applied to a surface, may provide temporary cover for the blemishes underneath. Much political use:

Then, in Hughes's opinion, the committee had produced a whitewash. (Colodny and Gettlin, 1991, writing about a report on the secret bombing of Vietnam)

To *whitewash* is to try to conceal or defuse such an event.

whizz an act of urination.

American onomatopoeic use:

'I just came down for a whizz.' He recoiled at the vulgarity. (Theroux, 1978)

whole can see HALF A CAN.

whole hog (the) copulation.

Usually after courtship involving exploratory sexual acts, and in the phrase *go the whole hog*:

She was disappointed. That I didn't go the whole hog. (Amis, 1980)

From the slang phrase *go the whole hog*, meaning 'to do something completely', which itself has a disputed derivation—either eating all of a roast male pig at a sitting, or drinking all of a hogshead of ale.

wholesome not suffering from venereal disease.

Literally, no more than healthy:

> The woman, endeed, is a most lovely woman; but I had no courege to meddle with her, for fear of her not being wholesome. (Pepys, 1664—perhaps too he was feeling weary, having already 'had his pleasure' of Mrs Lane twice that day)

wick the penis.

From the rhyming slang *Hampton Wick*, prick, and see also HAMPTON. This is a unique example of both parts of a rhyming slang phrase being used individually, although *hampton* and *wick* are not exact synonyms, *wick* alone being used figuratively as well as literally:

> It gets on my, you know, wick. (Bradbury, 1976)

wide-on (a) *American* female heterosexual lust.

I suppose from the inappropriateness of HARD-ON:

> That's the one thing about lady analysts... once in a while they fall in love with a stock, usually because they get a wide-on for the management. (M. Thomas, 1985—he was speaking figuratively)

wide parting (a) baldness.

Male humour, used only when speaking about others and unfunny when applied to ourselves.

wife[1] the senior prostitute of a pimp.

Prostitutes' jargon where the pimp has several females in his stable:

> Keep her as your 'wife', since she's more use to you than to me, and take me as your sweetheart. (Londres, 1928, in translation)

wife[2] a male homosexual taking the female role.

Homosexual use in a lasting domestic arrangement.

wife[3] a woman who copulates regularly with the same man outside marriage.

Often of a relationship with a male prostitute:

> Several of the other studs had regular customers—their 'wives'. (Sanders, 1983)

will a homosexual.

Widespread English dialect use of either sex, and of a hermaphrodite. Not from the standard English *will*, lust, but probably a shortened form of *Will-o'-the-wisp*, the *ignis fatuus*, of which the first appearance is deceptive.

will there be anything else? do you want any condoms?

The question asked by barbers of adult males when there was a taboo against buying condoms:

> ... the days when one's barber, hoping to sell a packet of Durex, used to murmur discreetly, 'Will there be anything else, sir?' (*Sunday Telegraph*, 27 March 1994—and they still called a customer 'Sir')

willie see WILLY.

willie-waught a drink of intoxicant.

The Scottish *good willie* means 'hospitable' and *waught* means 'to drink deeply':

> 'And we'll take a right guid willie-waught' was changed to, 'We'll give a right guid hearty shake', in deference to temperance principles. (Murray, 1977, writing of Sir James Murray, the creator of the OED and, domestically, the bowdlerizer of Burns)

willing prepared to engage in casual copulation.

Not just any old testatrix, and usually not of a prostitute:

> ... there might even be willing mountain women up there to warm his solitude. (Keneally, 1979)

willy the penis.

Both nursery and adult use:

> Does your willy rise like a snake out of a basket? (Theroux, 1978)

I can do no better than quote from *Man's Best Friend*:

> There are almost as many names for a man's most intimate possession as there are for man himself. Depending on the self-confidence of the owner, and the degree of esteem that exists between the two parties, these names vary from the optimistic (Big Steve, Oliver Twist) to the pessimistic (General Custer, The Sleeping Beauty); from formality (He Who Must Be Obeyed) to familiarity (Old Faithless); from Tom, Dick and Harry to Jean-Claude, Giorgio and Fritz. The villain of this book is called Willie. (Joliffe and Mayle, 1984)

wilted drunk.

The imagery is floral, though from too much liquid rather than too little.

win[1] to steal.

Very old general use:

> The cull has won a couple of rum glimsticks (Grose—a *glimstick* is a candlestick)

but still common in the British army:

> In the Army it is always considered more excusable to 'win' or 'borrow' things belonging to men from other companies. (Richards, 1936)

win[2] to copulate with.

In a bygone age, to *win* a woman was to do no more than secure her consent to marriage, but the modern use is of extramarital sexual conquest:

I resolved to win her altogether. (Harris, 1925)

win home to die.

Christian devout use of the death of others, although the speaker seldom appears anxious to secure a similar victory for himself:

Thro' a' life's troubles we'll win home at e'en. (Wright, 1897)

To *win your way* is obsolete:

Auld Jamie has giv'en up the ghost
And won his way.
(Hetrick, 1826)

wind[1] a belch or fart.

In genteel use only of belching, about which there are fewer taboos than farting:

Baked beans, which always give me terrible wind. (Matthew, 1978—it might apply to either form of expulsion)

Whence *windy*, so affected or, of an object, liable to cause such an effect:

...taters...es windy zorrt o grub. (Agrikler, 1872)

wind[2] **(the)** summary dismissal from employment.

An uncommon American variant of *the air* (see AIR (THE)).

winded incapacitated by a blow to the genitalia.

Supposedly, having received a disabling blow in the stomach. The jargon mainly of male sportsmen and those who comment on their activities but also some non-sporting use:

'Just winded,' groaned Harry, though in fact a flying brick had struck him a painful blow in the groin...he was holding his genitals cupped in his hand for they were too painful to massage. (Farrell, 1973)

windfall[1] *obsolete* a bastard.

Probably because the *fruits* arrived on the ground other than by design.

windfall[2] a bribe or other illegal benefit.

The fruit which fell to the ground used to be given to whomever wished to pick it up, whence in figurative use a *windfall* could be anything which you acquired of value without consideration, such as an unexpected legacy or a bribe:

The cop and those higher up share in the windfall. (Lavine, 1930, of bribery)

window blind *British* a towel worn during menstruation.

Punning on the periodical closing of a shop, indicated by the unrolling of a blind to conceal the goods normally on display.

window dressing falsely or fraudulently issuing figures or statements relating to a business.

Commercial and banking jargon, using imagery from retail trading. Of attempts to inflate profit, turnover, or reserves by improper adjustment, of any suppression or exaggeration designed to enhance the value or prospects of a business, and specifically of issuing cheques in your own favour which can only be met from their proceeds:

The cheques were part of the 'window dressing' of the balance sheet at London & County Securities. (*Private Eye*, September 1981)

windy[1] see WIND[1].

windy[2] cowardly.

From the First World War use, 'to get the wind up', 'to be afraid':

...he may be what the British soldier would call 'slightly windy'. (Moss, 1950)

In the First World War many danger spots in the battle zone were known by the punning *Windy Corner*.

wing your flight from this world *obsolete* to die.

How you made your way to heaven:

The Bonds of Life being gradually dissolved
She winged her Flight from this World in expectation of a better, the 15th January, 1810. (Memorial in Bath Abbey)

winged wounded.

Of humans, from the shooting of birds which, if hit in the wing, come to ground alive. Second World War and then general use.

winger a person protecting a criminal.

Not from his shooting abilities but a position to one side:

Couple of wingers, as the Boys called unobtrusive bodyguards. (M. Smith, 1993)

winkle the penis.

Nursery usage, perhaps from the *willie* in *Wee Willie Winkie* (see WILLY):

...unlikely to hurl himself diagonally across the polished walnut and snatch at his winkle. (Amis, 1978)

Winkie is rare:

Very butch, and he's got a gun trained on your winkie. (B. Forbes, 1986)

wipe off to kill.

Of an individual by a natural phenomenon:

What more useful bird can yer find, as wipes off worms an' grubs as they did? (Patterson, 1895)

or through human agency:

He'll wipe you off. (Chandler, 1939)

See also WIPE OUT[2]. The imagery of both is probably from chalk on a blackboard.

wipe out[1] to kill.

I worked with three gangs who got wiped out, all except me. (L. Thomas, 1979)

Usually of multiple deaths; see also WIPE OFF.

wipe out[2] to make bankrupt.

Again the shared imagery between death and bankruptcy:

... for was it fair to take a nice, dumb little guy like Lehman for such a ride, one that would inevitably wipe him out? (Erdman, 1987, of a dupe about to be cheated)

Also figuratively of discrediting:

It would wipe me out, of course. No one would employ me. (Deighton, 1988, of someone facing a criminal charge)

wire[1] see BEHIND THE WIRE.

wire[2] to bypass a tachometer on (a commercial vehicle).

The tachometer records the times when the vehicle is moving, thus providing evidence that statutory rest periods are taken by the drivers. An operator wishing to flout the regulations may deactivate the device by installing an electrical circuit which, when activated, will cut out the tachograph, called *wiring* a vehicle in the jargon. See also HOT-WIRE.

wire-pulling the covert use of influence or pressure.

From the actuation of a puppet:

... promises were also held out of 'wire-pulling tactics in high political circles'. (Foster, 1993, of the advance publicity for Mrs Parnell's 1914 autobiography)

Standard English.

wired[1] drunk or under the influence of narcotics.

Of the same tendency as LIT but now usually of the experience of illegal narcotics:

'Do you have to go to bed?' he asked. 'I'm wired. I can't sleep.' (Robbins, 1981, of narcotics)

wired[2] subject to clandestine surveillance.

This espionage jargon has survived the introduction of devices which are wireless:

Even the damn cats are wired, no exaggeration. (le Carré, 1980)

The device can be carried by the victim or by another knowingly; they may then also be said to be *wired up*:

... the defendant remained unaware ... that their interrogators were ... 'wired up'. (*Private Eye*, March 1981)

A *wireman* is an expert in such devices:

What we need is a first-class wireman. Somebody who can do it right. The apartment. The phone. (Diehl, 1978)

wise man a wizard.

And a *wise woman* was a witch:

Sure a wise woman came in from Finnaun ... and she said it's what ailed him he had the Fallen Palate. (Somerville and Ross, 1908)

with child pregnant.

Not just of a young woman holding a baby:

Once he had got a girl with child. (G. Greene, 1932)

with it in your hand always ready for extramarital copulation.

Northern English profligate males are said to *go about* in such a manner. *It* is the penis.

with Jesus dead.

Christian usage in various forms, from the posthumous heavenly gathering of the righteous and others. Other common companions are *the Lord* and *your Maker*:

If you make a wrong move, you're with your maker. (Fraser, 1970)

A dead fellow-worshipper or clubman may be described as *with us no more*.

with respect you are wrong.

Used in polite discussion and jargon of the British legal system where an advocate wants to contradict a judge without prejudicing his chances:

There is high authority for the view that [with respect] means 'You are wrong' ... just as 'with great respect' means 'you are utterly wrong' and 'with the utmost respect' equals 'send for the men in white coats'. (Mr Justice Staughton quoted in *Daily Telegraph*, February 1987)

withdraw your labour to go on strike.

Trade union jargon—it could simply mean to leave one job out of idleness, or for another. A *withdrawal of labour* is a strike.

withdrawal to prepared positions a retreat under pressure.

One way in which the vanquished seek to mitigate failure. A *withdrawal in good order* was probably a rout.

within-group norming see SCORE ADJUSTMENT.

without a head *Scottish, obsolete* unmarried.

This expression refers to the time when an unmarried woman had little security outside her parents' home, poor chances of maintaining herself, and almost no protection in law:

It's no easy thing ... for a woman to go through the world without a head. (Miller, 1879)

Males who are vexed by the antics of modern feminists should remember that this pendulum once swung the other way.

without baggage *obsolete* to execution.

Of the manner of departure of prisoners taken away by Russian Communists for summary execution:

> From time to time someone would depart from the camp 'without baggage'. Those were sinister words—we all knew what that meant. (Horrocks, 1960, of his 1920 Moscow imprisonment after serving with the White Russian forces)

woman a female viewed by a male for copulation.

The man who says *I feel like a woman tonight* does not postulate an incipient sex-change. Whence *womanizer*, a male profligate.

woman friend a man's sexual mistress.

As distinct from a friend who is a woman, and see also FRIEND:

> Samoza, his woman friend ... and four of his five children. (*Daily Telegraph*, August 1979)

woman in a gilded cage[1] a sexual mistress.

Denoting a female provided with separate accommodation, especially in luxury and in the 19th century:

> The companion of a girl's fall might himself be the unconscious utterer of a divine message ... the woman ... breaking away from her gilded cage. (H. Hunt, c. 1854)

woman in a gilded cage[2] *American* a young woman married to a rich old man.

Often coerced into the union by her greedy or ambitious family.

woman named a woman accused by the wife of copulating with her husband during the marriage.

British legal jargon. In a divorce suit, a male could be joined as a respondent along with the wife by a petitioning husband, and thereby made liable for damages and costs, but a woman would only be *named* by the wife, and liable for neither. See also CO-RESPONDENT.

woman of the town a prostitute.

Not merely a person whose sex comprises half the population:

> It is ordered that hereafter when any female shall ... show contempt for any officer or soldier of the United States, she shall be regarded and held liable as a woman of the town, plying her avocation. (G. C. Ward, 1990, quoting an order by the Yankee military governor of New Orleans in 1862)

Sometimes too as *woman of the world*, although to be a *man of the world* implies knowledge of, rather than participation in, shameful activities.

woman's thing (the) female homosexuality.

Not a brassière:

> The virago and her soulmate into, as they would say, the woman's thing. (Theroux, 1978)

woman's things see WOMEN'S THINGS.

women a lavatory for exclusive female use.

The companion of MEN, but no less salubrious than LADIES. Also as *women's room* etc.

women's disease (the) syphilis.

If British males could not attribute their misfortune to the French, the Neapolitans, or the Spanish, then it had to be the fault of women.

women's liberation aggressive feminism.

For most men and many women a dysphemism, especially when shortened to *women's lib.*:

> Women's lib meant more than burning your bra. It meant total commitment to the programme of women's superiority over men. (Sharpe, 1976)

An enthusiast may be called a *women's libber*:

> You make me sound like the worst sort of Women's Libber, an aggressive great lesbian with a foul placard (Pilcher, 1988)

or shortened to *libber*:

> She's gone to join some women friends. Libbers, you know. (Murdoch, 1980)

I note elsewhere the 19th-century social attitudes (see WITHOUT A HEAD) in the days when a *libber* was no more than a castrator of pigs (EDD).

women's movement an association of committed feminists.

And nothing to do with callisthenics.

women's rights the claim to or enjoyment of economic and social advantages historically exclusive to men.

An aspiration of the feminists, who may not necessarily wish to assume the corresponding responsibilities:

> ... extensive literature on Women's Rights and the Feminist movement. (Bradbury, 1976)

But before being too dismissive, see WITHOUT A HEAD.

women's things any taboo matter or article exclusive to women.

Occasionally denoting menstruation or illness, often towels worn during menstruation:

> For the curse—you know. Women's things. (W. Smith, 1979, of such towels)

Also as *woman's things*.

wonk a male homosexual.

From the slang *wonky*, askew. Formerly a *wonk* was no more than someone of limited intelligence.

wooden box a coffin.

In obsolete forms also as *wooden breeches, breeks, coat, overcoat,* etc.:

> A pair of wooden breeks
> Now him doth clede.
> (Sutherland, 1821)

Used figuratively, too, to refer to death:

> The Winston treatment when it finally comes to the wooden box. (*Private Eye*, June 1981 —Churchill had an elaborate state funeral)

wooden hill the staircase.

Nursery usage, for children urged to climb it when reluctant to go to bed (or, in the double pun, to *Bedfordshire*).

wooden log a human used involuntarily for dangerous medical research.

A Second World War usage, referring to both Russia and Japan, and attributing to human beings the characteristics of a tree which had been felled and cut up:

> White Russian Jews, nearly all living in Manchuria or Northern China, were already subject to appalling discrimination, not as Jews but as stateless White Russians, and potential 'wooden logs'. (Behr, 1989)

The practice was widespread and not confined to White Russian Jews. Unit 731 under General Ishii used prisoners for medical experimentation until the very end of the Second World War. He also tried to land plague-ridden fleas to infect the US forces in Saipan, being frustrated when the submarine involved was sunk. Neither Ishii, nor his master Hirohito, were charged as war criminals.

woolly bear *American* a policewoman.

Another extension of the BEAR[2] theme. The *wool* comes from the vulgar slang, a woman's pubic hair.

word from our sponsor (a) *American* a television advertisement.

Would that it were only one.

words an advertisement on television.

Another way of covering up the intrusion:

> We'll have a filmed report after these words. (Bryson, 1989, of an American advertisement —the British *after the break* at least concedes the loss of continuity)

work to copulate.

Of females, as in Iago's paradoxical satire:

> You rise to play, and go to bed to work. (Shakespeare, *Othello*)

And see also *bedwork* at BED[2].

work both sides of the street to serve people with conflicting interests.

> For years he'd been a Mr Fixit, working both sides of the street. (Deighton, 1988)

Usually dishonestly. In the days of common door-to-door selling, a salesman would be allocated one side of the street as his territory. To *work a street* is an attempt to sell from door to door, usually offering shoddy goods. See also WORK THE STREETS.

work of national importance exemption from military service.

British First and Second World War usage which, to the military, suggested that their own activities were of less importance. The timid tended to seek jobs which carried such exemption, while the braver or more robust went to any lengths to escape from a reserved occupation:

> Here they were doing 'Work of National Importance' and they were too windy even to join their own town guard. (Richards, 1933 —he spent the First World War in the trenches)

work on[1] to extract information from through violence.

Literally, to have an effect on physically, but often by rough treatment:

> 'Shellacking', 'massaging', 'breaking the news', 'working on the...', 'giving him the works',... express how they compel reluctant prisoners to refresh their memories. (Lavine, 1930, of the NYC police)

work on[2] to copulate with.

Of a male, from the concept of persistent rough handling rather than from the male posture:

> We could... give you an examination too, and see if you've been working on her tonight. (Mailer, 1965)

work the streets to be a prostitute.

From her public solicitation:

> She worked each side of the street with a skill shared... by the best of streetwalkers. (Mailer, 1965)

work to rule *British* to conduct yourself at work in a manner calculated to obstruct and cause loss.

Trade union jargon for a device which, if successful, enables the employee to be paid while effectively striking. The employee may purport to be following in all respects a *Rule Book*, which regulates his employment and duties. Also as the noun *work-to-rule*:

> Within months, he was asking me whether we ought not to be writing more about a work-to-rule on the Central Line of the Underground. (J. Cole, 1995)

See also GO-SLOW and SLOWDOWN.

work your ticket to contrive an early discharge from contracted service.

The British *ticket* was the certificate of honourable discharge. A soldier etc. unwilling to complete the period of his enlistment and unable to buy his release had to use great ingenuity to escape from his commitment with a clean record.

work yourself off to masturbate.

Of a male.

workers' control the oppressive rule of an oligarchy.

Communist jargon which tries to perpetuate the fiction that the populace controls the ruling and self-perpetuating oligarchs:

> Within the Leninist model... 'Workers' control' here means control of the workers. (*Sunday Telegraph*, August 1980, referring to Poland)

workhouse *obsolete* an institution for the homeless indigent.

The intention was that the unfortunates should work to pay for their keep, although the name outlived the concept:

> I was put in the workhouse when I was young ... I never knew my father or my mother. (Mayhew, 1862)

working class *British* in employment paid weekly.

More commonly used when a larger percentage of the population than in the 1990s was engaged in the manufacturing industry, in which employment it was normal to belong to a trade union. It did not necessarily imply that those not so described were unemployed. A variant in current use, especially among politicians, is *working people*:

> I doubt whether working people will be willing to go on making sacrifices of this nature for much longer. (*Daily Telegraph*, January 1977—the 'sacrifice' was not receiving a wage increase unsupported by an increase in productivity and much exceeding the rate of inflation)

The working class denoted some of those whose parents, or who themselves, were so employed or remunerated, as distinct (although less and less distinct) from the *middle*, *professional*, and *upper classes*.

working girl a prostitute.

Employed *in the business* (see at BUSINESS):

> ... lining up working girls for himself and for clients. (Wambaugh, 1983)

working people see WORKING CLASS.

workout the use of violence to extract information.

Literally, any exercise. Police and American underworld jargon:

> ... scream with fear when the 'workout' begins. (Lavine, 1930)

works (the) any taboo activity.

Usually inflicted on an unwilling participant with undue zeal, from the slang *give someone the works*. Of police harassment of or brutality to a prisoner; of injuring an adversary by shooting or hitting; of copulation by a male, although not necessarily with a reluctant partner.

workshop a non-industrial gathering with political or artistic objectives.

Denoting theatrical groups, discussion groups, etc., whose organizers wish to imply that something of value is being produced:

> You're a good cook; a friend asks you to cook lunch for twelve at an exercise workshop. (*Health and Fitness*, 1991)

World Peace Council an instrument of Soviet foreign policy.

A weapon of the Cold War:

> World Peace Council, see under FRONT ORGANIZATION. (FDMT—a magisterially dismissive comment)

worm-food see DIET OF WORMS.

worry persistently to attempt to copulate with (an unwilling partner).

Originally, of dogs and wild animals, to kill by gripping the throat, whence by transference mental distress in humans. Women may use the term more of the disturbance and discomfort than because of a concern about an unwanted pregnancy. To *worry at night* is specific:

> It is perfectly dreadful that Wifie should be so worried at night. (Kee, 1993—Parnell was writing to his sexual mistress, Katie O'Shea, whom he so addressed, commiserating with her about the fact that her husband sometimes wanted to copulate with her)

worse for wear (the) drunk.

Referring to mild drunkenness:

> Arrived home at four, rather the worse for wear. (Matthew, 1978)

See also THE WORSE.

worship at the shrine of to be unhealthily addicted to.

Usually of alcohol, illegal narcotics, or sexual excess:

> Among newspapermen, most of whom worshipped more frequently at the shrine of Bacchus than Ariadne... (Deighton, 1991—Bacchus was the god of wine; Ariadne's distinctions include being a granddaughter of Zeus, a daughter of Minos, a step-sister of the Minotaur, and the person who gave Theseus the ball of string to help him out of the maze devised by Daedalus. As punishment, Daedalus and his son Icarus were themselves confined to the maze, from which they made their aerial escape. Deighton implies that most journalists prefer the bar to the beat.)

wrack *obsolete* to copulate with (a female virgin).

Another form of the more common *wreck*, it means 'to destroy':

I fear'd he did but trifle,
And meant to wrack thee.
(Shakespeare, *Hamlet*)

The *wrack of maidenhead* was loss of virginity before marriage:

... the misery is, example, that so terrible shows in the wrack of maidenhood.
(Shakespeare, *All's Well*)

wreak your passion on to copulate with.

Of a male, usually extramaritally:

... overborne by desire, he had wreak'd his passion on a mere lifeless, spiritless body. (Cleland, 1749)

Passion, originally the suffering of pain as by Christ, has been used of lust, especially in males, since the 16th century.

wrecked drunk or under the influence of narcotics.

From the way you feel:

They were half blitzed, but both Dolly and Dilford were totally wrecked. (Wambaugh, 1983)

wretched calendar (the) I am menstruating.

From the female practice of noting the date of the onset. Used especially by apologetic females:

You must be kind. The wretched calendar. (Fowles, 1977)

wring out your socks to urinate.

British male usage, perhaps as a facetious explanation of the noise.

wrinkly an old person.

Used by the young, mindless of 'time's winged chariot':

... helping the wrinklies with their heating bills. (*Private Eye*, January 1987)

To qualify you do not have to reside in Wrinkle City (Miami).

wrist job (a) male masturbation.

Referring to the act and, as an insult, the actor:

Keen? In my book he's a wrist-job. (C. Forbes, 1983)

write yourself off to be killed in an accident.

To *write off* a vehicle etc. is to remove it from an inventory of serviceable equipment.

written out of the script killed or dismissed from employment.

Theatrical use in a serial play, soap opera, etc.:

... he had played a psychiatrist in a soap opera for seven years until he was written out of the script. (Sanders, 1983)

Not merely of actors:

I wouldn't write the D-G out of the script too early. (Deighton, 1988—or 'don't think he will give up his job soon or without a fight'; the *D-G* is the Director-General)

Whence the figurative use for death:

One jalopy like that in the flight could get us all written out of the script. (Deighton, 1982, of Second World War fliers)

wrong[1] *obsolete* to copulate with extramaritally.

Men *wrong* women, only marital copulation being *right*:

Vail your regard
Upon a wrong'd, I would fain have said, a maid.
(Shakespeare, *Measure for Measure*)

To copulate was bad enough, but to impregnate doubly so:

Ravish'd and wrong'd, as Philomela was.
(Shakespeare, *Titus Andronicus*)

wrong[2] homosexual.

When only heterosexuality was *right*:

Mildred genuinely suspected something 'wrong' with the girl, and 'wrong' with Barbie. (P. Scott, 1971)

And specifically as *wrong sexual preference* etc.:

Chris was a genuine Eastern aristocrat with the right name, right family, right connections, and the wrong sexual preference. (Sohmer, 1988)

wrong[3] pregnant.

Northern English usage, within or outside marriage, but probably referring to unplanned pregnancies only.

wrong side of the blanket an allusion to bastardy.

In addition to the evil of the act, the impregnation supposedly took place on or out of the marital bed, not in it:

Frank Kennedy, he said, was a gentleman though on the wrong side of the blanket.
(W. Scott, 1815)

wrong time of the month the period of menstruation.

Female usage, elaborating on TIME OF THE MONTH:

It's always the wrong time of the month.
(Weissman, quoted in Dickson, 1978)

yak a human carrier of illegal narcotics in bulk.

See also MULE2—different continent, same concept:

Maybe some of your yaks are mouthy guys. (Sanders, 1990—a *mouthy guy* cannot keep a secret)

yard1 the penis.

I hesitate to venture a derivation:

'Loves her by the foot'... 'He may not by the yard.' (Shakespeare, *Love's Labour's Lost*)

yard2 to copulate extramaritally while cohabiting with your spouse.

Perhaps you meet the third party just outside your home. Not from the English dialect *yarding*, the first step in courtship, in which the parties walked three feet apart, perhaps to be followed by *aiblen*, holding an elbow, and, if your suit prospered, with *waisting*.

yardbird a prisoner.

From the exercise yard in an American penitentiary:

The yardbirds ignored their chief and slacked off, shot craps while Ricco slept it off in the supply locker. (Adams, 1985)

year of progress a period of irreversible decline.

Look out for this in statements by politicians and company chairmen—it usually means things have gone badly:

In the year leading up to the Tet Offensive ('1967—Year of Progress' was the name of an official year-end report). (Herr, 1977)

years young old.

Journalese for the age of a spry geriatric, described perhaps as *74 years young* to avoid the taboo *old*.

yellow1 cowardly.

Probably from the paleness of fright; found in many compounds, such as *yellow belly*. Formerly *yellow stockings* were a sign of jealousy:

Remember who commended thy yellow stockings. (Shakespeare, *Twelfth Night*)

yellow2 (especially of a prostitute) of mixed black and white ancestry.

Originally describing a light-skinned American slave, of higher value than a pure-blooded black and often used as a house servant; whence a *yellow girl*, a prostitute:

The yellow girls stood around giggling. (Longstreet, 1956, of New Orleans)

Also as *high-yellow*:

... end up by being shot in the saloon by a high-yellow girl. (ibid.)

yes-girl a female known to copulate promiscuously.

American male usage, punning on the *yes-man*, or toady. She is not necessarily a prostitute.

yield to copulate with a man outside marriage.

Of a new partner if not a novel experience, from the sense 'to submit', with an implication of coercion (see also SURRENDER TO):

There is no woman, Euphues, but she will yeelde in time. (Lyly, 1579, quoted in ODEP)

I could not get her to yield. (Harris, 1925, in a rare admission of failure to persuade a woman to copulate)

The male may be identified:

My sisterly remorse confutes mine honour, And I did yield to him. (Shakespeare, *Measure for Measure*)

The female may also *yield to desire* or *solicitation*:

Without much demur I yielded to his desire. (Mayhew, 1862)

The pretty lady's maid will often yield to soft solicitation. (ibid.—the maid is the pretty one, not the lady)

Specifically she *yields* her *body*, *honour*, *person*, or *virginity*:

Yielding up thy body to my will. (Shakespeare, *Measure for Measure*)

If I would yield him my virginity. (ibid.)

... the innocent young woman, with full knowledge, usually yields, without remorse... her person to any man. (Pearsall, 1969, quoting Patmore, c. 1890)

yob a young lout.

Not really a euphemism, but an interesting back slang for *boy*. See also SLAG.

you-know-what any taboo subject within the context.

It may be copulation, as *a little of you-know-what*, or a lavatory:

'The you-know-what's in there,' she said helpfully. Frensic staggered into the bathroom and shut the door (Sharpe, 1977)

or parts of the body:

... scratching one another's you-know-whats. (le Carré, 1989)

you know what you can do with that a coarse rebuttal.

The reference is to figurative anal insertion: ' stick it. All euphemisms' (DAS). There are a number of similar variants.

young not over 45 years old.

Journalistic use to describe public figures who may have achieved prominence at an earlier age than most of their contemporaries:

> Nick was very young, still in his early thirties. (M. Thomas, 1982)

See also MIDDLE-AGED.

young lady a man's premarital female sexual partner.

> That marriage has been annulled by the papal courts and it would be very painful to me & my young lady to have it referred to. (A. Waugh, letter of January 1937, quoted in S. Hastings, 1994)

A genteel usage, implying no more than courtship, as does the more severe *young woman*.

your nose is bleeding your trouser zip is undone.

An oblique British warning to another male in mixed company.

youth (guidance) centre an institution for the punishment of young offenders.

It could be, as is the British *youth centre*, a place where the young can meet socially under mild supervision. For a young American, attendance may involve involuntary residence under a court order.

yo-yo a male homosexual.

Going up and down, it might seem, until he unwinds:

> I just can't see us going across to France with a load of yo-yoes as a crew. (L. Thomas, 1986)

Z

zap to kill violently.

Perhaps from American strip cartoon language, where it may mean no more than 'to hit':

> Clever bastards like us, we care about getting zapped. (Seymour, 1984—but Afghans fighting the Russian invader were braver or more reckless)

zero *obsolete* a female public display of nakedness.

Not directly from the meaning 'nothing' but a shortened form of *zerokini*, from *bikini* punning on *bi-*, or two-piece, via *monokini*, a bikini worn without the top by an adult female. Bikini was the atoll in the Marshall Islands used for testing atomic explosions and the use illustrates our slow realization of the long-term effects of nuclear pollution. Such an association with swimwear would be thought in deplorable taste and commercially disastrous, if introduced today.

zipper a male profligate.

> The quickest zipper in the West, someone had once called him. (Turow, 1990, of a philanderer)

Unzipper might seem more appropriate.

zoned out drunk or under the influence of illegal narcotics.

The imagery is from defensive play in American football and basketball.

zonked *American* drunk or under the influence of illegal narcotics.

From the slang meaning 'hit':

> ...he should be banging women zonked out of their gourds on high-quality coke. (Sanders, 1990—he invited females to take cocaine and then copulated with them)

zoo a brothel.

Particularly in America, from the choice offered to customers of prostitutes from differing racial groups.

Thematic Index

Classification under specific headings is necessarily inexact and is intended only to give the reader a quick guide to the most common areas of euphemism. It is not possible to avoid an overlap between such categories as *Copulation, Courtship and Marriage, Mistresses and Lovers*, and *Sexual Pursuit*; or between *Death, Funerals*, and *Killing and Suicide*. Please refer to *General and Miscellaneous*, located at the end of the index, for entries which are difficult to classify under the other headings.

A word or phrase which does not have its own entry but which is discussed within another entry may be listed in one of two ways. If its headword is listed in the index under the same subject heading and is alphabetically adjacent, it will appear indented beneath it:

blue hair
    blue rinse

If its headword is listed under a different subject heading or is at some remove alphabetically, it will be presented in this way:

temporary liquidity problem *at* temporary

Items are listed under the following subject headings:

Abortion and Miscarriage

birth quota
bring off[2]
criminal operation
D and C
drop a bundle *at* drop[4]
female pills
French renovating pills *at* French letter
hoovering
illegal operation
misgo *at* misfortune
mishap *at* misfortune
mis(s) *at* miss[2]
part with child
  part with Patrick
pick
planned parenthood
planned termination *at* planned
pregnancy interruption
pro-choice
procedure
reproductive freedom
slip[1]
slip a filly *at* filly
terminate[3]
VPI *at* pregnancy interruption

Age

active[1]
adult[2]
ageful
blue hair
  blue rinse
borrowed time *at* borrow
boy
certain age (a)
chair-days *at* chair (the)
convalescent home *at* rest home
crinkly
Darby and Joan
distinguished
evening of your days
  eventide home
forward at the knees *at* forward
get along
  get on
girl[2]
God's waiting room
golden ager
golden years (the)
home
kid
long in the tooth
longer-living
make old bones
matron
mature[1]
middle-aged
mutton dressed as lamb *at* mutton
no chicken
  no spring chicken
not as young as I was
  not in his first youth
of mature years
older woman (the)
resident *at* residential provision
rest home
Roman spring *at* Roman
senior citizen
  seniors
sheltered
somewhere where he can be looked after
state farm
  state home
  state hospital
sunset years
third age (the)
  University of the Third Age
twilight home
up along
wear down
white top
wrinkly
years young
young

Animals Live and as Meat

big animal
brute
cleanse[2]
cow brute
crower
dark meat[1]
  drumstick
fall[3]
gentleman cow
grunter
furry thing
he-biddy
he-cow
  he-thing
in use *at* in season
Johnny bum *at* arse
jolly
lady dog
male beast
  male brute
  male cow
  male hog
on heat
prairie oyster[1]
roach[2]
  rooster-roach
rooster
seed-ox *at* seed
sheep buck
stand[2]
stock beast
  stock animal
  stock brute
  stock cow
stoned-horse-man *at* stones
stony *at* stones
stunted hare
sweetbreads
take[4]
throw[2]
trotter *at* dark meat[1]
use[3]
variety meats
white meat[1]

Auctioneers and Estate Agents

bijou
boost[2]
blockbuster
character
colonial
commodious
convenient[2]
Dutch auction
East Village
eat-in kitchen
estate agent *at* agent
Georgian
gracious
handyman special
historic
home
ideal for modernization
immaculate
in the ring
knock-out
landscaped
lower ground floor
marine residence
negotiable
off the chandelier
old-fashioned
outstanding
part furnished
period[2]
Peter Funk
planning *at* planned
prestigious
realistic
ring[2]
secluded
select
semi-detached
snug
sought after
South Chelsea
starter home
studio
sweeten[2]
time-sharing
town house
unique
zoning *at* planned

Bankruptcy

arrangement with your creditors *at* arrange
bank

banker
belly up
bolt the moon *at* moonlight flit
bought and sold
bung[4]
bust
cash flow problem
close its doors
come to a sticky end[1]
corporate recovery
Deed of Arrangement *at* arrange
done for
drown the miller *at* drown your sorrows
fall at the staves *at* go[2]
fall out of bed
flit[1] (do a)
fly-by-night[1]
fold
get the shorts
go[2]
go Chapter Eleven
go Chapter Thirteen
go due north
go up Jackson's end
go up Johnson's end
go smash
go to staves
go bump *at* bump[1] (the)
go for a Burton
go to Bath *at* take a bath
go to the wall
go under
haircut
hammer[1]
in Carey Street
lame duck[2]
liquidator *at* liquidate
lose your shirt
lose your pants
lose your vest
moonlight flit
moonlight flight
moonlight march
moonlight touch
moonlight walk
on the go *at* go[2]
on the skids
on your ear
over-geared
pull the rug
put the skids under
refer to drawer
run[3]
shoot the moon *at* moonlight flit
stiff[4]
stiff out
swamped
take a bath
take a wheel off the cart
take someone's shirt off
take someone's pants off
temporary liquidity problem *at* temporary
turn south
up the creek
up the spout
wipe out[2]

Bastardy and Parentage

absent parents
basket[1]
bend sinister
born in the vestry *at* born in ...
break your elbow
break your leg (above the knee)
by(e)
by(e)-begit
by(e)-blow
by(e)-chap
by(e)-come
by(e)-scape
cast a (laggin) girth *at* cast
chance
chance-bairn
chance-begot
chance-born
chance-child
chance-come
chanceling
cheat the starter
child of sin
child of grief
come in at the window
come in at the back door
come in at the hatch
come in at the side door
come in at the wicket
come o' will
doorstep
force-put job
flyblow
hop-pole marriage *at* hop into bed
illegitimate
indiscretion
latchkey
left-handed[1]
lone parent
love child
love-bairn
love begotten
love bird
lover child
made in the bush *at* bush[1]
merry-begot(ten) *at* merry
midnight baby
misfortune
misbegot
mishap
natural[2]
natural father
natural son
nurse-child *at* nurse
one-parent family
parentally challenged *at* challenged
premature
SOB *at* son of a bachelor
single parent
slip a foot *at* slip[1]
slip a girth *at* slip[1]
son of a bachelor
son of a bitch
son of a gun *at* gunner's daughter
souvenir[1]
tender a fool
unlawful
unlawful issue
unlawfully born
windfall[1]
wrong side of the blanket

Bawds and Pimps

abbess
bawd
brother of the gusset *at* brother
Charlie
Charlie Ronce
Covent Garden abbess *at* Covent Garden
dab *at* dabble
double[2]
governess
handle a woman *at* handle[1]
husband[1]
joe[1]
Joe Ronce
live off *at* live as man and wife
live on *at* live as man and wife
mackerel
mac
mack
macko
madam
mother[1]
procure
procurer
procuress
take under protection
victualler
white slavery *at* white slave

Bisexuality

AC/DC
acey-deecy
all-rounder
ambidextrous
ambiguous
ambivalent
batting and bowling
bisexual
bi
both-way
double-gaited
jack of both sides *at* jack[1]
sexual ambiguity
swing both ways

Boasting, Flattery, and Lying

Bribery

kickback
kindness
lay pipes
lubricate
lunchtime engineering
massage[1]
negotiable
  neg.
oil
on the hill
on the pad
on the side
over-invoicing
palm[1]
  palm grease
  palm oil
  palm soap
  palmistry
payoff
piece off
political engineering
pourboire
promotion
questionable payment *at* questionable
rake-off
recognition
ride the gravy train *at* gravy[2]
sale preview *at* corporate entertainment
secondary distribution *at* distribution
see[2]
  see the cops
  see the doorman
sensitive payment
skim
slush
  slush fund
smear[1]
soft commission
straighten
street money
sugar[1]
supporters' club
sweeten[1]
  sweetener
take care of[2]
take, the *at* take[1]
take your end *at* take[1]
thank
third party payment
travel expenses
treat
  treating
under the table[2]
velvet[2]
windfall[2]

Brothels

abode of love
academy
accommodation house *at* accommodate
bag-shanty *at* baggage
bagnio
barrel-house
bat-house *at* bat[1]
bawdy house *at* bawd
bed-house *at* bed[2]
bird-cage *at* bird[1]
bitch
button-hole factory *at* button[1]
bum-shop *at* bum
call house *at* call girl
canhouse
case[1]
  casa
  casita
  caso
cat-house *at* cat[1]
chamber of commerce *at* chamber
cheap John
chicken ranch
chippie-joint *at* chippy[1]
common house[1]
coupling house *at* couple[1]
creep joint
crib
disorderly house
dress-house *at* dress for sale
doss-house
escort agency
fish market *at* fishmonger's daughter
fleshpot
fun-house *at* fun
garden house *at* Covent Garden
gay house *at* gay[1]
girlie bar *at* girl[1]
girlie club *at* girl[1]
girlie house *at* girl[1]
girlie parlour *at* girl[1]
goat-house *at* goat
grind-mill *at* grind
grinding-house *at* grind
hook-shop *at* hooker[2]
hot-house
hot-pillow motel
  hot-sheet business
hourly hotel
house[1]
  house in the suburbs
  house of accommodation
  house of assignation
  house of civil reception
  house of evil repute
  house of ill-fame
  house of ill-repute
  house of pleasure
  house of profession
  house of resort
  house of sale
  house of sin
  house of tolerance
ill-famed house *at* house[1]
immoral house *at* immoral
improper house *at* improper
introducing house
jag house
joy house *at* joy[1]
knocking-shop
  knocker's shop
  knocking-house
  knocking-joint
ladies' college *at* lady
Lahore house *at* bad
leaping academy *at* leap on
leaping house *at* leap on
loose house *at* loose
make-out joint *at* make[1]
massage parlour
meat-house *at* meat
microwave club
nanny-house *at* nanny
naughty-house *at* naughty
night club
  night house
nunnery
occupying house *at* occupy
panel-house
  panel-joint
parlor house
play house
pleasure house *at* pleasure
pushing academy *at* push[1]
pushing school *at* push[1]
pushing shop *at* push[1]
queen-house *at* queen
rag (the)
ramps *at* ramp
rap club
  rap parlor
  rap studio
red lamp
  red-light area
  red-light district
  red-light precinct
  red-lighted number
rib joint
sauna parlour
scalding-house *at* scald
seraglio
service station *at* service[1]
skivvie-house *at* skivvy
snake ranch
sport-trap *at* sport (the)
sporting-house *at* sport (the)
sporting section *at* sport (the)
stews (the)
touch-crib *at* touch[2]
vaulting-school *at* vault[1]
victualling-house *at* victualler
wang-house *at* wank
warm shop *at* warm
zoo

Charity - Private, National, and International

aid[2]
assistance
benefit
cargo
caring
concessional
  concessional export
  concessional fare
dole
  dole-bread
  dole-meats
  dole-money
feather-bed
financial assistance
fly a kite[2]
handout[1]
house[3]
  house of industry
income support
live on the high-fly *at* high[2]
national assistance *at* assistance
negative (income-) tax
old soldier (the)
on assistance *at* assistance
on the labour *at* labour[2]
on the parish
  on the parochial
  on the town
out of benefit *at* benefit
public assistance *at* assistance
relief[1]
remittance man
rock and roll
sheltered
social security
souper
tied aid *at* aid[2]
uncaring
welfare
  welfare state
workhouse

Childbirth

accouchement
bear[1]
bed[1]
  brought to bed
bond *at* confinement
cast
click[2]
come to your time
confinement
  confined
doorstep
drop[4]
  drop a bundle
facts (of life)
fall about *at* fall[3]
feed
fiddle
gooseberry bush *at* parsley bed
groper *at* grope
happy event
hatch
kid
labour[1]
lady in waiting[2]
  lady in the straw
lie in
  lay in
  lying-in
  lying-in wife
nurse
parsley bed
pup
slip[1]
time
trouble
upstairs[1]
whelp

Clothing

abandoned habits *at* abandoned
Abyssinian medal
appliance
at half mast
athletic supporter
BB *at* brassière
bags
bodice *at* brassière
body shaper
  body briefer
  body hugger
  body outline
boobytrap *at* brassière
box[2]
brassière
brothel-creeper *at* creep joint
bust bodice *at* brassière
canteen medal
catch a cold[2]
Charlie's dead *at* Charlie
cheaters
co-respondent's shoes *at* co-respondent
continuations
Cuban heels
cup
decent
don't-name-'ems
enhanced contouring *at* enhance
falsies
flag of distress *at* flag is up (my)
flapper
flying low
foundation garment
gay deceivers *at* gay[2]
gazelles are in the garden
indescribables
ineffables *at* unmentionables[1]
inexpressibles *at* unmentionables[1]
jock-strap *at* jock
Johnnie's out of jail *at* John Thomas
leg-bags *at* bags
lift[4]
medal showing (a)
nappy
one o'clock at the waterworks (it's)
sartorially challenged
sensible
shop door is open (the)
sides
sit-in-'ems *at* sit-upon
sit-upons *at* sit-upon
snowing down south (it's)
star in the east
step-ins
surgical appliance *at* appliance
Turkish medal
unmentionables[1]
  unexpressables
  unspeakables
  untalkaboutables
  unutterables
  unwhisperables
your nose is bleeding

Colour, Gypsies, and Slavery

affirmative action
African-American
  African-American worldview
African trade *at* triangular trade (the)
apartheid *at* separate development
Aryan
black *at* coloured[1]
blackbird
  black cattle
  black hides
  black ivory
  black pigs
  black sheep
  blackbird pie
  blackbird trade
  blackbirding
blockbuster
boy
brother
cattle[1]
chalkboard
change your luck
chi-chi
coloured[1]
community affairs
  community affairs correspondent
  community relations
dark[2]
  dark complected
  dark meat
  dark skinned
  darky
discrimination *at* reverse discrim-

Commerce and Industry

- product
- proposition selling
- provision
- public relations
  - public relations officer
- pull out of the air
  - pull figures out of a hat
- pull the pin[2]
- punter
- qualify accounts
  - qualification
- rainmaker
- RD *at* refer to drawer
- raise the wind
- redlining
- refer to drawer
- regular[2]
- remainder[2]
- representative *at* commerce
- restructured
- reverse engineering
- revision *at* price adjustment
- ring[2]
- rodent operator *at* exterminating engineer
- save
- scandal sheet
- Schwarzarbeit *at* black economy
- sculpted
- select
- self-catering
- send in your papers
- service[2]
  - service station
- shade[1]
- share pusher *at* push[4]
- sharp with the pen
- sharpen your pencil[1]
- sharpen your pencil[2]
- shave the gentry *at* slice
- shoe the colt *at* colt[2]
- silent copy *at* blind copy
- slack fill
- slice
- smooth
- snow[2]
  - snow-job
- sotto governo *at* black economy
- sovereign *at* tourist
- spike[2]
- sponsor
- standard
- steer
- structured
  - structured competition
- stuck
- subsidy publishing
- suggestion
- supporters' club
- swallow the anchor
- sweeten[2]
- sweetheart
  - sweetheart contract
  - sweetheart price
- switch-selling
- tail-pulling
- take a break
- take a powder
- take a view
- take-away
- tap[3]
- tap[4]
- technical adjustment
  - technical correction
  - technical reaction
- ticket punched *at* get your ticket punched
- tied up
- time-sharing
- top up
- touch signature
- tourist
- transfer pricing
- traveller *at* commerce
- triple entry
- turkey shoot
  - turkey farmer
- turn south
- twenty-four hour service
- tyre-kicking *at* kick the tyres
- unavailable[1]
- unbundling
  - unbundler
- uncertain
- under-invoicing
- underground production *at* black economy
- uneven
- unscheduled
- used
- vanity publishing
- velvet[2]
- visiting fireman[2]
- warehouse
- wash[2]
- water stock
- weekend
- weigh the thumb
- well rewarded
- what the traffic will bear
- white-knuckler
- white sale
- white tail
- window dressing
- wire[2]
- word from our sponsor (a)
- words
- work both sides of the street
  - work a street

Contraception

- armour
- bareback
  - bareback rider
- birth control
- bung[2]
- capote anglaise *at* English
- cardigan
- careful[2]
- circular protector
- coil (the)
- collapsible container
- conception prevention *at* birth control
- diaphragm
- Dutch cap *at* Dutch
- FL *at* French letter
- family planning
  - family planning requisites
- fight in armour
- French letter
  - French renovating pills
  - French tickler
  - Frenchie
  - froggie
- get fitted
- intrauterine device
- joy bag *at* joy[1]
- johnny
  - Johnnie
- loop[2]
  - on the loop
- Malthusianism
  - neo-Malthusianism
- pill[2] (the)
  - on the pill
- planned parenthood
- precautions
- preventative
- pro-pack *at* pro
- prophylactic
- protector[2]
- raincoat[1]
- rubber
  - rubber cookies
  - rubber goods
- safe sex
- safety
  - safe
- sheath
- skin
  - skin-diver
- something for the weekend
- tablet
- take care
  - take a risk
- take precautions *at* precautions
- tickler[2]
- unprotected sex
- Vatican roulette
- welly
- will there be anything else?

Copulation

- aboard[2]
- abuse a bed *at* abuse[1]
- accommodate
- act (the)
  - act of intercourse

  - act of generation
  - act of love
  - act of shame
- act like a husband
- active[2]
- all right
- all the way
- alley-cat *at* alley cat
- amateur
- amatory rites
- amorous business *at* business
- amorous favours
  - amorous sport
  - amorous tie
- amour
- antics in bed
- any
- ashes hauled *at* haul your ashes
- association[1]
  - associate intimately with
- astride
- athwart your hawse
- attend to
- attentions
- avail yourself of
- ball
- banana
- bang[1]
- banish your bed
- baser needs
- basket-making *at* basket[1]
- be nice to
- be with
- beard splitter *at* beard[1]
- beast with two backs
- beastliness
- beat the gun
- beat the mattress
- bed[2]
  - bed and breakfast
  - bed down
  - bed-hopping
  - bedwork
- beef
- been into *at* be with
- been there *at* be with
- beg a child of
- belly to belly
- belt[1]
- besomer *at* besom
- bestride
- betray
- between the sheets
- between the thighs of
- big prize (the)
- bit
  - bit of the other
  - bit on the side
- blanket drill
- block
  - block her passage
- blow[1]
  - blow off
  - blow the groundsels
  - blow the loose corns
  - blow through
- board
- board a train *at* pull a train
- boff[1]
- bonk
- boom-boom[2]
- bother
- bounce[1]
  - bouncy-bouncy
- bout
- break a commandment
- break a lance
- break the pale
- break your knee *at* break your elbow
- bring off[1]
- buckle to
- bull[1]
  - bullock
- bum-fighting *at* bum
- bump[3]
  - bump bones
- bundle
- bung up and bilge free
- bush patrol *at* bush[1]
- business
- buttock
  - buttock ball
  - buttock-mail
- button-hole *at* button[1]
- callisthenics in bed
- canned goods
- canoe
- capital act (the)
- carnal
  - carnal act
  - carnal knowledge
  - carnal necessities
  - carnal relations
- casting couch (the)
- catch
- cattle[2]
- chambering
- change your luck
- cheat
- clean-living *at* clean
- clean up[2]
- cleave *at* chopper[2]
- clicket
- climb
  - climb in with
  - climb into bed
- climb the ladder on her back *at* climb the ladder
- close the bedroom door
- cock
  - cock a leg across
  - cock a leg athwart
  - cock a leg over
- coition
- come about *at* come
- come across[2]
- come over *at* come
- come to
- come together *at* come
- commerce
- compound with
- compromise[1]
- congress
- connect[1]
  - connection
- connubial pleasures
- conquer a bed
- console
  - consolation
- consummate (a relationship)
  - consummate your desires
  - consummation
- contact with
- content your desire
- continence *at* incontinent[1]
- continent *at* incontinent[1]
- conversation *at* criminal conversation
- cop a cherry
- copulate
  - copulation
- corn[2]
- corrupt
- couple[1]
  - couple with
- cover[1]
- crack a Jane
  - crack a doll
  - crack a Judy
  - crack a pipkin
  - crack a pitcher
- crack your whip
- cracked in the ring *at* ring[1]
- cram
- crawl[1]
  - crawl in with
  - crawl into bed with
- criminal assault
- criminal connection
- criminal conversation
- cross
- cut the mustard
  - cut it
- Cythera
- dab it up *at* dabble
- damaged[2]
- dance the mattress quadrille
  - dance a Haymarket hornpipe
- debauch
- deceive (your regular sexual partner)
- deed (the)
- defend your virtue
- defile
  - defile a bed
  - defilement
  - defiler
- deflower

defloration
degraded
destruction
dick around *at* dick[1]
diddle[2]
dip Cecil in the hot grease *at* Cecil
dip your wick
dirty deed *at* dirty
dirty weekend *at* dirty
disgrace
dishonoured
disport amorously
dissolution[2]
distribute favours *at* favours
do[1]
do a perpendicular
do it
do what comes naturally
do the trick *at* trick
dock
double-header
double in stud
double time
droit de seigneur
drop your drawers
drop your pants
dry bob
dry run
East African activities
easy woman
eat flesh
embraces
embrace
employ
enjoy
enjoy favours
enjoy hospitality
enjoy someone's person
enjoyed
enter
entertain[2]
exchange flesh
exercise your marital rights *at* marital rights
experienced
extras
facts (of life)
faithful
fall[1]
false
fate worse than death
favour
favours
feed from home
feed your pussy
fidelity
filthy
filth
fire up *at* fire a shot
firk
flat on your back
flesh your will
fling (a)
flop
flower[1]
flutter[2]
foin
force your ardour upon
force favours from
force your attentions on
fork
foul desire
foul designs
foul lusts
foul way with
frail job *at* frail[2]
fraternize
free love
free of fumbler's hall *at* free[1]
free of your hips
freelance
freelancer
freeze[2]
frig
front door (the)
fuck around *at* dick[1]
fulfilment
full treatment (the)
fumble
fun
fun and games
gallant
gallantry
gallop
game[2] (the)
game fee
gasp and grunt *at* grumble
gentle art (the)
George
get a leg over
get your leg across
get your leg dressed
get down to business
get in her pants
get in the saddle *at* saddle up with
get into bed with
get into her bloomers
get it in, on, off, or up
get laid
get off with
get round
get stuffed *at* stuff[2]
get there
get through
get up
get your corner in *at* corner[3]
get your end away
get your end in
get your greens
get your hook into
get your muttons
get your nuts off
get your rocks off
get your share
get your way with
get your will of
gift of your body
give
give a little
give access to your body
give favours *at* favours
give it to
give out
give stick *at* stick[3]
give the ferret a run
give the time to
give up your treasure
give way
give your all
give your body
give yourself (to)
go all the way
go (any) further
go into
go on[2]
go short
go the length
go the whole hog *at* whole hog (the)
go the whole way
go through
go to bed with
go to it
go with
go wrong
good
good time
goose[3]
grant favours *at* favours
gratify your passion
gratification
gratify a man's desires
gratify your amorous works
grease the wheel *at* grease[1]
green fruit
green gown
grind
grumble
grummet
hang it on with *at* hang a few on
hanky-panky
haul your ashes
hauled
have
have a banana with *at* banana
have a man/woman
have at
have connection with *at* connect[1]
have it
have it away
have it off
have personal relations with *at* personal relations
have relations
have sexual relations
have sex
have something to do with
have your end away
have your end off

have your foul way with *at* foul desire
have your way with
  have your filthy way
have your will of
headache[2]
height of connubial bliss
hide the weenie
high jinks
hit[2]
hit on
hit the sack with
hochle
hoist your skirt
honest
honour
hop into bed
horizontal[1]
  horizontal aerobics
  horizontal conquest
  horizontal jogging
  horizontal position
horn of fidelity *at* horn[1]
hose[2]
hot-tailing
hot time (a)
how's your father
human relations
hump
  hump the mutton
hustle[2]
hymenal sweets
ill-used
illicit
  illicit commerce
  illicit connection
  illicit embraces
  illicit intercourse
illicit embraces *at* embraces
impale
impotent
improper
  improper suggestion
in[2]
in bed *at* bed[2]
in circulation
in flagrante delicto
  en flagrant délit
in full fling
in her mutton *at* mutton
in name only
in relation with *at* relate
in rut
in season
in the box
in the hay
in the sack
in the saddle
inconstancy
incontinent[1]
indecency
infidelity
initiation
  initiation into womanhood
intercourse
intimacy
  intimate
into the sack
intrigue (an)
invade
it[2]
itch
jack in the orchard *at* jack[1]
jam
jazz
jelly roll
jig-a-jig
  jig
  jig-jig
  jiggle
  jiggy-jig
jive
jock
join[1]
jolly[3]
joy[1]
  joy ride
juggle
jump[2]
keep your legs crossed
  keep your legs together
kind
kiss
knee-trembling *at* knee-trembler
knees up
knock off[2]
  knock
knot
know
  knowledge
know the score *at* score[1]
laid *at* lay
lance
last favour
  last intimacies
  last thing
lay
  lay a leg across
  lay a leg on
  lay a leg over
  lay down
  lay it
lay some pipe *at* lay pipes
lead apes in hell
leap on
  leap into
  leap into bed with
leave before the gospel
leave shoes under the bed
led astray
leg-over
  leg-over situation
leg-sliding
let in
lie with
  lie on
  lie together
lift a leg[1]
light the lamp
limit (the)
line
linked with
lose your cherry
  lose your character
  lose your innocence
  lose your reputation
  lose your snood
  lose your virtue
love
love affair
love-in *at* sit in
lovemaking
lubricate
lumber
main thing (the)
make[1]
  make a play for
  make it with
  make out
  make whoopee
make a suggestion
  make an improper suggestion
make arrangements *at* arrange
make babies together
make little of
make love to
make the beast with two backs *at* beast with two backs
make the (bed) springs creak
  make the (bed) springs squeak
make the chick scene
make time with
male congress *at* congress
managerial privileges
  management privileges
marital rights
marriage joys
massage[3]
mate
  mating
mattress
  mattress drill
  mattress extortion
mess[1]
mid-job *at* on the job
migraine
mingle bodies
misconduct
missionary position
misuse
momentary trick (the)
monkey business
mount
mutual joy *at* joy[1]
mutual pleasuring *at* pleasure
nail[1]
national indoor game *at* game[2]

(the)
natural vigours
naughty
nibble
nice time
nice-nice
night physic
night baseball
night exercise
night games
night's business *at* business
nightwork
no-tell
noble game *at* game[2] (the)
nocturnal exercise *at* night physic
non-penetrative sex
nookie
oats
oblige[1]
occupy
off[2]
offer yourself
offer kindness
on the couch
on the job
on top of
on your back
one-night stand
one-nighter
one thing
open your legs
open-legged
other, the *at* other woman (the)
outrage
over the broomstick
party[2]
pasture
peel a banana
penetrate
perform[2]
performer
personal relations
personal services
physical involvement
piece on a fork *at* piece[1]
pile into
plant a man *at* plant[1]
play
play around
play away
play games
play hookie
play in the hay
play mothers and fathers
play mums and dads
play on your back
play the ace against the jack
play the beast with two backs *at* beast with two backs
play the field
play the goat
play the organ
play tricks
play with
please yourself on
pleasure
pleasuring
plough[1]
pluck
pluck a rose
plug[2]
plumb
plunge away
pocket the red *at* pocket pool
poke[2]
pole
poontang
pop[2]
pork[3]
possess
pound
pound the keys
press conjugal rights on
press your attentions on
privy to a bed *at* bed[2]
probe
prod
prong
proposition
pull[2]
pull a train
pull his trigger
pump up
punch
pure
push[1]
pussy
put
put a man in your belly
put and take
put it about
put it in
put it up
put out
put to
put a move on
put the moves on
putter about *at* put
quickie *at* quick time
R and R *at* rest and recreation
racy
ram
rattle[1]
ravish
relate
relations *at* have relations
relief[3]
relieve of virginity
rest and recreation
ride
ride St George
rifle
rip off a piece of arse
rip off a bit of ass
ripple
rivet
rock *at* roll[1]
rod[1]
roger[1]
roll[1]
roll in the hay
roll the linen
romp
root[2]
root about
root rat
rootle
rub groins together *at* groin
rub off
rub the bacon
rub the pork
ruin
ruined (in character)
saddle up with
saddle-broken
sauce[2]
save it
score[1]
screw[1]
screw around
screwing
seduce
serve
serve your lust
service[1]
sex[1]
sex love
sexual act, the *at* act (the)
sexual intercourse
sexual commerce
sexual congress
sexual knowledge
sexual liaison
sexual relations
sexual relief
shaft[1]
shag[1]
shag like a rattlesnake
shame
share her embraces *at* embraces
share someone's bed
share someone's affections
share favours with
sharp and blunt
sheath the sword
sheets
shift[2]
short time
short session
shove[1]
sin
sinful commerce
skewer
slake your lust
slap and tickle
sleep around
sleep
sleep over
sleep together

Cosmetics

Courtship and Marriage

Cowardice

Crime (other than Stealing)

above ground
abuse[1]
action[1]
anti-social
    anti-social behaviour
Arkansas toothpick
assist the police *at* help the police (with their inquiries)
at it
badger game *at* badger
bend the rules
    bent
    bent copper
bird dog[2]
    bird-dog
bit
    biter
black market
black money
    black dollar
    black pounds
blow job *at* blow[1]
blow the gaff *at* blow[5]
blow the whistle *at* blow[5]
blue box
boning *at* bone[1]
bootleg
boys[2]
    boys upstairs
    boys uptown
bushwack
bust
carry the can
cherry-pick
chisel
claim responsibility for *at* claim
clean
clean up[1]
clock *at* put the clock back
cobbler
cocktail[1]
community alienation
con
    con-artist
    con-man
    confidence trick
cook[2]
cop out[2]
corner[1]
cosa nostra
cough[1]
cover-up *at* cover story
covert act
decontaminate
diddle[3]
dirt *at* dirty
do[5]
doctor
double entry
draw the king's picture
eat the Bible
fall money *at* fall[4]
family[2]
feed the bears
feed the meter
find[2]
finger[1]
    finger a job
    finger-man
    finger-mob
firm (the)
fit up
fix[1]
fleece
form
frame[1]
    frame-up
front[2]
fudge[1]
funny money
gentlemen *at* gentleman
goods (the)
gooseberry lay *at* gooseberry[2]
grass[1]
green goods
    green goods man
gunrunning *at* run[1]
hard case *at* hard
hatchet man
haul in *at* haul
Havana rider
heat[1]
help the police (with their inquiries)
hempshire gentleman *at* hemp[1]
hijack
    highjacker
    hijacker
hike[1]
hillside men
hook[3]
hose[1]
in need of supervision
in the black *at* black market
in the ring
in trouble
informal
    informal dealer
    informal market
inquiries *at* help the police (with their inquiries)
irregularity
jacket
jelly *at* soup
Jew's lightning
kangaroo court
King's evidence *at* Queen's evidence
known to the police
lard the books
launder
    laundry
lay paper
line your pocket
    line your coat
    line your vest
loaded[2]
lose[1]
low flying
lulu
Lydford law
mark[1]
Mickey Mouse
minder
mob
    mobster
moonlight[1]
    moonraker
nickel and dime
no show[1]
off the rails[2]
on the chisel *at* chisel
on the left
on the square
operator
organization (the)
penman
piece of the action *at* action[1]
pigeon[1]
    pigeon-drop
pigeon[3]
plant[2]
private enterprise
pull off[2]
put on
put the clock back
put the finger on *at* finger[1]
Queen's evidence
queer[4] (the)
questionable
    questionable act
    questionable motive
    questionable payment
ramp
rap
receiver
resolved without trial
restorer
revolving-door
rip off
    rip-off
    rip-off artist
    rip-off merchant
run[1]
    runner
run round the Horn
salt
score[2]
screw[3]
sell out
    sell-out
send to the cleaners *at* clean up[1]
set up[2]
shade[2]
shakedown *at* shake[1]
shanghai
share pusher *at* push[4]
sheet

shop[2]
sing
slice of the action *at* action[1]
smear[3]
smoking gun
soldier
soup
spill
  spill the beans
split on
squeal
sting
stitch up
stool pigeon
straight
stuck
swallow the Bible
switch the primer
syndicate
tagged[2]
take[5]
take for a ride[2]
take stock of the situation
take the can (back) *at* carry the can
take to the cleaners
throw the book at
tip[4]
  tip off
torpedo
toss[2]
tout
trunk
turn in
turn the clock back *at* put the clock back
turn up[2]
unavailable[3]
under the counter
underground *at* above ground
walk[5]
  walk free from court
walk penniless in Mark Lane *at* mark[1]
wash[3]
white money

Cuckoldry

Actaeon
blow the horn *at* horn[2]
forked plague *at* fork
freeman of Bucks *at* Freemans
graft
honour
horn[2]
  horn-maker
  horned
knight of Hornsey *at* knight
member for Horncastle *at* member[1]
prey to the bicorn *at* prey to (a)
put horns on *at* horn[2]
Vulcan's badge
wear a fork *at* fork
wind the horn *at* horn[2]

Death (other than Funerals, Killing, and Suicide)

afterlife
all up with
alleyed
anointed
answer the call[1]
asleep
  asleep in Jesus
at rest
  at peace
at your last *at* last call
auction of kit
away[1]
back-gate parole
better country
  better place
  better world
beyond salvage
big D
  big stand-easy
bite the dust
bonds of life being gradually dissolved
bone[3]
bought it
breathe your last
bring your heart to its final pause
brown
bung[4]
buy the farm *at* bought it
buzzed[1]
call (the)
  call it a day
  call souls
  called
  called away
  called home
  called to higher service
call off the bets
cardiac arrest *at* cardiac incident
cash in your checks
  cash in your chips
catch a packet
cease to be
check out
chop shot *at* chop[1]
chuck a seven *at* seven (chuck a)
church triumphant
clink off *at* clink
close your eyes
clunk
cold[1]
combat ineffective
come again
  come back
come home feet first
come to a sticky end[1]
come to your resting place
  come to the end of the road
  come to yourself
conk (out)
cool[1]
  cool one
cop a packet
cop it
  cop out
cough[2]
count (the)
  count the daisies
croak[1]
cross the Styx
  cross the River Jordan
curtains
cut off
cut the painter
  cut adrift
  cut your cable
dance a two-step in another world *at* dance[1]
Davy Jones' locker
depart this life
  dear departed
  departed
  departure
diet of worms
disappear[1]
dissolution[1]
done for
down for good
drop[3]
  drop dead
  drop in your tracks
  drop off the hooks
  drop off the perch
  drop your leaf
drop off
end of the road *at* end
end of watch *at* end
enter the next world
eternal life
everlasting life
exchange this life for a better
expended
expire
extremely ill
fade away
fall[3]
  fall asleep
  fall off the hooks
  fall off the perch
  fallen
fall out
feet first
flack out
flit *at* flit[1] (do a)
food for worms *at* diet of worms
gathered to your fathers
  gathered to God
  gathered to our ancestors
get away
  get it
  get the call

- get the chop *at* chop[1]
- give up the ghost
  - give up the spoon
- go[1]
- go aloft
- go away
- go corbie
- go down the nick
- go for a Burton
- go forth in your cerements
- go home
  - go home feet first
  - go home in a box
- go into the ground
- go off[1]
- go off the hooks
- go on[1]
- go out
- go over[1]
- go right
- go round land
- go the wrong way
- go to a better place
  - go to glory
  - go to life eternal
  - go to rest
- go to grass
- go to the wall
- go to your reward
- go to yourself
- go up[1]
  - go up the gate
- go west
- God called in the loan *at* call (the)
- gone[1]
- goner
  - gonner
- grave (the)
  - gravestone gentry
- great certainty
  - great leveller
  - great out
  - great perhaps
  - great secret
- great majority
- grim reaper *at* reaper (the)
- ground-sweat
- grounded for good *at* grounded
- had it
- hand in your dinner pail
- hang up your hat[2]
  - hang up your boots
  - hang up your harness
  - hang up your tackle
- happen to
- happy release
  - happy despatch
  - happy hunting grounds
- heels foremost
- hereafter (the)
- hole in the head *at* hole[1]
- hop off
  - hop the twig
- hump it
- in Abraham's bosom
- in heaven
  - in the arms of his Maker
  - in the arms of Jesus
  - in the arms of the Lord
- jack it in
  - jack it
- join[2]
  - join the great majority
  - join your Maker
- jump the last hurdle
- keel over
- kick[1]
  - kick in
  - kick it
  - kick off
  - kick the bucket
  - kick up
  - kick your heels
- kingdom come
- kiss the ground *at* kiss
- kiss off[1]
- laid to rest
  - laid in the lockers
- last call
  - last bow
  - last curtain
  - last debt
  - last end
  - last journey
  - last rattler
  - last resting place
  - last round-up
  - last trump
  - last voyage
- late[1]
  - latter end
- lay down your life
  - lay down the clay
  - lay down your burden
  - lay down your knife and fork
- leave the building
  - leave
  - leave the land of the living
  - leave the minority
  - left town
- lick the dust
- life everlasting *at* everlasting life
- life insurance
  - life
  - life cover
  - life office
- little gentleman in black velvet
- long count *at* count (the)
- long home
  - long day
  - long journey
- Lord has him (the)
- lose[3]
  - lose the number of the mess
  - lose the vital signs
  - lose the wind
- loss[1]
- lost[2]
  - lost at sea
- make old bones
- make the supreme sacrifice
- meet your Maker
  - meet the Prophet
- NYR
- napoo *at* nab
- narrow passageway to the unknown (the)
- negative patient care outcome
- next world *at* better country
- no longer with us
  - no more
- no right to correspondence (have)
- not dead but gone before
  - not lost but gone before
- not yet returned *at* NYR
- number is up (your)
- off the voting list
- on your shield
- on your way out
- other side (the)
- over Jordan
  - over the creek
- pack it in
  - pack up
- packet[1]
- pass[1]
  - pass away
  - pass in your checks
  - pass into the next world
  - pass off
  - pass on
  - pass out
  - pass over
  - passing
- pay nature's (last) debt
  - pay the extreme penalty
  - pay the final penalty
  - pay the supreme penalty
  - pay your debt to society
- peace at last
- peg out
- play your harp
- plucked from us
- pop off
  - pop your clogs
- popping up the daisies
- potted[1]
- promoted to Glory
- push up the daisies
- put away[1]
- put in your ticket
- put to bed with a shovel
  - put to bed with a mattock
  - put to bed with a spade
  - put under the sod
- put to rest
- quietus
- quit
  - quit breathing
  - quit the scene
- reaper (the)
- release[2]

released
remain above ground
removed *at* remove
repose
resign your spirit
ring eight bells
routine (nursing) care only
sale before the mast *at* auction of kit
say Kaddish for
screwed down
seven (chuck a)
shipped home in a box (be)
shovelled under
shuffle off this mortal coil
six feet of earth
six feet underground
sleep
sleep in Davy Jones' locker
sleep in your leaden hammock
sleep in your shoes
slip[2]
slip away
slip off
slip to Nod
slip your breath
slip your cable
slip your grip
slip your wind
snuff it *at* snuff (out)
stand before your Maker
stark
step off
step away
stoke Lucifer's fires
stop one
stop a slug
stop the big one
strike out
succumb[1]
suck daisy-roots
swing off
take a long sniff *at* sniff out
take a long walk off a short pier
take home
take leave of life
take off[2]
take your departal *at* depart this life
take your leave of
taken
taking
taking hence
talk to the old gentleman
taps (the)
throw a seven *at* seven (chuck a)
time
tip[2]
tip off
took *at* taken
turn it in
turn up your toes
turn up your heels
turn your face to the wall
under-arm bairn *at* under-arm
under hatches
under sailing orders
under the daisies
under the grass
under the sod
underground
undersod
undiscovered country (the)
Union Jack for
united
upstairs[1]
used up *at* used
wages of sin (the)
way of all flesh (the)
wear away
win home
win your way
wing your flight from this world
with Jesus
with the Lord
with us no more
with your Maker
worm-food *at* diet of worms
written out of the script

Defecation

accident[1]
alley apple
ammunition[1]
army form blank
be excused
been
behind a bush
big jobs
bind
bodily functions
bodily wastes
body wax *at* body rub
bog
Bombay milk-cart
Bombay oyster
boom-boom[1]
bowel movement
bum fodder *at* ammunition[1]
bury a quaker
business
call of nature
cast your pellet
caught short
CC pills
cement
change[3]
cleanliness training
clear-out (a)
confined *at* confinement
cowpat *at* horse apples
crap
crud
defecate
demands of nature
deposit
dirty your pants *at* dirty
do a dike *at* do a bunk
do a job *at* do a bunk
do a rural *at* do a bunk
drop your arse
droppings
drop a log
drop the crotte
drop the wax
dry waste *at* waste[2]
dump[1]
duty
ease your bowels *at* ease springs
evacuate
evacuation
feel the need
fertilizer
foul
foul yourself
George
go[3]
go about your business
go for a walk with a spade
go places
go to ground
go to the bathroom
go upstairs
gold-digger[2]
gold-dust
gold-finder
goldbrick
grunt
honey
honey barge
honey bucket
honey cart
honey-dipper
honey wagon
hooky
horse[2]
horse apples
house-trained
human waste
incontinent[2]
indiscretions[1]
irregular *at* regular[1]
job
loosen the bowels *at* loose[2]
make a mess
manure *at* horse apples
mess[2]
mistake[2]
morning George *at* George
motion (a)
move your bowels
movement
muck *at* horse apples
my word
nappy
natural functions (the)
natural purposes
nature's needs

necessities
need
nightsoil
number nine
number two(s)
on the seat
open your bowel(s)
opening medicine
ordure
pancake *at* pan
pellets
perform[1]
perform a natural function
physic
pony
poop[1]
pooper scooper
post a letter
potty training *at* pot[3]
prairie chips *at* horse apples
rear[2]
regular[1]
relief[2]
relieve your bowels *at* relieve yourself
Richard
road apples
run[4]
sausage
scour
see your aunt *at* see the rosebed
sewage *at* effluent
shift[1]
sit-down job
skidmarks
soil[2]
soil your pants
soil yourself
solid waste
souvenir[2]
squat[1]
stoppage[1]
tied up
tom[1]
top and tail
turn up your tail
use paper
visiting card
wander off
waste[2]

Diarrhoea

Adriatic tummy
Aztec two-step
Aztec hop
back-door trot
Basra belly
Bechuana tummy
bull-scutter *at* bull[2]
Cairo crud *at* crud
Delhi belly
Edgar Brits *at* Jimmy
flying handicap
gastric flu
gippy tummy
Hong Kong dog
Jimmy Brits *at* Jimmy
loose[2]
loose disease
Montezuma's revenge
Mexican fox-trot
Mexican toothache
Mexican two-step
Napoleon's revenge
on the trot *at* trots (the)
Rangoon runs
runny tummy *at* run[4]
runs *at* run[4]
scatters (the)
shoot[3]
Spanish tummy
squirt
skeet
skitters
squit
squitters
threepennies (the)
Tokyo trots
touristas (the)
trots (the)
tummy bug

Dismissal

air (the)
ax(e)
bag[2] (the)
beach
bench[2]
bobtail
boot (the)
boot out
bounce[3]
bowler hat
bucket, the *at* bucket[2]
bullet (the)
bump[1] (the)
bump off
California kiss-off *at* kiss-off[3] (the)
can[2]
cards (your)
chop[2] (the)
chuck (the)
consultancy
consultant
cop the bullet
DCM
dehired
demanning
deselect
dispense with someone's assistance
downsize
drop[6]
drop-dead list
early retirement
early release
excess
fire
flush down the drain
for the high jump
for the chop *at* chop[2] (the)
furlough
gardening leave
gate[2] (the)
general discharge
get on your bike
give someone the air
give time to other commitments
give time to other interests
given new responsibilities
golden bowler *at* bowler hat
golden handshake *at* golden
golden parachute *at* golden
goodbye
grand bounce *at* bounce[3]
graze on the common *at* graze
graze on the plain *at* graze
handshake
hatchet man
hatchet job
have the shout *at* shout[1] (the)
head-count reduction
hike[2]
hoist[3]
hoof (the)
house-cleaning[1]
in the barrel
kick[2] (the)
kiss-off[3] (the)
lay off
leave of absence
let go
let out
look after your other interests
lose[2]
measure for the drop
negative employee situation
New York kiss-off *at* kiss-off[3] (the)
notice
off the payroll
on the beach *at* beach
order of the boot (the)
order of the push
out to grass
pink slip
planned reduction *at* planned
poke[1] (the)
pursue other interests *at* look after your other intersts
push[2] (the)
put out to grass
railroad
rationalize
reduce your commitments
reduction in force *at* rif(f)
redundant
release[1]
relieve
relieve of duties

Drunkenness

fall among thieves
  fall among friends
fatigued
feel no pain
fired up *at* firewater
five or seven
flash[2]
flawed
floating
fly-by-night[2]
fly one wing low
foggy
  fogged
forward
fou *at* full
four sheets in the wind *at* sheet in the wind
foxed
fractured
fragile
frail[1]
fresh
  fresh in drink
fricasseed
fried
fuddled
full
  full as a boat
  full as a tick
full of liquor *at* liquor
fun-loving *at* fun
funny[2]
fuzzed
  fuzzled
  fuzzy
gaged *at* gage[1]
gargle
gassed
gay[1]
geared up
geezed up *at* geezer
given to drink *at* drink[1]
glass too many *at* glass
glassy-eyed
glazed
glow on (a)
gone[3]
grape-shot *at* grape (the)
greased
grog on board
  grogged
  grog-hound
growler-rushing
half and half
half-canned
  half-cooked
  half-corned
  half-cut
  half-foxed
  half-gone
  half in the bag
  half on
half-pint *at* half a can
half-screwed *at* screwed
half-seas over
  half-sea
half shot *at* shot[4]
half-sozzled *at* sozzled
half-sprung *at* sprung
half-stewed *at* stewed
half under *at* under the influence
hammered
hang a few on
  hang one on
hangover
happy
hard drink *at* hard
hard drinker *at* hard
have a thirst *at* thirst (a)
hazy
head[3]
  headache
heeled[1]
hen-drinking *at* hen-silver
high[2]
hit[1]
hit the bottle
  hit it
  hit the hooch
hoary-eyed
hoist[2]
  hoist a few
  hoist one
hold your liquor
hollow legs
honked
horizontal[2]
hot[5]
hung over *at* hangover
hunt the brass rail
hunt the fox down the red lane
ill[4]
illuminated
imbibe
impaired *at* mobility impaired
in drink *at* drink[1]
in liquor *at* liquor
in the bag[2]
in the Crown Office *at* in Carey street
in the down-pins *at* down among the dead men
in the rats
in the sun *at* sun has been hot today
in the tank
in your cups
incapable
indisposed[2]
indulge
intemperance
jag house
  jag
jagged
jet-lag
jolly[1]
jugged *at* jug[2]
juice *at* juice[1] (the)
juice head *at* juice[1] (the)
juiced *at* juice[1] (the)
jumbo
keelhauled
kiss the cap *at* kiss
knock it back
knock off[4]
Korsakow's Syndrome *at* syndrome
laid out *at* lay out
led astray
legless
lift your elbow
  lift your arm
  lift your little finger
  lift your wrist
like a drink *at* drink[1]
limp
liquored *at* liquor
lit
  lit up
load[1]
  loaded
look upon the wine when it is red *at* tight[1]
looped
lose your lunch
  lose your breakfast
  lose your dinner
  lose your doughnuts
lubricated
lumpy *at* lump
lush
  lushed
  lushy
malt
  malt above the meal
  malt above the water
  malt-sucker
  malt-worm
  malty
market-fresh *at* fresh
mellow
merry
migraine
morning after (the)
Mozart
muggy
muzzy
nasty[2]
non-drinker *at* drink[1]
off the wagon *at* on the wagon
oiled
  oil the wig
on[1]
on the bend *at* bend
on the booze *at* booze
on the bottle *at* bottle[1] (the)
on the piss
on the sauce *at* sauce[1] (the)
on the tiles

on the roof
on the town[1]
on the wagon
one over the eight
one too many
ossified
over-refreshed
over-excited
overcome
overtired
over the bat *at* bat[2]
overdo the Dionysian rites
overdone
overindulge *at* indulge
package on (a)
paint the town red
paralysed
paralytic
parboiled
pasted
peg
petrified
pickled
pie-eyed
piffled
pioneer[2]
piped
piran
pissed
plastered
pledge (the)
ploughed *at* plough[2]
polluted *at* pollute
pooped
pot[2]
pot-valour
pot-walloper
potted[2]
powder your hair *at* powder room
preserved
priest of Bacchus *at* bacchanalian
primed
problem drinker *at* problem
pruned *at* prune-juice
put away[3]
put it away
queer[1]
racked
ragged
ran-dan *at* randy
rattled
raunchy
reading Geneva print
ripe
ripped
rocky[2]
rosy
saturated
scorched
Scotch mist
screwed
see a man about a dog
sent
sewn up[2]
shakes (the)
sheet in the wind
shellacked
shoot the cat
shot[4]
shout[2]
shout yourself hoarse
sign the pledge *at* pledge (the)
siper *at* sip
skinful
slewed
slopped
sloshed
slugged
smashed
smash the teapot
smeared
snug
so-so[2]
soak
soaked
something aboard *at* aboard[1]
son of Bacchus *at* bacchanalian
sop
sot
souse
soused
sozzled
spiflicated
spiffed
spit feathers
splice the mainbrace
sponge
sprung
squashed
squiffy
stewed
stiff[2]
stinking
stinko
stitched
stitch in your wig
stoned
strong weakness *at* strong waters
stung
stunned
stupid
stupid-fou
suck the monkey
sun has been hot today
sun in your eyes
sun has gone over the yardarm
Sunday traveller
supercharged
swamped
swill
swill-pot
swilled
swiller
take a drop *at* take the drop
take something
take the cure *at* cure (the)
take the pledge *at* pledge (the)
take to the bottle *at* bottle[1] (the)
take too much
tanked (up)
tap[1]
tap the admiral
taste for the bottle
technicolor yawn
temperance *at* intemperance
the worse
thirst (a)
three sheets in the wind *at* sheet in the wind
tiddly
tie a bun on *at* bun on (have/get a)
tie one on
tight[1]
tip[3]
tip the bottle
tiper
tipped
tipper
tippler
tipsy
tired[2]
tired and emotional
too many sheets in the wind
toot
top-heavy
toss down
tot
touched[2]
translated
troubled with a malt sucker inside *at* malt
tuned
turn up your little finger
turn up your pinkie
twisted[2]
under the influence
under the table[1]
under the weather[1]
unwell[2]
uppish
visiting fireman[1]
wad-shifter
wall-eyed
wasted
waterlogged
weakness
well away
well bottled
well corned
well in the way
well oiled
well sprung
wet[2]
wet hand
wet the other eye
wet your beard
wet your mouth
wet your quill
wet your whistle

Education

Erection and Ejaculation

Espionage

- deep sleep
- denied area
- destabilize
- disposal
  - disposal facilities
- double[2]
  - double agent
- drop[5]
- ear
- electrical surveillance *at* surveillance
- electronic counter-measures
  - electronic penetration
- executive action *at* execute
- extreme prejudice *at* terminate[1]
- extremely sensitive source
- fan club[1]
- ferret
- firm (the)
- fishing expedition[2]
- honey trap
- hospital[2]
- human intelligence
- ill[3]
- illegal resident *at* legal resident
- in-depth study
- incompatible with diplomatic status
- intelligence
- joe[2]
- legal resident
- letter box
- liaison officer
- live source
- loose *at* tight[2]
- loose talk *at* tight[2]
- mail cover
- military intelligence
  - military attaché
- mole
- negotiation
  - negotiator
- no longer in service *at* no longer with us
- other side (the)
- overfly
- pavement artist
- penetrate
- persona non grata
- place-man
  - placed
- prejudice *at* terminate[1]
- safe house
- sanitized[2]
- secret agent *at* agent
- secure house *at* safe house
- security risk
- security service
- sleeper
- son et lumière
- source *at* live source
- special[1]
- sticky stranger
- surveillance
- tail[2]
  - tail job
- tap[2]
- technical surveillance *at* surveillance
- tick
- tight[2]
- turn[2]
  - turn around
  - turn round
- watcher
- wired[2]
  - wired up
  - wireman

Extortion, Violence, and Torture

- appropriate technology
- assault
- aversion therapy
- badger game *at* badger
- ball money *at* ball
- barker
- bederipe *at* droit de seigneur
- bell money
- bellyful of lead *at* lead
- benevolence
- biographic leverage
- blackmail
- bleed
- blood money
- boonwork *at* droit de seigneur
- bounce[4]
- break the news
- bunch of fives
- burn[3]
- cannon
- card
- carry[3]
- change someone's voice
- chopper[1]
- claim responsibility for *at* claim
- co-operate[2]
- come across[1]
- come through
- convince
- criminal assault
  - criminally used
- dance[2]
- debrief
- deep interrogation *at* interrogation (with prejudice)
- do[2]
  - do over
- electric methods
- enforcer
- entertain[1]
- extreme physical duress
- fill in
- filled with lead *at* lead
- force favours from *at* favours
- frightener
- gang-bang *at* bang[1]
- gang-shag *at* bang[1]
- gas
  - get the gas pipe
- get the shaft
- give someone the works *at* works (the)
- Glasgow kiss
- go abroad
- greenmailer
- gun[1]
  - gun down
- gunner's daughter
- handle[2]
- hatchet man
  - hatchet
- heat[1]
- heat[2]
  - heater
- heeled[2]
- heightened interrogation
- help the police (with their inquiries)
- hen-silver
- hook[2]
- horse-leech *at* horse collar
- in protection *at* protection
- indecent assault *at* assault
- inquisition
- interfere with
- interrogation (with prejudice)
- Irish hoist
- iron[1]
- joint[3]
- juice[2]
  - juice dealer
  - juice man
- junior jumper *at* jump[2]
- kiss the gunner's daughter *at* gunner's daughter
- kneecap
- knock around
- knuckle sandwich
- lay hands on
- lead
  - lead ballast
  - lead buttons
  - lead pill
  - lead poisoning
  - leaden fever
- lean on
- leather
- lift your hand to
- love-boonwork *at* droit de seigneur
- marry the gunner's daughter *at* gunner's daughter
- massage[2]
- mattress extortion *at* mattress
- molest
- moonlight[2]
  - moonlighter
- muscle

muscleman
on the take *at* take[1]
pack heat *at* heat[2]
persuade
persuader
piece[2]
post-war credit *at* benevolence
pressure
protection
put the arm on
put the black on
put the muscle on
put the scissors on
put the bite on *at* bite
put the burn on *at* burn[3]
put the scare into *at* scare (the)
rake-off
razor
refresh your memory
rent
retainer
ride the wooden horse
rod[2]
rodded
scare (the)
sexual assault
skim
slug[1]
something on you
squeeze
stick[1]
stick[5]
stick it into[1]
sweat it out of
sweat-box
take your end *at* take[1]
third degree (the)
tooled up
torpedo
treatment
voluntary
warm someone's backside
water cure
winger
wooden log
work on[1]
workout
works (the)

Farting

backfire
bad powder
blow[2]
blow off
boff[2]
break wind
break the sound barrier
Bronx cheer
buster
cut a cheese
cheeser
cut a leg
cut one
drop your guts
drop your lunch
let off
let fly
lift a gam *at* lift a leg[2]
pass air
poop[2]
raspberry
rude noise
wind[1]
windy

Female Genitalia and Breasts

amply endowed *at* well endowed
basket[2]
beard[1]
bearded clam
beaver
beaver-shot
between the legs
bird[3]
bird's nest
boobies *at* booby
boobs *at* booby
bouncers
box[3]
Bristols
bubbies *at* Charlie
busby
bush[1]
button[1]
cabbage[2]
cat[2]
charlies *at* Charlie
charms
cherry
cleavage
cleft
cock
cockpit
cotton
couple[2]
Cupid's arbour *at* Cupid's measles
Cupid's cave *at* Cupid's measles
Cupid's cloister *at* Cupid's measles
Cupid's corner *at* Cupid's measles
Cupid's cupboard *at* Cupid's measles
Cupid's hotel *at* Cupid's measles
dairies *at* Charlie
down below
down there
downstairs[2]
equipment
Eve's custom-house *at* Adam's arsenal
face between her forks *at* fork
fanny
feminine gender
finish off[2]
front door (the)
front parlour
gap
garden[1]
garden of Eden
gash
gasp and grunt *at* grumble
gentleman's pleasure
garden *at* pad[2]
glands
globes
grass[3]
grumble and grunt *at* grumble
grummet
hemispheres
hips
hole[2]
hole of contentment
holy of holies[1]
honeydew melons *at* melons
hypogastric cranny *at* hymenal sweets
intimate part
it[3]
jugs
keyhole *at* key
kitty
knobs
knockers
love-juice
lungs
maracas
marshmallows
Mary Jane[2]
Lady Jane
meat
melons
monosyllable
mousehole
muff
naughty bits (the)
nautch *at* nautch girl
nest
Netherlands (the)
nether parts
nether regions
nook *at* nookie
orbs
organ
oval office
pair
parts *at* private parts
pleasure-garden *at* garden[1]
plumbing[2]
private parts
privates
privities
privy parts
pussy
pussy lift
ring[1]
saddle *at* saddle up with
secret parts
sex[2]
shaft[2]
sharp and blunt

Funerals

Gambling and Cheating

Homosexuality and Sexual Variations

- unmarried
- unmentionable crime
- unnatural
  - unnatural crime
  - unnatural filth
  - unnatural practice
  - unnatural vice
- up the creek
- Uranian
- use the back door *at* front door (the)
- vibrator
- voyeur
  - voyeurize
- water sports
- weakness for boys *at* weakness
- wear Dick's hatband *at* dick[1]
- weirdie
- wife[2]
- will
- woman's thing (the)
- wonk
- wrong[2]
  - wrong sexual preference
- yo-yo

Illness and Injury

- ableism
- active[1]
- afflicted
- amenity
  - amenity bed
  - amenity ward
- American disease (the)
- Acquired Immune Deficiency Syndrome (AIDS) *at* syndrome
- aurally challenged *at* challenged
- aurally handicapped *at* handicap
- aurally inconvenienced *at* inconvenienced
- big C
- blighty
- bought it
- C
- cardiac incident
  - cardiac arrest
- case[2]
- cast for death *at* cast
- catch a packet
- challenged
- chuck up
- claret
  - claret-jug
- clip[2]
- cold[3]
- combat ineffective
- comfortable[3]
- complications
- condition[1]
- consumption
- cop a packet
- coronary inefficiency
- counsellor
- cream crackered
- crease
- cripple *at* hopping-Giles
- crippled *at* hopping-Giles
- cut[1]
- decline
- delicate
- dermatologically challenged *at* challenged
- devil disease (the)
- dicky
- differently abled *at* differently
- disabled
  - disability
- discomfort
- doctor
- Down's syndrome
- eliminate manhood *at* manhood
- executive flu *at* post-viral syndrome
- falling evil
  - falling sickness
- feed the fishes
- feel funny *at* funny[1]
- feminine complaint
- fix[2]
- follicularly challenged *at* challenged
- funny[1]
  - funny tummy
- gas
- geed up *at* G
- get a slug *at* slug[1]
- go on the box
- groper *at* grope
- growth
- handicap
  - handicapped
- Hansen's disease
- hard of hearing
- health
  - health care
  - health clinic
  - health insurance
- hearing-impaired
- heart condition
  - heart
- home
- hopping-Giles
  - Hopkins
- horizontally challenged *at* challenged
- human difference
- impaired hearing
- impairment *at* mobility impaired
- inconvenienced
- indisposed[2]
- Irish fever
- knackered *at* knackers
- lay a child *at* lay
- life preserver
- long illness (a)
- martyr to (a)
- mayday
- medical correctness
- misadventure
- mitotic disease
- mobility impaired
- motion discomfort
  - motion discomfort bag
- muster your bag
- mutilate
- mutt
- National Health Service *at* health
- neoplasm
- nick[4]
- nip[3]
- no mayday *at* mayday
- not very well
  - not at all well
  - not doing well
- off-colour[2]
- old man's friend
- old soldier (the)
- on the club
  - on the box
  - on the panel
- one foot in the grave
- optically challenged
  - optically inconvenienced
  - optically marginalized
- PWA *at* people of/with
- packet[1]
- panel, the *at* on the club
- partially sighted
- people with differing abilities *at* people of/with
- person with AIDS *at* people of/with
- physically challenged *at* challenged
- physically handicapped *at* handicap
- plug[1]
- poorly[1]
- post-viral syndrome
- prey to (a)
- private patient
- problem
- procedure
- put out for the count *at* count (the)
- raspberry[2]
- rather poorly *at* rather
- residential provision
- restricted growth
- routine (nursing) care only
- Royal Free disease *at* post-viral syndrome
- scratch[2]
- sight deprived
- smear[2]
- snip
  - snib
  - snicks
- so-so[1]

sore
spot[3]
staining
stone deaf *at* hard of hearing
stop one
stop a slug
surgical misadventure *at* misadventure
swing the lead
syndrome
TB *at* consumption
tagged[1]
temporarily abled *at* ableism
ten commandments (the)
therapeutic misadventure *at* misadventure
thick of hearing
throw up
throw a map
throw up your toenails
ticker
trouble
tumour (a)
turn[3]
Uncle Dick *at* Dicky
uniquely abled *at* uniquely
unmentionables[2]
used up *at* used
unsighted
unstoned
upstairs[1]
use a wheelchair
very poorly *at* poorly[1]
visually inconvenienced *at* inconvenienced
visually handicapped *at* handicap
wear a bullet
wearing *at* wear away
white plague (the)
winded
winged
yuppie flu *at* post-viral syndrome

Intoxicants

alcohol
amber fluid/liquid/nectar
ambrosia
angel foam *at* angel dust
anti-freeze
ardent spirits
auld kirk *at* auld
awful experiment
backhander
bar
belt[2]
beverage
beverage host
beverage room
blast[3]
blind pig
bloody Mary *at* bloody shame
blow me one
blue ruin *at* blue cheer
blue stone *at* blue cheer
booze
boozer
bracer
brother of the bung *at* brother
brownie[2]
burra peg *at* chota peg
bush house *at* bush[2]
cactus juice
chaser
chota peg
club[3]
club licence
cocktail[1]
cocktail bar
cocktail lounge
coffin varnish
cooler[2]
coquetel *at* cocktail[1]
cordial[1]
corn[1]
corn juice
corn mule
corn waters
cough medicine
cough syrup
creature (the)
cut[2]
dash[1]
dead soldier
dive[2]
doctor
double[1]
dram
drink[1]
drop[2]
drop of blood
dry[1]
Dutch cheer
embalming fluid *at* embalmed
eye-opener
fellow commoner
firewater
foot
footing
French article
French cream
French elixir
French lace
Frenchman
freshen a drink
G
gage[1]
gargle
gas *at* gassed
gas-house *at* gassed
gear
geezer
gentleman commoner
giggle water *at* giggle stick[1]
glass
glass of something
grape (the)
gravy[1]
groceries sundries
gypsy's warning
hair of the dog
hair of the dog that bit you
half a can
half a pint
half and half
happy hour *at* happy
hard drink *at* hard
hard liquor *at* hard
hard stuff *at* hard
harden a drink *at* hard
hardware[1]
heel-tap
highball
horn of the ox *at* hair of the dog
hospitality
hush-shop
Irish[1]
jar
John Barleycorn *at* barley cap
jolt
jug[2]
juice[1] (the)
juice joint
juice of the bear
juniper juice
kiddlywink
libation
lightning
liquid
liquid lunch
liquid refreshment
liquid restaurant
liquid supper
liquor
little something
livener
load[1]
loaded[3]
local
local pub
lotion
lush
malt
malt whisky
medicine
mercy
Micky Finn
Moll Thompson's mark *at* moll
moonlight[1]
moonshine
mother's ruin
mother's milk
mountain dew
mule[1]
nasty[1] (the)
nasty stuff
neck-oil
needle
Nelson's blood
nightcap

Killing and Suicide

- do[2]
    - do for
    - do in
    - do yourself in
- drill[1]
- drink milk
- drive a ball through
- drop[1]
    - drop down the shute
- dull
- dump[2]
- dust[2]
- Dutch act (the)
- Dutch (do the)
- East
- easy way out (the)
- eat a gun
- electric cure
- eliminate
    - elimination
- end
- erase
- execute
    - executive action
- executive measure
- exemplary punishment
- expedient demise
- fade *at* fade away
- feed a slug
    - feed a pill
- fill full of holes
    - filled with daylight
    - filled with lead
- finish[1]
    - finish off
- fog
    - fog away
- for the high jump
- foul play
- frag
- freeze off
- fry
- gaggler *at* crap
- game[1]
- gas
- get a bullet
- get a slug *at* slug[1]
- give someone the good news
- go down[1]
- go down the nick
- go to heaven in a string
- go up[1]
    - go up the chimney
- goof up
- grass[1]
- grease[2]
- gun down *at* gun[1]
- hang
    - hang-fair
    - hanging judge
- happy pill
- head[1]
    - heading
    - heading-hill
    - heading-man
- hemp[1]
    - hemp quinsy
    - hemp-string
    - hemp strung
    - hempen fever
    - hempen widow
    - hempshire gentleman
- hit[2]
    - hit man
- hoist[4]
- hole[1]
- hot seat[2] (the)
    - hot squat
- hummingbird
- hush
- ice[2]
- in the cart
- iron out
- justify
- kayo
- keep sheep by moonlight
- kick the wind *at* kick[1]
- King of Tyburn *at* Tyburn
- kiss St Giles' cup *at* kiss
- knock off[1]
    - knock down
- knock on the head
- knock over
- last drop *at* drop[3]
- last waltz
- lay hands on
- leap in the dark (a)
- liquidate
- long drop *at* drop[3]
- long walk off a short pier *at* long home
- look through the cottonwood leaves *at* cottonwood
- loop[1]
- make a hole in the water
- make away with
    - make an end of
- make bones *at* bone[3]
- make dead meat of *at* dead meat
- make off with
- measure for a necktie *at* necktie party
- meet with an accident
- mercy killing
    - mercy death
- necklace
- necktie party
    - necktie sociable
- neutralize
- nobble
- nullification
- off[1]
- one way ride
- overdose
    - OD
- Paddington
- paper[3]
- pay off
- pick off
- planned termination *at* planned
- plough under
- plug[1]
- pop[4]
- pot[1]
- pre-emptive self-defence *at* pre-emptive strike
- preach at Tyburn Cross *at* Tyburn
- pull the plug on
- push the button on
- put a person's lights out
- put away[1]
    - put hands in (or on) yourself
    - put yourself away
- put daylight through
- put down[1]
- put in a bag
- put off
- put on the spot
- put out[1]
- put out a contract *at* contract[2]
- put someone on ice *at* ice[2]
- put to sleep
- put to the sword
- put underground *at* put to bed with a shovel
- quit cold *at* quit
- relocation
- remainder[1]
- remove
    - removal
- resettlement
- retire[1]
- ride up Holborn Hill
    - ride backwards
    - ride the mare
- roll[3]
- rope[1] (the)
- roper *at* crap
- rub out
- run into a bullet
- sanction
- scalp
- scragger *at* crap
- scratch[3]
- scuppered
- self-deliverance
    - self-destruction
    - self-execution
    - self-immolation
    - self-termination
    - self-violence
- send home
    - send to heaven
    - send to the happy hunting ground
    - send to the happy land
    - send to the land of the lotus blossom
    - send to the skies

  - send to your long account
- service[3]
- settle[1]
- sheriff's journeyman *at* crap
- shoot[1]
- short illness, a *at* long illness (a)
- shove over
- showers[3]
- sizzle
- slotted *at* slot
- sluice[1]
- smear out
- smoke it
- sniff out
- snuff (out)
- solitaire
- spot[2]
- squash
- squib off
- stab with a Bridport dagger *at* stab
- stake (the)
- step on
- stick[1]
- stretch the hemp
  - stretch the neck
- string up
- suffer the supreme penalty
- sweep
- swing[1]
- take[3]
- take care of[1]
- take for a ride[1]
  - take for an outing
  - take out into the country
- take out[2]
- take the drop
- take with you
- take your life
  - take life
- terminate[1]
  - terminate with extreme prejudice
- top[2]
  - topping fellow
- topping-cove *at* crap
- triple tree *at* Tyburn
- tuck away in earth
  - tuck under the daisies
- turn in
- turn off[1]
  - turning tree
- twisted[1]
- Tyburn
  - Tyburn blossom
  - Tyburn dance
  - Tyburn hornpipe
  - Tyburn jig
  - Tyburn tippet
  - Tyburn tree
- walk the plank
- want out
- waste[1]
- wear a bullet
- wet-job
  - wet operations
- whack
- whiff
- wipe off
- wipe out[1]
- without baggage
- write yourself off
- upstairs out of this world *at* upstairs[1]
- zap

Lavatories

- ablutions
- Adam
- Ajax *at* jakes
- along the passage
- altar
  - altar room
- ammunition[1]
- army form blank
- arrangement *at* arrange
- article
- aunt[2]
  - Aunt Jones
- backside
- basement
- bathroom
  - bathroom paper
  - bathroom tissue
- bedpan *at* pan
- blue room
- bog
  - bog-house
- boys[1]
  - boys' room
- bucket[1]
- bum fodder *at* ammunition[1]
- buoys
- can[1]
- carsey
- chamber
  - chamber-pot
  - chamber of commerce
- chapel of ease
- chic sale
- cloakroom
- close stool
- closet[1]
- coffee shop *at* coffee grinder
- comfort station
- commode
- common house[2]
- convenience[1]
- corner[2]
- cottage *at* cottaging
- cousin John *at* john[1]
- crapper *at* crap
- crapping case *at* crap
- crapping-castle *at* crap
- crapping ken *at* crap
- dolls *at* guys
- duck
- EC
- earth closet
- Eve
- facility[1]
- female *at* male[1]
- fourth
- gang *at* go[3]
- gentlemen
  - gentlemen's convenience
  - gents
- geography
- gulls *at* buoys
- guys
- head[2]
  - heads
- holy of holies[2]
- hopper
- house[2]
  - house of commons
  - house of ease
  - house of lords
  - house of office
- jakes
- jane[2]
- jerry
  - Jericho
- jockum *at* jock
- joe[1]
- john[1]
- Jordan
- karsey *at* carsey
- karzey/karzy *at* carsey
- ladies
  - ladies' convenience
  - ladies' room
  - ladies' toilet
- lads
  - lassies
- latrine
- lavabo
- lavatory
- little boys' room
  - little girls' room
- little house
- loo
- looking glass
- male[1]
- men
  - men's room
  - men's toilet
- modern convenience
- Mrs Chant
- Mrs Jones *at* aunt[2]
- necessary
  - necessary house
  - necessary woman
- night bucket
- night stool
- outhouse
  - outdoor plumbing
- pan
- parliament
- personal hygiene station *at* per-

Low Intelligence

Male Genitalia

Masturbation

rub yourself
secret vice
secret indulgence
secret sin
self-abuse
self-gratification
self-indulgence
self-love
self-manipulation
self-pleasuring
self-pollution
shag[2]
solitary sex
solitary sin
solitary vice
Southern Comfort
stroke off
strop your beak
toss off
touch yourself *at* touch[1]
traffic with yourself
vibrator
wank
wank off
wanker
wankery
waste time
whack off
whip off
whip your wire
work yourself off
wrist job (a)

Menstruation

ammunition[2]
Aunt Flo *at* aunt[2]
bad news
bad week
baker flying *at* baker
bends (the)
bloody[1]
the bloody flag is up
bunny[2]
buns on *at* bun on (have/get a)
caller (a)
captain is at home (the)
cardinal is at home (the)
cease to be
change[1] (the)
Charlie's come *at* Charlie
clout
come on[1]
come around
country cousins *at* relations have come (my)
courses
cramps (the)
curse (the)
the curse of Eve
danger signal is up (the)
domestic afflictions
facts (of life)
fall off the roof
female physiology
feminine hygiene *at* personal hygiene
flag is up (my)
flag of defiance
flowers
flux
fly the red flag *at* flag is up (my)
friend has come (my)
(little) friend to stay
grandmother to stay
have the painters in
holy week
hygiene *at* personal hygiene
ill[1]
ill of those
in purdah
indisposed[1]
irregular *at* regular[1]
Kit has come
late[2]
leaky[1]
little visitor
little friend
little sister
mense(s)
miss[2]
monthly flowers *at* flowers
monthly period
month's
monthlies
monthly blues
monthly courses
number one, London
off duty
off games
old faithful
out of circulation
over your time
pad[2]
padlock
painters are in (the)
pause[1]
period[1]
personal hygiene
pleasure-garden padlock *at* pleasure
poorly[2]
problem days *at* problem
rag(s) on
ragging *at* ragged
ragtime *at* rag(s) on
red rag (the)
the red flag is up
Red Sea is in (the)
redhaired visitor (a)
reds (the)
regular[1]
relations have come (my)
ride the red horse
road is up for repair (the)
roses (your)
run on (a)
sanitary towel
sanitary napkin
show[1]
sick
so-so[3]
so
ST
start bleeding
stomach cramps *at* cramps (the)
Tampax time
term
time of the month
trouble
tummy ache
turn of life
unavailable[2]
under the weather[2]
unstanched
unwell[1]
vapours (the)
visitor (a)
visitor from Redbank
wallflower week
well
wet weekend (a)
window blind
woman's problem *at* problem
women's things
wretched calendar (the)
wrong time of the month

Mental Illness

acorn academy
adjustment[3]
afflicted[1]
ape
asylum
balance of the mind disturbed
barmy *at* balmy
bananas
bats (in the belfry)
batty
bin
blue-devil factory *at* blue cheer
bonkers *at* bonk
booby
booby hatch/hutch
both oars in the water
bughouse
by yourself
camisole
certified
certifiable
change[2]
changeling
coco
cocoa
commit
content
counsellor
cracked
crack-brained
crackers

crackpot
cuckoo[2]
dateless
Deolalic tap *at* do-lally-tap
derailed
devil's mark (the)
disabled
disability
distressed
disturbed[2]
do-lally-tap
dotty
eccentric
fatigue
flake[1]
flip your lid
fruitcake
fruit
funny[4]
funny farm
funny house
funny place
gears have slipped
God's child
half-deck
harpic
have a screw loose
have a slate loose
head-case
head-shrinker
hospital[1]
ill-adjusted
in left field
institutionalize
laughing academy
left-field
loco
loopy *at* up the loop
loose in the attic
loose in the head
lose your marbles *at* marbles
lose your grip
lose your hold
maladjustment *at* maladjusted
mental
mentally handicapped
mental fatigue *at* fatigue
meshugga
natural[1]
nervous breakdown
nut
nut-box
nut college
nut farm
nut house
nut hutch
nutter
nuts
nutty
off[3]
off at the side
off the rails
off the wall
off your chump
off your gourd
off your head
off your napper
off your nut
off your onion
off your rocker
off your tree
off your trolley
off your turnip
out of your mind
out of the envelope
out of your head
out of your senses
out of your skull
out to lunch
potty[1]
psycho
put away[2]
queer[2]
residential provision
rocky[1]
round the bend
screwy
screw factory
seclusion
sectioned
send over the edge
shrink *at* head-shrinker
slate off *at* have a slate loose
slippage
snake pit
squirrel
squirrel tank
state farm
state home
state hospital
state (training) school
tap *at* do-lally-tap
touched[1]
unbalanced
unglued
unhinged
unscrewed
unslated
up the loop
voluntary patient
wandered
whiff of march hare *at* whiff of

Mistresses and Lovers

à trois
admirer
adult[3]
adventure[2]
adventuress
affair
affaire
affinity
amour
assignation
association[1]
back-door man
beard[2]
beau
bedfellow *at* bed[2]
bit of meat, a *at* meat
bit on the side *at* bit
boyfriend
brother starling *at* brother
camp down with
canary
carry on with
chère amie
close[2]
close relationship
close friend *at* friend
cohabit
commit misconduct
companion
company[1]
consort with
constant companion
cookie
cruise[1]
daddy
dalliance *at* dally
dear friend *at* chère amie
dirty weekend *at* dirty
err
errant
escort
extra-curricular activity
extra-curricular sex
extra-mural
extramarital excursion
fair lady
fallen woman
familiar with
fancy bit *at* fancy
fancy man *at* fancy
fancy piece *at* fancy
fancy woman *at* fancy
friend
gallant
garçonnière
gentleman friend
girlfriend *at* boyfriend
good friends *at* friend
grass-widow
hand-fast
homework[1]
honey man
housekeeper
inamorata
inamorato
involved with
item (an)
john[2]
Judy
just good friends
keep
keeper
keep company with
keep old boots *at* old boots
kept mistress *at* keep

kept wench *at* keep
kept woman *at* keep
lad
lady friend
lady of intrigue
lass
lay
learn on the pillow
liaison
light housekeeping
little woman
live as man and wife
live-in
live in sin
live off
live on
live tally
live together
live with
long-term friend
love nest
lover
man[1]
man friend
meaningful
meaningful relationship
ménage à trois
mistress
miss
more than a good friend
native comfort *at* native
new cookie *at* cookie
old boots
old man[3]
on the side
other woman (the)
parallel parking
paramour
partner
patron[1]
peculiar
pet[2]
petite amie
petit ami
petite femme
pillow partner
playmate
playfellow
protector[1]
relationship
rich friend *at* friend
ride in old boots *at* old boots
riding master *at* ride
romance
romantic entanglement
run around with
St Colman's girdle has lost its virtue
secretary
see[1]
see company
set up[1]
shack up
shack up with
share embraces *at* embraces
share someone's affections
sin
sinful
skin off all dead horses
sleeping dictionary *at* sleep with
sleeping partner
spoken for
steady company
sugar daddy
sweet momma
sweet man
take up with
thing going
toy boy
travel
triangular situation
turn off[2]
two-time
warm up old porridge *at* warm
wife[3]
woman friend
woman in a gilded cage[1]

Nakedness

as Allah made him
as God made him
au naturel
be in your naturals *at* natural functions (the)
birthday suit
birthday attire
birthday finery
birthday gear
bollocky *at* bollocks
bottomless *at* bottom
buff
decent
garb of Eden
in the altogether
in the skin
in the buff
in the raw
nature's garb
naturism
naturist
raw
skin-
skinny-dip
stark
streak
streaker
state of nature
wear a smile
zero

Narcotics

A
A bomb
abuse[2]
Acapulco gold
acid
acid fascism
acid-freak
additional means
angel dust
anti-freeze
Arctic explorer
artillery
B-bomb *at* A
B-pill *at* B
bag[4]
bagman
balloon room
bang[2]
beat the gong
beat pad
bee
belt[1]
black beauty
black stuff
black pills
black smoke
blast[3]
blast party
blocked
bloke
blow[6]
blow horse
blow snow
blow a stick
blow your cool
blue cheer
blue devil
blue flags
blue heaven
blue jay
blue velvet
bombed out
bomb
bomber
bombita
brick
brownie[3]
Buddha stick
burn[3]
burn out
bush[2]
business
bust
bust a cap
buyer
buzz on (a)
C
California sunshine
can on (a)
candy
candy man
canned *at* can on (a)
cap
carry[2]
cement
channel
charge[2]

  - charged up
- chef
- chemically inconvenienced
  - chemically affected
- Chinese tobacco
  - China white
- chippy[2]
- chuck horrors *at* chucked
- clean
- clear up
- co-pilot
- coast
- cocktail[2]
- coke
  - coke-hound
  - coked
- cold turkey
- Columbian gold
- come down
  - come off
- connect[2]
  - connection
- controlled substance
- cook[3]
- cookie *at* cookie pusher
- cool[2]
- cop[3]
- crack *at* crystal
- crash
- cruise[2]
- crystal
- cube
  - cube head
  - cube juice
- cure (the)
- cut[2]
- dabble
- deal
- deck
  - deck up
- dependency[2]
- dirty
- dissolution[2]
- doctor
- doll[2]
- dope
- dope-fiend *at* fiend
- downs
  - downers
- drag[2]
- dream dust
  - dream stick
- drop acid *at* acid
- drug
  - drug abuse
  - drug dealer
  - drug habit
  - drug pusher
- dust[1]
- dynamite
- Eastern substances
- eat[2]
- ecstasy
- equipment
- eye-opener
- feed your nose
- fiend
- fix[3]
- flake[2]
- floating
- fly[2]
  - flight
  - flying
- foil
- forwards *at* forward
- freak
  - freak out
- French blue
- fruit salad[1]
- gage[2]
- gay[1]
- gear
- geezer
  - geezed up
- get off[3]
  - get on
- giggle stick[2]
- go up[2]
- God's (own) medicine
- golden triangle
  - gold dust
- gom *at* God's (own) medicine
- gone[3]
- goof
  - goofball
  - goofed
- gow
- grass[2]
  - grass-weed
  - green grass
- Guatemala (go to)
- gun[2]
- H
  - H and C
- habit
- happenings
- happy dust
- hard
- hash
  - hash-head
- hay
- head[5]
  - head-kit
- headache-wine
- heaven dust
  - heaven and hell
  - heavenly blue
- hemp[2]
- herb (the)
- high[2]
- hit[4]
  - hit the pipe
- hold
- hook[4]
- hooked
- hop
  - hophead
  - hop joint
  - hopped
- horner
- horse[4]
  - horsed
- hot shot
- hustle[1]
  - hustler
- ice[3]
- ice cream
  - ice cream habit
  - ice cream man
  - ice creamer
- J
  - J smoke
  - J stick
- jab a vein
  - jab off
- jerk off[2]
- joint[1]
- jolt
- joy[2]
  - joy flakes
  - joy popper
  - joy powder
  - joy rider
  - joy smoke
  - joy stick
- juju
- junk
  - junked up
  - junker
  - junkie
  - junkman
- kick stick
  - kick party
  - kick the gong around
- kick the habit *at* habit
- kilo connection
- lamp habit *at* chef
- lid
- life (the)
- lift[2]
- lit
- loaded *at* load[1]
- loco weed *at* loco
- low
- Lucy in the sky with diamonds
- M
- MJ *at* Mary[2]
- mainline
- Mary[2]
  - Mary Anne
  - Mary Jane
- merchandise
- Mexican brown
  - Mexican green
  - Mexican mushroom
  - Mexican red
- Miss Emma
- monkey[2] (the)
- mood freshener

mother's blessing
mule[2]
nail[2]
nose-candy *at* candy
nose habit *at* habit
O
on[3]
on a cloud
on the needle
operator
ounce man
paper[2]
    paper acid
pharmaceuticals
pharmacy
piece[3]
polluted *at* pollute
pop[1]
    popper
pot[4]
potted[2]
powder
psychologically disadvantaged
punk
    punk pills
purple heart
push[4]
    pusher
racked
recreational drug
red devil
    red cross
    red dirt marijuana
reefer
roach[1]
rope[2]
scorched
score[2]
sent
shoot[4]
    shoot gravy
    shooting gallery
shot[3]
sleighride
smack
smashed
smeared
smell the stuff
smoke
snapper
sniff
snort[2]
snow[1]
    snow-storm
    snowball
    snowbird
    snowed in
    snowed under
    snowed up
snow-head *at* head[5]
soft[2]
spaced out
    spaced
speed
    speedball
spike[1]
spike[3]
sports medicine
stardust
stash
stewed
stick[4]
    stick of tea
stimulant
stoned
street drugs
strung out
stuff[1]
substance
    substance abuse
sugar[3]
supercharged
sweet tooth
swing[2]
take needle
take off[1]
take something
tall
tar
tea
    tea head
    tea party
    tea stick
ten two thousand *at* ten one hundred
Texas tea *at* tea
toss[2]
toss out
tracks
trip
turn on
twisted[2]
up[2]
    uppers
    uppies
    ups
use[2]
    user
viper
    viper's weed
way out
weed (the)
    weedhead
white lady
    white powder
    white stuff
white lightning
wired[1]
wrecked
yak
zoned out
zonked

Obesity

a bit of a stomach *at* stomach (a)
ample
battle of the bulge
bay window
big-boned
big girl
brewer's goitre
buxom
calorie counter
chubby
classic proportions
contour
corn-fed
cure (the)
differently-weighted *at* differently
full-figured
    full-bodied
    fuller figure
larger
mature[2]
middle-aged spread *at* middle-aged
puppy fat
quantitatively challenged
reduce your contour *at* contour
rubber tire *at* spare tyre
shortening the front line[2]
spare tyre
stomach (a)
weight problem *at* problem
well built
    well fleshed

Parts of the Body (other than Genitalia and Breasts)

after part
back passage
    back garden
    back way
backside
    backseat
basement
behind
benders *at* bend
bottom
bronze eye
butt
can *at* canhouse
cornhole
derrière
double jug *at* jug
duff[2]
elephant and castle
eye
fanny
fleshy part of the thigh
hindside
Khyber
latter end[1]
    latter part
limb[1]
little Mary
moon
plumbing[2]
posterior(s)

rear[1]
  rear end
ring[1]
seat
second eye
sit-upon
  sit-down-upon
sitting
stern
tassel *at* pencil[1]
under-arm

Police

accommodation collar *at* collar[2]
arm
B specials *at* special[3]
badge
  badge bandit
bear[2]
  bear bite
  bear cage
  bear food
  bear in the air
  bear trap
  bearded buddy
  bears are crawling
bent copper *at* bend the rules
big brother
bill
bird in the air
black and whites
blow the whistle on
blue[1]
  blue and white
  blue-belly
  blue jeans
  blue lamp
  blue police
  bluebird
  bluebottle
  bluecoat
bobby
bogy
boned *at* bone[1]
boy scouts (the)
bracelets *at* brace
Bridewell
brown paper bag
bull[3]
bust
busy
button[3]
camera
canary
Charlie
chirp
collar[2]
cop[2]
  cop house
  cop shop
copper
country Joe
cuff
decoy
dick[2]
  dickless tracy
do a number
do your paperwork *at* paper-hanger[1]
drop the hook on
eye in the sky
field associate
filth (the)
fireman[3]
flash tin *at* flash[1]
flatfoot
  flat
fly[1]
  fly ball
  fly bob
  fly bull
  fly cop
  fly dick
flying squad
folding camera *at* camera
frog
front office
fuzz
  fuzzy bear
G-man *at* G
Gestapo *at* secret (state) police
goldilocks
goon squad
grasshopper
green stamp collector
gumshoe
harness bull
hawk
headhunter[2]
helmet
horny[1]
house man
informer
internal affairs
jack[2]
  jack rabbit
john[4]
  John Law
KGB *at* Committee of State Security
lady bear *at* bear[2]
lift[3]
limb[2]
  limb of the law
little bears
local bear
  local boy
  local yokel
lower the boom on
mama bear
  mama smokey
man[2]
  man with a gun
men in blue *at* blue[1]
Morals Squad *at* moral
mug-shot
nick[3]
nightingale
noddy
old bill
paddy wagon
paper-hanger[1]
peeler
  peel
peeper
penny
pig
pinch[2]
pink panther
plod
polar bear
pull in
question[1]
Radical Squad *at* radical
raincoat[2]
red-squad (the)
roach[2]
rubber heel
run round the Horn
runner
running bear
Sam[1]
secret (state) police
shake[2]
shield
slops
smoke screen
smokey
  smokey bear
  smokey beaver
  smokey on four legs
  smokey on rubber
  smokey with camera
  smokey with ears
snatch[4]
  snatch squad
snout
snowdrop
special[3]
Special Branch
special fuzz
special police
  special patrol group
stake-out
state bear *at* bear[2]
stool pigeon
Sweeney *at* flying squad
Tijuana taxi
use your tin
woolly bear

Politics

(*C* or *N* indicate Communist or Nazi)
action[2] *N*
activist
adviser
alternative
America first

Poverty and Parsimony

remittance man
seen better days (that has)
set back
shoot the moon *at* moonlight flit
shorts (the)
social worker
south (the)
stroller
tapped
third world
tied aid *at* aid[2]
tightwad
to the knuckle
uncle
underdeveloped
underprivileged
union[2]
urban renewal
vulnerable
whistling
white nigger
workhouse

Pregnancy

accident[2]
afterthought
anticipating
away the trip *at* away[2]
awkward
bagged
bear[1]
beat the gun
belly plea
big
  big belly
bump[2] (the)
bun in the oven (a)
carry[1]
  carrying
caught[1]
certain condition *at* condition[2]
cheat the starter
club *at* in the club
colt[1]
come to a sticky end[1]
condition[2]
costume wedding
delicate condition *at* condition
do the right thing
do your duty by
eat for two
enceinte
expectant
  expecting
facts (of life)
fall[2]
  fall for a child
  fall in the family way
  fall wrong to
fill a pannier
force-put job
free of fumbler's hall *at* free[1]
full in the belly
get with child
gone[2]
grass-widow
great
have a watermelon on the vine *at* watermelon
heavy *at* heavy date
high in the belly
how's your father
in calf
  in foal
  in for it
  in kindle
  in pig
  in pod
  in pup
in season
in the club
  in the pudding club
  in the plum(p) pudding club
in the family way
in trouble
interesting condition *at* condition[2]
Irish toothache
join the club *at* in the club
kid
knock up
labour[1]
lady in waiting[2]
large[1]
lined *at* line
little stranger
lumpy
make a child *at* make babies together
make a decent woman of
make an honest woman of
mistake[1]
off-white wedding
on[2]
on her way
on the nest
overdue[1]
plum(p) pudding club *at* in the club
pudding club *at* in the club
pup
quick
raise a belly
rank
riding time *at* ride
ring the bell
rush job
sewn up[1]
she has given Lawton Gate a clap *at* away[2]
shot in the tail
shotgun marriage
  shotgun wedding
so
  so-and-so[1]
split a woman's shape *at* split
stung by a serpent *at* stung
swallow a watermelon seed *at* watermelon
swell
that way[2]
trouble
up the pole
up the spout
up the stick
upstairs[1]
watermelon
wear the belly high *at* belly plea
with child
wrong[3]

Prisons and Prisoners

adjustment[3]
approved school
assembly centre
at Her Majesty's pleasure
attendance centre
away[2]
back-gate parole
behind the wire
big house
  big pasture
  big school
bird[2]
  birdcage
black hole
board school *at* residential provision
boat *at* boat people
book
boom-passenger *at* boom-boom[2]
both hands
Bridewell
brig
bucket[2]
bullpen *at* bull[3]
buried
cage
camp *at* concentration camp
can[3]
canary
  canary-bird
chokey
chuck horrors *at* chuck
clink
cockchafer[1]
come to a sticky end[1]
Community Treatment Centre *at* community alienation
concentration camp
control unit
cooler[1]
coop
cop[2]
correctional
  correctional facility
  correctional officer
corrective training
  Corrective Labour Camp

cross bar hotel
custody suite
dance-hall *at* dance[1]
deep freeze
detainee
down the line[1]
drunk tank *at* in the tank
enjoy her Majesty's hospitality
eat porridge
entertain[1]
everlasting staircase
fall[4]
Fanny Hill *at* fanny
fistful
five fingers
flowery
fly a kite[3]
forced labour
freezer
G
glass house
go down[2]
go over the hill *at* go over[2]
go over the wall *at* go over[2]
go to the Bay
go up the river
grind the wind *at* grind
guest
guest of her Majesty
guest of Uncle Sam
gulag *at* corrective training
handful[1]
Hanoi Hilton
hard room
hit the bricks[2]
    hit the hump
hole *at* black hole
holiday
home
hoosegow
horse[3]
house of correction
    house of detention
hulk
ice box[1]
    icehouse
ill[3]
in[1]
in need of supervision
in the bag[1]
individual behavior adjustment unit
indoctrination camp
inside
jolt
jug[1]
kiss the counter *at* kiss
kiss the clink *at* kiss
kitty
labour education
last shame (the)
length
limbo *at* limb[2]
little school *at* big house
man[2]
municipal farm
nab the stoop *at* nab
Newgate
nick[3]
on ice *at* ice box[3]
on the rockpile
on the trot[1]
out of town
over the wall
pacification camp *at* pacify
pacification centre *at* pacify
periodic rest
place of safety
    place of correction
poke[3]
political education
porridge
Potsdam
preventive detention
protective custody
put away[2]
quod
railroad
residential provision
    resident
resisting arrest
rock crusher
room and board with Uncle Sam
school
screw[2]
seclusion
segregation unit
send down *at* down the line[1]
send up
sheriff's hotel
slam
slammer
sneezer
special education
special regime
sponging-house *at* sponge
spring
state farm
    state home
    state (training) school
stir
    stir-wise
stockade
stretch
sweat-box *at* sweat it out of
take to the hills
tank
tempered in the forge of labour
time
tolbooth
transported
trying to escape
unauthorized departure *at* depart this life
under the screw
University of ...
up the river
vacation
yardbird
youth (guidance) centre

Prostitution

AMW
abandoned
accost
available casual indigenous female companion *at* available
academician *at* academy
actress
all-nighter
amateur
ammunition wife *at* ammunition[2]
arse-pedlar *at* arse
article
aunt[1]
B girl *at* bar
bachelor's wife
bad
    bad girls
badger
bag[5]
baggage
bang-tail *at* bang[1]
banger *at* bang[1]
bar girl *at* bar
bash
bat[1]
bed-faggot *at* bed[2]
beef
belly-piece *at* belly to belly
belter *at* belt[1]
besom
bibi
    bidi
biddy
bint
bird[1]
bit
bitch
black-eyed Susan
black velvet
blow[1]
    blowen
board lodger *at* board
bobtail
bona-roba
bottle[3]
bottom woman *at* bottom
brass
break luck *at* break a lance
broad
bum
bun
business
    business woman
buttered bun *at* bun
buttock
    buttock and file
    buttock and twang

buy love
call-button girl
call girl
call boy
call the tricks *at* trick
camp follower
can[4]
cannibal
case vrow *at* case[1]
cat[1]
cattle[2]
cavalry
champagne trick
Charlie
chick
chickie
child of Venus
chippy[1]
cockatrice *at* cocktail
cockchafer[2]
cocktail[1]
coffee grinder
comfort women
commercial sex worker
common customer
common jack
common maid
common sewer
common tart
commoner o' the camp
convenient[1]
country-club girls
courtesan
Covent Garden
Covent Garden goddess
creature of sale *at* creature (the)
Cressida
cross girl
cruiser *at* cruise[1]
Cyprian
dance a Haymarket hornpipe *at* dance the mattress quadrille
dance-hall hostess
dasher *at* dash[2]
daughter of the game
dead meat[2]
demi-mondaine
demi-rep
doe
dolly
dolly-common
dolly mop
doxy
dress for sale
dress-lodger
Drury Lane vestal *at* Drury Lane ague
Dutch widow
edie
escort
ewe mutton
faggot *at* fag
fallen woman
feather-bed soldier *at* feather bed
femme du monde *at* femme fatale
filth *at* filthy
fish[1]
fish[2]
fishmonger's daughter
fix up
flapper
flash girl *at* flash[1]
flash panney *at* flash[1]
flash-tail *at* flash[1]
flash woman *at* flash[1]
flat-backer *at* flat on your back
flutter a skirt *at* flutter[2]
forty-four
frail sister *at* frail[2]
fresh and sweet *at* dead meat[2]
fresh meat *at* dead meat[2]
fruit fly *at* fruit[1]
game[2] (the)
game pullet
gamester[1]
gay girl *at* gay[1]
gay it *at* gay[1]
gay lady *at* gay[1]
gay life *at* gay[1]
girl[1]
girl of the streets
girlie
go case
go into the streets
go to Paul's for a wife
goat-milker *at* goat
good-time girl *at* good time
goose[1]
grande horizontale *at* horizontal[1]
grass bibi/bidi *at* bibi
guinea-hen
gull
harlot
hawk your mutton
hawk your pearly
head chick *at* head job (a)
high-yellow girl *at* yellow
hobby-horse
hold-door trade (the)
hooker[2]
horizontal life *at* horizontal[1]
hostess
husband[1]
hustle[2]
hustler
immoral
immoral earnings
immoral girls
Immorality Act
importune
in circulation
in the business *at* business
in the game *at* game[2] (the)
in the trade
infantry *at* cavalry
jam tart *at* tart
jane[1]
jerker *at* jerk off[1]
Jezebel
john[5]
joy girl *at* joy[1]
joy sister *at* joy[1]
Judy
kerb crawling
knee-trembler
lady
lady boarder
lady in waiting[1]
lady of a certain description
lady of easy virtue
lady of no virtue
lady of pleasure
lady of the night
lady of the stage
ladybird
liberated *at* liberate[3]
lie backwards and let out your forerooms *at* lie with
life (the)
life of infamy
life of shame
light ladies *at* light[1]
light-skirts *at* light-fingered
light wenches *at* light[1]
little bit
living by trade *at* trade (the)
lost *at* lose your cherry
low girls *at* low-budget
Magdalene
masseuse
messer *at* mess[1]
mixer
mob
model
moll
moose
mud-kicker
Murphy
mutton
nanny
naughty girl *at* naughty
naughty lady *at* naughty
nautch girl
nice time
night job
nightingale
nocturne
nun
nymph
nymph of darkness
nymph of delight
nymph of the pavement
oldest profession *at* profession (the)
on the bash *at* bash
on the game *at* game[2] (the)
on the grind
on the street(s) *at* street (the)
on the stroll

Religion and Superstition

Sexual Pursuit

  - town
- maul
- meat
- mouse *at* mousehole
- mouth
- muslin
- mutton
  - mutton-monger
- neck
- no better than she ought to be
  - no better than she should be
- nonsense
- nose open
- not inconsolable
- old Adam (the)
- on[4]
- on heat[2]
- on the make
- on the pull
- over-familiar *at* familiar with
- over-gallant *at* gallant
- park[1]
- pash
- pass[2]
- paw
- pet[1]
- pick up
  - pick-up
- piece[1]
  - piece of arse/ass
  - piece of buttered bun
  - piece of crackling
  - piece of crumpet
  - piece of gash goods
  - piece of muslin
  - piece of rump
  - piece of skirt
- piece of tail *at* tail
- pin-up
- play the field
- play the goat
- popsy
  - poppet
- pussy-whipped *at* pussy
- put a move on
  - put the moves on
- put yourself about
- randy
- raunchy
- roundheels
- roving eye
  - rover
- SA *at* sex appeal
- salute upon the lips
- San Quentin jail bait *at* jail bait
- scarlet fever
- seat cover
- sex appeal
- sexual variety
- sheila
- shoot the agate
- shoot the breeze *at* shoot a line
- skirt
- slag
- Spanish fly
- spoon with
- stern-chaser *at* stern
- stoat
- stuck on
- stud
- sure thing
- switch on
- swordsman
- tail[1]
- take liberties
  - take a liberty
- talent
  - talent-spotting
- turn on
- venereal *at* Venus
- venery *at* Venus
- want[2]
  - want a body
  - want intercourse
  - want love
  - want relations
  - want sex
- warm
- weakness
- wear your heart upon your sleeve
- wet for
- wide-on (a)
- will
- willing
- with it in your hand
- woman
  - womanizer
- worry
  - worry at night
- yes-girl

Stealing

- acquire
  - acquisition
- adjustment[5]
- alienate
- appropriate[1]
- bag[1]
  - bag job
- black-bag
- black George *at* black gentleman
- black fishing *at* black gentleman
- bleed the monkey *at* bleed
- bone[1]
- boost[1]
  - booster
  - booster bag
  - booster bloomers
- bootleg
- borrow
- browse
- cabbage[1]
- cadge
  - cadger
- canary
- cannon
- carpetbagger
- chisel
- cleaners (the)
- click[1]
  - Clickem Fair
  - Clickem Inn
  - clicker
- clip[1]
  - clip-artist
  - clip-joint
- clout
- collar[1]
- collector *at* collect
- convey
  - conveyance
  - conveyancing
  - conveyor
- cop[1]
- crack[1]
  - cracksman
- crib man *at* crib
- cut out
- Davy Jones' natural children *at* Davy Jones' locker
- dip[1]
  - dipper
- dive[1]
  - diver
- divert
  - diverted
- do[3]
  - do over
  - do the dirty
- dodgy night *at* dodgy
- drop[5]
- earn
- fair trader
- fall off the back of a lorry
  - fall off the back of a half-track
  - fall off the back of a truck
  - fall off the back of a wagon
- fence
- fetch[1]
- fiddle
- fillet
- find[1]
  - finder
- finger-blight
- fish hook *at* hook[1]
- five-fingered discount
- flash panney *at* flash[1]
- fly the blue pigeon
- footpad *at* highwayman
- forage
- freeloader
  - freebie
  - freeload
- free trade
- Freemans
- freeze on to
- G
- gain

Sweat

Trade Unions

Unemployment

unwaged
waiting for employment

Urination

accident[1]
accommodate yourself
adjourn
answer the call[2]
arrangement *at* arrange
back teeth afloat
bale out[1]
be excused
beat a retreat *at* retreat
bedwetting
been
bodily functions
bodily wastes
bottle[2]
break your neck
burst
business
call of nature
cash a check *at* cash in your checks
caught short
chamber-lye
choke your chicken
cleanliness training
clout
cock the leg
comfort break
comfortable[1]
  comfy
commit a nuisance
consulting Mrs Jones
continent *at* incontinent[2]
contribution[1]
cover your feet *at* cover[1]
decant
demands of nature
diddle
  Dicky Diddle
dirty your pants *at* dirty
disappear[2]
do a bunk
  do a dike
  do a shift
drain off
ease springs
  ease your bladder
  ease yourself
empty your bladder
feed a dog
  feed a goldfish
  feed a horse
  feed a parrot
feel the need
forget yourself
freshen up
gather a daisy/pea/rose *at* pick a daisy
go[3]
go on the coal
go over the heap
go places
go round a corner
go to Cannes
go to the bathroom
go upstairs
gold water
grow your greens
house-trained
incontinent[2]
indiscretions[1]
Jimmy
  Jerry Riddle
  Jimmy Riddle
kill a snake
leak[1]
leave the class
  leave the room
lift a leg[2]
little jobs *at* big jobs
littles *at* big jobs
look at the garden
  look at the compost heap
  look at the crops
  look at the flowers
  look at the lawn
  look at the roses
  look at the vegetable garden
mail a letter
make a call
make a mess
make room for tea
make water
mess[2]
Micky
  Micky Bliss
  Mike Bliss
minor function (the)
mistake[2]
natural functions (the)
  natural purposes
nature's needs
necessities
need
night water *at* nightsoil
nuisance *at* commit a nuisance
number one(s)[1]
P
pass water
pay a visit
  pay a visit to the old soldiers' home
pee
  pee-pee
perform[1]
  perform a natural function
pick a daisy
  pick a pea
  pick a rose
piss *at* P
pit-stop
pluck a daisy/pea/rose *at* pick a daisy
point Percy at the porcelain
polish the mahogany
potty training *at* pot[3]
powder your nose *at* powder room
powder your puff *at* powder room
puddle
pull a daisy/pea/rose *at* pick a daisy
pump ship
  pump bilges
rattle[2]
relief[2]
relieve yourself
retire[2]
run[4]
run off the bathwater
  run-off
sample
scatter *at* scatters
see a man about a dog
see the rosebed
  see the compost heap
  see the vegetable garden
  see the view
  see your aunt
shake hands with the bishop
  shake hands with the unemployed
  shake hands with your best friend
  shake hands with your wife's best friend
shed a tear
shoot a lion
  shoot a dog
slack
  slack off
slash
specimen
spend a penny
splash your boots
squat[1]
stale
stretch your legs
syphon the python
take a leak *at* leak[1]
take the air
take your snake for a gallop
taken short
tap a kidney
ten one hundred
that and this
tinkle
turn your bike around
uncomfortable
unlimber your joint *at* joint[2]
upstairs[1]
use the facilities
visit *at* pay a visit
visiting card
void water
wash[1]
  wash-mug

Venereal Disease

Vulgarisms, Pornography, and Swearing

gollin
golls
gom
gommy
goms
gomz
goom
gull
gum
Gordon Bennet(t)
gow
H
hail Columbia
hard core *at* hard
hell *at* H
horn-emporium *at* horn[1]
horse collar
jerk *at* jerk off[1]
joe[1]
language
men's magazine
mercy
micky
Micky Bliss
Mike Bliss
Mrs Duckett
monkey's (a)
mother[2]
muck
naff off
off-colour[1]
P
P off
pillock *at* pill[1]
pissed off *at* pissed
P.O.'d *at* pissed
potty mouth *at* pot[3]
pound salt
pound sand
poxy *at* pox (the)
prick
rat
rabbit
rose-coloured *at* roses (your)
round objects
ruddy
sanguinary
silly B *at* B
skin-
skin-flick
skin-house
skin-magazine
smut house
so-and-so[2]
sod *at* sodomite
something-something
spicy
squirt
stag
stiff[1]
stripper
sugar[2]
sweet FA *at* Fanny Adams
sweet Fanny Adams *at* Fanny Adams
take the mickey
take the Michael
take the Mick
take the micky
take the Mike
Tijuana bible
tinpot
topless
toss[1]
tosser *at* toss off
two fingers
wanker *at* wank
what the H? *at* H
you know what you can do with that

Warfare

active (air) defence *at* defence
adjustment[4]
adventure[1]
aid[1]
air support
alternative
Anschluss
anti-personnel
appeasement
army of occupation *at* occupied
bale out[2]
blue-on-blue
bogey *at* bogy
bomb[1]
border incident *at* incident
boys in the bush *at* boys[2]
brew
brushfire war
bug out
ceasefire
Charlie
chemical warfare agent *at* agent
chopper[1]
circular error probability
civilian impacting
clean
clean locality
cleanse[1]
co-belligerent
co-operate[1]
collaborator
collaborationist
collateral damage
come up with the rations
coming of peace
conflict
confrontation
constructed
conventional
counter-force capability
counter-force attack
counter-force weapon
counter-value capability
counterattack
counterinsurgency
cruise[3]
defence
D Notice
defence budget
defence debate
defence procurement
defence strategy
defence votes
defensive victory
degrade
deliver
delivery vehicle
device
ditch
do[4]
done for
dove
draw the enemy into a trap
drill[2]
drink[3] (the)
drop your flag *at* drop your drawers
duration
egg
emergency
enhanced radiation weapon *at* enhance
ethnic cleansing
expendable *at* expended
fire has gone out
first strike
first strike capability
fish[3]
fizzer
flap
frag
fratricide
free a man for duty
freedom fighters
French leave
friendly fire
frontier guards
garden[2]
go over the side *at* go over[2]
go over the top *at* over the top
good voyage
groceries
hardware[2]
hawk
hit the silk *at* silk (the)
hot[4]
incident
incontinent ordnance
incursion
intervention
intruder
late disturbances
late nastiness
late unpleasantness
liberate[1]
limited action
limited covert war

General and Miscellaneous